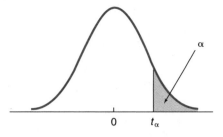

TABLE B The t-Distribution

df	$t_{0.050}$	$t_{0.025}$	$t_{0.010}$	$t_{0.005}$	df
1	6.314	12.706	31.821	63.657	1
2	2.920	4.303	6.965	9.925	2
3	2.353	3.182	4.541	5.841	3
4	2.132	2.776	3.747	4.604	4
5	2.015	2.571	3.365	4.032	5
6	1.943	2.447	3.143	3.707	6
7	1.895	2.365	2.998	3.499	7
8	1.860	2.306	2.896	3.355	8
9	1.833	2.262	2.821	3.250	9
10	1.812	2.228	2.764	3.169	10
11	1.796	2.201	2.718	3.106	11
12	1.782	2.179	2.681	3.055	12
13	1.771	2.160	2.650	3.012	13
14	1.761	2.145	2.624	2.977	14
15	1.753	2.131	2.602	2.947	15
16	1.746	2.120	2.583	2.921	16
17	1.740	2.110	2.567	2.898	17
18	1.734	2.101	2.552	2.878	18
19	1.729	2.093	2.539	2.861	19
20	1.725	2.086	2.528	2.845	20
21	1.721	2.080	2.518	2.831	21
22	1.717	2.074	2.508	2.819	22
23	1.714	2.069	2.500	2.807	23
24	1.711	2.064	2.492	2.797	24
25	1.708	2.060	2.485	2.787	25
26	1.706	2.056	2.479	2.779	26
27	1.703	2.052	2.473	2.771	27
28	1.701	2.048	2.467	2.763	28
29	1.699	2.045	2.462	2.756	29
inf.	1.645	1.960	2.326	2.576	inf.

This table is abridged from Table A of R. A. Fisher and F. Yates: *Statistical Tables for Biological, Agricultural, and Medical Research*, published by Longman Group, Ltd., London (previously published by Oliver & Boyd, Edinburgh), by permission of the authors and publishers.

STATISTICS AND PROBABILITY IN MODERN LIFE

Sixth Edition

Joseph Newmark

The College of Staten Island of the City University of New York

SAUNDERS COLLEGE PUBLISHING

Harcourt Brace College Publishers

Fort Worth   Philadelphia   San Diego   New York   Orlando   Austin   San Antonio
Toronto   Montreal   London   Sydney   Tokyo

Text typeface: New Caledonia
Compositor: University Graphics, Inc.
Executive Editor: Bill Hoffman
Developmental Editor: Patrick Farace
Managing Editor and Project Editor: Sally Kusch
Copy Editor: Charlotte Nelson
Manager of Art and Design: Carol Bleistine
Senior Art Director: Joan S. Wendt
Illustration Supervisor: Sue Kinney
Text Designer: Rebecca Lemna
Cover Designer: Joan S. Wendt
Text Artwork: Tech Graphics
Photo Researcher: George Semple
Manager of Production: Joanne Cassetti
Production Manager: Alicia Jackson
Marketing Manager: Nick Agnew

Cover Credit: Michael Howell/Photonica

Printed in the United States of America

Statistics and Probability in Modern Life, Sixth edition

ISBN 0-03-006393-0

Library of Congress Catalog Card Number: 96-70011

890123456 039 1098765

To
Trudy
and our
children

PREFACE

Today, more than ever, statistics plays a very important role in our lives. This book has been designed to provide the reader with a comprehensive treatment of introductory statistics and probability in such areas as sociology, business, ecology, economics, education, medicine, psychology, and mathematics. Students in these fields must frequently demonstrate a knowledge of the language and methods of statistics. Methodology and applications have been integrated throughout the text.

A special effort has been made in this edition to make the concepts of statistics available to students who are not prepared for elaborate symbolisms or complex arithmetic. Although the mathematical content is complete and correct, the language is elementary and easy to understand. Expressions that students have difficulty comprehending have been avoided, and the introduction of new terminology has been held to a minimum. In frequent **COMMENTS**, various points that students often misunderstand or miss completely are carefully discussed. Yet mathematical rigor has not been sacrificed. Introductory high school algebra is a sufficient prerequisite to using this text satisfactorily.

Each chapter starts off with a chapter objectives section and a newspaper clipping to set the stage for the material to be covered, and concludes with a summary and study guide to reinforce the ideas covered in the chapter. Additionally, each idea introduced is first explained in terms of an example chosen from everyday real-life situations with which the student can identify.

CHANGES IN THE SIXTH EDITION

This revised edition of the text reflects the many suggestions and recommendations of users of the earlier editions; the changes are outlined below.

- All exercises and examples have been updated.
- Each chapter now starts off with a "Did You Know" section containing some interesting facts that are discussed later in the chapter.
- The end of each chapter has been streamlined into three groups of exercises; Testing Your Understanding of This Chapter's Concepts, Chapter Test, and Thinking Critically.
- Many new newspaper articles and magazine clippings have been added throughout the book.
- The sources of the data for numerous exercises are now clearly indicated.
- Each chapter now concludes with one or more case studies which illustrate how the concepts discussed within the chapter are actually applied in real-life situations.
- Each chapter has a historical note which presents some interesting sidelights about the mathematicians discussed within that chapter.
- The Using Computer Packages section has been considerably expanded. Each chapter now has such a section. MINITAB exercises have been added at the end of each such section.
- The number of exercises in the book has been greatly increased. Some are purely mechanical, some are applications, and some are theoretical. There are now well over 1500 exercises (of all types) throughout the book. All exercises have been checked for accuracy.

SUPPLEMENTS

The following supplements are available for use with this book:

1. A MINITAB for Windows supplement has been prepared by Lloyd Jaisingh of Morehead State University to teach the reader how to use the Windows version of the popular statistical package MINITAB. No prior computer knowledge is assumed.
2. A TI-83 Graphing Calculator Manual prepared by Stephen Kokoska, of Bloomsburg University, has been prepared to show the reader how graphics calculators can be used to solve typical introductory statistics problems.
3. An *Instructor's Resource Manual*, written by Joseph Newmark, gives detailed line-by-line solutions to *all* of the exercises, as well as to the questions in the Testing Your Understanding and Thinking Critically sections.
4. A *Student Solutions Manual*, written by Joseph Newmark, includes worked-out solutions to every other odd-numbered problem in the exercises and end-of-chapter tests.
5. A *Printed Test Bank* and *Computerized Test Banks* for Mac, IBM, and Windows computers, prepared by Don Staake of Jackson Community College, contains from 50 to 80 questions per chapter. It is free to all adopters.

ACKNOWLEDGMENTS

I would like to thank the many instructors and students who received the earlier editions of this book so warmly. I am grateful to those who took the time to send comments, suggestions, and corrections. I am also grateful to the following people who reviewed this edition of the text and who made valuable suggestions for its improvement:

Robert Baldree, Campbell University
Darrell F. Clevidence, Carl Sandburg College
Carol Freeman, Finger Lakes Community College
Charles Klein, De Anza College
Marlene Kovaly, Florida Community College
Ethel Muter, Raritan Valley Community College
Ronald Schrader, University of New Mexico
Linn Stranak, Union University

Special thanks to the following instructor for his accuracy reviews of all examples and exercises: Robert Smidt, California State Polytechnic University at San Luis Obispo.

I am greatly indebted to the staff of Saunders College Publishing. I especially want to thank Barry Fetterolf (Vice President) for his enthusiastic support, Patrick Farace and Marc Sherman for their expertise and skillful guidance, and Sally Kusch for her knowledge and dedication in bringing this project to a successful completion.

Many thanks also go to the authors and publishers who granted permission to use the statistical tables so necessary for this work. Among others, I am indebted to the Literary Executor of the late Sir Ronald A. Fisher, F.R.S., and to Oliver and Boyd, Ltd., Edinburgh for their permission to reprint tables from *Statistical Methods for Biological, Agricultural, and Medical Research*.

Finally, and most important, I wish to thank my wife Trudy and our children for their understanding and patience as they endured the enormous strain associated with completing this project. Without their encouragement it could not have been undertaken or completed.

Joseph Newmark
November 1996

TO THE INSTRUCTOR

In order to write a statistics and probability book that is easily understood by readers with varying backgrounds and help the student become actively involved in the learning process, the text contains the following features:

1. ***Did You Know That***   At the beginning of each chapter, some interesting statistical facts are presented. These are later discussed within the chapter.
2. ***Statistics in Action***   Each chapter begins with an article taken from the daily press that presents the ideas discussed in the chapter in applied context. This is intended to motivate the student by showing how the ideas mentioned in the chapter are applied in real-life situations. Moreover, there are numerous clippings and illustrations throughout the book to further clarify the material.
3. ***Chapter Objectives***   Keyed to the appropriate section of the text, chapter objectives highlight the key ideas to be discussed.
4. ***Introduction***   Each chapter's introductory section sets the stage for the ideas contained in the chapter.
5. ***Comments***   Throughout the text "Comments" are included to emphasize ideas that students often miss or find confusing.
6. ***Historical Notes***   Most chapters contain brief historical notes on mathematicians who contributed to the ideas discussed within the chapter. These should help humanize mathematics.
7. ***Use of Boldface and Italic Type***   Certain words or definitions are set in italic type for emphasis; marginal notes are set in boldface type in text as well as in color in the margin. Formulas and rules are highlighted for the student by a box.
8. ***Examples and Exercises***   All of the applied examples and exercises have been carefully written to capture student interest. The exercises (graded from easy to more challenging) are grouped into two categories: (1) routine exercises that help build skills and understanding of concepts and (2) problem-solving exercises. As suggested by the NCTM, the use of calculators or computers is highly recommended. Throughout the book, some of the examples or exercises are specifically designed to be worked out using a hand-held calculator or computer.

9. ***End-of-Chapter Material*** Each chapter concludes with a Study Guide and Formulas to Remember. The Study Guide highlights and summarizes the main points of the chapter and includes page numbers for easy reference. Students should find these extremely useful when studying and preparing for exams.

10. ***Testing Your Understanding of This Chapter's Concepts*** Each chapter contains a section which presents questions that are often asked in class and that are designed to see if the student truly understands the concepts.

11. ***Thinking Critically*** Each chapter contains a section designed to make the reader think very carefully. These questions are somewhat more challenging.

12. ***Chapter Tests*** Each chapter concludes with two comprehensive sets of exercises that can be used both as a review and as a student self-test. Some of the exercises are multiple choice while others are of the completion type.

13. ***Answers*** Answers are provided at the end of the book for selected exercises as well as for all of the questions in the Testing Your Understanding and Thinking Critically sections, all of the Chapter Test Exercises, and the Case Studies.

14. ***Ancillary Package*** A complete set of ancillaries is available to supplement this text, including an Instructor's Resource Manual, a Student Solutions Manual, a MINITAB for Windows supplement, a TI-83 Graphing Calculator Manual, a Printed Test Bank, and Computerized Test Banks for Mac, IBM and Windows computers.

Saunders College Publishing may provide complimentary instructional aids and supplements or supplement packages to those adopters qualified under our adoption policy. Please contact your sales representative for more information. If as an adopter or potential user you receive supplements you do not need, please return them to your sales representative or send them to: Attn: Returns Department
Troy Warehouse
465 South Lincoln Drive
Troy, MO 63379

Suggested Course Outline An instructor should have no difficulty in selecting material for a one- or two-semester course. The following outlines indicate how this text can be used:

One-Semester Course (meets 40 times per semester, 40 min. per session)

Text material	Approximate amount of time	Prerequisite needed for each chapter
Chapter 1	1 lesson	none
Chapter 2	6 lessons	none
Chapter 3	6 lessons	Chapter 2, Section 2.2 or the equivalent
Chapter 4	5 lessons	none
Chapter 5 (skip 5.5)	4 lessons	Chapter 4
Chapter 6 (skip 6.7 & 6.8)	5 lessons	Chapter 4
Chapter 7	5 lessons	Chapter 2
Chapter 8	4 lessons	Chapter 7
	36°	

°The remaining meetings can be devoted to exams and review.

Two-Semester Course—Semester 1

Text material	Approximate amount of time	Prerequisite needed for each chapter
Chapter 1	1 lesson	none
Chapter 2	6 lessons	none
Chapter 3	6 lessons	Chapter 2, Section 2.2 or the equivalent
Chapter 4	5 lessons	none
Chapter 5	5 lessons	Chapter 4
Chapter 6	7 lessons	Chapter 4
Chapter 7	5 lessons	Chapter 2
Chapter 8	4 lessons	Chapter 7
	39°	

°The remaining meetings can be devoted to exams and review.

Two-Semester Course—Semester 2

Text material	Approximate amount of time	Prerequisite needed for each chapter
Chapter 9	5 lessons	Chapter 7
Chapter 10	7 lessons	Chapter 7
Chapter 11	8 lessons	Chapter 10
Chapter 12	4 lessons	Chapter 10
Chapter 13	7 lessons	Chapter 10
Chapter 14	7 lessons	Chapters 8, 10, 11, and 12
	38°	

°The remaining meetings can be devoted to exams and review.

Since comments and suggestions from instructors using previous editions of this text proved invaluable in completing this revision, I would be grateful to continue receiving them. Please address all correspondence to me directly at The College of Staten Island, 2800 Victory Blvd., Staten Island, N.Y. 10314. My e-mail address is NEWMARK@POSTBOX.CSI.CUNY.EDU

TO THE STUDENT

This is a nonrigorous mathematical text on elementary probability and statistics. If you are afraid that your mathematical background is rusty, don't worry. The only math background needed to use this book is introductory high school algebra or the equivalent; however, those of you who are fairly knowledgeable in math will see how statistics can be used in interesting and challenging problems.

The examples are plentiful and are chosen from everyday situations. Also, the ideas of statistics are applied to a variety of subject areas. This variety indicates the general applicability of statistical methods. In frequent COMMENTS, appropriate explanations of statistical theory are given and the "why's" of statistical methods answered.

Occasionally a section or exercise is starred. This means that it is slightly more difficult and may require some time and thought.

Some of the exercises have been specifically designed to be worked out using a hand-held calculator. However, they can also be easily worked out without such a calculator.

Each chapter concludes with a summary and a review of the formulas and major terms and concepts introduced in the chapter; page numbers are included for easy reference. In addition, there are Chapter Tests which will help in preparing for exams. Answers to selected exercises and to all Chapter Test questions are given in the back of the book.

I hope that you will find reading and using this book an enjoyable and rewarding experience, since a basic knowledge of statistics is essential. Good luck.

CONTENTS

APPENDIX   A.1

STATISTICAL TABLES

ANSWERS TO SELECTED EXERCISES   A-25

INDEX   I.1

1

THE NATURE OF STATISTICS—
WHAT IS STATISTICS?

DID YOU KNOW THAT

a   approximately 25% of the American population will be over 65 years of age by the end of this century? (*Source:* U.S. Bureau of the Census, 1995)

b   the number of bank failures has been increasing over the past few years, with more than 100 such failures predicted in each of the next few years? (*Source:* Federal Reserve Bank of New York)

c   more than 800,000 children are reported missing every year in the United States and that the trend is on the rise? (*Source:* National Center for Missing and Exploited Children, 1995)

d   giant pandas will become extinct within the next 30 years unless efforts succeed in improving their low birth rates and saving their habitats from encroachment by humans and that by outfitting pandas with electronic radio transmitters statisticians are able to predict the total population of such endangered species?

Pictured above is a game warden attaching an electronic radio transmitter to an animal so that it can be used to predict the total population of an endangered species. See the discussion on page 15. (*LuRay Parker/U.S. Fish and Wildlife Service*)

When you hear the word "statistics" what do you think of? The word statistics as defined in many dictionaries and encyclopedias has several meanings.

In this introductory chapter we describe the general nature of statistics and discuss some of its uses (and abuses). We present some basic terminology that will be used throughout the book.

Chapter Objectives

- **To discuss** the nature of statistics and numerous examples of how they are used. (Section 1.1)

- **To identify** the two major areas of statistics—descriptive statistics and inferential statistics. Descriptive statistics involves collecting data and tabulating it in a meaningful way. Inferential statistics involves making predictions based on the sample data. (Section 1.2)

- **To distinguish** between part of a group, called a sample, and the whole group, called the population. (Section 1.2)

- **To present** a brief discussion of the historical development of statistics and probability. (Section 1.3)

Consider the first newspaper article. When the 20th century began, only one person in ten lived in an urban area. Today nearly one half of all people are urban dwellers. It is projected that in the coming century more people will dwell in cities than live on the entire planet today. The sheer force of such rapid growth has important implications for many urban areas that are already plagued by overloaded infrastructures, pollution, and conspicuous human suffering.

Now consider the second clipping. The United States uses more fresh water per capita

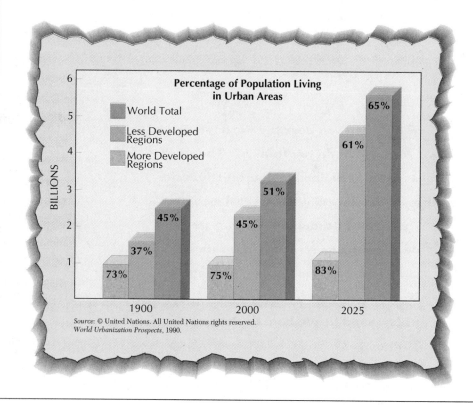

Source: © United Nations. All United Nations rights reserved.
World Urbanization Prospects, 1990.

than any other industrialized country in the world. In 1992, for example, enough water was used to provide every American with the equivalent of 37 baths each day. As the population grows, groundwater supplies are being pumped out faster than they can be replenished. Today, government officials claim that parts of the United States already face constant water shortages because of this growth phenomenon. Such statistics have important implications for our society.

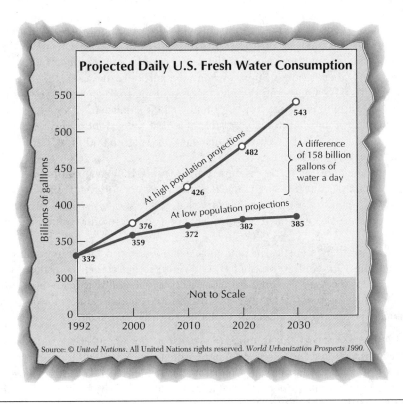

Projected Daily U.S. Fresh Water Consumption

Billions of gallons

At high population projections

A difference of 158 billion gallons of water a day

At low population projections

543

482

426

376

359 372 382 385

332

Not to Scale

1992 2000 2010 2020 2030

1.1   INTRODUCTION

Statistics   What is **statistics**? Most of us tend to think of statistics as having something to do with charts or tables of numbers. While this idea is not wrong, mathematicians and statisticians use the word "statistics" in a more general sense. Roughly speaking, the term

Data   statistics, as used by the statistician, involves collecting any information called **data**, analyzing it, and making meaningful decisions based on the data. Collected data, which represent observations or measurements of something of interest, can be classified into

Qualitative data   two general types: qualitative and quantitative. **Qualitative data** refer to observations that are attributes. For example, a person's sex, eye color, or blood type, or the brand of whiskey preferred by an individual, and so on. Each person can be placed in one and only one category depending on the situation and how the categories are defined.

Quantitative data   On the other hand, **quantitative data** represent observations or measurements that are numerical, for example, the weight of a student, the percentage of homes that are contaminated with radon gas in New Jersey, or the number of college students receiving financial aid at your college. In each of these cases an individual observation measures some quantity. This gives us quantitative data.

Discrete numerical   Numerical or quantitative data can be further subdivided into two categories, **dis-**
data   **crete numerical data** and **continuous numerical data**. When data result from a
Continuous   *count*, we have discrete numerical data. The number of fans attending a football game
numerical data   in the New Jersey Meadowlands Sports Arena on a particular day is 0, 1, 2, 3, . . . , but it cannot be 2.4 or 6.95. On the other hand, data resulting from a *measure* of a quantity will usually be continuous. Thus, if a bag of sugar in a supermarket weighs 5 pounds (lb) (to the nearest pound), all we can be sure of is that the weight of the sugar is some value between 4.5 and 5.5 lb. Actually the sugar can weigh 5.000 lb, 4.998 lb, or any value in the interval 4.5 to 5.5.

Often data may appear to be discrete but in reality they are continuous. For example, Jennifer may claim that she is married for 12 years. Although she was married for 12 years on her last wedding anniversary day, today she is married 12 years plus some part of a year. We will have more to say about discrete and continuous data in later chapters.

The role played by statistics in our daily activities is constantly increasing. As a matter of fact, the 19th-century prophet H. G. Wells predicted that "statistical thinking will one day be as necessary for efficient citizenship as the ability to read and write."

The following examples on the uses of statistics indicate that a knowledge of statistics today is quickly becoming an important tool, even for the layperson.

1. Statistics show that the number of American households with video cassette recorders (VCRs) increased from approximately 4% in 1980 to more than 75% in 1994. Also, the number of American households that subscribe to cable TV tripled in the same period and that approximately 63% of American households subscribed to cable TV in 1994. (*Source:* U.S. Bureau of the Census, 1995)

2. Statistics show that there may be an oversupply of doctors in certain areas of the United States. (*Source:* Statistical Abstract of the U.S. (112th Edition), Washington, D.C., 1992)

3. Statistics collected by the Social Security Administration show that people, on the average, are living longer today than did their parents. This has important implications for predicting the future fiscal soundness of our social security system. (See the following article.)

AMERICA GROWING OLDER

Baltimore: According to the survey conducted by Balcy and Smith for the Social Security Administration, the number of Americans over 65 years of age increased by 6.3% over last year at this time. The survey further indicated that if current trends continue then by the end of this century, approximately 25% of the American population will be over 65 years of age.

These findings have far reaching implications for the financial solvency of the Social Security System.

Monday–January 7, 1991

4. Statistics show that the number of hours in the American workweek is constantly shrinking and that if the present trend continues, the average workweek will decline to 34 hours. (See Figure 1.1.)
5. Statistics show that if the air we breathe is excessively polluted, then undoubtedly some people will become ill.
6. Statistics show that the United States is facing a serious energy shortage and that alternative sources of energy must be found to meet the ever-increasing demand. As a matter of fact, energy consumption in the United States reached a record 81.9 quadrillion BTU in 1992. Only 7.9% of this consumption was from "renewable" energy sources. (*Source:* U.S. Department of Energy's Energy Information Administration *Annual Energy Review*, 1993)
7. Statistics show that the world's population is growing at a faster rate than the availability of food. (See Figure 1.2.)
8. Statistics show that it is often impossible to obtain accurate records on the prevalence of crime because many victims simply do not report the crime to police. This makes apprehension of the offenders impossible. (See Table 1.1.)

FIGURE 1.1   *Note:* Figures for 1910 and 1930 are for workers in manufacturing.   *(Source: U.S. Department of Labor)*

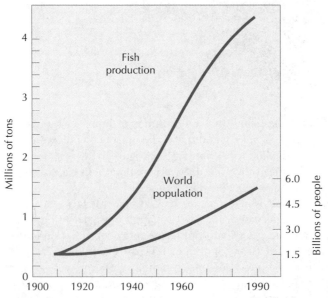

FIGURE 1.2   Over the years both the world population and the production of fish have increased considerably. Will the world's fish production continue to increase indefinitely, or will we reach a maximum point beyond which production will decrease?   *(Source: Fisheries of North America)*

TABLE 1.1   Personal Crimes of Violence: Percent Distribution of Reasons for Not Reporting Victimizations to the Police, by Victim-Offender Relationship and Type of Crime, 1990*

Victim-Offender Relationship and Type of Crime	Total	Nothing Could Be Done; Lack of Proof	Not Important Enough	Police Would Not Want to Be Bothered	Too Inconvenient or Time Consuming	Private or Personal Matter	Fear of Reprisal	Reported to Someone Else	Other and Not Given
Involving strangers									
Crimes of violence	100.0	24.2	23.6	8.1	3.7	9.7	3.9	8.2	18.6
Rape	100.0	34.0	[1]1.8	[1]13.3	[1]0.0	[1]9.9	[1]6.9	[1]10.2	24.0
Robbery	100.0	31.0	17.1	10.3	4.6	7.5	5.5	7.6	16.2
Assault	100.0	21.4	26.7	7.1	3.6	10.4	3.2	8.3	19.2
Involving nonstrangers									
Crimes of violence	100.0	8.7	17.6	6.0	1.4	29.3	6.2	14.5	16.3
Rape	100.0	[1]12.1	[1]2.3	[1]9.3	[1]2.8	22.8	[1]10.7	[1]5.3	34.9
Robbery	100.0	12.5	17.8	[1]2.9	[1]0.0	21.6	10.4	17.3	17.5
Assault	100.0	8.2	18.2	6.2	1.5	30.4	5.5	14.6	15.4

Note: Detail may not add to total shown because of rounding.

*Rate per 1000 population age 12 and over

[1]Estimate, based on zero or on about ten or fewer sample cases, is statistically unreliable.

(*Source: Criminal Victimization in the United States—A National Crime Survey Report.* U.S. Department of Justice, Law Enforcement Assistance Administration, National Criminal Justice Information and Statistics Service)

9. Statistics show that any student, no matter what his or her high school background, will succeed in college if properly motivated.

10. Statistics show that chewing sugarless gum appears to be a cheap and simple way to relieve heartburn. Apparently, gum chomping boosts saliva production that helps counteract stomach acids. (*Source: USA Today*, 1995)

11. Statistics given in Table 1.2 show the victimization rates for persons age 12 and over by race, annual family income, and type of crime.

12. Federal Reserve Bank statistics indicate that the number of bank failures has been increasing over the past few years, with more than 100 such failures predicted in each of the next few years. Such statistics have important implications for the integrity of our banking and monetary system.

Very little formal mathematical knowledge is needed to collect and tabulate data. However, the interpretation of the data in a meaningful way requires careful analysis. If this is not done by a statistician or mathematician who has been trained to interpret data, statistics can be misused. The following example indicates how statistics can be misused.

TABLE 1.2   Personal Crimes: Victimization Rates for Persons Age 12 and Over, by Race, Annual Family Income of Victims, Type of Crime, 1990*

Race and Income	Crimes of Violence	Rape	Robbery			Assault			Crimes of Theft	Personal Larceny	
			Total	With Injury	Without Injury	Total	Aggravated	Simple		With Contact	Without Contact
White											
Less than $3,000 (10,907,000)	54.4	3.1	10.7	4.4	6.3	40.5	15.9	24.6	83.9	4.3	79.6
$3,000–$7,499 (30,186,000)	35.3	1.0	7.2	2.8	4.4	27.1	11.3	15.8	80.3	3.6	76.7
$7,500–$9,999 (16,700,000)	33.8	[1]0.6	6.6	1.9	4.7	26.6	11.6	15.0	95.3	3.0	92.3
$10,000–$14,999 (38,650,000)	27.2	0.5	4.3	1.2	3.1	22.4	9.2	13.1	93.8	1.7	92.1
$15,000–$24,999 (29,168,000)	27.3	0.4	5.2	1.6	3.6	21.7	8.6	13.2	115.6	2.5	113.1
$25,000 or more (9,888,000)	25.6	[1]0.6	5.7	1.5	4.3	19.3	5.8	13.5	128.3	2.7	125.6
Black											
Less than $3,000 (3,377,000)	48.0	4.1	15.9	7.8	8.1	27.9	15.9	12.0	65.5	7.6	58.0
$3,000–$7,499 (6,469,000)	39.9	1.9	14.8	3.8	11.0	23.1	13.7	9.4	73.0	5.5	67.5
$7,500–$9,999 (1,926,000)	46.5	[1]1.3	17.2	8.8	8.4	28.0	15.4	12.6	91.2	8.2	83.0
$10,000–$14,999 (2,914,000)	32.9	[1]0.5	10.6	[1]2.5	8.2	21.8	11.7	10.1	93.3	4.5	88.4
$15,000–$24,999 (1,533,000)	42.0	[1]2.5	15.0	[1]5.1	9.9	24.5	10.8	13.6	129.5	[1]6.1	123.5
$25,000 or more (242,000)	10.7	[1]0.0	[1]10.7	[1]5.1	[1]5.5	[1]0.0	[1]0.0	[1]0.0	110.7	[1]4.5	106.1

Note: Detail may not add to total shown because of rounding. Numbers in parentheses refer to population in the group; excludes data on persons whose income level was not ascertained.

*Rate per 1000 population age 12 and over

[1]Estimate, based on zero or on about ten or fewer sample cases, is statistically unreliable.

(*Source: Criminal Victimization in the United States—A National Crime Survey Report.* U.S. Department of Justice, Law Enforcement Assistance Administration, National Criminal Justice Information and Statistics Service)

In the city of Bushtown the occurrence of Lyme disease, thought to be nonexistent, increased by 100% from the year 1993 to 1994. Such a statistic would horrify any parent. However, upon careful analysis it was found that in 1993 there were two reported cases of Lyme disease out of a population of five million and in 1994 there were four cases.

1.2   CHOICE OF ACTIONS SUGGESTED BY STATISTICAL STUDIES

Since the word statistics is often used by many people in different ways, statisticians have divided the field of statistics into two major areas called *descriptive statistics* and *inferential statistics*.

Descriptive statistics

Descriptive statistics involves collecting data and tabulating results using, for example, tables, charts, or graphs to make the data more manageable and meaningful. Very often, certain numerical computations are made that enable us to analyze data more intelligently. Most of us are concerned with this branch of statistics. For example, even if we know the income of every family in California, we are still unable to analyze the figures because there are too many to consider. These figures somehow must be condensed so that meaningful statements can be made.

Inferential statistics or **statistical inference**, on the other hand, is much more involved. To understand why, imagine that we are interested in the average height of students at the University of California. Since there are so many students attending the university, it would require an enormous amount of work to interview each student and gather all the data. Furthermore, the procedure would undoubtedly be costly and could take too much time. Possibly we could obtain the necessary information from a sample of sufficient size that would be accurate for our needs. We could then use the data based

Population

on this sample to make predictions about the entire student body, called the **population**. This is exactly what statistical inference involves. We have the following definitions.

Definition 1.1

Sample
Parameter
Statistic

A **sample** is any small group of individuals or objects selected to represent the entire group called the **population**, where the population is the set of *all* measurements of interest to the sample collector. A **parameter** is a numerical measurement describing some characteristic of a *population*. On the other hand, a **statistic** is a numerical measurement describing some characteristic of a *sample*.

The relationship between a population and a sample is shown in Figure 1.3.

Of the 1800 families residing in the Marcy Plaza Houses, it was found that 648 of them owned two TV sets. Since 648 is 36% of 1800, we can conclude that for the residents of Marcy Plaza Houses 36% of them owned two TV sets. In this case, 36% is a *parameter* (not a *statistic*) because it is based upon a survey of *all* households in this housing development. If we assume that the households in this development are fairly

FIGURE 1.3 The relationship between a population and a sample. The highlighted cars in the foreground were selected as a sample of the entire population of cars in the parking lot.

representative of all households in the United States, or that they represent a sample drawn from a larger population, then the 36% becomes a statistic. As a matter of fact, according to 1993 figures released by *Nielsen Media Research*, 98% of U.S. homes (some 93.1 million households) have a color TV, with 36% of the households having two TV sets. (*Newsweek*, August 2, 1993)

Definition 1.2 *Inferential statistics*	**Inferential statistics** is the study of procedures by which we draw conclusions and make decisions or predictions about a population on the basis of a sample.

Of course, we would like to make the best possible decisions about the population. To do this successfully, we will need some ideas from the theory of probability. Statisticians must therefore be familiar with both statistics and probability. Exactly how such decisions are made will be discussed in later chapters.

1.3 STATISTICS IN MODERN LIFE

Although statistics is one of the oldest branches of mathematics, its use did not become widespread until the 20th century. Originally, it involved summarizing data by means of charts and tables. Historically, the use of statistics can be traced back to the ancient

Egyptians and Chinese who used statistics for keeping state records. The Chinese under the Chou Dynasty, 2000 B.C., maintained extensive lists of revenue collection and government expenditures. They also maintained records on the availability of warriors.

The study of statistics was really begun by the Englishman John Graunt (1620–1674). In 1662 he published his book, *Natural and Political Observations upon the Bills of Mortality* (Figure 1.4). Graunt studied the causes of death in different cities and noticed that the percentage of deaths from different causes was about the same and did not change considerably from year to year. For example, deaths from suicide, accidents, and certain diseases occurred not only with surprising regularity but with approximately the same percentage from year to year. Furthermore, Graunt's statistical analysis led him to discover that there were more male than female births. But since men were more subject to death from occupational hazards, diseases, and war, it turned out that at marriageable age the number of men and women was about equal. Graunt believed that this was nature's way of assuring monogamy.

After Graunt published his *Bills of Mortality*, other mathematicians became interested in statistics and made important contributions. Pierre-Simon Laplace (1749–

JOHN GRAUNT (1620–1674)

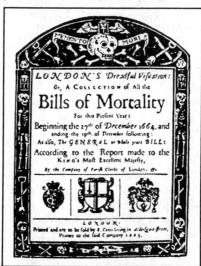

FIGURE 1.4 An illustration of John Graunt's "Bills of Mortality," redrawn from *Devils, Drugs, and Doctors* by Howard W. Haggard, M.D. *(Copyright 1929 by Harper and Row, Publishers, Inc.; renewed 1957 by Howard W. Haggard. Reprinted by permission of the publisher)*

1827), Abraham De Moivre (1667–1754), and Carl Friedrich Gauss (1777–1855) studied and applied the **normal distribution** (see page 403). Karl Pearson (1857–1936) and Sir Francis Galton (1822–1911) studied the **correlation coefficient** (see page 589). These are but a few of the many mathematicians who made valuable contributions to statistical theory. In later chapters we will further discuss their works.

Although a great deal of modern statistical theory was known before 1930, it was not commonly used, simply because the accumulation and analysis of statistical data involved time-consuming, complicated computations. However, all this changed with the invention of the computer and its ability to perform long and difficult calculations in a relatively short period of time. Statistics soon began to be used for **inference**, that is, in making generalizations on the basis of samples. Also, probability theory was soon applied to the statistical analysis of data. The use of statistics for inference resulted in the discovery of new techniques for treating data.

HISTORICAL NOTE

The mathematical study of probability can be traced back to the mathematician Jerome Cardan (1501–1576). The illegitimate child of a distinguished lawyer, Cardan became a famous doctor, who treated many prominent people throughout Europe. On various occasions he was also a professor of medicine at several Italian universities. While practicing as a doctor, he also studied, taught, and wrote mathematics.

Although he was extremely talented, Cardan's personality and personal life appear to have been less than perfect. He was very hot tempered. In fact, he is said to have cut off one of his son's ears in a fit of rage. (His sons seem to have followed their father's example. One of them poisoned his own wife.)

Cardan was also an astrologer. There is a legend that claims that he predicted the date of his death astrologically and, to guarantee its accuracy, he drank poison on that day. That's one way of being right!

Cardan suffered from many illnesses that prevented him from enjoying life. To forget his troubles, he gambled daily for many years. His intense interest in gambling led him to write a book on the subject. This work, called *The Book on Games of Chance*, is really a textbook for gamblers, complete with tips on how to succeed in cheating. In this book we find the beginnings of the study of probability.

The development of mathematical probability was further helped along its way by the Chevalier de Méré. Like Cardan, he was a gambler. He was also an amateur mathematician and was interested in the following problem: Suppose a gambling game must be interrupted before it is finished. How should the players divide up the money that is on the table? He sent the problem to his friend, the mathematical genius Blaise Pascal (1623–1662).

When Pascal received the Chevalier de Méré's gambling problem, he sent it to his friend, the great amateur mathematician Pierre Fermat (1602–1665). The two men wrote to each other on this subject. These correspondences, during the year 1654, were the starting point for the modern theory of probability. Many other gifted mathematicians were attracted by the work that Pascal and Fermat had begun.

Statistics is so important to our way of living that many of us often use statistical analysis in making decisions without even realizing it. Today statistics are used by the nonmathematician as well as by the mathematician in such areas as psychology, ecology, sports, insurance, education, biology, business, agriculture, music, and sociology, to name but a few. The fields of study to which statistics and probability are being applied is constantly increasing. Their usefulness in the fields of biology, economics, and psychology are so enormous that the subjects of biometrics, econometrics, and psychometrics have come into being.

By tagging birds, statisticians are able to predict the population of endangered species of birds.
(© Degginger, Fran Heyl Associates)

As statistics developed, probability began to assume more importance because of its wide range of applications. Today the application of probability in gambling is but one of its minor uses. In recent years, statistics has even been applied to determine the total population of various species of living things. In particular, by using very simple procedures, statisticians have been able to predict the total population of such endangered species as the whooping crane, the giant panda, and various fish. In each case a number of birds, pandas, or fish are caught, tagged with a label or some other form of identification, and then released for breeding. When they are recaptured or sighted at a later date, the proportion of tagged fish, pandas, or birds out of the total catch is calculated and used to predict the total population of the species.

Statistics and probability have recently been used to answer such questions as "Did Shakespeare author a newly discovered poem?" (*Science*, January 24, 1986, p. 335) or "What is the probability that a randomly selected person who has tested positive in a drug screening actually uses drugs?" (*Newsweek*, November 29, 1986, p. 18)

In the future many new and interesting applications of statistics and probability are likely to be found.

1.4   WHAT LIES AHEAD

In the preceding sections we mentioned several uses of statistics to convince you that the development of statistics and probability is not static. It is constantly changing. Who knows what a beginning course in statistics will be like by the year 2000? Undoubtedly, different things will be stressed and new applications for statistics and probability will be found. Yet certain basic ideas of probability and its uses in statistical studies will not be changed. Such ideas are too fundamental. It is with these ideas that we will concern ourselves in this text.

As mentioned earlier, with the advent and widespread use of both the personal computer and the large mainframe computer in the 1980s, many time-consuming and tedious calculations can now be performed quickly and efficiently by the computer. Very little background in computer programming is actually required since the programs are usually incorporated as part of routine statistical tasks. There are many statistical computer packages currently on the market such as COMP-STAT, MINITAB, and SPSS (Statistical Package for the Social Sciences) Batch System. One of the most widely used packages is MINITAB. All the statistical computer programs given in this book will work whether you are using the DOS version of MINITAB° (Release 8) or the Windows version of MINITAB (Release 10XTRA). Occasionally, an alternate command may be needed when using the Windows version. These will be noted as needed.

Along with the widespread acceptance and use of statistical concepts has come the unfortunate use of statistics for deception. Statistics can often be manipulated so that they show what a person wants them to show. In the next chapter we will indicate how statistical graphs can distort information presented by data. Statistical mistakes can be honest, also. In its January 2, 1967, issue, *Newsweek* reported on page 10 that Mao-Tse Tung cut the salaries of certain Chinese government officials by 300%. Is it possible for a salary to be cut by 300%?

Also, consider the advertisement by the Morgan Nursing Home chain, which claims that the medical care rendered by its medical staff is so superior that the nursing home has not had any stroke victim deaths in its 25 years of operation. What the advertisement

---

°MINITAB is a registered trademark of Minitab, Inc., 3081 Enterprise Drive, State College, PA 16801. Telephone: (814) 238-3280; Fax: (814) 238-4383.

fails to mention is that all seriously ill patients are transferred to a nearby hospital. The patient deaths are then recorded on the hospital records where the deaths actually occurred, rather than on the nursing home records.

These are only two examples of the misuses (intentionally or not) of statistics. Read the book *How to Lie with Statistics* by Darrell Huff for a number of interesting examples on the misuses of statistics. Robert Reichard's *The Figure Finaglers* also indicates how statistics can be used deceptively.

In the following chapters we will develop the techniques used in all applications of statistics and discuss the role played by probability in these applications. If you follow the well-defined statistical rules that are given, you will avoid falling into some of the aforementioned traps.

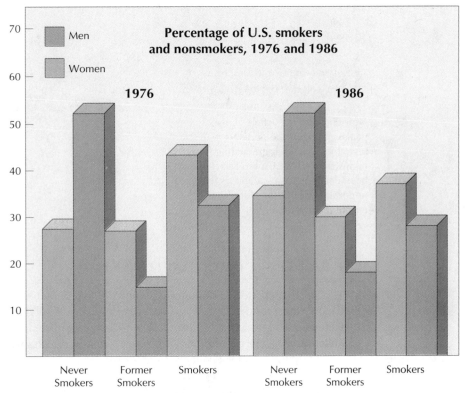

Source: National Health Interview Survey (U.S.DHHS NCHS 1984).
Reported in "Facts and Figures On Smoking, 1976-1986" American Cancer Society.

The graph above indicates that new statistics have been gathered regarding the percentage of smokers and nonsmokers. Obviously, the cigarette companies interpret such statistics in a different manner. It is apparent that such statistical data must be analyzed very carefully before arriving at any conclusions.

HIGH SCHOOL DEGREES

Percentage of seventeen-year-olds graduating from high school in the United States:

1870	2%
1900	6%
1920	17%
1930	29%
1940	51%
1950	59%
1960	65%
1970	76%
1980	72%
1990	70%

Source: U.S. Department of Education, National Center for Education Statistics.

Now consider the article above. Since peaking in 1970 the percentage of seventeen-year-olds graduating from high school in the United States has been declining. Such statistics have important implications for our educational system.

EXERCISES FOR SECTION 1.4

1. The word "statistics" is often used by newscasters in radio and television newscasts. Classify these as descriptive or inferential statistics. Explain your answer.
2. Determine whether each of the following variables is qualitative or quantitative. If quantitative, state whether it is discrete or continuous. Explain your answer.
   a. Amount of money a college student spends on textbooks per semester.
   b. Number of telephones per household.
   c. Amount of time spent by a college student applying for financial aid.
   d. Method of payment for tuition: cash, check, or credit card.
   e. Religion of college student: Catholic, Protestant, Jewish, other.
   f. Height of your mother.
3. A counselor at Dorby College is interested in determining the composition of the student body at the college. Which variables of interest measure characteristics that result in

a. attribute data
b. discrete numerical data
c. continuous numerical data

4. Several of the questions asked by the counselor mentioned in Exercise 3 are as follows:
   a. How do you get to school—walk, drive, use public transportation, or ride a bicycle?
   b. How many days per week do you go to school?
   c. How long does it take you to get from your home to school?
   Classify the responses to parts (a), (b), and (c) as attribute data, discrete numerical data, or continuous numerical data.

5. Consider the newspaper article given on page 7.
   a. What part of the survey is an example of descriptive statistics?
   b. What part of the survey is an example of inferential statistics?

6. A student who is registered for a math course claims that she will not do well in the course since math has never been an easy subject for her. This conclusion involves
   a. descriptive statistics only
   b. inferential statistics only
   c. both descriptive and inferential statistics
   d. no statistics at all
   e. none of these

7. After carefully analyzing 10,000 pregnancies, a medical research team concludes that cigarette smoking or drug use by *any* pregnant woman is harmful to the unborn child. This conclusion involves
   a. quantitative data   b. qualitative data
   c. descriptive statistics   d. inferential statistics

INTERNATIONAL LEFT-HANDERS DAY

New York: August 13 is International Left Handers Day. The likelihood that a left-hander compared with a right-hander will be accidentally injured while playing a sport is 20% greater; while working it is 25% greater, while in the home it is 49% greater and while driving the likelihood is 85% greater.

Are the odds stacked against lefties?

Adapted from *US News and World Report* Monday –August 9, 1993

8. Consider the newspaper article on the preceding page. What part of the article involves descriptive and/or inferential statistics?
9. What type of statistics is presented in the financial section of a newspaper, inferential or descriptive? Explain your answer.

ARE YOU OVERWEIGHT?

Washington: The latest medical research indicates that women who carry their extra weight around the waist rather than hips and thighs tend to have higher levels of "bad" cholesterol which increases their risk of heart trouble.

Adapted from *Health* magazine, p. 12
May/June, 1993

10. Consider the information contained in the article from *Health* magazine. What part of this article involves descriptive and/or inferential statistics?
11. After carefully analyzing individual 1040 tax returns covering the past seven years, an IRS agent claims that 7% of the rejected returns involve mathematical errors. This conclusion involves what type of statistics, descriptive or inferential?

For Questions 12–14, use the following information that was obtained from a local video movie rental store:°

Year	Price Charged to Rent Newly Released Movies	Usual Number of Newly Released Movies Rented Weekly
1991	$2.00	127
1992	1.90	186
1993	1.80	211
1994	1.65	287
1995	1.50	346

°*Source:* Videorific, Inc., New York City, 1995.

12. The conclusion that "There is a considerable difference between the number of newly released movies rented weekly when the price is $2.00 instead of $1.50" involves what type of statistics: statistical inference or descriptive statistics?

13. The conclusion that "The usual number of newly released movies rented weekly has been increasing over the years" involves what type of statistics: statistical inference or descriptive statistics?

14. The conclusion that "Any future decrease in the price charged to rent a newly released movie will result in an increase in the number of such movies rented weekly" involves what kind of statistics: statistical inference or descriptive statistics?

15. A safety engineer is interested in estimating the proportion of cars that have tires with unsafe treads. Can the engineer select the collection of cars in a shopping center as a sample? Explain your answer.

16. Can you find examples of how you use statistics in your daily experiences?

17. Imagine that you are interested in determining the number of people who cheat on their federal income tax return. How could you go about doing this; that is, how would you select a sample? Discuss the difficulties in obtaining accurate information on a question of this type.

18. Read the article "Statistics-Watching: A Guide for the Perplexed" in the *New York Times* (November 23, 1975, page 45). This article discusses how economic statistics can often be confusing.

19. Read the book *How to Lie with Statistics* by Darell Huff (New York: Norton, 1954) for a number of interesting examples on the misuses of statistics.

20. In 1995, only 25% of the registered Democrats actually voted in local primary elections. Can we conclude that 75% of the registered Democrats did not care who was nominated? Explain your answer.

21. Due to economic conditions and to avoid bankruptcy, a large airline company required all of its pilots to take a 10% cut in their 1994 salary. In 1995, the economic health of the company had improved drastically, and the company gave each pilot a 10% raise in salary. Are the pilots now earning as much as they were before? Explain your answer.

22. To decrease the number of accidents, the Highway Patrol issued 79,869 speeding tickets in 1994 to motorists on the Riverview Expressway. In 1995, only 41,602 speeding tickets were issued. Can we conclude that motorists are obeying the posted speed limits to decrease the number of accidents?

Exercises 23–27 indicate how statistics can be misused. After carefully analyzing them, try to find any possible errors in the reasoning.

23. A real estate broker made the following claim: Between 1970 and 1995 the salary of the average American rose by the same rate as did the cost for a new home. The broker concluded that the more money an American worker earns, the higher the price will be for a new home. Do the facts support this claim? Explain your answer.

24. A television commercial claims that "our tiles are manufactured to such high standards that they will keep their shine 50% longer." If you are considering purchasing tiles, should you buy this brand of tiles because of its superior "shining" ability?

25. After analyzing the police records of Sun Valley, a young and inexperienced statistician comes to the following conclusion: "Since two thirds of all rape and murder victims on the town's college campus were relatives or former friends of their assailants, a female college student is much safer going out at night with strangers than remaining in her room." Do you agree with this conclusion? Explain your answer.

26. A national health food magazine claims that "95% of its subscribers who follow the magazine's recommendation and take megadoses of vitamin C are healthy and vigorous." Should you begin taking the vitamin C supplement?

27. During 1995 there were 234 accidents involving drunken drivers and 15,897 accidents involving drunken pedestrians reported in Danville. Can we conclude that it is safer in Danville to be a drunken driver than a drunken pedestrian? Explain your answer.

1.5   USING COMPUTER PACKAGES

We live in a computer age. The computer is a very important work machine. At home, as well as at the office or in school, personal computers are appearing in greater numbers and being used for a variety of purposes, such as word processing, record keeping, accounting, and other workaday tasks. The use of the computer in the field of statistics is no exception. The widespread availability of computers today (both mainframe and personal) and the supporting software allow the user to perform many detailed statistical calculations.

As with learning to drive a car, operating a computer (mainframe or personal) is an easy task once you learn how to navigate the keyboard, how to negotiate the disk operating system (DOS), what each applications program can do, and which key gets it to do these things. Fortunately, it is rarely necessary for you to write your own computer program to perform statistical analysis. An ample supply of computer software is currently on the market to perform most statistical tasks. These programs are ideally suited to perform the tedious and time-consuming statistical calculations. Many mainframe computers already have such programs stored in memory and can be called upon for use with a few simple commands.

Although there are many readily available statistical computer programs, the most popular of these are MINITAB, SAS (Statistical Analysis System), SPSS (Statistical Package for the Social Sciences), and BMDP.° Check with your local computer center to see which programs are available. (These centers have assistants who can show you how to use these programs if you need help.) Throughout this book you will see examples of computer printouts illustrating particular ideas.

°For a detailed discussion of some of these packages, see J. Lefkowitz, *Introduction to Statistical Computer Packages* (Boston: Duxbury Press, 1985).

At this point we merely mention the existence of statistical packages. We will present the details of the packages at the end of most chapters of this text. In this way you can see how the packages can be used to do the calculations discussed within the chapter.

The MINITAB statistical package was developed in 1972 at Pennsylvania State University. Today it is used at more than 2000 colleges and universities and at 70% of the Fortune 500 companies. Its impressive popularity and widespread usage can be attributed to its simplicity. No programming background is needed and by using a few simple commands, one can have MINITAB perform sophisticated calculations and analyses. Although all computer statistical packages have certain features in common, they do differ in certain respects. In the next few paragraphs we indicate how to get started in MINITAB.

The first step in getting started is to load the MINITAB program into the computer. If you are using a personal computer (PC), you simply load the program diskettes into the disk drive(s) or load the program from a hard drive. Then you type MINITAB and press the **Enter/Return** button. If you are using a mainframe computer, you probably have to enter an identification code or account number before you can get onto the system. You should consult with your instructor or the computer center director who can provide you with the exact details on how to access the system.

Throughout the text, we will assume that the reader has access to the standard version of MINITAB (mainframe or microcomputer). All of the commands and instructions reflect this assumption. MINITAB Release 10XTRA is now available in a Windows version. The input/output commands differ slightly. The specifics are discussed in a supplementary MINITAB for WINDOWS pamphlet that accompanies this text.

After the MINITAB program is loaded and you have typed MINITAB, the following ready prompt MTB> will be displayed on the screen indicating that MINITAB is awaiting further instructions. Any information that is entered on the computer terminal is called *input* and any solution provided by MINITAB as a result of our commands is called *output*.

If you are working on a microcomputer (IBM PC or compatible, or Macintosh) that is connected to a printer, then you may wish to print everything from the very beginning of your MINITAB work session. To facilitate this, type PAPER at the very first MINITAB prompt as shown: MTB> PAPER, and hit the **Return/Enter** key. The printer must be turned on when you type the PAPER command. All of your subsequent work, whether it appears currently on the screen or not, will be printed by MINITAB when so instructed. To stop printing from some point onward, type NOPAPER and hit the **Return/Enter** key as shown: MTB> NOPAPER. Any current screen can also be printed by pressing the key marked **Print Screen**.

It is advisable, especially when your output screen will be printed with other computer outputs, to have your name printed on the output. This can be accomplished by typing NOTE: your name at the MTB prompt as shown:

```
MTB> NOTE: your name
```

A MINITAB command that begins with the word "NOTE" is not processed by MINITAB. It is merely for the benefit of the user. It also can be used to enter any comments for your information.

Data are stored in MINITAB in a worksheet that consists of rows and columns. The rows are designated 1, 2, 3, . . . and the columns are labeled C1, C2, C3, ... However, as we shall see shortly, the data do not have to be entered in column format.

First we type the SET C1 command and then press the **Return/Enter** key. This instructs MINITAB to put the data into column 1. MINITAB will then display the DATA> prompt. The user now enters the data.

To illustrate, suppose we want to form a column of data called C1 that will contain the integers 1 through 10. This can be accomplished as follows:

MTB> SET C1	This command instructs MINITAB that you are to enter data in C1.
DATA> 1 2 3 4 5 6 7 8 9 10	These data values can be entered in as many rows as we want.
DATA> END	This command indicates the end of data entry.
MTB> NAME C1 'INTEGERS'	This command names the data entered in C1 as "integers."

When we reach the number 10, there are no more data so we type END and the **Return/Enter** key is pressed.

Data can be entered onto the worksheet either from the keyboard or from previously stored files or it can be generated randomly by MINITAB as in the following:

MTB> RANDOM 30 C2;	Note the semicolon at the end of this command.
SUBC> INTEGER 0 30.	Note the period at the end of this command.
MTB> NAME C2 'RANDOM'	

A semicolon at the end of a MINITAB command instructs MINITAB that a subcommand (SUBC) containing some additional information is to follow. A period at the end of a subcommand instructs MINITAB that all MINITAB commands and subcommands have already been entered.

In this case, the RANDOM command causes MINITAB to randomly generate a large list of numbers. The subcommand INTEGER tells MINITAB to use an algebraic formula to convert each of these numbers into integers between 0 and 30 and then to place the results in C2.

The contents of any of the columns in the worksheet can be examined at any time by using the PRINT command. In our case when we type PRINT C1-C2, MINITAB will respond with the following screen display.

```
MTB> PRINT C1-C2
```

ROW	INTEGERS	RANDOM
1	1	16
2	2	21
3	3	19
4	4	17
5	5	2
6	6	19
7	7	9
8	8	7
9	9	12
10	10	25
11		21
12		13
13		4
14		23
15		19
16		24
17		24
18		8
19		11
20		0
21		1
22		30
23		16
24		30
25		10
26		28
27		13
28		1
29		18
30		30

The contents of columns 1 and 2 would be printed by the printer and at the same time would be shown on the screen. To disengage the printer after a printout, you must give the command NOPAPER.

Even if your computer is not connected to a printer, you may want to save your results so that they can be retrieved for later work. To accomplish this, choose an appropriate file name (up to eight characters) and type SAVE as shown below. If you are using a personal computer and have a disk in a particular drive, for example, if you want to save the data in the worksheet by the name EXAMPLE 1 on your hard drive or on a disk in drive A, you would type

```
MTB> Save 'EXAMPLE1'
```
This command will save the data file as EXAMPLE1.MTW (which stands for

MINITAB WORKSHEET) on the hard disk.

MTB> SAVE 'A : EXAMPLE1'   This command will save the data file on a floppy disk in drive A.

Be careful to use the single quotation marks and the colon exactly as indicated.
    To recall data, type

MTB> RETRIEVE 'EXAMPLE1'   This command is used if the saved file is on the hard disk.

or if the data has been stored on a floppy disk, type:

MTB> RETRIEVE 'A : EXAMPLE1'   This command is used if the saved file is on a floppy disk in drive A.

    After completing your work, you must inform MINITAB that you want to exit it. This is accomplished by typing STOP. The computer will now exit MINITAB and return to its operating system.
    Throughout this book, we will present specific MINITAB computer programs that can be used to analyze concepts discussed within the chapter.

EXERCISES 1.5

The following table gives the sales figures of the Albe Supermarket for the first six days of June.

June	Sales (in dollars)
1	157,802
2	181,362
3	159,576
4	172,887
5	193,216
6	138,076

Perform each of the following MINITAB commands and discuss the output for each command.

1. MTB > SET THE FOLLOWING IN C1
2. DATA> 1 2 3 4 5 6
3. DATA> END
4. MTB > SET THE FOLLOWING IN C2
5. DATA> 157802 181362 159576 172887 193216 138076

```
 6. DATA> END
 7. MTB > NAME C1 'JUNE'
 8. MTB > NAME C2 'SALES'
 9. MTB > LET C3 = SUM(C2)
10. MTB > NAME C3 'TOTAL'
11. MTB > PRINT C1 C2 C3
12. MTB > SAVE 'ASSIGN1'
13. MTB > SAVE 'A : ASSIGN1'
14. MTB > RETRIEVE 'A : ASSIGN1'
15. MTB > STOP
```

1.6 SUMMARY

In this chapter we discussed the nature of statistics and how they are used. We distinguished between descriptive statistics and statistical inference. A brief discussion on the origins of statistics and probability was given. Finally, we pointed out how statistics and probability are constantly gaining in importance because of their ever-increasing wide range of applications. We also mentioned some basic facts about the MINITAB statistical package.

Study Guide

The following is a chapter outline in capsule form. You should now be able to demonstrate your knowledge of the ideas mentioned by giving definitions, descriptions, or specific examples. Page references are given in parentheses.

Statistics involve collecting numerical information called **data**, analyzing it, and making meaningful decisions based on the data. (page 6)

Qualitative data refer to observations that are attributes. (page 6)

Numerical or **quantitative data** represent observations or measurements that are numerical. When data result from a *count*, we have **discrete numerical data**. On the other hand, data resulting from a *measure* of a quantity will usually be **continuous numerical data**. (page 6)

Descriptive statistics involves collecting data and tabulating results using, for example, charts, tables, or graphs to make the data more manageable and meaningful. (page 11)

A **sample** is any small group of individuals or objects selected to represent the entire group called the **population**, where the population is the set of all measurements of interest to the sample collector. (page 11)

A **parameter** is a numerical measurement describing some characteristic of a population, whereas a **statistic** is a numerical measurement describing some characteristic of a sample. (page 11)

Inferential statistics or **statistical inference** is the study of procedures by which we draw conclusions and make decisions or predictions on the basis of a sample. (page 12)

Testing Your Understanding of This Chapter's Concepts

1. A researcher claims that "Now that women have entered the job market and assumed stress-ridden managerial jobs, they too will be subject to the same type of illnesses that until now primarily afflicted men only." Do you agree? Explain your answer.

2. When applying for admission to college, a student must also submit a high school transcript that gives the high school average. Explain how the high school average is both descriptive and inferential.

3. A recent survey by Haber and Sullivan (1995) concluded that 80% of all juvenile delinquents arrested in Parksville for various crimes during the years 1990–1994 were from broken homes. Can one conclude that a juvenile from a broken home in Parksville is likely to be arrested for committing various crimes? Explain your answer.

4. In a study° of the incidence of heart attacks in men, a researcher noted that men who had only one job, on the average, had somewhat fewer heart attacks than men who had two jobs. The researcher concluded that having more than one job increases a man's chance of having a heart attack. Do you think that the facts support the conclusion? Explain your answer.

5. *Is Smoking Dangerous to Your Health?* The American Cancer Society claims that statistics overwhelmingly support this claim. Yet the tobacco industry claims that the statistics are not conclusive. Explain how both sides in the controversy interpret the results. (See newspaper article on page 17.)

6. A cardiologist claims, "Statistics show that the average blood serum cholesterol level of Americans is dropping because the foods we eat contain less fatty items." Based on your own research, would you say that the facts support this claim? Explain your answer.

7. An insurance company official claims that 90% of the company's policy holders renew their car insurance with the company when their policy expires. If we randomly select 100 of the company's car insurance policy holders, would we expect exactly 90 of them to renew their car insurance policy with the company when it expires? Explain your answer.

8. The Stapleton Medical Association claims that 8 of every 11 of its doctors carry medical malpractice insurance. If you select 11 Stapleton doctors at random, would you expect 8 of them to carry this type of insurance? Explain your answer.

Chapter Test

Multiple Choice

1. After carefully analyzing meteorological data, several scientists have concluded that the Earth has been gradually experiencing warmer temperatures over the past 100 years. This conclusion involves

   a. parametric statistics   b. inferential statistics
   c. descriptive statistics   d. no statistics at all

°Briggs and Rogers, 1994.

2. Refer back to the previous question. Several of the scientists have concluded that during the 21st century our winters will become milder and milder. This conclusion involves
   a. parametric statistics   b. inferential statistics
   c. descriptive statistics   d. no statistics at all

3. The set of *all* measurements of interest to the sample collector is referred to as the
   a. sample   b. population   c. survey
   d. hypothesis   e. none of these

4. A numerical measurement describing some characteristic of a population is referred to as a
   a. parameter   b. statistic   c. sample
   d. variable   e. qualitative data

5. Statistical data that result from a count are often called
   a. continuous numerical data   b. qualitative data
   c. approximate data   d. discrete numerical data
   e. none of these

Supplementary Exercises

6. Consider the newspaper article below. What part of the article is an example of descriptive statistics?

BLOOD SUPPLY AT DANGEROUSLY LOW LEVELS

Nov. 22: According to the greater Metropolitan Blood Council, the blood levels at various hospitals throughout the city stood at 20% of their normal level. Only 837 pints of blood were donated last week. This represents an all-time low record. Administrator Halsey Davis claimed that the drop in blood donations is due to an erroneous fear of the AIDS virus.

These findings have far-reaching implications for those planning to have major surgery in the future and for the general health of our citizens.

Wednesday –November 22, 1989

7. In the previous exercise, what part of the article is an example of inferential statistics?

8. A researcher decided to investigate the reason for the drop in blood donations. The researcher interviewed 5000 college students and determined that the reason was indeed the fear of contracting the AIDS virus. The researcher concluded that this reason seems to be true for the entire population. This conclusion involves
   a. quantitative data only
   b. qualitative data only
   c. descriptive statistics
   d. statistical inference
   e. none of these

9. Which of the following statements most likely involve statistical inference and which most likely involve descriptive statistics?
   a. The warning label that appears on the packages of cigarettes: "Warning: The Surgeon General has determined that cigarette smoking is dangerous to your health."
   b. The average annual cost of malpractice insurance for a neurosurgeon is $239 more in New York City than in Chicago.
   c. The world consumption of oil is increasing so rapidly that new sources of energy will have to be found or we will not have an adequate supply of energy in the near future.
   d. After carefully analyzing the available food supply and the world population, several environmentalists have concluded that the world will experience a world-wide food shortage by the year 2010.

For Test Questions 10 and 11, use the following information: A consumer's group collected data on the prices charged by two large supermarket chains for a dozen large eggs over a four-week period as indicated in the accompanying chart:

Week	Prices Charged by George's Supermarket (cents)	Prices Charged by Bandy's Supermarket (cents)
January 1–7	79	88
January 8–14	84	85
January 15–21	73	72
January 22–28	81	80

10. The conclusion "The average price charged by George's Supermarket for a dozen large eggs is less than the average price charged by Bandy's Supermarket" involves descriptive statistics. True or false?

11. The conclusion "George's Supermarket will generally give you more for your money than Bandy's Supermarket" involves statistical inference. True or false?

12. What part of the following article involves descriptive statistics?

RADON GAS A SILENT KILLER

Dec. 2: Agents from the state's Environmental Protection Agency (EPA) inspected 527 houses yesterday and found 14 of them to be contaminated with the deadly and naturally occurring gas, radon. The gas is entering the houses from underground sources.

Bill Class estimates that about 15% of the homes in the region are contaminated with the radon gases, and called upon the governor to investigate the matter further.

Sunday--December 2, 1990

13. What part of the article involves inferential statistics?

14. A nurse in the Emergency Room at Maimonides Hospital has obtained the following information about Bill McKinley upon his admittance:
    a. Sex: Male
    b. Age: 67 years
    c. Type of medical coverage: Medicare
    d. Blood pressure: 140/70 msg/mm
    e. Pulse: 80 beats per minute
    f. Temperature: 101.8°F (oral)
    g. Medication currently being taken by patient: Cardizan
    h. Religion: Catholic

    Classify each of these variables as qualitative or quantitative. If it is quantitative, determine whether the variable is discrete or continuous.

15. Over the years numerous claims have been made that fluoridation of our water supply is associated with a higher incidence of cancer than occurs with use of non-fluoridated water. After a careful and detailed analysis of the water supply systems of 20 American cities, P.D. Oldham* concluded that there is no significant difference in the cancer rates after adjustments for the nonhomogeneity of the populations are made. Does the accompanying newspaper article indicate that Oldham's analysis of the water supply systems and his conclusions are wrong? Explain your answer.

> ## CANCER AND FLUORIDATION OF OUR WATER SUPPLY
>
> *June 1*: Medical researchers reported yesterday that preliminary studies of the fluoridation of our water system showed an apparent increase in the death rate from cancer per 100,000 of population. Further studies will be undertaken before any definite conclusions can be drawn.
>
> June 1, 1990

16. Give three examples of variables that are actually discrete but might be considered continuous.
17. Explain the difference between discrete and continuous random variables and give three examples of each. Do not use any of the examples given in the text.

Thinking Critically

Some of the following questions involve statistical ideas to be discussed in greater detail in later chapters. Nevertheless, they are presented here to alert you to the type of

---

*Oldham, P.D. "Fluoridation of Water Supplies and Cancer: A Possible Association?" *Applied Statistics*:26, (1977), pp. 125–135.

thinking that is necessary in the statistical analysis of a problem. Based on your own experiences, you should be able to discuss and answer most, if not all, of them.

1. Consider the accompanying newspaper article. Do you agree that the new net helps to protect dolphins or is it possible that the number of dolphins in the sea has already been reduced from earlier catches?

NEW TUNA FISH NET PROVES SUCCESSFUL

Los Angeles (Sept. 2): Responding to claims that tuna fish nets have been inadvertently capturing and killing dolphins also, the tuna fish industry has developed a new kind of net for capturing tuna fish. The new net has been in use for one year. Statistics show that the percentage of dolphins out of the total tonnage of tuna fish caught has dropped radically when compared to the average percentage over the preceding twenty-five years.

Said a spokesperson for the industry "We think that we have the problem under control."

Tuesday – September 3, 1988

2. Suppose we are interested in determining the percentage of women who have at least as many children as their mothers did. Can we obtain this percentage by selecting all of our friends and relatives as a random sample? Explain your answer.

3. Suppose you are interested in determining who are safer drivers in Florida, men or women. Can you go to the Motor Vehicle Bureau and simply analyze the number of accidents involving men or women drivers? Explain your answer.

4. Suppose you are interested in determining the percentage of married men who get along well with their mother-in-law. Can you obtain this percentage by selecting all your relatives and your friends at work as a valid sample? Explain your answer.

5. After analyzing accident statistics from an insurance company, an analyst concludes that "since there are more automobile accidents when the roads are dry than when the roads are covered with ice, it is safer to drive when there is ice on the roads." Do you agree with this conclusion?

6. The Cuban National Commission for the Propaganda and Defense of Havana Tobacco once noted that the human life span has more than doubled since the Tobacco plant was discovered. Is it true that increased use of tobacco will result in a longer life? Explain your answer.

7. In its annual report, the Aviation Council reported that there were three separate incidents in which a plane was struck by lightning during a blinding rainstorm. Furthermore, on clear sunny days there were 25 incidents in which engine failure resulted from birds being sucked into the plane's engines. In view of these findings, the Council concluded that it is less dangerous to fly an airplane during a blinding rainstorm than on a clear sunny day when birds are present. Do you agree with this conclusion? Explain your answer.

8. To decrease the number of accidents, the Highway Patrol randomly stops cars driving on the Clearpark Expressway and subjects the drivers to sobriety tests to determine whether the driver is under the influence of alcohol. In 1994 the Highway Patrol arrested 782 motorists for drunken driving. During 1995 only 645 motorists were arrested for drunken driving. Can we conclude that more people are obeying the law and not driving while intoxicated? Explain your answer.

9. In 1984 there were 23 licensed day-care centers in Dover. In 1994 there were 49 licensed day-care centers. Can we conclude that the reason for the increase in the number of licensed day-care centers in Dover is because a greater percentage of the population is now sending their children to these centers? Explain your answer.

Case Study   Mouthwash and Plaque

In addition to freshening their breath, many people use mouthwash in the belief that it will help them fight plaque—the germ-laden film that coats teeth and gums. Although plaque can be brushed and flossed away quite effectively, some mouth rinses with plaque-fighting chemicals can help finish the job.

A few years ago, Pfizer Inc. ran advertisements claiming that its *PLAX* brand of mouthwash was capable of "removing 300% more plaque than brushing alone." Several competing companies complained to the National Advertisement Division of the Better Business Bureau about the 300% claim. Shortly thereafter, Pfizer Inc. replaced its ad with the new claim, "Removes more plaque than brushing alone."

Soon the Attorneys General of ten states sued the company and its advertising agency for misleading advertisements and unsubstantiated and deceptive product health claims. Although admitting no wrongdoing, the company agreed to halt its claims and to reimburse the state $70,000 in legal costs.

(Adapted from *Consumer Reports*, September 1992, p. 610)

1. Explain what, if anything, is wrong with the claim "**PLAX** is capable of removing 300% more plaque than brushing alone."

2. Explain what, if anything, is wrong with the claim "**PLAX** removes more plaque than brushing alone."

2

THE DESCRIPTION
OF SAMPLE DATA

Chapter Outline

DID YOU KNOW THAT

a a person eats, on average, about 5 gallons of frozen desserts per year: 63% from ice cream, 24% from ice milk, 10% from frozen yogurt, and 3% from sherbet? (*Source:* U.S. Department of Agriculture, 1992)

b the number of cars on the road in America dropped in 1991, the first time since the end of World War II, as can be seen in the following graph:

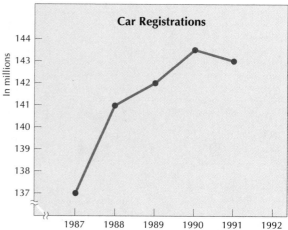

Source: American Automobile Manufacturers' Association, 1992.

c only 16% of the public trusts lawyers? As a matter of fact, in a recent survey (*USA TODAY*, July 29, 1993) it was found that 68% of the

The lead content in our drinking water must be tested frequently since it often exceeds the safe level of 15 ppb. See the discussion on page 120. (© *David R. Frazier Photolibrary*)

public trusts pharmacists, 50% trusts police, 19% trusts local officials, 18% trusts senators, and only 14% trusts congressmen and state officials.

d   19.5 million Americans participated in some type of water sport in 1992? (*Source:* Sporting Goods Manufacturing Association)

As indicated in Chapter 1, descriptive statistics involves collecting data and tabulating them in a meaningful way. In this chapter we indicate how to organize the data and then present the results graphically.

Chapter Objectives

- **To discuss** what is meant by a frequency distribution. This is a convenient way of grouping data so that meaningful patterns can be found. (Section 2.2)

- **To draw** a histogram, which is nothing more than a graphical representation of the data in a frequency distribution. (Section 2.2)

- **To analyze** circle graphs or pie charts, which are often used when discussing distributions of money. (Section 2.3)

- **To apply** frequency polygons and bar graphs, which are used when we wish to emphasize changes in frequency, for example, in business and economic situations. (Section 2.3)

- **To introduce** the graph of a frequency distribution, which resembles what we call a normal curve. (Section 2.3)

- **To analyze** how the area under a curve is related to relative frequency and hence to probability. (Section 2.3)

- **To present** an alternate way of analyzing data graphically by means of stem-and-leaf diagrams, which are particularly well suited for computer sorting techniques. (Section 2.4)

- **To show how** graphs can be misused. (Section 2.5)

- **To explain** the use of index numbers, which are used to show changes over a period of time. (Section 2.6)

- **To show how** we use such words as random sample, frequency, class mark, relative frequency, area under a curve, and pictograph.

S T A T I S T I C S   I N   A C T I O N

Many unions negotiate an escalator clause in their contract with management. When the Consumer Price Index (CPI) or some other index rises, then so will an employee's salary. Thus, it is hoped that the employee's purchasing power can keep up with inflation. How are such index numbers computed? In this chapter we indicate how index numbers are calculated.

Now consider the second newspaper article. The information that is given in grouped data form has important implications

DOLLAR'S BUYING POWER DECLINED MORE THAN 60% SINCE 1970

Washington (June 18): Despite slowing inflation, the purchasing power of the dollar has eroded nearly 60% since 1970, the Conference Board said. The business information firm said an average family of four that earned $10,000 in 1970 now needs $26,450 to maintain the same after-tax buying power. While inflation was cited as the major cause of the change, higher federal income and Social Security taxes were said to account for about one-third of the drop.

June 19, 1989

IT WILL COST YOU $128,730 TO RAISE A CHILD BORN TODAY

Washington (July 28): According to the Family Economic Research Group of the U.S. Department of Agriculture, middle income couples who had babies in 1995 will spend an average of $128,730 by the time the baby is 18 years old as follows:

Age	Avg. Cost per year
0–5	$6820
6–11	6865
12–14	7540
15–17	8000

Friday–July 28, 1995

for middle income couples. Can the same information be presented in a graph?

The information contained in the bar graph below is very important for parents and health officials.

In this chapter we indicate how to draw and interpret graphs properly.

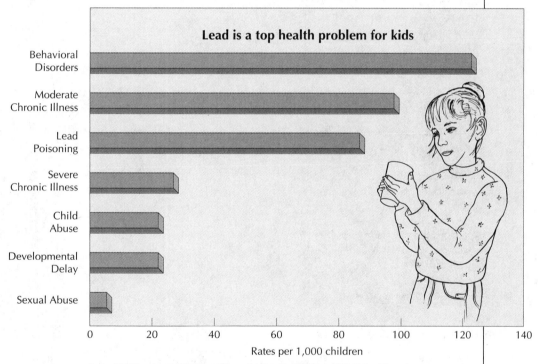

Lead is a top health problem for kids

Rates per 1,000 children

Source: U.S. Department of Health and Human Services. Healthy People 2000. *National Promotion and Disease Prevention Objectives*. Washington, D.C., U.S. Printing Office

2.1   INTRODUCTION

Suppose that John, a student in a sociology class, is interested in determining the average age at which women in New York City marry for the first time (see the accompanying newspaper article). He could go to the Marriage License Bureau and obtain the necessary data. (We will assume that all marriages are reported to the Bureau.) Since there are literally thousands of numbers to analyze, he would want to organize and condense the data so that meaningful interpretations can be drawn from them. How should he proceed?

NEWLYWEDS GET OLDER

Washington (July 19): According to statistics released yesterday by the U.S. Census Bureau, the average age at which Americans marry is on the rise. Fifteen years ago, the median age was 23.2 for men and 20.8 for women.

But in 1984, the age for first marriage rose to 25.4 for men and 23 for women. The highest ever median age at marriage for men was 25.9 years in 1900. For women the previous high was 22 in 1890.

Wednesday–July 20, 1988

One of the first things to do when given a mass of data is to group it in some meaningful way and then construct a frequency distribution for the data. After doing this, we can use various forms or graphs of the distribution so that different distributions can be discussed and compared.

In this chapter we consider some of the common methods of describing data graphically. In the next chapter we will discuss the numerical methods for analyzing data. In each case we will use examples from everyday situations so that you can see how and where statistics is applied.

2.2 FREQUENCY DISTRIBUTIONS

Rather than analyze thousands of numbers, John decides to take a sample. He will examine the records of the Marriage Bureau on a given day and then use the information obtained from this sample to make inferences about the ages of *all* women who marry for the first time in New York City.

Random sample

The most important requirement in taking a sample of this sort is that it be a **random sample**. This means that each individual of the population, a woman who applies for a marriage license in our case, must have an equally likely chance of being selected. If this requirement is not satisfied, one cannot make meaningful inferences about the population based on the sample. Since we will say more about random samples in Chapter 8, we will not analyze here whether or not John's sampling procedure is random.

John has obtained the ages of 150 women for a given day from the Marriage Bureau. Table 2.1 lists the ages in the order in which they were obtained from the Bureau records.

TABLE 2.1   Age at Which 150 Women in New York City Married for the First time

20	17	24	18	16	23	21	17	21	27	23	17	16	19	14
23	26	38	33	20	33	26	22	26	33	26	35	28	35	24
21	18	21	23	22	21	19	23	19	18	19	15	23	21	18
27	40	27	26	22	25	27	25	25	25	30	19	26	32	22
22	19	22	19	24	15	20	22	23	26	23	18	21	20	21
34	41	35	20	29	20	29	27	29	32	29	29	32	28	31
19	22	23	25	23	23	18	19	24	24	21	20	24	22	20
30	31	39	43	38	37	30	37	33	30	36	34	36	32	26
25	23	17	24	18	24	24	21	16	20	18	22	25	24	17
28	29	34	31	25	34	25	36	28	27	31	27	28	30	28

The only thing we can say for sure is that most women were in their 20s and that one was as young as 14 while another woman was as old as 43 years when they first married. However, since the ages are not arranged in any particular order, it is somewhat difficult to conclude anything else. Clearly, the data must be reorganized. We will use a frequency distribution to do so.

Definition 2.1
Frequency distribution
Frequency

A **frequency distribution** is a convenient way of grouping data so that meaningful patterns can be found. The word **frequency** will mean how often some number occurs.

A frequency distribution is easy to construct. We first make a list of numbers starting at 14, the youngest age, and going up to 43, the oldest age, to indicate the age of each woman. This list is the first column. In the second column we indicate tally marks for each age; that is, we go through the list of original numbers and put a mark in the appropriate space for each age. Finally, in the third column we enter the total number of tally marks for each particular age. The sum for each age gives us the frequency column. When applied to our example, we get the distribution shown in Table 2.2.

TABLE 2.2   Construction of a Frequency Distribution for Age at Which 150 Women in New York City Married for the First Time

Column 1	Column 2	Column 3
Age	Tally	Frequency
14		1
15		2
16		3
17		5
18		8
19		9
20		9
21		10
22		10
23		12
24		10
25		9
26		8
27		7
28		6
29		6
30		5
31		4
32		4
33		4
34		4
35		3
36		3
37		2
38		2
39		1
40		1
41		1
42		0
43		1

A number of interesting facts can be seen at a glance from Table 2.2. Since 23 occurred most often, this is the age at which most women married. Also, the frequency began to decrease as age increased beyond the age of 23.

Classes or intervals

The data in Table 2.2 can also be arranged into a more compact form as shown in Table 2.3. In this table we have arranged the data by age groups, also called **classes** or **intervals**. We select a group size of 3 years since this will result in 10 different groups. Although any number of intervals can be used, for this example we have chosen to work with 10. With the possible exception of the first and last interval, classes are usually of equal size. Generally speaking, if we subtract the smallest age from the largest age and divide the results by 10, we will get a number, rounded off if necessary, that can be used as the size of each group.

TABLE 2.3   Age at Which 150 Women in New York City Married for the First Time

Class Number	Ages	Class Mark	Tally	Class Frequency	Relative Frequency
1	14–16	15		6	6/150
2	17–19	18		22	22/150
3	20–22	21		29	29/150
4	23–25	24		31	31/150
5	26–28	27		21	21/150
6	29–31	30		15	15/150
7	32–34	33		12	12/150
8	35–37	36		8	8/150
9	38–40	39		4	4/150
10	41–43	42		2	2/150
			Total frequency =	150	

If we let L equal the largest number and S equal the smallest number, then a guideline to obtain **class width** or size is given by $\dfrac{L - S}{\text{number of classes}}$. In our case, we have

$$\frac{43 - 14}{10} = \frac{29}{10} = 2.9 \quad \text{or} \quad 3 \text{ when rounded}$$

COMMENT   Strictly speaking, the class width represents the difference between the lower class limit of any class and the lower class limit of the next higher class.

COMMENT   When setting up class intervals make sure that no data are deleted. For example, if the largest number is 10 and the smallest number is 1 and we wish to have three classes, the previous formula would give a class width of $\dfrac{10 - 1}{3} = 3$. If we keep the class width as 3 then the classes would be 1–3, 4–6, and 7–9 and the number 10 would have to be placed in a fourth class or be lost completely. If the class width is 4, then the classes would be 1–4, 5–8, and 9–12, thereby avoiding any problems.

Class number

We have labeled the first column of Table 2.3 **class number** because we will need to refer to the various age groups in our later discussions. Thus, we will refer to age group 26 to 28 as class 5. Notice also that each class has an *upper limit*, the oldest age, and a *lower limit*, the youngest age. A **class mark** represents the point that is midway between the limits of a class; that is, it is the midpoint of a class. Thus, 18 years is the class mark of class 2. We have indicated the class mark for each class in the third column of Table 2.3.

Class mark

COMMENT   To obtain the class mark, we have

$$\text{Class mark} = \frac{\text{lower limit} + \text{upper limit}}{2}$$

Thus, the class mark does not necessarily need to be a whole number.

COMMENT   Although we mentioned that equal class intervals are commonly used, this does not necessarily include the first and last classes if we do not wish to lose any data. We can always use larger class widths and have equal class intervals.

Grouped data

Table 2.3, containing a series of intervals and the corresponding frequency for each interval, is an example of **grouped data**. (It should be noted that we can group data using classes so that each represents a single numerical piece of data.) When using grouped data, any piece of data can belong to one and only one class.

Class frequency

In the fifth column of Table 2.3 we have the **class frequency**, that is, the total tally for each class. Finally, the last column gives the relative frequency of each class. Formally stated, we have the following definition.

Definition 2.2

Relative frequency

The **relative frequency** of a class is defined as the frequency of that class divided by the total number of measurements (the total frequency). Symbolically, if we let f_i denote the frequency of class i where i represents any of the classes, and we let n represent the total number of measurements, then

$$\text{Relative frequency} = \frac{f_i}{n} \qquad \text{for class } i$$

Since in our example the total frequency is 150, the relative frequency of class 6, for example, is 15 divided by 150, or $\frac{15}{150}$. Here $i = 6$ and $f_i = 15$. Similarly, the relative frequency of class 7 where $i = 7$ and $f_i = 12$ is $\frac{12}{150}$. The relative frequency of class 10 is $\frac{2}{150}$.

Histogram

Once we have a frequency distribution, we can present the information it contains in the form of a graph called a **histogram**. To do this, we first draw two lines, one horizontal (across) and one vertical (up-down). We mark the class boundaries along the horizontal line and indicate frequencies along the vertical line. We draw rectangles over each interval, with the height of each rectangle equal to the frequency of that class. All of our rectangles have equal widths. Generally, the areas of the rectangles should be proportional to the frequencies.

The histogram for the data of Table 2.3 is shown in Figure 2.1. The area under the histogram for any particular rectangle or combination of rectangles is proportional to the relative frequency. Thus, the rectangle for class 4 will contain $\frac{31}{150}$ of the total area under the histogram. The rectangle for class 7 will contain $\frac{12}{150}$ of the total area under the histogram. The rectangles for classes 9 and 10 together will contain $\frac{6}{150}$ of the total area under the histogram since

$$\frac{4}{150} + \frac{2}{150} = \frac{4 + 2}{150} = \frac{6}{150}$$

The area under a histogram is important in statistical inference and we will discuss it in detail in later chapters.

COMMENT By changing the number of intervals used, we can change the appearance of a histogram and hence the information it gives. (See Exercise 15 at the end of this section.)

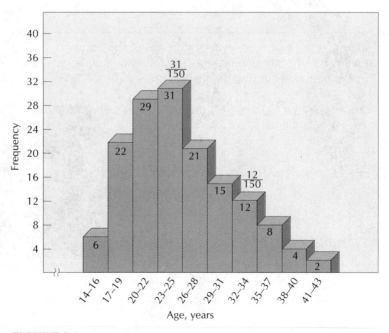

FIGURE 2.1

We can summarize the procedure to be used in the following rules.

RULES FOR GROUPING DATA AND FOR DRAWING HISTOGRAMS

When unorganized data are grouped into intervals and histograms are drawn, the following guidelines should be observed.

1. The intervals should be of equal size (with the possible exception of the end intervals).
2. The number of intervals should be between 5 and 15. (Using too many intervals or too few intervals will result in much of the data not being effectively presented.)
3. The intervals can never overlap each other. If an interval ends with a counting number, then the following interval must begin with the next counting number.
4. Any score to be tallied can fall into one and only one interval.
5. The histogram is then drawn by using rectangles placed next to each other. The rectangles are placed together to indicate that the next interval begins as soon as one interval ends. The height of each rectangle represents its frequency (assuming the widths of the rectangles are equal). There are no gaps between the rectangles drawn in a histogram (with the exception of an interval having a frequency of zero).

To illustrate further the concept of a frequency distribution and its histogram, let us consider the following example.

EXAMPLE 1

Stock Transactions

Manya, a stock broker, keeps a daily list of the number of transactions when her customers buy the stock on margin. The following number of stocks was bought on margin by her customers during the first ten weeks of 1995.

30	28	16	23	22	10	8	15	16	23
29	30	22	15	24	9	5	4	15	24
21	24	16	14	21	13	6	2	23	22
24	21	27	13	20	8	4	3	24	20
22	18	23	23	17	7	6	1	16	17

Graphical analysis often helps a stockbroker analyze transactions. *(Courtesy of UNISYS)*

Construct a frequency distribution for the preceding data and then draw its histogram.

SOLUTION   We will use 10 classes. The largest number is 30 and the smallest number is 1. To determine the class size, we subtract 1 from 30 and divide the result by 10, getting

$$\frac{30 - 1}{10} = \frac{29}{10} = 2.9 \quad \text{or} \quad 3 \text{ when rounded}$$

Thus, our group size will be 3. We now construct the frequency distribution. It will contain six columns as indicated.

Class Number	Number of Stocks	Class Mark	Tally	Class Frequency	Relative Frequency
1	1–3	2	|||	3	3/50
2	4–6	5	ℍℍ	5	5/50
3	7–9	8	||||	4	4/50
4	10–12	11			1	1/50
5	13–15	14	ℍℍ		6	6/50
6	16–18	17	ℍℍ ||	7	7/50
7	19–21	20	ℍℍ	5	5/50
8	22–24	23	ℍℍ ℍℍ ||||	14	14/50
9	25–27	26			1	1/50
10	28–30	29	||||	4	4/50

Total frequency = 50

The relative frequency column is obtained by dividing each class frequency by the total frequency, which is 50. We draw the following histogram, which has the frequency on the vertical line and the number of stocks on the horizontal line.

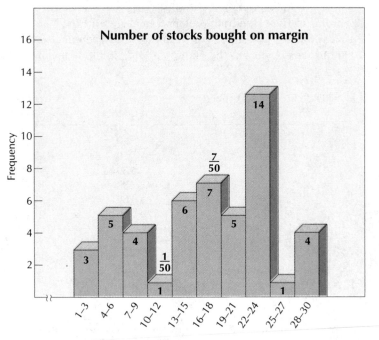

The rectangle for class 6 will contain $\dfrac{7}{50}$ of the total area under the histogram.

Similarly, the rectangle for class 4 contains $\dfrac{1}{50}$ of that area. Finally, the rectangles for classes 1 and 2 contain

$$\frac{3}{50} + \frac{5}{50} = \frac{3+5}{50} = \frac{8}{50}$$

of the total area under the histogram. What part of the area is contained in *all* the rectangles under the histogram?

Relative frequency histogram

Often we may be interested in picturing data that have been arranged in relative frequency form. We can then draw a **relative frequency histogram** by constructing a histogram in which the rectangle heights are relative frequencies of each class.

In both the frequency histogram and the relative frequency histogram, the classes are displayed on the horizontal axes. However, in the frequency histogram, the frequency of each class is represented by a vertical bar whose height equals the relative frequency of the class.

The preceding ideas are illustrated in the following example.

EXAMPLE   2

Young Business Owners

According to Rick Hendricks, regional manager of First National Bank, the age at which business owners first start their businesses has been decreasing. In a random survey of 1000 business owners in the Hampton District the following results were obtained.

Age at Which Owner Started Business	Frequency	Relative Frequency (in decimal form)
16–20	20	20/1000 = 0.02 or 2%
21–25	120	120/1000 = 0.12 or 12%
26–30	230	230/1000 = 0.23 or 23%
31–35	200	200/1000 = 0.20 or 20%
36–40	160	160/1000 = 0.16 or 16%
41–45	80	80/1000 = 0.08 or 8%
46–50	70	70/1000 = 0.07 or 7%
51–55	60	60/1000 = 0.06 or 6%
56–60	50	50/1000 = 0.05 or 5%
61–65	10	10/1000 = 0.01 or 1%
Total Frequency = 1000		*Total* = 100%

The frequency histogram as well as the relative frequency histogram for the preceding data are shown here.

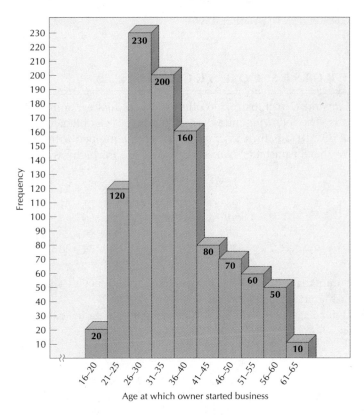

Frequency histogram for age at which owner started business.

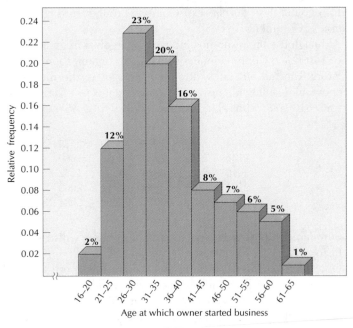

Relative frequency histogram for age at which owner started business.

EXERCISES FOR SECTION 2.2

1. *Car and Driver* magazine often provides fuel-economy information about various cars. The following miles per gallon data were obtained by the EPA from a sample of fifty 1993 Saturn SW2 station wagons equipped with a 1.9-liter engine with dual overhead camshafts. (*Note:* MPG data are for highway driving.)

35	27	30	31	32	36	32	34	39	34
38	41	32	32	37	33	31	37	43	38
31	39	35	33	34	35	38	27	26	36
36	27	29	37	33	31	29	28	33	30
30	29	30	29	31	30	34	33	31	32

   Summarize the data by constructing a frequency distribution and then drawing its histogram. (Use 10 classes.)

2. Refer back to Exercise 1. Miles-per-gallon data for city driving with the cars were as follows:

21	26	24	25	29	25	25	24	21	24
23	24	23	26	20	20	29	20	28	28
29	25	29	23	23	26	21	27	21	25
20	21	20	22	24	23	24	29	24	26
39	27	22	20	25	29	23	22	22	27

   Summarize the data by constructing a frequency distribution and then drawing its histogram. (Use 10 classes.) Compare your results with those obtained in Exercise 1. Comment.

3. Although the life insurance industry is largely unregulated, the National Association of Insurance Commissioners (NAIC), an organization made up of insurance regulators from all 50 states, attempts to regulate the industry. A 1995 survey of 35 companies selling life insurance in most states revealed the following premium costs for a $100,000 whole-life insurance policy for a 45-year-old male nonsmoker.

$3504	$2055	$1858	$2053	$2075	$1600	$1800
2156	2151	2130	2059	1880	2035	1778
2016	2135	2020	1898	1652	1708	1955
2101	2061	2082	1672	2063	1514	1960
2113	2425	2073	2002	2094	1743	1719

   Summarize the data by constructing a frequency distribution and then drawing its histogram. (Use 10 classes.)

4. Officials from the paint industry estimate that more than half of America's home owners paint or stain their houses by themselves (thereby saving thousands of dollars). Since a paint brush uses less paint than a roller, and a spray gun uses the least paint of all, the manufacturer must label each gallon with its approximate spreading

rate depending on the method of application and the type or condition of the surface. The research department of one company painted 60 homes with a particular type of paint and reported the following spreading rate per U.S. gallon.

421	495	388	407	451	440	427	427	428	437	402	409
408	376	461	412	426	427	433	436	439	446	423	419
462	388	422	418	431	437	417	445	426	402	427	423
427	401	437	401	428	418	423	447	483	412	436	432
403	412	444	447	433	401	432	423	401	417	407	409

Summarize the data by constructing a frequency distribution and then drawing its histogram. (Use 10 classes.)

5. The Federal Communications Commission oversees the cable-TV industry and the nation's 11,000 cable franchises. Since the industry was deregulated in 1986, the average monthly bill has risen to $18.85, which is considerably higher than the $11.09 monthly bill in 1988. A survey of the monthly bill of 40 viewers indicated the following charges:

$17.49	$22.79	$21.17	$20.18	$18.76	$24.79	$20.62	$20.57
18.76	18.99	15.26	21.23	19.21	21.18	19.82	21.43
19.23	19.46	16.02	18.48	18.84	20.71	18.69	19.85
14.69	17.69	18.28	17.62	19.81	23.79	17.93	18.48
15.21	16.86	17.68	19.06	20.72	18.71	19.41	20.29

Summarize the data by constructing a frequency distribution and then drawing its histogram. (Use 10 classes.)

6. Credit card purchases and returns are costly to retail establishments. During the first 100 days of 1996, the Marcy Department Store chain reported the following number of daily credit card returns of merchandise:

36	57	75	73	37	60	56	24	48	28
32	49	63	35	48	57	57	25	29	79
35	35	47	39	59	68	71	33	77	35
28	42	41	24	35	75	68	29	68	74
42	38	52	25	32	77	76	51	34	38
40	26	33	29	37	37	34	48	48	75
35	25	37	62	46	29	45	45	44	36
52	34	74	50	56	48	77	37	41	28
37	31	79	44	67	36	63	36	37	28
35	24	62	29	80	38	39	48	29	59

Summarize the data by constructing a frequency distribution and then drawing its histogram. (Use 10 classes.)

7. The number of pounds of recyclable glass collected weekly from 60 auto crushing machines located in various cities of the United States is as follows:

880	1370	1390	760	940	790	1110	1030	1110	1210
860	880	680	120	620	1210	810	1080	480	950
950	640	1790	610	790	1060	610	960	690	1500
290	910	1520	1530	1540	990	370	850	280	460
770	630	1090	1950	730	640	210	780	1450	790
1310	650	1140	1690	1460	430	1400	620	2000	1200

a. Construct a frequency distribution and its histogram using 10 classes.
b. Construct a frequency distribution and its histogram using only 5 classes.
c. Construct a frequency distribution and its histogram using 15 classes.
d. Compare the histograms in parts (a), (b), and (c). What information, if any, is lost by using fewer classes? by using too many classes?

8. A dress company employs 40 workers, each of whom completes similar products in the company's two factories. In each factory there are 20 employees. The number of products completed by each employee during a given time period in each factory is as follows:

	Factory A						Factory B			
17	24	16	28	11		7	26	11	9	24
25	5	7	10	30		17	14	32	5	28
9	4	31	21	13		19	25	28	31	12
35	26	21	19	28		31	27	25	36	19

a. Construct frequency distributions and histograms for each factory. (Use 8 intervals.)
b. Combine the data for both factories and construct a frequency distribution and its histogram. (Use 8 intervals.)
c. Compare the histograms in parts (a) and (b). Comment.

9. Refer back to Exercise 8. The personnel manager decides to change the working conditions in Factory A to determine what effect this will have on production. She installs new lighting facilities, new air conditioning, carpeting, piped music, and a new coffee machine. She now notices that the production of the employees in Factory A is:

11	21	33	15	17	18	25	19	28	28
14	21	29	17	21	16	34	16	28	21

a. Construct the frequency distribution and its histogram for the new data. (Use 8 intervals.)
b. Compare the new histogram with the original histogram for Factory A and the histogram for the combined data. Comment.

10. Thirty entering freshmen at Bologna University received the following scores on the mathematics portion and the verbal portion of the Scholastic Aptitude Test:

Mathematics Scores					Verbal Scores				
631	531	478	612	686	560	689	582	578	568
438	523	593	676	583	577	628	646	557	589
383	613	601	511	612	574	676	598	643	574
449	486	626	676	610	600	591	557	571	589
573	503	612	628	633	604	574	540	584	594
486	517	572	692	563	594	590	559	587	673

a. Construct histograms for both sets of scores. (Use 10 intervals.)
b. By looking at the histograms can we tell if the scores are approximately consistent?

11. Why is it important that classes be of equal width?

12. Juanita Rodriguez and Lloyd Black are employed by the Gale Media Corporation. Their job is to promote newspaper home delivery. Each day they telephone prospective customers. During the month of September 1995 each was asked to promote a particular new deal. The number of new accounts opened daily by each worker is as follows:

Juanita Rodriguez					Lloyd Black				
38	27	27	61	47	39	41	58	44	37
41	55	38	47	73	53	59	60	66	29
57	53	43	38	51	58	38	53	36	43
61	41	52	63	29	53	53	55	27	41
33	27	27	42	38	47	35	39	38	53
43	63	32	71	52	56	45	37	45	33

a. Construct a histogram for each worker. (Use 10 intervals.)
b. Who is a better employee (in terms of new accounts opened)? Explain your answer.

13. The ages of 200 runners in a marathon were recorded. The results are given here:

Age (in years)	Frequency
Over 78	11
70–77 inclusive	14
62–69 inclusive	19
54–61 inclusive	21
46–53 inclusive	32
38–45 inclusive	24
30–37 inclusive	27
23–29 inclusive	34
15–22 inclusive	18

Draw the histogram for the data and answer the following:

a. What part of the area is below, that is, to the left of the 46 to 53 years category?

b. What part of the area is in the 38 or above category?

c. What part of the area is in the 30 to 53 years category?

d. What part of the area is in the 70 to 77 years category?

e. What part of the area is *not* in the 70 to 77 years category?

14. There are many companies (in addition to the U.S. Postal Service) that offer next-day delivery service for packages. To determine the popularity of such services, a survey° of 50 randomly selected businesses in Chicago indicated that these companies used next-day delivery services the following number of times on July 12, 1995:

11	37	10	2	25	3	14	5	17	1
21	6	16	12	11	10	10	26	15	14
29	23	12	7	28	9	3	0	7	4
15	17	14	16	2	8	8	12	11	10
10	27	14	17	7	33	3	22	15	10

Draw the histogram for the data (using 10 intervals, each of length 3) and then answer the following:

a. What part of the area is below, that is, to the left of 12?

b. What part of the area is 16 or above?

c. What part of the area is between 4 and 27 inclusive?

d. What part of the area is either below 3 or above 36?

e. What part of the area is *not* between 12 and 19?

f. What part of the area is above 28?

15. a. Refer back to the data of Example 1 on page 46. Construct a frequency distribution and histogram using only (i) 5 classes (ii) 15 classes.

b. Compare the histograms obtained in part (a) with the histogram given on page 47. Comment.

16. Although there are many potent medications for arthritis on the market today, patients taking these drugs often experience some sort of reaction, ranging from mild to severe. Medical researchers at ULAC administered a new experimental drug to 50 subjects. They recorded the following number of minutes elapsed before the drug showed any effect:

77	59	10	70	54	79	67	38	42	46
75	55	60	38	45	56	28	96	74	86
79	41	45	68	23	47	65	82	78	54
84	83	53	75	61	23	31	38	68	32
7	57	82	46	38	51	42	45	21	89

°Bailey and Brokaw, 1995.

a. Construct a relative frequency distribution for this data.

b. Draw a relative frequency histogram for this data.

2.3   OTHER GRAPHICAL TECHNIQUES

In the preceding section we saw how frequency distributions and histograms are often used to picture information graphically. In this section we discuss some other forms of graphs that are often of great help in picturing information contained in data.

Bar Graphs and Pictographs

Bar graph
Pictograph

The **bar graph** and a simplified version, the **pictograph**, are commonly used to describe data graphically. In such graphs *vertical bars* are usually (but not necessarily) used. The height of each bar represents the number of members, that is, the frequency, of that class. The bars are often also drawn horizontally. We illustrate the use of such graphs with the following examples.

E X A M P L E   1   The number of people calling the police emergency number in New York City for assistance during a 24-hour period on a particular day was as follows:

	Time	
Starting at	Ending at (up to but not including)	Number of Calls Received
12 midnight–	2 A.M.	138
2 A.M.–	4 A.M.	127
4 A.M.–	6 A.M.	119
6 A.M.–	8 A.M.	120
8 A.M.–	10 A.M.	122
10 A.M.–	12 noon	124
12 noon–	2 P.M.	125
2 P.M.–	4 P.M.	128
4 P.M.–	6 P.M.	131
6 P.M.–	8 P.M.	139
8 P.M.–	10 P.M.	141
10 P.M.–	12 midnight	140

The bar graph for these data is shown in Figure 2.2.

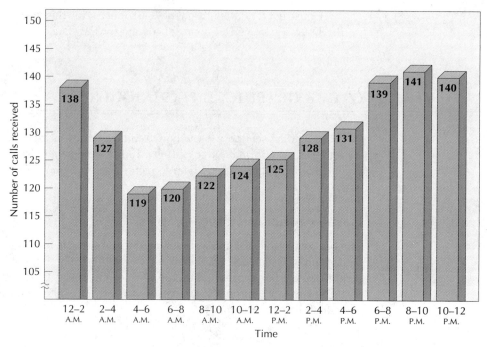

FIGURE 2.2

From the bar graph we see that the least number of calls was received during the hours of 4 A.M. to 6 A.M. The number of calls received after that period steadily increased until a maximum occurred during the 8 P.M. to 10 P.M. period.

When such statistical information is presented in a bar graph rather than in a table, it can be readily used by police officials to determine the number of police officers needed for each time period.

COMMENT   Although the time periods in Example 1 are consecutive, the bars in the bar graph are drawn apart from each other for emphasis and ease in interpretation.

COMMENT   Note that the beginning of the vertical line in Figure 2.2 is drawn with a jagged edge. We should always start a vertical scale with 0. However, when all the frequencies are large, we insert this jagged edge in the beginning to indicate a break in the scale.

Often the bars of the bar graph are placed side by side or superimposed one upon the other for easy comparison. This is shown in the bar graphs in Figures 2.3 through 2.5. Also, note that the bars in Figure 2.3 have been drawn horizontally.

Several variations of the bar graph are commonly used. In such modifications, columns of coins, pictures, or symbols are used in place of bars. When symbols or pictures are used, the bars are sometimes drawn horizontally. We call the resulting graph a *Pictograph*   **pictograph**. Pictographs do not necessarily have to be drawn in the form of a bar graph. Several pictographs are shown in Figures 2.6 through 2.8.

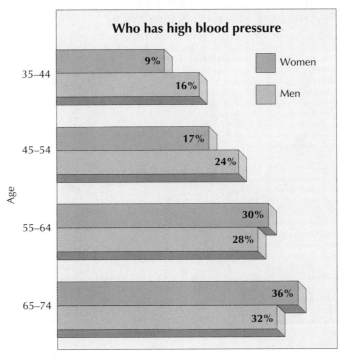

Source: U.S. Bureau of the Census.

FIGURE 2.3

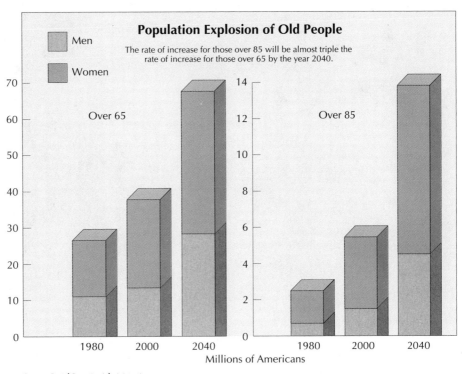

Source: Social Security Administration

FIGURE 2.4

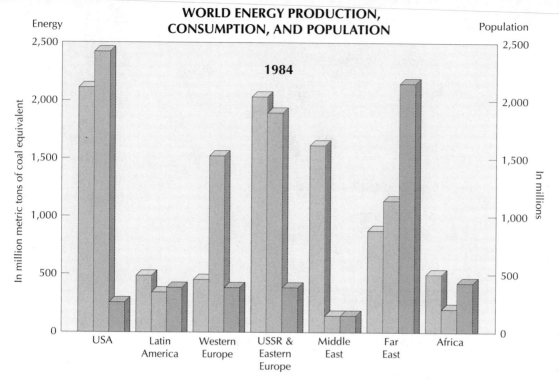

WORLD ENERGY PRODUCTION, CONSUMPTION, AND POPULATION

Energy
In million metric tons of coal equivalent

Population
In millions

1984

USA Latin America Western Europe USSR & Eastern Europe Middle East Far East Africa

FIGURE 2.5

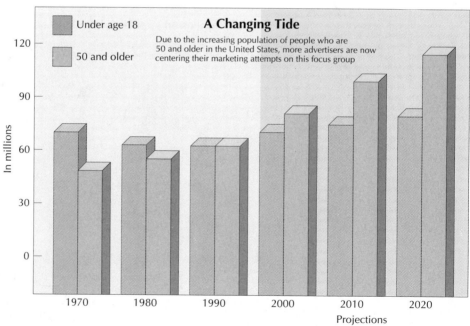

A Changing Tide

Under age 18

50 and older

Due to the increasing population of people who are 50 and older in the United States, more advertisers are now centering their marketing attempts on this focus group

In millions

1970   1980   1990   2000   2010   2020

Projections

Source: U.S. Bureau of the Census.

A Changing Tide
Because of a shift in the numbers of people who are 50 and older in the United States, more advertisers are focusing on this group in their marketing efforts.

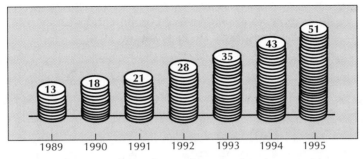

FIGURE 2.6 Expenditure for social services by a large northeastern city government.

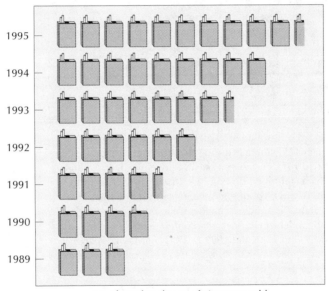

Number of packages of cigarettes sold

FIGURE 2.7 The number of packs of cigarettes sold weekly in the Magway Bus Terminal. Each symbol represents 200 packages.

Summarizing

Sometimes we use a bar graph to picture numerical facts, whereas other times we use a histogram for the same purpose. In a bar graph we use a series of bars in the same direction, either all vertically or all horizontally. The length of a bar that represents some

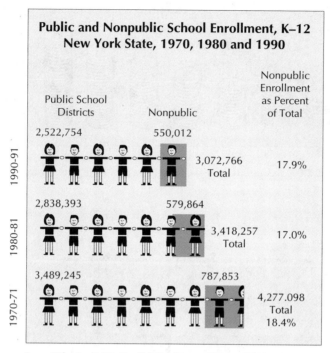

Source: "Education Statistics New York State," January 1993, State Education Department

FIGURE 2.8   Enrollment in elementary and secondary schools increased annually during the 1960s. However, beginning in 1971 enrollment began to decline. This declining trend, which continued through the 1980s, was due both to reductions in the birth rate in recent years and outmigration from the state. Nonpublic school enrollment was expected to decline at a much slower rate than public school enrollment. Between 1983 to 1984 and 1990 to 1991, nonpublic school enrollment declined by 2.6%, whereas public school enrollment decreased by 4.4%.

numerical fact depends on the size of the number that it represents. We must start the scale at zero and use equally spaced intervals so that the approximate size of a number can be read. A histogram is simply a bar graph in which the bars are placed next to each other in order to show that when one interval ends the other begins.

Line graphs
Broken line graph

COMMENT   Often, to emphasize changes in a particular item such as the price of a stock, or temperature changes, we use a **line graph** (sometimes called a **broken line graph**) where a point is used to represent each numerical fact. The consecutive points on the graph are joined together with line segments. If the line segment rises, this indicates that the item being analyzed is increasing. The use of such line graphs will be illustrated in the exercises. It should be noted that line graphs are not used to compare different items.

Frequency Polygons

Another alternate graphical representation of the data of a frequency distribution is a *frequency polygon*. Here again, the vertical line represents the frequency and the horizontal line represents the class boundaries.

Definition 2.3 *Frequency polygon*	If the midpoints (class marks) of the tops of the bars in a bar graph or histogram are joined together by straight lines, then the resulting figure, without the bars, is a **frequency polygon**.

E X A M P L E   2   Draw the frequency polygon for the data of Example 1 on page 46.

SOLUTION   Since the histogram has already been drawn (page 47), we place dots on the midpoints of the top of each rectangle and then join these dots. The result is the frequency polygon shown in Figure 2.9.

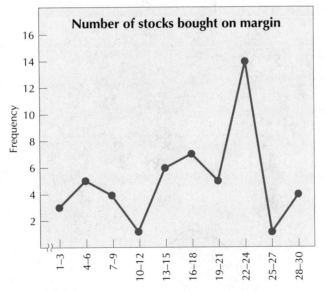

FIGURE 2.9

COMMENT   Since frequency polygons emphasize changes (rise and fall) in frequency more clearly than any other graphical representation, they are often used to display business and economic data.

E X A M P L E   3   Consider the frequency polygon showing the distribution of the speeds of cars as they passed through a particular speed enforcement station (radar trap). See Figure 2.10.

Normal curve
Normal distribution

There are many frequency distributions whose graphs resemble a bell-shaped curve as shown in Figure 2.11. Such graphs are called **normal curves** and their distributions are known as **normal distributions**. Since many things that occur in nature are normally distributed, it is no surprise that they are studied in great detail by mathematicians. We will discuss this distribution in detail in a later chapter.

It should be noted that histograms can also be uniform. This will occur when every value appears with equal frequency. Histograms can be skewed to the left or right where one tail is stretched out longer than the other and the direction of skewness is on the side of the longer tail. See the graphs on pages 45, 47, and 49.

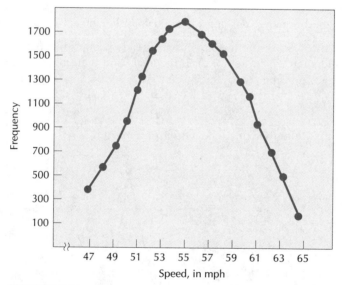

FIGURE 2.10

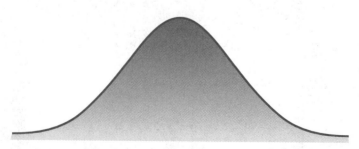

FIGURE 2.11   A normal curve.

Cumulative Frequency Histograms and Cumulative Frequency Polygons

Suppose we analyze the data presented in the following table that represents the weights of 250 army recruits admitted to a particular training base.

Weight (rounded to the nearest lb)	Frequency (number)
131–140	23
141–150	43
151–160	59
161–170	69
171–180	56
	250

A histogram representing the grouped data is shown in Figure 2.12. Based on the frequency distribution and the histogram, we can say that the weights of 23 of the

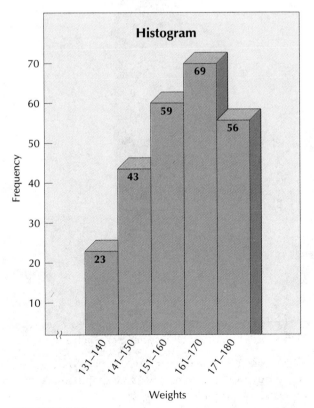

FIGURE 2.12

recruits were in the interval 131 to 140 pounds, that the weights of 43 of the recruits were in the interval 141 to 150 pounds, and so forth.

Often we are interested in answering questions of the type, "How many of the army recruits weighed less than or equal to a certain weight?" For example, suppose we wanted to know, "How many of the recruits weighed less than or equal to 170 pounds?" We can answer this question by adding or "accumulating" the frequencies in the grouped data. Thus, by adding the frequencies for the four lowest intervals, 23 + 43 + 59 + 69, we find that 194 of the recruits weighed 170 pounds or less. A histogram that displays these "accumulated" figures is called a **cumulative frequency histogram**. For our example, the cumulative frequency histogram is given in Figure 2.13.

Cumulative frequency histogram

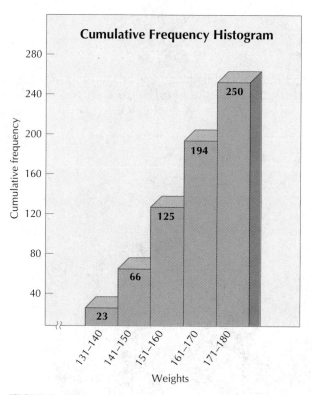

FIGURE 2.13

COMMENT The frequency scale for our cumulative frequency histogram will go from 0 to 250, which represents the total frequency for our data.

Cumulative frequency polygon

We can also draw a cumulative frequency polygon for the preceding data. A **cumulative frequency polygon** is simply a line graph connecting a series of points that answer the question mentioned earlier, namely, "How many of the army recruits weighed less than or equal to a certain weight?" Such a graph is shown on the next page.

We construct the cumulative frequency polygon as follows: For the interval 171 to 180, a point is placed at the upper right of the bar to show that 250 of the recruits weighed 180 pounds or less. For the interval 161 to 170, a point is placed at the upper right of the bar to show that 194 of the recruits weighed 170 pounds or less.

We continue this process of placing dots for all five intervals. You will notice that the last dot is placed at the bottom left of the interval 131 to 140 since "0 of the recruits weighed 130 pounds or less." The line graph connecting these six points represents the cumulative frequency polygon.

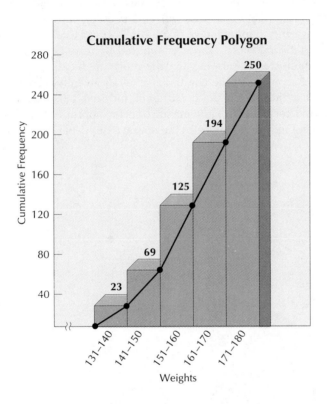

COMMENT   It is also possible to label the vertical scale of a cumulative frequency histogram (or polygon) in a slightly different manner, namely, one involving percents. This will be done in Chapter 3 when we discuss percentiles.

Ogives   Cumulative frequency polygons are also known as **ogives** (pronounced "oh-jive"). Any frequency polygon can be converted into an ogive by simply replacing the frequency with cumulative frequency. As indicated earlier, the cumulative frequency for any given class is obtained by adding the frequency for that class with the frequencies for *all* classes

Cumulative relative   of smaller values. (A **cumulative relative frequency distribution** combines the cu-
frequency   mulative frequency with the relative frequency.) The vertical scale is used to indicate
distribution   frequency. However, we must adjust the scale so that the total of all individual

frequencies will fit. The horizontal scale gives the class boundaries. Ogives are often used when we want to know how many scores are above or below some level.

Circle Charts or Pie Charts

Circle chart
Pie chart

A common method for graphically describing qualitative data is the **circle chart** or **pie chart**. A circle (which contains 360°, or 360 degrees) is broken up into various categories of interest in the same way as one might slice a pie. Each category is assigned a certain percentage of the 360° of the total circle, depending on the data.

The magazine clipping shown in Figure 2.14 is an example of a pie chart that reveals interesting statistics. Since crimes committed by juveniles are on the increase, many people are interested in knowing how the courts are handling such cases. Are the courts too liberal? According to the chart, 36.6% of all juveniles taken into police custody in 1988 were handled within the department and released. Also, 55.9% of the juveniles were referred to juvenile court jurisdiction for appropriate action. Similarly, the pie chart given in Figure 2.15 indicates the reason that people gave for possessing a handgun or a pistol.

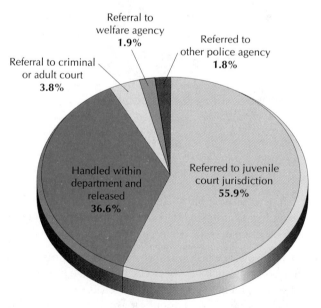

Source: U.S. Department of Justice, Federal Bureau of Investigation, *Uniform Crime Reports for the United States.* 1988 (Washington, D.C.) U.S. Government Printing Office, 1989.

FIGURE 2.14   Percent distribution of juveniles taken into police custody by method of disposition, 1988.

Source: *Sourcebook of Criminal Justice Statistics.* U.S. Government
Printing Office, Washington, D.C., p. 204.

FIGURE 2.15   Gun owners' reasons for possessing a handgun
or pistol in the United States, 1995.

The information contained in both of these charts has important implications for
our criminal justice system in particular and our society in general.

To illustrate the procedure for drawing pie charts, consider the data given in Table
2.4, which indicates the monthly living expenses of a college student at a state university
during 1995.

TABLE 2.4   Monthly Living Expenses in 1995 for a
College Student at a State University

Item	Amount (in dollars)
Food	100
Car and Transportation	75
Utilities	40
Entertainment	85
Laundry	20
Miscellaneous	40
	Total = 360

Since the total expenditure was $360 and there are 360° in a circle, we can construct the pie chart directly (without any conversions). We draw a circle and partition it in such a way that each category will contain the appropriate number of degrees, as shown in Figure 2.16.

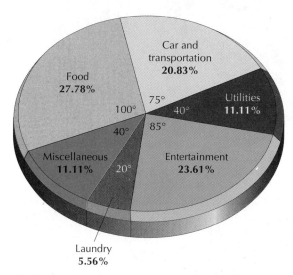

FIGURE 2.16   Pie chart showing the monthly living expenses in 1995 for a college student at a state university.

Since it is more convenient to work with percentages than with amounts of money, we may want to convert the amount of money spent in each category into percentages. Thus, the student's money was spent as follows:

Relative Frequency

$$\frac{100}{360} = 0.2778, \text{ or } 27.78\%, \text{ for food}$$

$$\frac{75}{360} = 0.2083, \text{ or } 20.83\%, \text{ for the car and transportation}$$

$$\frac{40}{360} = 0.1111, \text{ or } 11.11\%, \text{ for utilities}$$

$$\frac{85}{360} = 0.2361, \text{ or } 23.61\%, \text{ for entertainment}$$

$$\frac{20}{360} = 0.0556, \text{ or } 5.56\%, \text{ for laundry}$$

$$\frac{40}{360} = 0.1111, \text{ or } 11.11\%, \text{ for miscellaneous items}$$

We have indicated these percentages in the pie chart of Figure 2.16.

COMMENT   We convert a fraction into a decimal by dividing the denominator, that is, the bottom number, into the numerator, that is, the top number. Thus, $\frac{100}{360}$ becomes

$$
\begin{array}{r}
0.27777 \\
360\overline{)100.00000} \\
\underline{720} \\
2800 \\
\underline{2520} \\
2800 \\
\underline{2520} \\
2800
\end{array}
$$

When rounded, this becomes 0.2778. This number is written in percentage form as 27.78%.

To round decimals, we use the following rule.

RULE • Rounding Decimals

1. Underline the digit that appears in the position to which the number is to be rounded.
2. Examine the first digit to the right of the underlined position.
   a. If the digit is 0, 1, 2, 3, or 4, replace all digits to the right of the underlined position by zeros.
   b. If the digit is 5, 6, 7, 8 or 9, add 1 to the digit in the underlined position and replace all digits to the right of the underlined position by zeros.
3. If any of these zeros (from step 2) are to the right of the decimal point, omit them.

Let us see how this rule is used.

E X A M P L E   4   1. Round 61.379 to the nearest hundredth.

We underline the digit that appears in the position to which the number is to be rounded:

$$61.3\underline{7}9$$

Since the digit to the right of the underlined position is 5 or more, we add 1 to the digit in the underlined position. Thus, 61.379 rounded to the nearest hundredth is 61.38.

2. 0.0792 rounded to the nearest thousandth is 0.079.
3. 36.746 rounded to the nearest whole number is 37.

COMMENT The sum of the percentages may not necessarily be 100%. This discrepancy is due to the rounding off of numbers.

We further illustrate the technique of drawing pie charts by working several examples.

EXAMPLE 5 During 1995 a nationwide auto-leasing company sold 15,000 cars to the individuals who had originally leased them. The types of cars involved are shown in the following frequency distribution.

Type of Car	Number Sold
Manufactured by Japanese companies	3100
Manufactured by General Motors	4800
Manufactured by Chrysler Corp.	2000
Manufactured by Ford Motor Co.	1150
Manufactured by American Motors	850
Manufactured by German companies	2330
Manufactured by other companies	770
	15,000 = *Total sold*

Draw the pie chart for these data.

SOLUTION We first convert the numbers into percentages by dividing each number by the total 15,000. Thus, we have the following results.

Type of Car	Number Sold	Percentage of Total
Japanese companies	3100	$\frac{3100}{15000} = 0.2067$, or 20.67%
General Motors	4800	$\frac{4800}{15000} = 0.32$, or 32%
Chrysler Corp.	2000	$\frac{2000}{15000} = 0.1333$, or 13.33%
Ford Motor Co.	1150	$\frac{1150}{15000} = 0.0767$, or 7.67%

American Motors	850	$\frac{850}{15000} = 0.0567$, or 5.67%
German companies	2330	$\frac{2330}{15000} = 0.1553$, or 15.53%
Other companies	770	$\frac{770}{15000} = 0.0513$, or 5.13%

Now we multiply each percentage by 360° (the number of degrees in a circle) to determine the number of degrees to assign to each part. We get

$$0.2067 \times 360° = 74.41°, \quad \text{or} \quad 74°, \quad \text{for Japanese companies}$$
$$0.32 \quad \times 360° = 115.20°, \quad \text{or } 115°, \quad \text{for General Motors}$$
$$0.1333 \times 360° = 47.99°, \quad \text{or} \quad 48°, \quad \text{for Chrysler Corp.}$$
$$0.0767 \times 360° = 27.61°, \quad \text{or} \quad 28°, \quad \text{for Ford Motor Co.}$$
$$0.0567 \times 360° = 20.41°, \quad \text{or} \quad 20°, \quad \text{for American Motors}$$
$$0.1553 \times 360° = 55.91°, \quad \text{or} \quad 56°, \quad \text{for German companies}$$
$$0.0513 \times 360° = 18.47°, \quad \text{or} \quad 18°, \quad \text{for other companies}$$

Then we use a protractor and compass to draw each part in order, using the appropriate number of degrees. In our case we obtain the pie chart represented in Figure 2.17.

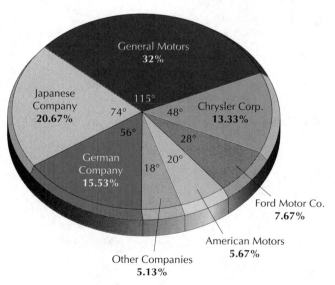

FIGURE 2.17

E X A M P L E   6

Oil Reserves

It has been estimated that the number of billions of barrels of oil in the Western Hemisphere, excluding Alaska, is given by the pie chart shown in Figure 2.18. Assuming that there are 130 billion barrels of oil in reserve, answer the following:

  a. How many barrels are in reserve in the United States?
  b. How many barrels are there in Mexico?
  c. How many barrels are there in Canada?

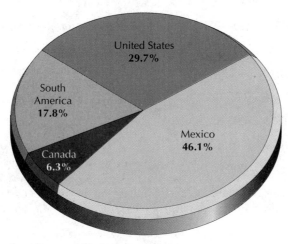

Source: Department of the Interior

FIGURE 2.18

SOLUTION

  a. The United States has 29.7% of the 130 billion barrels. Thus, since

$$0.297 \times 130 = 38.61$$

the United States has 38.61 billion barrels of oil. (In decimal form 29.7% is written as 0.297.)

  b. Mexico has 59.93 billion barrels of oil since

$$0.461 \times 130 = 59.93$$

  c. Canada has 8.19 billion barrels of oil since

$$0.063 \times 130 = 8.19$$

We summarize the procedure to be used in constructing pie charts:

RULES FOR CONSTRUCTING A PIE CHART OR CIRCLE GRAPH

1. Determine all the categories that are of interest from the data.
2. For each category determined in step 1, calculate its relative frequency.
3. Draw a circle and assign a slice of the circle to each category. The size of each slice should be proportional to the fraction of observations in that category. Also, the central angle should be an angle whose measure is 360° times the relative frequency for that category. The sum of the measures of all the central angles should be 360° (except for possible rounding errors).
4. Place an appropriate label in each category and indicate the percentage of the total number of observations in the category. This can be found by using the fact that for each category,

$$\text{Percentage} = \text{relative frequency} \times 100$$

The sum of all the percentages must always be 100% (except for possible rounding errors).

COMMENT   Again, we wish to point out that pie charts are commonly used to summarize qualitative (or categorical) data. Some people believe that pie charts are more difficult to read than other graphs; however, such charts are particularly useful when discussing distributions of money.

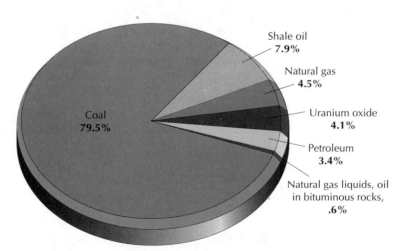

Statistics show that America has been using its natural resources at an increasing rate. As a result, government officials are looking for alternate sources of energy. The circle graph indicates some possible sources.

With modern computer graphics technology, pie charts can easily be drawn using many of the available statistical packages. Pictured above is a person analyzing a computer drawn pie chart. *(Courtesy of UNISYS)*

EXERCISES FOR SECTION 2.3

1. Consider the newspaper article on the next page. Draw a bar graph to picture the information about the number of confirmed rabid animal cases.

2. Different animals react differently to strangers. In a survey of how dogs and cats react to strangers, *USA Today* (Aug. 10 and Aug. 11, 1993) obtained the following results:

Type of Reaction	Dogs (%)	Cats (%)
Friendly	70	51
Aloof	13	35
Hostile	14	6
Neurotic	3	8

a. Draw a bar graph and a pie chart to picture this information for both dogs and cats.

b. Which graph (bar graph or pie chart) makes comparisons easier? Explain your answer.

FATAL RABIES DISEASE RETURNS

New York (Aug. 15): The rabies disease in New York is re-emerging after 40 years of relative dormancy in some upstate counties. The affliction attacks the body's central nervous system causing fever, headaches, muscle throat spasms and eventually paralysis and death. The number of verified rabid animal cases in 1992 and 1993 was 98 and 25, respectively, in Westchester, 62 and 4 in Rockland, 69 and 9 in Orange, 117 and 5 in Ulster, 34 and 639 in Albany, 143 and 16 in Duchess, 1 and 3 in Sullivan, and 15 and 1 in Putnam counties.

Sunday–August 15, 1993

3. According to the U.S. Department of Energy's Energy Information Administration *Annual Energy Review, 1991*, Americans consumed a record 81.4 quadrillion BTUs in 1990. The chief source of this energy for the years 1960 and 1990 are as follows:

Year	Coal (%)	Petroleum (%)	Natural Gas (%)	Other (%)
1960	26.1	36.0	34.0	3.9
1990	33.5	22.9	29.9	13.7

Draw a bar graph to picture this information.

4. **Are airplanes safe?** The number of worldwide airline fatalities from 1981 to 1990 was as follows:

Year	Number of Passenger Deaths
1981	362
1982	764
1983	809
1984	223
1985	1066

continued

Year	Number of Passenger Deaths
1986	546
1987	901
1988	729
1989	817
1990	495

(*Source:* International Civil Aviation Organization, Montreal, Canada, *Civil Aviation Statistics of the World, 1993*)

Draw a bar graph to picture this information.

5. Consider the following bar graph that indicates how the percentage of Americans with the various communication devices has changed from 1980 to 1990:

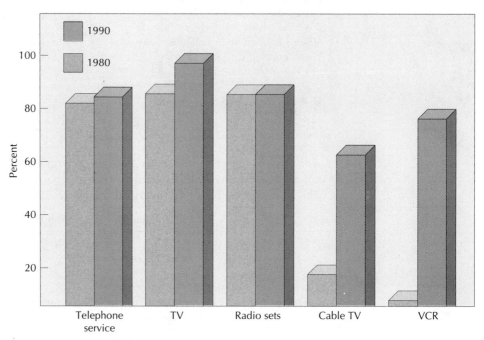

Source: U.S. Bureau of the Census.

a. Estimate the percent of Americans with the various devices in 1980 and in 1990.

b. By what percent did the number of Americans with VCRs increase over this period?

c. By what percent did the number of Americans with radio sets increase over this period?
6. Each of the world's six largest suspension bridges is longer than 1000 meters. Use the information given in the graph in the following figure to answer the questions.
    a. Find the length of each bridge to the nearest 100 meters.
    b. What is the approximate difference in length between the Humber Bridge and the Verrazano Narrows Bridge?
    c. What is the approximate difference in length between the Golden Gate Bridge and the Bosporus Bridge?

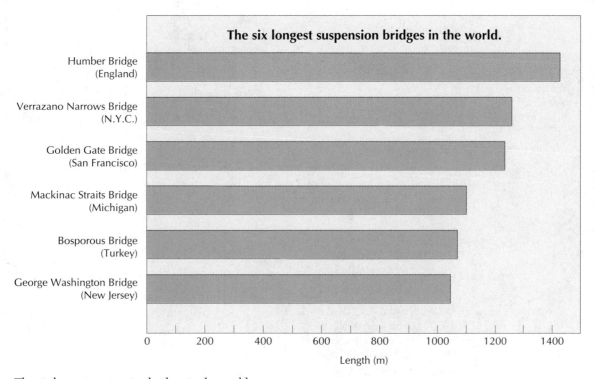

The six longest suspension bridges in the world.

7. The Bookerville chapter of the American Red Cross compiles statistics on the number of pints of blood that it collects during the year. For the first six months of 1995 it has constructed the following line graph:

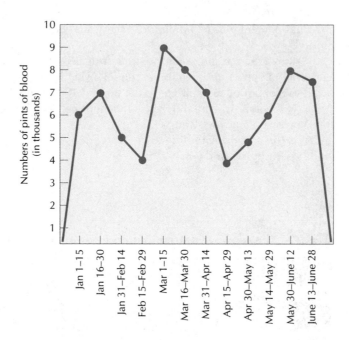

Answer the following:

a. During which period of the year were there at least 8000 pints of blood collected?

b. During which period(s) was the most blood collected?

8. Consider the accompanying newspaper article. The following list gives the total

MISSISSIPPI WATERS BEGIN TO RECEDE AS RAIN STOPS

St. Louis (July 28): After two days of continuous rainfall, the skies finally cleared yesterday afternoon and the sun reappeared. The rampaging waters from the Mississippi River which crested far above flood levels are beginning to recede slowly. The weekend rainfall dumped 6.1 inches of rain into an already overburdened river system. So far this year's rainfall is running far ahead of last year's total.

Wednesday–July 28, 1993

monthly rainfall (in inches) for the region over the past year. Draw a line graph for these data:

Aug. 12.1	Nov. 13.7	Feb. 19.9	May 14.6
Sept. 10.3	Dec. 15.6	Mar. 19.3	June 18.8
Oct. 12.7	Jan. 17.6	Apr. 11.4	July 10.3

9. The five leading causes of death in the United States during 1991 were as follows:

Leading Cause of Death, 1991

Heart disease	725,000
Cancer	506,000
Stroke	145,340
Accidents	93,550
Chronic lung diseases	88,989

(*Source: Vital Statistics of the U.S.* Published by the National Center for Health Statistics (NCHS), 1992)

Draw a bar graph to picture this information.

10. The United States imports huge amounts of crude oil as shown in the following table:

Crude Oil Imports into the U.S. by Country of Origin, 1990 (In millions of barrels— where each barrel contains 42 gallons)

Canada	235
Norway	35
Trinidad-Tobago	28
United Kingdom	56
OPEC	1286
Other	264

(*Source:* U.S. Energy Information Administration *Petroleum Supply Annual*, 1991, Volume 1)

Draw a pie chart to picture this information.

11. The following pie chart indicates the nature of the most prevalent U.S. disasters:

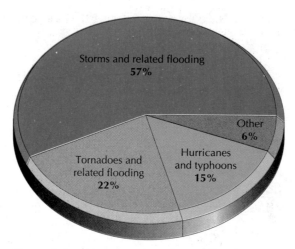

(Source: Federal Emergency Management Agency (FEMA), 1992)

According to FEMA, there were 312 major U.S. disasters in the last ten years.
a. How many disasters were from hurricanes and typhoons?
b. How many disasters were from tornadoes and related flooding?
c. How many disasters were not from storms and related flooding?

12. How important is TV to Americans? Consider the following information:

Percent of Homes with TV

98% have color TV
36% have two TV sets
77% have a VCR
62% have basic cable
28% have pay cable

(Source: Nielsen Media Research,
1992)

a. Draw a bar graph to picture the above information.
b. *Newsweek* (August 2, 1993, p. 6) estimated that there were 93.1 million households in America with televisions. How many households were in each of the above categories in 1993?

13. Is there too much violence on TV? Between the hours of 6 A.M. and midnight on April 2, 1992, ABC, CBS, NBC, PBS, FOX, WDCA-Washington, TURNER, USA, MTV, and HBO combined broadcast scenes involving the following acts of violence (adapted from *Newsweek* July 12, 1993):

Acts of Violence	Number of Scenes
Serious assaults (without guns)	389
Gunplay	302
Isolated punches	273
Pushing, dragging	272
Menacing threat with a weapon	226

Acts of Violence	Number of Scenes
Slapping	128
Deliberate destruction of property	95
Simple assault	73
All other types	28

(*Source:* Center for Media and Public Affairs, June 1992)

a. Draw a pie chart *and* a bar graph to picture the above information.
b. Which graph is more useful?

14. Water sports have become very popular in the United States as the following table indicates:

Type of Water Sport	Number of People Participating
Water skiing	10.5 million
Sailing	4.2
Scuba diving	2.9
Surfing	1.2
Board sailing	0.7

(*Source:* Sporting Goods Manufacturing Association, 1992)

a. Draw a pie chart *and* a bar graph to picture the above information.
b. Which graph is more useful?

15. *How do we pay for things?* On January 2, 1996 a major U.S. airline reported that customers paid for their plane tickets as follows.

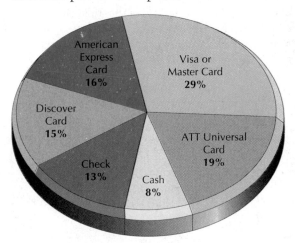

If the airline received payment for 4276 tickets on that day, how many customers paid for their tickets
a. by using the American Express card?
b. by using the Discover card or the ATT Universal card?
c. by not using a check or cash?

16. In an effort to promote conservation, government officials encourage car pooling. The McKormack Bridge and Tunnel authority offers motorists crossing its bridge or tunnel a 60% discount from the toll when the car contains four or more people. The following pictograph indicates the number of motorists participating in the car pooling program over a seven-month period. (Each symbol represents 450 cars.)

a. How many motorists participated in the program in January 1995?
b. What is the difference in the number of motorists who car pooled in July 1995 when compared with January 1995?
c. What is the total number of motorists who participated in the program from January 1995 to July 1995?

17. The following pictograph indicates the 1980 population of the world's most populated countries. (Each symbol represents 100 million people.)
a. Approximately how many people did each of these countries have in 1980?
b. Approximately how many more people did China have than did the Soviet Union in 1980?

18. Draw a cumulative frequency histogram and cumulative frequency polygon for the data given in each of the following distributions:
    a. The heights of all the entering freshmen at Polytechnic University:

Height (inches)	Frequency
62–63	57
64–65	160
66–67	360
68–69	462
70–71	298
72–73	179
74–75	32

    b. The systolic blood pressure (at rest) of the 700 female staff members at Polytechnic University:

Blood Pressure	Frequency
100–110	68
111–120	187
121–130	209
131–140	88
141–150	76
Over 150	72

19. Based on your own experience, which of the following do you think is (are) likely to be normally distributed:
    a. Age at which women give birth to their first child.
    b. Amount of money that an individual claims as a charity deduction on a 1040 U.S. Individual Income Tax return.
    c. Age of the students in your statistics class.
    d. Length of stay by mother in hospital for routine delivery.
    e. Weight of individuals in your college.
    f. Number of hours of sleep needed daily by a college student.
20. In a survey conducted by Hadley and Kolb, people were asked whether they favored building additional nuclear energy–generating facilities or were opposed. The results were then compared with earlier such surveys. Their findings are given below:

Year	Opposed	In Favor	Not Sure
1988	940	770	260
1989	760	980	240
1990	840	950	210
1991	630	1110	200
1992	520	1240	150
1993	380	1280	90

Draw a bar graph to picture this information.

BEWARE OF NATIONWIDE GAS RIPOFFS

Washington (Sept. 1): Many motorists purchase premium gas for the smooth operation of their vehicles. Authentic premium gas should have an octane rating of 90 or higher. This grade of gas has been most often implicated in retail shenanigans. There are 22 states today where motorists have little or no protection against octane fraud and contaminated fuel.

As much as 50 percent of the gasoline tested in certain states was found to be 2 to 4 octane ratings below what was posted on the pump. A spokesperson for the AAA noted that gas samples in those states with vigorous testing programs showed octane problems in the 3.9 percent, on average, as opposed to those states with little or no testing programs. According to Joe Bartfai, director of Weights and Measures for New York State, only 1.5 percent of the 1159 samples tested during the first quarter of 1993 failed octane tests. Declining failure rates have been occurring over the years throughout the state.

Source: New York State Department of Weights and Measures, 1993

(Source: New York State Department of Weights and Measures, 1993)

21. Consider the above newspaper article. Note the use of three bar graphs where one has been superimposed on the other.
    a. Of the 1158 samples tested during the first quarter of 1993, how many failed the octane tests?
    b. Is there anything wrong with these bar graphs? (*Hint:* Consider the 1993 data.)

2.4 STEM-AND-LEAF DIAGRAMS

Stem-and-leaf diagrams

The graphical techniques discussed to this point are well suited to handle most situations. In recent years, however, a new technique known as **stem-and-leaf diagrams** has become very popular. It represents a combination of the sorting techniques often used by computers and a graphical technique.

To see how this new method works, let us analyze some information on the number of people using the cash machines daily at an automated banking facility. The following data are available for 30 business days:

$$
\begin{array}{cccccc}
162 & 146 & 110 & 219 & 174 & 165 \\
128 & 159 & 197 & 205 & 152 & 166 \\
151 & 142 & 188 & 212 & 162 & 123 \\
203 & 137 & 167 & 178 & 183 & 153 \\
178 & 198 & 143 & 179 & 189 & 138 \\
\end{array}
$$

Stem
Leaf

Using stem-and-leaf diagrams, we can group the data and at the same time obtain a display that looks like a histogram. This is done as follows. The first number on the list is 162. We designate the first two leading digits (16) as its **stem**. We call the last (or trailing) digit its **leaf** as illustrated here.

Stem (first or leading digits)	Leaf (last or trailing digit)
16	2

The stem and leaf of the number 128 are 12 and 8, respectively. Also, the stem and leaf of the number 151 are 15 and 1, respectively.

To form a stem-and-leaf display for the preceding data, we first list all stem possibilities in a column starting with the smallest stem (11, which corresponds to the number 110) and ending with the largest stem (21, which corresponds to the number 219). Then we place the leaf of each number from the original data in the row of the display corresponding to the number's stem. This is accomplished by placing the last (or trailing) digit on the right side of the vertical line opposite its corresponding leading digit or stem. For example, our first data value is 162. The leaf 2 is placed in the stem row 16. Similarly, for the number 128, the leaf 8 is placed in the stem row 12. We continue in this manner until each of the leaves is placed in the appropriate stem rows. The completed stem-and-leaf display will appear as shown in Figure 2.19.

The stem-and-leaf diagram arranges the data in a convenient form since we can now count the number of leaves for each stem. We then obtain the frequency distribution. From this it is very easy to draw the histogram or bar graph. If we turn the preceding stem-and-leaf display on its side, we obtain the same type of bar graph provided by the frequency distribution. This is shown in Figure 2.20.

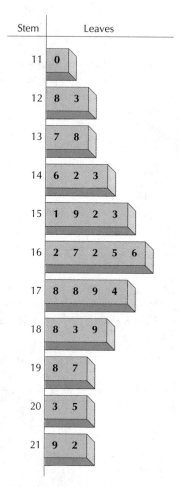

FIGURE 2.19

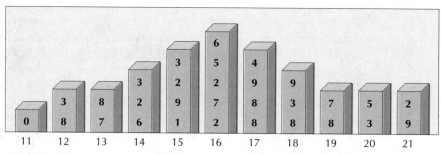

FIGURE 2.20

COMMENT   One major advantage of a stem-and-leaf diagram over a frequency distribution is that the original data are preserved. The stem-and-leaf diagram displays the value of each individual score as well as the size of each data class. This is not possible with a bar graph.

COMMENT   A stem-and-leaf diagram presents the data in a more convenient form. This will make it easy to perform various arithmetic calculations to be studied in the next chapter.

We summarize the procedure to be used in constructing stem-and-leaf diagrams in the following rules.

RULES FOR CONSTRUCTING STEM-AND-LEAF DIAGRAMS

To construct a stem-and-leaf diagram, proceed as follows:

1. Determine how the stems and leaves will be identified.
2. Arrange the stems in order in a vertical column, starting with the smallest stem and ending with the largest.
3. Go through the original data and place a leaf for each observation in the appropriate stem row.
4. If the display looks too cramped and narrow, we can stretch the display by using two lines (or more) per stem so that we can place leaf digits 0, 1, 2, 3, and 4 on one line of the stem and leaf digits 5, 6, 7, 8, and 9 on the other line of the stem.

Let us illustrate the preceding rule with another example.

EXAMPLE   1   A new drug treatment clinic recently opened in a city, and the number of addicts treated with methadone per day during the first month was as follows:

37	95	34	26	45
88	89	24	61	28
42	78	67	32	79
67	29	28	24	35
68	72	91	86	78

Draw a stem-and-leaf diagram for these data.

SOLUTION   Let us use the first digit of each of the numbers as the stem and the second digit as the leaf. Then we arrange the stems in order in a vertical column. Although they can be arranged horizontally, the stems are usually arranged vertically. We get the following table.

Stem
_____

2
3
4
5
6
7
8
9

Now we go through the original data and place a leaf for each observation in the appropriate stem row. The stem-and-leaf diagram is shown in Figure 2.21.

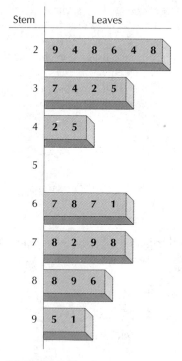

FIGURE 2.21

Dotplot   Another commonly used graphical display of numerical data is the *dotplot*. **Dotplots** are very helpful in showing the relative positions of numbers in a set of data or for comparing several sets of data. We illustrate the construction of dotplots in the following example.

EXAMPLE 2 Consider the accompanying newspaper article. The following list gives the rainfall (in inches) for the region over the past year. Construct a dotplot for the data.

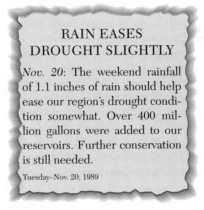

RAIN EASES
DROUGHT SLIGHTLY

Nov. 20: The weekend rainfall of 1.1 inches of rain should help ease our region's drought condition somewhat. Over 400 million gallons were added to our reservoirs. Further conservation is still needed.

Tuesday–Nov. 20, 1989

Jan.	7	Apr.	8	Jul.	4	Oct.	11
Feb.	6	May	6	Aug.	4	Nov.	8
Mar.	10	Jun.	5	Sept.	3	Dec.	9

SOLUTION We first draw a horizontal line that indicates the possible number of inches of rain. Then we go through the data and place a dot over the appropriate value on the horizontal axis. Thus, for example, we place a dot over the "7" on the horizontal axis to correspond to the rainfall in January. The complete dotplot for these data is pictured below.

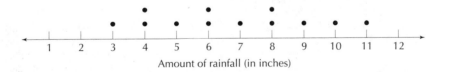

Amount of rainfall (in inches)

COMMENT Although dotplots are quite similar to histograms, they are generally more convenient to work with when dealing with single-value grouped data that involve decimals.

EXERCISES FOR SECTION 2.4

1. In the 1992 *Gas Mileage Guide* published by the U.S. Department of Energy, an estimate of 27 miles per gallon (mpg) is given for the Ford Ranger pickup truck equipped with a 2.3-liter engine and a two-wheel drive, five-speed transmission. These estimates are based on repeated tests of the vehicles. In one survey of 50 Ford Ranger pickup trucks, the following mpg results were obtained.

```
31   24   29   30   31   26   26   29   31
25   23   32   27   32   28   29   27   30
27   27   33   26   28   27   27   28
28   26   25   28   26   29   24   27
33   28   28   24   24   30   29   26
24   30   31   34   27   32   33   28
```

    a. Draw a stem-and-leaf diagram for the above data.

    b. Draw a dotplot for the above data.

2. In an effort to better serve the public and to have the appropriate number of personnel on duty as needed, many governmental agencies analyze the number of clients handled daily. Gordon Blaisley is director of the Motor Vehicle Bureau in his state. He supervises the issuance of driver's licenses. During the first 60 working days of 1995, his agency issued the following number of driver's licenses (on a daily basis):

```
181   224   250   296   278   281   200   286   294   257
302   302   275   283   189   245   199   199   265   225
186   193   258   210   198   258   263   249   201   300
253   199   196   299   275   193   293   200   275   293
279   276   189   187   260   290   187   300   253   232
201   292   294   225   236   238   260   245   289   248
```

Draw a stem-and-leaf diagram for the data.

3. In *Savings and Home Financing Service*, published annually by the Federal Home Loan Bank, the median purchase prices for existing single-family or two-family houses for various metropolitan areas are given. The following are the purchase prices for 42 two-family houses in the San Francisco-Oakland-San Jose, California, metropolitan area during July 1993:

```
$172,000   $270,000   $254,000   $170,000   $229,000   $175,000   $220,000
 185,000    198,000    189,000    270,000    180,000    160,000    180,000
 251,000    215,000    194,000    268,000    265,000    185,000    205,000
 208,000    140,000    142,000    220,000    248,000    169,000    250,000
 198,000    158,000    165,000    238,000    225,000    181,000    175,000
 137,000    143,000    215,000    245,000    176,000    200,000    225,000
```

Draw a stem-and-leaf diagram for the data.

4. The ages of the 50 participants in a recent jogging marathon were as follows:

```
53   32   23   27   42   45   33   60   32   31
21   69   31   62   30   23   56   62   31   51
29   28   35   71   31   27   51   55   42   62
37   21   38   23   39   19   39   41   32   41
69   22   41   42   42   18   28   40   23   34
```

Draw a stem-and-leaf diagram for the data.

5. Randy is interested in buying an elaborate and complete computer system for his child. Numerous dealers have quoted him the following prices, in dollars, for the

same system. Group the numbers so that 50 values fall on each line. Thus, the stems
will be 1850–1899, 1900–1949, and so on.

$2149	$2050	$1875	$2000	$2000
2249	1945	2249	1998	2198
2150	1989	1976	1987	1945
1901	2149	2004	2049	1998
1945	2200	1984	2145	2100

Draw a stem-and-leaf diagram for the data.

6. A scientist from the Environmental Protection Agency took samples of the toxic
   substance polychlorinated biphenyl (PCB) levels from the soil at 60 different waste
   disposal facilities located throughout the United States. The following results (in
   0.0001 grams per kilogram of soil) were obtained:

38.8	35.6	31.8	32.8	36.3	40.2	39.7	33.9	34.4	33.1	39.3	34.8
35.3	38.1	35.7	39.1	37.8	39.5	36.4	38.6	37.6	37.8	31.7	35.7
39.1	35.8	38.4	34.5	37.9	38.2	38.3	40.1	38.8	33.9	30.8	37.6
31.8	32.4	35.9	36.1	38.1	37.6	36.7	30.8	37.8	35.5	39.8	36.9
40.2	33.8	34.7	39.0	36.0	37.3	31.4	31.7	32.9	30.7	37.5	31.8

Draw a stem-and-leaf diagram for the data.

7. Many airlines require that passengers check in at least 45 minutes before the scheduled departure time. A recent survey by an FAA (Federal Aviation Administration)
   official at Los Angeles International Airport of 50 passengers traveling overseas indicated that these passengers had arrived the following number of minutes before
   their scheduled departure times:

57	53	51	55	54	47	47	45	58	54
46	45	48	48	50	42	53	53	46	50
54	53	47	56	41	58	51	44	53	53
41	58	48	54	52	48	47	48	45	47
53	52	54	46	46	55	42	49	42	49

a. Draw a dotplot for the above data.
b. Draw a stem-and-leaf diagram for the data.
c. Which is more useful, the dotplot or the stem-and-leaf diagram?

2.5 HOW TO LIE WITH GRAPHS

In the previous sections we indicated how to analyze a list of numbers by graphical
techniques. Nevertheless, it is possible to use these techniques to present the data in a
misleading way. The following examples indicate some misuses of statistics and the
reason for the incorrect use.

   Since frequency polygons emphasize changes (rise and fall) in frequency more
clearly than any other graphical representation, they are used often to display business

and economic data. However, this must be done with great care. Figures 2.22 and 2.23 both represent the same idea, that is, the number of cases of malpractice insurance filed against neurosurgeons in a large northeastern city in the United States during the years 1991–1995. Figure 2.23 seems to indicate that the number of malpractice insurance cases filed increased significantly over this five-year period. Figure 2.22 also indicates that the number of such cases increased, although not so dramatically. How can we have two *different* graphs representing the same situation? Which graph better displays the situation?

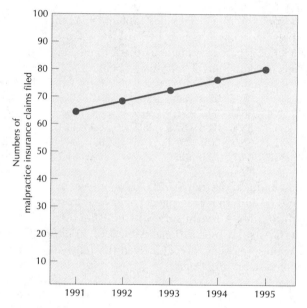

FIGURE 2.22

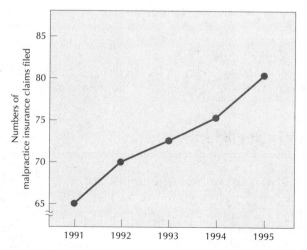

FIGURE 2.23

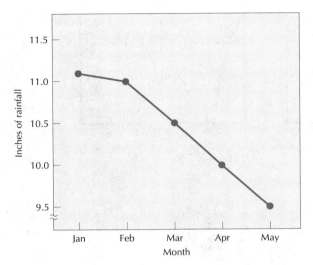

FIGURE 2.24

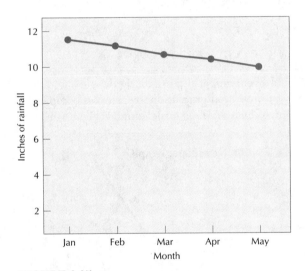

FIGURE 2.25

Truncated graphs

Notice that the vertical scales in the two graphs are not the same. The vertical scale of Figure 2.23 has been truncated or cut off. **Truncated graphs** often tend to distort the information presented. Consider the graphs on a drought situation shown in Figures 2.24 and 2.25. Again, notice the truncated graph. How bad is the drought situation?

Scaling

Another misuse of statistics involves **scaling**. Consider the graphs given in Figures 2.26 and 2.27 to show how the circulation of a computer magazine has doubled. Anything wrong? The graph in Figure 2.27 is twice as tall as the graph in Figure 2.26. But it is also twice as wide, so it is four times as large. In Figure 2.27 each of the four rectangles is exactly the same size as the one in Figure 2.26. Do people draw graphs in such a misleading way? Unfortunately, the answer is yes.

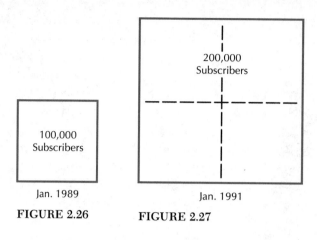

100,000
Subscribers

Jan. 1989

FIGURE 2.26

200,000
Subscribers

Jan. 1991

FIGURE 2.27

Up to this point we have merely indicated two ways that statistical graphs can be misused. Many other ways are discussed in Darrell Huff's book *How to Lie with Statistics* (New York: W. W. Norton, 1954).

COMMENT   By now the point should be obvious. Read and construct statistical graphs carefully to avoid the common pitfalls.

EXERCISES FOR SECTION 2.5

1. The number of accidents at a particularly dangerous railroad crossing over a five-year period was presented graphically by railroad company officials and in the local city newspaper. Which graph is misleading? Explain your answer.

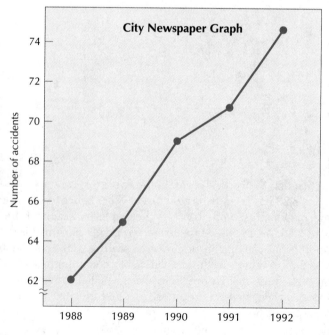

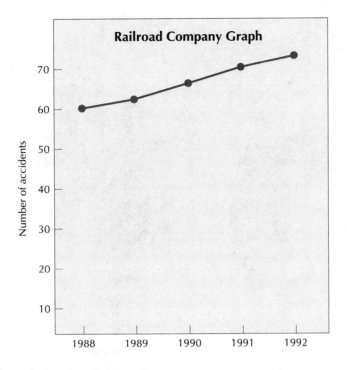

2. Consider the following advertisement promoted by a swimming pool construction company. Anything wrong with the claim?

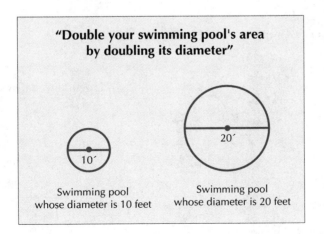

3. The following graph concerning the unemployment rate in a large metropolitan city recently appeared in the city's newspaper indicating that unemployment was declining considerably. Anything wrong with this graph?

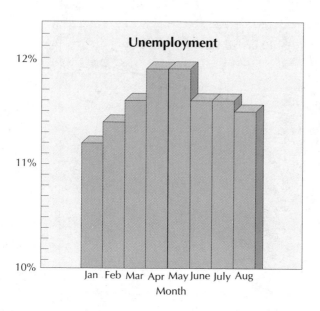

4. The following graphs, one prepared by a company statistician and the other prepared by an independent statistician, indicate how the Arjon Tool Company's share of the market has increased from 1987 to 1991. Which graph is the correct one?

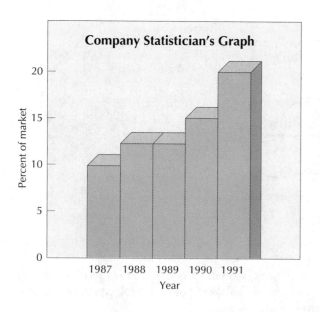

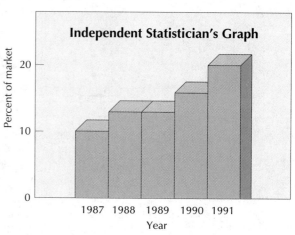

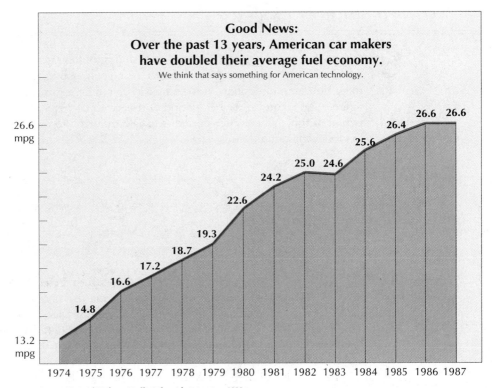

Good News:
Over the past 13 years, American car makers
have doubled their average fuel economy.
We think that says something for American technology.

Source: National Highway Traffic Saftey Administration, 1989.

5. Consider the article shown above concerning American cars and fuel economy. Is the picture presented by this graph fair? What should the height of the 1987 bar be as compared to the height of the 1979 bar? From these data, can we conclude that American car makers have doubled their average fuel economy?

2.6   INDEX NUMBERS

Base period

Often when analyzing data we are interested in obtaining a clear picture of trend. We could then use index numbers that give us a good analysis of changes over a period of time. Index numbers allow us to compare such things as prices, production figures, sales figures, and so on for a given period of time with corresponding values in some earlier period of time. The earlier period is usually referred to as the **base period**. We have the following definition.

Definition 2.4
Index number

An **index number** is a special form of ratio that is used to show percentage changes over a period of time.

Consumer Price Index (CPI)

The best known index number is the **Consumer Price Index (CPI)** published by the U.S. Bureau of Labor Statistics. In the media this is often called the cost-of-living index. It reflects the changes in prices of goods and services purchased by a typical wage earner in a large city. Since the labor contracts negotiated by some unions often tie increases in salary to changes in the CPI through cost-of-living escalation clauses, a knowledge of how to compute index numbers is important.

The CPI is published each month and includes the results of surveys for the prices of over 400 goods and services including among other things food, clothing, medical costs, automobile repair, college tuition, and so on. Thus, it can be used as a measure of inflation. When the CPI goes up, wages, pensions, and Social Security payments go up.

COMMENT   Great care must be exercised when using the CPI. The prices of the items included in the computation must be weighted according to their importance.

To calculate an index number, we use the following formula.

FORMULA 2.1

$$\text{Index number} = \frac{\text{given year's values}}{\text{base year's value}} \times 100$$

We illustrate the use of this formula with several examples.

E X A M P L E   1 The Musicktone Corporation sells various stereo components. The annual sales for these components over the years 1988 to 1993 is as follows.

Year	Sales
1988	$400,000
1989	425,000
1990	491,000
1991	350,000
1992	410,000
1993	380,000

Using 1988 as a base year, calculate the sales index for each year.

SOLUTION   Since we are using 1988 as our base year, we must divide each year's value by the base year's value and multiply the result by 100. We have the following table.

Year	Ratio	Ratio × 100 = Index Number
1988	$\dfrac{400,000}{400,000} = 1$	$1 \times 100 = 100$
1989	$\dfrac{425,000}{400,000} = 1.0625$	$1.0625 \times 100 \times 106.25$
1990	$\dfrac{491,000}{400,000} = 1.2275$	$1.2275 \times 100 = 122.75$
1991	$\dfrac{350,000}{400,000} = 0.875$	$0.875 \times 100 = 87.5$
1992	$\dfrac{410,000}{400,000} = 1.025$	$1.025 \times 100 = 102.5$
1993	$\dfrac{380,000}{400,000} = 0.95$	$0.95 \times 100 = 95$

The index numbers calculated in the previous example show us the percent change from the base year. This percent change is the difference from 100. We have the following:

a. The sales index for 1989 indicates an increase of 6.25% when compared with 1988 since $106.25 - 100 = 6.25$.
b. The sales index for 1990 indicates an increase of 22.75% when compared with 1988 since $122.75 - 100 = 22.75$.
c. The sales index for 1991 indicates a decrease of 12.5% when compared with 1988 since $87.5 - 100 = -12.5$.

Similar conclusions can be arrived at for the other years.

EXAMPLE 2
*Oil Dependence
of the United
States*

The number of thousands of barrels of oil imported by the United States from OPEC (Organization of Petroleum Exporting Countries) sources for the years 1981 to 1984 is as follows.

Year	Oil Imported
1981	1,849,017
1982	2,260,482
1983	2,057,468
1984	2,023,341

(*Source:* American Petroleum
Institute, 1987)

Using 1981 as a base year, compute the import index for each of the years given and interpret the results.

SOLUTION   Using the same tabular arrangement as we did in the previous example, we have the following table.

Year	Ratio	Ratio × 100 = Index Number
1981	$\dfrac{1,849,017}{1,849,017} = 1$	$1 \times 100 = 100$
1982	$\dfrac{2,260,482}{1,849,017} = 1.2225$	$1.2225 \times 100 = 122.25$
1983	$\dfrac{2,057,468}{1,849,017} = 1.1127$	$1.1127 \times 100 = 111.27$
1984	$\dfrac{2,023,341}{1,849,017} = 1.0943$	$1.0943 \times 100 = 109.43$

The import indexes for these years indicate

  a. for 1982 an increase of 22.25% when compared with 1981.
  b. for 1983 an increase of 11.27% when compared with 1981.
  c. for 1984 an increase of 9.43% when compared with 1981.

COMMENT   In the previous example we can use a different base year and arrive at different conclusions. Thus, if 1983 is used as a base year, we get index numbers of approximately 90, 110, 100, and 98, respectively, for the years 1981, 1982, 1983, and 1984. These index numbers give us different interpretations than when we used 1981 as a base year.

COMMENT   Often when using index numbers an important decision involves deciding which base year is best to use.

EXERCISES FOR SECTION 2.6

For Exercises 1–4, use the following information: Memorial Day heralds the unofficial arrival of summer. Each year the AAA (Automobile Association of America) surveys many gas stations nationwide to determine the availability of gas for the coming summer driving season and its price. At one particular gas station on Broadway and Melville Road, the price of a gallon of unleaded premium gas over the past seven years was as follows:

Year	Price per Gallon
1989	$1.26
1990	1.37
1991	1.39
1992	1.42
1993	1.37
1994	1.35
1995	1.38

(*Source:* AAA reproduced by permission)

1. Using 1989 as a base year, what was the price index for 1995?
2. Using 1989 as a base year, what does the price index for 1992 tell us?
3. Using 1995 as a base year, what does the price index for 1995 tell us?
4. Using 1995 as a base year, what does the price index for 1992 tell us?
5. Medical costs are constantly rising. According to insurance company records the average charge for delivering a baby (not by caesarean section) in Brooklyn over the years has been increasing by leaps and bounds as can be seen in the following table.

Year	Average Charge
1990	$3300
1991	3700
1992	4000
1993	4400
1994	4600
1995	5000

(*Source:* Acme Insurance Group, 1995)

Using 1990 as a base year, compute cost indexes for each of the years given and interpret the results.

6. The charge for malpractice insurance to doctors depends among other things on the area of specialization in which the doctor practices and also the region in which

the doctor practices. The average charge for malpractice insurance to doctors specializing in internal medicine in one particular city over the years was as follows:

Year	Charge for Malpractice Insurance
1989	$10,200
1990	11,100
1991	12,000
1992	13,000
1993	14,500
1994	17,000
1995	20,000

Using 1989 as a base year, compute cost indexes for each of the years given and interpret the results.

7. Refer back to Exercise 6. Using 1992 as a base year, compute cost indexes for each of the years given and interpret the results. Compare the results obtained for Exercises 6 and 7. Comment.

8. The Budget Director of the Beck Corporation has computed cost indexes for several items directly related to its employees. The following is available:

Index of Costs (January 1990 = 100)

Month	Labor Costs	Health Insurance Costs	Pension Costs
January 1993	106	115	103
January 1994	108	111	94
January 1995	109	113	100

a. Which cost showed the smallest increase between January 1993 and January 1995?

b. Which cost showed the greatest increase between January 1993 and January 1995?

9. Refer back to Exercise 8. For the month of January 1994, pension costs had a cost index of 94. Explain the cost index.

10. Refer back to the newspaper article at the beginning of this chapter on page 38. In 1991 Ellen Ingram earned $65,000 and in 1992 she earned $66,000. According to government records, the CPI for Ellen's region was 106.7 in 1991 and 108.1 in 1992. Did Ellen's purchasing power increase or decrease over the year?

2.7 USING COMPUTER PACKAGES

To illustrate how we can use the MINITAB computer package to draw a histogram, let us consider the following data representing the number of credit card sales reported by a large department store chain at all of its branches over a 50-day period.

Day	Number of Sales	Day	Number of Sales	Day	Number of Sales
Mon.	52	Mon.	60	Mon.	53
Tues.	68	Tues.	62	Tues.	58
Wed.	51	Wed.	58	Wed.	62
Thurs.	62	Thurs.	48	Thurs.	61
Fri.	49	Fri.	61	Fri.	60
Mon.	37	Mon.	45	Mon.	45
Tues.	74	Tues.	49	Tues.	61
Wed.	64	Wed.	53	Wed.	53
Thurs.	48	Thurs.	67	Thurs.	42
Fri.	63	Fri.	49	Fri.	37

Day	Number of Sales	Day	Number of Sales
Mon.	34	Mon.	47
Tues.	68	Tues.	59
Wed.	51	Wed.	38
Thurs.	47	Thurs.	32
Fri.	56	Fri.	76
Mon.	59	Mon.	57
Tues.	69	Tues.	41
Wed.	42	Wed.	39
Thurs.	38	Thurs.	31
Fri.	36	Fri.	37

After logging in on the computer, we can construct the histogram° for the preceding data by typing in the following instructions:

```
MTB  > SET THE FOLLOWING DATA INTO C1
DATA > 52 68 51 62 49 37 74 64 48 63
DATA > 60 62 58 48 61 45 49 53 67 49
DATA > 53 58 62 61 60 45 61 53 42 37
DATA > 34 68 51 47 56 59 69 42 38 36
DATA > 47 59 38 32 76 57 41 39 31 37
DATA > END

MTB  > HISTOGRAM OF C1;
SUBC > Start 30;
SUBC > Increment 5.
```

The first line tells the computer that the data should be placed in column C1 of the worksheet that MINITAB maintains in the computer. The next few lines contain the

°In this text, we are assuming that the standard DOS version of MINITAB is being used. If you are using the Windows version of MINITAB, then you must type GSTD at the MTB prompt to obtain the results shown here.

data themselves. The command after that instructs the computer to construct the histogram. The semicolon at the end of the line informs MINITAB that another command (subcommand) will follow on the next line. The subcommand START is used to indicate where the histogram is to begin. In our case the starting value is the first class mark (the midpoint of the first class). Again there is a semicolon at the end of this line indicating that another subcommand will appear on the next line.

This line contains the subcommand INCREMENT that is used to indicate the class width. In our case it will be 5. A period at the end informs MINITAB that there will be no more subcommands.

After typing the above information, the computer will automatically print out the frequency distribution and histogram shown below. Notice that the histogram is printed sideways with the bars (actually they are asterisks) drawn horizontally rather than vertically.

```
Histogram of C1    N = 50

Midpoint      Count
   30.00        2      **
   35.00        5      *****
   40.00        6      ******
   45.00        4      ****
   50.00        8      ********
   55.00        5      *****
   60.00       12      ************
   65.00        3      ***
   70.00        3      ***
   75.00        2      **
```

The subcommands START and INCREMENT are optional with the HISTOGRAM command. We specified them so as to obtain a frequency distribution and a frequency histogram that are based on the intervals 28–32, 33–37, 38–42, 43–47, . . . and whose midpoints are 30.00, 35.00, 40.00, 45.00, ... If the first midpoint and interval width are not specified and one just types HISTOGRAM C1, then MINITAB will automatically select its own class.

We can use MINITAB to obtain a dotplot of the data in C1. We simply type the command DOTPLOT followed by the storage location of the data (C1). In our case we have

MTB > **DOTPLOT OF C1**

We can also have MINITAB draw a stem-and-leaf diagram for the data in C1. We have the following:

```
MTB > STEM-AND-LEAF OF C1

Stem-and-leaf of C1          N = 50
Leaf Unit = 1.0
      3      3 124
     10      3 6777889
     13      4 122
     22      4 557788999
     (6)     5 112333
     22      5 678899
     16      6 0011122234
      6      6 7889
      2      7 4
      1      7 6
```

```
MTB > STOP
```

The first line of this output gives us a verbal description of what MINITAB is doing—in our case "stem-and-leaf of C1." It also tells us that N = 50 or that we have 50 pieces of data. The second line of the output (Leaf Unit = 1.0) tells us where the decimal point goes—in our case directly after each leaf digit.

There are three columns in this stem-and-leaf printout. The second column of numbers gives us the stems. The third column of numbers gives us the leaves. The leaves are ordered so that we now have an ordered stem-and-leaf diagram. The first

Depth column of numbers gives us **depth**. The depths are used to display cumulative frequencies. Starting from the top, the depths tell us the number of leaves (pieces of data) that lie on a given row or on earlier rows. Thus, the "22" in the fourth row tells us that there is a total of 22 leaves in the first four rows.

When we reach the row in which the middle observations lie, the cumulative frequency is replaced by the number of leaves in that row enclosed by parentheses. In our case the middle observations lie in the 50 to 54 interval and there are six data values in this interval. Finally, the depths following the 50 to 54 interval indicate the number of leaves that lie in the given row and in subsequent rows. Thus, the "16" in the seventh row indicates that there is a total of 16 leaves in the last four rows.

COMMENT   Our objective here is not to teach you how to become an expert in the MINITAB statistical package but merely to familiarize you with its general nature and usefulness. The details can be obtained from manuals. For a manual specifically dealing with MINITAB, see T. Ryan, B. Joiner, and B. Ryan, *MINITAB STUDENT HANDBOOK* (Boston: Duxbury Press, 1985).

EXERCISES FOR SECTION 2.7

1. The U.S. Bureau of Labor Statistics obtains data on earnings of U.S. workers in numerous industries and publishes its findings in *Employment and Earnings*. A survey of 50 workers in the trucking industry revealed the following weekly earnings (in dollars):

430	441	460	420	437	450	426	460	438	456
400	425	409	428	416	442	422	475	453	443
450	410	422	406	404	460	429	420	446	405
375	408	430	411	417	420	435	403	422	412
493	430	401	429	406	409	442	427	409	459

    a. Using MINITAB construct a histogram for the data using increments of 12.
    b. Using MINITAB construct a stem-and-leaf diagram for the data.

2. The U.S. Census Bureau obtains data on the price of new mobile homes and publishes its findings in *Construction Reports*. A survey revealed the following prices (in dollars) for 35 new mobile homes:

25,000	29,000	27,000	24,500	22,500	27,000	24,000
26,000	27,000	34,000	26,000	24,000	28,000	22,000
22,500	33,000	32,000	26,500	27,500	22,000	25,000
28,000	31,000	20,000	25,500	26,000	29,000	26,000
23,000	28,500	22,000	28,500	28,000	31,000	29,000

    a. Using MINITAB construct a histogram for the data using increments of 1500.
    b. Using MINITAB construct a stem-and-leaf diagram for the data.

3. Driving while under the influence of alcohol can be deadly. Police officers often administer a sobriety test to any driver suspected of being drunk and will arrest the driver when the blood test results exceed the legal limits. The number of arrests of drunken drivers at different ages is often analyzed by Justice Department officials. The ages of 61 drivers arrested and charged with DWI (Driving While Intoxicated) are as follows:

31	29	26	27	46	37	43	37	30	30	29	25	25
25	32	28	26	45	38	41	36	32	32	26	29	
27	34	40	31	49	36	42	38	33	33	28	26	
26	33	35	38	47	39	43	39	32	31	27	27	
28	27	38	36	42	32	45	34	31	25	25	25	

(*Source:* U.S. Bureau of Statistics: *Drunk Driving, Special Report*)

a. Using MINITAB construct a histogram for the data.

b. Using MINITAB construct a stem-and-leaf diagram for the data.

c. Using MINITAB construct a dotplot for the data.

4. Do you drive with your seat belt buckled up properly? In its May 25, 1990, issue, *USA Today* gave the following percentages (state by state) of motorists who use their seat belts:

State	Percentage of Drivers Using Seat Belts	State	Percentage of Drivers Using Seat Belts
AL	31.0	MT	63.0
AK	45.0	NE	32.0
AZ	48.0	NV	38.4
AR	30.1	NH	50.1
CA	66.0	NJ	44.1
CO	47.0	NM	51.7
CT	54.8	NY	60.0
DE	43.0	NC	62.0
FL	55.2	ND	28.0
GA	38.8	OH	44.4
HI	80.5	OK	36.3
ID	33.7	OR	48.0
IL	40.5	PA	49.5
IN	47.4	RI	23.9
IA	59.0	SC	37.9
KS	52.0	SD	26.0
KY	20.5	TN	41.0
LA	40.6	TX	63.0
ME	34.4	UT	44.2
MD	67.0	VT	35.0
MA	28.0	VA	55.0
MI	45.6	WA	55.4
MN	44.1	WV	42.0
MS	17.0	WI	50.3
MO	54.1	WY	35.5

(*Source:* Copyright 1990, USA TODAY. Reprinted with permission)

a. Using MINITAB construct a histogram for the data.

b. Using MINITAB construct a stem-and-leaf diagram for the data.

5. Refer back to Exercise 1 of Section 2.2 on page 50. Using MINITAB, construct a dotplot for the data.

6. Refer back to Exercise 2 of Section 2.2 on page 50. Using MINITAB construct a dotplot for the data.

2.8   SUMMARY

In this chapter we discussed the different graphical methods that can be used to picture a mass of data so that meaningful statements can be made. When given a large quantity of numbers to analyze, it is recommended that frequency tables be constructed. Some forms of graphical representation should then be used to serve as visual aids for thinking about and discussing statistical problems in a clear and easily understood manner. We discussed the different graphical techniques that can be used and also pointed out in Figures 2.22 and 2.24 on pages 92 and 93 how they can be misused. We studied an alternate and relatively new way of analyzing data. This is by means of stem-and-leaf diagrams or dotplots. We also analyzed index numbers, which are special ratios that can be used to study percentage changes over a period of time. We pointed out that great care must be exercised when interpreting such numbers. The best known of these index numbers is the CPI published by the U.S. Bureau of Labor Statistics. There are many other indexes that can be computed. Some of these measure changes in regional and national retail and wholesale prices, industrial and agricultural production, and so on. Each of these is computed in a manner similar to the way we computed index numbers.

Study Guide

The following is a chapter outline in capsule form. You should now be able to demonstrate your knowledge of the ideas mentioned by giving definitions, descriptions, or specific examples. Page references are given in parentheses.

All sample data in this book are the results of **random samples**. This means that each individual of the population must have an equally likely chance of being selected. (page 41)

A **frequency distribution** is a convenient way of grouping data in order to find meaningful patterns. The word **frequency** means how often some number occurs. (page 41)

Data are often arranged by groups called **classes** or **intervals**. (page 43)

When data are arranged so that we have a series of intervals with the corresponding frequency for each interval, we have **grouped data**. (page 44)

A **class mark** represents the point that is midway between the limits of a class; that is, it is the midpoint of a class. (page 44)

The **class width** is defined as the difference between the lower class limit of any given class and the lower class limit of the next higher class. (page 43)

To determine the size of an interval we use

$$\frac{\text{Largest measurement} - \text{smallest measurement}}{\text{Number of intervals}}$$

We usually choose between 5 and 15 measurement classes. (page 44)

The tally for each class is called the **class frequency**. The class frequency for class i is denoted by the symbol f_i. (page 44)

The **relative frequency** for a given class is defined as the frequency of that class divided by the total number of measurements (the total frequency). (page 44)

A graphical portrayal of a frequency distribution is called a **histogram**. The height of each rectangle of a histogram represents its frequency. (page 44)

A **relative frequency histogram** is drawn by constructing a histogram in which the rectangle heights are relative frequencies of each class. (page 48)

The **bar graph** is commonly used to describe data graphically. The height of each bar represents the number of members, that is, the frequency of that class. (page 55)

When symbols and pictures are used in place of bars, the resulting graph is called a **pictograph**. (page 59)

If the midpoints (class marks) of the tops of the bars in a bar graph or histogram are joined together by straight lines, then the resulting figure (without the bars) is called a **frequency polygon**. (page 61)

A frequency distribution whose graph resembles a bell-shaped curve is known as a **normal distribution** and its graph is called a **normal curve**. (page 62)

A histogram that displays "accumulated" frequencies that are obtained by adding frequencies for the lower intervals is called a **cumulative frequency histogram**. (page 64)

A **cumulative frequency polygon** is simply a line graph connecting a series of points that answers the question, "How many of the scores are less than or equal to a given score?" (page 64)

Cumulative frequency polygons are also known as **ogives**. (page 65)

A **cumulative relative frequency distribution** combines the cumulative frequency with the relative frequency. (page 65)

Circle charts or **pie charts** are commonly used to describe qualitative data. A circle is broken up into various categories of interest in the same way that a pie is sliced, where each category is assigned a certain percentage of the 360° of the total circle. (page 66)

A useful graphical description of quantitative date is the **stem-and-leaf diagram**, where each measurement is partitioned into two components—a **stem** and a **leaf**. (page 85)

Dotplots are also used to display numerical data. They are helpful in showing the relative position of numbers in a set of data or for comparing several sets of data. (page 88)

Statistical graphs can be used to misrepresent the true picture. This is done by using **truncated graphs**, where the vertical scales of two "comparable" graphs are not the same. Sometimes the vertical scale is truncated or cut off. (page 93)

Another misuse of statistics involves **scaling**, where the scales of two comparable graphs are not the same. (page 93)

An **index number** is a special form of ratio that is used to show percentage changes over a period of time. The earlier period is usually referred to as the **base period**. The best known index number is the **Consumer Price Index (CPI)** published by the U.S. Bureau of Labor Statistics. (page 98)

When the MINITAB stem-and-leaf program is used, the **depths** indicate the number of leaves (pieces of data) that lie on a given row or on earlier rows. (page 105)

Formulas to Remember

The following list summarizes the formulas given in this chapter:

1. Rules for grouping data (page 46)

2. Class interval width $= \dfrac{\text{largest measurement} - \text{smallest measurement}}{\text{number of intervals}}$ (page 43)

3. Relative frequency for class $i = \dfrac{f_i}{n}$ (page 44)

4. Rules for drawing histograms (page 46)

5. Rules for rounding decimals (page 69)

6. Rules for drawing bar graphs, frequency polygons, pie charts (pages 61, 73)

7. Rules for constructing stem-and-leaf diagrams (page 87)

8. Index number $= \dfrac{\text{given year's values}}{\text{base year's values}} \times 100$ (page 98)

Testing Your Understanding of This Chapter's Concepts

1. When frequency tables and histograms are constructed, must we include a class if its frequency is zero?
2. *True or false.* Every ogive starts on the left with a relative frequency of zero at the lower class boundary of the first class and ends on the right with a relative frequency of 100% at the upper class boundary of the last class. Explain your answer.
3. The MINITAB statistical package was used to generate the following computer print-out of a histogram:

```
Midpoint     Count
      30        4      ****
      35        6      ******
      40        9      *********
      45        8      ********
      50        4      ****
```

```
Midpoint    Count
       55       3     ***
       60       7     *******
       65       5     *****
       71       1     *
```

a. How many pieces of data are described by this histogram?
b. What type of histogram is displayed—relative, frequency, or cumulative histogram? Explain your answer.

4. An FAA official is interested in determining the class-interval widths for constructing a frequency distribution on the number of late arriving planes. The following information is available:

Number of Minutes Late	
Largest value	112 minutes
Smallest value	0 minutes
No. of intervals	12

What should the class width of each interval be?

Chapter Test

Multiple Choice

For Questions 1–3, refer to the following circle graph that gives the place of origin of the 150,000 immigrants living in one section of a large city in the southeast.

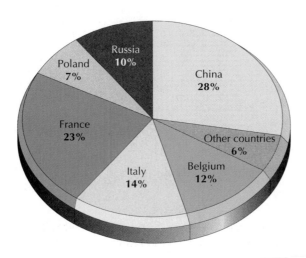

1. How many of the immigrants originated in Italy?
   a. 21,028   b. 28,000   c. 2,100   d. 21,000   e. none of these
2. How many of the immigrants originated in France or Poland?
   a. 34,500   b. 10,500   c. 45,000   d. 4,500   e. none of these
3. How many of the immigrants did not originate in Russia, China, or Belgium?
   a. 72,000   b. 75,000   c. 42,000   d. 135,000   e. none of these
4. Consider the following frequency distribution. What is the relative frequency of the score of 61?

Score	Frequency
12	89
20	71
36	64
43	53
61	29
69	12
74	6
	Total = 324

   a. 29   b. $\dfrac{29}{61}$ c. $\dfrac{61}{324}$ d. $\dfrac{29}{324}$ e. none of these

5. Refer back to Question 4. What is the relative frequency of the score of 29?
   a. $\dfrac{29}{324}$ b. 0   c. $\dfrac{29}{61}$ d. $\dfrac{61}{324}$ e. none of these

6. Consider the following frequency distribution. Find the class mark for interval 5.

Interval	Score	Frequency
1	70–89	33
2	90–109	39
3	110–129	84
4	130–149	48
5	150–169	36
6	170–189	18
		Total = 258

   a. 150   b. 169   c. 159.5   d. 36   e. $\dfrac{36}{258}$

Use the following information to answer Questions 7 and 8. The graph on the next page shows how many applications for admission were received by a particular medical school over the period 1983–1993.

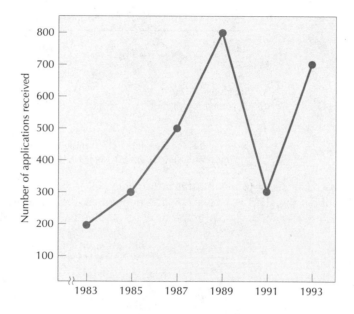

7. In which time period did the number of applications received increase the most?
   a. 1983–1985   b. 1985–1987   c. 1987–1989   d. 1989–1991   e. 1991–1993
8. In which time period did the number of applications received increase the least?
   a. 1983–1985   b. 1985–1987   c. 1987–1989   d. 1989–1991   e. 1991–1993

Supplementary Exercises

9. In recent years attendance at air shows has really soared as indicated in the following table. Draw a frequency polygon to illustrate the data.

Year	Attendance (in millions)
1987	14.1
1988	18.3
1989	22.0
1990	23.1
1991	23.9
1992	24.4

(*Source:* International Council of Airshows, 1993)

10. There were 2,162,000 deaths reported in the United States during 1990. The leading causes of these deaths are shown in the following chart:

Cause of Death	Number of People
Heart disease	725,000
Cancer	506,000
Stroke	145,340
Accidents	93,550
Chronic lung disease	88,989
Other	603,121

(*Source: Vital Statistics of the U.S.* published by the National Center for Health Statistics (NCHS), 1992)

Draw a bar graph to picture this information.

11. The following chart indicates how the population of the United States has changed over the years. Draw a bar graph to illustrate the data.

Year	Population (in millions)
1900	157,553
1910	163,026
1920	168,903
1930	174,882
1940	180,671
1950	188,543
1960	196,308
1970	205,052
1980	227,722
1990	249,924
1991	252,688

(*Source:* U.S. Bureau of the Census, *Current Population Reports*, Series P-24, Nos. 311, 1045, and 1083)

12. Many health insurers will reimburse their clients for claims on the basis of a schedule of allowances. This schedule is often determined by the prevailing rate charged by doctors in a region for the same service. According to the Metropolitan Health Insurance Company, the fee (in dollars) charged by 50 doctors in 1995 for a routine delivery of a child was as follows:

```
4500  5250  4600  4600  4600  4900  4600  5100  4600  4500
4850  5200  4950  4900  5200  5100  4900  5000  5100  4600
5300  5150  4600  4600  5100  5000  5300  4600  4600  5400
5000  4800  4800  4800  4600  4600  5100  4700  5400  5100
5100  5499  5400  4600  4800  5000  4600  4900  4800  4600
```

Construct a frequency distribution and then draw its histogram. (Use 10 intervals.)

13. Draw a dotplot for the following data:

```
76  73  75  76  69  70
72  71  72  73  67  71
```

```
71  77  74  61  67  62
67  71  72  71  79  78
74  77  75  73  67  70
70  74  72  69  67  71
```

14. Construct a stem-and-leaf diagram for the following data, which represent the monthly energy use (in billions of kilowatt hours) for several cities.

```
 7.9    4.4   13.8   12.3
 8.4    4.9   12.9    6.6
10.0    7.2   10.3    8.6
 5.0   11.9    6.8    9.5
```

(*Source:* U.S. Department of Energy's Energy Information Administration, *Annual Energy Review*, 1991)

15. Tuition costs have been increasing. At one particular college, the tuition charge to a full-time student registering for 12 credits has been as follows:

Year	Tuition Cost
1988	$2500
1989	2600
1990	2750
1991	3000
1992	3200
1993	3500

Using 1988 as a base year, find index numbers for the tuition for each year. Interpret the results.

16. The pictograph below is intended to illustrate the fact that the amount of money collected from sales tax in Greenville has doubled between 1985 and 1995. How should it be modified so that it will convey a fair impression of the actual change?

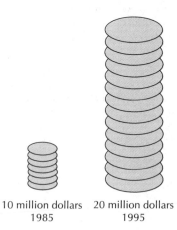

10 million dollars   20 million dollars
       1985   1995

17. The Medical Director of the emergency room at Billings Hospital has been collecting data regarding the blood serum cholesterol level of patients. The following data are available for one particular day:

240	267	287	251	324	281	229	327	267	281
228	296	303	244	309	245	239	291	278	259
237	229	312	258	262	261	262	279	289	256
237	229	312	258	262	261	262	279	289	256
224	241	242	272	278	243	251	261	245	267
238	321	239	261	285	272	241	248	248	236

Summarize the data by constructing
a. a frequency distribution. (Use 10 intervals.)
b. a percent frequency distribution
c. a cumulative percent frequency distribution
d. a histogram
e. a stem-and-leaf diagram

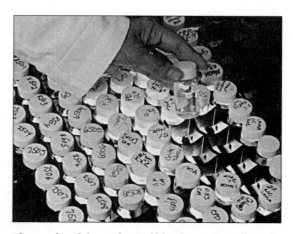

The results of the analysis of blood samples collected from large groups of individuals can be analyzed using stem-and-leaf diagrams. (© *DiMaggio/Kalish, Fran Heyl Associates*)

18. Consider the following stem-and-leaf display.

Stem	Leaf
1	2 2 5
2	3 6 8 2 9
3	5 4 7 6 3
4	7 5 3 2 0 1

Stem	Leaf
5	1 7 5
6	3 6 6 8 2
7	1 2

How many items are in the data set?

19. The average cost of a cab ride from the airport to the center of a large city over the years has changed as follows:

Year	Cost
1990	$12.45
1991	13.16
1992	15.67
1993	16.49
1994	18.11
1995	19.95

Using 1990 as a base year, compute cost indexes for the years 1991–1995.

20. Each of the 33 students in a day-care center was asked to indicate the number of brothers or sisters in his or her family. The following table presents the results of their answers:

Number of Brothers or Sisters	Frequency
0	4
1	5
2	11
3	7
4	4
5	2

Draw a frequency histogram for the data.

21. Based upon your own experience, which of the following do you think is (are) likely to be normally distributed?
    a. The height of fellow students at your college.
    b. The time needed by many students to drive to school.
    c. The amount of money deducted for medical expenses from an individual's 1040 tax form.
    d. IQ scores of math teachers.

22. Refer back to the Did You Know That section at the beginning of this chapter (page 36).
    a. Draw a circle graph to illustrate the information presented in part (a).
    b. Draw a bar graph to illustrate the information presented in part (b).
    c. Draw a bar graph to illustrate the information presented in part (c).

23. A total of 473,977 immigrants entered the United States legally during 1992. Their reasons for entering the United States are as follows:

Reason for Entering U.S.	Percentage
Employment based	12
Relatives of U.S. citizen	24
Sponsored by their family	22
1986 Amnesty legislation	17
Refugee/asylum status	12
Other	13

(*Source:* Immigration and Naturalization Service, 1993)

a. How many immigrants entered the United States as a result of the 1986 amnesty legislation?
b. How many immigrants entered the United States as relatives of a U.S. citizen or with refugee/asylum status?

24. The *National Commission for Prevention of Child Abuse* reported that 2.9 million children were maltreated in 1992 as opposed to 1.9 million maltreated children in 1985. Do family values play any role? As a matter of fact, the following statistics are available:

72% of American children live with both parents.
23% of American children live with the mother only.
3% of American children live with the father only.
2% of American children live with neither parent.

Draw a pie chart to picture this information.

Thinking Critically

1. An inexperienced statistician constructed the following frequency distribution of the salary of 6229 people. Is there anything wrong with the designations for class intervals?

Interval	Frequency
Under $10,000	269
$11,000–$15,000	1384
$16,000–$25,000	2469
$26,000–$35,000	1201
$36,000–$45,000	861
$46,000–$55,000	45

2. After surveying numerous families in Breyerville to determine their monthly telephone bills, several students set up the following class intervals. Is anything wrong?

Monthly Bill	Frequency
$10–15	22
15–25	39
25–35	57
35–45	61
Over $45	89

3. Thirty students in a statistics class received the following scores on the mathematics section of the Scholastic Aptitude Test (SAT). Draw a stem-and-leaf diagram for the data by grouping the numbers so that 20 values fall on each line. Thus the stems will be 4 and 5 where each stem will have five lines (leaves). Values falling between 400 and 419 (inclusive) are on the first leaf, values falling between 420 and 439 (inclusive) are on the second leaf, and so on.

586	512	478	531	532
483	576	493	523	428
512	419	501	513	488
510	576	526	486	449
433	528	512	503	573
463	593	572	517	586

4. The following table presents information on recent fatal aircraft accidents and the number of passenger deaths involved.

Year	Number of Fatal Aircraft Accidents	Number of Passenger Deaths
1981	21	362
1982	26	764
1983	20	809
1984	16	223
1985	22	1066
1986	22	546
1987	26	901
1988	28	729
1989	27	817
1990	25	495

(*Source:* International Civil Aviation Organization, Montreal, Canada; Annual Report of the Council, 1990)

    a. Draw a bar graph to picture this information.

    b. Draw a frequency polygon to picture this information.

    c. Which graph (bar graph or frequency polygon) is more helpful in analyzing trends in fatal aircraft accidents? Explain your answer.

5. Refer back to the second newspaper article given at the beginning of this chapter on (page 38). Draw a bar graph to picture the information contained in the article.

Case Study

1. Consider the following newspaper article:

IS YOUR WATER SAFE?

New York (Aug. 11): Nearly two decades after Congress passed the Safe Drinking Water Act, the water coming from a faucet may contain contaminants and impurities. One in six people drink water with excessive amounts of lead, a heavy metal that impairs children's I.Q., attention span, and raises blood pressure levels.

    Recently, the EPA stated that lead levels in the tap in many Northeastern cities (New York, Boston) are excessive. In New York alone, the water quality of 20 ppb (parts per billion) of lead far exceeds the safe action level of 15 ppb. Drinking water contributes about 10 to 20% of the total lead exposure of children according to the EPA.

Source: Environmental Protection Agency, 1995

A July 1995 survey conducted by the EPA of 60 systems in the New York City and Long Island areas disclosed the following lead levels (in ppb):

22	21	24	29	23	22	23	19	20	22	22	23
17	19	20	26	22	20	24	18	18	19	23	18
18	24	18	31	27	19	28	15	17	16	18	19
25	22	15	29	21	28	20	14	26	18	23	22
32	30	19	27	20	21	18	16	25	21	21	17

Summarize the above data using a
  a. frequency distribution (10 classes)
  b. histogram
  c. stem-and-leaf diagram
  d. dotplot diagram

3

NUMERICAL METHODS FOR ANALYZING DATA

Chapter Outline

DID YOU KNOW THAT

a   the *median* United States income in 1993 was $32,073. (*Source:* U.S. Bureau of the Census)

b   Americans take their dogs to the vet an *average* of 2.4 times a year at a cost of $82.86 and that they take their cats to the vet an *average* of 1.6 times a year at a cost of $54.26. (*Source: Statistical Abstract of the U.S.,* 1993)

c   tall people earn more money than shorter people with an *average* salary benefit of $600 to a tall man for each extra inch (in height). (*Source: Health,* May/June 1993, p. 12)

d   the *average* size of a mortgage applied for dropped to $116,991 in 1993 as opposed to $119,199 in 1992. (*Source:* Federal Reserve Bank of New York)

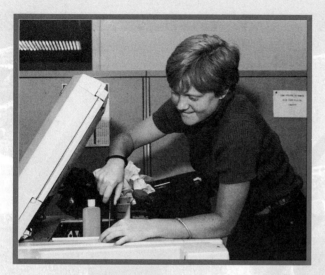

When repairing photocopying machines, a technician must keep accurate records on the performance of certain components to facilitate comparisons among various brands of products. See further discussion on page 159. *(George Semple)*

In this chapter we discuss arithmetic methods of summarizing data. We compute various numbers such as the mean, standard deviation, and so on. In turn, these descriptive measures enable us to analyze a data set adequately.

Chapter Objectives

- **To work with** summation notation, which is used to denote the addition of numbers. (Section 3.2)

- **To discuss** the mean, median, and mode as three different ways of measuring some general trend or location of the data. These will be called measures of central tendency. (Section 3.3)

- **To calculate** the range, standard deviation, variance, and average deviation. Even when we know the mean, median, or mode, we might still want to know whether the numbers are close to each other or whether they are spread out. These are called measures of dispersion or measures of variation. (Section 3.4)

- **To apply** several shortcut formulas for calculating the mean, variance, and standard deviation for grouped or ungrouped data. (Section 3.5)

- **To find** an interpretation for the standard deviation. (Section 3.6)

- **To understand** Chebyshev's theorem, which gives us some information about where many terms of a distribution will lie. (Section 3.6)

- **To draw** the graphs of cumulative frequency distributions and the graphs of cumulative relative frequency distributions. These are called ogives. (Section 3.7)

- **To analyze** the idea of percentiles, which give us the relative standing of one score when compared with the rest. Specifically, percentiles tell us what percent of the scores are above or below a given score. (Section 3.7)

- **To compute** z-scores. These give us the relative position of a score with respect to the mean and are expressed in terms of standard deviations. (Section 3.8)

STATISTICS IN ACTION

Note the use of the word "average" in both newspaper articles. Is the word "average" used in the same manner in both instances? How do we calculate such averages?

MUST WE SPEED?

Jan. 10: A new study conducted by the Department of Transportation found that more and more drivers are ignoring the states' new legal 65 mph speed limit on the highways and freeways. A survey of 500 cars found them travelling at an average speed of 73 mph. Stricter enforcement of existing speed limits was urged.

January 10, 1996

CREDIT CARD INTEREST STILL AN AVERAGE 18%

New York (Dec. 2): Despite the recent sharp drop in interest rates, many of our country's major lending institutions have still not lowered the annual interest rates charged to their credit card customers. A survey of 100 banks nationwide disclosed that the average annual interest rate charged was 18%. Several U.S. senators have called this legal usury.

December 2, 1995

Now consider the newspaper article shown below. In this case the word "median" is used. Exactly what is meant by a median duration? Is median duration the same as average duration? In this chapter we analyze the meaning of these words.

FIRST MARRIAGES FOR WOMEN LAST ONLY 3.4 YEARS BEFORE DIVORCE

Washington (April 2): In its latest *Marriage, Divorce and Remarriage in the 1990s* report, the U.S. Census Bureau reports that the median duration of first marriages that end in divorce is approximately 6.3 years. For women in their late 20s and early 30s—the prime ages for divorce—their average is 3.4 and 4.9 years. Females in the 30–44 age group have the highest divorce rate (about 41%). Those under 30, 38% and around 33% for 45–54 year olds. College graduates are least likely to divorce.

The median duration between a first divorce and remarriage is 2.5 years. 3/4 of remarriages occur within 5 years of divorce.

Monday–April 4, 1993

3.1 INTRODUCTION

In the preceding chapter we learned how to summarize data and present it graphically. Although the techniques discussed there are quite useful in describing the features of a distribution, statistical inference usually requires more precise analysis of the data. In particular, we will discuss many measures that locate the *center* of a distribution of a set of data and analyze how dispersed or spread out the distribution is. In this chapter we examine measures of central tendency and measures of variation (dispersion).

Summation notation

Most analyses involve various arithmetic computations that must be performed on the data. In each case the operation of addition plays a key role in these calculations. It is for this reason that we introduce a special shorthand notation called **summation notation**.

Now consider the card from the Emerson Medical Laboratory shown below.

<table>
<tr><td colspan="4" align="center">**Emerson Medical Laboratory**
Staten Island, N.Y.</td></tr>
<tr><td>Smith, Mary Ellen</td><td>F</td><td>32</td><td>Nov 1, 1995</td></tr>
<tr><td>patient's name</td><td>sex</td><td>age</td><td>date</td></tr>
<tr><td>Results</td><td colspan="2" align="center">109</td><td>266</td></tr>
<tr><td>Percentile Rank</td><td colspan="2" align="center">79</td><td>68</td></tr>
<tr><td></td><td colspan="2" align="center">Triglycerides</td><td>Cholesterol</td></tr>
</table>

The card gives Mary Ellen Smith her percentile rank for triglycerides and cholesterol. How should she interpret this result? Is it better to have a higher or lower percentile rank? It should be apparent that we cannot answer this unless we understand the meaning of percentiles and learn how to calculate them. This will be done in this chapter.

After the data have been analyzed numerically, the techniques of statistical inference can be applied. These will be discussed in later chapters.

3.2 SUMMATION NOTATION

Often when working with a distribution of many numbers, we use letters with subscripts, that is, small numbers attached to them. Thus, we write $x_1, x_2, x_3, \ldots, x_n$, which is read "x sub one, x sub two, x sub three, . . . , x sub n." To be specific, consider the following set of numbers:

$$7, \ 15, \ 5, \ 3, \ 9, \ 8, \ 14, \ 21, \ 10$$

Here we let x_1 denote the first number; that is, $x_1 = 7$, and let x_2 denote the second number; that is, $x_2 = 15$. Similarly, $x_3 = 5$, $x_4 = 3$, $x_5 = 9$, $x_6 = 8$, $x_7 = 14$, $x_8 = 21$, and $x_9 = 10$.

To indicate the operation of taking the sum of a sequence of numbers, we use the Greek symbol Σ, which is read as sigma. To add all the x's, we write

$$\sum_{i=1}^{9} x_i$$

This tells us to add all consecutive values of x starting with x_1 and preceeding to x_9. Thus,

$$\sum_{i=1}^{9} x_i = x_1 + x_2 + x_3 + x_4 + x_5 + x_6 + x_7 + x_8 + x_9$$
$$= 7 + 15 + 5 + 3 + 9 + 8 + 14 + 21 + 10$$
$$= 92$$

If we only wanted to add $x_1 + x_2 + x_3 + x_4 + x_5$, we would write $\sum_{i=1}^{5} x_i$. The i (or j) in

Index the summation symbol $\sum_{i=1}^{n}$ is referred to as the **index**.

Lower limit of summation The **lower limit of summation** is the value of the index placed below the sum-
Upper limit of summation mation symbol and the **upper limit of summation** is the value of the index placed above the summation symbol. In $\sum_{i=1}^{4} i$ the lower limit is 1 and the upper limit is 4. Of

course, $\sum_{i=1}^{4} i$ means the sum of the integers designated by i, from $i = 1$ to $i = 4$, so that

$$\sum_{i=1}^{4} i = 1 + 2 + 3 + 4 = 10.$$

Throughout our study of statistics we will have need to work with various applications of the summation Σ symbol. Great care must be exercised when using this symbol. For example, using the Σ notation, we have the following:

$$\sum_{i=1}^{n} x_i^2 = x_1^2 + x_2^2 + x_3^2 + \cdots + x_n^2$$

$$\sum_{i=1}^{n} x_i y_i = x_1 y_1 + x_2 y_2 + \cdots + x_n y_n$$

$$\sum_{i=1}^{n} x_i^2 f_i = x_1^2 f_1 + x_2^2 f_2 + \cdots + x_n^2 f_n$$

COMMENT The symbols $\sum_{i=1}^{n} x_i^2$ and $\left(\sum_{i=1}^{n} x_i\right)^2$ are quite different. The symbol $\sum_{i=1}^{n} x_i^2$ means that we first square the numbers and add them together, whereas the symbol

$$\left(\sum_{i=1}^{n} x_i \right)^2$$ means that we add all the x_i's together to obtain a sum and then square the sum.

Let us illustrate the use of the Σ symbols with several examples.

EXAMPLE 1

Find $\sum_{i=1}^{5} x_i^2$ and $\left(\sum_{i=1}^{5} x_i \right)^2$ for the following data:

$$x_1 = 3, \qquad x_2 = 7, \qquad x_3 = 6, \qquad x_4 = 3, \qquad x_5 = 9$$

SOLUTION $\sum_{i=1}^{5} x_i^2$ means $x_1^2 + x_2^2 + x_3^2 + x_4^2 + x_5^2$

Thus,

$$\begin{aligned}
\sum_{i=1}^{5} x_i^2 &= 3^2 + 7^2 + 6^2 + 3^2 + 9^2 \\
&= 9 + 49 + 36 + 9 + 81 \\
&= 184
\end{aligned}$$

Also $\left(\sum_{i=1}^{5} x_i \right)^2$ means $(x_1 + x_2 + x_3 + x_4 + x_5)^2$.

Thus,

$$\begin{aligned}
\left(\sum_{i=1}^{5} x_i \right)^2 &= (3 + 7 + 6 + 3 + 9)^2 \\
&= 28^2 \\
&= 784
\end{aligned}$$

Therefore, $\sum_{i=1}^{5} x_i^2$ and $\left(\sum_{i=1}^{n} x_i \right)^2$ are quite different.

NOTATION   Throughout this text, whenever we use the Σ notation, we will want to add *all* available data. Therefore, to simplify the formulas, we will sometimes write the summation symbol without any index. *Thus, when no index is indicated, it is understood that all the data are being used.*

COMMENT   In Example 1 we can write $\sum_{i=1}^{5} x_i^2$ and $\left(\sum_{i=1}^{5} x_i \right)^2$ as Σx^2 and $(\Sigma x)^2$ since we are using all available data.

E X A M P L E   2   Using the data given below, find (a) $\Sigma\, x$, (b) $\Sigma\, y$, (c) $\Sigma\, xy$, (d) $\Sigma\, x \cdot \Sigma\, y$.

x	3	5	6	8	11
y	8	7	9	10	12

   a. $\Sigma\, x$ means $3 + 5 + 6 + 8 + 11 = 33$. Thus, $\Sigma\, x = 33$.

   b. $\Sigma\, y$ means $8 + 7 + 9 + 10 + 12 = 46$. Thus, $\Sigma\, y = 46$.

   c. To find $\Sigma\, xy$, we must first find all the products of the corresponding x and y values and then add these products together. Thus, we have

$$\Sigma\, xy = 3(8) + 5(7) + 6(9) + 8(10) + 11(12)$$
$$= 24 + 35 + 54 + 80 + 132$$
$$= 325$$

     Therefore, $\Sigma\, xy = 325$.

   d. The symbol $\Sigma\, x \cdot \Sigma\, y$ means the product of the two summations $\Sigma\, x$ and $\Sigma\, y$. From parts (a) and (b) we already know that $\Sigma\, x = 33$ and $\Sigma\, y = 46$ so that

$$\Sigma\, x \cdot \Sigma\, y = 33(46) = 1518$$

E X A M P L E   3   If $x_1 = 7$, $x_2 = 3$, $x_3 = 9$, $x_4 = 4$, $f_1 = 5$, $f_2 = 1$, $f_3 = 8$, and $f_4 = 2$, find

   a. $\displaystyle\sum_{i=1}^{4} x_i$ b. $\displaystyle\sum_{i=1}^{4} f_i$ c. $\displaystyle\sum_{i=1}^{4} x_i f_i$

SOLUTION

   a. $\displaystyle\sum_{i=1}^{4} x_i$ means $x_1 + x_2 + x_3 + x_4$ so that

$$\sum_{i=1}^{4} x_i = 7 + 3 + 9 + 4 = 23$$

   b. $\displaystyle\sum_{i=1}^{4} f_i = f_1 + f_2 + f_3 + f_4 = 5 + 1 + 8 + 2 = 16$

   c. $\displaystyle\sum_{i=1}^{4} x_i f_i = x_1 f_1 + x_2 f_2 + x_3 f_3 + x_4 f_4$
$$= 7(5) + 3(1) + 9(8) + 4(2) = 118$$

When we work with summations, there are certain rules that we use. These rules are easily verified by simply writing out in full what each of the summations represents.

RULE 1

$$\sum_{i=1}^{n} (x_i + y_i) = \sum_{i=1}^{n} x_i + \sum_{i=1}^{n} y_i$$

RULE 2

$$\sum_{i=1}^{n} kx_i = k \cdot \sum_{i=1}^{n} x_i$$

RULE 3

$$\sum_{i=1}^{n} k = n \cdot k$$

In words, the *first rule* says that the summation of a sum of two terms equals the sum of the individual summations. The *second rule* says that we can "factor" a constant out from under the operation of a summation. Thus,

$$\sum_{i=1}^{n} kx_i = kx_1 + kx_2 + kx_3 + \cdots + kx_n$$
$$= k(x_1 + x_2 + \cdots + x_n)$$
$$= k \cdot \sum_{i=1}^{n} x_i$$

The *third rule* says that the summation of a constant is simply n times that constant or the constant times the number of indicated terms in the summation.

These rules are easy to use and will be applied throughout the book.

EXAMPLE 4

If $x_1 = 4$, $x_2 = 5$, $x_3 = 7$, and $x_4 = 9$, find $\sum_{i=1}^{4} (3x_i - 1)$.

SOLUTION

$$\sum_{i=1}^{4} (3x_i - 1) = (3x_1 - 1) + (3x_2 - 1) + (3x_3 - 1) + (3x_4 - 1)$$
$$= (3 \cdot 4 - 1) + (3 \cdot 5 - 1) + (3 \cdot 7 - 1) + (3 \cdot 9 - 1)$$
$$= 11 + 14 + 20 + 26$$
$$= 71$$

Thus, $\sum_{i=1}^{4} (3x_i - 1) = 71$ for the previous values of x_i.

COMMENT Example 4 can also be evaluated by using Rules 1 to 3. Can you see how?

EXERCISES FOR SECTION 3.2

1. Rewrite each of the following using summation notation:
   a. $x_1 + x_2 + x_3 + x_4 + x_5 + x_6 + x_7$
   b. $x_1^2 + x_2^2 + x_3^2 + x_4^2 + x_5^2 + x_6^2 + x_7^2$
   c. $x_1 f_1 + x_2 f_2 + x_3 f_3 + x_4 f_4 + x_5 f_5 + x_6 f_6 + x_7 f_7$
   d. $10x_1 + 10x_2 + 10x_3 + 10x_4 + 10x_5 + 10x_6 + 10x_7$
   e. $(2y_1 + x_1) + (2y_2 + x_2) + (2y_3 + x_3) + (2y_4 + x_4) + (2y_5 + x_5)$
      $+ \cdots + (2y_n + x_n)$

2. Rewrite each of the following without summation notation.
   a. $\displaystyle\sum_{i=1}^{5} x_i$
   b. $\displaystyle\sum_{i=5}^{10} x_i y_i$
   c. $\displaystyle\sum_{i=3}^{10} (5x_i + 3)$
   d. $\displaystyle\sum_{i=1}^{6} 2y_i f_i$

3. If $x_1 = 4$, $x_2 = 17$, $x_3 = 7$, $x_4 = 19$, and $x_5 = 21$, find each of the following:
   a. Σx
   b. $(\Sigma x)^2$
   c. Σx^2
   d. $\Sigma(x + 5)$
   e. $\Sigma(x + 5)^2$
   f. $\Sigma(2x + 3)$
   g. $\Sigma(2x + 3)^2$

4. If $x_1 = 4$, $x_2 = 7$, $x_3 = 13$, $x_4 = 31$, $f_1 = 2$, $f_2 = 6$, $f_3 = 8$, $f_4 = 10$, $y_1 = 0$, $y_2 = 3$, $y_3 = -2$, and $y_4 = -5$, find each of the following:
   a. Σx^2
   b. $\Sigma x \cdot f$
   c. $\Sigma x \cdot \Sigma y$
   d. Σxy
   e. $\Sigma x \cdot y \cdot f$
   f. $\Sigma(x - y)f$

5. If $\displaystyle\sum_{i=1}^{10} x_i = 18$ and $\displaystyle\sum_{i=1}^{10} x_i^2 = 37$, find each of the following:
   a. $\displaystyle\sum_{i=1}^{10} (x_i + 8)$
   b. $\displaystyle\sum_{i=1}^{10} (x_i + 8)^2$

6. Show that each of the following are true:
   a. $\displaystyle\sum_{i=1}^{n} (x_i - y_i) = \sum_{i=1}^{n} x_i - \sum_{i=1}^{n} y_i$
   b. $\displaystyle\sum_{i=1}^{n} (x_i + k) = \sum_{i=1}^{n} x_i + nk$

7. Professor Lorenz requires each of the students in her English composition course to submit five different essays during the semester. In the following frequency distribution she has recorded the number of students who have completed 1, 2, 3, 4, or 5 of these assignments.

Number of Essays Completed x	Frequency f	xf
1	7	7
2	11	22
3	8	24
4	6	24
5	4	20
	Total = 36	97

   a. The sum $\Sigma f = 36$. What does this sum represent?
   b. The sum $\Sigma xf = 97$. What does this sum represent?

For Exercises 8–10, find:

    a. $\Sum x$ b. $(\Sum x)^2$ c. $\Sum x^2$

8. According to the 1989 *World Almanac*, the annual salaries of the governors of the Great Lakes States were as follows: (All data are rounded to the nearest thousand dollars.)

        82   60   64   58   76

9. A certain brand of sugar comes in packages marked "net weight 5 lb." To ensure that the amount of sugar being packaged does indeed weigh 5 lb on the average, a sample of 10 bags were randomly selected and then their contents carefully weighed. The results were:

    4.9   5.2   5.3   5.1   4.8   4.7   4.9   5.0   5.1   5.2

10. The blood serum cholesterol level of ten male patients in the cardiac unit at University Hospital was as follows:

    223   240   212   193   276   253   248   239   278   269

3.3   MEASURES OF CENTRAL TENDENCY

To help us understand what we mean by measures of central tendency, consider the Metropolis Police Department, which recently purchased tires from two different manufacturers. The police department is interested in determining which tire is superior and has compiled a list on the number of miles each set of tires lasted before replacement was needed for 14 of its identical police cars. Seven cars were fitted with Brand X tires and seven with Brand Y. The number of miles each set lasted before replacement is indicated in the following chart.

Brand X	Brand Y
14,000	10,000
12,000	8,000
12,000	14,000
14,000	10,000
14,000	8,000
11,000	40,000
14,000	8,000
Total = 91,000	98,000

It would appear that Brand Y is the better tire since the seven police cars were driven a combined total of 98,000 miles using Brand Y tires but only 91,000 miles with Brand X tires. Let us, however, analyze the data by computing the average number of miles driven with each brand of tires. We will divide each total by the number of police cars used for each brand. We have the following.

	Average	
Brand X		**Brand Y**
$\dfrac{91,000}{7} = 13,000$		$\dfrac{98,000}{7} = 14,000$

Since Brand Y tires lasted on the average 14,000 miles and Brand X on the average 13,000 miles, it again appears that Brand Y is superior. If we look at the data more carefully, however, we find that Brand X tires consistently lasted around 13,000 miles. As a matter of fact, they lasted 14,000 miles most often, four times. Brand Y, on the other hand, lasted 14,000 miles only once. They lasted 8000 miles most often. Thus, in terms of consistency of performance one might say that Brand X is more consistent. Let us arrange the data for each brand in order from smallest to largest. We have the following.

Brand X	Brand Y
11,000	8,000
12,000	8,000
12,000	8,000
(14,000)	(10,000)
14,000	10,000
14,000	14,000
14,000	40,000

In this chart we have circled two numbers. These are the numbers that are in the middle for each brand: 14,000 for Brand X and 10,000 for Brand Y.

All the preceding ideas are summarized in the following definitions.

Definition 3.1
Mean or average

The **mean**, or **average**, of a set of numbers is obtained by adding the numbers together and dividing the sum by the number of numbers added. We denote the sample mean by the symbol $\bar{x}$, which is read x bar, and the population mean by the Greek letter μ, which is read mu.

Definition 3.2
Mode

The **mode** of a set of numbers is the number (or numbers) that occurs most often. If no number occurs more than once, there is no mode.

Definition 3.3
Median

The **median** of a set of numbers is the number that is in the middle when the numbers are arranged in order from the smallest to the largest. The median is easy to calculate when we have an odd number of numbers. If we have an even number of them, the median is defined as the average of the middle two numbers when arranged in increasing order of size. We denote the median by the symbol $\tilde{x}$, read "x tilde."

COMMENT A set of numbers may have more than one mode. (See Example 1 of this section.)

When the preceding definitions are applied to our example, we get the following table.

	Brand X	Brand Y
Mean	13,000	14,000
Median	14,000	10,000
Mode	14,000	8,000

Which number should the police department use to determine the superior tire: the mean, median, or mode? You might say that the mean is not particularly helpful since one set of Brand Y tires lasted 40,000 miles and this instance had the effect of increasing the average for Brand Y considerably. In terms of consistency Brand X appears to be superior to Brand Y.

Notice that in calculating the mean we had to add numbers. Since the operation of addition plays a key role in our calculations, we use the summation notation discussed in the last section.

Consider the following set of numbers:

$$7, \ 15, \ 5, \ 3, \ 9, \ 8, \ 14, \ 21, \ 10$$

We have

$$\Sigma x = x_1 + x_2 + x_3 + x_4 + x_5 + x_6 + x_7 + x_8 + x_9$$

$$\Sigma x = 7 + 15 + 5 + 3 + 9 + 8 + 14 + 21 + 10$$

$$\Sigma x = 92$$

If we divide this sum by n, the number of terms added, the result will be the mean for these numbers.

COMMENT   Remember that usually no indexes will be shown in the summation formulas in this book: The summations are understood to be over all available data.

FORMULA 3.1

Sample Mean

Sample mean

The mean of a set of sample values $x_1, x_2, x_3, \ldots, x_n$ is given by

$$\textbf{Sample mean} = \bar{x} = \frac{\Sigma x}{n}$$

$$= \frac{x_1 + x_2 + x_3 + \cdots + x_n}{n} \quad \begin{array}{l} \leftarrow \text{ sum of the data} \\ \leftarrow \text{ number of pieces of data} \end{array}$$

For our set of numbers the mean is

$$\bar{x} = \frac{\Sigma x}{n} = \frac{92}{9} = 10.22$$

EXAMPLE   1
The Better Worker

The district office of a state unemployment insurance department recently hired two new employees, Rochelle and Sharon, to interview prospective aid recipients. Their supervisor is interested in determining who is the better worker. The following chart indicates the number of clients interviewed daily by each on seven randomly selected days.

Rochelle	Sharon
54	38
67	51
46	46
52	49
45	46
39	38
41	44
344	312

Calculate the sample mean, median, and mode for each employee.

SOLUTION   Let x represent the number of clients interviewed daily by Rochelle and let y represent the number of clients interviewed daily by Sharon. For Rochelle the sample mean is

$$\bar{x} = \frac{\Sigma x}{n} = \frac{x_1 + x_2 + x_3 + x_4 + x_5 + x_6 + x_7}{7}$$

$$= \frac{54 + 67 + 46 + 52 + 45 + 39 + 41}{7}$$

$$= \frac{344}{7} = 49.14$$

For Sharon the sample mean is

$$\bar{y} = \frac{\Sigma y}{n} = \frac{y_1 + y_2 + y_3 + y_4 + y_5 + y_6 + y_7}{7}$$

$$= \frac{38 + 51 + 46 + 49 + 46 + 38 + 44}{7}$$

$$= \frac{312}{7} = 44.57$$

Let us now arrange the numbers for each employee in order from the lowest to the highest. We get the following.

Rochelle	39	41	45	(46)	52	54	67
Sharon	38	38	44	(46)	46	49	51

Notice that one number has been circled for each worker. This number is in the middle and it represents the median. For both workers the median is 46.

The mode is the number that occurs most often. For Rochelle there is no mode since no number occurs more than once. For Sharon there are two modes, 38 and 46.

Which statistic, the mean, mode, or median, should the supervisor use in determining who is the better worker?

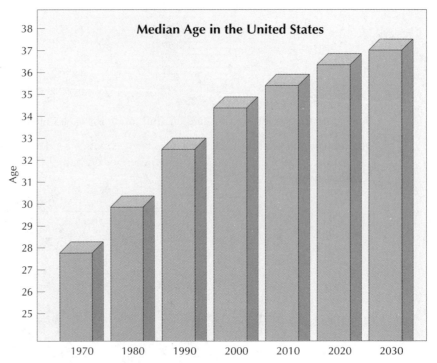

Note the use of the word "median" in this newspaper clipping. The information contained in it has important implications for our future.

E X A M P L E 2   Calculate the sample mean of the following numbers:

28	19	25	17	28	19	26	17	28	25
17	28	31	22	31	17	31	28	31	14
31	14	17	28	24	31	17	14	26	24
24	24	19	24	14	28	22	31	17	22
25	12	26	19	26	12	19	26	19	28

SOLUTION One way of calculating the mean is to add the numbers and divide the result by 50, which is the number of terms. This takes a considerable amount of time, but nevertheless we have

$$\bar{x} = \frac{\Sigma x}{n} = \frac{x_1 + x_2 + \cdots + x_{50}}{n}$$

$$= \frac{28 + 19 + 25 + \cdots + 28}{50}$$

$$= \frac{1145}{50} = 22.9$$

Thus, the sample mean is 22.9.

We can also group the data (not necessarily in the usual interval format) as follows.

Column 1	Column 2	Column 3	Column 4
Number (x)	Tally	Frequency (f)	$x \cdot f$
12	||	2	24
14	||||	4	56
17	||||| ||	7	119
19	||||| |	6	114
22	|||	3	66
24							5	120
25	|||	3	75
26							5	130
28	||||| |||	8	224
31	||||| ||	7	217
		Total = 50	1145

Column 4 was obtained by multiplying each number by its frequency. If we now sum column 3 and column 4 individually and divide the column 4 total by the column 3 total, our answer will be 22.9, which is the mean. Although we get the same result as before, it was considerably easier to obtain it by grouping the data as we did.

FORMULA 3.2

Sample mean of a distribution of grouped data

The **sample mean of a distribution of grouped data** is given by

$$\bar{x} = \frac{\Sigma xf}{\Sigma f}$$

where xf represents the product of each class mark and its frequency and Σf represents the total number of items in the distribution.

COMMENT It may seem that Formula 3.2 is considerably different from Formula 3.1 for calculating the sample mean. In reality it is not. In ungrouped data the frequency of each observation is 1 and the total number of terms, Σf, is n. So, Formulas 3.1 and 3.2 are really the same.

To illustrate further the use of Formula 3.2, we consider the following example (where we have grouped the data in interval form).

EXAMPLE 3   George is a maintenance person in the Auburn Shopping Center. He keeps accurate records on the life of the special security light bulbs that he services. He has recorded the life of 60 light bulbs in the following chart.

Life of Bulb (hours)	Class Mark x	Frequency f	Product $x \cdot f$
40–49	44.5	5	222.5
50–59	54.5	7	381.5
60–69	64.5	8	516.0
70–79	74.5	9	670.5
80–89	84.5	12	1014.0
90–99	94.5	9	850.5
100–109	104.5	6	627.0
110–119	114.5	4	458.0
		Total = 60	4740

Calculate the sample mean life of a light bulb.

SOLUTION   In using Formula 3.2, we use the class mark (see page 44) for each interval. Thus, we assume that any bulb that lasted between 40 and 49 hours actually lasted 44.5 hours. Although this may introduce a slight error, the error will be minimal when the number of bulbs is large. Thus, we use the class mark. Applying Formula 3.2, we have

$$\bar{x} = \frac{\Sigma xf}{\Sigma f} = \frac{222.5 + 381.5 + \cdots + 458.0}{5 + 7 + \cdots + 4} = \frac{4740}{60} = 79$$

In the previous example we assumed that any bulb that lasted between 40 and 49 hours actually lasted 44.5 hours, which is the class mark. Some statisticians prefer to compute medians (and percentiles, to be discussed later) for grouped data by assuming that the entries in a class are evenly distributed in that class. For example, suppose we were interested in calculating the median of a distribution for grouped data. These statisticians would then use the following formula:

Statistician's
formula for median

$$\text{Median} \qquad \tilde{x} = L + \frac{j}{f} \cdot c$$

where L is the lower boundary or lower limit of the category into which the median must fall, f is its frequency, c is its interval length, and j is the number of items still to be counted after reaching L.

To illustrate, suppose we were interested in computing the median for the grouped data given below.

	Interval	
(from)	(up to but not including)	Frequency
9.5	19.5	12
19.5	29.5	10
29.5	39.5	13
39.5	49.5	11
49.5	59.5	7
	Total =	53

When the numbers are arranged in order of size (from the smallest to the largest), the median will be the 27th term. This falls in the 29.5 to 39.5 interval. Since there are $12 + 10$ or 22 below this category, we need another five items in addition to the 22 that fall below this class. We add $\frac{5}{13}$ of the class interval to the lower bound of that class. Here we have $L = 29.5$, $f = 13$, $c = 10$, and $j = 5$ so that

$$\tilde{x} = 29.5 + \frac{5}{13} \cdot 10$$

$$= 33.35 \text{ when rounded}$$

Now let us consider the following examples in which all values are not of equal importance. To solve such examples we must use Formula 3.3.

FORMULA 3.3

Weighted sample
mean

If $w_1, w_2, \ldots, w_n$ are the weights assigned to the numbers $x_1, x_2, \ldots, x_n$, then the **weighted sample mean** denoted by the symbol $\bar{x}_w$ is given by

$$\bar{x}_w = \frac{\Sigma\, xw}{\Sigma\, w} = \frac{x_1 w_1 + x_2 w_2 + \cdots + x_n w_n}{w_1 + w_2 + \cdots + w_n}$$

This formula indicates that we multiply each number by its weight and divide the sum of these products by the sum of the weights.

E X A M P L E   4

Calculating
Term Grades

The grades that Liz received on her exams in a statistics course and the weight assigned to each are as follows.

	Grade	Weight Assigned
Test 1	84	1
Test 2	73	2
Test 3	62	5
Test 4	91	4
Final exam	96	3

Find Liz's average term grade.

SOLUTION   Since each test did not have the same weight, that is, count as much, Formula 3.1 or 3.2 has to be modified. This change is necessary because Formula 3.1 assumes that all numbers are of equal importance, which is not the case in this example. To calculate a weighted mean, we use Formula 3.3. When this formula is applied to our example, we have

$$\bar{x}_w = \frac{x_1 w_1 + x_2 w_2 + \cdots + x_5 w_5}{w_1 + w_2 + \cdots + w_5}$$

$$= \frac{(84 \cdot 1) + (73 \cdot 2) + (62 \cdot 5) + (91 \cdot 4) + (96 \cdot 3)}{1 + 2 + 5 + 4 + 3}$$

$$= \frac{1192}{15} = 79.47 \qquad \text{when rounded off}$$

Thus, the weighted mean is 79.47, not 81.2, which is obtained by adding the numbers together and dividing the sum by n, which is 5.

E X A M P L E   5

Average Cost
of Gasoline

On a recent vacation trip Bill Hunt kept the following record of his gasoline purchases:

	Price per Gallon	Number of Gallons Purchased
Town 1	$1.53	17
Town 2	1.46	21
Town 3	1.49	16
Town 4	1.35	11
Town 5	1.51	19

What is the average cost per gallon of gasoline for the entire trip?

SOLUTION   Since Bill did not purchase an identical amount of gasoline in each town, we cannot use Formula 3.1. We must use Formula 3.3 instead. We have

$$\bar{x}_w = \frac{(1.53)(17) + (1.46)(21) + (1.49)(16) + (1.35)(11) + (1.51)(19)}{17 + 21 + 16 + 11 + 19}$$

$$= \frac{124.05}{84} = \$1.48 \qquad \text{rounded off}$$

Thus, the weighted average cost per gallon of gasoline for the entire trip was $1.48.

Midrange

There are other measures that describe in some way the middle or center of a set of numbers. One of these is known as the **midrange**, which is found by taking the average of the lowest value L and the highest value H. Thus,

$$\text{Midrange} = \frac{L + H}{2}$$

For the sample 3, 7, 11, 15, 16, 18, 21, 22, and 23, the lowest value L is 3 and the highest value H is 23. Therefore, the midrange is

$$\frac{3 + 23}{2} = 13$$

Geometric mean
Harmonic mean

Other measures that are sometimes used are the **geometric mean** and the **harmonic mean**. (See Exercise 17 at the end of this section.)

We mentioned earlier that both $\bar{x}$ and μ will be used to represent the mean. The symbol $\bar{x}$ is used to represent the sample mean or the mean of a sample. Thus, the mean $\bar{x}$ of the sample values $x_1, x_2, x_3, \ldots, x_n$ is given by the formula

$$\text{Sample mean} \qquad \bar{x} = \frac{\Sigma x}{n}$$

The symbol μ is used to represent the mean of the entire population. Thus, the mean of a population of N values, $x_1, x_2, x_3, \ldots, x_N$ is given by the formula

$$\text{Population mean} \qquad \mu = \frac{\Sigma x}{N}$$

where the N values of x constitute the entire population.

COMMENT Generally speaking, Greek letters are used when we are referring to a description of the population as opposed to English letters, which are used when we are referring to a description of a sample.

COMMENT Throughout this section we used $\bar{x}$ since we were calculating sample means. From a given problem we can almost always tell whether we are referring to only part of the population or to the entire population.

Measures of central tendency   **COMMENT**   The mean, median, and mode are known as **measures of central tendency** since each measures some central or general trend, that is, location of the data. In any particular situation one measure will usually be more helpful than the others.

EXERCISES FOR SECTION 3.3

1. The American Hospital Association publishes *Hospital Statistics* that contains information on hospital costs, length of patient stays, and so forth. For one hospital, the age of ten randomly selected mothers at the time of the birth of their first child in 1995 was as follows:

Patient	Age of Mother
Sally	17
Sue	31
Ellen	28
Jennifer	32
Pat	16
Priscilla	19
Deidre	23
Sherry	24
Holly	27
Gwendolyn	15

   Find the mean, median, and modal age of the mother at the time of the birth of their first child. Which is more useful?

2. The United States Commerce Department publishes data on the incomes of people in different occupations. For one trucking company that employs 21 drivers, the following were their 1995 incomes:

$41,000	$39,000	$42,000	$42,000	$45,500	$35,000
57,000	45,000	37,000	96,000	50,500	
49,000	49,000	52,000	52,000	53,000	
37,000	38,000	49,000	47,000	64,000	

   Calculate the mean, median, and modal 1995 incomes for this particular trucking company.

3. Refer back to Exercise 2. Neglecting the one driver who earned $96,000, calculate the mean, median, and mode for the remaining truck drivers. Compare the answers obtained here with the answers obtained in Exercise 2. Comment.

4. According to the American Manufacturer's Association, very few shoppers actually use manufacturer's cents-off coupons. The total value of coupons redeemed by 12 people observed shopping at a local supermarket was as follows:

| 45¢ | 55¢ | $1.05 | 25¢ | 35¢ | 55¢ |
| 25¢ | 25¢ | 40¢ | 60¢ | 80¢ | 55¢ |

Calculate the mean, median, and modal redemption value of the coupons. Which is more useful to a manufacturer?

5. We mentioned earlier that the average size of a mortgage loan dropped to $116,991 in 1993 as opposed to $119,199 in 1992. (*Source:* Federal Reserve Bank of New York) During April 1993 the National Commercial Bank received 20 mortgage loan applications. The amount of money applied for was as follows:

$100,000	$125,000	$180,000	$ 95,000	$155,000
228,000	110,000	140,000	88,000	185,000
175,000	115,000	90,000	105,000	145,000
150,000	80,000	50,000	112,000	130,000

Calculate the mean, median, and mode for these data.

6. There are six photocopying machines in the back office of one Wall Street brokerage firm. During the month of January 1996, these machines produced 5716, 8249, 6283, 7956, 8116, and 7223 copies.
   a. Find the mean, median and modal number of copies produced by these machines.
   b. Which measure of central location do you believe more accurately describes the "typical" number of copies produced?

7. Refer back to Exercise 6. Due to an increase in business, each of these machines will be required to produce an additional 500 copies. Find the mean, median, and modal number of copies under these new conditions.

8. Although the Automobile Manufacturer's Association recommends that the oil levels in cars be checked periodically, nevertheless, drivers frequently neglect to check the oil in their cars. Mary Aquilla owns a gas station on heavily traveled U.S. Highway 1. For the first 60 days in 1995 the number of drivers who asked to have the oil checked while buying gas was as follows:

13	17	8	23	16	18	18	13	8	3
19	16	16	12	14	16	23	8	7	12
11	14	12	14	12	14	17	18	14	8
16	10	12	12	11	22	19	21	16	8
10	8	10	3	19	19	18	17	21	10
3	3	14	8	16	13	18	19	11	8

By grouping the data, find the mean, median, and mode. Which statistic is more useful in supporting the Automobile Manufacturer's claim?

9. *Cost Averaging.* Bill purchased 80 shares of ABC stock on January 2 at a price of $60 per share. On February 2, he purchased 100 shares of the same stock at a price of $46 per share. On March 2, he purchased 180 shares of the same stock at a price of $48 per share. What is Bill's average cost of a share of stock of this company?

10. Are drivers obeying the posted speed limits? The speed of 60 cars as they passed a posted "Maximum 40 MPH Speed Limit" sign was accurately measured. The results are as follows:

Speed (MPH)	Frequency
31–35	4
36–40	7
41–45	14
46–50	16
51–55	9
56–60	8
61–65	2

Find the mean, median, and mode for the data.

11. Important information on the United States prison population is contained in *Profile of Jail Inmates* published annually by the U.S. Bureau of Justice Statistics. One particular correctional facility averaged 138 white-female inmates during 1993. For the months of January through November 1993 the number of white-female inmates was as follows:

109   162   151   123   108   176   161   143   151   102   111

How many white-female inmates were housed in this facility in December 1993?

12. The Consumer's Affairs Department purchased a particular dosage of a popular drug in 88 different pharmacies located throughout the city to determine the average price. The results are as follows:

Number of Pharmacies	Price Charged by These Pharmacies
27	$19.95
32	21.95
16	23.75
13	24.50

Find the average price charged by these pharmacies.

13. During August 1993, the First National State Bank approved 22 home equity loans. If the combined amount loaned was $1,716,000, find the average amount of a loan.

14. Consider the following newspaper article. Find the average price of an admission ticket paid by these 100 people.

STATE TO INVESTIGATE TICKET SCALPING

Dover: The Attorney General announced yesterday that he would launch an immediate investigation into the widespread practice of ticket scalping for tickets to rock concerts. A survey of 100 ticket holders for last night's rock concert indicated that these people paid exorbitant prices for the tickets. The prices paid for the same general admission ticket were as follows:

Number of people	Price paid
36	$30
28	35
19	40
17	45

The consumer is being victimized by this unlawful practice.

December 1, 1985

15. Professor Rodriguez teaches two sections of an introductory English class. In one class the average final exam grade was 84 and in the second class the average final exam grade was 74. Is it safe to assume that the average final exam grade for both classes combined is 79? Explain your answer.

16. An important number to hay fever sufferers is the pollen count. Health officials monitor the situation by taking daily pollen counts. The following data for Stevensville is available concerning the 1992 hay fever season.

Pollen Count Number	Frequency
1–5	8
6–10	18
11–15	19
16–20	23
21–25	16
26–30	12
31–35	4
36–40	3
41–45	2

(*Source:* Lutheran Medical Center, Stevensville, 1993)

Find the mean, median, and mode for the data.

17. The **harmonic mean** of n numbers $x_1, x_2, \ldots, x_n$ is defined as n divided by the sum of the reciprocals of the numbers; that is,

$$\text{Harmonic mean} = \frac{n}{\displaystyle\sum_{i=1}^{n} 1/x_i}$$

Also, the **geometric mean** of a set of n positive numbers $x_1, x_2, \ldots, x_n$ is defined as the nth root of their product; that is,

$$\text{Geometric mean} = \sqrt[n]{x_1 \cdot x_2 \cdots x_n}$$

Find the harmonic mean and the geometric mean of the numbers 8, 5, 9, and 6. (Both the harmonic mean and the geometric mean are used in certain applications.)

18. *Bank failures.* To insure a depositor's money, many banks belong to the Federal Deposit Insurance Corporation (FDIC). In the event of a bank failure the FDIC will insure each customer's deposits (currently) up to $100,000. According to government records, the number of banks that failed and were subsequently taken over by the FDIC in a certain region over the years is as follows.

Year	Number of Bank Failures
1983	2
1984	7
1985	6
1986	4
1987	9
1988	8
1989	6
1990	7
1991	5
1992	14

(*Source:* Federal Reserve Bank of New York, 1993)

Outliers

Trimmed mean

We notice that the preceding set of numbers contains two extremes (or **outliers**) that really should not be included in the data set for one reason or another. For situations such as these, statisticians will compute a **trimmed mean** where high and low values are excluded or "TRIMMED OFF" before calculating the mean. In our case if we exclude *both* the top 10% (the number 14) and the bottom 10% (the number 2) and then calculate the mean of the remaining data, we get

$$10\% \text{ trimmed mean} = \frac{7 + 6 + 4 + 9 + 8 + 6 + 7 + 5}{8} = 6.5$$

The following are the scores of 20 students on the mathematics part of the Scholastic Aptitude Test (SAT):

604	523	551	525	528	498	587	578	611	557
587	499	574	586	591	580	528	593	602	582

  a. Calculate the arithmetic mean for the data.
  b. Calculate the 5% trimmed mean for the data.
  c. Calculate the 10% trimmed mean for the data.
  d. Compare the three means that you obtained in parts (a) to (c). Which of the three means do you think is the best measure of central tendency for the data? Explain your answer.

19. Refer back to Exercise 18. Calculate the midrange for
  a. the number of bank failures
  b. the mathematics SAT scores

20. Three insurance companies claim that the average time that it takes their company to process a claim is five days. A consumer's group decides to test each insurance company's claim. It obtains the following list on the number of days needed by each company to process a claim.

Company A	*Company B*	*Company C*
Days Needed to Process 8 Different Claims	**Days Needed to Process 11 Different Claims**	**Days Needed to Process 8 Different Claims**
1	4	2
2	4	3
3	5	4
4	5	4
5	5	6
7	6	13
8	11	14
10	13	15
	14	
	15	
	16	

  a. Which measurement of average was each insurance company using to support its claim?

  b. From which insurance company would you buy insurance?

21. Using summation notation, verify that the sum of the differences from the mean is always zero, that is verify that

$$\sum_{i=1}^{n} (x_i - \bar{x}) = 0$$

22. Deidre Vasilious is a student at the College of Staten Island. A transcript of her record is shown below. Compute Deidre's grade point average (GPA).

DATE OF ISSUE 9/12/95	THE COLLEGE OF STATEN ISLAND OF THE CITY UNIVERSITY OF NEW YORK				PAGE 1

CURRICULUM		CANDIDATE FOR:		IDENTIFICATION NUMBER
9/01/90 LIB ARTS AND SCI		AA		123-45-6789
2/01/92 PSYCHOLOGY		BA		

STUDENT NAME AND ADDRESS

DATE OF BIRTH
06/23/73

DEIDRE VASILIOUS
2800 VICTORY BOULEVARD
STATEN ISLAND, N.Y. 10314

DATE OF H.S. GRAD. HIGH SCHOOL OF GRADUATION

06/90 PORT RICHMOND HIGH SCHOOL

END OF SESSION OR DATE OF ACTION MO. DAY YR.	COURSE NUMBER	COURSE TITLE	GRADE	CREDITS	QUALITY POINTS
12/23/94		FALL SEMESTER 1994			
	EDE 200	EDUC SOC FOUNDATIONS	B	4.0	12.00
	EDP 420	CURRICULUM I	C+	3.0	6.90
	PSY 202	PSYCHOPATHOLOGY	A-	4.0	14.80
	SOC 210	HEALTH & MEDICINE SOC	B	4.0	12.00
5/28/95		SPRING SEMESTER 1995			
	MTH 123	COLL ALG & TRIG	A	4.0	16.00
	GEO 100	PHYSICAL GEOLOGY	A-	3.0	11.10
	GEO 101	PHYSICAL GEO LAB	A-	1.0	3.70
	ENG 111	COMM WORKSHOP	B-	4.0	8.10
7/12/95		SUMMER SESSION 1995			
	PED 150	AEROBIC MOVEMENT	B+	1.0	3.30
	PSY 332	TESTS/MEASURMENTS	D	4.0	4.00
	MTH 112	PROBABILITY STATISTICS	A	3.0	12.00

GPA CREDITS ATTEMPTED	TOTALCREDITS COMPLETED	QUALITY POINTS	GRADE POINT AVERAGE
35.0	35.0		

3.4   MEASURES OF VARIATION

Although the mean, median, or mode are very useful in analyzing a distribution, there are some disadvantages in using them alone. These measures only locate the center of the distribution. In certain situations location of the center may not be adequate. We need some method of analyzing variation, that is, the difference among the terms of a distribution. In this section we discuss some of the most commonly used methods for analyzing variation.

First let us consider Christina, who is interested in determining the best route to drive to work. During one week she drove to work on the Brooks Expressway and during a second week she drove on the Kingston Expressway. The number of minutes needed to drive to work each day was

Brooks Expressway	15	26	30	39	45
Kingston Expressway	29	30	31	32	33

In each case the average time that it took her to drive to work was 31 minutes. Which way is better?

When she used the Brooks Expressway, the time varied from 15 to 45 minutes. We then say that for the Brooks Expressway the *range* is $45 - 15 = 30$ minutes.

On the Kingston Expressway the time varied from 29 to 33 minutes. Thus, the range is $33 - 29 = 4$ minutes.

Definition 3.4
Range

The **range** of a set of numbers is the difference between the largest number in the distribution and the smallest number in the distribution.

The range is frequently used by manufacturers as a measure of dispersion (spread) in specifying the variation in the quality of a product. So, although the average diameter of a drill bit may be $\frac{15}{32}$ inches, in reality the range in size may be enormous. The manufacturer usually specifies the range to prospective customers.

The range is also used frequently by stock brokers to describe the prices of certain stock. One often hears such statements as "Stock X had a price range of 15 to 75 dollars, or 60 dollars, during the year."

The range is by far the simplest measure of variation to calculate since only two numbers are needed to calculate it; however, it does not tell us anything about how the other terms vary. Furthermore, if there is one extreme value in a distribution, the dispersion or the range will appear very large. If we remove the extreme term, the dispersion may become quite small. Because of this, other measures of variation such as variance, standard deviation, or average deviation are used.

To calculate the sample variance of a set of numbers, we first calculate the sample mean of the numbers. We then subtract the sample mean from each number and square the result. Finally, we divide the result by $n - 1$, where n is the number of items in the sample. The result is called the **sample variance** of the numbers. If we now take the square root of the sample variance, we get the **sample standard deviation** for the numbers.[*] If instead of squaring the differences from the mean we take the absolute value (that is, we neglect any negative signs) of these differences and find the average of these absolute values, the resulting number is called the **average deviation**. The symbol for absolute value is two vertical lines. Thus, $|+8|$ is read as "the absolute value of $+8$."

Let us illustrate the preceding ideas by calculating the sample variance, sample standard deviation, and average deviation for the two routes that Christina uses to drive to work. Since the sample mean, $\bar{x}$, is 31 we can arrange our calculations as shown in the following chart.

	Time (x)	Difference from Mean, $(x - \bar{x})$	Square of Difference $(x - \bar{x})^2$	Absolute Value of Difference $|x - \bar{x}|$
	15	$15 - 31 = -16$	$(-16)^2 = 256$	16
By Way of	26	$26 - 31 = -5$	$(-5)^2 = 25$	5
Brooks	30	$30 - 31 = -1$	$(-1)^2 = 1$	1
Expressway	39	$39 - 31 = 8$	$8^2 = 64$	8
	45	$45 - 31 = \underline{14}$	$14^2 = \underline{196}$	$\underline{14}$
		Sum = 0	Sum = 542	Sum = 44

Therefore, if Christina travels to work by way of the Brooks Expressway, the sample variance is $\dfrac{542}{5 - 1}$, or 135.5; the sample standard deviation is $\sqrt{135.5}$ or approximately 11.64; and the average deviation is $\dfrac{44}{5}$, or 8.8.

Notice that in computing these measures of variation, we used symbols. Thus, $\bar{x}$ represents the sample mean, $x - \bar{x}$ represents the difference of any number from the mean, $(x - \bar{x})^2$ represents the square of the difference, and $|x - \bar{x}|$ represents the absolute value of the difference from the sample mean. Furthermore, the sum of the differences from the sample mean, $\Sigma(x - \bar{x})$, is 0. Can you see why?

Let us now compute the sample variance, sample standard deviation, and average deviation for traveling to work by way of the Kingston Expressway. Again we will use symbols.

---

[*]A knowledge of how to compute square roots is not needed. Such values can be obtained by using a calculator or a square root table.

	Time (x)	Difference from Mean $(x - \bar{x})$	Square of Difference $(x - \bar{x})^2$	Absolute Value of Difference $|x - \bar{x}|$
	29	$29 - 31 = -2$	$(-2)^2 = 4$	2
By Way of	30	$30 - 31 = -1$	$(-1)^2 = 1$	1
Kingston	31	$31 - 31 = 0$	$0^2 = 0$	0
Expressway	32	$32 - 31 = 1$	$1^2 = 1$	1
	33	$33 - 31 = 2$	$2^2 = 4$	2
		Sum = 0	Sum = 10	Sum = 6

In this case the sample variance is $\dfrac{10}{5 - 1}$, or 2.5; the sample standard deviation is $\sqrt{2.5}$, or approximately 1.58; and the average deviation is $\dfrac{6}{5}$, or 1.2. Here again, the sum of the differences from the mean, $\Sigma(x - \bar{x})$, is 0. This is always the case.

We now formally define sample variance, sample standard deviation, and average deviation.

Definition 3.5 | *Sample Variance*
The sample variance of a sample of n numbers is a measure of the spread of the numbers about the sample mean and is given by

Sample variance

$$\textbf{Sample variance} = s^2 = \frac{\Sigma(x - \bar{x})^2}{n - 1}$$

Definition 3.6 | *Sample Standard Deviation*
The sample standard deviation of a sample of n numbers is the positive square root of the sample variance. Thus,

Sample standard deviation

$$\textbf{Sample standard deviation} = s = \sqrt{s^2} = \sqrt{\frac{\Sigma(x - \bar{x})^2}{n - 1}}$$

Definition 3.7 | The **average deviation** of a sample of numbers is the average of the absolute value of the differences from the sample mean. Symbolically,

Average deviation

$$\textbf{Average deviation} = \frac{\Sigma|(x - \bar{x})|}{n}$$

If we are working with the entire population rather than with a sample, then we have the following:

$$\text{Population variance} = \sigma^2 = \frac{\Sigma(x - \mu)^2}{N}$$

$$\text{Population standard deviation} = \sigma = \sqrt{\frac{\Sigma(x - \mu)^2}{N}}$$

COMMENT   When calculating the sample variance (or sample standard deviation), we use $n - 1$ instead of n in the denominators of the two formulas given in Definitions 3.5 and 3.6. However, when calculating the population variance (or standard deviation), we use N instead of $N - 1$ in the denominators of the formulas. There is a sound statistical reason for doing this, but we will not concern ourselves with the reason at this point. To summarize, the difference between σ and s is whether we divide by N or by $n - 1$ and whether we use the population mean, μ, or the sample mean, $\bar{x}$. When calculating the population standard deviation, we use μ and divide by N and denote our result by σ, whereas when calculating a sample standard deviation, we use $\bar{x}$ and divide by $n - 1$ and denote our result by s. Thus, s is really an estimate of σ, the population standard deviation. Very often statisticians will refer to s as *the* standard deviation, even though it is only an estimate.

COMMENT   It may seem that the standard deviation is a complicated and useless number to calculate. At the moment let us say that it is a useful number to the statistician. Just as the measures of central tendency help us locate the "center" of a relative frequency distribution, the standard deviation helps us measure its "spread." It tells us how much the observations differ from the mean. Notice that most of the observations in Figure 3.1 deviate very little from the mean of the distribution. As opposed to this, most of the observations in Figure 3.2 deviate substantially from the mean of the distribution. When we discuss the normal distribution in later chapters, you will understand

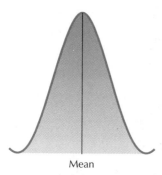

Mean

FIGURE 3.1   Most of the observations deviate very little from the mean.

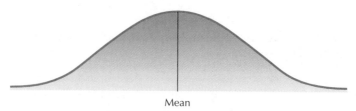

Mean

FIGURE 3.2   Most of the observations deviate substantially from the mean.

the significance and usefulness of the standard deviation. Statisticians prefer to work with the standard deviation as opposed to the variance, since the standard deviation is usually smaller than the variance.

In most statistical problems we do not have all the data for the population. Instead, we have only a small part, that is, a sample, of the population. It is for this reason that we use the sample variance and sample standard deviation, which we restate below using a formula number for later reference.

FORMULA 3.4	
Sample Variance	$$s^2 = \frac{\Sigma(x - \bar{x})^2}{n - 1}$$
Sample Standard Deviation	$$s = \sqrt{\frac{\Sigma(x - \bar{x})^2}{n - 1}}$$

We will illustrate the preceding ideas with another example.

EXAMPLE   1   The number of hours per day that several technicians spent adjusting the timing mechanism on CAT scan machines during the past week is 5, 3, 2, 6, 4, 2, and 6. Find the sample variance, sample standard deviation, and average deviation.

SOLUTION   We arrange the data in order as shown in the following chart and perform the indicated calculations. The sample mean is

$$\bar{x} = \frac{\Sigma x}{n} = \frac{2 + 2 + 3 + 4 + 5 + 6 + 6}{7}$$

$$= \frac{28}{7} = 4$$

Number of Hours x	Difference from Mean $(x - \bar{x})$	Square of Difference $(x - \bar{x})^2$	Absolute Value of Difference $\lvert x - \bar{x} \rvert$
2	$2 - 4 = -2$	$(-2)^2 = 4$	2
2	$2 - 4 = -2$	$(-2)^2 = 4$	2
3	$3 - 4 = -1$	$(-1)^2 = 1$	1
4	$4 - 4 = 0$	$0^2 = 0$	0
5	$5 - 4 = 1$	$1^2 = 1$	1
6	$6 - 4 = 2$	$2^2 = 4$	2
6	$6 - 4 = 2$	$2^2 = 4$	2
	Sum = 0	*Sum* = 18	*Sum* = 10

Our answers then are

$$\text{Sample variance} = \frac{\Sigma(x - \bar{x})^2}{n - 1} = \frac{18}{7 - 1} = 3$$

$$\text{Sample standard deviation} = \sqrt{\text{variance}} = \sqrt{3}, \qquad \text{or approximately } 1.73$$

$$\text{Average deviation} = \frac{\Sigma|x - \bar{x}|}{n} = \frac{10}{7} = 1.43 \qquad \text{when rounded}$$

3.5   COMPUTATIONAL FORMULAS FOR CALCULATING THE VARIANCE

Although the formulas in Definitions 3.5 and 3.6 of the preceding section can always be used for calculating the sample variance, in practice, it turns out that the calculations become quite tedious. For this reason we use more convenient formulas, which follow.

FORMULA 3.5

Computational Formulas

$$\text{Population variance} = \sigma^2 = \frac{\Sigma x^2}{N} - \frac{(\Sigma x)^2}{N^2} \qquad \text{or} \qquad \frac{N \Sigma x^2 - (\Sigma x)^2}{N^2}$$

$$\text{Sample variance} = s^2 = \frac{n \Sigma x^2 - (\Sigma x)^2}{n(n - 1)}$$

$$\text{Population standard deviation} = \sigma = \sqrt{\frac{\Sigma x^2}{N} - \frac{(\Sigma x)^2}{N^2}}$$

$$\text{Sample standard deviation} = s = \sqrt{\frac{n \Sigma x^2 - (\Sigma x)^2}{n(n - 1)}}$$

In Formula 3.5, Σx^2 means that we square each number and add the squares together, whereas $(\Sigma x)^2$ means we first sum the numbers and then square the sum.

Using summation notation for the data in Table 3.1, we have

$$\Sigma x = 1 + 2 + 3 + 4 + 5 = 15$$

$$\Sigma x^2 = 1^2 + 2^2 + 3^2 + 4^2 + 5^2$$

$$= 1 + 4 + 9 + 16 + 25 = 55$$

We now use Formula 3.5 to calculate the sample variance for the data of Table 3.1:

TABLE 3.1	**Squares of Integers**	
	(x^2 Means x Times x)	

x	x^2
1	1
2	4
3	9
4	16
$\underline{5}$	$\underline{25}$
15	55
$\Sigma x = 15$	$\Sigma x^2 = 55$

$$s^2 = \frac{n\,\Sigma\,x^2 - (\Sigma\,x)^2}{n(n-1)}$$

$$= \frac{5(55) - (15)^2}{5(5-1)}$$

$$= \frac{275 - 225}{5(4)}$$

$$= \frac{50}{20} = 2.5$$

If you now calculate the sample variance by using the formula in Definition 3.5 of the preceding section and compare the results, your answer will be the same. It is considerably simpler, however, to get the answer by using Formula 3.5. The sample standard deviation is obtained by taking the square root of the variance. Thus,

$$\text{Standard deviation} = \sqrt{\text{variance}}$$
$$= \sqrt{2.5} \approx 1.58 \qquad \text{(The symbol} \approx \text{stands for approximately.)}$$

COMMENT The advantage of computing the sample variance by Formula 3.5 is that we do not have to subtract the sample mean from each term of the distribution.

BEWARE Do not confuse the symbols $\Sigma\,x^2$ and $(\Sigma\,x)^2$. The symbol $\Sigma\,x^2$ represents the sum of the squares of each number, whereas the symbol $(\Sigma\,x)^2$ represents the square of the sum of the numbers. If your calculation of the variance results in a negative number, you probably have confused the two symbols.

How are the mean, variance, and standard deviation affected if we *multiply* each term of a distribution by some number? To see what happens, multiply each number of the distribution given in Table 3.1 by, say, 10, and compute the mean, variance, and standard deviation of the new distribution. (You will be asked to do this in one of the exercises.)

How are the mean, variance, and standard deviation affected if we *add* the same constant to each term of a distribution? (Again, you will be asked to do this in one of the exercises.)

To calculate the variance and standard deviation of sample data that is presented in frequency distribution form, as is often the case with published data, we need a computational formula for determining the sample variance and sample standard deviation for grouped data. We can always use the definitions. Thus, we have

Sample Variance for Grouped Data

$$s^2 = \frac{\Sigma(x - \bar{x})^2 \cdot f}{n - 1}$$

Sample Standard Deviation for Grouped Data

$$s = \sqrt{\frac{\Sigma(x - \bar{x})^2 \cdot f}{n - 1}}$$

However, the computations involved in using these formulas can be very tedious. Formula 3.6 is a shortcut formula that gives the same results as if we had used the definition.

FORMULA 3.6

Computational Formulas for Grouped Data

Sample Variance for Grouped Data

$$s^2 = \frac{n(\Sigma x^2 \cdot f) - (\Sigma x \cdot f)^2}{n(n - 1)}$$

Sample Standard Deviation for Grouped Data

$$s = \sqrt{\frac{n(\Sigma x^2 \cdot f) - (\Sigma x \cdot f)^2}{n(n - 1)}}$$

EXAMPLE 1 The IQ of 50 students in the fifth grade of a special school was measured. The results are as follows:

IQ Score	Frequency
71–79	3
80–88	6
89–97	12
98–106	14
107–115	8
116–124	5
125–133	2

Compute the sample variance and sample standard deviation.

SOLUTION We rearrange the data as shown in the following table. Note that we use the class marks for each category. In grouped data we assume that all the numbers in each group are at the class mark for that group.

IQ Score	Class Mark x	Frequency f	$x \cdot f$	$x^2 \cdot f$
71–79	75	3	225	16,875
80–88	84	6	504	42,336
89–97	93	12	1116	103,788
98–106	102	14	1428	145,656
107–115	111	8	888	98,568
116–124	120	5	600	72,000
125–133	129	2	258	33,282
		50	5019	512,505

Here $n = \Sigma f = 50$ since the sample consisted of 50 students. Then using Formula 3.6, we get

$$s^2 = \frac{n(\Sigma x^2 \cdot f) - (\Sigma x \cdot f)^2}{n(n-1)}$$

$$= \frac{50(512,505) - (5019)^2}{50(50-1)}$$

$$= 177.5057$$

and

$$s = \sqrt{s^2} = \sqrt{177.5057} \approx 13.32$$

Thus, the sample variance is 177.51 and the sample standard deviation is approximately 13.32. (Do not use a rounded variance to get a standard deviation. Round all answers at the end of all calculations.)

EXERCISES FOR SECTION 3.5

1. Each year, many television stations conduct fund raising drives through telethons for various charities. For the past six years one television station kept records on the time that passed before the first $25,000 was pledged. The following information is available:

Year	Minutes Passed Before First $25,000 Was Pledged
1990	14
1991	11
1992	10
1993	17
1994	12
1995	8

Find the range, population variance, and population standard deviation.

2. Is cancer more prevalent in states on the eastern coast of the United States? The following table represents the reported incidence of cancer (per 100,000 population) for ten eastern states during 1991.

State	Incidence of Cancer (per 100,000 population)
Connecticut	434
Delaware	500
Maine	391
Massachusetts	443
New Hampshire	403
New Jersey	464
New York	329
Pennsylvania	442
Rhode Island	445
Vermont	376

(*Source:* National Cancer Institute, 1993)

Find the range, population variance, and population standard deviation.

3. The U.S. National Center for Health Statistics reports in *Vital and Health Statistics* that a 6-foot-tall male between 18 and 24 years of age weighs about 175 pounds. A survey of ten such males found that their weights were 183, 162, 171, 157, 197, 203, 159, 167, 183, and 188 pounds. Find the range, sample variation, sample standard deviation, and average deviation for these data.

4. A large publishing company in Philadelphia published ten new math books in 1995. The number of chapters in these books is as follows: 13, 9, 12, 11, 8, 11, 17, 14, 15, and 10. Find the range, population mean, population variance, population standard deviation, and average deviation for the number of chapters in a math book published by this company.

5. The A.C. Nielsen Company reports in *Nielsen Report on Television* that children aged two to five years watch about 27 hours of television weekly. A sample of nine such children found that they watched 21, 33, 16, 29, 20, 32, 25, 27, and 22 hours of television weekly. Find the sample mean, sample variance, sample standard deviation, and average deviation for the number of hours of television watched.

6. The ages of 162 prospective jurors considered for a controversial case are shown below:

Age (years)	Frequency
19–27	25
28–36	38
37–45	27
46–54	36
55–63	21
64–72	15

Find the sample variance and sample standard deviation.

7. The time spent by patients in a doctor's office before receiving medical attention directly affects the efficient operation of the doctor's practice. The number of minutes (after their scheduled appointment time) that each of 56 patients had to wait before seeing an obstetrician at the Bayview Medical Center is shown below:

Waiting Time (minutes)	Number of Patients
0–5	15
6–11	12
12–17	11
18–23	10
24–29	8

Find the sample variance and sample standard deviation.

8. The population of the United States is aging. Nevertheless, many senior citizens must take a variety of medications that for many of these people represent life's necessities. A recent congressional investigation into the prices charged for these drugs found that some senior citizens take as many as 12 different prescription drugs daily. In a random sample of nine senior citizens in the Palm Gardens Nursing Home, a government fact-finding commission found that these individuals were taking 4, 2, 8, 3, 10, 9, 2, 6, and 1 different kinds of medication. Find the range, sample variance, sample standard deviation, and average deviation for the number of medications taken by the senior citizens.

9. Multiply each number in the preceding exercise by four and then compute the sample variance, sample standard deviation, and average deviation for the new distribution. How do these results compare with those of the preceding exercise? Can you generalize?

10. The National Flood Insurance Program administered by the Federal Emergency Management Agency (FEMA) offers flood insurance to home owners through the Federal Government and some 85 private insurers. Annual rates vary according to the structure of the house and its vulnerability to flooding, with an average premium of about $308 and a standard deviation of $27. As a result of the summer 1993 floods, the annual premium charged to each homeowner in a particular region was increased by $40. What will the new average premium and new standard deviation be?

11. Refer back to Exercise 10. If the premium charged to each homeowner is increased by 20% instead of $40, what will the new mean and new standard deviation be?

12. Priscilla Hodges is responsible for maintaining the photocopying machines in a large office complex. Her crews periodically replace key mechanical components of the machines to guard against unwanted breakdowns. She has two potential suppliers for one particular mechanical part. To help her decide which supplier to use, she installs 20 of those components, 10 from each supplier, into the machines and subjects them to normal operating conditions. The number of copies (in thousands) provided by these machines is as follows:

Supplier A					Supplier B				
462	432	412	486	503	283	495	589	460	390
414	495	498	427	481	492	376	705	453	487

Average = 461
Standard deviation = 36.396

Average = 473
Standard deviation = 116.088

From which of these suppliers should Priscilla order the mechanical parts? Explain your answer.

13. A trucking company tested two different brands of tires on its fleet of trucks to determine the useful life of each brand of tire. The following results are available:

	Brand A	Brand B
Average useful life	32,000 miles	33,000 miles
Standard deviation	900 miles	1,100 miles

Which brand of tire would you buy?

14. The price of a 2-liter bottle of a popular soft drink charged by seven vendors in the Brighton section of town during 1995 was $1.29, $1.10, $1.14, $0.99, $1.06, $1.25, and $1.07.
    a. Calculate the population mean and population standard deviation by subtracting $1.07 from each price.
    b. Calculate the population mean and population standard deviation by subtracting $1.14 from each price.
    c. How are the population mean and population standard deviation affected if we subtract a different number from each price?

15. There are 140 male convicts and 88 female convicts in the Greenborough Correctional Facilities System. The number of years remaining to be served from the convicts' original sentences before being eligible for parole is as follows:

Number of Years	Males	Females
0–3	47	26
4–7	32	21
8–11	22	16
12–15	17	11
16–19	12	9
20–23	10	5

(*Source:* Greenborough Correctional Facilities System, 1993)

a. Find the sample standard deviation for the number of years remaining to be served by the males.
b. Find the sample standard deviation for the number of years remaining to be served by the females.

c. Compare the answer obtained in part (a) with the answer obtained in part (b). Comment.

3.6 INTERPRETATION OF THE STANDARD DEVIATION: CHEBYSHEV'S THEOREM

Chebyshev's Theorem

Up to this point we have been discussing formulas for calculating the standard deviation of a set of numbers. If the standard deviation turns out to be small, then we can conclude that all the data values are concentrated around the mean. On the other hand, if the standard deviation is large, then the data values will be widely scattered about the mean.

In a later chapter when we discuss the normal distribution, we will notice that a substantial number of the data is bunched within 1, 2, or 3 standard deviations above or below the mean. Nevertheless, a more general result, which is true for any set of measurements—population or sample—and regardless of the shape of the frequency distribution, is known as **Chebyshev's Theorem**. We state it now.

CHEBYSHEV'S THEOREM:

Let k be any number equal to or greater than 1. Then the proportion of any distribution that lies within k standard deviations of the mean is at least $1 - \dfrac{1}{k^2}$.

To see what Chebyshev's Theorem means, let us compute some values of $1 - \dfrac{1}{k^2}$ as shown here:

Value of k	Value of $1 - \dfrac{1}{k^2}$
1	0
2	$\dfrac{3}{4}$ or 75%
3	$\dfrac{8}{9}$ or 89%

When $k = 2$, then if μ is the population mean and σ the population standard deviation, Chebyshev's Theorem says that you will always find at least $\dfrac{3}{4}$, that is, 75%

or more, of the measurements will lie within the interval $\mu - 2\sigma$ and $\mu + 2\sigma$, that is, within 2 standard deviations of the mean on either side. This can be seen in Figure 3.3.

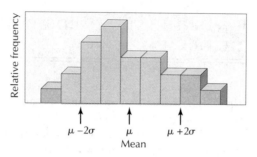

FIGURE 3.3 At least $\dfrac{3}{4}$ of the measurements will fall within the shaded portion.

Similarly, when $k = 3$, Chebyshev's Theorem says that at least $\dfrac{8}{9}$ of the measurements, or 89% of the data, will lie within the interval $\mu - 3\sigma$ and $\mu + 3\sigma$, that is, within 3 standard deviations of the mean on either side. This can be seen in Figure 3.4.

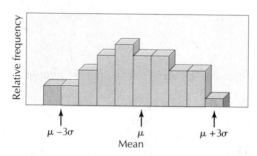

FIGURE 3.4 At least $\dfrac{8}{9}$ of the measurements will fall within the shaded portion.

E X A M P L E   1   A statistician is analyzing the claims filed with an auto insurance company. A sample of 100 claims discloses that the average claim filed was $831, with a standard deviation of $150. If we let $k = 2$, then at least $1 - \dfrac{1}{2^2}$ or $\dfrac{3}{4}$, that is, 75% or more, of the claims will be within

$$831 - 2(150) \text{ and } 831 + 2(150) \quad \text{or} \quad \text{between } \$531 \text{ and } \$1131$$

Similarly, at least $1 - \dfrac{1}{3^2}$ or $\dfrac{8}{9}$, or 89% or more, of the claims will be within

$$\$831 - 3(150) \text{ and } \$831 + 3(150) \quad \text{or} \quad \text{between } \$381 \text{ and } \$1281$$

E X A M P L E   2 If all the light bulbs manufactured by a certain company have a mean life of 1600 hours with a standard deviation of 100 hours, at least what percentage of the bulbs will have a mean life of between 1450 and 1750 hours?

SOLUTION We first find $1750 - 1600 = 1600 - 1450 = 150$. Using Chebyshev's Theorem, we know that k standard deviations or $k(100)$ will equal 150. That is,

$$100k = 150$$

so that

$$k = 1.5$$

Thus, at least $1 - \dfrac{1}{(1.5)^2} = 1 - \dfrac{1}{2.25} = 0.5556$, or at least 55.56% of the light bulbs will have a mean life between 1450 and 1750 hours.

For a specific set of data, we can always answer the question, "How many measurements are within 1, 2, or 3 standard deviations of the mean?" We can simply count the number of measurements in each of the intervals. Based on Chebyshev's Theorem, we can say the following:

FACT I For any sample of measurements (regardless of the shape of the frequency distribution):

1. Possibly none of the measurements fall within 1 standard deviation of the mean, that is, within $\bar{x} \pm s$.

2. At least $\dfrac{3}{4}$ of the measurements will fall within 2 standard deviations of the mean, that is, within $\bar{x} \pm 2s$.

3. At least $\dfrac{8}{9}$ of the measurements will fall within 3 standard deviations of the mean, that is, within $\bar{x} \pm 3s$.

4. At least $(1 - 1/k^2)$ will fall within k standard deviations of the mean, that is, within $\bar{x} \pm ks$, where $k > 1$.

FACT II If the frequency distribution of a set of sample data is mound-shaped (normally distributed), then

1. Approximately 68% of the measurements will fall within 1 standard deviation of the mean, that is, within $\bar{x} \pm s$.

2. Approximately 95% of the measurements will fall within 2 standard deviations of the mean, that is, within $\bar{x} \pm 2s$.

3. Approximately 99.7% of the measurements will fall within 3 standard deviations of the mean, that is, within $\bar{x} \pm 3s$.

Empirical Rule   Fact I is often called Chebyshev's Theorem and Fact II is often called the **Empirical Rule**. We will discuss the normal distribution in considerable detail in a later chapter.

EXERCISES FOR SECTION 3.6

1. The American Hospital Association publishes *Hospital Statistics* that contains information on hospital costs, length of patient stays, and so forth. The cost per day of six randomly selected patients was

$$\$584 \quad \$681 \quad \$500 \quad \$525 \quad \$586 \quad \$532$$

Verify Chebyshev's Theorem with $k = 2$ and $k = 3$ for the above data.

2. As an economy move, many colleges and universities hire adjunct personnel rather than full-time people to teach various courses. At Boylan University the number of adjuncts employed and the subject matter taught are as follows:

Department	Number of Adjuncts
Mathematics	38
English	27
History	35
Computer Science	29
Biology	29
Business	42

Verify Chebyshev's Theorem with $k = 2$ and $k = 3$ for the above data.

3. If all the pilots of Lakeville Airlines have been flying an average of 12 years with a standard deviation of 2.7 years, at least what percentage of pilots will have been flying for the company between 8 and 16 years?

4. For a variety of health reasons, many people must restrict their intake of sodium (salt). Often when eating in a restaurant, diners find it difficult to control sodium intake. The average sodium content of all the dinners at Luigi's Restaurant is 975 milligrams (mg), with a standard deviation of 35 mg. At least what percentage of the dinners will contain between 935 and 1015 mg of sodium?

5. Radon, a radioactive gas produced by the natural decay of radium in the ground, often finds its way into homes and buildings through openings in the foundation. Officials of the EPA recommend that corrective measures be taken when levels reach 4 or more picocuries per liter (pc/1). Readings were taken at 20 different schools in one city and produced the following results:

$$4.2 \quad 1.8 \quad 3.2 \quad 2.9 \quad 5.7 \quad 3.4 \quad 2.8 \quad 3.2 \quad 5.1 \quad 6.1$$
$$1.3 \quad 2.1 \quad 3.7 \quad 3.5 \quad 4.1 \quad 3.8 \quad 2.9 \quad 3.1 \quad 2.3 \quad 1.1$$

a. Find the mean reading $\bar{x}$ for these sample readings. Also find s^2 and s.

b. According to Chebyshev's Theorem, what percentage of the measurements would you expect to fall within the interval $\bar{x} \pm 1.75s$.

c. What percentage of measurements actually fall in the intervals of part (b)?

d. Compare the results obtained in part (c) with the result of part (b). Comment.

3.7 PERCENTILES AND PERCENTILE RANK

Consider the following newspaper article. The article indicates that percentile ranks are

NEW ADMISSIONS GUIDELINES ADOPTED
Trudy Hoffman

March 20: As a result of further budget cutbacks the Chancellor's office announced new admission guidelines. Effective this fall, no students will be admitted to a senior college unless his or her high school average is 80% or higher or is in an least the 75th percentile in his or her graduating class. To be admitted to a community college, a student must have a 70% high school average or have a percentile rank of 40 or higher.

Sunday–March 20, 1987

used by many colleges in determining which students will be admitted. Also, if we analyze the card from the Emerson Medical Laboratory shown in Figure 3.5, we again notice the use of percentiles. How are percentiles calculated? How do we interpret them?

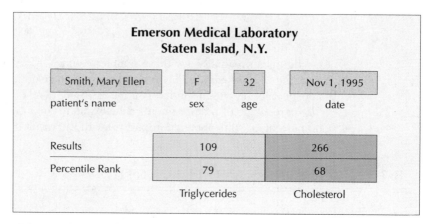

FIGURE 3.5

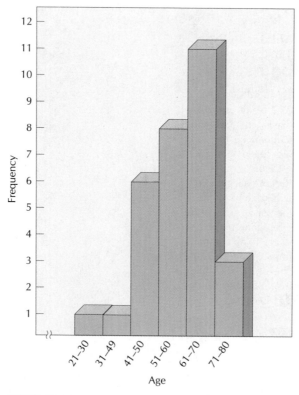

FIGURE 3.6

Now consider the table shown on top of the next page, which shows the distribution of the ages of 30 runners in a Florida jogging marathon. The histogram for the data is shown in Figure 3.6.

Age (in years)	Frequency
21–30	1
31–40	1
41–50	6
51–60	8
61–70	11
71–80	3

Suppose we were interested in determining how many runners were 70 years old or younger. We would add the frequencies in the five lowest categories. In our case we would add $1 + 1 + 6 + 8 + 11 = 27$. We can obtain the same answers by drawing a
Cumulative   **cumulative frequency histogram**. In essence, we accumulate the frequencies by add-
frequency   ing the frequencies in each interval of the grouped data. The resulting cumulative fre-
histogram   quency histogram is shown in Figure 3.7. Such cumulative frequency or cumulative
Ogives   relative frequency graphs are called **ogives**.

Instead of using the cumulative frequency as the vertical scale, we can use percents where 100% corresponds to 30 runners, and 0% corresponds to 0 runners. Fifty percent would correspond to 15 runners, and so on. By having the percent on the vertical scale

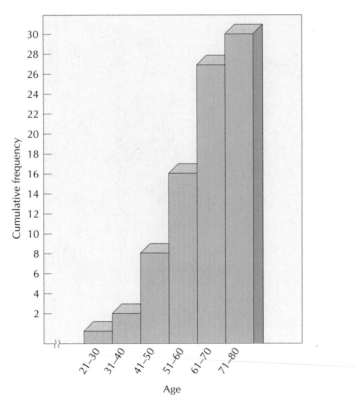

FIGURE 3.7   A cumulative frequency histogram.

Cumulative relative frequency histogram

we can answer such questions as, "What percent of the runners were 70 years old or younger?" We can read our answer from the graph, called a **cumulative relative frequency histogram**, given in Figure 3.8. The answer is 90%. More generally, a score that tells us what percent of the total population scored at or below that measure is called the *percentile*. How do we determine such percentiles?

To get us started, let us consider Lorraine, who received a 76 on her midterm psychology examination. There are 150 students, including her, in the class. She knows that 60% of the class got below 76, 10% of the class got 76, and the remaining 30% got above 76.

Since 60% of the class got below her grade of 76 and 30% got above her grade, her percentile rank should be between 60 and 70. We will use 65, which is midway between 60 and 70. What we do is find the percent of scores that are below the given score and add one-half the percent of the scores that are the same as the given score. In our case 60% of the class grades were below Lorraine's and 10% were the same as Lorraine's. Thus, the percentile rank of Lorraine's grade is

$$60 + \left(\frac{1}{2}\right)(10) = 65$$

Figure 3.9 illustrates the situation.

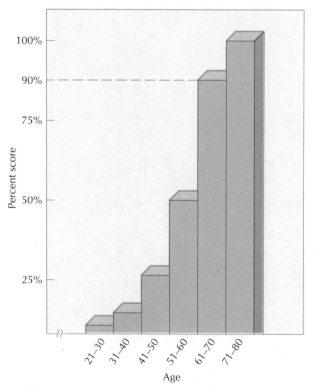

FIGURE 3.8   A cumulative relative frequency histogram.

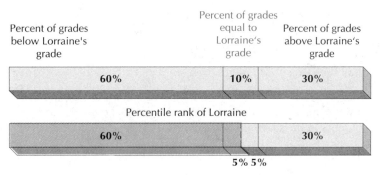

FIGURE 3.9

We now say that Lorraine's percentile rank is 65. This means that approximately 35% of the class did better than Lorraine on the exam and that she did better than 65% of the class. Essentially, the percentile rank of a score tells us the percentage of the distribution that is below that score. Formally, we have the following definition.

Definition 3.8
Percentile rank
Percentile

The **percentile rank** or **percentile** of a term in a distribution is found by adding the percentage of terms below it with one-half the percentage of terms equal to the given term.

Let X be a given score, let B represent the *number* of terms below the given score X, and let E represent the *number* of terms equal to the given score X. If there are N terms altogether (that is, if the entire population consists of N terms together), then the percentile rank of X is given by Formula 3.7.

FORMULA 3.7

$$\text{Percentile rank of } X = \frac{B + \frac{1}{2}E}{N} \cdot 100$$

We now illustrate the use of Formula 3.7.

EXAMPLE 1 Bill and Jill are twins, but they are in different classes. Recently, they both got 80 on a math test. The grades of the other students in their classes were as follows:

Jill's class

64 67 73 73 73 74 77 77 78 78 79 80 80 82 91 94 100

Bill's class

43 65 68 73 75 76 76 77 79 80 80 80 80 85
86 87 88 90 92 96

Find the percentile rank of each student.

SOLUTION   We use Formula 3.7. Jill's grade is 80. There were two 80s (including Jill's) in the class, so $E = 2$. There were 11 grades below 80, so $B = 11$. Since there are 17 students in the class altogether, $N = 17$. Thus,

$$\text{Jill's percentile rank} = \frac{11 + \frac{1}{2}(2)}{17} \cdot 100$$

$$= \frac{11 + 1}{17} \cdot 100 = \frac{12}{17} \cdot 100$$

$$= \frac{1200}{17} = 70.59$$

Jill's percentile rank is 70.59.

Using a similar procedure for Bill's class, we find that $B = 9$, $E = 4$, and $N = 20$. Thus,

$$\text{Bill's percentile rank} = \frac{9 + \frac{1}{2}(4)}{20} \cdot 100$$

$$= \frac{9 + 2}{20} \cdot 100 = \frac{11}{20} \cdot 100$$

$$= \frac{1100}{20} = 55$$

Bill's percentile rank is 55.

COMMENT   The percentile rank of an individual score is often more helpful than the particular score value. Although both Bill and Jill had grades of 80, Jill's percentile rank is considerably higher. If we assume that the levels of competition are equivalent in both classes, this may indicate that Jill's performance is superior to Bill's performance when compared with the rest of their respective classes.

We often use the word **percentile** to refer directly to a score in a distribution. So, instead of saying that the percentile rank of Lorraine's grade is 65, we say that her grade is in the 65th percentile. Similarly, if a term has a percentile rank of 40, we say that it is in the 40th percentile.

Percentiles are used quite frequently to describe the results of achievement tests and the subsequent ranking of people taking those tests. This is especially true when applying for many civil service jobs. If there are more applicants than available jobs, candidates are often ranked according to percentiles. Many colleges use only percentile ranks, rather than the numerical high school average, to determine which candidates to admit. The reason is that percentile ranks of a student's high school average reflect how they did with respect to their classmates, whereas numerical averages only indicate an individual student's performance.

Since percentiles are numbers that divide the set of data into 100 equal parts, we can easily compare percentiles. Thus, in Example 1 we were able to find the percentile rank of Jill and Bill, even though they both were in different classes.

During World War II the United States Army administered the Army General Classification Tests (AGCT) to thousands of enlisted men. The results showed important differences in the average IQ of men in various jobs, ranging from 93 for miners and farmhands to around 120 for accountants, lawyers, and engineers. Figure 3.10 shows the IQ range between the 10th and 90th percentiles for workers in various occupations. Furthermore, the vertical bars represent the 50th percentile or median scores. Very often, when such tests are administered to large groups of people, the results are given in terms of percentile bands, as shown in Figure 3.11. Since percentiles are used quite

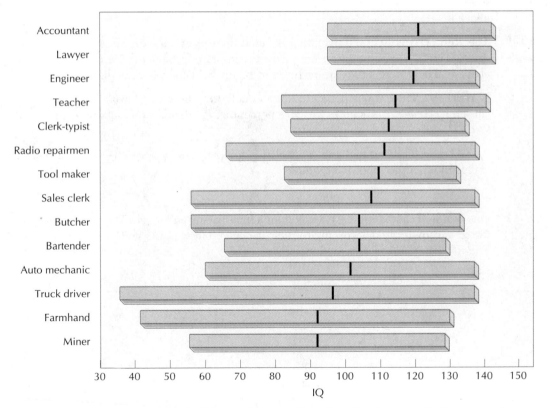

FIGURE 3.10 Each bar shows the IQ range between the 10th and 90th percentiles for men in that occupation. The vertical bars represent the 50th percentiles. Note that although the average IQ score of accountants was 121 and that of miners 93, some miners had higher IQ scores than some accountants.

often, special names are given to the 25th and 75th percentiles of a distribution. Of course, the *50th percentile is the median of the distribution*. See Figure 3.11.

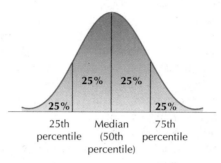

25th
percentile

Median
(50th
percentile)

75th
percentile

FIGURE 3.11   Some of the frequently used percentiles.

Definition 3.9
Lower quartile

The 25th percentile of a distribution is called the **lower quartile**. It is denoted by P_{25} or Q_1. Thus, 25% of the terms are below the lower quartile and 75% of the terms are above it.

Definition 3.10
Upper quartile

The 75th percentile of a distribution is called the **upper quartile**. It is denoted by P_{75} or Q_3. Thus, 75% of the terms are below the upper quartile and 25% of the terms are above it. These percentiles are pictured in Figure 3.11 for a normal distribution.

NOTATION   The various percentiles are denoted by the letter P with the appropriate subscript. Hence, P_{37} denotes the 37th percentile, P_{99} denotes the 99th percentile, and so on.

When calculating percentiles involving grouped data, we use the class mark. This is illustrated in the following example.

EXAMPLE   2   One hundred candidates have applied for a high-paying acting job. The candidates were tested and then rated according to their dancing ability, poise, and overall acting skill. The following ratings were obtained.

Rating	Frequency
10–19	17
20–29	8
30–39	6
40–49	18
50–59	28
60–69	16
70–79	4
80–89	3
Total =	100

Calculate the percentile rank of Heather McAllister, who scored 44.5 in the ratings.

SOLUTION   Since Heather scored 44.5 in the ratings, she is in the 40 to 49 category. We know that there are 17 + 8 + 6 or 31 below the 40 to 49 category. We now assume that all the 18 candidates who scored in the 40 to 49 category scored 44.5, which is the class mark. Thus, there are 18 people including Heather who scored 44.5. Using Formula 3.7 with $N = 100$, $B = 31$, and $E = 18$, we have

$$\text{Heather's percentile rank} = \frac{31 + \frac{1}{2}(18)}{100} \times 100$$

$$= \frac{31 + 9}{100} \times 100$$

$$= 40$$

Heather is in the 40th percentile.

COMMENT   In the previous example we assumed that all of the candidates who scored in the 40 to 49 category scored 44.5, which is the class mark. Some statisticians prefer to compute percentiles for grouped data by assuming that the entries in a class are evenly distributed in that class. These statisticians then use the formula $\tilde{x} = L + \frac{j}{f} \cdot c$. Refer back to the discussion of this approach given on page 139.

Statisticians often use percentiles to group data into different classes. Two popular percentile groupings are **deciles** and **quartiles**. When a set of data is divided by deciles then approximately 10% of the data are put into each of the ten classes. Similarly, when a set of data is divided by quartiles, then approximately 25% of the data will fall into each of the four classes. There is a general procedure that can be used to find the quartiles for a set of data.

GENERAL PROCEDURE FOR OBTAINING Q_1, Q_2, AND Q_3

---

1. Arrange the data in increasing order.
2. If there are n pieces of data, then the median or Q_2 will be at position $\frac{n+1}{2}$.

3. The lower quartile or Q_1 is the median of the smallest $\frac{n}{2}$ data points when n is even or the median of the smallest $\frac{n-1}{2}$ data points when n is odd.

4. The upper quartile or Q_3 is the median of the largest $\frac{n}{2}$ data points when n is even or the median of the largest $\frac{n-1}{2}$ data points when n is odd.

COMMENT   **Quartiles** divide the data into quarters, or four equal parts; **deciles** divide the data into tenths, or ten equal parts; and **percentiles** divide the data into hundredths, or 100 equal parts.

Box plots
Box-and-whisker
diagram
Interquartile range

A relatively new concept in the descriptive analysis of data involves **box plots** or **box-and-whisker diagrams**. This method, invented by Professor John Tukey, is based on the quartiles of a data set and the **interquartile range**, which is the distance between the upper and lower quartile. We illustrate the procedure for constructing box plots in the following example.

HISTORICAL NOTE

The author of numerous books and several hundred articles on mathematics, statistics, and other sciences, John Wilder Tukey was born in New Bedford, Massachusetts, on June 16, 1915. He attended Brown University where he majored in chemistry. He received his bachelor's degree in 1936 and his master's degree in 1937, both from Brown University. He then enrolled in the mathematics program at Princeton University, from which he received his master's degree in 1938 and his doctorate in 1939.

Upon graduation, he was appointed to the mathematics staff at Princeton University. When Princeton created a Department of Statistics in 1965, Tukey was immediately named its first chairperson. Today he remains on the staff of Princeton as Donner Professor of Science, Emeritus, and as Professor of Statistics, Emeritus.

Tukey has been the recipient of numerous honors and awards and has also served on a Presidential commission, specifically President Eisenhower's Science Advisory Commission. As a research statistician at Princeton and also as a member of the technical staff at AT & T Bell Laboratories, Tukey has pioneered new statistical techniques such as stem-and-leaf diagrams, box plots, and box-and-whisker diagrams. He has made valuable contributions in the field of exploratory data analysis.

E X A M P L E   3   Mary Aquilla is the telephone switchboard operator for the Hewlit Corporation. The number of "wrong number" telephone calls that she received over the past 20 days is as follows:

$$3 \quad 4 \quad 13 \quad 31 \quad 16 \quad 24 \quad 9 \quad 17 \quad 13 \quad 33$$
$$7 \quad 12 \quad 8 \quad 23 \quad 21 \quad 18 \quad 14 \quad 19 \quad 21 \quad 35$$

Construct a box plot for these data.

SOLUTION

Step 1.   We first find the smallest and the largest data value. In our case we have

Smallest value $= 3$ and   largest value $= 35$

Step 2. We arrange the data in increasing order to find the median (middle quartile). In our case when the data are arranged in order, we get the following:

3 4 7 8 9 12 13 13 14 16 17 18 19 21 21 23 24 31 33 35

so that the median is between 16 and 17 or that

$$\text{Median or middle quartile} = \frac{16 + 17}{2} = 16.5$$

Step 3. Now we find the median of the top half of the data and also the median of the bottom half of the data. In our case we get the following:

Lower half of the data: 3   4   7   8   9   12   13   13   14   16
Median of lower half or lower quartile = 10.5
Upper half of the data: 17   18   19   21   21   23   24   31   33   35
Median of upper half or upper quartile = 22

Hinges   The lower and upper quartiles are called the **hinges**.

Step 4. Locate the values obtained in steps 1 to 3 on a horizontal axis. Now we draw a rectangle (the box) with the ends of the rectangle (the hinges) drawn at the lower and upper quartiles. We connect the hinges to each other to get a box. Now we mark the median (middle quartile), the low value, and the high value on the graph above the axis as shown in Figure 3.12. Finally, we connect the low and high values to the hinges by *Whiskers*   means of lines called **whiskers**. The resulting graph is called the box-and-whisker diagram.

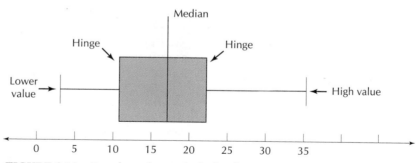

FIGURE 3.12   Box plot or box-and-whisker diagram.

In Figure 3.12 we have labeled everything on the graph so that we can interpret correctly the information that it provides. We note that the two boxes in the box-and-whisker diagram tell us how spread out the two middle quartiles are. Also, the two whiskers indicate how spread out the lower and upper quartiles are. By visually examining the lengths of the whiskers, we can safely say that if one whisker is clearly longer, then the distribution of the data is probably skewed in the direction of the longer whisker.

COMMENT   From the preceding box-and-whisker diagram, we can conclude that since the right whisker is longer than the left whisker, the data are skewed to the right. There is more variation in the upper quarter than in any other quarter.

EXERCISES FOR SECTION 3.7

1. *Ecology.* Twenty students in an ecology class have been collecting aluminum soda cans so that they can be used for recycling. The number of cans collected by each of these students in one month is shown in the following chart:

Jennifer	427	John	426	Jimmy	327
Heather	408	Michele	402	Lisa	453
Ike	381	Christine	387	Ricardo	442
Moses	307	Rufus	352	Eddie	406
Cassandra	297	Michael	391	Terrie	423
Vincent	365	Robert	402	Carol	430
Paul	428	Patrick	462		

   Find the percentile rank of Michele and the percentile rank of Ricardo.

2. A study[*] of the blood serum cholesterol levels of 100 people was undertaken by a medical research team to determine the relationship between high-cholesterol levels and the incidence of heart attacks. The following results were obtained.

Cholesterol Level (mg)	Frequency
240–249	6
250–259	9
260–269	11
270–279	29
280–289	25
290–299	10
300–309	9
310–319	1

   a. Find the percentile rank of Melissa, whose cholesterol level was 274.5.
   b. Find the percentile rank of Maurice, whose cholesterol level was 294.5.

3. Refer back to Figure 3.5. On November 1, 1995, the Emerson Medical Laboratories received blood samples from 112 patients. Ed and seven other patients had a blood serum cholesterol level of 281, whereas 76 others had a cholesterol level below 281. Find Ed's percentile rank.

[*]Grieves and Johnson, 1993.

4. An analysis of the family income of the 900 members of the Bergenfield Golf and Country Club reveals the following:

$$P_{25} = \$28,000$$

$$P_{50} = \$39,000$$

$$P_{75} = \$53,000$$

$$P_{85} = \$61,000$$

What percent of the members have family incomes that are
a. less than $28,000?
b. more than $61,000?
c. more than $39,000?
d. between $28,000 and $61,000?

5. Two hundred students in a driver education program volunteered to participate in an obstacle-avoidance driving test. The following scores were obtained:

Score	Frequency
10–19	16
20–29	19
30–39	22
40–49	39
50–59	34
60–69	28
70–79	29
80–89	13
	Total = 200

a. Draw the cumulative frequency histogram for the above data.
b. Draw the cumulative relative frequency histogram (in percent form) for the above data.

6. As an incentive to attract additional customers, one Caribbean hotel recently installed a toll-free 800 phone number. During the first three weeks of its operation, the hotel received the following number of requests for additional information on a daily basis:

```
22  28  23  25  29  28  24  29  19  30  28
27  26  18  20  24  31  27  21  19  27
```

a. Find the lower quartile for the data.
b. Find the upper quartile for the data.
c. Find the interquartile range.
d. Draw the box-and-whisker diagram for the data.

7. Some hospitals operate advance life support ambulances. The American Hospital Association reports in *Hospital Statistics* that such ambulances are very costly to maintain. Hollow Hill Hospital operates one such ambulance. During the first 20 days of 1995, this ambulance made the following number of runs (on a daily basis):

$$
\begin{array}{ccccccc}
23 & 19 & 25 & 19 & 23 & 31 & 28 \\
31 & 27 & 29 & 24 & 21 & 26 & 22 \\
16 & 18 & 30 & 27 & 25 & 20 &
\end{array}
$$

a. Find the upper quartile for the data.
b. Find the lower quartile for the data.
c. Find the interquartile range.
d. Draw the box-and-whisker diagram for the data.

8. The following relative frequencies were obtained by entering freshmen college students from a certain school on the Scholastic Aptitude Test administered by the Educational Testing Service.

Test Scores	Relative Frequency
200–249	0.01
250–299	0.02
300–349	0.03
350–399	0.08
400–449	0.14
450–499	0.18
500–549	0.23
550–599	0.16
600–649	0.11
650–699	0.03
700–749	0.01
750–799	0

Approximately what score must a student have in order to be in the
a. 75th percentile   b. 90th percentile

9. The Food and Drug Administration (FDA) monitors the effectiveness of new drugs. Some medications cause serious side effects. One particular antihistamine drug can lead to drowsiness or adversely affect the user's ability to operate a motor vehicle. The following is a list of the frequency of reaction and the age of the person using the medication.

Age of Person	No. Who Experienced a Reaction
16–25	12
26–35	18
36–45	22
46–55	19
56–65	17
66–75	8
76–85	4
	Total = 100

a. Draw the cumulative frequency histogram for the data.

b. Draw the cumulative relative frequency histogram (in percent form) for the data.

10. Many students enroll in college courses but withdraw from them for a variety of reasons. This affects other students who need particular courses but who are "closed out" of the classes. An analysis of 25 Fall 1995 lecture classes at Boyerville College indicates the following number of students who withdrew from the classes by the end of the semester.

$$
\begin{array}{cccccccccccccc}
10 & 22 & 30 & 16 & 20 & 11 & 9 & 14 & 21 & 27 & 24 & 12 & 13 \\
8 & 15 & 18 & 7 & 18 & 2 & 14 & 16 & 23 & 31 & 9 & 17 \\
\end{array}
$$

a. Find the lower quartile for the data.

b. Find the upper quartile for the data.

c. Find the interquartile range.

d. Draw the box-and-whisker diagram for the data.

11. All newborn children are measured and weighed at birth. During the week of Sept. 1–Sept. 5, 22 babies were born. Their weights (in ounces) at birth were as follows:

$$
\begin{array}{cccccccccccc}
112 & 96 & 105 & 110 & 100 & 108 & 85 & 132 & 147 & 127 & 125 \\
103 & 91 & 108 & 99 & 107 & 116 & 92 & 125 & 140 & 131 & 136 \\
\end{array}
$$

a. Find the percentile rank of Dawn, who weighed 108 ounces.

b. Find the percentile rank of Spector, who weighed 147 ounces.

• c. Find the lowest decile, D_{10}, for the data.

3.8 z-SCORES

As we saw in the preceding section, one way of measuring the performance of an individual score in a population is by determining its percentile rank. Using percentile rank alone, however, can sometimes be misleading. For example, two students in different classes may have the same percentile rank. Yet one student may be far superior to his or her competitors, whereas the second student may only slightly surpass the others in his or her class.

Statisticians have another very important way of measuring the performance of an individual score in a population. This measure is called the **z-score**. The z-score measures how many standard deviations an individual score is away from the mean. We define it formally as follows:

z-Score

Definition 3.11

FORMULA 3.8

The z-score of any number x in a distribution whose mean is μ and whose standard deviation is σ is given by

$$
z = \frac{x - \mu}{\sigma} \quad \text{or} \quad z = \frac{x - \bar{x}}{s}
$$

where x = value of number in original units

μ = population mean

σ = population standard deviation

$\bar{x}$ = sample mean

s = sample standard deviation

z-Value
Measurement in
standard units

COMMENT   The z-score of a number in a population is sometimes called the **z-value** or **measurement in standard units**.

COMMENT   Since σ is always a positive number, z will be a negative number whenever x is less than μ, as $x - \mu$ is then a negative number. A z-score of 0 implies that the term has the same value as the mean.

We now illustrate how to calculate z-scores with several examples.

E X A M P L E   1   A certain brand of flashlight battery has a mean life, μ, of 40 hours and a standard deviation of 5 hours. Find the z-score of a battery that lasts

a. 50 hours   b. 35 hours   c. 40 hours

SOLUTION   Since $\mu = 40$ and $\sigma = 5$, we use Formula 3.8.

a. The z-score of 50 is

$$\frac{50 - 40}{5} = \frac{10}{5} = 2$$

b. The z-score of 35 is

$$\frac{35 - 40}{5} = \frac{-5}{5} = -1$$

c. The z-score of 40 is

$$\frac{40 - 40}{5} = 0$$

E X A M P L E   2

Testing Tuna Fish

Two consumer's groups, one in New York and one in California, recently tested at numerous local colleges a number of different brands of canned tuna fish for taste appeal. Each consumer group used a different rating system. The following results were obtained:

Brand	New York Rating
A	1
B	10
C	15
D	21
E	28

Brand	California Rating
M	25
N	35
P	45
Q	50
R	70

Which brand has the greatest taste appeal?

SOLUTION   At first glance it would appear that Brand R is superior since its California rating was 70. However, we see from the rating and from the given information that the two consumer's groups awarded their points differently so that the point value alone is not enough of a basis for deciding among the different brands. We therefore convert each of the ratings into standard scores. These calculations are shown in Tables 3.2 and 3.3.

We can now use the z-scores as a basis for comparison of the different brands. Clearly, Brand R for which $z = 1.65$ is superior to Brand E for which $z = 1.41$.

Notice that the sum of the z-scores for the New York ratings of the brands is 0:

$$(-1.52) + (-0.54) + (0) + (0.65) + (1.41) = 0$$

This means that the average of the z-scores is 0 since 0 divided by 5, the number of z-scores, is 0. Also, the sum of the z-scores for the California rating of the brands is 0:

$$(-1.32) + (-0.66) + (0) + (0.33) + (1.65) = 0$$

TABLE 3.2   **New York Rating of Tuna Fish ($\mu = 15$, $\sigma = 9.23$)**

Brand	Rating x	Mean μ	Difference from Mean $(x - \mu)$	z-Score $z = \dfrac{x - \mu}{\sigma}$
A	1	15	$1 - 15 = -14$	$\dfrac{-14}{9.23} = -1.52$
B	10	15	$10 - 15 = -5$	$\dfrac{-5}{9.23} = -0.54$
C	15	15	$15 - 15 = 0$	$\dfrac{0}{9.23} = 0$
D	21	15	$21 - 15 = 6$	$\dfrac{6}{9.23} = 0.65$
E	28	15	$28 - 15 = 13$	$\dfrac{13}{9.23} = 1.41$

TABLE 3.3   California Rating of Tuna Fish ($\mu = 45$, $\sigma = 15.17$)

Brand	Rating x	Mean μ	Difference from Mean $(x - \mu)$	z-Score $z = \dfrac{x - \mu}{\sigma}$
M	25	45	$25 - 45 = -20$	$\dfrac{-20}{15.17} = -1.32$
N	35	45	$35 - 45 = -10$	$\dfrac{-10}{15.17} = -0.66$
P	45	45	$45 - 45 = \quad 0$	$\dfrac{0}{15.17} = \quad 0$
Q	50	45	$50 - 45 = \quad 5$	$\dfrac{5}{15.17} = \quad 0.33$
R	70	45	$70 - 45 = \quad 25$	$\dfrac{25}{15.17} = \quad 1.65$

Therefore, the mean is 0. If you now compute the standard deviations of the z-scores in Tables 3.2 and 3.3, you will find that the standard deviation in each case is 1. We summarize these facts in the following rule.

RULE

In any distribution the mean of the z-scores is 0 and the standard deviation of the z-scores is 1.

Formula 3.8 can be changed so that if we are given a particular z-score, we can calculate the corresponding original score. The changed formula is as follows.

FORMULA 3.9

$$x = \mu + z\sigma$$

EXAMPLE   3   In a recent swimming contest the mean score was 40 and the standard deviation was 4. If Carlos had a z-score of -1.2, how many points did he score?

SOLUTION   Since $\mu = 40$, $\sigma = 4$, and $z = -1.2$, we can use Formula 3.9. Thus, we have

$$x = 40 + (-1.2)(4)$$
$$= 40 - 4.8$$
$$= 35.2$$

Carlos's score was 35.2.

EXERCISES FOR SECTION 3.8

1. In a recent weight-lifting contest, the average weight lifted was 210 pounds with a standard deviation of 20 pounds. Find the z-score of
   a. José, who lifted 250 pounds
   b. Bob, who lifted 180 pounds
   c. Mark, who lifted 270 pounds
   d. Mike, who lifted 210 pounds
2. Refer back to Exercise 1. The z-scores of five other contestants participating in the contest were

   Gwen, -1; Shelly, $+1.6$; Maurine, -3; Roger, -0.1;
   Sandy, $+1.01$

   a. Rank these five people from lowest to highest.
   b. Which of these five contestants were above average?
   c. Which of these five contestants were below average?
3. A computer literacy test was given to 100 secretaries. The mean score was 47 with a standard deviation of 4.7.
   a. If Heather had a z-score of -0.89, what was her actual score?
   b. If Harley had a z-score of 2.46, what was his actual score?
   c. If Lucille had a z-score of -3.79, what was her actual score?
4. A statistics professor posts the following test results on the departmental bulletin board.

Test Grade	z-Score	Percentile Rank
55	-2	2
65	-1	16
75	0	30
85	1	50
95	2	85

   Note: $\sigma = 10$.

   a. What was the average test grade?
   b. What percent of the class grades were between 65 and 85?
   c. If Al's z-score on the exam was -3.04, what was his actual test score?
5. Verify that the standard deviation of the z-scores of Tables 3.2 and 3.3 is 1.

6. Vera and Judy have both applied for the same job. Vera scored 80 on the state aptitude test where the mean was 70 and the standard deviation was 4.2. Judy scored 510 on the company exam where the mean was 490 and the standard deviation was 10.3. Assuming that the company uses these test results as the only criterion for hiring new employees and that both tests are considered as equal by company officials, who will get the job? Explain your answer.

7. Alumni records indicate that the average starting salary of an accounting major graduating from one particular college is $31,000 with a standard deviation of $2700. Find the actual starting salaries of Claudette, Tom, and Gina if their z-scores are 2.6, -0.98, and 1.37, respectively.

8. The estimated life expectancy for males and females in several Asian countries is as follows:

Life Expectancy at Birth for Ten Asian Countries

Country	Male (years)	Female (years)
China	66.0	69.0
Hong Kong	71.9	77.6
India	53.9	52.9
Indonesia	51.2	53.9
Israel	72.1	75.7
Japan	74.2	79.7
Kuwait	68.1	72.9
Pakistan	54.4	54.2
Taiwan	69.7	74.6
Thailand	59.5	65.1

(*Source:* © United Nations. All United Nations rights reserved)

For these countries $\bar{x} = 64.10$ and $s = 8.59$ for males and $\bar{x} = 67.56$ and $s = 10.44$ for females.

a. Find the z-score for the life expectancy for females in Japan and Pakistan.
b. Find the z-score for the life expectancy for males in Japan and Pakistan.
c. Compare the answers obtained in parts (a) and (b). Comment.

9. *Pollution.* There are eight particularly polluted lakes in New York and six particularly polluted lakes in neighboring New Jersey that were analyzed by environmentalists from both states. (Each state has its own method for measuring the pollution level of its waterways.) The following results on the amount of pollution were obtained by these two groups.

Garbage and industrial pollutants dumped into our nation's waterways are adversely affecting our environment.

New York Lake	Rating
A	22
B	15
C	29
D	33
E	24
F	17
G	37
H	28

New Jersey Lake	Rating
Q	68
R	75
S	61
T	70
U	70
V	76

Assume that these lakes represent the entire population for both states.
  a. Which of these 14 lakes tested has the greatest amount of pollution?
  b. Which of these 14 lakes tested has the least amount of pollution?
10. The distribution shown on the next page indicates the range of grades that can be expected on many intelligence tests. Notice that the scores are normally distributed.
  a. If someone scores 300 on the ETS exam, what is the corresponding percentile rank?
  b. What percent of the children taking the WISC exam will score higher than 130?
  c. Find the percentile rank of a score that has a z-value of $+2$.

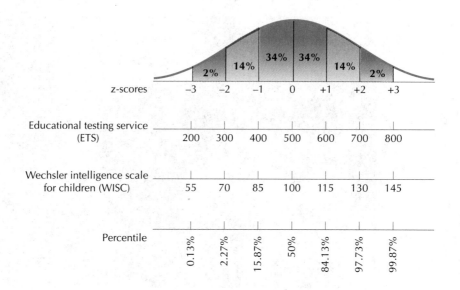

3.9   USING COMPUTER PACKAGES

To illustrate how to use the MINITAB computer package to compute the various statistical measures discussed in this chapter, let us consider the following data on the ages of ten patients who were admitted to the 1995 Rainville Drug Detoxification Program because of their addiction to "Crack." Their ages were 16, 24, 32, 19, 21, 14, 18, 19, 17, and 25. The following MINITAB program gets the data into column C1:

```
MTB  > SET THE FOLLOWING DATA INTO C1
DATA > 16   24   32   19   21
DATA > 14   18   19   17   25
DATA > END
```

The DESCRIBE statement in MINITAB yields the sample size (n), the mean, median, trimmed mean (TRMEAN); the standard deviation; the standard error of the mean (to be discussed in a later chapter); the maximum and minimum values; the upper (75th) percentile Q_3; and the lower (25th) percentile Q_1. The DESCRIBE statement for the preceding data produces the following computer printout:

```
MTB  > DESCRIBE DATA IN C1
```

	N	MEAN	MEDIAN	TRMEAN	STDEV	SEMEAN
C1	10	20.50	19.00	19.88	5.28	1.67

```
     MIN    MAX     Q1     Q3
C1 14.00  32.00  16.75  24.25
```

NOTE The value given in the preceding printout for the first quartile (Q1) is 16.75, which differs slightly from the value of 16.5 that we obtain when we calculate Q1 in the usual manner. This small discrepancy occurs because MINITAB computes quartiles in a slightly different way.

MINITAB can also be used to generate box-and-whisker diagrams. This is accomplished by using the instruction BOXPLOT followed by the location of the data (C1). When applied to the previous data, the BOXPLOT instruction produces the following printout:

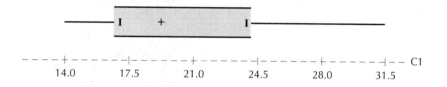

In this printout the median is indicated by the + symbol. Note the length of the whiskers. There is less variation in the bottom two quarters of the drug-age data than in the top two quarters, and the top quarter has the greatest variation of all. MINITAB's box-and-whisker diagram is slightly different from the following boxplot for the same data that we obtain when constructing it by hand. You should compare both diagrams.

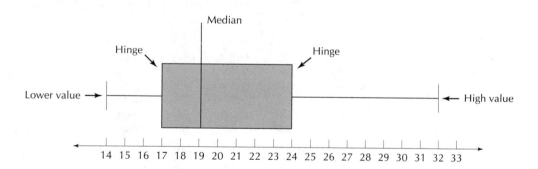

EXERCISES FOR SECTION 3.9

1. According to *Construction Reports*, published by the U.S. Census Bureau, the average price of a new mobile home is approximately $25,000. A survey of the price of 34 new mobile homes in Monticello gave the following results:

$29,500	$28,700	$25,500	$20,500	$23,000	$29,750
23,000	22,000	22,000	26,100	24,000	21,000
28,500	20,500	21,000	23,000	21,750	23,000
27,400	22,750	27,500	22,750	25,000	28,000
26,000	19,700	26,750	22,000	26,000	21,500
25,500	27,500	25,000	24,000		

a. Using MINITAB, compute the mean, median, sample standard deviation, and lower and upper quartiles.

b. Draw the box plot for the data.

c. Find the interquartile range.

2. The Motor Vehicle Manufacturer's Association surveys the age of cars in use and publishes its findings in *Motor Vehicle Facts and Figures*. In one survey of 36 cars in use, their age (in years) was as follows:

8.6	5.7	4.5	6.1	7.1	6.4
5.2	8.8	7.6	8.8	8.4	8.3
4.8	7.6	8.4	9.9	8.6	9.1
3.9	7.9	9.3	8.4	7.4	4.2
9.7	8.1	4.8	6.8	5.3	2.7
10.2	2.3	5.7	7.3	8.4	1.2

a. Using MINITAB, compute the mean, median, sample standard deviation, and lower and upper quartiles.

b. Draw the box plot for the data.

c. Find the interquartile range.

3. The life expectancy (in years) in 30 countries is as follows:

Argentina	65	France	73	Pakistan	48	Turkey	57
Australia	74	Greece	72	Poland	71	United Kingdom	73
Bangladesh	46	India	47	Soviet Union	70	United States	73
Belgium	71	Indonesia	46	Spain	73	Venezuela	63
Brazil	60	Italy	73	Sri Lanka	64	West Germany	72
Canada	73	Japan	76	Sweden	75	Yugoslavia	70
Chile	62	Mexico	60	Thailand	61	Zaire	39
Egypt	54	Nigeria	41				

(*Source:* © United Nations. All United Nations rights reserved)

a. Using MINITAB, compute the mean, median, sample standard deviation, and lower and upper quartiles.

b. Draw the box plot for the data.

c. Find the interquartile range.

4. The following data represent the reported incidence of cancer per 100,000 population in all 50 states during the year 1991.

Incidence of Cancer per 100,000 Population

State	Number	State	Number
Alabama	433	Montana	372
Alaska	442	Nebraska	336
Arizona	360	Nevada	422
Arkansas	383	New Mexico	375
California	366	New Jersey	464
Colorado	282	New York	329
Connecticut	434	New Hampshire	403
Delaware	500	North Carolina	355
Florida	367	North Dakota	408
Georgia	406	Ohio	463
Hawaii	371	Oklahoma	326
Idaho	307	Oregon	396
Illinois	402	Pennsylvania	442
Indiana	438	Rhode Island	445
Iowa	377	South Carolina	348
Kansas	345	South Dakota	418
Kentucky	414	Tennessee	408
Louisiana	422	Texas	313
Maine	391	Utah	229
Maryland	491	Vermont	376
Massachusetts	443	Virginia	440
Michigan	454	Washington	364
Minnesota	366	West Virginia	409
Mississippi	438	Wisconsin	398
Missouri	390	Wyoming	238

(*Source:* National Cancer Institute)

a. Using MINITAB, compute the mean, median, sample standard deviation, and lower and upper quartile.
b. Draw the box plot for the data.
c. Find the interquartile range.

3.10 SUMMARY

In this chapter we discussed various numerical methods for analyzing data. In particular, we calculated and compared three measures of central tendency: the mean, median, and mode. We pointed out that each has its advantages and disadvantages. In addition, various properties of each measure were discussed. Thus, we mentioned that the mean is affected by extreme values and that the sum of the differences from the mean is zero. The mean is the most frequently used measure of central tendency. We also demonstrated how to calculate a weighted mean when the terms of a distribution are not of equal weight. In the process we introduced summation notation.

We then discussed four measures of variation that tell us how dispersed, that is, how spread out, the terms of the distribution are around the *center* of the distribution. These were the range, variance, standard deviation, and average deviation. Various short-cuts for computing the standard deviation and variance were introduced.

We also saw that an individual score is sometimes meaningless unless it is accom-panied by a percentile rank or z-score. When scores are converted into percentile ranks or z-values, we can then make meaningful statements about them and compare them with other scores.

We discussed and demonstrated how to calculate percentile ranks as well as z-scores. The latter play an important role in the normal distribution to be discussed in a later chapter.

Study Guide

The following is a chapter outline in capsule form. You should now be able to demon-strate your knowledge of the ideas mentioned by giving definitions, descriptions, or specific examples. Page references are given in parentheses.

To indicate the operation of taking the sum of a sequence of numbers, we use **sum-mation notation**. (page 126)

The i (or j) in the summation symbol $\sum_{i=1}^{n}$ is referred to as the **index**. (page 127)

The **lower limit of summation** is the value of the index placed below the summation symbol and the **upper limit of summation** is the value of the index placed above the summation symbol. (page 127)

The **mean** or **average** of a set of numbers, denoted by $\bar{x}$ or μ, is obtained by adding the numbers together and dividing the sum by the number of numbers added. (page 133)

The **mode** of a set of numbers is the number (or numbers) that occurs most often. (page 133)

The **median** of a set of numbers, denoted by $\tilde{x}$, is the number that is in the middle when the numbers are arranged in order from the lowest to the highest. (page 133)

The mean of a sample of numbers is called the **sample mean**. (page 134)

When all values in a set of numbers are not of equal importance, we must multiply each number by its weight and divide the sum of these products by the sum of the weights to obtain the **weighted sample mean**. (page 139)

The **midrange** of a set of numbers is found by taking the average of the lowest value L and the highest value H. (page 141)

The **harmonic mean** of n numbers $x_1, x_2, \ldots, x_n$ is defined as n divided by the sum of the reciprocals. (page 141)

The **geometric mean** of a set of n positive numbers $x_1, x_2, \ldots, x_n$ is defined as the nth root of their product. (page 141)

The mean, median, and mode are known as **measures of central tendency** since each measures some central or general trend of the data. (page 142)

Any extremes in a set of numbers are called **outliers**. (page 147)

A **trimmed mean** is calculated by deleting high and low values from a set of numbers and then calculating the mean of the remaining numbers. (page 147)

The **range** of a set of numbers is the difference between the largest number in the distribution and the smallest number in the distribution. (page 149)

The **sample variance** of a sample of numbers is a measure of the spread of the numbers about the sample mean and is given by $s^2 = \dfrac{\Sigma(x - \bar{x})^2}{n - 1}$. (page 151)

The **sample standard deviation** of a sample of numbers, denoted by s, is the positive square root of the sample variance. (page 151)

The **average deviation** of a sample of numbers is the average of the absolute value of the differences from the mean. (page 151)

For any set of measurements—population or sample—and regardless of the shape of the distribution, **Chebyshev's Theorem** specifies the following: Let k be any number ≥ 1. Then the proportion of the distribution that lies within k standard deviations of the mean is at least $1 - \dfrac{1}{k^2}$. (page 161)

If the frequency distribution of a set of sample data is mound-shaped (normally distributed), then the **Empirical Rule** states that approximately 68% of the measurements will fall within 1 standard deviation of the mean, 95% of the measurements will fall within 2 standard deviations of the mean, and 99.7% of the measurements will fall within 3 standard deviations of the mean. (page 164)

If we add together all the frequencies in each interval of grouped data that are less than or equal to a given interval and graph, the results are a **cumulative frequency histogram** or **ogive**. (page 167)

A score that indicates what percent of the population scored at or below a given measure is called the **percentile**. (page 168)

The **percentile rank** of a term in a distribution is found by adding the percentage of terms below it with one-half the percentage of terms equal to the given term. (page 169)

Percentiles divide the data into hundredths or 100 equal parts. (page 174)

Quartiles divide the data into quarters, or four equal parts, so that approximately 25% of the data will fall into each of the four classes. (page 174)

Deciles divide the data into tenths, or ten equal parts, so that approximately 10% of the data are put into each of ten classes. (page 174)

The 25th percentile of a distribution, denoted as P_{25} or Q_1, is called the **lower** or **first quartile**. (page 172)

The 50th percentile or middle quartile of a distribution is its **median**. (page 172)

The 75th percentile of a distribution, denoted as P_{75} or Q_3, is called the **upper** or **third quartile**. (page 172)

The **interquartile range** is the difference between the upper and lower quartiles. (page 174)

A **box plot** or **box-and-whisker diagram** is a way of analyzing data that is based on the quartiles of the data set and the interquartile range. The lower and upper quartiles are called **hinges**. They are connected to the low and high values by means of **whiskers**. (page 174)

The **z-score** of any number x in a distribution whose mean is μ and whose standard deviation is σ is given by $z = \dfrac{x - \mu}{\sigma}$ or $z = \dfrac{x - \bar{x}}{s}$. (page 179)

The z-score of a number in a population is sometimes called the **z-value** or **measurement in standard units**. (page 180)

Formulas to Remember

At this point you have learned some of the common terms used in statistical analysis and some of the graphic techniques. The formulas are important too.

The following list is a summary of all formulas given in the chapter. You should be able to identify each symbol, understand the relationships among the symbols expressed in each formula, understand the significance of each formula, and use the formulas in solving problems.

1. When using summation notation the following are true:

$$\sum_{i=1}^{n} (x_i + y_i) = \sum_{i=1}^{n} x_i + \sum_{i=1}^{n} y_i$$

$$\sum_{i=1}^{n} kx_i = k \cdot \sum_{i=1}^{n} x_i$$

$$\sum_{i=1}^{n} k = n \cdot k$$

2. Sample mean: $\bar{x} = \dfrac{\Sigma x}{n}$

   Population mean: $\mu = \dfrac{\Sigma x}{N}$

3. Sample mean for grouped data: $\bar{x} = \dfrac{\Sigma xf}{\Sigma f}$

4. Weighted sample mean: $\bar{x}_w = \dfrac{\Sigma xw}{\Sigma w}$

5. Sample variance: $s^2 = \dfrac{\Sigma(x - \bar{x})^2}{n - 1}$

6. Sample standard deviation: $s = \sqrt{\dfrac{\Sigma(x - \bar{x})^2}{n - 1}}$

7. Average deviation: $\dfrac{\Sigma|x - \bar{x}|}{n}$

8. Population variance: $\sigma^2 = \dfrac{\Sigma(x - \mu)^2}{N}$

9. Population standard deviation: $\sigma = \sqrt{\dfrac{\Sigma(x - \mu)^2}{N}}$

10. Computational formula for population variance: $\sigma^2 = \dfrac{\Sigma x^2}{N} - \dfrac{(\Sigma x)^2}{N^2}$

11. Computational formula for population standard deviation: $\sigma = \sqrt{\dfrac{\Sigma x^2}{N} - \dfrac{(\Sigma x)^2}{N^2}}$

12. Computational formula for sample variance: $s^2 = \dfrac{n(\Sigma x^2) - (\Sigma x)^2}{n(n - 1)}$

13. Computational formula for sample standard deviation: $s = \sqrt{\dfrac{n(\Sigma x^2) - (\Sigma x)^2}{n(n - 1)}}$

14. Computational formulas for grouped data:

    Sample variance: $s^2 = \dfrac{n(\Sigma x^2 \cdot f) - (\Sigma x \cdot f)^2}{n(n - 1)}$

    Sample standard deviation: $s = \sqrt{\dfrac{n(\Sigma x^2 \cdot f) - (\Sigma x \cdot f)^2}{n(n - 1)}}$

15. *Chebyshev's Theorem:* At least $1 - \dfrac{1}{k^2}$ of a set of measurements will lie within k standard deviation units of the mean (assuming $k \geq 1$).

16. Percentile rank of $X = \dfrac{B + \frac{1}{2} E}{N} \cdot 100$

17. z-score: $z = \dfrac{x - \mu}{\sigma}$ or $z = \dfrac{x - \bar{x}}{s}$

18. Original score: $x = \mu + z\sigma$

Testing Your Understanding of This Chapter's Concepts

1. Locate the approximate location of the mean, median, and mode for the frequency distributions given in parts (b) and (c). We have already done it for the diagram of part (a).

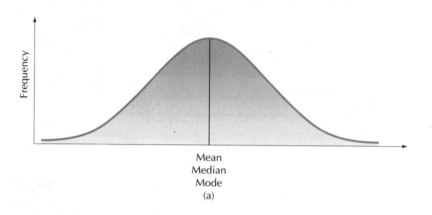

Mean
Median
Mode
(a)

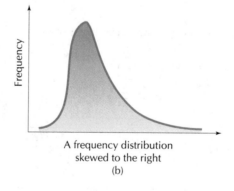

A frequency distribution
skewed to the right
(b)

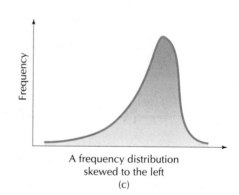

A frequency distribution
skewed to the left
(c)

2. If the third quartile for scores on a test is 68, what percent of the scores are below 68?

3. Is it safe for a man who is 6 feet tall and a nonswimmer to walk in a swimming pool that has an *average* depth of 3 feet? Explain your answer.

4. In a certain distribution $\mu = 69$ and $\sigma = 0$. What is the z-score of any term in this distribution? Explain your answer.

5. The average age of a student at Blake University is 24 years. The average age of a student at Collin's School of Engineering is 20 years. Can we conclude that the average age for both schools is 22 years? Explain your answer.

6. Joe has a high school average of 93. The college that he wishes to attend will not accept any applicant with a percentile rank below 85. Is Joe sure he will be accepted by this college or is it possible that he will be denied admission? Explain your answer.

7. Can a percentile rank of 56 have a negative z-score? Explain your answer.

8. The Printex Corporation pays its workers the following salaries depending on title:

Title	Number of Employees	Weekly Salary
Manager	1	$500
Supervisor	3	300
Machinists	8	275
Other Workers	14	200
Secretaries	4	150

Two students were asked to calculate the average. One did as follows:

$$\mu = \frac{500 + 300 + 275 + 200 + 150}{5} = \frac{1425}{5} = \$285$$

The student then claimed that the average salary was $285. A second student computed a weighted arithmetic mean as follows:

$$\mu = \frac{1 \cdot 500 + 3 \cdot 300 + 8 \cdot 275 + 14 \cdot 200 + 4 \cdot 150}{30} = \frac{7000}{30} = \$233.33$$

This student then claimed that the average salary was $233.33. Which student is correct and why?

9. Owing to numerous boiler breakdowns, a new heating system was recently installed at Skytown Towers. Several tenants are now complaining that it is too warm, whereas others are complaining that it is now too cold. In an attempt to satisfy everyone, management has decided to install an energy-saving thermostat and to poll each of the residents as to what temperature it should be set. When this was done, the following data were obtained.

Mean	71°
Median	70°
Mode	69°

At what temperature should the thermostat be set so as to satisfy as many residents as possible? Explain your answer.

Chapter Test

Multiple Choice Questions

1. If $x_1 = 5$, $x_2 = 8$, $x_3 = 9$, $x_4 = 10$, and $x_5 = 13$, find Σx^2.
   a. 2025   b. 439   c. 45   d. 25   e. none of these

For Questions 2–7, use the following information that gives the weights (in pounds) of all of the people who joined the Brighton Health Club on January 17.

George	153	Heather	123	Peter	165	Tom	147
Martha	102	Gail	119	Priscilla	123		
Bill	137	Jim	145	Robert	156		

2. Find the (population) mean weight of these people.
   a. 137   b. 123   c. 141   d. 145   e. none of these
3. Find the median weight for these people.
   a. 137   b. 123   c. 141   d. 145   e. none of these
4. Find the modal weight for these people.
   a. 137   b. 123   c. 141   d. 145   e. none of these
5. Find Heather's percentile rank.
   a. 20th percentile   b. 25th percentile   c. 80th percentile
   d. 30th percentile   e. none of these
6. Find the population standard deviation for this data.
   a. 350.6   b. 18.724   c. 389.556   d. 19.737   e. none of these
7. Find Heather's z-score.
   a. -0.709 b. $+0.709$ c. -0.7477 d. $+0.7477$
   e. none of these
8. All applicants for a prestigious office managerial job are required to take four different exams. Arlene's results, as well as the results of the other applicants for the job, are shown below.

Type of Exam	Arlene's Score	Average Score	Standard Deviation
Personality	89	76	14.7
Business skills	74	67	7.3
Medical	56	49	9.6
Computer skills	107	95	11.2

By transforming each of Arlene's results into z-scores, determine in which test Arlene performed the best (on a comparative basis).
   a. Personality   b. Business skills   c. Medical   d. Computer skills
9. If the variance of a set of numbers is 0.81, find the standard deviation.
   a. 0.9   b. 9   c. 0.81   d. 0.09   e. none of these
10. On February 14, the ten busses on the Rawley route carried an average of 40 riders whereas the 16 busses on the Piedro route carried an average of 46 riders. Find the average number of riders carried by the busses on both of these routes on February 14.
   a. 13   b. 43   c. 40   d. 43.692   e. none of these

Supplementary Exercises

11. A certain set of ten measurements has a mean equal to 55 and a population variance equal to 25. Find   a. Σx b. Σx^2.

12. In any distribution, what proportion of terms lies within 2.3 standard deviations of the mean?

13. MINITAB was used to generate the following box plot:

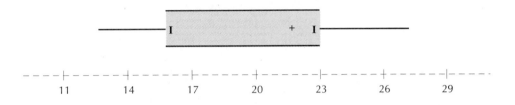

a. Find the median for the data (approximately).
b. Find the upper and lower quartile for the data (approximately).
c. Find the interquartile range for the data set.

14. The number of express-mail next-day delivery packages received daily by the U.S. Postal Service Fox Street Station during the first 30 business days of January 1993 was as follows:

39	38	30	40	43	34	39	36	42	44
30	45	31	32	46	44	30	29	33	40
41	32	36	35	29	30	33	35	34	41

a. Construct a relative frequency histogram for the data.
b. Find the midrange for the data.

15. Refer back to Exercise 14. Find the lower and upper quartile for the data.

16. A Wall Street brokerage company has a large back office that processes all the transactions. The average age of the 18 women in the back office is 27 years, whereas the average age of the 17 men in the back office is 33 years. What is the average age for the entire group?

17. Eighty-eight people have applied for a prestigious civil service job. Maureen is one of the applicants. She scored better than 32 of the applicants on the qualifying exam and four other people scored the same as she did. Find her percentile rank.

18. Refer to the newspaper article on the next page. The company agreed to retrain some of the workers so that they could find other jobs. Each of the workers was given an aptitude test to determine suitability for particular jobs. Laura Snyder's results as well as the results of the other workers who took the test are shown in the table.

ACME CLOTHING TO CLOSE

Feb. 6: Acme Clothing Company officials announced yesterday that they would close the Pachogue plant on March 1, 1996, with the resulting loss of 1400 jobs. The closing is a direct result of rising foreign imports with which the company cannot compete.

Tuesday–February 6, 1996

Skill	Average Test Score	Standard Deviation	Laura's Score
Marketing/sales	61	7.8	71
Plant maintenance	103	6.2	101
Inventory/shipping	42	3.1	46
Personnel	83	5.9	81

  a. Transform each of Laura's scores into a z-score.
  b. In which skill does she have the most talent?
  c. In which skill does she have the least talent?

19. A survey of the families living in the Martin Luther King Housing Complex was conducted to determine the number of children per household. The following table lists the results of the survey:

Number of Children per Household	Frequency
1	30
2	45
3	120
4	80
5	60
6	20
7	5

Find: a. Q_1 b. Q_3 c. P_{80} d. P_{95}.

20. The average time required for runners to complete a racing marathon is 45 minutes with a standard deviation of 10.3 minutes. Find the time required by Lori, Coty, and Walter to complete the marathon if their z-scores are 2.6, -0.53, and 1.87, respectively.

21. Senior citizens residing in nursing homes frequently complain that their children often forget to call or visit them. The 415 residents of the Palm View Nursing Home were polled to determine the number of times during the week of Jan 1–7 that their children called or visited them. The results are shown below:

Number of Times/Week That Children Called	Frequency
0–1	125
2–3	85
4–5	65
6–7	57
8–9	50
10–11	15
12–13	14
14–15	4

   a. Find the (sample) mean and standard deviation.
   b. Find P_{55}.

22. John, who is 23 years old, wants to date a girl who is approximately 20 years old. He is considering two computer dating services in the Los Angeles area, both of which claim that the average age of the girls they have on file is 20 years. John does not know which dating service to select as he believes that both are equally good. His friend persuades him to look at the individual ages of the girls and not at the average. The individual ages are as follows.

Dating Service A	Dating Service B
15	18
17	19
18	20
23	21
27	22

   Is his friend right? Explain your answer.

23. The following relative frequencies were obtained by entering college students from Roosevelt High School on the Scholastic Aptitude Test administered by the Educational Testing Service:

Test Score	Relative Frequency
200–249	0.01
250–299	0.02
300–349	0.06
350–399	0.10
400–449	0.17
450–499	0.19
500–549	0.18
550–599	0.15
600–649	0.08
650–699	0.03
700–740	0.01
750–799	0

Approximately what score must a student have in order to be in the   (a) 60th percentile,   (b) 80th percentile?

24. The 20 girls of the Alpha Sorority contributed an average of $7.50 toward a holiday party. What was the average amount contributed by the 30 girls of the Beta Sorority if the average amount contributed by both sororities was $7.80?

25. Two identical computer components manufactured by two different companies were thoroughly tested to determine the useful life of each component. The following results were obtained:

Company	Average Life	Standard Deviation
A	1400 hr	70 hr
B	1350 hr	25 hr

Which computer component would you buy? Explain your answer.

26. In its *Vital and Health Statistics*, the U.S. National Center for Health Statistics publishes statistical data pertaining to health. A survey of 20,000 female smokers disclosed a mean age of 31.6 years with a standard deviation of 8.6 years. Using Chebyshev's Theorem, what conclusions can be drawn?

27. Bill receives a grade of 85 on a statistics midterm given to 200 students. The average grade was 82 with a standard deviation of 4.3. Did Bill do well on the exam? Explain by using the z-score and Chebyshev's Theorem.

Thinking Critically

1. Is it ever possible for the average deviation, standard deviation, and variance of a set of numbers to be equal? Explain your answer.
2. Verify, by expanding, that $\Sigma(x - \mu) = 0$.

3. Verify that the following two formulas for calculating the sample standard deviation are the same:

$$s = \sqrt{\frac{\Sigma(x - \bar{x})^2}{n - 1}} = \sqrt{\frac{n(\Sigma x^2) - (\Sigma x)^2}{n(n - 1)}}$$

4. What is the z-score of the mean (in any distribution)?

5. Is it true that the median (the 50th percentile) is the average of the lower quartile (the 25th percentile) and the upper quartile (the 75th percentile)? In other words, is it true that $\dfrac{P_{25} + P_{75}}{2} = P_{50}$? Explain your answer.

6. It is true that negative terms can never have positive z-scores? Explain your answer.

7. When working with a normal distribution, z-scores will usually be between -3 and $+3$. Can you explain why?

8. The union representing 812 textile workers is negotiating a new labor contract with the company. The union claims that the *average* hourly rate of pay is $6.11, which is much lower than the pay of workers in competing companies. Management claims that the *average* hourly rate of pay is $9.69, which is much higher than the pay of workers in competing companies. A labor negotiator believes that both the union claim and the management claim are accurate. How can this be? Explain your answer.

Case Studies

1. Drugstore prices for prescription drugs are not uniform. A 1992 General Accounting Office comparison of prices for 121 drugs found that often the most expensive price for a drug can be 25% to 40% more than the cheapest price charged for the same drug. The following prices were obtained by a consumer's group in New York for sixty 500-mg tablets of the arthritis drug Naprosyn. These prices were obtained by sampling 12 local pharmacies, including several independents, several chains, and a drug store inside a chain supermarket.°

Naprosyn
(Sixty 500-mg tablets)

$60.49 (national chain)	$69.23	$68.32	$70.59
61.17	68.61	66.88	71.26
76.00 (independent)	74.19	72.28	63.14

Find the (sample) mean and (sample) variance for the above data.

2. As an incentive to attract additional customers, one Caribbean hotel recently installed a toll-free 800 number. During the first three weeks of operation, the hotel received

°Adapted from "How to Buy Drugs for Less." In *Consumer Reports*, published by Consumer's Union, Mount Vernon, N.Y., October 1993, pp. 675–676.

the following number (on a daily basis) of requests for additional information about the hotel's programs.[**]

$$
\begin{array}{ccccccc}
46 & 62 & 44 & 53 & 58 & 57 & 55 \\
53 & 29 & 20 & 46 & 49 & 48 & 40 \\
28 & 32 & 35 & 37 & 38 & 43 & 28
\end{array}
$$

a. Find the (population) mean and population standard deviation for the above data.
b. Using MINITAB, draw a box plot for the data.

3. Despite governmental supervision, many companies often dispose of industrial wastes by dumping them into nearby rivers, lakes, or streams, thereby polluting the water supply for the animals and fish that feed in these lakes. In one such study an environmental group documented the devastating effect of such activities for a particular river.[†] They collected data on the number of fish and deer that died (based on pathological studies) as a result of industrial dumping. The data that follow are for a six-month period.

Month	No. Dead Deer Found
April	28
May	37
June	26
July	31
August	45
September	21

a. Find the (sample) mean and standard deviation for the preceding data.
b. Using MINITAB, draw a box plot for the data.

4. To cover operating costs and losses due to theft, some supermarket chains charge different prices for the same item, particularly if the store is located in poorer neighborhoods. This type of discrimination is illegal. Recently, the Attorney General began an investigation into such practices. Samples of the milk prices charged in stores in various parts of the city were taken. The following data were obtained:

| *Prices Charged for a Quart of Milk* | |
Poorer Neighborhoods	Middle Class Neighborhoods
63¢	58¢
58	62
71	60
57	59
51	61

[**]DeMadrid and Rodriguez, Puerto Rico, 1993.

[†]Boggs and Jones, New York, 1993.

The supermarket chain claims that the average cost for a quart of milk in either type of neighborhood is 60¢. What is your reaction? Explain your answer.

5. In 1990 there were approximately 25 million Americans, mostly women, who had osteoporosis. This is a disorder in which the bones deteriorate due to the excessive loss of bone tissue. Abnormally low bone mass results in an increase in susceptibility to bone fractures. While the precise cause of osteoporosis is not fully understood, some researchers believe that calcium is a key to preventing osteoporosis. The recommended daily allowance (RDA) of calcium varies with age and gender. For women of child-bearing age, the RDA is 800 milligrams daily. A research team[*] randomly sampled 100 women suffering from osteoporosis to determine their daily intakes of calcium. The following frequency distribution gives the results.

Daily Intake of Calcium (in milligrams)	Frequency
Under 200	6
200–under 400	7
400–under 600	16
600–under 800	32
800–under 1000	21
1000–under 1200	15
1200–under 1400	3

Calculate the sample mean, sample variance, and sample standard deviation for these women.

[*]Todd & Rodgers, New York, 1990.

4

PROBABILITY

Chapter Outline

DID YOU KNOW THAT

a   women at least 40 years old were 41% more likely to be referred for a mammogram if their doctors were female? (*Source: Wall Street Journal*, Thursday, August 12, 1993)

b   65% of professional football players suffer serious injuries as a result of their career? (*Source: Injuries*, NFL Players Association, Washington, D.C.)

c   2,000,000 children will receive a bicycle as a gift this year and that 50,000 children will be treated in hospital emergency rooms for bicycle related head injuries? (*Source: Health*, March/April, 1993, p. 12)

What is the probability of winning the jackpot on a slot machine showing three items on three separate wheels? See further discussion on page 226. *(Tropicana Casino And Resort)*

See further discussion on page 226.

Chapter Objectives

- **To define** what is meant by the word "probability." (Section 4.2)

- **To explain** the use of the word "probability" in such statements as "the probability of a particular thing happening is 0.78." (Section 4.2)

- **To discuss** a formula for determining the number of possible outcomes when an experiment is performed. This is done so that we can find the total number of possible outcomes, which we use for probability calculations. (Section 4.3)

- **To analyze** the number of different ways of arranging things, depending on whether or not order counts. Thus, we will analyze permutations and combinations. (Sections 4.4 and 4.5)

- **To work** with a convenient notation used to represent a special type of multiplication. This is the factorial notation. (Section 4.4)

- **To apply** a computational device for calculating the number of possible combinations. This is Pascal's triangle. (Section 4.5)

- **To talk** about "odds" and "mathematical expectation." These words, often used by gamblers, represent the payoff for a situation and the likelihood of obtaining it. (Section 4.6)

Until now we have been discussing methods for organizing and summarizing data, that is, descriptive statistics. Since we often use results of a sample to make inferences about the unknown population from which the sample was drawn, we can never be certain that our inferences are correct. Thus, before studying the techniques of statistical inference, we need to be familiar with probability theory—the science of uncertainty.

In this chapter we introduce some of the basic ideas of probability that will enable us to evaluate the likelihood that our statistical inferences are correct.

STATISTICS IN ACTION

The first newspaper article indicates that a committee of four governors would be selected to go to Washington to plead for state aid. How do we select such committees? How many four-

GOVERNORS ASK FOR MORE STATE AID

May 10: It was decided yesterday at a governor's meeting that a committee of four governors would be selected to go to Washington to plead for more state aid. Many governor's expressed the view that the current level of federal aid to the states was appalling, and that it was the obligation of the federal government to share the state's financial burden.

May 10, 1993

SMOKING AND YOUR HEALTH—THE LATEST STATISTICS

Washington: Smoking is a powerful addiction. The latest statistics released by the American Cancer Society indicate that young women under age 23 make up the fastest growing group of new smokers in the United States. One study of high school seniors found that an average 31.6% of girls smoked compared to 28% for boys. Most of the kids began the habit before the age of 19. According to a report from the Surgeon General Antonia Novello, after just one year off cigarettes the probability of the added cancer risk caused by previous smoking is reduced by 50%. Furthermore, after about 15 years of nonsmoking, the probability of lung cancer is about the same as that of individuals who never smoked.

Friday—January 4, 1991

governor committees can possibly be formed from among all the governors?

Note the use of the word "probability" in the second newspaper article. How are such probabilities determined? Are they reliable?

In the third newspaper article the word "likelihood" is used. Is likelihood the same as probability? How do we determine likelihoods?

In this chapter we analyze what we mean by probability and how we determine such probabilities.

WHAT IS THE LIKELIHOOD THAT YOU WILL BE INJURED ON THE JOB?

Washington: According to a survey conducted by the *National Center for Statistics*, 1 in 15 U.S. workers had one or more job-related injuries this past year. Injuries on the job were twice as common among men as women and more likely among workers aged 1–29 years than among workers aged 45–64 years.

The most common job related injuries reported by the agency were:

26% back strains or sprains
21% cuts or punctures
13% bruises or abrasions

Adapted from *USA Today*—July 28, 1993

4.1   INTRODUCTION

Although the word "probability" may sound strange to you, it is not so unfamiliar as you may think. In everyday situations we frequently make decisions and take action as a result of the probability of certain events. Thus, if the weather forecaster predicts rain with a probability of 80%, we undoubtedly would prepare ourselves accordingly.

Let us, however, analyze the weather forecaster's prediction. What the forecaster really means is that based on past records, 80% of the time when the weather conditions have been the same as they are today, rain has followed. Thus, the probability calculations and resultant forecasts are based on past records. They are based on the assumption that since in the *past* rain has occurred a certain percentage of the time, it will occur the same percentage of times in the *future*. This is but one usage of probability. It is based on **relative frequency**. We will explain this idea in greater detail shortly.

Probability is also used in statements that express a personal judgment or conviction. This can be best illustrated by the following statements: "If the United States had not dropped the atomic bomb on Japan, World War II would *probably* have lasted several more years" or "If all the New York Mets players had been healthy the entire season, they *probably* would have won the pennant last year."

Probability can also be used in other situations. For example, if a fair coin° is tossed, we would all agree that the probability is $\frac{1}{2}$ that heads comes up. This is because there are only two possible outcomes when we flip a coin, heads or tails.

Now consider the following conversation overheard in a student cafeteria.

Bill: I am going to cut math today.
Eric: Why?
Bill: I didn't do my homework.
Eric: So what? I didn't either.
Bill: So, since the teacher calls on at least half of the class each day for answers, he will *probably* call on me today and find out that I am not prepared.
Eric: The teacher called on me yesterday, so he *probably* won't get me today. I am going to class.

In the preceding situation, each student is making a decision based on probability.

Basically, the theory of probability deals with the study of uncertainties. Thus, it has been found to have wide applications in the following situations:

1. It is used by insurance companies when they calculate insurance premiums and the *probable* life expectancies of their policyholders.
2. It is used (formally or informally) by a gambler who decides to bet 10 to 1 on a particular horse.
3. It is used by industry officials in determining the reliability of certain equipment.

°A *fair coin* is a coin that has the same chance of landing on heads as on tails. Throughout this book we will always assume that we have fair coins unless told otherwise.

4. It is used by medical researchers who claim that smoking increases your *chance* of getting lung cancer.
5. It is used by biologists in their study of genetics.
6. It is used by pollsters in such polls as the Harris poll, the Gallup poll, and the Nielsen ratings to determine the reliability of their polls.
7. It is used by an investor who decides that a particular stock has a greater chance for future growth than any other stock.
8. It is used by business managers in determining which products to manufacture, which products to advertise, and which media to use in advertising: television, radio, magazines, newspapers, subway and bus advertisements, and so on.
9. It is used by psychologists in predicting reactions or behavioral patterns under certain stimuli.
10. It is used by government economists in predicting that the inflation rate will increase or decrease in the future.

Since probability has so many possible meanings and uses, we will first analyze the nature of probability and how to calculate it. This will be done in this chapter. In the

HISTORICAL NOTE

Historically, probability had its origin in the gambling room. The Chevalier de Méré, a professional French gambler, had asked his friend Blaise Pascal (1623–1662) to solve the following problem: In what proportion should two players of equal skill divide the stakes remaining on the gambling table if they are forced to stop before finishing the game? Pascal wrote to and began an active correspondence with Pierre Fermat (1601–1665) concerning the problem. Although Pascal and Fermat agreed on the answer, they gave different proofs. It was in this series of correspondences during the year 1652 that they developed the modern theory of probability.

A century earlier the Italian mathematician and gambler Girolomo Cardan (1501–1576) wrote *The Book on Games of Chance*. For a further discussion of Cardan, see p. 14.

Another famous mathematician who contributed to the theory of probability was Abraham De Moivre (1667–1754). Like de Méré, De Moivre spent many hours with London gamblers and wrote a manual for gamblers entitled *Doctrine of Chances*. Like Cardan, De Moivre predicted the day of his death. A rather interesting story is told of De Moivre's death. De Moivre was ill and each day he noticed that he was sleeping 15 minutes longer than he did on the preceding day. Using progressions, he computed that he would die in his sleep on the day after he slept 23 hours and 45 minutes. On the day following a sleep of 24 hours, De Moivre died.

The French mathematician Pierre Simon de Laplace (1749–1827) also contributed much to the historical development of probability. In 1812 he published *Théorie Analytique des Probabilités*, in which he referred to probability as a science that began with games but that had wide-ranging applications. In particular, he applied probability theory not to gambling situations but as an aid in astronomy.

Over the course of many years probability theory has left the gambling rooms and has grown to be an important and ever-expanding branch of mathematics.

next chapter we will discuss various rules that allow us to calculate probabilities for many different situations.

4.2   DEFINITION OF PROBABILITY

Probability theory can be thought of as that branch of mathematics that is concerned with calculating the probability of outcomes of experiments.

Since many ideas of probability were derived from gambling situations, let us consider the following experiment. An honest die (the plural is dice) was rolled many times and the number of 1's that came up was recorded. The results are

Number of 1's that came up	1	11	18	99	1001	10,001
Number of rolls of the die	6	60	120	600	6000	60,000

Notice that in each case the number of 1's that appeared is approximately $\frac{1}{6}$ of the total number of tosses of the die. It would then be reasonable to conclude that the probability of a 1 appearing is $\frac{1}{6}$.

Although when a die is rolled there are six equally likely possible outcomes if it is an honest die (see Figure 4.1), we are concerned with the number of 1's appearing.

FIGURE 4.1   Possible outcomes when a die is rolled once.

Favorable outcome Each time a 1 appears, we call it a **favorable outcome**. There are six possible outcomes of which only one is favorable. The probability is thus the number of favorable outcomes divided by the total number of possible outcomes, which is 1 divided by 6, or $\frac{1}{6}$. The preceding chart indicates that our guess, that the probability is approximately $\frac{1}{6}$, is correct.

Similarly, if a coin is tossed once, we would say that the probability of getting heads is $\frac{1}{2}$ since there are two possible outcomes, heads and tails, and only one is favorable.

Sample space These are the only two possible outcomes in this case. All the possible outcomes of an experiment are referred to as the **sample space** of the experiment. We will usually be interested in only some outcomes of the experiment. The outcomes that are of interest *Event* to us will be referred to as an **event**. Thus, in flipping a coin once the sample space is heads or tails, abbreviated as H, T. The event of interest is H.

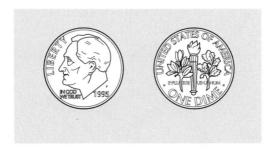

Possible outcomes when a coin is flipped.

In the rolling of a die the sample space consists of six possible outcomes: 1, 2, 3, 4, 5, and 6. We may be interested in the event "getting a 1."

If we toss a coin twice, the sample space is HH, HT, TH, and TT. There are four possibilities. In this abbreviated notation HT means heads on the first toss and tails on the second toss, whereas TH means tails on the first toss and heads on the second toss. The event "getting a head on both tosses" is denoted by HH. The event "no head" is TT.

To illustrate further the idea of sample space and event, consider the following examples.

EXAMPLE 1 Two dice are rolled at the same time. Find the sample space.

SOLUTION   There are 36 possible outcomes as pictured in Figure 4.2.

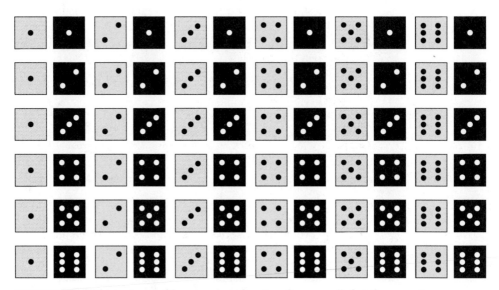

FIGURE 4.2   Thirty-six possible outcomes when two dice are rolled at the same time.

The possible outcomes are summarized as follows:

Die 1	Die 2	Die 1	Die 2	Die 1	Die 2
1	1	3	1	5	1
1	2	3	2	5	2
1	3	3	3	5	3
1	4	3	4	5	4
1	5	3	5	5	5
1	6	3	6	5	6
2	1	4	1	6	1
2	2	4	2	6	2
2	3	4	3	6	3
2	4	4	4	6	4
2	5	4	5	6	5
2	6	4	6	6	6

The event "sum of 7 on both dice together" can happen in six ways. These are

Die 1	Die 2	Die 1	Die 2
1	6	3	4
6	1	5	2
4	3	2	5

Similarly, the event "sum of 9 on both dice together" can happen in four ways. The event "sum of 13 together" can happen in zero ways.

EXAMPLE   2   Two contestants, Jack and Jill, are on a quiz show. Each is in a soundproof booth and cannot hear what the other is saying. They are each asked to select a number from 1 to 3. They each win 10,000 if they select the same number. In how many different ways can they win the prize?

SOLUTION   They will win the prize if they both say 1, 2, or 3. The sample space for this experiment is

Jill Guesses	Jack Guesses	Jill Guesses	Jack Guesses	Jill Guesses	Jack Guesses
1	1	2	1	3	1
1	2	2	2	3	2
1	3	2	3	3	3

Thus, the event "winning the prize" can occur in three possible ways.

We now define formally what we mean by probability.

Definition 4.1
Probability

If an event can occur in n equally likely ways and if f of these ways are considered favorable, then the **probability** of getting a favorable outcome is

$$\frac{\text{Number of favorable outcomes}}{\text{Total number of outcomes}} = \frac{f}{n}$$

Thus, the probability of any event equals the number of favorable outcomes divided by the total number of possible outcomes.

We use the symbol $p(A)$ to stand for "the probability of event A."

Classical interpretation of probability

Definition 4.1 is often called the **classical interpretation of probability**. It assumes that all outcomes of an experiment are equally likely. Thus, if we say that the probability of getting heads when flipping an honest coin is $\frac{1}{2}$, we are basing this on the following facts: When a coin is flipped once, there are 2 possible outcomes, heads and tails. Therefore, the probability of getting heads is $\frac{1}{2}$ (1 out of a possible 2 outcomes).

Similarly, the probability of getting a 1 when an honest die is rolled is $\frac{1}{6}$ since there are 6 possible outcomes, only 1 of which is favorable.

Relative frequency concept of probability

An alternate interpretation of probability is called the **relative frequency concept of probability**. Suppose we tossed a coin 100 times and it landed heads 50 of the times. It would seem reasonable to claim that the probability of landing heads is approximately $\frac{50}{100}$ or $\frac{1}{2}$. We can think of the probability as the relative frequency of the event. Of course, to convince ourselves that our answer is reasonable, we could toss the coin many, many additional times. Since it is much easier to determine relative frequencies, this interpretation is very easy to understand and is commonly used. The relative-frequency interpretation represents the percentage of times that the event will happen in repeated experiments.

Probability can also be defined from a strictly mathematical, that is, axiomatic point of view; however, this is beyond the scope of this text. Let us now illustrate the concept of probability with several examples.

EXAMPLE 3

A family plans to have three children. What is the probability that all three children will be girls? (Assume that the probability of a girl being born in a given instance is $\frac{1}{2}$.)

SOLUTION   We will use Definition 4.1. We first find the total number of ways of having three children, that is, the sample space. There are eight possibilities as shown in the following table:

Child 1	Child 2	Child 3
Boy	Boy	Boy
Boy	Boy	Girl
Boy	Girl	Boy
Boy	Girl	Girl
Girl	Boy	Boy
Girl	Girl	Boy
Girl	Boy	Girl
Girl	Girl	Girl

Of these, only one is favorable, namely, the outcome Girl, Girl, Girl. Thus,

$$p(3 \text{ girls}) = \frac{1}{8}$$

EXAMPLE   4
Playing Cards

A card is selected from an ordinary deck of 52 cards. What is the probability of getting

   a. a queen?
   b. a diamond?
   c. a black card?
   d. a picture card?
   e. the king of clubs?

SOLUTION   A deck of playing cards consists of 52 cards, as shown in Figure 4.3. When we perform the experiment of randomly selecting a card from the deck, there are 52 possible outcomes, namely, the 52 cards pictured in Figure 4.3.

   a. As shown in Figure 4.3, there are 4 queens in the deck, so there are 4 favorable outcomes. Definition 4.1 tells us that

$$p(\text{queen}) = \frac{4}{52} = \frac{1}{13}$$

   b. There are 13 diamonds in the deck, so there are 13 favorable outcomes. Therefore,

$$p(\text{diamonds}) = \frac{13}{52} = \frac{1}{4}$$

c. Since a black card can be either a spade or a club, there are 26 black cards in the deck as shown in Figure 4.3. Therefore,

$$p(\text{black card}) = \frac{26}{52} = \frac{1}{2}$$

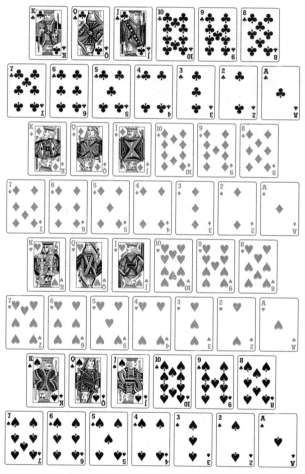

FIGURE 4.3 The sample space when a card is drawn from an ordinary deck of 52 cards.

d. There are 12 picture cards (4 jacks, 4 queens, and 4 kings), so there are 12 favorable outcomes. Therefore,

$$p(\text{picture card}) = \frac{12}{52} = \frac{3}{13}$$

e. There is only one king of clubs in a deck of 52 cards. Thus,

$$p(\text{king of clubs}) = \frac{1}{52}$$

EXAMPLE 5   Mary and her friend Gwendolyn are visiting the Sears' Tower in Chicago. Each of them enters a different elevator in the main lobby that is going up and that can let them off on any floor from 1 to 6. Assuming that Mary and Gwendolyn are as likely to get off at one floor as another, what is the probability that they both get off on the same floor?

SOLUTION   Since both Mary and Gwendolyn can get off at any floor between 1 and 6, there are 36 possible outcomes as shown in the following table. A favorable outcome occurs if both get off on floor 1, floor 2, and so on as shown. Thus, there are 6 favorable outcomes out of 36 possible outcomes, so that

$$p(\text{both get off on same floor}) = \frac{6}{36} = \frac{1}{6}$$

Floor Where Mary Gets Off	Floor Where Gwendolyn Gets Off	Floor Where Mary Gets Off	Floor Where Gwendolyn Gets Off
1	1 ← Favorable outcome	4	1
1	2	4	2
1	3	4	3
1	4	4	4 ← Favorable outcome
1	5	4	5
1	6	4	6
2	1	5	1
2	2 ← Favorable outcome	5	2
2	3	5	3
2	4	5	4
2	5	5	5 ← Favorable outcome
2	6	5	6
3	1	6	1
3	2	6	2
3	3 ← Favorable outcome	6	3
3	4	6	4
3	5	6	5
3	6	6	6 ← Favorable outcome

EXAMPLE 6 A wheel of fortune has the numbers 1 through 50 painted on it (see the diagram below). Tickets numbered 1 through 50 have been sold. The wheel will be rotated and the number on which the pointer lands will be the winning number. If the pointer stops on the line between two numbers, the wheel must be turned again. A prize of a new car will be awarded to the person whose ticket number matches the winning number. What is the probability that

    a. ticket number 42 wins?
    b. a ticket between the numbers 26 and 39 (not including these numbers) wins?
    c. ticket number 51 wins?

SOLUTION Since 50 tickets were sold, the total number of possible outcomes, that is, the sample space, is 50.

    a. Only 1 ticket numbered 42 was sold. There is only one favorable outcome. Thus,

$$p(\text{ticket number 42 wins}) = \frac{1}{50}$$

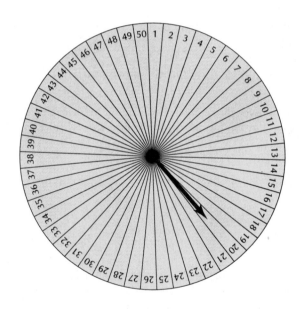

    b. There are 12 ticket numbers between 26 and 39, not including 26 and 39. Thus,

$$p(\text{a ticket between the numbers 26 and 39 wins}) = \frac{12}{50} = \frac{6}{25}$$

c. Since tickets numbered up to 50 were sold, ticket number 51 can never be a winning number. Thus,

$$p(\text{ticket number 51 wins}) = \frac{0}{50} = 0$$

Null event   An event that can never happen is called a **null event** and its probability is 0.

E X A M P L E   7   Two fair dice are rolled at the same time and the number of dots appearing on both dice is counted. Find the probability that this sum

    a. is 7.
    b. is an odd number larger than 6.
    c. is less than 2.
    d. is more than 12.
    e. is between 2 and 12, including these two numbers.

SOLUTION   When two dice are rolled, there are 36 possible outcomes; that is, the sample space has 36 possibilities. These were listed in Example 1 on page 211.

    a. The sum of 7 on both dice together can happen in 6 ways so that

$$p(\text{sum of 7}) = \frac{6}{36} \text{ or } \frac{1}{6}$$

    b. The statement "a sum that is an odd number larger than 6" means a sum of 7, a sum of 9, or a sum of 11. A sum of 7 on both dice together can happen in 6 ways. Similarly, a sum of 9 on both dice together can happen in 4 ways, and a sum of 11 can happen in 2 ways. There are then 12 favorable outcomes out of 36 possibilities. Thus,

$$p(\text{a sum that is an odd number larger than 6}) = \frac{12}{36} = \frac{1}{3}$$

c. When two dice are rolled, the minimum sum is 2, and we cannot obtain a sum less than 2. There are *no* favorable events. This is the null event. Hence,

$$p(\text{a sum that is less than 2}) = 0$$

d. When two dice are rolled, the maximum sum is 12. We cannot obtain a sum that is more than 12. This is the null event. Thus,

$$p(\text{a sum that is more than 12}) = 0$$

e. When two dice are rolled we *must* obtain a sum that is between 2 and 12, including the numbers 2 and 12. There are 36 possible outcomes and *all* these are favorable. Thus,

$$p(\text{a sum between 2 and 12, including 2 and 12}) = \frac{36}{36} = 1$$

Therefore, a favorable outcome *must* occur in this case.

Certain event or Definite event   An event that is certain to occur is called the **certain event** or **definite event** and its probability is 1.

COMMENT   Any event, call it A, may or may not occur. If it is sure to occur, we have the certain event and its probability is 1. If it will never occur, we have the null event and its probability is 0. Thus, if we are given any event A, then we know that its probability *must* be between 0 and 1 and possibly equal to 0 or 1. This is because the event may or may not occur. Probability can *never* be a negative number.

COMMENT   We mentioned earlier that probability can be thought of as the fraction of times that an outcome will occur in a long series of repetitions of an experiment. However, there may be certain experiments that cannot be repeated. For example, if Gary's kidney has to be removed surgically, we cannot think of this as an experiment that can be repeated over and over again, at least as far as Gary is concerned. How do we assign probabilities in this case? This is not an easy task. Calculating the probability in such situations requires the judgment and experience of a doctor familiar with *many* experiments of a similar type. Thus, if doctors tell you that you have an 80% chance of surviving the operation, they mean that based upon their previous experiences with such situations, 80% of the patients with similar operations have survived. Usually an experienced surgeon can assign a fairly reasonable probability to the success of a nonrepeatable operation.

1970 Draft Lottery

January		February		March		April		May		June	
Birthday	Draft-Priority Number	Birthday	Draft-Priority Number	Birthday	Draft-Priority Number	Birthday	Draft-Priority Number	Birthday	Draft-Priority Number	Birthday	Draft-Priority Number
1	■305	1	86■	1	108■	1	32■	1	■330	1	■249
2	159■	2	144■	2	29■	2	■271	2	■298	2	■228
3	■251	3	■297	3	■267	3	83■	3	40■	3	■301
4	■215	4	■210	4	■275	4	81■	4	■276	4	20■
5	101■	5	■214	5	■293	5	■269	5	■364	5	28■
6	■224	6	■347	6	139■	6	■253	6	155■	6	110■
7	■306	7	91■	7	122■	7	147■	7	35■	7	85■
8	■199	8	181■	8	■213	8	■312	8	■321	8	■366
9	■194	9	■338	9	■317	9	■219	9	■197	9	■335
10	■325	10	■216	10	■323	10	■218	10	65■	10	■206
11	■329	11	150■	11	136■	11	14■	11	37■	11	134■
12	■221	12	68■	12	■300	12	■346	12	133■	12	■272
13	■318	13	152■	13	■259	13	124■	13	■295	13	69■
14	■238	14	4■	14	■354	14	■231	14	178■	14	■356
15	17■	15	89■	15	169■	15	■273	15	130■	15	180■
16	121■	16	■212	16	166■	16	148■	16	55■	16	■274
17	■235	17	■189	17	33■	17	■260	17	112■	17	73■
18	140■	18	■292	18	■332	18	90■	18	■278	18	■341
19	58■	19	25■	19	■200	19	■336	19	75■	19	104■
20	■280	20	■302	20	■239	20	■345	20	183■	20	■360
21	■186	21	■363	21	■334	21	62■	21	■250	21	60■
22	■337	22	■290	22	■265	22	■316	22	■326	22	■247
23	118■	23	57■	23	■256	23	■252	23	■319	23	109■
24	59■	24	■236	24	■258	24	2■	24	31■	24	■358
25	52■	25	179■	25	■343	25	■351	25	■361	25	137■
26	92■	26	■365	26	170■	26	■340	26	■357	26	22■
27	■355	27	■205	27	■268	27	74■	27	■296	27	64■
28	77■	28	■299	28	■223	28	■262	28	■308	28	■222
29	■349	29	■285	29	■362	29	■191	29	■226	29	■353
30	164■			30	■217	30	■208	30	103■	30	■209
31	■211			31	30■			31	■313		

In December 1969 the United States Selective Service established a priority system for determining which young men would be drafted into the army (see chart above). Capsules representing each birth date were placed in a drum and selected at random. Those men whose numbers were selected first were almost certain that they would be drafted. The different birthdays were given draft priority numbers as indicated in the clipping on the draft lottery. Supposedly, each birthday had an equally likely probability of being selected. However, by analyzing the numbers carefully, we find that the majority of those birthdays that occurred later in the year had a higher priority number than those that occurred earlier in the year. Did each birthday have an equal probability of being selected?

EXERCISES FOR SECTION 4.2

1. The FBI publishes the book *Crime in the United States* that presents information on arrests for violent crimes. The table below presents information on 1032 prisoners in one particular city as to type of crime committed and the arrested person's age.

1970 Draft Lottery

July		*August*		*September*		*October*		*November*		*December*	
Birthday	Draft-Priority Number	Birthday	Draft-Priority Number	Birthday	Draft-Priority Number	Birthday	Draft-Priority Number	Birthday	Draft-Priority Number	Birthday	Draft-Priority Number
1	93 ■	1	111 ■	1	■ 225	1	■ 359	1	19 ■	1	129 ■
2	■ 350	2	45 ■	2	161 ■	2	125 ■	2	34 ■	2	■ 328
3	115 ■	3	■ 261	3	49 ■	3	■ 244	3	■ 348	3	157 ■
4	■ 279	4	145 ■	4	■ 322	4	■ 202	4	■ 266	4	165 ■
5	■ 188	5	54 ■	5	82 ■	5	24 ■	5	■ 310	5	56 ■
6	■ 327	6	114 ■	6	6 ■	6	87 ■	6	76 ■	6	10 ■
7	50 ■	7	168 ■	7	8 ■	7	■ 234	7	51 ■	7	12 ■
8	13 ■	8	48 ■	8	■ 184	8	■ 283	8	97 ■	8	105 ■
9	■ 277	9	106 ■	9	■ 263	9	■ 342	9	80 ■	9	43 ■
10	■ 284	10	21 ■	10	71 ■	10	■ 220	10	■ 282	10	41 ■
11	■ 248	11	■ 324	11	158 ■	11	■ 237	11	46 ■	11	39 ■
12	15 ■	12	142 ■	12	■ 242	12	72 ■	12	66 ■	12	■ 314
13	42 ■	13	■ 307	13	175 ■	13	138 ■	13	126 ■	13	163 ■
14	■ 331	14	■ 198	14	1 ■	14	■ 294	14	127 ■	14	26 ■
15	■ 322	15	102 ■	15	113 ■	15	171 ■	15	131 ■	15	■ 320
16	120 ■	16	44 ■	16	■ 207	16	■ 254	16	107 ■	16	96 ■
17	98 ■	17	154 ■	17	■ 255	17	■ 288	17	143 ■	17	■ 304
18	■ 190	18	141 ■	18	■ 246	18	5 ■	18	146 ■	18	128 ■
19	■ 227	19	■ 311	19	177 ■	19	■ 241	19	■ 203	19	■ 240
20	■ 187	20	■ 344	20	63 ■	20	■ 192	20	■ 185	20	135 ■
21	27 ■	21	■ 291	21	■ 204	21	■ 243	21	156 ■	21	70 ■
22	153 ■	22	■ 339	22	160 ■	22	117 ■	22	9 ■	22	53 ■
23	172 ■	23	116 ■	23	119 ■	23	■ 201	23	182 ■	23	162 ■
24	23 ■	24	36 ■	24	■ 195	24	■ 196	24	■ 230	24	95 ■
25	67 ■	25	■ 286	25	149 ■	25	176 ■	25	132 ■	25	84 ■
26	■ 303	26	■ 245	26	18 ■	26	7 ■	26	■ 309	26	173 ■
27	■ 289	27	■ 352	27	■ 233	27	■ 264	27	47 ■	27	78 ■
28	88 ■	28	167 ■	28	■ 257	28	94 ■	28	■ 281	28	123 ■
29	■ 270	29	61 ■	29	151 ■	29	■ 229	29	99 ■	29	16 ■
30	■ 287	30	■ 333	30	■ 315	30	38 ■	30	174 ■	30	3 ■
31	■ 193	31	11 ■			31	79 ■			31	100 ■

Type of Crime Committed	Age of Person Arrested			
	16–22	23–30	31–38	45+
Murder	19	17	16	12
Forcible rape	49	28	4	1
Robbery	137	102	98	81
Aggravated assault	157	206	59	46

Find the probability that a randomly selected prisoner from this group
a. was arrested for robbery.
b. was between 16–22 years of age at the time of arrest.
c. was at least 23 years of age at the time of arrest.
d. was arrested for murder or forcible rape.

2. A charity box contains 17 nickels, 13 dimes, 8 quarters, and 5 half-dollars. When the box is shaken a coin falls out. (Assume that different-sized coins are equally likely to fall out of the box.) Find the probability that the coin is

   a. a nickel.

   b. a penny.

   c. a quarter.

3. A man and a woman who do not know each other board a plane in Paris that is bound for St. Louis, with stop offs in London, New York, and Chicago. What is the probability that they both get off at the same airport? (Assume that people are equally likely to get off at each of these cities.)

4. According to the *Insurance Institute*, the premium charged for life insurance is determined by many factors such as whether the proposed insured is a smoker, drinker, and so forth. The following information about the 316 employees of the Printex Corporation who applied for insurance is available.

	Heavy Smoker and Drinker	Heavy Smoker and Nondrinker	Nonsmoker and Heavy Drinker	Nonsmoker and Nondrinker
Male	81	49	57	3
Female	32	68	21	5

Find the probability that a randomly selected person from this group is

   a. a female.

   b. a female who is a nonsmoker and nondrinker.

   c. a nonsmoker and a nondrinker.

5. The FBI in its *Uniform Crime Statistics* reports that car theft is so prevalent in some cities that a car is reported stolen once every 30 seconds. To combat auto theft, various antitheft devices are available. The Acme Insurance Company surveyed 300 cars equipped with some antitheft device and reported the following:

Type of Car	*Antitheft Device Used*		
	Ignition Shutoff	Steering Wheel Lock	Burglar Alarm
Compact	48	27	53
Intermediate size	32	19	46
Large size	17	22	36

Find the probability that a randomly selected car from this group is

   a. a compact car.

   b. equipped with a steering wheel lock.

   c. a compact car that is equipped with a steering wheel lock.

   d. a subcompact car.

6. Which of the following cannot be the probability of some event?

   a. 9.97   b. 1.36   c. −0.01   d. 0   e. $\dfrac{5003}{5004}$

7. The Aluminum Association of America reports that 17.5 billion pounds of aluminum were used in the year of 1989 for a variety of purposes as shown below:

Aluminum Use	Percentage
Building and construction	16
Consumer durables	7
Containers and packaging	27
Electrical	9
Exports	13
Machinery and equipment	6
Transportation	19
Other	3
Total =	100

(*Source: New York Times*, July 22, 1994)

What is the probability that the aluminum was used
a. for electrical purposes?
b. for consumer durables, exports, or transportation purposes?

8. At a recent New Year's party, four of the guests became drunk and were asked to leave. As each of the drunk guests left the party, he or she picked up a coat that definitely belonged to one of these four people. Since the guests were drunk, however, they did not know whose coat they were taking. Find the probability that each person got his or her own coat.

9. During the spring semester, Professor Carrington taught a calculus and a statistics course. The following is the grade distribution for these classes:

Grade Distribution

Class	Calculus Course	Statistics Course
A	7	8
B	9	10
C	11	12
D	6	9
F	5	8
	Total = 38	Total = 47

If a student that was in one of Professor Carrington's classes during the spring semester is randomly selected, what is the probability that the student
a. received an A?
b. was in the calculus class
c. was in the calculus class and received an A?

10. An emergency room nurse has just taken blood samples from three different patients. Before attaching the appropriate identifying label to the three tubes contain-

ing the blood samples, the nurse drops all three labels on the floor. If the nurse picks up all the labels and randomly attaches the labels to the blood samples, what is the probability that each tube will be labeled with its correct identifying label? (*Hint:* Find the sample space.)

11. Consider the accompanying newspaper article.

> ### HOW DO YOU GET TO WORK?
>
> *January 5*: According to the lat- est U.S. Bureau of the Census statistics, most people prefer to use their automobiles or car pools to travel to work. Many Americans simply refuse to use public transportation to get to work. Among the reasons given for this refusal are unreliability, lack of safety, and convenience.
>
> January 5, 1991

The following statistics are available on how 702,000 Americans get to work.

	By Car	By Public Transportation
Urban Worker	470,000	157,000
Rural Worker	70,000	5,000

(*Source:* U.S. Bureau of the Census, 1991)

If one of these 702,000 American workers is selected at random, what is the probability that the individual
a. is an urban worker?
b. comes to work by car?
c. is a rural worker who travels to work by means of public transportation?

12. An appliance company manufactured 10,000 microwave ovens during the first three months of 1996, each stamped with a number from 1 to 10,000. A few months later, the company decides to recall those microwave ovens whose last digit is 5, 6, or 7 to check for possible radiation leakage. Helen Hoffman owns a microwave oven that was manufactured during the first three months of 1996. What is the probability that she owns one of the ovens that is being recalled?

13. A pharmaceutical research company is conducting research with gene-splicing tech- niques that produce four different kinds of viruses, A, B, C, and D, in the ratio 7:6:4:3, respectively. A medical researcher randomly selects a virus that has been produced by the new process. Find the probability that it is a virus

a. of type A.

b. of type B or C.

c. of type B, C, or D.

d. not of type A or B.

14. MRCA Information Services reports that 68.6% of all men's apparel is purchased by women. If a men's apparel purchase is randomly selected, what is the probability that it was bought by the man himself?

15. Consider the accompanying newspaper article. If a lawyer that graduated from law school between 1990 and 1992 is randomly selected, what is the probability that the lawyer graduated in 1991?

> **MORE LAWSUITS ON THE HORIZON**
>
> *Washington*: (Sept. 29) As if America doesn't already have enough lawyers, there are plenty more to come. The American Bar Association reports that the number of law school graduates continues to rise each year, from 36,985 in 1990, to 38,800 in 1991 to 42,037 in 1992.
>
> September 20, 1994

16. Boris is taking a true-false test and has no idea of the answers to the last four questions. He decides to guess at the answers.

    a. If C stands for a correct answer and W stands for a wrong answer, list all possible outcomes.

    b. Find the probability that he guesses all four answers correctly.

    c. Find the probability that at least one of his answers is correct.

17. Refer back to the previous question. Find the probability that exactly two of the four answers are correct.

18. According to *Consumer Reports* (January 1993, pp. 9–12), the number of cellular phone users grew by an average of 7300 per day during most of 1992, bringing the total to 10 million by December. Fifty-five percent of *Consumer Reports* readers stated that they were "very" or "completely" satisfied with cellular phone service. If a *Consumer Reports* reader that has cellular phone service is randomly selected, what is the probability that the reader is *not* "very or completely" satisfied with the service?

19. An individual's genetic makeup is determined by the genes obtained from each parent. For every genetic trait, each parent possesses a gene pair; and each contributes one half of this gene pair, with equal probability, to their offspring, forming

a new gene pair. The offspring's traits come from this new gene pair, where each gene in this pair possesses some characteristic. It is known that in humans, the eye color brown is dominant and the color blue is recessive. Let B represent a gene for brown eyes and let b represent a gene for blue eyes. Thus a person who has BB genes (that is, brown from each parent) or Bb genes (brown from one parent and blue from the other) will have brown eyes, since brown is dominant. On the other hand, a person who has bb genes (blue from both parents) will have blue eyes. Arlene, who is known to have Bb genes, marries Tom, who is also known to have Bb genes. What is the probability that their child will have blue eyes?

20. Refer back to Exercise 19. If Arlene has brown eyes, type BB, what is the probability that her child will have blue eyes?

21. For security reasons, many companies require their workers to wear their photo identification badges on their clothing at all times while at work. Typically, a worker's photo is taken and then laminated onto a card that has the worker's signature and department on it. Because of a mechanical malfunction of the laminating machine, four photos have been found on the floor next to four signature cards. If a clerk randomly attaches one photo to a signature card, find the probability that each photo is attached to the proper signature card.

22. A slot machine in a gambling casino has three wheels, and each wheel has a picture of a lemon, a cherry, and an apple on it. When the appropriate amount of money is deposited and the button is pushed, each wheel rotates and then displays a picture of one of the items mentioned. Each wheel operates independently of the other. When all three wheels show the same item, then the player wins $5000.
   a. List all possible outcomes for this machine.
   b. Find the probability of a player winning $5000 when playing this slot machine.

4.3 COUNTING PROBLEMS

In determining the probability of an event, we must first know the total number of possible outcomes. In many situations it is a rather simple task to list all possible outcomes and then to determine how many of these are favorable. In other situations there may be so many possible outcomes that it would be too time consuming to list all of them. Thus, when two dice are rolled, there are 36 possible outcomes. These are listed on page 211. Exercise 22 of Section 4.2 has 27 possible outcomes. When there are too many possibilities to list, we can use rules, which will be given shortly, to determine the actual number.

Tree diagram   One technique that is sometimes used to determine the number of possible outcomes is to construct a **tree diagram**. The following examples will illustrate how this is done.

EXAMPLE 1 By means of a tree diagram we can determine the number of possible outcomes when a coin is repeatedly tossed. If one coin is tossed, there are two possible outcomes, heads or tails, as shown in the tree diagram in Figure 4.4. If two coins are tossed, there are

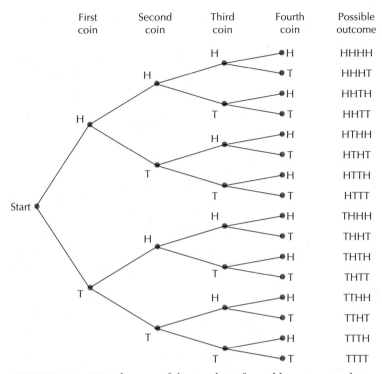

	First coin	Second coin	Third coin	Fourth coin	Possible outcome

FIGURE 4.4 Tree diagram of the number of possible outcomes when a coin is repeatedly tossed.

now four possible outcomes. These are HH, HT, TH, and TT, since each of the possible outcomes on the first toss can occur with each of the two possibilities on the second toss. So, if heads appeared on the first toss, we may get heads or tails on the second toss. The same is true if tails appeared on the first toss. The tree diagram shows that there are four possibilities. When three coins are tossed, the diagram shows that there are eight possible outcomes. Also, there are 16 possible outcomes when four coins are tossed.

E X A M P L E   2   Hazel Brown is about to order dinner in a restaurant. She can choose any one of three main courses and any one of four desserts.

Main Course	Dessert
Hamburger	Hot-fudge sundae
Steak	Jello
Southern-fried chicken	Cake
	Fruit

Using a tree diagram, find all the possible dinners that Hazel can order.

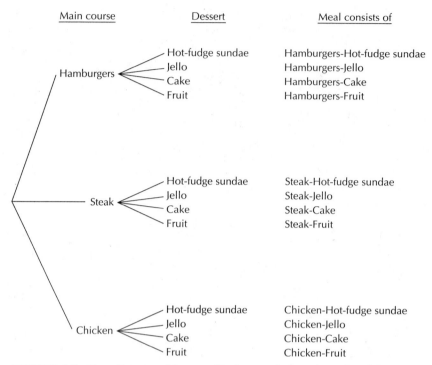

| Main course | Dessert | Meal consists of |

Hamburgers
- Hot-fudge sundae → Hamburgers-Hot-fudge sundae
- Jello → Hamburgers-Jello
- Cake → Hamburgers-Cake
- Fruit → Hamburgers-Fruit

Steak
- Hot-fudge sundae → Steak-Hot-fudge sundae
- Jello → Steak-Jello
- Cake → Steak-Cake
- Fruit → Steak-Fruit

Chicken
- Hot-fudge sundae → Chicken-Hot-fudge sundae
- Jello → Chicken-Jello
- Cake → Chicken-Cake
- Fruit → Chicken-Fruit

FIGURE 4.5 Tree diagram of the possible dinners that can be ordered from a choice of any one of three main courses and any one of four desserts.

SOLUTION Hazel may order any one of the three dishes as a main course. With each main course she may order any one of four desserts. These possibilities are pictured in Figure 4.5. The diagram shows that there are 12 possible meals that Hazel can order.

COMMENT Although counting the number of possible outcomes by using a tree diagram is not difficult when there are only several possibilities, it becomes very impractical to construct a tree when there are many possibilities. For example, if ten coins are tossed, there are 1024 different outcomes. Similarly, if a die is rolled four times, there are 1296 different outcomes. For situations such as these we need a rule to help us determine the number of possible outcomes.

Before stating the rule, however, let us analyze the following situations. When one die is rolled, there are six possible outcomes, 1, 2, 3, 4, 5, 6. When a second die is rolled, there are 6×6, or 36, possible outcomes for both dice together. These are listed on page 211.

If again we analyze Exercise 22 of Section 4.2, we find that the first wheel can show a lemon, cherry, or apple. This gives us 3 possibilities. Similarly, the second wheel can also show a lemon, cherry, or apple. The same is true for the third wheel. This gives us

3 possibilities for wheel 1, 3 possibilities for wheel 2, and 3 possibilities for wheel 3. We then have a total of 27 possible outcomes since

$$3 \times 3 \times 3 = 27$$

This leads us to the following rule.

COUNTING RULE

If one thing can be done in m ways, and if after this is done, something else can be done in n ways, then both things can be done in a total of $m \cdot n$ different ways in the stated order. (The same rule extends to more than two things done in sequence.)

EXAMPLE 3 A geology teacher plans to travel from New York to Florida and then on to Mexico to collect rock specimens for her class. From New York to Florida she can travel by train, airplane, boat, or car. However, from Florida to Mexico she cannot travel by train. In how many different ways can she make the trip?

SOLUTION Since she can travel from New York to Florida in four different ways and from Florida to Mexico in three different ways, she can make the trip in 4×3, or 12, different ways.

EXAMPLE 4 In a certain state license plates have three letters followed by two digits. If the first digit cannot be 0, how many different license plates can be made if

   a. repetitions of letters or numbers are allowed?
   b. repetitions of letters are not allowed?

SOLUTION There are 26 letters and 10 possible digits (0, 1, 2, . . . , 9).

   a. If repetitions are allowed, the same letter can be used again. Since 0 cannot be used as the first digit, the total number of different license plates is

$$26 \times 26 \times 26 \times ⑨ \times 10 = 1{,}581{,}840$$

   Thus, 1,581,840 different license plates are possible. Note that the circled position has only 9 possibilities. Why?
   b. If repetition of letters is not allowed, there are 26 possibilities for the first letter, but only 25 possibilities for the second letter since once a letter is used it may not be used again. For the third letter there are only 24 possibilities. There are then a total of 1,404,000 different license plates since

$$26 \times 25 \times 24 \times 9 \times 10 = 1{,}404{,}000$$

E X A M P L E   5   In Example 1 (see page 226) there are two possible outcomes for the first toss, two possibilities for the second toss, and two possibilities for the third toss. Thus, there are a total of eight possible outcomes when three coins are tossed or when one coin is tossed three times since

$$2 \times 2 \times 2 = 8$$

This is the same result we obtained using tree diagrams. It is considerably easier to do it this way.

E X E R C I S E S   F O R   S E C T I O N   4 . 3

1. A hospital administrator is interested in up-grading the hospital's computers. The particular model chosen has numerous options for storing information as follows: three possible floppy drives, seven possible hard drives, and two possible tape options. How many different computer setups with these options are possible?

2. There are seven elevators in a museum that connect the main floor to the third floor exhibits. In how many different ways can a visitor to the museum take an elevator up to the third floor to see the exhibits and take a different elevator down from the third floor?

3. There are seven pitchers and three catchers on a baseball team. In how many different ways can the manager select a starting battery (pitcher and catcher) for a game? (Use tree diagrams.)

4. At a recent reception party guests were asked to select a meal from the following items:

Entrée	Soup	Main Course	Beverage
Slice of melon	Vegetable	Fish	Wine
Fruit cup	Cream of chicken	Chicken	Champagne
	Mushroom	Duck	Soft drink
		Meat	Coffee
			Tea

If a guest selects one item from each category, how many possible meals can be obtained?

5. A gambler selects three cards (without replacement) from a deck of 52 cards. Assuming order counts, in how many different ways can the three cards be selected?

6. All students at Amerville College are issued student identification tags as follows: Each student is assigned a five-digit number, with repetition allowed, followed by the letter F (fall semester) or S (spring semester). Additionally, each identification number is preceded by a D (day student) or N (night student). How many different identification tags are possible?

7. The *New York Daily News* and *Chicago Tribune* have as a daily feature, a five- or six-letter nonsensical word that the reader must unscramble to make a meaningful

word. In how many different ways can the letters of the word "FACTOR" be arranged? (*Note:* Each arrangement does not necessarily have to form a meaningful word.)

8. In how many different ways can the letters of the word "MAGIC" be arranged if each arrangement is to consist of five letters and
   a. repetition is not allowed?
   b. repetition is allowed?

9. How many different seven-digit telephone numbers are possible if the first digit cannot be a zero or a one? (Assume no other restrictions and that repetitions are allowed.)

10. Refer back to the previous question. Using the same restrictions, how many telephone numbers are possible if there are 40 different area codes?

11. An amateur auto mechanic is repairing an old car. After replacing the six spark plugs, the mechanic must reconnect each of the six ignition cables with the appropriate spark plug. In how many different ways can these cables be reconnected?

12. In a certain state, license plates consist of three numbers followed by three letters. How many different license plates can be formed if
    a. repetition of numbers or letters is allowed?
    b. repetition of numbers or letters is not allowed?
    c. only numbers can be repeated?
    d. only letters can be repeated?

13. How many different four-digit numbers greater than 4000 can be formed from the digits 3, 4, 8, and 9 if
    a. repetition is allowed?
    b. repetition is not allowed?

14. Consider the accompanying newspaper article. Eight depositors are anxiously waiting in line to withdraw their money. In how many different ways can they stand in line?

POLICE QUELL DEPOSITOR UNREST

New York: Local police had to be called in yesterday to restore law and order at the Golden Pacific Bank following the announcement by state banking officials that the bank was insolvent and that the depositors would be allowed to withdraw only $100 per person pending further action by the Federal Deposit Insurance Corporation.

June 10, 1985

15. There are seven members on the student governing board at Palisades College as follows: Manya Rubashin, Wallace McKenzie, Fil Washington, Sandy Johnson, Heather Peterson, Eugene Brooks, and April Rizzo. At any meeting, these members are seated at a (straight) head table.

   a. In how many different ways can these members be seated?

   • b. If Wallace McKenzie and Eugene Brooks are bitter enemies and cannot be seated together, how many different seating arrangements are possible?

16. Heather has been advised by her counselor that in order to graduate in June she must complete one course from each of the three categories shown below:

	Category	
A	B	C
Math 113	Eng 111	Educ 176
Bio 153	Hist 212	Music 053
Chem 101	Soc 153	

Neglecting any other considerations and assuming these courses will be offered in the spring semester, how many *different* programs can Heather select? (Use tree diagrams.)

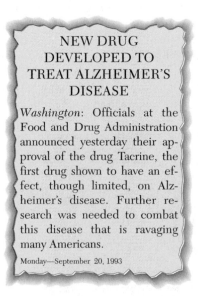

NEW DRUG DEVELOPED TO TREAT ALZHEIMER'S DISEASE

Washington: Officials at the Food and Drug Administration announced yesterday their approval of the drug Tacrine, the first drug shown to have an effect, though limited, on Alzheimer's disease. Further research was needed to combat this disease that is ravaging many Americans.

Monday—September 20, 1993

17. Refer to the accompanying article. Doctors at Caledony Hospital have selected four patients for the testing program. Some of the patients will be given the new drug while others will be given a placebo. In how many different ways can the patients be assigned to receive either the new drug or the placebo? (Use tree diagrams.)

18. The Felquois Indians sent a message from one tribe to another by using the following different types of smoke signals:

> Long, black smoke cloud
> Short, black smoke cloud
> Long, white smoke cloud
> Short, white smoke cloud

Each of these clouds was followed by a two-minute pause, after which a second smoke cloud was created. Again a two-minute pause followed, after which a third smoke cloud was created. Each message consisted of three smoke clouds.

a. Using this scheme, how many different messages could the Felquois Indians transmit from one tribe to another?

b. Assuming that all messages are equally likely to be transmitted, what is the probability that a transmitted message will consist of three short, white smoke clouds one after another?

4.4 PERMUTATIONS

Consider the following situation: Three vacationing students, Mel, Carl, and Rhoda, have purchased standby tickets on a transatlantic flight that will take them from London back to New York. There are three aisle seats remaining vacant in the plane: one in the forward section, one in the midsection, and one in the tail section of the airplane. In how many different ways can the three travelers line up to board the plane and pick the best of the remaining seats?

There are six ways as shown in the accompanying table. In this case you will notice that in each arrangement order is important.

First Person to Board Plane	Second Person to Board Plane	Last Person to board Plane
Mel	Carl	Rhoda
Mel	Rhoda	Carl
Carl	Mel	Rhoda
Carl	Rhoda	Mel
Rhoda	Carl	Mel
Rhoda	Mel	Carl

Thus, the arrangement Mel, Carl, Rhoda means that Mel boards first, then Carl, and finally Rhoda. If the number of seats is limited, then the person who boards first has first choice of selecting the best of the remaining seats as compared with the person who boards last.

This important idea in mathematics, in which a number of objects can be arranged in a particular order, is known as a *permutation*.

Definition 4.2 | A **permutation** is any arrangement of distinct objects in a particular order.
Permutation

Let us now examine another such problem.

E X A M P L E   1   Four young ladies, Stephanie Marie Gallagher, Liz Armstrong, Ann Sullivan, and Patricia Beth O'Connell, are finalists in a college scholarship contest. The judges must select a winner and an alternate from among these four contestants. In how many different ways can this be done?

SOLUTION   The winner and the alternate can be selected in 12 different ways. We list these possibilities in a table.

Winner	Alternate	Winner	Alternate
Stephanie	Liz	Ann	Stephanie
Stephanie	Ann	Ann	Liz
Stephanie	Patricia	Ann	Patricia
Liz	Stephanie	Patricia	Stephanie
Liz	Ann	Patricia	Liz
Liz	Patricia	Patricia	Ann

COMMENT   Perhaps you are wondering whether there are any formulas that can be used to determine the number of possible permutations. The answer is yes and this will be done shortly.

E X A M P L E   2   A doctor has five examination rooms. There are five patients in the waiting room. In how many different ways can the patients be assigned to the examination rooms?

SOLUTION   We can solve this problem by numbering the waiting rooms 1 through 5 as shown here and then considering them in sequence.

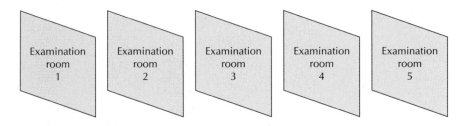

Examination room 1 can be used by any one of the five patients. Once a patient has been assigned to examination room 1, there are four patients who can use examination room 2, so there are (using the counting rule)

$$5 \cdot 4 = 20 \text{ ways}$$

of using examination room 1 and examination room 2. Similarly, once examination rooms 1 and 2 are occupied, there are three patients remaining who can use examination room 3, so there are

$$5 \cdot 4 \cdot 3 = 60 \text{ ways}$$

of using examination rooms 1, 2, and 3. Continuing in this manner we find that there are

$$5 \cdot 4 \cdot 3 \cdot 2 \cdot 1 = 120 \text{ ways}$$

of using examination rooms 1, 2, 3, 4, and 5. Thus, there are 120 different ways in which the five patients can be assigned to the five examination rooms.

Notice in Example 2 that we had to multiply $5 \cdot 4 \cdot 3 \cdot 2 \cdot 1$. Often in mathematics we have to multiply a series of numbers together starting from a given whole number and multiplying this by the number that is 1 less than that, and so on until we get to the number 1, which is where we stop. For this type of multiplication we introduce *Factorial notation* **factorial notation** and use the symbol $n!$ read as "n factorial." In this case we would write the product as $5!$. Thus,

$$5! = 5 \cdot 4 \cdot 3 \cdot 2 \cdot 1 = 120$$

Accordingly,

$$1! = 1$$
$$2! = 2 \cdot 1 = 2$$
$$3! = 3 \cdot 2 \cdot 1 = 6$$
$$4! = 4 \cdot 3 \cdot 2 \cdot 1 = 24$$
$$5! = 5 \cdot 4 \cdot 3 \cdot 2 \cdot 1 = 120$$
$$6! = 6 \cdot 5 \cdot 4 \cdot 3 \cdot 2 \cdot 1 = 720$$
$$7! = 7 \cdot 6 \cdot 5 \cdot 4 \cdot 3 \cdot 2 \cdot 1 = 5040$$
$$8! = 8 \cdot 7 \cdot 6 \cdot 5 \cdot 4 \cdot 3 \cdot 2 \cdot 1 = 40{,}320$$
$$9! = 9 \cdot 8 \cdot 7 \cdot 6 \cdot 5 \cdot 4 \cdot 3 \cdot 2 \cdot 1 = 362{,}880$$
$$10! = 10 \cdot 9 \cdot 8 \cdot 7 \cdot 6 \cdot 5 \cdot 4 \cdot 3 \cdot 2 \cdot 1 = 3{,}628{,}800$$

Also, we define $0!$ to be 1. This makes formulas involving $0!$ meaningful.

Let us now return to the college scholarship contest example (Example 1) discussed earlier. In that example the judges were interested in selecting two winners from among the four finalists. This actually represents the number of permutations of four things taken two at a time, or the number of possible permutations of two things that can be formed out of a possible four things. We abbreviate this by writing $_4P_2$. This is read as the number of permutations of four things taken two at a time.

NOTATION $_nP_r$ represents the number of permutations of n things taken r at a time, and $_nP_n$ represents the number of permutations of n things taken n at a time.

To determine the number of possible permutations, we use the following formula:

FORMULA 4.1

$$_nP_r = \frac{n!}{(n-r)!}$$

EXAMPLE 3

a. Find $_6P_4$.
b. Find $_7P_5$.
c. Find $_5P_5$.

SOLUTION

a. $_6P_4$ means the number of possible permutations of six things taken four at a time. Using Formula 4.1 we have $n = 6$, $r = 4$, and $n - r = 6 - 4 = 2$. Thus,

$$_6P_4 = \frac{6!}{(6-4)!} = \frac{6!}{2!} = \frac{6 \cdot 5 \cdot 4 \cdot 3 \cdot 2 \cdot 1}{2 \cdot 1} = \frac{6 \cdot 5 \cdot 4 \cdot 3 \cdot \cancel{2} \cdot \cancel{1}}{\cancel{2} \cdot \cancel{1}}$$

$$= 6 \cdot 5 \cdot 4 \cdot 3 = 360$$

Thus, $_6P_4 = 360$.

b. $_7P_5$ means the number of permutations of seven things taken five at a time, so that $n = 7$ and $r = 5$. Using Formula 4.1 we have

$$_7P_5 = \frac{7!}{(7-5)!} = \frac{7!}{2!} = \frac{7 \cdot 6 \cdot 5 \cdot 4 \cdot 3 \cdot 2 \cdot 1}{2 \cdot 1}$$

$$= \frac{7 \cdot 6 \cdot 5 \cdot 4 \cdot 3 \cdot \cancel{2} \cdot \cancel{1}}{\cancel{2} \cdot \cancel{1}}$$

$$= 7 \cdot 6 \cdot 5 \cdot 4 \cdot 3 = 2520$$

Thus, $_7P_5 = 2520$.

c. $_5P_5$ means the number of permutations of five things taken five at a time, so that $n = 5$ and $r = 5$. Using Formula 4.1 we have

$$_5P_5 = \frac{5!}{(5-5)!} = \frac{5!}{0!}$$

$$= \frac{5!}{1} \quad \text{(Since 0! is 1)}$$

$$= 5! = 5 \cdot 4 \cdot 3 \cdot 2 \cdot 1$$

$$= 120$$

Thus, $_5P_5 = 5! = 120$.

EXAMPLE 4 A baseball scout has received a list of 15 promising prospects for the team's consideration. The scout is asked to list, in order of preference, the five most outstanding of these prospects. In how many different ways can the scout select the five best players?

SOLUTION As there are 15 promising prospects, $n = 15$. Also only the top five are to be considered, so that $r = 5$ and $n - r = 15 - 5 = 10$. Using Formula 4.1, we have

$$_nP_r = \frac{n!}{(n-r)!}$$

$$= \frac{15!}{10!}$$

$$= \frac{15 \cdot 14 \cdot 13 \cdot 12 \cdot 11 \cdot \cancel{10} \cdot \cancel{9} \cdot \cancel{8} \cdot \cancel{7} \cdot \cancel{6} \cdot \cancel{5} \cdot \cancel{4} \cdot \cancel{3} \cdot \cancel{2} \cdot \cancel{1}}{\cancel{10} \cdot \cancel{9} \cdot \cancel{8} \cdot \cancel{7} \cdot \cancel{6} \cdot \cancel{5} \cdot \cancel{4} \cdot \cancel{3} \cdot \cancel{2} \cdot \cancel{1}}$$

$$= 15 \cdot 14 \cdot 13 \cdot 12 \cdot 11 = 360{,}360$$

Thus, the scout can select the top five players in 360,360 different ways.

COMMENT In Example 3 we calculated $_5P_5$, which represents the number of permutations of five things taken five at a time. This really means the number of different ways of arranging five things, where order counts. Our answer turned out to be 5! This leads us to Formula 4.2.

FORMULA 4.2

The number of possible permutations of n things taken n at a time, denoted as $_nP_n$, is

$$_nP_n = n!$$

EXAMPLE 5 How many different permutations can be formed from the letters of the word CAT?

SOLUTION There are six permutations. These are

CAT   CTA   TAC   TCA   ACT   ATC

This actually represents the number of possible permutations of three letters taken three at a time. Formula 4.2 tells us that there are 3!, or 6, possible permutations since

$$3! = 3 \cdot 2 \cdot 1 = 6.$$

These are listed above.

Suppose a librarian has two identical algebra books and three identical geometry books to be shelved. In how many different ways can this be done? Since there are five books altogether and all have to be shelved, we are tempted to use Formula 4.2 or Formula 4.1. Unfortunately, since the two algebra books are identical, we cannot tell them apart. There are actually ten possible permutations. These are

algebra	algebra	geometry	geometry	geometry
algebra	geometry	geometry	geometry	algebra
algebra	geometry	algebra	geometry	geometry
algebra	geometry	geometry	algebra	geometry
geometry	geometry	geometry	algebra	algebra
geometry	geometry	algebra	algebra	geometry
geometry	geometry	algebra	geometry	algebra
geometry	algebra	algebra	geometry	geometry
geometry	algebra	geometry	algebra	geometry
geometry	algebra	geometry	geometry	algebra

Of course, if we label the algebra books as copy 1 and copy 2, which is sometimes done by some libraries, we can use Formula 4.1. Thus,

Algebra copy 1	Algebra copy 2	Geometry copy 1	Geometry copy 2	Geometry copy 3

would be a different permutation than

Algebra copy 2	Algebra copy 1	Geometry copy 1	Geometry copy 2	Geometry copy 3

However, this is usually not done. Thus, these two permutations have to be counted as the same or as only one permutation. Formulas 4.1 and 4.2 have to be revised somewhat to allow for the possibility of repetitions. This leads us to the following formula:

FORMULA 4.3

The number of different permutations of n things of which p are alike, q are alike, or r are alike, and so on, is

$$\frac{n!}{p!q!r!} \cdots$$

It is understood that $p + q + r + \cdots = n$.

In our example we have five books to be shelved, so $n = 5$. Since the two algebra books are identical and the three geometry books are also identical, $p = 2$ and $q = 3$. Formula 4.3 tells us that the number of permutations is

$$\frac{5!}{2! \cdot 3!} = \frac{5 \cdot 4 \cdot 3 \cdot 2 \cdot 1}{2 \cdot 1 \cdot 3 \cdot 2 \cdot 1}$$

$$= \frac{5 \cdot \overset{2}{\cancel{4}} \cdot \cancel{3} \cdot \cancel{2} \cdot \cancel{1}}{\cancel{2} \cdot 1 \cdot \cancel{3} \cdot \cancel{2} \cdot \cancel{1}}$$

$$= 5 \cdot 2 = 10$$

Thus, there are ten permutations, which were listed previously.

EXAMPLE 6 How many different permutations are there of the letters in the word

a. "IDIOT"?
b. "STATISTICS"?

SOLUTION

a. The word "IDIOT" has five letters, so $n = 5$. There are 2 I's, so $p = 2$. Formula 4.3 tells us that the number of permutations is

$$\frac{5!}{2!1!1!1!} = \frac{5 \cdot 4 \cdot 3 \cdot \cancel{2} \cdot \cancel{1}}{\cancel{2} \cdot \cancel{1} \cdot 1 \cdot 1 \cdot 1}$$

$$= 5 \cdot 4 \cdot 3 = 60$$

There are 60 permutations.

b. Since the word "STATISTICS" has ten letters, $n = 10$. The letter "S" is repeated three times, "T" is repeated three times, and "I" is repeated twice, so $p = 3$, $q = 3$, and $r = 2$. Formula 4.3 tells us that the number of permutations is

$$\frac{10!}{3!3!2!1!1!} = \frac{10 \cdot 9 \cdot 8 \cdot 7 \cdot \cancel{6} \cdot 5 \cdot \overset{2}{\cancel{4}} \cdot \cancel{3} \cdot \cancel{2} \cdot \cancel{1}}{\cancel{3} \cdot \cancel{2} \cdot \cancel{1} \cdot \cancel{3} \cdot \cancel{2} \cdot \cancel{1} \cdot \cancel{2} \cdot \cancel{1} \cdot 1 \cdot 1}$$

$$= 10 \cdot 9 \cdot 8 \cdot 7 \cdot 5 \cdot 2 = 50,400$$

There are 50,400 permutations.

Calculations involving factorials can easily be done using a hand-held calculator. However, if you are familiar with the MINITAB statistical computer package, then you can use MINITAB to evaluate factorials quite easily. For example, to evaluate 1! to 10! using MINITAB, we do the following:

```
MTB > SET THE FOLLOWING INTO C1
DATA> 1:10
DATA> END
MTB > PARPRODUCTS OF C1 PUT INTO C2
MTB > PRINT C1 C2
```

```
Data Display

Row     C1          C2

 1       1           1
 2       2           2
 3       3           6
 4       4          24
 5       5         120
 6       6         720
 7       7        5040
 8       8       40320
 9       9      362880
10      10     3628800
```

MTB > STOP

In a similar manner, we can have MINITAB evaluate any factorial.

EXERCISES FOR SECTION 4.4

1. Evaluate each of the following symbols:

   a. $6!$ b. $7!$ c. $2!$ d. $\dfrac{7!}{6!}$ e. $\dfrac{0!}{3}$ f. $\dfrac{6!}{4!2!}$

   g. $\dfrac{8!}{6!2!}$ h. $\dfrac{6!}{3!3!}$ i. $_7P_5$ j. $_6P_4$ k. $_7P_3$ l. $_5P_4$

   m. $_6P_6$ n. $_4P_4$ o. $_0P_0$ p. $_5P_0$

2. There are ten runners entered in tomorrow's marathon. In how many different ways can four of them finish first, second, third, and fourth?

3. A biochemist mixed six chemicals together in a solution and created a new synthetic drug. Unfortunately, the biochemist does not recall the order in which the chemicals were introduced into the solution. It is decided to repeat the experiment. How many possibilities are there?

4. Each year, readers of a certain magazine are asked to rank the top three best-dressed men from among a list of 12 candidates. In how many different ways can this be done?

5. Mrs. Gibbs goes to the playground where seven of her children are playing. She needs three volunteers: one to go to the grocery store, one to do the dishes, and one to do the family laundry. In how many different ways can she select the three volunteers?

6. The student governing board at Milton University consists of nine members. A new president, vice president, secretary, and treasurer are to be elected for the coming semester.

    a. How many permutations of the list of these officers is possible from the nine governing members?

    b. If Maureen is a member of the governing board, what is the probability that she will be elected president?

    c. What is the probability that Maureen will be elected to any office?

7. Four math books, three ecology books, two music books, and three economics books are to be arranged on a bookshelf. None of the books are identical.

    a. How many different permutations of these books are there?

    b. How many different permutations of these books are there if books on the same subject are to be grouped together?

8. In how many different ways can the letters of the word "SMOKING" be arranged if a vowel must be the first letter of each permutation? (Repetitions are not allowed.)

9. Six couples are to be honored for their outstanding humanitarian and philanthropic efforts. In how many different ways can these twelve people be seated at a straight head table if

    a. any person can sit in any seat?

    b. men are to be seated on one side of the master of ceremonies and women on the other side of the master of ceremonies?

    c. each man will sit with his wife? (*Note:* The master of ceremonies will not be seated.)

10. How many different permutations are there of the letters in these words?

    a. MATHEMATICS   b. MISSISSIPPI

    c. RECONSTRUCTION   d. PHOTOGRAPHY

11. Priscilla asks her kid brother (who is too young to read) to place her five-volume Bible on a shelf.

    a. In how many different ways can the books be placed on the shelf?

    b. Find the probability that the little boy places the books on the shelf in proper sequential order: Volume I, Volume II, and so on.

12. Bill Clinton is about to enter his six-digit personal identification number into an automatic teller machine (ATM) in a shopping mall. However, he does not recall the sequence of the digits 4, 6, 2, 8, 3, and 9 of his code. How many possibilities are there?

13. In how many different ways can the letters of the word "TRIGONOMETRY" be arranged if a vowel must be the first letter of each permutation?

14. In a certain state, license plates consist of five digits followed by a letter of the alphabet. How many different license plates can be formed if

    a. the first digit cannot be a zero but repetitions are allowed?

    b. no digit can be repeated?

    c. there are no restrictions at all?

15. A baseball manager has selected his nine starting players.

    a. How many different batting orders are possible?

    b. Assuming that all batting orders are equally likely, what is the probability that the pitcher bats last?

    c. Is the assumption given in part (b) reasonable?

> **WINNING TICKET EXPECTED TO TOP 90 MILLION DOLLARS**
>
> *Washington* (Dec 20): Tickets in this week's lottery were selling very briskly. With dreams of becoming the instant winner of the expected $90 million payoff, many people were reportedly waiting in line for hours to buy their tickets.
>
> Wednesday—December 20, 1995

16. Facing severe cash shortfalls, many states now conduct weekly lotteries. Typically, in such lotteries a player selects a number from 1 to 40 in each of six columns.
    a. Determine the total number of possible lottery tickets. (See accompanying article.)
    b. What is the probability of becoming an instant winner? (Assume that each person buys only one ticket.)
17. The Eastern Division of the American Baseball League consists of seven teams as follows: New York, Toronto, Milwaukee, Detroit, Cleveland, Baltimore, and Boston. How many different orders of finish are there in which the teams can place?
18. Many companies often sponsor "employee efficiency and increased worker productivity recommendations" by using a suggestion box. Bonuses are usually rewarded for any suggestions that are implemented. Recently, workers at the Apex Wire Company submitted nine different recommendations for increasing worker productivity and employee efficiency. Company officials plan to rank these suggestions and to reward the top five suggestions with prizes of $1000, $500, $250, $100, and $50, respectively. In how many different ways can the winning suggestions be selected?
19. The number of different ways in which n distinct objects can be arranged in a circle is $(n - 1)!$.
    a. Explain why this formula is valid.
    b. In how many distinct ways can a baker display six different cakes in a circular arrangement in a showcase window?
20. In how many ways can seven people be seated around a circular table to play cards?

4.5   COMBINATIONS

Imagine that a six-person rescue party is climbing a mountain in search of survivors of an airplane crash. They suddenly spot the plane on a ledge but the passageway to the ledge is very narrow and only four people can proceed. The remaining two people will have to return. How many different four-person rescue groups can be formed to reach the ledge?

In this situation we are obviously interested in selecting four out of six people. However, since the order in which the selection is to be made is not important, Formula 4.1 of Section 4.4 (see page 236) has to be changed.

Any selection of things in which the order is not important is called a *combination*. Let us determine how many possible combinations there actually are. If the names of the people of the six-person rescue party are Alice, Betty, Calvin, Drew, Ellen, and Frank, denoted as A, B, C, D, E, and F, then Formula 4.1 of Section 4.4 tells us that there are $_6P_4$ possible ways of selecting the four people out of a possible six. This yields

$$_6P_4 = \frac{6!}{(6-4)!}$$

$$= \frac{6!}{2!}$$

$$= 360 \text{ possibilities}$$

However, we know that permutations take order into account. Since we are not interested in order, there cannot be 360 possible combinations. Thus, if the four-person rescue party consists of A, B, C, and D, there would be the following 24 different permutations:

A B C D	B A C D	C A B D	D A B C
A B D C	B A D C	C A D B	D A C B
A D C B	B D A C	C B A D	D B A C
A D B C	B D C A	C B D A	D B C A
A C B D	B C A D	C D A B	D C B A
A C D B	B C D A	C D B A	D C A B

Since these permutations consist of the same four people, A, B, C, and D, we consider them as only one combination of these people.

Similarly, for any other combination of four people there are 24 permutations of these people. Thus, it would seem reasonable to divide the 360 by 24 getting 15 and to conclude that there are only 15 different combinations.

Notice that $24 = 4!$. Therefore, the number of combinations of six things taken four at a time is

$$\frac{_6P_4}{4!} = \frac{6!}{4!2!}$$

$$= 15$$

In general, consider the problem of selecting r objects from a possible n objects. We have the following definitions and formula.

Definition 4.3
Combination

A **combination** is a selection from a collection of distinct objects where order is not important.

Definition 4.4

The number of different ways of selecting r objects from a possible n distinct objects, where the order is not important, is called the **number of combinations of n things taken r at a time** and is denoted as $_nC_r$. Some books use the symbol $\binom{n}{r}$ instead of

Binomial coefficient

$_nC_r$. The symbol $\binom{n}{r}$ is called a **binomial coefficient**.

FORMULA 4.4

The number of combinations of n things taken r at a time is

$$\binom{n}{r} = {}_nC_r = \frac{n!}{r!(n-r)!}$$

Formula 4.4 is especially useful when calculating the probability of certain events. We illustrate the use of this formula with several examples.

EXAMPLE   1   Consider any five people, whom we shall name A, B, C, D, and E.

a. In how many ways can a committee of three be selected from among them?
b. What is the probability of selecting the three-person committee consisting of A, B, and C? (Assume that each committee is equally likely to be selected.)
c. In how many ways can a committee of five be selected from among them?

SOLUTION

a. We must select any three people from a possible five, and order does not matter. Since this is the number of combinations of five things taken three at a time, we want $_5C_3$. Using Formula 4.4 with $n = 5$ and $r = 3$, we have

$$\binom{5}{3} = {}_5C_3 = \frac{5!}{3!(5-3)!} = \frac{5!}{3!2!} = \frac{5 \cdot \overset{2}{\cancel{4}} \cdot \cancel{3} \cdot \cancel{2} \cdot \cancel{1}}{\cancel{3} \cdot \cancel{2} \cdot \cancel{1} \cdot \cancel{2} \cdot \cancel{1}} = 10$$

There are ten possible three-person committees that can be formed. We can verify this answer by listing them:

A B C	A B E	A C E	B C D	B D E
A B D	A C D	A D E	B C E	C D E

b. There are ten possible three-person committees that can be formed, as listed in part (a). Of these only one consists of A, B, and C. Thus,

$$p(\text{committee consists of A, B, and C}) = \frac{1}{10}$$

c. We are interested in selecting any five people from a possible five, and order does not matter. This is $_5C_5$. Thus,

$$\binom{5}{5} = {_5C_5} = \frac{5!}{5!(5-5)!}$$

$$= \frac{5!}{5!0!} \qquad (\text{Remember } 0! = 1)$$

$$= \frac{5 \cdot 4 \cdot 3 \cdot 2 \cdot 1}{5 \cdot 4 \cdot 3 \cdot 2 \cdot 1 \cdot 1}$$

$$= 1$$

So, there is only one combination containing all five people.

EXAMPLE 2
Nuclear Accident

On Wednesday, March 27, 1979, the nuclear generating facility on Three Mile Island near Middletown, Pennsylvania, malfunctioned and began discharging radiation into the air. In an attempt to prevent a nuclear "meltdown," ten nuclear physicists were

The four cooling towers at the Three Mile Island Nuclear Power Plant. *(Photo: UPI/ Bettmann Newsphotos)*

contacted to shed some light on the problem. (Later, additional physicists were called in.) Because the ten scientists were busily engaged on another project, they agreed that only seven of them would come to the crippled nuclear plant. (Any three of them could remain to oversee their other existing project.) In how many different ways could the seven scientists have been selected?

SOLUTION   Order was not important, so the seven scientists had to be selected out of a possible ten. This could be done in $_{10}C_7$ ways. Using Formula 4.4, we have

$$\binom{10}{7} = {}_{10}C_7 = \frac{10!}{7!(10-7)!} = \frac{10!}{7!3!} = 120$$

Thus, the seven scientists could have been selected in 120 different ways.

EXAMPLE   3
Football Teams

How many different 11-member football teams can be formed from a possible 20 players if any player can play any position?

SOLUTION   We are interested in the number of combinations of 20 players taken 11 at a time. So, $n = 20$ and $r = 11$. Thus,

$$\binom{20}{11} = {}_{20}C_{11} = \frac{20!}{11!(20-11)!}$$

$$= \frac{20!}{11!9!}$$

$$= 167{,}960$$

There are 167,960 possible 11-member football teams.

EXAMPLE   4

a. How many different poker hands consisting of five cards can be dealt from a deck of 52 cards?
b. What is the probability of being dealt a royal flush in five-card poker? (A royal flush consists of the ten, jack, queen, king, and ace of the same suit.)

SOLUTION

a. Since order is not important, we are interested in the number of combinations of 5 things out of a possible 52. So, $n = 52$ and $r = 5$. Thus,

$$\binom{52}{5} = {}_{52}C_5 = \frac{52!}{5!(52-5)!}$$

$$= \frac{52!}{5!47!}$$

$$= \frac{52 \cdot 51 \cdot 50 \cdot 49 \cdot 48 \cdot 47!}{5 \cdot 4 \cdot 3 \cdot 2 \cdot 1 \cdot 47!}$$

$$= 2{,}598{,}960$$

There are 2,598,960 possible poker hands.

b. Out of the possible 2,598,960 different poker hands only four are favorable. These are the ten, jack, queen, king, and ace of hearts; the ten, jack, queen, king, and ace of clubs; the ten, jack, queen, king, and ace of diamonds; and the ten, jack, queen, king, and ace of spades. Thus,

$$p(\text{royal flush}) = \frac{4}{2{,}598{,}960} = \frac{1}{649{,}740}$$

COMMENT   For those familiar with poker, we present certain probabilities in Table 4.1.

TABLE 4.1   **Different Poker Hands and Their Probabilities**

Type of Hand	Probability of It Being Dealt to You
Royal flush (ace, king, queen, jack, 10 in the same suit)	0.0000015
Four of a kind (four of a kind, like four 4's or four queens)	0.00024
Flush (five cards in a single suit but not straight)	0.0020
Two pairs	0.0475
Nothing of interest	0.5012

E X A M P L E   5   John has ten single dollar bills of which three are counterfeit. If he selects four of them at random, what is the probability of getting two good bills and two counterfeit bills?

SOLUTION   We are interested in selecting four bills from a possible ten. Thus, the number of possible outcomes is $_{10}C_4$. The two good bills to be drawn must be drawn from the seven good ones. This can happen in $_7C_2$ ways. Also, the two counterfeit bills to be drawn must be drawn from the three counterfeit ones. This can happen in $_3C_2$ ways. Thus,

$$p(\text{two good bills and two counterfeit bills}) = \frac{_7C_2 \cdot {}_3C_2}{_{10}C_4}$$

Now

$$_7C_2 = \frac{7!}{2!(7-2)!}$$

$$= \frac{7!}{2!5!} = 21$$

and

$$_3C_2 = \frac{3!}{2!(3-2)!}$$

$$= \frac{3!}{2!1!} = 3$$

Also,

$$_{10}C_4 = \frac{10!}{4!(10-4)!}$$

$$= \frac{10!}{4!6!} = 210$$

Therefore,

$$p(\text{two good bills and two counterfeit bills}) = \frac{21 \cdot 3}{210}$$

$$= \frac{3}{10}$$

E X A M P L E   6   Find the probability of selecting all good bills in the preceding example.

SOLUTION   Since we are interested in selecting only good bills, we will select four good bills from a possible seven and no counterfeit bills. Thus,

$$p(\text{all good bills}) = \frac{_7C_4 \cdot {}_3C_0}{_{10}C_4}$$

$$= \frac{35 \cdot 1}{210} = \frac{1}{6}$$

The probability of selecting all good bills is $\frac{1}{6}$.

There is an alternate method for computing the number of possible combinations of n things taken r at a time, $_nC_r$, which completely avoids the factorial notation. This can be accomplished by using *Pascal's triangle*. Such a triangle is shown in Figure 4.6. How do we construct such a triangle?

Each row has a 1 on either end. All in-between entries are obtained by adding the numbers immediately above and directly to the right and left of them as shown by the arrows in the diagram of Figure 4.7. For example, to obtain the entries for the eighth row we first put a 1 on each end (moving over slightly). Then we add the 1 and 7 from row 7, getting 8 as shown. We then add 7 and 21, getting 28; then we add 21 to 35, getting 56, and so on. Remember to place the 1's on each end. The numbers must be

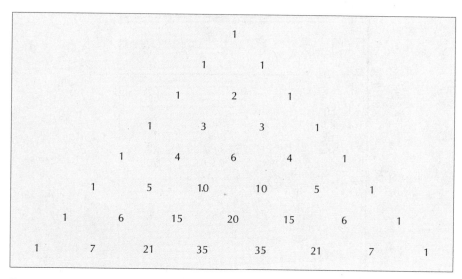

FIGURE 4.6   Pascal's triangle.

Pascal's triangle   lined up as shown in the diagram. Such a triangle of numbers is known as **Pascal's triangle** in honor of the French mathematician who found many applications for it. Although the Chinese knew of this triangle several centuries earlier (Figure 4.8), it is named for Pascal.

Let us now apply Pascal's triangle to solve problems in combinations.

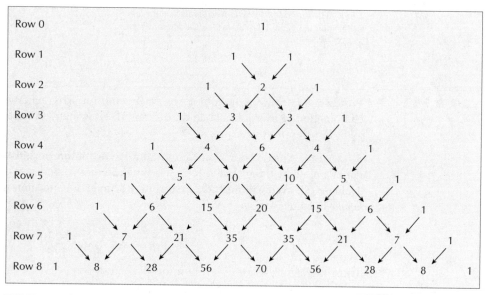

FIGURE 4.7   Construction of Pascal's triangle.

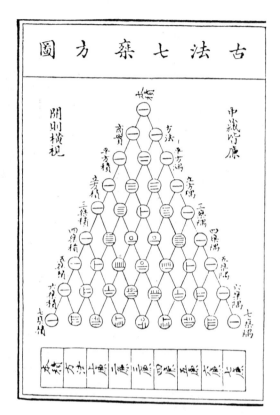

FIGURE 4.8   The Pascal triangle as used in
1303.   *(Redrawn from Joseph Needham,* Science and
Civilization in China, III, *Cambridge University Press,
p. 135)*

E X A M P L E   7   Professor Matthews tells four jokes each semester. His policy is never to repeat any combination of four jokes once they are used. How many semesters will eight different jokes last the professor?

SOLUTION   Since order is not important, we are interested in the number of combinations of eight things taken four at a time. So, $n = 8$ and $r = 4$. Therefore, we must evaluate $_8C_4$. We will find $_8C_4$ first using Formula 4.4 and then using Pascal's triangle. Using Formula 4.4, we have

$$_8C_4 = \frac{8!}{4!(8-4)!} = \frac{8!}{4!4!} = 70$$

Thus, the eight jokes will last him for 70 semesters.

Let us now evaluate $_8C_4$ using Pascal's triangle. Since $n = 8$, we must write down the first nine rows of Pascal's triangle as shown in Figure 4.9.

Row 0								1								
Row 1							1		1							
Row 2						1		2		1						
Row 3					1		3		3		1					
Row 4				1		4		6		4		1				
Row 5			1		5		10		10		5		1			
Row 6		1		6		15		20		15		6		1		
Row 7	1		7		21		35		35		21		7		1	
Row 8	1	8		28		56		70		56		28		8		1

FIGURE 4.9   First nine rows of Pascal's triangle.

Now we look at row 8. Since we are interested in selecting four things from a possible eight things, we skip the 1 and move to the fourth entry appearing after the 1. It is 70. This number represents the value of $_8C_4$. Thus, $_8C_4 = 70$. Similarly, to find $_8C_6$ we move to the sixth entry after the end 1 on row 8. It is 28. Thus, $_8C_6 = 28$. Also, the first entry after the end 1 is 8 so that $_8C_1 = 8$. If we wanted $_8C_8$, we would move to the eighth entry after the end 1. It is 1, so that $_8C_8 = 1$. To find $_8C_0$ we must move to the zeroth entry after the end 1. This means we do not go anywhere; we just stay at the 1. Thus, $_8C_0 = 1$.

COMMENT   When you use Pascal's triangle, remember that the rows are numbered from 0 rather than from 1 so that the rows are labeled row 0, row 1, row 2, and so on.

E X A M P L E   8   Using Pascal's triangle find the following:

   a. $_5C_2$ b. $_5C_3$ c. $_6C_6$ d. $_7C_0$

SOLUTION   We will use the Pascal triangle shown in Figure 4.9.

   a. To find $_5C_2$, go to row 5 and move across to the second number from the left after the end 1. The entry is 10. Thus, $_5C_2 = 10$.
   b. To find $_5C_3$, go to row 5 and move across to the third number from the left after the end 1. The entry is 10. Thus, $_5C_3 = 10$.
   c. To find $_6C_6$, go to row 6 and move across to the sixth number from the left after the end 1. The entry is 1. Thus, $_6C_6 = 1$.

d. To find $_7C_0$, go to row 7 and do not move anywhere after the end 1. Remain there. Thus, $_7C_0 = 1$.

EXERCISES FOR SECTION 4.5

1. Evaluate each of the following:
   a. $_7C_6$
   b. $_6C_4$
   c. $_8C_3$
   d. $_9C_0$
   e. $_6C_1$

   f. $_8C_5$
   g. $\binom{9}{6}$
   h. $\binom{8}{6}$
   i. $\binom{7}{8}$
   j. $\binom{6}{6}$

2. Nine police officers arrive at Victory Hospital to give blood for a wounded fellow cop. However, the hospital can accept only five volunteers. In how many different ways can these volunteers be selected?

3. As presented in *Proceedings of the National Academy of Sciences* (Vol. 86, No. 16) and reported in *Science News* (Vol. 136, August 19, 1989, p. 116), scientists have demonstrated that a newly developed simian immunodeficiency virus (SIV) vaccine can protect rhesus monkeys from infection by a virus closely related to the Aids-causing human immunodeficiency HIV virus. In how many different ways can researchers select three out of a possible nine rhesus monkeys to be injected with this new vaccine to analyze the effectiveness of this new vaccine?

4. The Argavon Corp. believes that it is in the company's best interest to maintain the physical fitness of its 30 employees. It recently purchased six exercise machines to be used by its employees during their lunch break or after work. In how many different ways can 6 of the 30 employees be assigned to these machines?

5. The Metro Automobile Insurance Company received ten auto theft claims on June 12, 1995. Management decides to randomly select four of these claims and to investigate these claims thoroughly for the possibility of fraud. In how many different ways can this be done?

6. A student has eight textbooks that she would like to place in her attaché case. However, regardless of the arrangement, only five books fit into the attaché case. In how many different ways can the five books to be placed in the attaché case be selected?

7. Professor Bergen, a psychology teacher, needs six student volunteers to be subjects for her latest personality tests. If 14 students from her class have volunteered to be subjects, in how many different ways can she select the six subjects for the experiment?

8. See the newpaper article on the next page. A four-person Presidential Search Committee is to be set up to interview numerous candidates and to present their recommendations for President to the University Board. The members of the committee are to be selected from seven full professors, four students, three deans, and two university officials. In how many different ways can the committee be selected if
   a. one member must be from each of the above mentioned categories?
   •b. at least one full professor and one student must be on the committee?

> **BOARD AUTHORIZES PRESIDENTIAL SEARCH COMMITTEE**
>
> *New York*, Nov. 23: At its November 22, 1993 meeting the Board of Higher Education of the City University of New York authorized the creation of a Presidential Search committee to conduct a nationwide search for a replacement for the retiring President Edmond Volpe of the College of Staten Island and to present their recommendations to the University.
>
> Tuesday—November 23, 1993

9. The Marco Trucking Company employs 13 mechanics on its morning shift and 11 mechanics on its evening shift. As part of its down-sizing policy, management decides to retrench two mechanics. What is the probability that
   a. both retrenched mechanics will be from the morning shift?
   b. both retrenched mechanics will be from the evening shift?
   c. one of the retrenched mechanics will be from the morning shift and one will be from the evening shift?

10. How many different worker-management health committees can be formed from a total of 15 workers and seven management personnel if each committee is to consist of five workers and four management personnel?

11. The law firm of Bellen & Co. plans to hire seven new lawyers for the coming fiscal year. After interviewing prospective employees, it is determined that 13 men and 11 women qualify for the job. In how many different ways can the new lawyer job openings be filled if
    a. the new lawyers must be four men and three women?
    b. the new lawyers must be more women than men?

12. A tour operator is arranging a winter travel package from the East Coast to the West coast. Although there are 12 airlines that could be used for the travel package, the tour operator wishes to consider only five of the airlines. This will cut down on the paperwork involved.
    a. In how many different ways can the five airlines to be considered be selected?
    b. In how many different ways can the seven airlines not to be considered be selected?
    c. Compare the answers obtained in parts (a) and (b). They should be the same. Can you explain why?

13. It is known that two of the 20 hundred-dollar bills submitted to a teller at the Monticello Savings Bank are counterfeit. The branch manager of the bank randomly selects three of the bills from the collection and carefully checks each one. In how many different ways can the branch manager select three of the hundred-dollar bills
    a. so that none of the counterfeit bills are selected?
    b. so that one of the counterfeit bills is selected?
    c. so that both of the counterfeit bills are selected?

14. Refer back to Exercise 13. Find the probability that the branch manager selects both counterfeit bills (so that no counterfeit bill is recirculated).

15. The Emergency Room at George Washington Hospital employs eight male and ten female nurses. A special triage unit of six nurses is being formed. In how many different ways can this be done if
    a. the unit must have three male nurses and three female nurses on it?
    b. the unit must have at least one female nurse on it?
    c. the unit must have at least one male and one female nurse on it?

16. All day-care centers that receive government funding are inspected periodically to verify compliance with government regulations. In one particular city, there are 22 day-care centers. If it is decided to check six of these centers, in how many different ways can the centers to be inspected be selected?

17. Refer back to the newspaper article given at the beginning of this chapter, on page 206. Assume that there are 36 governors that are Democrats and 14 governors that are Republicans. Find the probability that the committee will consist of
    a. all Republicans.
    b. at least two Republicans.
    c. no Republicans.

4.6 ODDS AND MATHEMATICAL EXPECTATION

Gamblers are frequently interested in determining which games are profitable to them. What they would really like to know is how much money can be earned in the long run from a particular game. If in the long run nothing can be won, that is, if as much money that is won will be lost, then why play at all? Similarly, if in the long run no money can be won but money can be lost, then the gambler will not play such a game. He or she will play any game only if money can be won. When applied to gambling or business situations, the **mathematical expectation** of an event is the amount of money to be won or lost in the long run.

COMMENT In Chapter 6, when we discuss random variables and their probability distributions in detail, we will present a more general definition of mathematical expectation.

Since many applied examples of mathematical expectations involve gambling or business situations, where money can be won or lost in the long run, we have the following convenient formula for such cases.

Gamblers frequently speak of odds when betting on horses.
(*© Bancroft, Fran Heyl Associates*)

FORMULA 4.5

Mathematical expectation

Consider an event that has probability p_1 of occurring and that has a payoff, that is, the amount won, m_1. Consider also a second event with probability p_2 and payoff m_2, a third event with probability p_3 and payoff m_3, and so on. The **mathematical expectation** of the event is

$$m_1 p_1 + m_2 p_2 + m_3 p_3 + \cdots + m_n p_n$$

when the event has n different payoffs.

The numbers attached to the letters are called subscripts. They have no special significance. We use them only to avoid using too many different letters.)

We illustrate these ideas with several examples.

EXAMPLE 1   A large company is considering opening two new factories in different towns. If it opens in town A, it can expect to make $63,000 profit per year with a probability of $\frac{4}{7}$. However, if it opens in town B, it can expect to make a profit of $77,000 with a probability of only $\frac{3}{7}$. What is the company's mathematical expectation?

SOLUTION   We use Formula 4.5. We have

$$(63,000)\left(\frac{4}{7}\right) + (77,000)\left(\frac{3}{7}\right) = 36,000 + 33,000 = 69,000$$

Thus, the company's mathematical expectation is $69,000.

EXAMPLE 2   A contractor is bidding on a road construction job that promises a profit of \$200,000 with a probability of $\frac{7}{10}$ and a loss, due to strikes, weather conditions, late arrival of building materials, and so on, of \$40,000 with a probability of $\frac{3}{10}$. What is the contractor's mathematical expectation?

SOLUTION   A "+" sign will denote a gain and a "−" sign will denote a loss. Using Formula 4.5, we find that the contractor's mathematical expectation is

$$(+\$200{,}000)\left(\frac{7}{10}\right) \; + \; (-\$40{,}000)\left(\frac{3}{10}\right)$$

$$= \$140{,}000 - \$12{,}000 = \$128{,}000$$

The contractor's mathematical expectation is \$128,000.

CAN YOUR CHANCES OF WINNING IN A LOTTERY BE IMPROVED?

Sept 17: Because winning lottery numbers are generated randomly, no set of numbers has a higher probability of winning than any other, according to Dr. Jim Maxwell of the American Mathematics Society. "It is my understanding that lotteries are designed so that no one will have an advantage over anyone else," he said. "You could play the same number over and over again and have the same chance of winning as you would if you changed the number each time." Nevertheless, lottery players can decrease the number of players with whom they share their winnings by decreasing the odds that other players will choose the same numbers. "This is done by staying away from common number combinations such as months of the year, birthdays, holidays, et cetera," Dr. Maxwell said, since numbers drawn from the calendar are usually 31 or less. Because many lotteries call for numbers up to 48, players may increase their winnings, though not their chances of winning, by selecting some numbers above 31.

September 17, 1988

This newspaper article indicates that we can often make decisions (and select numbers accordingly) that affect our mathematical expectations.

EXAMPLE 3   Peter selects one card from a deck of 52 cards. If it is an ace, he wins \$5. If it is a club, he wins only \$1. However, if it is the ace of clubs, then he wins an extra \$10. What is his mathematical expectation?

SOLUTION When one card is selected from a deck of cards, we have the following probabilities:

$$p(\text{ace}) = \frac{4}{52}$$

$$p(\text{clubs}) = \frac{13}{52}$$

$$p(\text{ace of clubs}) = \frac{1}{52}$$

Thus, Peter's mathematical expectation is

$$5\left(\frac{4}{52}\right) + 1\left(\frac{13}{52}\right) + 10\left(\frac{1}{52}\right) = \frac{20}{52} + \frac{13}{52} + \frac{10}{52}$$

$$= \frac{20 + 13 + 10}{52} = \frac{43}{52} = 0.83$$

His mathematical expectation is 83 cents.

COMMENT Eighty-three cents is the fair price to pay to play the game, or is the break-even point. If he pays more than 83¢ per game to play, then in the long run he will lose money. If he pays less, then in the long run he will win.

Another interesting application of probability is concerned with betting. Gamblers frequently speak of the odds of a game. To understand this idea best, consider George, who believes that whenever he washes his car, it usually rains the following day. If George has just washed his car and if the probability of it raining tomorrow is $\frac{3}{10}$, gamblers would say that the odds in favor of it raining are 3 to 7 and the odds against it raining are 7 to 3. The 7 represents the 7 chances of it not raining. Formally we have the following definitions.

Definition 4.5
Odds in favor of an event
| The **odds in favor** of an event occurring are p to q, where p is the number of favorable outcomes and q is the number of unfavorable outcomes.

Definition 4.6
Odds against an event
| If p and q are the same as in Definition 4.5, the **odds against** an event happening are q to p.

We now illustrate these definitions with several examples.

E X A M P L E   4 What are the odds in favor of the New York Mets winning the World Series if the probability of their winning is $\frac{4}{7}$ and the probability of their losing is $\frac{3}{7}$?

SOLUTION   Since the probability of their winning is $\frac{4}{7}$, this means that out of 7 possibilities 4 are favorable and 3 are unfavorable. Thus, the odds in favor of the New York Mets winning the series are 4 to 3.

METS ARE FAVORED TO WIN SERIES

New York: (Sept. 20)— The New York Mets are favored to win the upcoming baseball World Series. Local oddsmakers were betting 7-to-4 odds that the Mets would win the series in six games.

September 20, 1990

EXAMPLE 5   Leon is in a restaurant. He decides to give the waiter a tip consisting of only 1 coin selected randomly from among the 6 that he has in his pocket. What are the odds against him giving the waiter a penny tip, if he has a penny, a nickel, a dime, a quarter, a half-dollar, or a dollar piece?

SOLUTION   There are 6 coins; 5 are favorable and 1 is unfavorable. Thus, the odds against giving the waiter a penny tip are 5 to 1.

EXAMPLE 6   What are the odds in favor of getting a face card when selecting a card at random from a deck of 52 cards?

SOLUTION   There are 52 possibilities; 12 are favorable and 40 are unfavorable. Thus, the odds in favor of getting a face card are 12 to 40.

EXERCISES FOR SECTION 4.6

1. A caterer has just prepared numerous specialty dishes to be sold. There is a $\frac{5}{11}$ chance that they will be sold today, in which case the profit will be $76. There is a $\frac{6}{11}$ chance that they will not be sold today, in which case the caterer will lose $31

since the caterer does not have refrigerator space to store the dishes. Find the caterer's mathematical expectation.

2. Juanita Johnson has just invested $175,000 to open a new drive-up food store. If successful, she can expect an annual income of $75,000. If unsuccessful, she will lose $95,000. (The remaining $80,000 can be recovered by selling the equipment.) If the probability of success is 0.85, find her mathematical expectation for the first year.

3. Roger is in a gambling casino that charges $4 per chance to roll an honest die. He will be paid, in dollars, the number of dots on the face of the die. Find his mathematical expectation.

4. Margaret Billingsley has just sent some construction plans to Europe via one of the overnight delivery services. There is a 95% chance that they will arrive at their destination by 9:00 A.M. the next morning, in which case her company will be awarded a $80,000 contract. If the plans do not arrive on time, then the contract will be awarded to a competing company, in which case her company will lose the $10,000 already invested to draw up the plans. Find her company's mathematical expectation.

5. A card is drawn at random from an ordinary deck of cards. If the card selected is a face card, then Trudy wins $15. If the card selected is an ace, then Trudy wins $20. Otherwise, she loses $10. What is her mathematical expectation?

6. On a television show, a blindfolded contestant is asked to select one bill from a box containing money as follows:

Denomination of Bill	Number of Bills in Box
$20	16
$50	10
$100	7
$500	5
$1000	2

If the contestant gets to keep the bill selected, find the contestant's mathematical expectation.

7. Beverly Washington has just written a letter to the president of a cosmetics company complaining about the quality of a product. If there is a $\frac{7}{13}$ chance that the president will respond to her letter, what are the odds against the president responding to her letter?

8. According to the *Ladies's Home Journal* as reported in *Vitality* (September 1993, Vol. 7, No. 9, p. 16), 81% of cat owners and 63% of dog owners kiss their pets. If a cat owner is randomly selected,
   a. what is the probability that the cat is *not* kissed by its owner?
   b. what are the odds in favor of the cat being kissed by its owner?

9. Consider the accompanying newspaper article. If an American family is randomly selected,

> ## WHAT DO AMERICAN PARENTS WORRY ABOUT?
>
> *Washington*: (Sept. 20)—According to the latest survey, 43% of American parents worry most about their children becoming victims of violence (especially kidnapping). Furthermore, 39% worry about them becoming addicted to drugs and alcohol, and 12% are concerned about their becoming involved in gangs.
>
> Wednesday—September 20, 1995

a. what is the probability that the parents are not concerned about their children becoming involved with gangs?

b. what are the odds in favor of the parents not being concerned about their children joining or becoming involved with a gang?

10. Antonia Brier is a salesperson for a wholesale grocer. She is analyzing her five best accounts and comes up with the following information:

Account	Weekly Sales	Estimated Probability That Sales Will Be Realized
I	$4698	0.61
II	4203	0.82
III	4506	0.59
IV	4475	0.51
V	4388	0.46

Find the expected sales from each account.

11. Refer back to Exercise 10. Antonia's car is in the repair shop and as a result Antonia finds that by using her friend's car she has enough time to visit only two of the accounts. Using mathematical expectation, help Antonia decide which accounts to visit.

12. Along with seven other guests, Bill places his hat on the table. All the hats look alike. The hostess puts the hats in a closet. After the party, Bill returns and selects a hat at random. What are the odds in favor of his getting his own hat?

13. According to *Time* magazine (Fall 1993, p. 14), 21% of all Los Angeles residents under 18 years of age are foreign born. If a Los Angeles resident under 18 years of age is randomly selected, what are the odds against the resident being foreign born?
14. Consider the accompanying newspaper clipping. Based on the information contained in it,

> ## EXPLORATION FOR OIL TO BEGIN TODAY
>
> *Prudhoe Bay: (March 17)* A conglomeration of oil companies will begin drilling for oil in an area which is 70 miles north of area A. Geologists estimate that the probability of finding oil in the area is 0.85. If oil is discovered in large enough quantities to make it commercially feasible to drill for, the oil companies are expected to make a profit of $7 million. On the other hand, if the quantity of oil discovered is not large enough, then the oil companies will lose 2.5 million dollars in exploration costs.
>
> Thursday—March 17, 1981

    a. find the oil companies' mathematical expectation.
    b. what are the odds in favor of finding oil in the area?

4.7   USING COMPUTER PACKAGES

The term **simulation** refers to a computer's ability to duplicate a theoretical experiment. MINITAB can be used to simulate a variety of experiments. This is accomplished by using MINITAB's **RANDOM** command along with several subcommands. For example, suppose we wish to perform an experiment that consists of randomly rolling a fair die 60 times. This can be accomplished by having MINITAB generate a sequence of 60 equally likely integers 1 through 6, where 1 corresponds to 1 dot showing on the die, 2 corresponds to 2 dots showing on the die, and so on. The following are the commands and the resulting table from a MINITAB simulation of this experiment.

```
MTB  >  RANDOM  60  C1;    (We must have a semicolon since a subcommand
                            follows.)
SUBC >  INTEGER  1  TO  6.    (We must have a period since this is the last sub-
                               command.)
MTB  >  PRINT  C1

C1
    1   3   6   4   6   3   3   4   4   2   3   4   1   4   1
    1   4   6   4   3   4   3   6   2   5   2   2   4   1   4
    2   1   3   4   6   2   2   6   3   4   3   6   6   6   1
    3   4   2   5   6   2   1   5   2   4   2   3   5   3   4

MTB  >  TALLY  C1

    C1      COUNT
    1          8
    2         11
    3         12
    4         15
    5          4
    6         10
    N=        60
```

The first two commands tell **MINITAB** to generate 60 integers (1 to 6) and to place the results in C1. We use the command **TALLY** to make a frequency table summarizing the results of this experiment.

Now let us consider an experiment in which a fair coin is tossed 50 times. This can be accomplished by having **MINITAB** generate a sequence of 50 equally likely digits, where each digit is either a 0 (a tail) or a 1 (a head). The following are the commands and the resulting table for a **MINITAB** simulation of this experiment.

```
MTB  >  RANDOM  50  C2;
SUBC >  INTEGER  0  TO  1.
MTB  >  PRINT  C2

C2
    0   1   1   1   0   1   0   1   1   1   1   0   0   1   0
    0   1   0   1   0   0   0   1   1   0   0   0   0   1   1
    0   1   0   1   0   1   0   1   0   0   1   1   0   1   1
    0   0   0   0   0

MTB  >  TALLY  C2

    C2      COUNT
    0         27
    1         23
    N=        50
```

To illustrate the relative frequency definition of probability, we had MINITAB simulate the tossing of a fair coin 500 times, 5000 times, 10,000 times, and so on. The following results were obtained:

C3	COUNT	C4	COUNT	C5	COUNT
0	253	0	2496	0	4993
1	247	1	2504	1	5007
N	500	N	5000	N	10000

We summarize these findings as follows:

Number of Tosses of Coin (n)	Number of Heads Appearing (f)	Relative Frequency (f/n)
50	23	0.460000
500	247	0.49400
5,000	2,504	0.50080
10,000	5,007	0.50070

The relative frequency is close to 0.5. Does this contradict the fact that the probability that a fair coin will show heads is exactly 0.5? Can you explain why not?

EXERCISES FOR SECTION 4.7

1. Using MINITAB, simulate the rolling of a fair die a total of
   a. 180 times   b. 300 times   c. 600 times   d. 6000 times
   e. Are the relative frequencies of the number of dots appearing approximately $\frac{1}{6}$?

2. Using MINITAB, simulate the tossing of a fair coin 180 times.
3. MINITAB was used to simulate the experiment of rolling a die 120 times. Each time the number of dots showing was recorded. The following results were obtained:

```
MTB > TALLY C1
C1    COUNT
 1      24
 2      18
 3       8
 4      23
 5      22
 6      25
```

Does it appear that the die is biased? Explain your answer.
4. Using MINITAB, simulate the tossing of a fair coin ten times and record the number of heads obtained. Repeat the experiment 50 times.

4.8   SUMMARY

In this chapter we discussed various aspects of probability. We noticed that probability is concerned with the total number of possible outcomes and experiments. Among the different ways of defining probability, we mentioned both the classical and the relative-frequency interpretation of probability. In the classical definition we have

$$p = \frac{\text{number of outcomes favoring an event}}{\text{total number of equally possible outcomes}}$$

We noticed that the probability of an event was between 0 and 1, the null event and the definite event, respectively.

To enable us to determine the total number of possible outcomes, we analyzed various counting techniques. Tree diagrams, permutations, and combinations were introduced and discussed in detail. Permutations represent arrangements of objects where order *is* important, whereas combinations represent selections of objects where order is *not* important. Applications of permutations and combinations to many different situations were given, in addition to the usual gambling problems.

Finally, probability was applied to determine the amount of money to be won in the long run in various situations. This was called the mathematical expectation of the event. We also discussed what is meant by statements such as odds in favor of an event and odds against an event. Definitions were given that allow us to calculate these odds.

Study Guide

The following is a chapter outline in capsule form. You should now be able to demonstrate your knowledge of the ideas mentioned by giving definitions, descriptions, or specific examples. Page references are given in parentheses.

The outcomes that are of interest to us represent a **favorable event**. (page 210)

All possible outcomes of an experiment are referred to as the **sample space** of the experiment. (page 210)

In the **classical interpretation of probability** we define the probability of any event as the number of favorable outcomes divided by the total number of possible outcomes, assuming that all outcomes of an experiment are equally likely. (page 213)

The **relative frequency interpretation of probability** represents the percentage of times that an event will happen in repeated experiments. (page 213)

An event that can never happen is called a **null event** and its probability is 0. (page 218)

An event that is certain to occur is called the **certain event** or **definite event** and its probability is 1. (page 219)

The probability of any event must be between 0 and 1. (page 219)

One technique that is sometimes used to determine the number of possible outcomes for an experiment is to construct a **tree diagram** where each branch represents a different possible outcome. (page 226)

The **counting rule** states that if one thing can be done in m ways, and if after this is done, something else can be done in n ways, then both things can be done in a total of $m \cdot n$ different ways in the stated order. (page 229)

A **permutation** is any arrangement of distinct objects in a particular order. (page 234)

We use **factorial notation** to represent a special type of multiplication. Thus, $n!$ means that we start with the whole number n and multiply this by the number that is 1 less than n, and so on until we get to the number 1, which is where we stop. (page 235)

$_nP_r$ represents the **number of permutations of n things taken r at a time**. (page 236)

The **number of permutations of n things taken n at a time** is denoted by $_nP_n$ and is equal to $n!$. (page 237)

A **combination** is a selection from a collection of distinct objects where order is not important. (page 244)

The **number of combinations of n things taken r at a time** is denoted by $_nC_r$ or $\binom{n}{r}$. The symbol $\binom{n}{r}$ or $_nC_r$ is also called a **binomial coefficient**. (page 244)

Pascal's triangle can be used for computing the number of possible combinations of n things taken r at a time. This technique avoids the factorial notation. (page 249)

If one event has probability p_1 of occurring with a payoff of m_1, a second event has probability p_2 of occurring with payoff of m_2, a third event has probability p_3 of occurring with payoff m_3, and so on, then the **mathematical expectation** of the events is $m_1p_1 + m_2p_2 + m_3p_3 + \cdots + m_np_n$. When applied to gambling or business situations, the mathematical expectation of an event is the amount of money to be won or lost in the long run. (page 255)

The **odds in favor** of an event occurring are p to q, where p is the number of favorable outcomes and q is the number of unfavorable outcomes. (page 257)

The **odds against** an event happening are q to p, where p and q are the same as in the previous definition. (page 257)

The term **simulation** refers to a computer's ability to duplicate a theoretical experiment. (page 261)

Formulas to Remember

You should be able to identify each symbol in the following formulas, understand the relationships among the symbols expressed in each formula, understand the significance of each formula, and use the formulas in solving problems.

1. $p(A) = \dfrac{\text{number of favorable outcomes}}{\text{total number of equally possible outcomes}} = \dfrac{f}{n}$

2. $_nP_r = \dfrac{n!}{(n-r)!}$ The number of permutations of n things taken r at a time

3. $_nP_n = n!$ The number of permutations of n things taken n at a time

4. $\dfrac{n!}{p!q!r!}\cdots$ The number of permutations of n things where p are alike, q are alike, and so on

5. $_nC_r = \dfrac{n!}{r!(n-r)!}$ The number of combinations of n things taken r at a time

6. $m_1p_1 + m_2p_2 + m_3p_3 + \cdots$ Mathematical expectation of an event.

7. Odds in favor of an event are p to q, and odds against an event are q to p,

   where $\begin{cases} p = \text{the number of favorable outcomes} \\ q = \text{the number of unfavorable outcomes.} \end{cases}$

Testing Your Understanding of This Chapter's Concepts

1. Jennifer is a clerk in a department store. A customer purchases an item for $86 and pays for it with one $50 bill, one $20 bill, one $10 bill, one $5 bill, and one $1 bill. Jennifer's cash register drawer has five different compartments, one for each of the bills mentioned. If Jennifer picks up the bills and, without looking, randomly places the bills in the drawer (one per compartment), find the probability that she inserts each bill in its proper compartment.

2. A jewelry salesman carries all of his samples in his attaché case. There is a combination lock on either side of the case. Each lock has four dials that have to be rotated independently so that any number from 0 to 9 inclusive shows on each dial. Both locks can be opened only if the correct number shows on each dial displayed. How many combinations does a thief have to try before the locks open and the thief can steal the sample jewels from the attaché case?

3. Is a combination lock actually a combination lock or should it be named a permutation lock?

4. Six couples are to be honored for their outstanding humanitarian and philanthropic efforts. Each husband will sit with his wife at a circular head table. In how many different ways can these 12 people be seated?

5. How many different social security numbers are there? If a tenth digit is added, how many different social security numbers will there be? (Assume that there are no restrictions at all.)

6. Many states conduct weekly or daily lotteries. (See the accompanying ticket.) In the Pick 10 lottery, a player must select ten different numbers from 1 to 40.

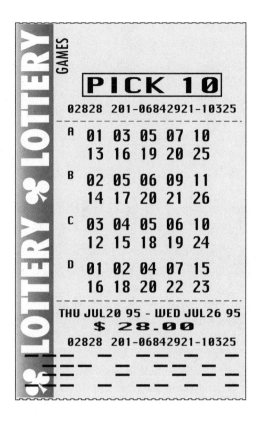

a. Determine the number of possible lottery tickets.

b. What are the odds in favor of a person winning in this lottery?

7. Kelly, Noel, and Patricia work as part-time nurses at Maimonides Hospital. Each week Noel works three days, Kelly works two days, and Patricia works two days. In how many different orders can they be assigned to work for one week at the hospital? (*Hint:* This is an arrangement of seven things with repetition.)

Chapter Test

Multiple Choice Questions

1. A pair of dice is rolled. Find the probability that the sum of the number of dots appearing is larger than 9.

   a. $\dfrac{3}{36}$ b. $\dfrac{6}{36}$ c. $\dfrac{10}{36}$ d. $\dfrac{5}{36}$ e. none of these

2. A family is known to have four children. Find the probability that the family consists of at least two boys.

   a. $\dfrac{6}{16}$ b. $\dfrac{4}{16}$ c. $\dfrac{1}{16}$ d. $\dfrac{11}{16}$ e. none of these

3. Which of the events shown below has the greatest likelihood of occurring?

Event	Probability That This Event Will Occur
A	0.18
B	0.81
C	0.97
D	0.097

   a. event A   b. event B   c. event C   d. event D
   e. none of these

4. Each license plate in a certain state consists of three letters followed by two numbers with repetition allowed. How many different license plates are possible in this state?
   a. 1,757,600   b. 328,536   c. 1,404,000   d. 365,040
   e. none of these

5. A teacher receives three test papers from three students. Unfortunately, the three students forgot to write their names on the papers. The teacher decides to randomly assign each of the test papers to one of these students. Find the probability that each student receives his or her own test paper.

   a. $\dfrac{1}{6}$ b. $\dfrac{1}{3}$ c. $\dfrac{2}{3}$ d. $\dfrac{1}{2}$ e. none of these

6. How many different male-female committees can be formed from among nine males and ten females to investigate discrimination charges if each committee is to consist of five males and six females?
   a. 39,916,800   b. 126   c. 210   d. 26,460   e. none of these

7. There is a 90% chance that a certain active volcano will erupt this year. What are the odds in favor that the volcano will erupt this year?
   a. 9 to 10   b. 9 to 1   c. 1 to 9   d. 10 to 9   e. none of these

8. In how many different ways can the letters of the word RESEARCH be arranged?
   a. 40,320   b. 20,160   c. 10,080   d. 24   e. none of these

9. At a local fund-raising drive, tickets numbered consecutively from 1 to 100 were sold. One winning ticket will be selected. What is the probability that the winning number is a multiple of 9?

   a. $\dfrac{11}{100}$ b. $\dfrac{10}{100}$ c. $\dfrac{9}{100}$ d. $\dfrac{1}{9}$ e. none of these

10. The probability that Nicole will go on to medical school after graduating college is 0.89. What is the probability that she will *not* go on to medical school?
    a. 0.011   b. 0.11   c. 1.1   d. 0.089   e. none of these

Supplementary Exercises

11. Carl Berenson remembers the first three digits of an auto claim that he filed with his insurance company. However, he does not recall the sequence of the remaining five digits: 3, 1, 2, 8, and 0. How many possibilities are there?

12. In how many different ways can the letters in the word MONEY be arranged? (*Note:* Each arrangement does not necessarily have to form a word.)

13. The lock on the vault at Empire Savings Bank consists of three dials, each with 30 positions. In order for the lock to be opened, each dial must be in the correct position. How many different "dial positions" are there for this lock?

14. According to a recent survey, 71% of the state's residents are in favor of the state's proposed anti-gun legislation. If a state resident is randomly selected, what are the odds that the resident is opposed to the new legislation?

15. In a special lottery, 50,000 $1-tickets are sold. The first prize is $15,000 and the five second-prize winners will share $2000 equally. Manya buys one ticket for this lottery.
    a. What is her mathematical expectation?
    b. What are the odds against her being a first-prize winner?
    c. What are the odds against her being a winner?

16. A child is playing with a telephone and dials a long-distance (ten-digit) number.
    a. How many possible long-distance numbers are there? (Assume that there are no restrictions.)
    b. What is the probability that the child dials his or her own number?

17. The Daily Double at the local race track consists of picking the winners of the first two races. If eight horses are entered in the first race, and 10 horses are entered in the second race, how many Daily Double possibilities are there?

18. Joanna is interested in purchasing a computer system. The following list presents some of the possibilities.

Computer Type	Monitor	Printer
IBM or compatible	EGA	Laser Jet
Macintosh	CGA	Dot Matrix
	VGA	
	Super VGA	

Using tree diagrams, find how many possible computer systems can be purchased.

19. How many different permutations are there of the letters in the word
    a. TRIGONOMETRIC   c. COOPERATION
    b. DIFFERENT   d. INTRODUCTION

20. The manager of a clothing store wishes to analyze the relationship between the type of customer and the form of payment. The following data have been collected for 1000 customers:

Customer	*Method of Payment*		
	Credit Card	Check	Cash
Male	293	102	205
Female	186	172	42

If a customer is selected at random from this group, find the probability that

a. the customer is a male.

b. the customer paid by credit card.

c. the customer is a male who paid by credit card.

d. the customer did not pay by credit card.

21. Consider the accompanying newspaper article. If a child is randomly selected, what are the odds that the child has not been vaccinated against such deadly diseases?

DEATHS FROM CHILDHOOD DISEASES DOWN

New York (Jan 3): According to the U.N., fewer youngsters worldwide are dying of childhood diseases now than at any time in history. In 1993, measles killed 1.1 million children, down from 2.5 million a year just a decade ago. Polio crippled 140,000 children last year, down from 500,000 in 1980. About 80% of children today are vaccinated against such deadly diseases as measles and polio, compared with 20% in the early 1980s.

Monday—January 3, 1994

22. How many four-digit numbers larger than 6000 can be formed using the digits 3, 4, 6, 8, 9, and 0? (No repetition allowed.)

23. Steve and Joan (who do not know each other) are each registering for next semester's courses at their college. Each student must register for one of four different courses. Assuming that there is only one open section of each of these courses and that they are all offered at the same time, what is the probability that Steve and Joan both register for the same course?

24. Each radio manufactured by a large electronics company is labeled with a serial number that consists of two letters followed by four numbers. How many different serial identification numbers are possible?

25. A nurse can select any one of five skirts, four blouses, and three hats to wear for work. How many different outfits are possible?

26. How many different three-person paramedic crews can be formed to drive a special ambulance if ten equally qualified paramedics are available for the job?

27. A baseball scout is ranking the top four promising baseball prospects from a possible list of 11 candidates. In how many different ways can this be done?

28. Dr. Marilyn Briggs and Dr. Juanita Rodriguez both share the same medical office. Each week Dr. Briggs works four days and Dr. Rodriguez works three days. In how many different orders (one doctor per day) can they choose to work in the medical office? (*Hint:* This is an arrangement of seven things with repetition.)

29. If the letters of the word "VALUE" are rearranged at random, what is the probability that the first letter of the new "word" is a vowel?

30. A dairy company is considering using one of two identification codes to place on each container of yogurt that it sells; four letters followed by three numbers or four numbers followed by three letters. Which of these schemes will result in more identification codes?

31. An employee in a supermarket discovers three cans of tuna fish on the floor from which the identifying labels have fallen off. The labels are nearby on the floor. One label reads Bumble Bee Tuna Fish, one label reads Chicken of the Sea Tuna Fish, and one label reads Star Kist Tuna Fish. It is definitely known that the labels are for these cans. If the employee randomly attaches a label to a can, what is the probability that each can gets labeled properly?

32. Construct a tree diagram to determine the number of possible three-digit numbers larger than 600 that can be formed from the digits 5, 6, 7, and 8 if repetition is not allowed?

33. Consider the following newspaper clipping. What are the odds against a particular business being checked?

STATE TO CHECK ON SALES TAX CHEATS

Marlowe: (*April 17*):—The State Sales Tax Bureau announced yesterday that it would begin checking on merchants who consistently charge the wrong sales tax. Said a spokesman for the bureau, the State is losing an estimated 5 million dollars annually from merchants who do not collect sales tax.

It is expected that approximately 15% of all business establishments within the state will be checked.

April 17, 1990

34. Each of the 17,225 students at a state university commutes to school by bicycle. In an attempt to discourage theft, the administration is considering a plan to carve identification codes on each bike.
    a. Will a three-digit code carved on each bike be sufficient for all bikes? Explain your answer.
    b. Will a three-letter code carved on each bike be sufficient for all the bikes? Explain your answer.

35. The Social Security Administration never reuses any Social Security number; that is, after an individual dies, that individual's Social Security number is no longer used.
    a. How many different nine-digit Social Security numbers are there?
    b. If a tenth digit is added, how many different Social Security numbers will there be?
    c. If a letter is added to everybody's Social Security number, how many different Social Security numbers will there be?

36. In an effort to protect the consumer, the agriculture department of a certain state requires that all canned goods manufactured within the state be stamped with numbers or letters as follows:

    i. The company numbers as assigned by the department: There are nine companies within the state.
    ii. The plant location within the state: The state is divided into eight geographical areas labeled A, B, C, D, E, F, G, or H.
    iii. The date of manufacture: This consists of the month and last two digits of the year. For example, a can with the code 1A0288 on it means that it was manufactured by company 1 located in area A during the month of February 1988.

    Using the preceding scheme, how many different codes are possible for cans manufactured during the years 1989 through 1994?

37. A wheel of fortune is equally divided into six colored areas: black, green, blue, yellow, red, and brown. If the wheel stops on black or blue after one spin, then the prize is $7. If it stops on green, the prize is $8. However, if it stops on red or brown, then $3 will be lost. There is no prize when the wheel stops on yellow.
    a. What are the odds of winning a prize in any one spin of the wheel?
    b. What is the mathematical expectation for someone who plays this game?

Thinking Critically

1. José tosses a fair coin twice and observes the number of heads obtained. He reasons that since the number of heads obtained can be either 0, 1, or 2, the probability of obtaining no heads is $\frac{1}{3}$. Do you agree with this conclusion? Explain your answer.

2. In the baseball World Series, two teams play against each other in a contest of up to seven games. The first team to win four games is the winner. Assume that the teams are labeled A and B and that the probability that team A wins any game is $\frac{1}{2}$. Find the probability that the World Series will end in

   a. 4 games.   b. 5 games.   c. 6 games.   d. 7 games.

3. There are three cards on a table as shown in the following figure. Under one of these cards there is a one thousand dollar bill. Under the second card there is a one dollar bill and under the third card there is nothing. A player must guess under which card the $1000 bill appears.

   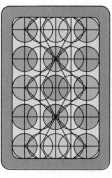   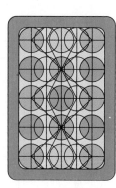

Card 1   Card 2   Card 3

   If the player guesses correctly, then the player keeps the $1000. After the player selects a card, the card dealer reveals one of the remaining cards under which there is nothing. The dealer now offers the player the opportunity of switching choices among the remaining cards since the probability of winning $1000 is now $\frac{1}{2}$ instead of $\frac{1}{3}$. Do you agree with this reasoning? Explain your answer.

4. Determine whether each of the following pairs of "counts" are equal.

   a. $_nP_1$ and $_nC_1$

   b. $_nP_n$ and $_nP_{n-1}$

5. Verify that the probabilities given in Table 4.1 are accurate.

6. Explain how the relationship $\binom{n+1}{r} = \binom{n}{r} + \binom{n}{r-1}$ enables us to construct Pascal's triangle. (*Hint:* First verify that the equation is true.)

7. Using Pascal's triangle, show that $\binom{n}{r} = \binom{n}{n-r}$, that is, numbers in a particular row read from left to right are the same as read from right to left.

8. Using Pascal's triangle, show that the sum of the numbers in any row is 2^n where n is the row number.

9. A pair of dice is rolled and the sum of the number of dots appearing on both dice is recorded. Since the sum is either 2, 3, 4, 5, 6, 7, 8, 9, 10, 11, or 12, the probability of obtaining a sum of 12 is $\frac{1}{11}$. Do you agree with this reasoning? Explain your answer.

10. Grunwall and Co., a Wall Street brokerage firm, has decided to replace the key lock entrance to its high security computer room with a keyless plastic card system where a plastic card has to be inserted into a slot above the door knob for the door to open. Each employee authorized to enter the room will be issued a plastic card that has been engraved with notches arranged in a grid of 10 rows, each with seven notches as shown below.

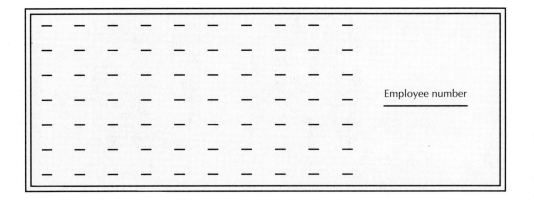

The notches can be molded into the card in either a vertical position ▮ or a horizontal position ▬. Each authorized employee will be issued a card that has been coded uniquely for that employee.

Since the company is in the midst of a major expansion campaign, company officials would like to know how many different coded cards can be made. Using the above coding scheme, how many different cards are actually possible?

Case Studies

1. Many oil companies sponsor contests so as to encourage customers to buy their brands of gasoline. Between September 13, 1993, and October 24, 1993, the American Oil Company sponsored one such contest at participating AMOCO gas stations in the states of AL, DE, FL, GA, MS, NC, OH, PA, SC, TN, CT, NY, CO, IL, IA, KS, MI, MN, NE, ND, SD, MO, WV, WI, and DC. With each purchase of gasoline, a customer was given a card (as shown on the next page) that had five covered boxes. The first two boxes contained the letter of the states listed above and the last three boxes contained digits with repetition allowed. (000 was not a possibility.) The cus-

tomer was required to scratch all five boxes on the card to reveal a license plate number. If the revealed number matched an official list displayed at any participating AMOCO station, then the card holder would win a new 1994 Dodge Caravan car.

If Jennifer scratches the boxes from one such card,
a. what are her chances of winning the Dodge Caravan?
b. what are the odds against her winning the Dodge Caravan?
2. An article in *Newsday* indicated that more than 100 million cases of beer were distributed in New York State during the year 1989.

Brewing Company	Percentage of Market Share
Anheuser-Busch Co.	46
Miller Brewing Co.	20
Imported Beers	11
Malt Liquor	9
Adolph Coors Co.	6
G. Heileman Brewing Co.	3
Latrobe Brewing Co.	1
Stroh Brewing Co.	1
Others	3
	Total = 100

(*Source: Newsday* May 2, 1990 (Extracted from *Beverage World*))

If a can of beer distributed in New York State in 1989 is randomly selected, what is the probability that it was brewed by
a. one of the two companies that have the largest market share?
b. one of the two companies that have the smallest market share?

5

RULES OF PROBABILITY

DID YOU KNOW THAT

a   a woman's risk of breast cancer rises dramatically with her intake of meat, eggs, and butter? (*Source: Conference on Breast Cancer and Diet*–U.S.-Japan Cooperative Cancer Research Program, Fred Hutchison Cancer Center, Seattle, Wash.)

b   despite EPA rules limiting the use of dangerous pesticides, children are still at risk from pesticide residues in food? (*Source: Pesticides Monitoring Journal*, 24, 1991, and *Time*, July 12, 1993, p. 18)

c   there were 67 million handguns owned by private citizens in the United States in 1993 and that 1.5 million handguns were made in the United States in 1992? (*Source:* Bureau of Alcohol, Tobacco and Firearms, *FBI Uniform Crime Reports*, U.S. Justice Dept.)

d   55% of patients who had no emotional support died within one year of their heart attacks compared with only 27% who had two or more people giving them attention? (*Source: Annals of Internal Medicine*, as quoted in *Vitality*, September 1993, p. 24)

e   when spouses argue, the wife's blood pressure rises an average of 6% as compared with the 14% for the husband? (*Source: Harper's*)

f   since 1901, 30% of the U.S. Nobel prizewinners have been immigrants?

g   more than 100 languages are spoken in the school systems of New York City, Chicago, Los Angeles, and Fairfax County, Va.? (*Source: Newcomers in American Schools*, Rand)

What is the probability that the car in the above photograph won't start because of a dead battery? See discussion on page 289. *(Pamela Farace)*

Although probability was defined in the last chapter, nevertheless, we need rules for determining probabilities of events involving "and" or events involving "or." Also, we may be interested in determining probabilities given that other events have occurred. In this chapter we analyze various rules of probability.

Chapter Objectives

- **To use** two addition rules that allow us to determine the probability of either event *A*, event *B*, or both happening. (Section 5.2)

- **To discuss** whether particular events are or are not mutually exclusive and how this determines which addition rule to use. (Section 5.2)

- **To determine** when one event is affected by the occurrence or nonoccurrence of another and how this affects probability calculations. This is known as conditional probability. (Section 5.3)

- **To understand** that when one event is not affected at all by the occurrence of another, we have independent events and a simplified multiplication rule. (Section 5.4)

- **To apply** Bayes' rule. Thus, if we know the outcome of some experiment, we might be interested in determining the probability that it occurred because of some specific event. (Section 5.5)

STATISTICS IN ACTION

BEWARE (A LITTLE) FALLING METEORITES

A falling meteorite may not be one of the hazards of modern life, but a person can be struck by one. Scientists in Canada have even calculated the magnitude of the risk.

The scientists, at Herzberg Institute of Astrophysics in Ottowa, used a network of 60 cameras in Western Canada for the past nine years to study meteorite falls. From the observations, they said in a recent letter to *Nature*, it can be calculated that one human should be hit in North America every 180 years.

The researchers, I. Halliday, A. T. Blackwell and A. A. Griffin, based their calculations on the number of meteorite falls of a size large enough to be detected, the number of humans in the total Canadian and United States populations, the average human size, the time a person could be expected to be out of doors and other factors. On these assumptions they calculated an annual rate of human meteorite hits at 0.0055 per year, or one every 180 years.

Rare as these events should be, they noted, one such case actually occurred only 31 years

November 14, 1986

ago. It is believed to be the only well-documented case of a collision between a meteorite and a human.

On Nov. 30, 1954, a nine-pound stony meteorite plunged through the roof of a home in Sylacauga, Ala., bounced off a large radio and struck Mrs. E. H. Hodges, inflicting painful bruises but causing her no serious injury.

"At first glance, it would appear unlikely that there would be even one known event only 31 years ago," the Canadian scientists said, "but the fact that there are no other verified cases elsewhere in the world indicates that the impacts on people are extremely rare."

Meteorite impacts on buildings are much less rare, they said, noting that there have been seven documented reports in North America during the past 20 years.

Worldwide, the Canadian scientists said, one could expect a human to be struck by a meteorite once in every nine years, while 16 buildings a year would be expected to sustain some meteorite damage.

Consider the newspaper article on the previous page. What is the likelihood (probability) of a meteorite falling on a human being? The newspaper article indicates that given the average human size, the time that a person can be expected to be outdoors, and other factors, the probability of a human hit is still 0.0055 per year. How are such conditional probabilities calculated?

As mentioned in the following article, many airports are increasing their security. What is the probability that someone can avoid being detected given all the additional security checks? How do we calculate such conditional probabilities?

AIRPORTS TO TIGHTEN SECURITY

Washington (Jan. 10)—In the wake of the recent threats by international terrorists to bomb American facilities and the reports by many organizations on the easiness by which security can be breached at our airports, the Federal Aeronautics Administration (FAA) yesterday ordered all airports throughout the country to tighten security and to check all passengers thoroughly before they board planes. Several airports have already installed highly technological screening machines which can detect plastic explosives. Other airports have discontinued curbside check-ins.

January 10, 1991

5.1   INTRODUCTION

In Chapter 4 we discussed the nature of probability and how to calculate its value. However, in order to determine the probability of an event in many situations, we must often first calculate the probability of other related events and then combine these probabilities. In this chapter we discuss several rules for combining probabilities, including rules for addition, multiplication, conditional probability, and Bayes' rule. Depending on the situation, these rules enable us to combine probabilities so that we can determine the probability of some event of interest.

5.2   ADDITION RULES

Mutually Exclusive Events

Let us look in on Charlie, who is playing cards. He is about to select one card from an ordinary deck of 52 playing cards. His opponent, Dick, will pay him $50 if the card selected is a face card (that is, a jack, queen, or king) *or* an ace. What is the probability that Charlie wins the $50?

Again, we note that a deck of playing cards consists of 52 playing cards as shown in Figure 5.1.

Now, to answer the question we first notice that a card selected cannot be a face card and an ace at the same time. Mathematically, we say that the events of "drawing a face card" and of "drawing an ace" are **mutually exclusive**.

Since there are 12 face cards in a deck (4 jacks, 4 queens, and 4 kings), the probability of getting a face card is $\frac{12}{52}$, or $\frac{3}{13}$. Similarly, the probability of getting an ace is $\frac{4}{52}$, or $\frac{1}{13}$, since there are 4 aces in the deck. There are then 12 face cards and 4 aces. Thus, there are 16 favorable outcomes out of a possible 52 cards in the deck. Applying the definition of probability, we get

$$p(\text{face card or ace}) = \frac{16}{52} = \frac{4}{13}$$

Let us now add the probability of getting a face card with the probability of getting an ace. We have

$$p(\text{face card}) + p(\text{ace}) = \frac{12}{52} + \frac{4}{52}$$

$$= \frac{12 + 4}{52} \qquad \text{(since the denominators are the same)}$$

$$= \frac{16}{52} = \frac{4}{13}$$

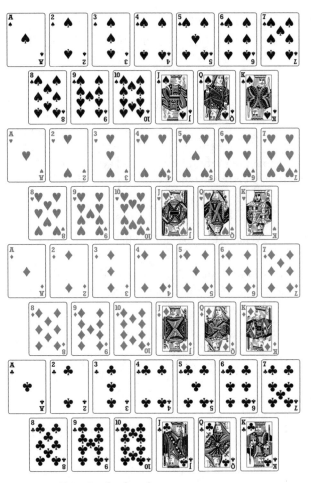

FIGURE 5.1 Deck of cards.

This indicates that

$$p(\text{face card or ace}) = p(\text{face card}) + p(\text{ace})$$

The same reasoning can be applied for any mutually exclusive events. First we have the following definition and formula.

Definition 5.1
Mutually exclusive events

Consider any two events A and B. If both events cannot occur at the same time, we say that the events A and B are **mutually exclusive**.

FORMULA 5.1

Addition Rule for Mutually Exclusive Events

If A and B are mutually exclusive, then

$$p(A \text{ or } B) = p(A) + p(B)$$

We illustrate the use of Formula 5.1 with several examples.

E X A M P L E   1   Louis has been shopping around for a calculator and has decided to buy the scientific model graphics calculator. It has been estimated that the probability a customer will buy the Casio model is $\frac{1}{9}$ and the probability that a customer will buy the Texas Instruments model is $\frac{4}{9}$. What is the probability that Louis will buy either of these two models?

SOLUTION   Since Louis will buy only one calculator, the events "buys the Texas Instruments model" and "buys the Casio model" are mutually exclusive. Thus, Formula 5.1 can be used. We have

$$
\begin{aligned}
p(\text{buys either model}) &= p(\text{buys Texas Instruments model}) \\
&\quad + p(\text{buys Casio model}) \\
&= \frac{4}{9} + \frac{1}{9} \\
&= \frac{4+1}{9} = \frac{5}{9}
\end{aligned}
$$

Therefore, $p(\text{Louis buys either model}) = \frac{5}{9}$.

E X A M P L E   2 Mary turns on the television set. Based upon her viewing habits, it is estimated that the probability that the television is on Channel 2 is $\frac{1}{3}$, and the probability that it is on Channel 7 is $\frac{4}{13}$. What is the probability that it is either on Channel 2 or Channel 7?

SOLUTION   Since the same television set cannot be on Channel 2 and on Channel 7 at the same time, the events "television set on Channel 2" and "television set on Channel 7" are mutually exclusive; thus, Formula 5.1 can be used. We have

$$p(\text{television set on Channel 2 or 7}) = p(\text{television set on Channel 2})$$
$$+ \; p(\text{television set on Channel 7})$$
$$= \frac{1}{3} + \frac{4}{13}$$
$$= \frac{13}{39} + \frac{12}{39} = \frac{25}{39}$$

Therefore,

$$p(\text{TV on Channel 2 or 7}) = \frac{25}{39}$$

(*Note:* We cannot add the fractions $\frac{1}{3}$ and $\frac{4}{13}$ together as they are since they do not have the same denominators. We must change both fractions to fractions with the same denominators. Thus, $\frac{1}{3}$ becomes $\frac{13}{39}$ and $\frac{4}{13}$ becomes $\frac{12}{39}$. Hence, the probability that the television is on Channel 2 or 7 is $\frac{25}{39}$.)

E X A M P L E   3 Two dice are rolled. What is the probability that the sum of the dots appearing on both dice together is 9 or 11?

SOLUTION   Since the events "getting a sum of 9" and "getting a sum of 11" are mutually exclusive, Formula 5.1 can be used. When 2 dice are rolled there are 36 possible outcomes. These were listed on page 211. There are 4 possible ways of getting a sum of 9. Thus,

$$p(\text{sum of 9}) = \frac{4}{36}$$

Also, there are only two possible ways of getting a sum of 11. Thus,

$$p(\text{sum of 11}) = \frac{2}{36}$$

Therefore,

$$p(\text{sum of 9 or 11}) = p(\text{sum of 9}) + p(\text{sum of 11})$$

$$= \frac{4}{36} + \frac{2}{36}$$

$$= \frac{6}{36} = \frac{1}{6}$$

Hence, the probability that the sum is 9 or 11 is $\frac{1}{6}$.

E X A M P L E   4   Doris and her friends plan to travel to Florida during the winter intersession period. Based upon past experience they know that the probability that they go by car is $\frac{2}{3}$ and the probability that they go by plane is $\frac{1}{5}$. What is the probability that they travel to Florida by car or plane only?

SOLUTION   Since they plan to travel to Florida either by car or by plane, not by both, we are dealing with mutually exclusive events. Formula 5.1 can be used. Therefore,

$$p(\text{go by car or plane}) = p(\text{go by car}) + p(\text{go by plane})$$

$$= \frac{2}{3} + \frac{1}{5}$$

$$= \frac{10}{15} + \frac{3}{15}$$

$$= \frac{13}{15}$$

Complementary Events

E X A M P L E   5   Rosemary buys a ticket in the state lottery. The probability that she will win the grand prize of one million dollars is $\frac{1}{50,000}$. What is the probability that she does not win the one million dollars?

SOLUTION   Since the events "Rosemary wins the million dollars" and "Rosemary does not win the million dollars" are mutually exclusive, we can use Formula 5.1. One of those events must occur so that the event "Rosemary wins the million dollars or does not win the million dollars" is the definite event. We know that the definite event has probability 1 (see page 219). Thus,

p(Rosemary wins million dollars or does not win) = p(wins million dollars)
$\qquad\qquad\qquad\qquad\qquad\qquad\qquad\qquad$ + p(does not win million dollars)

$$1 = \frac{1}{50,000} + p(\text{does not win million dollars})$$

$$1 - \frac{1}{50,000} = p(\text{does not win million dollars})$$

$$\frac{50,000}{50,000} - \frac{1}{50,000} = p(\text{does not win million dollars})$$

$$\frac{49,999}{50,000} = p(\text{does not win million dollars})$$

Therefore, the probability that Rosemary does not win the million dollars is $\dfrac{49,999}{50,000}$.

More generally, consider any event A. Let $p(A)$ be the probability that A happens and let $p(A')$ read as the probability of A prime, be the probability that A does not happen. Since either A happens or does not happen, we can use Formula 5.1. Thus,

$$p(A \text{ happens or does not happen}) = p(A \text{ happens}) + p(A \text{ does not happen})$$
$$1 = p(A) + p(A')$$
$$1 - p(A) = p(A') \qquad (\text{We subtract } p(A) \text{ from both sides.})$$

Complement of an event

Therefore, the probability of A not happening is $1 - p(A)$. (*Note:* Some books refer to the event A' as the **complement** of event A.)

Addition Rule—General Case

Now consider the following problem. One card is drawn from a deck of cards. What is the probability of getting a king or a red card? At first we might say that since there are 4 kings and 26 red cards, then

$$p(\text{king or red card}) = p(\text{king}) + p(\text{red card})$$
$$= \frac{4}{52} + \frac{26}{52}$$
$$= \frac{30}{52} = \frac{15}{26}$$

Thus, we would say that the probability of getting a king or a red card is $\dfrac{15}{26}$. Notice, however, that in arriving at this answer we have counted some cards twice. The 2 red kings have been counted as both kings and red cards. Obviously, we must count them only once in probability calculations. The events "getting a king" and "getting a red card" are not mutually exclusive. We therefore have to revise our original estimate of the total number of favorable outcomes by deducting the number of cards that have been counted twice. We will subtract 2. When this is done, we get

$$p(\text{king or red card}) = p(\text{king}) + p(\text{red card}) - p(\text{king also a red card})$$

$$= \frac{4}{52} + \frac{26}{52} - \frac{2}{52}$$

$$= \frac{4 + 26 - 2}{52}$$

$$= \frac{28}{52} = \frac{7}{13}$$

Thus, the probability of getting a king or a red card is $\frac{7}{13}$. This leads us to a more general formula.

Addition rule (general case)

ADDITION RULE • (General Case)

If A and B are any events, the probability of obtaining either of them is equal to the probability of A plus the probability of B minus the probability of both occurring at the same time.

Symbolically, the addition rule is as follows.

FORMULA 5.2

If A and B are any events, then

$$p(A \text{ or } B) = p(A) + p(B) - p(A \text{ and } B)$$

We now apply Formula 5.2 in several examples.

E X A M P L E   6

The probability that any member of the Hamilton Bay Ensemble plays a guitar is $\frac{1}{4}$, and the probability that a member plays a clarinet is $\frac{5}{8}$. If the probability that a member plays both instruments is $\frac{5}{24}$, what is the probability that the member plays the guitar or that the member plays the clarinet?

SOLUTION   Since it is possible that a member plays both these instruments, these events are not mutually exclusive. Thus, we must use Formula 5.2. We have

$$p(\text{plays guitar or clarinet}) = p(\text{plays guitar}) + p(\text{plays clarinet})$$
$$- p(\text{plays guitar and clarinet})$$
$$= \frac{1}{4} + \frac{5}{8} - \frac{5}{24}$$
$$= \frac{6}{24} + \frac{15}{24} - \frac{5}{24}$$
$$= \frac{6 + 15 - 5}{24}$$
$$= \frac{16}{24} = \frac{2}{3}$$

Thus, the probability that a member plays either instrument is $\frac{2}{3}$.

EXAMPLE 7

Voting

Consider the accompanying newspaper article given below. What is the probability that either a husband will vote or his wife will vote in the coming mayoral election?

LOW VOTER TURNOUT EXPECTED

Greensburgh (Oct. 20)—According to a poll released yesterday, voter turnout for the coming mayoral election is expected to be at an all-time low. One out of eleven married men said that they would vote. For the women, the figure was one out of nine. Only one out of 28 couples said that they would both vote. Both candidates indicated that they were mobilizing their forces to get out more votes.

November 1, 1994

SOLUTION   The events husband votes and wife votes are not mutually exclusive events, since both events can occur. Thus, we use Formula 5.2. We have

$$p(\text{husband or wife votes}) = p(\text{husband votes}) + p(\text{wife votes}) - p(\text{both vote})$$

Based on the newspaper article,

$$p(\text{husband votes}) = \frac{1}{11}$$

$$p(\text{wife votes}) = \frac{1}{9}$$

$$p(\text{both vote}) = \frac{1}{28}$$

Therefore,

$$p(\text{husband or wife votes}) = \frac{1}{11} + \frac{1}{9} - \frac{1}{28}$$

$$= \frac{252}{2772} + \frac{308}{2772} - \frac{99}{2772} = \frac{461}{2772}$$

Thus, $p(\text{husband or wife votes}) = \dfrac{461}{2772}$.

E X A M P L E   8
Controlling
Pollution

Environmentalists have accused a large company in the eastern United States of dumping nuclear waste material into a local river. The probability that *either* the fish in the river or the animals that drink from the river will die is $\dfrac{11}{21}$. The probability that only the fish will die is $\dfrac{1}{3}$, and the probability that only the animals that drink from the river will die is $\dfrac{2}{7}$. What is the probability that both the fish *and* the animals that drink from the river will die?

SOLUTION   Since both the fish and animals may die, the events are not mutually exclusive. We then use Formula 5.2. We have

$$p(\text{fish or animals die}) = p(\text{fish die}) + p(\text{animals die}) - p(\text{both fish and animals die})$$

$$\frac{11}{21} = \frac{1}{3} + \frac{2}{7} - p(\text{both die})$$

$$\frac{11}{21} = \frac{7}{21} + \frac{6}{21} - p(\text{both die})$$

$$\frac{11}{21} = \frac{13}{21} - p(\text{both die})$$

$$\frac{11}{21} + p(\text{both die}) = \frac{13}{21}$$

$$p(\text{both die}) = \frac{13}{21} - \frac{11}{21} = \frac{2}{21}$$

Thus, the probability that *both* the fish and animals that drink from the river will die is $\frac{2}{21}$.

COMMENT   Although you may think that Formulas 5.1 and 5.2 are different, this is not the case. Formula 5.1 is just a special case of Formula 5.2. Formula 5.2 can always be used since if the events A and B are mutually exclusive, the probability of them happening together is 0. In this case Formula 5.2 becomes

$$p(A \text{ or } B) = p(A) + p(B) - 0$$

which is exactly Formula 5.1.

COMMENT   Formula 5.2 can be used when we have only two events, A and B. For any three events A, B, and C, the probability of A or B or C is given by the formula

$$p(A \text{ or } B \text{ or } C) = p(A) + p(B) + p(C) - p(A \text{ and } B)$$
$$- p(A \text{ and } C) - p(B \text{ and } C) + p(A \text{ and } B \text{ and } C)$$

The use of this formula will be illustrated in the exercises.

EXERCISES FOR SECTION 5.2

1. Determine which of the following events are mutually exclusive:
   a. being overweight and having high blood pressure.
   b. being a U.S. Senator and being a member of the U.S. House of Representatives at the same time.
   c. winning first prize and winning the second prize in the state lottery.
   d. taking out a life insurance policy with Company A and taking out a life insurance policy with Company B.
   e. the N.Y. Yankees and the N.Y. Mets both winning the same world series.
   f. being pregnant and giving birth to twins and triplets.
   g. heating your home with gas and heating your home with oil.
   h. being a registered Democrat and a registered Republican.
2. The R.R. Bowker Company of New York collects data on annual subscription rates to periodicals. Results are published in *Library Journal*. In a recent independent study, it was found that 63% of all students at Mill University read *Time* magazine, 51% read *U.S. News and World Report*, and 24% read both magazines. If a student at Mill University is randomly selected, what is the probability that the student reads either magazine?
3. According to *The Motorist*, a publication of the American Automobile Association, 90% of all calls for emergency service on a cold winter day in Tottenville are for

cars that won't start because of a dead battery, 10% of the calls are for cars that won't start because they have no gas, and 4% of the calls are for cars that won't start because of a dead battery *and* no gas. If a typical call is randomly selected, what is the probability that it is for a car that won't start because of a dead battery *or* because of no gas?

4. The Gable Construction Corp. has placed bids for installing electronic message boards on two different highways. Because of competition from other companies, Gable Construction Corp. officials estimate that the probability that they will be awarded the Macon Expressway contract is 0.65 and the probability that they will be awarded the Seaview Expressway contract is 0.43. Furthermore, the probability that they will be awarded both contracts is 0.22. What is the probability that they will be awarded the contract for either the Macon Expressway or the Seaview Expressway?

5. The National Center for Health Statistics publishes information on the educational level of married Americans in *Vital Statistics of the United States*. For one community, the probability that a married man is a college graduate is 0.35. The probability that the wife is a college graduate is 0.41, and the probability that both the husband and the wife are college graduates is 0.14. If a married couple from this community is randomly selected, what is the probability that either the man *or* his wife is a college graduate?

6. The Metropolitan Transportation Authority (MTA) is one of the many regional quasi-governmental agencies responsible for mass transit covering the New York, New Jersey, and Connecticut tri-state area. Other regions have similar agencies. Regional transportation officials in one area claim that the probability that a commuter bus will break down is 0.12 and the probability that a commuter train will break down is 0.07. If the probability that both a bus and a train will break down is 0.02, what is the probability that there will be *no* breakdowns on either system?

7. According to the *Journal of the American Veterinary Medical Association*, there were an estimated 57 million pet cats in the United States in 1991, an increase of two million since 1987. On the other hand, the dog population remained constant. A survey found that 74% of the families in Morganville Lake have either a pet dog or pet cat in their home. Furthermore, 50% have a pet dog and 38% have a pet cat in their home. If a family in Morganville is randomly selected, what is the probability that the family has *both* a pet cat and pet dog at home?

8. In *Birthday Gifts*, Brier and Rogers found that 71% of married men will get flowers *or* buy jewelry for their wives on their birthdays. Fifty-eight percent of married men will buy jewelry and 39% of married men will buy flowers for their wives. What is the probability that a married man will buy flowers *and* jewelry for his wife's birthday?

9. University officials estimate that 96% of the degrees to be awarded at this year's graduation ceremonies will be for the bachelor's or master's degree. (The remainder will be A.A. degrees.) No student is allowed to receive two degrees at the same time. If 70% of the graduates will receive the bachelor's degree, what is the probability that a randomly selected student will be receiving a master's degree?

10. Various environmental studies indicate that the probability that the Croton River contains toxic chemical wastes at any time is 0.62 and the probability that it contains coliform bacteria at any time is 0.86. If the probability that it contains either coliform bacteria or toxic chemical wastes is 0.92, what is the probability that it contains both pollutants?

WOMEN PUTTING OFF HAVING THEIR FIRST CHILD

New York (Sept. 25): Insurance company officials disclosed the results of an exhaustive 5-year study. Due to career considerations, many women are postponing having their first child much later than ever, until they are in their 30s. These results have far-reaching implications for our society.

Saturday—September 25, 1993

11. Consider the accompanying newspaper article. Based upon thousands of claims, the following facts are available:

Fact	Probability
Mother over 35 years of age	0.27
First child for mother	0.21
Mother had a well-paying job	0.42
First child for mother and over 35 years of age	0.17
First child for mother and mother had a well-paying career job	0.09
Mother over 35 years of age and had a well-paying career job	0.16
First child for mother and mother over 35 years of age who had a well-paying career job	0.07

If a mother from this group is randomly selected, what is the probability that she is either over 35 years of age or had a well-paying career job or that the child is a first child for the mother?

12. Marilyn Schaefer is a manager of the local McDonald's. Over the past few years, she has determined the following probabilities on the items that a customer will order.

Item(s)	Probability
Cheeseburger	0.70
French fries	0.42
Drink	0.44
Cheeseburger and drink	0.26
French fries and drink	0.14
Cheeseburger and french fries	0.25
Cheeseburger, french fries, and drink	0.09

What is the probability that a randomly selected customer will order either a cheeseburger, french fries, or a drink?

13. To maintain their physical fitness, Americans are exercising more than ever in a variety of ways. Bradley and Robbins interviewed many Americans to determine how they were exercising. In *How Americans Exercise*, they present the following: 53% jog, 44% swim, 46% cycle, 18% jog and swim, 15% swim and cycle, 17% jog and cycle, and 7% jog, swim, and cycle. If an American who exercises regularly is randomly selected, what is the probability that the person either jogs, swims or cycles to maintain physical fitness?

14. A flight attendant for one of the major airline companies has been studying the reading habits of business-class passengers who fly with the company. The accompanying table gives the probability that a business-class passenger will ask for some of the various magazines available.

Magazine	Probability
Cosmopolitan	0.42
Business Week	0.49
U.S. News and World Report	0.46
Cosmopolitan and *Business Week*	0.16
Cosmopolitan and *U.S. News and World Report*	0.11
Business Week and *U.S. News and World Report*	0.14
Any of these three magazines	0.98
None of these three magazines	0.02

Assuming that a passenger will definitely ask for one of these magazines, find the probability that a passenger will ask for all three of these magazines.

5.3   CONDITIONAL PROBABILITY

Although the addition rule given in Section 5.2 applies to many different situations, there are still other problems that cannot be solved by that formula. It is for this reason that we introduce conditional probability. Let us first consider the following problem.

E X A M P L E   1

Antismoking Campaign

In an effort to reduce the amount of smoking, the administration of Podunk University is considering the establishment of a smoking clinic to help students "break the habit." However, not all the students at the school favor the proposal. As a result, a survey of the 1000 students at the school was conducted to determine student opinion about the proposal. The table below summarizes the results of the survey.

	Against Smoking Clinic	For Smoking Clinic	No Opinion	Total
Freshmen	23	122	18	163
Sophomores	39	165	27	231
Juniors	58	238	46	342
Seniors	71	127	66	264
Total	191	652	157	1000

a. What is the probability that a student selected at random voted against the establishment of the smoking clinic?

b. If a student is a freshman, what is the probability that the student voted for the smoking clinic?

c. If a senior is selected at random, what is the probability that the senior has no opinion about the clinic?

SOLUTION

a. Since there was a total of 191 students who voted against the establishment of the smoking clinic out of a possible 1000 students, we apply the definition of probability and get

$$p(\text{student voted against the smoking clinic}) = \frac{191}{1000}$$

Thus, the probability that a student selected at random voted against the establishment of the smoking clinic is $\frac{191}{1000}$.

b. There are 163 freshmen in the school. One hundred twenty-two of them voted for the smoking clinic. Since we are concerned with freshmen only, the number of possible outcomes of interest to us is 163, not 1000. Out of these, 122 are favorable. Thus, the probability that a student voted for the smoking clinic given that the student is a freshman is $\frac{122}{163}$.

c. In this case the information given narrows the sample space to the 264 seniors, 66 of which had no opinion. Thus, the probability that a student has no opinion given that the student is a senior is $\frac{66}{264}$, or $\frac{1}{4}$.

Conditional probability The situation of part (b) or that of part (c) in Example 1 is called a **conditional probability** because we are interested in the probability of a student voting in favor of the establishment of the smoking clinic given that, or conditional upon the fact that, the student is a freshman. We express this condition mathematically by using a vertical line "$|$" to stand for the words "given that" or "if we know that." We then write

$$p(\text{student voted in favor of smoking clinic} \mid \text{student in a freshman}) = \frac{122}{163}$$

Similarly, for part (c) we write

$$p(\text{student had no opinion} \mid \text{student is a senior}) = \frac{66}{264} = \frac{1}{4}$$

EXAMPLE 2 Sherman is repairing his car. He has removed the six spark plugs. Four are good and two are defective. He now selects one plug and then, without replacing it, selects a second plug. What is the probability that both spark plugs selected are good?

SOLUTION   We list the possible outcomes and then count all the favorable ones. To do this, we label the good spark plugs as g_1, g_2, g_3, and g_4, and the defective ones as d_1 and d_2. The possible outcomes are

g_1, g_2	g_2, g_1	g_3, g_1	g_4, g_1	d_1, g_1	d_2, g_1
g_1, g_3	g_2, g_3	g_3, g_2	g_4, g_2	d_1, g_2	d_2, g_2
g_1, g_4	g_2, g_4	g_3, g_4	g_4, g_3	d_1, g_3	d_2, g_3
g_1, d_1	g_2, d_1	g_3, d_1	g_4, d_1	d_1, g_4	d_2, g_4
g_1, d_2	g_2, d_2	g_3, d_2	g_4, d_2	d_1, d_2	d_2, d_1

There are 30 possible outcomes. Twelve of these are favorable. These are the circled ones, which represent the outcome that both spark plugs are good. Thus,

$$p(\text{both spark plugs selected are good}) = \frac{12}{30} = \frac{2}{5}$$

EXAMPLE 3 In Example 2, what is the probability that both spark plugs selected are good if we know that the first plug selected is good?

SOLUTION   Again, we list all possible outcomes and count the number of favorable ones.

(g_1, g_2)	(g_2, g_1)	(g_3, g_1)	(g_4, g_1)
(g_1, g_3)	(g_2, g_3)	(g_3, g_2)	(g_4, g_2)
(g_1, g_4)	(g_2, g_4)	(g_3, g_4)	(g_4, g_3)
g_1, d_1	g_2, d_1	g_3, d_1	g_4, d_1
g_1, d_2	g_2, d_2	g_3, d_2	g_4, d_2

Since we know that the first plug selected is good, there are only 20 possible outcomes. Of these, 12 are favorable. These are the circled ones. Thus, the probability that both spark plugs are good if we know that the first plug is good is

$$\frac{12}{20} = \frac{3}{5}$$

Using the conditional probability notation, we can write this result as

$$p(\text{both spark plugs are good} \mid \text{first spark plug is good}) = \frac{3}{5}$$

COMMENT Example 3 differs from Example 2 since in Example 3 we are interested in determining the probability of getting two good spark plugs once we know that the first one selected is good. On the other hand, in Example 2 we were interested in determining the probability of getting two good plugs without knowing whether or not the first plug is defective.

Let us analyze the problem discussed at the beginning of this section in detail. There are a total of 163 freshmen out of a possible 1000 students in the school. Thus,

$$p(\text{freshman}) = \frac{163}{1000}$$

Also, there were 122 freshmen who voted in favor of the clinic. Thus,

$$p(\text{freshman and voted in favor of clinic}) = \frac{122}{1000}$$

Summarizing these results we have

$$p(\text{freshman}) = \frac{163}{1000} \quad \text{and}$$

$$p(\text{freshman and voted in favor of clinic}) = \frac{122}{1000}$$

Let us now divide p(freshman *and* voted in favor of clinic) by p(freshman). We get

$$\frac{p(\text{freshman } and \text{ voted in favor of clinic})}{p(\text{freshman})} = \frac{122/1000}{163/1000}$$

$$= \frac{122}{1000} \div \frac{163}{1000}$$

$$= \frac{122}{\cancel{1000}} \cdot \frac{\cancel{1000}}{163}$$

$$= \frac{122}{163}$$

This is the same result as p(student voted in favor of smoking clinic | student is a freshman). In both cases the answer is $\frac{122}{163}$.

If we let A stand for "student voted in favor of smoking clinic" and B stand for "student is a freshman," then the previous result suggests that

$$p(A \mid B) = \frac{p(A \text{ and } B)}{p(B)}$$

We can apply the same analysis for part (c) of the problem. We have

$$p(\text{senior}) = \frac{264}{1000} \quad \text{and} \quad p(\text{senior and no opinion}) = \frac{66}{1000}$$

If we divide p(senior and no opinion) by p(senior), we get

$$\frac{p(\text{senior and no opinion})}{p(\text{senior})} = \frac{66/1000}{264/1000}$$

$$= \frac{66}{1000} \div \frac{264}{1000}$$

$$= \frac{66}{\cancel{1000}} \cdot \frac{\cancel{1000}}{264} = \frac{66}{264} = \frac{1}{4}$$

Thus,

$$p(\text{student had no opinion} \mid \text{student is a senior}) = \frac{p(\text{senior and no opinion})}{p(\text{senior})}$$

Conditional probability formula

We can generalize our discussion by using a formula that is called the **conditional probability formula**.

FORMULA 5.3

Conditional Probability Formula

If A and B are any events, then

$$p(A \mid B) = \frac{p(A \text{ and } B)}{p(B)}, \qquad \text{provided } p(B) \neq 0$$

We illustrate the use of Formula 5.3 with several examples.

EXAMPLE 4 In Ashville the probability that a married man drives is 0.90. If the probability that a married man *and* his wife both drive is 0.85, what is the probability that his wife drives given that he drives?

SOLUTION   We use Formula 5.3. We are told that

$$p(\text{husband drives}) = 0.90$$

and

$$p(\text{husband and wife drive}) = 0.85$$

Thus,

$$p(\text{wife drives} \mid \text{husband drives}) = \frac{p(\text{husband and wife drive})}{p(\text{husband drives})}$$

$$= \frac{0.85}{0.90}$$

$$= \frac{85}{90} \qquad \text{(We multiply numerator and denominator by 100)}$$

$$= \frac{17}{18}$$

Thus, $p(\text{wife drives} \mid \text{husband drives}) = \dfrac{17}{18}$

EXAMPLE 5 Joe often speeds while driving to school in order to arrive on time. The probability that he will speed to school is 0.75. If the probability that he speeds and gets stopped by a police officer is 0.25, find the probability that he is stopped, given that he is speeding.

SOLUTION   We use Formula 5.3. We are told that $p(\text{Joe speeds})$ is 0.75 and $p(\text{speeds and is stopped})$ is 0.25. Thus,

$$p(\text{he is stopped} \mid \text{he speeds}) = \frac{p(\text{speeds and is stopped})}{p(\text{speeds})}$$

$$= \frac{0.25}{0.75}$$

$$= \frac{25}{75}$$

$$= \frac{1}{3}$$

Thus, $p(\text{he is stopped} \mid \text{he speeds}) = \dfrac{1}{3}$.

E X A M P L E   6   Janet likes to study. The probability that she studies *and* passes her math test is 0.80. If the probability that she studies is 0.83, what is the probability that she passes the math test, given that she has studied?

SOLUTION   We use Formula 5.3. We have

$$p(\text{passes math test} \mid \text{she studied}) = \frac{p(\text{studies and passes math test})}{p(\text{she studies})}$$

$$= \frac{0.80}{0.83}$$

$$= \frac{80}{83}$$

Thus, $p(\text{she passes math test} \mid \text{she has studied}) = \dfrac{80}{83}$.

EXERCISES FOR SECTION 5.3

1. The U.S. Bureau of Prisons compiles data on the prisoner population as well as the time served by prisoners released from federal institutions for the first time. The data are published in the document *Statistical Report*. For one prison, the following information is available for five different offense categories. The data represents the number of prisoners in jail in 1994 for the specified crime.

Crime Committed

	Fraud	Forgery	Drug Laws	Counterfeiting	Robbery
Male	26	61	89	59	46
Female	22	49	12	37	32

Find the probability that a randomly selected prisoner from this group
a. is a male.
b. is in jail because of counterfeiting given that he is a male.
c. is a male given that the prisoner is in jail because of counterfeiting.

2. Emergency triage centers at most hospitals have the facilities and personnel to deal with a variety of medical emergencies. During the last six months of 1994, the Emergency Room at Lincoln Hospital provided medical attention to people in various age-groups. Some of the conditions treated are shown below.

Age of Patients	Cardiac Problems	Breathing Difficulties (Asthma)	Kidney Failure	Drug Overdose
20–35	2	13	3	39
36–50	18	16	8	12
51–70	47	11	14	5
Over 70	32	10	37	1

(*Source:* Lincoln Hospital Patient Analysis)

Find the probability that a randomly selected individual in the group
a. was between 36–50 years of age.
b. was between 36–50 years of age given that the patient received treatment for drug overdose.
c. received treatment for drug overdose given that the patient was between 36–50 years of age.

3. State banking officials claim that the probability that a person in Newberg has a checking account is 0.86 and the probability that the person has a checking account as well as overdraft privileges is 0.35. (Overdraft privileges allow customers to write checks for amounts that exceed their current balance.) If a resident in Newberg who has a checking account is randomly selected, what is the probability that the customer also has overdraft privileges?

4. Red Cross officials estimate that for the white American population, 37% have type A blood, 13% have type B blood, 6% have type AB blood, and 44% have type O blood. Moreover, only 8% of the white population of Middletown have type A blood and participate in the annual blood drive. If a white individual of Middletown who has type A blood is randomly selected, find the probability that the individual will participate in the blood drive. (*Source: American Red Cross Association, 1995*)

5. Seventy-seven percent of the medium-sized companies doing business in Brooklyn provide their CEO's with desktop computers. Moreover, 23% of these companies also provide their CEO's with laptop computers. If a company that provides its CEO with a desktop computer is randomly selected, what is the probability that it also provides the CEO with a laptop computer?

6. According to a survey conducted by a local affiliate of the American Automobile Association, attendants at 42% of the stations surveyed will check your oil when purchasing gas. Additionally, 24% of the attendants will *also* wash your windows. If

an attendant is observed checking the oil in Deidre's car when she buys gas, what is the probability that the attendant will also wash the windows?

7. Consider the accompanying newspaper article. The probability of a motorist having a mechanical breakdown on the city's highways is 0.03. What is the probability that a motorist's car will break down and that the motorist will *not* be able to summon help because of a call box that is not functioning properly?

> ## MAJORITY OF POLICE CALL BOXES NOT FUNCTIONING
>
> *New York*: A random survey by reporters for the local auto club found that 57% of the emergency call boxes on the city's highways were not functioning properly because of vandalism or were missing telephones completely. The phones in these strategically placed call boxes allow a motorist to summon help in the event of a mechanical breakdown.
>
> May 17, 1994

8. Twenty-four percent of the employees of a large publishing company are female sales personnel. Forty-seven percent of the workers are sales personnel. If a salesperson comes to a particular school promoting a book, what is the probability that this person is female?

9. Many colleges require all students to pass an English competency exam before being allowed to enroll in any credit-bearing college English course. Based upon past experience, it has been determined that the probability is 0.75 that a student will pass the competency exam (on the first attempt). If a student passes the competency exam, then the probability is 0.88 that he or she will pass any credit-bearing English course. Find the probability that a randomly selected student will pass *both* the competency exam and any credit-bearing English course.

10. Seventy-five percent of all the cars insured by the Metro Insurance company for theft have both a hood lock *and* a passive-type ignition shutoff alarm system to discourage car theft. Furthermore, 88% of the insured cars have a passive-type ignition shutoff alarm system. If a car insured by this company that is known to have a passive-type ignition shutoff alarm system is randomly selected, what is the probability that it also has a hood lock?

11. A newspaper reporter conducted a nationwide survey of 1638 people to determine what they thought about the proposed requirement that all handguns be registered with the government similar to car registrations. The results of the survey are shown in the accompanying table.

Region of Country in Which Respondent Lives	Against Proposal	In Favor of Proposal
East	164	204
Midwest	110	358
South	128	276
Far west	146	252

Find the probability that a randomly selected individual in the group
a. lives in the East given that he or she is against the proposal.
b. is against the proposal given that he or she lives in the East.
c. is against the proposal.

12. Fifty-four percent of the Bayview Health Club members exercise regularly (at least three times a week) *and* watch their diets. Seventy-eight percent of the club members watch their diet. If a club member who is known to watch his or her diet is randomly selected, find the probability that the member exercises regularly.

5.4   MULTIPLICATION RULES

In Section 5.3 we discussed the conditional probability formula and how it is used. In this section we discuss a variation of the conditional probability formula known as the multiplication rule.

Consider a large electric company in the northeastern United States. In recent years it has been unable to meet the demand for electricity. To prevent any cable damage and blackouts as a result of overload, that is, too much electrical demand, it has installed two special switching devices to shut off the flow of electricity automatically and thus prevent cable damage when an overload occurs. The probability that the first switch will not work properly, is 0.4, and the probability that the second switch will not work properly given that the first switch fails is 0.3. What is the probability that both switches will fail?

Let us look at Formula 5.3 in Section 5.3. It says that for any events A and B

$$p(A \mid B) = \frac{p(A \text{ and } B)}{p(B)}$$

If we multiply both sides of this equation by $p(B)$, we get

$$p(A \mid B) \cdot p(B) = p(A \text{ and } B)$$

Multiplication rule   This equation is called the **multiplication rule**. We state this formally as follows:

FORMULA 5.4

Multiplication Rule

> If A and B are any events, then
> $$p(A \text{ and } B) = p(A \mid B) \cdot p(B)$$

If we now apply Formula 5.4 to our example, we get

$p(\text{both switches fail})$
$\quad = p(\text{switch 2 fails} \mid \text{switch 1 has failed}) \cdot p(\text{switch 1 fails})$
$\quad = (0.3)(0.4)$
$\quad = 0.12$

Thus, the probability that both switches fail is 0.12.

EXAMPLE 1   In a certain community the probability that a man over 40 years old is overweight is 0.42. The probability that his blood pressure is high given that he is overweight is 0.67. If a man over 40 years of age is selected at random, what is the probability that he is overweight and that he has high blood pressure?

SOLUTION   We use Formula 5.4. We have

$p(\text{overweight and high blood pressure})$
$\quad = p(\text{high blood pressure} \mid \text{overweight}) \cdot p(\text{overweight})$
$\quad = (0.67)(0.42)$
$\quad = 0.2814$

Thus, the probability that a man over 40 is overweight and has high blood pressure is approximately 0.28.

EXAMPLE 2
TV Commercials

A new cleansing product has recently been introduced and is being advertised on television as having remarkable cleansing qualities. The manufacturer believes that if a homemaker is selected at random, the probability that the homemaker watches television and sees the commercial between the hours of 12 noon and 4 P.M. is $\frac{4}{11}$. Furthermore, if the homemaker sees the commercial, then the probability that the homemaker buys the cleanser is $\frac{22}{36}$. What is the probability that a homemaker selected at random will watch television *and* buy the product?

SOLUTION   We use Formula 5.4. We have

$$p(\text{watches TV and buys product})$$
$$= p(\text{buys product} \mid \text{watches TV}) \cdot p(\text{watches TV})$$
$$= \frac{22}{36} \cdot \frac{4}{11}$$
$$= \frac{88}{396} = \frac{2}{9}$$

Thus, the probability that a homemaker selected at random watches television and buys the cleanser is $\frac{2}{9}$.

Independent Events

In many cases it turns out that whether or not one event happens does not affect whether another will happen. For example, if two cards are drawn from a deck and the first card is replaced before the second card is drawn, the outcome on the first draw has nothing to do with the outcome on the second draw. Also, if two dice are rolled, the outcome for one die has nothing to do with the outcome for the second die. Such events are *Independent events*   called **independent events**.

Definition 5.2 | Two events A and B are said to be **independent** if the likelihood of the occurrence of event B is in no way affected by the occurrence or non-occurrence of event A.

When dealing with independent events, we can simplify Formula 5.4. The following example shows this.

E X A M P L E   3   Two cards are drawn from a deck of 52 cards. Find the probability that both cards drawn are aces if the first card

    a. is *not* replaced before the second card is drawn.
    b. is replaced before the second card is drawn.

SOLUTION

    a. Since the first card is not replaced, we use the multiplication rule. We have

$$p(\text{both cards are aces}) =$$
$$p(\text{2nd card is ace} \mid \text{1st card is ace}) \cdot p(\text{1st card is ace})$$

Notice that since the first card is not replaced, there are only three aces remaining out of a possible 51 cards. This is because the first card removed was an ace. Thus,

$$p(\text{both cards are aces}) = \frac{3}{51} \cdot \frac{4}{52}$$

$$= \frac{12}{2652} = \frac{1}{221}$$

Thus, the probability that both cards are aces is $\frac{1}{221}$.

b. Since the first card is replaced before the second card is drawn, then whether or not an ace appeared on the first card in no way affects what happens on the second draw. The events "ace on second draw" and "ace on first draw" are independent. Thus, $p(\text{2nd card is ace} \mid \text{1st card is ace})$ is exactly the same as $p(\text{2nd card is ace})$. Therefore,

$$p(\text{both cards are aces}) = p(\text{2nd card is ace} \mid \text{1st card is ace})$$
$$\cdot \, p(\text{1st card is ace})$$
$$= p(\text{2nd card is ace}) \cdot p(\text{1st card is ace})$$
$$= \frac{4}{52} \cdot \frac{4}{52}$$
$$= \frac{16}{2704} = \frac{1}{169}$$

Hence, the probability that both cards are aces in this case is $\frac{1}{169}$.

Example 3 suggests that if two events A and B are independent, we can substitute $p(B)$ for $p(B \mid A)$ since B is in no way affected by what happens with A. We then get a special multiplication rule for independent events.

FORMULA 5.5

If A and B are independent events, then

$$p(A \text{ and } B) = p(A) \cdot p(B)$$

EXAMPLE 4   Two randomly selected travelers, Carlos and Pedro, who do not know each other, are at the information desk at Kennedy International Airport. The probability that Carlos speaks Spanish is 0.86, and the probability that Pedro speaks Spanish is 0.73. What is the probability that they both speak Spanish?

SOLUTION   Since both travelers do not know each other, the events "Carlos speaks Spanish" and "Pedro speaks Spanish" are independent. We therefore use Formula 5.5. We have

$$p(\text{both speak Spanish}) = p(\text{Carlos speaks Spanish}) \cdot p(\text{Pedro speaks Spanish})$$
$$= (0.86)(0.73)$$
$$= 0.6278$$

Thus, the probability that they both speak Spanish is approximately 0.63.

EXAMPLE 5   If the probability of a skin diver in a certain community having untreated diabetes is 0.15, what is the probability that two totally unrelated skin divers from the community do *not* have untreated diabetes? (Assume independence.)

SOLUTION   These are independent events, so we use Formula 5.5. The probability of a skin diver having untreated diabetes is 0.15. Thus, the probability that the diver does not have untreated diabetes is $1 - 0.15$, or 0.85. Therefore,

$$p(\text{both divers do not have diabetes}) = p(\text{diver 1 does not have diabetes})$$
$$\cdot p(\text{diver 2 does not have diabetes})$$
$$= (0.85)(0.85)$$
$$= 0.7225$$

Hence, the probability that neither of two totally unrelated skin divers have untreated diabetes is approximately 0.72.

COMMENT   The multiplication rule for independent events can be generalized for more than two independent events. We simply multiply all the respective probabilities. Thus, if event A has probability 0.7 of occurring, event B has probability 0.6 of occurring, and event C has probability 0.5 of occurring, and if these events are independent, then the probability that all three occur is

$$p(A \text{ and } B \text{ and } C) = p(A) \cdot p(B) \cdot p(C)$$
$$= (0.7)(0.6)(0.5)$$
$$= 0.21$$

Therefore, the probability that all three occur is 0.21.

EXERCISES FOR SECTION 5.4

1. Due to the rising costs of auto insurance, the probability that a randomly selected driver in one particular city drives an uninsured motor vehicle is 0.13. Moreover, the probability that the car's driver is under 30 years of age, given that the car is uninsured, is 0.39. If a driver is randomly selected, find the probability that the driver is under 30 years of age *and* that the car driven is uninsured.

2. Medical records indicate that the probability that a man over 40 years of age in a mining community has black lung disease is 0.35. The probability that the individual has high blood pressure given that the individual has black lung disease is 0.21. If an individual in this mining community over 40 years of age is randomly selected, find the probability that the individual has high blood pressure *and* black lung disease.

3. The Board of Trustees of Merck College is considering increasing tuition for the coming school year by 9%. The probability that they vote to raise tuition is 0.83. Past experience has shown that whenever tuition is increased by 9%, student enrollment drops by approximately 5% with a probability of 0.95. Find the probability that the board votes to raise tuition by 9% *and* that the enrollment drops by 5%.

4. A recent survey in Alexville showed that 81% of the population have some form of health insurance. Furthermore, the survey showed that 22% of the people have some form of life insurance when they have health insurance. If an adult resident of Alexville is randomly selected, find the probability that the individual has both health *and* life insurance.

5. Morgan Electronics assembles computers by using disk drives manufactured by several different companies. Seventy-one percent of the disk drives used are manufactured by the Stanton Corp. The probability that the disk drive will function properly given that it was manufactured by the Stanton Corp. is 0.93. If a computer that was assembled by Morgan Electronics is randomly selected, find the probability that it contains a disk manufactured by the Stanton Corp. *and* that it will function properly.

6. Marilyn McIntyre is driving on the highway with her children when she notices two service stations up ahead on the road. The probability that the first service station has a functioning public rest room is 0.62. The probability that the second gas station has a functioning public rest room is 0.49. What is the probability that neither has a functioning public rest room? (Assume independence.)

7. As the tax laws are becoming more complicated, many people are turning to professional tax advisors to help them complete their tax returns. According to the Internal Revenue Service, the probability that a randomly selected individual in Woodridge has his or her annual federal income tax form prepared by a professional tax consultant is 0.61. Find the probability that three totally unrelated people in Woodridge have their annual federal income tax forms prepared by a professional tax consultant.

8. With the high cost of car insurance and depending upon the state involved, many automobile owners frequently do not have collision and comprehensive insurance for their car. (Some companies will not issue collision insurance without comprehensive insurance.) According to *Insurance Facts*, for one state the probability is 0.62 that an automobile owner has collision insurance given that the owner has comprehensive insurance. Furthermore, the probability is 0.78 that a car owner has comprehensive insurance. What is the probability that an automobile owner in this state has both collision *and* comprehensive car insurance?

9. At Los Gables Airport, two kinds of security checks are used to prevent any passenger from taking an explosive device and/or a gun on to an airplane. One is a visual check by a security guard coupled with an explosive sniff test by a specially

trained dog and the second is a screening by a metal detector. The probability that the guard or dog stops a person carrying a gun or bomb is 0.74, and the probability that the person is caught by the metal detector given that he or she was not stopped by the guard is 0.99. What is the probability that a person will *not* be stopped by the guard nor caught by the metal detector? (See the newspaper article on page 279.)

10. Consider the following newspaper article. The probability that the printer's demands will be met is 0.19. Also, the probability that the price of a newspaper will go up given that the printer's demands are met is 0.89. Find the probability that the printer's demands are met *and* that the price of a newspaper goes up.

NEWSPAPER WORKERS STILL ON STRIKE

New York (Jan. 7): The strike by workers of the *New York Daily News* enters its fourth month today with no end in sight. The workers are demanding more job security and increased contributions by management to the pension fund. Management claims that if all of the worker's demands are met, then the cost of a newspaper will rise to over a dollar or the company will be forced to go out of business.

No new talks were scheduled for this week.

January 7, 1991

11. According to the *National Safety Council*, many drivers frequently do not use seat belts when they drive. In recent years, however, seat-belt use has been gradually increasing. It is estimated that the probability that a driver uses a seat belt is 0.68. Find the probability that three randomly selected drivers passing a particular check point will be using their seat belts.

12. The Casco Corp. uses many delivery services. The probability that any given parcel will be sent with the ABC Speedy Delivery Service is 0.72. The probability that the parcel will arrive on time given that the ABC Speedy Delivery service was used is 0.94. If a parcel is randomly selected, find the probability that it will be sent with the ABC Speedy Delivery Service *and* that it will arrive on time.

13. As a cost-cutting measure, many companies hire part-time employees. Thus, they avoid having to pay for costly medical insurance and pension benefits. The probability that a male employee of the Calvington Corporation is a part-time worker is 0.46 and the probability that a female employee of the Calvington Corp. is a part-

time worker is 0.53. Find the probability that two randomly selected and totally unrelated employees (one male and one female) are *not* part-time workers.

14. A particular nuclear reactor has two safety valves that are designed to prevent a nuclear accident similar to the one that occurred at the Three Mile Island facility in Pennsylvania. The probability that the first safety valve will function satisfactorily in an emergency is 0.985, and the probability that the second safety valve will function satisfactorily in an emergency if the first fails is 0.99. Find the probability that both valves fail and that a nuclear mishap occurs.

*5.5 BAYES' FORMULA

Let us look in on Dr. Carey, who has two bottles of sample pills on his desk for the treatment of arthritic pain. One day he gives Madeline a few pills from one of the bottles. (All other treatments have failed.) However, he does not remember from which bottle

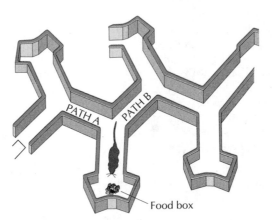

FIGURE 5.2 Consider the situation pictured above. Often we read about psychologists experimenting with rats to determine how quickly they learn maze patterns. Mazes used to study human learning are similar, in principle, to those used with animals. We notice that the rat in the picture has already arrived at the food box. What is the probability that it came there from Path A and not from Path B? By using Bayes' formula, we will be able to answer this and similar questions.

*An asterisk indicates that the section requires more time and thought than other sections.

he took the pills. The pills in bottle B_1 are effective 70% of the time, with no known side effects. The pills in bottle B_2 are effective 90% of the time, with some possible side effects. Bottle B_1 is closer to Dr. Carey on his desk and the probability is $\frac{2}{3}$ that he selected the pills from this bottle. On the other hand, bottle B_2 is farther away from Dr. Carey and the probability is therefore $\frac{1}{3}$ that he selected the pills from this bottle. The problem is to determine the bottle from which the pills were taken.

In many problems we are given situations such as this one, where we know the outcome of the experiment and are interested in concluding that the outcome happened because of a particular event. Figure 5.2 is an example of this. For these situations we need a formula.

EXAMPLE   1

Medicine

In the situation we are discussing, find the probability that the pills are effective in relieving Madeline's pain.

SOLUTION   Madeline can be relieved of her pain by taking the pills from either bottle B_1 or bottle B_2. Let A represent the event *Madeline's pain is relieved*, let B_1 represent the event that *the pills were taken from bottle B_1*, and let B_2 represent the event that *the pills were taken from bottle B_2*. The tree diagram in Figure 5.3 illustrates how Madeline's pain can be relieved. From the given information

$$\text{Prob}(A \mid B_1) = 0.70 \qquad \text{and} \qquad \text{Prob}(B_1) = \frac{2}{3}$$

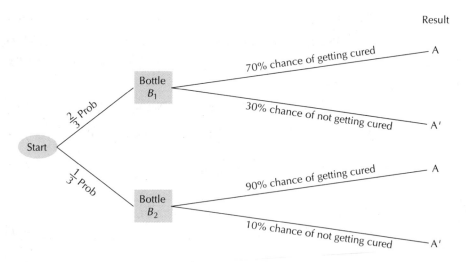

FIGURE 5.3

Applying the multiplication rule (Formula 5.4) gives

$$\text{Prob(selects } B_1 \text{ } and \text{ pill from } B_1 \text{ is effective)}$$
$$= \text{Prob(Madeline is cured } | \text{ pill from } B_1 \text{ was selected)}$$
$$\cdot \text{ Prob(selecting } B_1)$$
$$= (0.70)\frac{2}{3} = \frac{7}{10} \cdot \frac{2}{3}$$
$$= \frac{14}{30} = \frac{7}{15}$$

Symbolically,

$$\text{Prob}(A \text{ and } B_1) = \text{Prob}(A \mid B_1) \cdot \text{Prob}(B_1) = \frac{7}{15}$$

Also, from the given information

$$\text{Prob}(A \mid B_2) = \frac{9}{10} \quad \text{and} \quad \text{Prob}(B_2) = \frac{1}{3}$$

so that

$$\text{Prob}(A \text{ and } B_2) = \text{Prob}(A \mid B_2) \cdot \text{Prob}(B_2)$$
$$= \frac{9}{10} \cdot \frac{1}{3} = \frac{3}{10}$$

Madeline can be cured in one of two mutually exclusive ways:

1. Bottle B_1 is selected and a pill from it is effective.
2. Bottle B_2 is selected and a pill from it is effective.

Thus, using the addition rule for mutually exclusive events (Formula 5.1), we have

$$\text{Prob(pills are effective)} = \text{Prob(selects } B_1 \text{ and pills are effective)}$$
$$+ \text{Prob(selects } B_2 \text{ and pills are effective)}$$

Symbolically,

$$\text{Prob(pills are effective)} = \text{Prob}(A \text{ and } B_1) + \text{Prob}(A \text{ and } B_2)$$
$$\text{Prob}(A) = \text{Prob}(A \mid B_1) \cdot \text{Prob}(B_1) + \text{Prob}(A \mid B_2) \cdot \text{Prob}(B_2)$$
$$= \frac{7}{15} + \frac{3}{10}$$
$$= \frac{14}{30} + \frac{9}{30} = \frac{23}{30}$$

Therefore, the probability that the pills are effective in relieving Madeline of her pain is $\frac{23}{30}$.

EXAMPLE 2   Two weeks later Madeline returns to Dr. Carey and reports that the pills were extremely effective. Dr. Carey would now like to recommend the same medicine for his other patients who suffer from the same pain. What is the probability that the pills came from

    a. Bottle B_1?
    b. Bottle B_2?

SOLUTION

    a. Using the conditional probability formula (Formula 5.3), we have

$$\text{Prob(pills came from } B_1 \mid \text{pills are effective)}$$
$$= \frac{\text{Prob(selects } B_1 \text{ and pills are effective)}}{\text{Prob(pills are effective)}}$$

Symbolically,

$$\text{Prob}(B_1 \mid A) = \frac{\text{Prob}(A \text{ and } B_1)}{\text{Prob}(A)}$$

Substituting the values obtained in Example 1 gives

$$\text{Prob}(B_1 \mid A) = \frac{7/15}{23/30} = \frac{7}{15} \div \frac{23}{30}$$
$$= \frac{7}{15} \cdot \frac{30}{23} = \frac{14}{23}$$

Thus, the probability that the pills came from bottle B_1 is $\frac{14}{23}$.

    b. To find the probability that the pills came from bottle B_2, we use a procedure similar to the one used in part (a). We have

$$\text{Prob(pills came from } B_2 \mid \text{pills are effective)}$$
$$= \frac{\text{Prob(selects } B_2 \text{ and pills are effective)}}{\text{Prob(pills are effective)}}$$

In symbols,

$$\text{Prob}(B_2 \mid A) = \frac{\text{Prob}(A \text{ and } B_2)}{\text{Prob}(A)}$$
$$= \frac{3/10}{23/30} = \frac{3}{10} \div \frac{23}{30}$$
$$= \frac{3}{10} \cdot \frac{30}{23} = \frac{9}{23}$$

Thus, the probability that the pills came from bottle B_2 is $\frac{9}{23}$.

We can combine the results of the preceding two examples as follows:

$$\text{Prob}(B_1 \mid A) = \frac{\text{Prob}(A \text{ and } B_1)}{\text{Prob}(A)}$$

$$= \frac{\text{Prob}(A \text{ and } B_1)}{\text{Prob}(A \text{ and } B_1) + \text{Prob}(A \text{ and } B_2)}$$

$$= \frac{\text{Prob}(A \mid B_1) \cdot \text{Prob}(B_1)}{\text{Prob}(A \mid B_1) \cdot \text{Prob}(B_1) + \text{Prob}(A \mid B_2) \cdot \text{Prob}(B_2)}$$

and

$$\text{Prob}(B_2 \mid A) = \frac{\text{Prob}(A \text{ and } B_2)}{\text{Prob}(A)}$$

$$= \frac{\text{Prob}(A \text{ and } B_2)}{\text{Prob}(A \text{ and } B_1) + \text{Prob}(A \text{ and } B_2)}$$

$$= \frac{\text{Prob}(A \mid B_2) \cdot \text{Prob}(B_2)}{\text{Prob}(A \mid B_1) \cdot \text{Prob}(B_1) + \text{Prob}(A \mid B_2) \cdot \text{Prob}(B_2)}$$

When these results are generalized, we have Bayes' rule.

FORMULA 5.6

Bayes' Rule

Consider a sample space that is composed of the mutually exclusive events A_1, A_2, A_3, . . . , A_n. Suppose each event has a nonzero probability of occurring and that one must definitely occur. If B is any event in the sample space, then

$$\text{Prob}(A_1 \mid B) = \frac{\text{Prob}(B \mid A_1) \cdot \text{Prob}(A_1)}{\text{Prob}(B \mid A_1) \cdot \text{Prob}(A_1) + \text{Prob}(B \mid A_2) \cdot \text{Prob}(A_2) + \cdots + \text{Prob}(B \mid A_n) \cdot \text{Prob}(A_n)}$$

$$\text{Prob}(A_2 \mid B) = \frac{\text{Prob}(B \mid A_2) \cdot \text{Prob}(A_2)}{\text{Prob}(B \mid A_1) \cdot \text{Prob}(A_1) + \text{Prob}(B \mid A_2) \cdot \text{Prob}(A_2) + \cdots + \text{Prob}(B \mid A_n) \cdot \text{Prob}(A_n)}$$

$$\vdots \qquad\qquad\qquad \vdots$$

$$\text{Prob}(A_n \mid B) = \frac{\text{Prob}(B \mid A_n) \cdot \text{Prob}(A_n)}{\text{Prob}(B \mid A_1) \cdot \text{Prob}(A_1) + \text{Prob}(B \mid A_2) \cdot \text{Prob}(A_2) + \cdots + \text{Prob}(B \mid A_n) \cdot \text{Prob}(A_n)}$$

Bayes' rule may seem rather complicated but it is easy to use, as the following examples illustrate.

EXAMPLE 3

If $p(A \mid B) = \dfrac{1}{5}$, $p(A \mid C) = \dfrac{2}{7}$, $p(B) = \dfrac{1}{2}$, and $p(C) = \dfrac{1}{2}$, find

a. $p(B \mid A)$.
b. $p(C \mid A)$.

SOLUTION

a. Using Bayes' rule, we have

$$p(B \mid A) = \frac{p(A \mid B) \cdot p(B)}{p(A \mid B) \cdot p(B) + p(A \mid C) \cdot p(C)}$$

$$= \frac{\left(\dfrac{1}{5}\right)\left(\dfrac{1}{2}\right)}{\left(\dfrac{1}{5}\right)\left(\dfrac{1}{2}\right) + \left(\dfrac{2}{7}\right)\left(\dfrac{1}{2}\right)}$$

$$= \frac{\dfrac{1}{10}}{\left(\dfrac{1}{10}\right) + \left(\dfrac{1}{7}\right)}$$

$$= \frac{1/10}{17/70} = \frac{1}{10} \div \frac{17}{70}$$

$$= \frac{1}{10} \cdot \frac{70}{17} = \frac{7}{17}$$

Thus, $p(B \mid A) = \dfrac{7}{17}$

b. Again, we use Bayes' formula. We have

$$p(C \mid A) = \frac{p(A \mid C) \cdot p(C)}{p(A \mid B) \cdot p(B) + p(A \mid C) \cdot p(C)}$$

$$= \frac{\left(\dfrac{2}{7}\right)\left(\dfrac{1}{2}\right)}{\left(\dfrac{1}{5}\right)\left(\dfrac{1}{2}\right) + \left(\dfrac{2}{7}\right)\left(\dfrac{1}{2}\right)} = \frac{\dfrac{1}{7}}{\left(\dfrac{1}{10}\right) + \left(\dfrac{1}{7}\right)} = \frac{\dfrac{1}{7}}{\dfrac{17}{70}}$$

$$= \frac{10}{17}$$

Thus, $p(C \mid A) = \dfrac{10}{17}$

E X A M P L E   4 A prisoner has just escaped from jail. There are three roads leading away from the jail. If the prisoner selects road A to make good her escape, the probability that she succeeds is $\dfrac{1}{4}$. If she selects road B, the probability that she succeeds is $\dfrac{1}{5}$. If she selects road C, the probability that she succeeds is $\dfrac{1}{6}$. Furthermore, the probability that she selects each of these roads is the same. It is $\frac{1}{3}$. If the prisoner succeeds in her escape, what is the probability that she made good her escape by using road B?

SOLUTION We use Bayes' rule. We have

$$p(\text{uses road B} \mid \text{succeeds}) = \frac{p(\text{succeeds} \mid \text{uses road B}) \cdot p(\text{uses road B})}{\begin{array}{c}p(\text{succeeds} \mid \text{uses road A}) \cdot p(\text{uses road A}) + p(\text{succeeds} \mid \text{uses road B}) \\ \cdot p(\text{uses road B}) + p(\text{succeeds} \mid \text{uses road C}) \cdot p(\text{uses road C})\end{array}}$$

$$= \frac{\left(\frac{1}{5}\right)\left(\frac{1}{3}\right)}{\left(\frac{1}{4}\right)\left(\frac{1}{3}\right) + \left(\frac{1}{5}\right)\left(\frac{1}{3}\right) + \left(\frac{1}{6}\right)\left(\frac{1}{3}\right)}$$

$$= \frac{\frac{1}{15}}{\left(\frac{1}{12}\right) + \left(\frac{1}{15}\right) + \left(\frac{1}{18}\right)}$$

$$= \frac{12}{37}$$

Thus, the probability that she made good her escape by using road B is $\frac{12}{37}$.

EXAMPLE 5

Smoke Detectors

A large real estate manager purchased 50,000 smoke detectors to comply with new city ordinances. She purchased 25,000 from company A, 15,000 from company B, and 10,000 from company C. It is known that some of the smoke detectors malfunction and go off spontaneously. It is also known that 4% of the detectors produced by company A are defective, 5% of the detectors produced by company B are defective, and 6% of the detectors produced by company C are defective. A call is received by the management office that one of the detectors is malfunctioning. Find the probability that it was produced by company A.

SOLUTION From the given information,

$$\text{Prob(produced by company A)} = \frac{25,000}{50,000} = 0.5$$

$$\text{Prob(produced by company B)} = \frac{15,000}{50,000} = 0.3$$

$$\text{Prob(produced by company C)} = \frac{10,000}{50,000} = 0.2$$

$$\text{Prob(defective} \mid \text{produced by company A)} = 0.04$$

Prob(defective | produced by company B) = 0.05
Prob(defective | produced by company C) = 0.06

Here we must determine Prob(produced by company A | defective). We apply Bayes' rule:

Prob(produced by A | defective)

$$= \frac{\text{Prob(defective | prod. by A)} \cdot \text{Prob(prod. by A)}}{\text{Prob(def. | prod. by A)} \cdot \text{Prob(prod. by A)} + \text{Prob(def. | prod. by B)} \cdot \text{Prob(prod. by B)} + \text{Prob(def. | prod. by C)} \cdot \text{Prob(prod. by C)}}$$

$$= \frac{(0.04)(0.5)}{(0.04)(0.5) + (0.05)(0.3) + (0.06)(0.2)}$$

$$= \frac{0.02}{0.02 + 0.015 + 0.012} = \frac{0.02}{0.047}$$

$$= \frac{20}{47}$$

Thus, the probability that the defective smoke detector was produced by company A is $\frac{20}{47}$.

EXAMPLE 6   There are four photocopying machines I, II, III, and IV on the third floor of a large office building. The probabilities that the copies produced from each of these machines will be blurred are 0.2, 0.5, 0.3, and 0.1, respectively. Furthermore, because of the location of these machines, management estimates that the probabilities that a worker will use any one of these machines are 0.6, 0.2, 0.1, and 0.1, respectively. The president of the company receives a blurred memo that was photocopied on one of these machines. What is the probability that it was photocopied on machine I?

SOLUTION   We use Bayes' formula. Let A represent the event "copy is blurred," B_1, represent the event "machine I is used," B_2 represent the event "machine II is used," B_3 represent the event "machine III is used," and B_4 represent the event "machine IV is used." Then we are interested in $p(B_1 \mid A)$.

Using Bayes' rule, we have

$$p(B_1 \mid A) = \frac{p(A \mid B_1) \cdot p(B_1)}{p(A \mid B_1) \cdot p(B_1) + p(A \mid B_2) \cdot p(B_2) + p(A \mid B_3) \cdot p(B_3) + p(A \mid B_4) \cdot p(B_4)}$$

$$= \frac{(0.6)(0.2)}{(0.6)(0.2) + (0.2)(0.5) + (0.1)(0.3) + (0.1)(0.1)}$$

$$= \frac{0.12}{0.26} = \frac{12}{26} = \frac{6}{13}$$

Thus the probability that the blurred memo was photocopied on machine I is $\frac{6}{13}$.

HISTORICAL NOTE

Bayes' Theorem is named for the English clergyman Thomas Bayes who did early work in probability and decision theory. Bayes was born in Tunbridge Wells, Kent, England, in 1702 and died on April 17, 1761. He was the first to use probability inductively and to establish a mathematical basis for probability inference. In his Bayesian Estimator Theorem he established a statistical technique for determining probabilities on the basis of an *a priori* estimate of its probability.

Although Bayes' Theorem is in no sense controversial, the question of its appropriate use has, in past years, been a focal point in a controversy between those who favor a strict relative-frequency interpretation of probability and those who would prefer a subjective interpretation as well. Nevertheless, modern statistical theory and decision making clearly embrace Bayes' Theorem.

EXERCISES FOR SECTION 5.5

1. A computer company sells computers that are assembled in one of its three factories. The factories vary in quality due to the experience levels of the workers. Factory 1, which accounts for 60% of all computers assembled, employs workers with little experience and has a 7% defect rate. Factory 2, which accounts for 30% of all computers assembled, employs workers with much experience and has a 3% defect rate. Factory 3, which accounts for 10% of all computers assembled, employs workers with average experience and has a 6% defect rate. If a randomly selected computer assembled by this company is inspected and found to be defective, what is the probability that it was assembled by workers in Factory 3?

2. Refer back to Exercise 1. If the inspected computer is *not* defective, what is the probability that it was assembled by workers in Factory 2?

3. One large Wall Street brokerage firm reports that its work force consists of 40% females and 60% males. If 20% of the females and 40% of the males participate in the company's stock option program, what is the probability that a randomly selected employee participating in the stock option program is a female?

4. Twenty-five percent of the employees of the brokerage firm mentioned in Exercise 3 have MBA degrees, and 75% of these are senior account executives. Of those who do not have MBA degrees, 20% are senior account executives. What is the probability that a senior account executive of this firm has an MBA degree?

5. According to *Current Population Reports* published by the U.S. Bureau of the Census, 40% of the male population of one particular city is 60 or older in age. Furthermore, 60% of the population is female and 42% of the females are 60 years or older. If a man of this city is randomly selected, what is the probability that he is 60 years or older?

6. One nationwide fast-food restaurant chain reports that 45% of its order takers are female. Moreover, 30% of its male order takers and 20% of its female order takers

attend college. If an order taker that attends college is randomly selected, what is the probability that the order taker is a female?

7. The American Cancer Society as well as the medical profession recommends that people have themselves checked annually for any cancerous growths. If a person has cancer, then the probability is 0.99 that it will be detected by a test. Furthermore, the probability that the test results will be positive (meaning that cancer is possible) when no cancer actually exists is 0.10. Government records indicate that 8% of the population in the vicinity of a chemical corporation that produces asbestos have some form of cancer. Donald Williams takes the test and the results are positive. What is the probability that he does *not* have cancer?

8. Focal Electronics reports that cordless telephones were a popular selling item at its stores in 1995. Of all the cordless phones sold, 10% were manufactured by company A, 20% were manufactured by company B, 40% were manufactured by company C, and 30% were manufactured by company D. The proportion of these phones requiring major adjustments within the first six months of use are 0.001 for those from company A, 0.005 for those from company B, 0.003 for those from company C, and 0.002 for those from company D. A cordless telephone that requires major adjustments within the first six months of use is randomly selected. What is the probability that it was manufactured by company C?

9. Refer back to Exercise 8. What is the probability that a randomly selected cordless telephone that requires major adjustments within the first six months was manufactured by company A or B?

10. Many airline companies offer "fly and drive" deals whereby they provide ticket buyers with rental cars. One airline company provides its customers with cars from one of three car rental agencies, with 28% from agency A, 32% from agency B, and 40% from agency C. If 9% of the cars from A, 7% of the cars from B, and 4% of the cars from C need tune-ups, what is the probability that a car needing a tune-up that is delivered to an airline customer came from car rental agency A?

11. The Environmental Protection Agency has found that the quality of air we breathe is often unhealthy because of automobile pollution. As mandated by law, many states require that all automobiles be tested periodically for emission of excessive amounts of pollutants. When tested properly, 99% of the cars that emit excessive amounts of pollutants will fail. However, 12% of the cars that do not emit excessive amounts of pollutants will also fail. In one state it is known that 15% of all cars emit excessive amounts of pollutants. What is the probability that a car that fails the test actually emits excessive amounts of pollutants?

12. *Ecology.* Environmentalists have developed a test for determining when the mercury level in fish is above the permissible levels. If the fish actually contain an excessive amount of mercury, then the test is 99% effective in determining this, and only 1% will escape undetected. On the other hand, if the mercury content is within permissible limits, then the test will correctly indicate this 96% of the time. Only 4% of the time will the test incorrectly indicate that the mercury content is not within permissible limits. The test is to be used on fish from a river into which a chemical company has been dumping its wastes. It is estimated that 30% of the fish in the river contain excessive amounts of mercury. A fish is caught and tested by

this procedure. The test indicates that the mercury level is within permissible limits. Find the probability that the mercury content is actually greater than the permissible level.

5.6   USING COMPUTER PACKAGES

MINITAB can be used to simulate independent trials of an experiment where we have equally or unequally likely outcomes. For example, suppose a doctor received hypodermic needles from four different suppliers whom we shall name 1, 2, 3, or 4. Based upon past experience, the doctor knows that 30% of the needles used were supplied by company 1, 40% were supplied by company 2, 20% were supplied by company 3, and 10% were supplied by company 4. Thus, each needle selected by the doctor can result in four possible outcomes (company 1, 2, 3, or 4) with four unequal probabilities 0.30, 0.40, 0.20, and 0.10. The following is the MINITAB simulation of the doctor selecting 100 needles.

We first read the possible values for the companies 1, 2, 3, or 4 into column C1 and their respective probabilities into column C2. This is accomplished as follows:

```
MTB   > READ C1 C2
DATA  > 1 0.30
DATA  > 2 0.40
DATA  > 3 0.20
DATA  > 4 0:10
DATA  > END


        4 ROWS READ
```

Then we use the MINITAB RANDOM command with the subcommand DISCRETE. We get

```
MTB   > RANDOM 100 VALUES AND PLACE IN C3;
SUBC  > DISCRETE SAMPLE VALUES FROM C1 WITH PROBABILITIES IN C2.
```

To see the results, we can set up a histogram° with single-value classes for the contents of C3. We get

```
MTB   > HISTOGRAM OF C3;

SUBC  > START 1;
```

°Again we are assuming that the DOS Version of MINITAB is being used. If you are using the WINDOWS version of MINITAB, then type GSTD at the MTB prompt to obtain the results given here.

```
SUBC  >  INCREMENT 1.

HISTOGRAM OF C3    N=100

MIDPOINT     COUNT
    1.00       27     * * * * * * * * * * * * * * * * * * * * * * * * * * *
    2.00       41     * * * * * * * * * * * * * * * * * * * * * * * * * * * * * * * * * * * * * * * * *
    3.00       15     * * * * * * * * * * * * * * *
    4.00       17     * * * * * * * * * * * * * * * * *

MTB  >  STOP
```

EXERCISES FOR SECTION 5.6

1. All students at Podunk University must register for any one and only one of the following courses in their freshman year: Math, Physics, Chemistry, or Biology. Previous experience has shown that the probabilities that students will register for these courses are 0.35, 0.25, 0.22, and 0.18 respectively. Use MINITAB to simulate for which of these courses 100 entering freshmen will register.
2. Refer back to the previous exercise. Use MINITAB to simulate for which of these courses 150 entering freshmen will register.
3. The students of Exercise 1 must also register for any one and only one of the following courses: History, Economics, and Sociology. The probabilities that students will register for these courses are 0.45, 0.35, and 0.20. Use MINITAB to simulate for which of these courses the 100 entering freshmen will register.

5.7 SUMMARY

In this chapter we discussed many different rules concerning the calculation of probabilities. Each formula given applies to different situations. Thus, the addition rule allows us to determine the probability of event A or event B or both events happening. We distinguished between mutually exclusive and non-mutually exclusive events and their effect on the addition rule.

We then discussed conditional probability and how the probability of one event is affected by the occurrence or nonoccurrence of a second event. This led us to the multiplication rule. When one event is in no way affected by the occurrence or non-occurrence of a second event, we have independent events and a simplified multiplication rule.

Finally we discussed Bayes' formula, which is used when we know the outcome of some experiment and are interested in determining the probability that it was caused by or is the result of some other event. In each case many applications of all the formulas introduced were given.

Study Guide

The following is a chapter outline in capsule form. You should now be able to demonstrate your knowledge of the ideas mentioned by giving definitions, descriptions, or specific examples. Page references are given in parentheses.

Two events A and B that cannot occur at the same time are said to be **mutually exclusive**. (page 282)

If $p(A)$ is the probability of event A happening, then $p(A')$ is the probability that A will not happen. Thus, $p(A') = 1 - p(A)$. The event A' is called the **complement** of event A. (page 285)

If A and B are any events, then the **addition rule** states

$$p(A \text{ or } B) = p(A) + p(B) - p(A \text{ and } B). \text{ (page 286)}$$

We use the symbol $p(A \mid B)$ to represent the **conditional probability** of A given that B has occurred or conditional upon the fact that A has occurred. (page 294)

The **conditional probability formula** allows us to calculate $p(A \mid B)$ in terms of $p(A \text{ and } B)$ and $p(B)$. (page 297)

If A and B are any events, then the **multiplication rule** states that

$$p(A \text{ and } B) = p(A \mid B) \cdot p(B) \text{ (page 302)}$$

Two events are said to be **independent** if the likelihood of the occurrence of event B is in no way affected by the occurrence or nonoccurrence of event A. (page 303)

If we know the outcome of an experiment and are interested in determining the probability that the outcome happened because of a particular event, then we use **Bayes' formula**. (page 312)

Formulas to Remember

You should be able to identify each symbol in the following formulas, understand the relationships among the symbols expressed in each formula, understand the significance of each formula, and use the formulas in solving problems.

1. Addition rule, for mutually exclusive events:

   $$p(A \text{ or } B) = p(A) + p(B)$$

2. Addition rule, general case: $p(A \text{ or } B) = p(A) + p(B) - p(A \text{ and } B)$

3. Addition rule, for three events: $p(A \text{ or } B \text{ or } C) = p(A) + p(B) + p(C) - p(A \text{ and } B) - p(A \text{ and } C) - p(B \text{ and } C) + p(A \text{ and } B \text{ and } C)$

4. Complement of event A: $p(A') = 1 - p(A)$

5. Conditional probability formula: $p(A \mid B) = \dfrac{p(A \text{ and } B)}{p(B)}$

6. Multiplication rule, general: $p(A \text{ and } B) = p(A \mid B) \cdot p(B)$

7. Multiplication rule, for independent events: $p(A \text{ and } B) = p(A) \cdot p(B)$

8. Bayes' rule:

$$p(A_i \mid B) = \frac{p(B \mid A_i) \cdot p(A_i)}{p(B \mid A_1) \cdot p(A_1) + p(B \mid A_2) \cdot p(A_2) + \cdots + p(B \mid A_n) \cdot p(A_n)}$$

Testing Your Understanding of This Chapter's Concepts

1. Explain the difference between mutually exclusive events and independent events.
2. *True or False:* If two events are mutually exclusive, their probabilities must add up to 1. Explain.
3. If two events, A and B, each have nonzero probability and are mutually exclusive, must they also be independent? Explain your answer.
4. When surveyed, 40% of American women said they had read a book last weekend, compared with 31% of men. (*Source: Vitality*, September 1989, p. 16) Therefore, the probability that they both read a book last weekend is $(0.40)(0.31) = 0.124$, which would mean that the probability that they *both* did not read a book last weekend is 0.876. Do you agree with this reasoning? Explain your answer.
5. Mr. and Mrs. Pascal have two girls and would like to have a boy. After taking a course in probability they decide to have a third child, reasoning that the probability that the third child is a boy is $\dfrac{7}{8}$. There are eight possible ways in which a family can consist of three children, as shown in the following chart.

	Child 1	Child 2	Child 3
Possibility 1	Boy	Boy	Boy
Possibility 2	Boy	Boy	Girl
Possibility 3	Boy	Girl	Boy
Possibility 4	Boy	Girl	Girl
Possibility 5	Girl	Boy	Boy
Possibility 6	Girl	Boy	Girl
Possibility 7	Girl	Girl	Boy
Possibility 8	Girl	Girl	Girl

Of these, seven of them have at least one boy. Thus, they reason that the probability that the third child is a boy is $\dfrac{7}{8}$. Do you agree with this reasoning? Explain your answer.

6. In an attempt to attract new listeners, a radio station announces that it will randomly select a name from the local residential phone directory. The name will be announced "on the air." If the named person calls the station within 10 minutes, then that person will be awarded $1000. A curious listener whose last name begins with the letter "'Z'" calls the station and inquires about her chances of winning. She is informed that since there are 26 letters in the alphabet, then the probability of her winning is $\frac{1}{26}$. Do you agree? Explain your answer.

7. A family is known to have two children. A student claims that the probability that both children are boys is $\frac{1}{3}$, since there are three possible outcomes: two boys, one boy and one girl, or two girls. Do you agree? Explain your answer.

Chapter Test

Multiple Choice Questions

For questions 1–4, use the following information: Two hundred seventy people were asked to comment on the assembly instructions supplied with a particular computer. The results are shown below.

	Very Easy to Follow	Average Difficulty to Follow	Very Difficult to Follow	Impossible to Follow
Male	51	39	27	19
Female	47	45	22	20

1. One of these people is randomly selected. Find the probability that this individual is *not* a male.
   a. $\frac{136}{270}$ b. $\frac{134}{270}$ c. $\frac{134}{136}$ d. $\frac{47}{270}$ e. none of these

2. One of these people is randomly selected. If it is known that this person found the instructions very difficult to follow, find the probability that this individual is a male.
   a. $\frac{27}{136}$ b. $\frac{22}{134}$ c. $\frac{27}{270}$ d. $\frac{27}{49}$ e. none of these

3. One of these people is randomly selected. If it is known that this person is a male, find the probability that this person found the instructions very difficult to follow.
   a. $\frac{27}{136}$ b. $\frac{22}{134}$ c. $\frac{27}{270}$ d. $\frac{27}{49}$ e. none of these

4. Find the probability of selecting an individual from this group who is a male and who found the instructions very difficult to follow.
   a. $\frac{27}{136}$ b. $\frac{22}{134}$ c. $\frac{27}{270}$ d. $\frac{27}{49}$ e. none of these

5. A card is randomly selected from an ordinary deck of cards. Find the probability that it is a picture card or a red card.

   a. $\dfrac{12}{52}$ b. $\dfrac{26}{52}$ c. $\dfrac{38}{52}$ d. $\dfrac{32}{52}$ e. none of these

6. A pair of dice is rolled. What is the probability that the sum will be an odd number *and* that a 6 will appear on at least one die?

   a. $\dfrac{1}{6}$ b. $\dfrac{5}{36}$ c. $\dfrac{1}{9}$ d. $\dfrac{18}{36}$ e. none of these

7. Two cards are drawn at random from a 52-card deck of cards, with the first card replaced before the second one is drawn. What is the probability of selecting a diamond *and* a queen in the specified order?

   a. $\dfrac{52}{2652}$ b. $\dfrac{39}{2704}$ c. $\dfrac{16}{52}$ d. $\dfrac{39}{2652}$ e. $\dfrac{52}{2704}$

8. Mutually exclusive events
   a. can be dependent   b. can be independent   c. are always related
   d. all of the above   e. none of these

Supplementary Exercises

9. Two students at a college are randomly selected. The probability that the first student is majoring in mathematics is 0.72. The probability that the second student is majoring in art is 0.51. Assuming independence, find the probability that the first student is a math major and that the second student is an art major.

10. In a particular unit of a girls scout group, the probability that a girl scout is a good mountain climber is 0.83, and the probability that a girl scout is a good mountain climber as well as a good hunter is 0.52. If a girl scout that is a good mountain climber is randomly selected, what is the probability that she is also a good hunter?

11. Refer back to the previous question. The probability that a girl scout in this group knows how to treat a snake bite is 0.67. Find the probability that three independently selected and totally unrelated girl scouts in this group do *not* know how to treat a snake bite.

12. If $p(A \mid B) = 0.2$, $p(A \mid C) = 0.4$, $p(A \mid D) = 0.3$, $p(B) = 0.31$, $p(C) = 0.48$, and $p(D) = 0.21$, find $p(C \mid A)$.

13. A fish tank contains seven guppies, six algae-eaters, and four catfish. A storekeeper randomly selects two fish (one after another without replacement) from the tank for a customer. Find the probability that both fish selected are catfish.

14. A restaurant owner receives food supplies from one of four different vendors, A, B, C, or D, which the owner selects on a random basis. The probabilities that the food received from these vendors is fresh are 0.93, 0.90, 0.89, and 0.87, respectively. The restaurant owner randomly selects a food supply received from one of these vendors. If the food is fresh, what is the probability that it was supplied by vendor C?

15. The Marleton Elementary School uses school busses supplied by companies A, B, and C. The probability that the busses supplied by company A arrive on time is 0.95, the probability that the busses supplied by company B arrive on time is 0.93,

and the probability that the busses supplied by company C arrive on time is 0.97. If a day is randomly selected, find the probability that the busses supplied by all three companies to the school will be late. (Assume independence.)

16. A large soap company is planning to introduce a new dish-washing detergent into the market. In the past, 35% of the products introduced by the company have been successful and 65% of the products have not been successful. Before the product is actually marketed, market research is conducted and a report (favorable or unfavorable) must be prepared. In the past, 75% of the successful products received favorable reports and 25% of the unsuccessful products also received favorable reports. If the new dish-washing detergent receives a favorable marketing report, what is the probability that it will indeed be successful?

17. Charlie drives up to a parking meter and notices a piece of paper on the ground that reads "Meter out of order—No parking." Is this an old sign? Should he deposit money in the meter and park there? The probability that the meter is out of order is $\frac{4}{9}$. The probability that the meter is out of order *and* that he will get a ticket for parking there is $\frac{2}{13}$. If the meter is actually out of order, what is the probability that he will get a ticket?

18. A nationwide poll[°] was taken of 800 mathematicians to determine their professional responsibilities. The following is the distribution of their professional responsibilities:

Responsibility[†]	Distribution (%)
R	39
T	69
C	29
R and C	21
R and T	12
R, C, and T	5
R or C or T	Total = 100%

[†]R = research; T = teaching; C = consultation and textbook writing.

Find the percentage that are involved in T and C (teaching and consultation with textbook writing).

19. A Japanese video cassette recorder manufacturer shipped 100 VCRs to an American store. Due to a packer's error, 20 of these VCRs were packed with operating instructions in Japanese only. The manufacturer also knows that the store will choose

[°]Yates and Piscato, 1994.

two of these VCRs at random, check them thoroughly, and accept the entire shipment if neither is defective in any way (including operating instructions since the store's customers do not read or understand Japanese).

a. What is the probability that the first VCR chosen by the store will pass inspection?

b. If the first VCR selected passes inspection, what is the probability that the second will *not* pass inspection?

c. What is the probability that the entire shipment of VCRs is accepted by the store?

20. A hospital administrator is arranging emergency plans. The weather bureau has forecast freezing rain with a probability of 0.90. The administration knows that if freezing rain does occur, then there is a 0.75 probability that all hospital personnel will not be able to get to the hospital. Find the probability that freezing rain does occur *and* that all hospital personnel cannot get to the hospital.

21. Despite the numerous articles that have been written on the subject, many Americans have a blood serum cholesterol level that is considerably higher than the 200 milligram level recommended by the American Medical Association. According to the American Heart Association, 42% of the people in Boonton have a high blood serum cholesterol level, 30% of the people are overweight, and 15% of the people have a high blood cholesterol level and are also overweight. If a resident of Boonton is randomly selected, what is the probability that this individual has a high blood serum cholesterol level *or* is overweight?

22. A survey (Hodges and Kiddey, *Information About New York City Children*, New York, 1991) of 10,000 children found that 69% of them had brothers or sisters that graduated high school. Also, 33% of these children needed remediation with their school work. Moreover, 12% of the children needed remediation *and* had brothers or sisters that graduated high school. If a child from this group is randomly selected, what is the probability that the child needs remediation *or* has a brother or sister that graduated high school?

23. In some large cities, tow truck operators monitor police emergency calls so as to be the first at an accident scene. Midview Collision Co. responds to accident scenes 75% of the time. Crossland Collision Co. responds to accident scenes only 53% of the time. Each tow company operates independently of the other. What is the probability that exactly one of them will respond to a police report of a major accident on Broadway?

24. There are many computer viruses around today that can wreak havoc on a computer once it becomes infected. To forestall such possibilities, many businesses or private individual users install virus scanners to detect the presence of any virus on a data disk. One company has installed three different types of virus scanners, each capable of detecting different types of viruses. Each scanner operates independently of the other. The probabilities of these scanners being successful in detecting viruses are 0.8, 0.72, and 0.6. What is the probability that

a. one of them succeeds in detecting a contaminated disk?

b. none of them succeed in detecting a contaminated disk?

25. A new blood test has been developed that can correctly identify 97% of the people with a particular disease and 92% of the people who do not have the disease. If it is known that 4% of the population have this disease, what is the probability that
    a. a person that is classified as positive by the blood test (meaning that the person has the disease) does indeed have the disease?
    b. a person that is classified as negative by the blood test (meaning that the person does not have the disease) does indeed not have the disease?
26. If a fair coin is tossed 50 times, what is the probability that it will land on "heads" all 50 times?

If a fair coin is tossed 50 times, how likely is it that it will land on "heads" 50 times?   (*© Rapho, Photo Researchers, Inc.*)

Thinking Critically

1. There are only two people in Dr. Plotkin's waiting room. What is the probability that their birthdays are different?
2. If only three people are in a beauty salon, what is the probability that at least two of them have the same birthday?

3. Four people who do not know each other are standing on a line at a movie theater waiting for the box office to open. What is the probability that their birthdays (month and day) are different?

Would you believe that we need only 23 people in a crowd to have a 50% probability that at least two of these people will have the same birthday? The probability increases to about 1, almost certainly, when we have a crowd of 63 people.

4. Joan Hartman was born in January. What is the probability that she will die in January?
5. A fair die is rolled five times. If it is known that on exactly one of these rolls, one dot showed, what is the probability that it was on the third roll of the die?
6. A family is known to have six children. If we know that one child is definitely a girl, what is the probability that the family consists of *at least* two girls?

Case Studies

1. In June of 1964 an elderly woman was mugged in San Pedro, California. In the vicinity of the crime a bearded black man sat waiting in a yellow car. Shortly after the crime was committed, a young white woman, wearing her blonde hair in a ponytail, was seen running from the scene of the crime and getting into the car, which sped off. The police broadcast a description of the suspected muggers. Soon afterward, a couple fitting the description was arrested and convicted of the crime. Although the evidence in the case was largely circumstantial, the prosecutor based his case on probability and the unlikeliness of another couple having such characteristics. He assumed the following probabilities.

Characteristic	Assumed Probability
Drives yellow car	$\dfrac{1}{10}$
Black-white couple	$\dfrac{1}{1000}$
Black man	$\dfrac{1}{3}$
Man with beard	$\dfrac{1}{10}$
Blonde woman	$\dfrac{1}{4}$
Woman wears her hair in ponytail	$\dfrac{1}{10}$

He then multiplied the individual probabilities:

$$\frac{1}{10} \cdot \frac{1}{1000} \cdot \frac{1}{3} \cdot \frac{1}{10} \cdot \frac{1}{4} \cdot \frac{1}{10} = \frac{1}{12,000,000}$$

He claimed that the probability is $\dfrac{1}{12,000,000}$ that another couple has such characteristics. The jury agreed and convicted the couple. The conviction was overturned
by the California Supreme Court in 1968. The defense attorneys got some professional advice on probability. Serious errors were found in the prosecutor's probability
calculations. Some of these involved assumptions about independent events. As a
matter of fact it was demonstrated that the probability is 0.41 that another couple
with the same characteristics existed in the area once it was known that there was at
least one such couple.

For a complete discussion of this probability case, read "Trial by Mathematics,"
which appeared in *Time*, January 8, 1965, p. 42, and April 26, 1968, p. 41.

2. A TV host informs a contestant that behind one of three doors there is a new car
and behind the other doors are worthless prizes. All of the doors are equally likely
to be selected. The contestant selects door 1. The host, who knows behind which
door the car is, opens door 3 to reveal a worthless prize. The host then offers the
contestant the opportunity to switch the choice of doors. Should the contestant
switch? This problem generated much discussion in the *Ask Marilyn* column of *Parade*. Marilyn recommends that the contestant should switch arguing that the probability of winning, originally $\dfrac{1}{3}$, has now increased to $\dfrac{2}{3}$. Other mathematicians disagree, arguing that with door 3 eliminated, doors 1 and 2 were equally likely, so that
the probability of winning is $\dfrac{1}{2}$. Assume that you are the contestant. Given that the

host opened door 3, what is the probability that you win if you switch? (For a more detailed discussion see Marilyn vos Savant "Ask Marilyn," *Parade*, September 9, 1990; December 2, 1990; and February 17, 1991, and "The Car and the Greats," by Leonard Gillman in *The American Mathematical Monthly*, Vol. 99, No. 1, January 1992, pp. 3–7.)

3. It has been estimated that when a person buys one ticket for exactly two different New Jersey lotteries, then the probability that the person will hit the lottery twice in a lifetime is 1 in 17 trillion or 1/17,000,000,000,000. Despite these odds, in 1986, Evelyn Marie Adams won the New Jersey lottery twice within 4 months. If a person is a big winner in one of the many state lotteries that are run nationwide, what is the probability that the person will be the big winner in that same state for a second time? *The Wall Street Journal* on February 27, 1990, discussed how such probabilities are calculated.

6

SOME DISCRETE PROBABILITY DISTRIBUTIONS AND THEIR PROPERTIES

DID YOU KNOW THAT·

a   a pregnant woman has a greater chance of having a boy than a girl, that is, more boys are born than girls? (*Source: Bureau of Vital Statistics*, Washington, D.C.)

b   30% of American parents allow their children under age 13 to stay home alone after school? (*Source: U.S. Department of Labor*, 1994)

c   95% of Americans grind their teeth while asleep at some point in their life? (*Source: Vitality*, December 1993, p. 8)

d   88% of American women wear shoes that are too small for their feet? (*Source: American Orthopaedic Foot and Ankle Society*, 1993)

e   although 49 states prohibit the sale of cigarettes to minors, one study found that 63% of underage children could buy them? (*Redbook*, 1993)

f   31% of the adult U.S. population has borderline high blood cholesterol levels of between 200 and 239 mg/dl and 20% has high blood cholesterol levels of over 240 mg/dl? (*Source: JAMA* 269(23):3009–3014, June 16, 1993)

Laptop computers are fast becoming an important part of a business executive's tools. See discussion on page 340. (*Patrick Farace*)

As random variables and probability distributions play a key role in our later studies of statistical inferences, in this chapter we analyze the nature of and applications involving discrete random variables and their associated probability distributions.

Chapter Objectives

- **To discuss** the concept of a random variable, where the outcome of some experiment is of interest to us. (Section 6.2)

- **To see** how probability functions assign probabilities to the different values of the random variable. (Section 6.2)

- **To understand** how the mean and variance of a probability function give us the expected value of a probability function as well as the spread of the distribution. (Sections 6.3 and 6.4)

- **To study** Bernoulli or binomial experiments, which are experiments having only two possible outcomes, success or failure. (Section 6.5)

- **To apply** the binomial probability distribution, which gives us the probability of obtaining a specified number of successes when an experiment is performed n times. (Section 6.5)

- **To work with** several important binomial distribution properties. (Section 6.6)

- **To analyze** the Poisson probability distribution, which can be applied in very specific cases. (Section 6.7)

- **To use** the hypergeometric probability distribution when the absence of certain assumptions makes the binomial distribution inappropriate. (Section 6.8)

According to this article, about one out of every eight women in America is expected to develop breast cancer. For the American population as a whole, can we determine the expected number of women that will develop breast cancer? Such medical statistics have important implications for our society.

In this chapter we will discuss how such statistics are calculated.

ONE OUT OF EIGHT WOMEN WILL DEVELOP BREAST CANCER

Washington (Jan. 15):—According to statistics released yesterday, it is expected that one out of every 8 American women will develop breast cancer this year. This represents an increase in the estimated 1 out of 9 women projected only last year. The American Cancer Society as well as the American Medical Association urges all women over 40 to take a mammography test. When detected early, breast cancer can be treated.

Tuesday—January 15, 1993

6.1   INTRODUCTION

In Chapter 2 we discussed frequency distributions of sets of data. Using these distributions we were able to analyze data more intelligently to determine which outcomes occurred most often, least often, and so on. In Chapters 4 and 5 we discussed the various rules of probability and how they can be applied to many different situations. These rules enable us to predict how often something will happen in the long run. In this chapter we combine these ideas.

In any given experiment there may be many different things of interest. For example, if a scientist decides to mate a white rat with a black rat, she may be interested in the number of offspring that are white, black, gray, and so on. We will therefore have to define what is meant by a random variable and then discuss its probability function. Special emphasis will be given in this chapter to the binomial random variable and its probability distribution.

6.2   DISCRETE PROBABILITY FUNCTIONS

To understand what is meant by a discrete probability function, let us analyze the following examples.

Valerie is a dentist and keeps accurate records on the number of cavities of each of her patients. Her records indicate that each patient has anywhere from zero to five cavities. Based on past experience, she has compiled the data given in Table 6.1.

TABLE 6.1   The Number of Cavities and Their Probabilities

Number of Cavities	Probability
0	$\frac{1}{16}$
1	$\frac{4}{16}$
2	$\frac{5}{16}$
3	$\frac{3}{16}$
4	$\frac{2}{16}$
5	$\frac{1}{16}$
	$Total = \frac{16}{16} = 1$

TABLE 6.2   **The Number of People Entering the Elevator and Their Probabilities**

Number of People Entering Elevator	Probability	Number of People Entering Elevator	Probability
0	$\dfrac{1}{50}$	6	$\dfrac{10}{50}$
1	$\dfrac{3}{50}$	7	$\dfrac{6}{50}$
2	$\dfrac{4}{50}$	8	$\dfrac{3}{50}$
3	$\dfrac{5}{50}$	9	$\dfrac{2}{50}$
4	$\dfrac{7}{50}$	10	$\dfrac{1}{50}$
5	$\dfrac{8}{50}$		$Total = \dfrac{50}{50} = 1$

Notice that each patient has anywhere from zero to five cavities. Thus, the number of cavities that each patient has is somehow dependent on chance, as the probabilities in the table indicate.

Now consider Eric who is an elevator operator. The number of people who enter the elevator at exactly 9:00 A.M. varies from zero to ten. The capacity of the elevator is ten people. From past experience Eric has been able to set up the chart shown in Table 6.2.

In each of the two previous examples the values assumed by the item of interest, that is, the number of cavities or the number of people entering the elevator, were whole numbers and were somehow dependent on chance. We refer to such a quantity as a **discrete random variable**.

The term discrete random variable applies to many different situations. Thus, it may represent the number of people buying tickets to a movie, the number of mistakes made by a secretary in typing a letter, the number of telephone calls received by the school switchboard during the month of September, the number of games that the Green Bay Packers will win next season, or the number of students that will enroll in a particular course, Music 161, to be offered for the first time in the spring.

Random variable

Basically, if an experiment is performed and some quantitative variable, denoted by x, is measured or observed, then the quantitative variable x is called a **random variable** since the values that x may assume in the given experiment depend on chance. It is a random outcome. Whenever all possible values that a random variable may assume can be listed (or counted), then the random variable is said to be **discrete**. As opposed to

Discrete random variable

this, the time required to complete a transaction at a bank is a continuous random variable because it could theoretically assume any one of an infinite number of values—namely, any value 0 seconds or more. Random variables that can assume values corresponding to any of the points contained in one or more intervals on a line are called

Continuous
random variable

continuous random variables.

Other examples of continuous random variables are

1. the time it takes for a drug to take effect.
2. the height (in cm) of a player on a basketball team.
3. the blood serum cholesterol level of a person.
4. the weight of a bag of sugar or of a large jar of coffee.
5. the length of time between births in the maternity ward of a hospital.

In this chapter we discuss discrete random variables and their probability distributions. In the next chapter we will discuss continuous random variables and their distributions. *To summarize:* A **random variable** is a *numerical* quantity, the *value* of which is determined by an experiment. In other words, its value is determined by chance.

The following examples further illustrate the idea of a random variable.

EXAMPLE   1   Three cards are selected, without replacement, from a deck of 52 cards. The random variable may be the number of aces obtained. It would then have values of 0, 1, 2, or 3, depending on the number of aces actually obtained. This is a discrete random variable.

EXAMPLE   2   Calvin drives his car over some nails. The random variable is the number of flat tires that Calvin gets. The values of the random variable are 0, 1, 2, 3, 4, corresponding to zero flats, one flat, two flats, three flats, and four flats, respectively. This is a discrete random variable.

EXAMPLE   3   Let the number of people who will attend the next concert at the Hollywood Bowl be the random variable of interest. Then this random variable can assume values ranging from 0 to the seating capacity of the Hollywood Bowl. This is a discrete random variable.

EXAMPLE   4   A nurse is taking Chuck's blood pressure. Let the random variable be Chuck's systolic blood pressure. What values can the random variable assume? This represents a continuous random variable.

EXAMPLE   5   Get on a scale and weigh yourself. Let the random variable be your weight. What values can the random variable assume? This represents a continuous random variable.

Let us now return to the two examples discussed at the beginning of this section. You will notice that the probabilities associated with the different values of the random variable are indicated. Thus, Table 6.1 tells us that the probability of having one cavity is $\frac{4}{16}$ and that the probability of having three cavities is $\frac{3}{16}$. Similarly, Table 6.2 tells us that the probability of three people entering the elevator is $\frac{5}{50}$.

When discussing a random variable, we are almost always interested in assigning probabilities to the various values of the random variable. For this reason we now discuss probability functions.

Definition 6.1 *Probability function* *Probability distribution*	A **probability function** or **probability distribution** is a correspondence that assigns probabilities to the values of a random variable.

EXAMPLE 6   If a pair of dice is rolled, the random variable that may be of interest to us is the number of dots appearing on both dice together. When a pair of dice is rolled, there are 36 possible outcomes. These are shown in Figure 6.1.

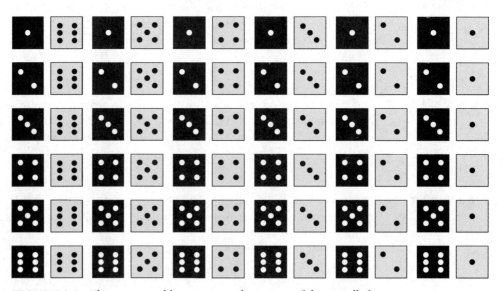

FIGURE 6.1   Thirty-six possible outcomes when a pair of dice is rolled.

We can then set up the following chart.

Sum on Both Dice	Number of Different Ways in Which Sum Can Be Obtained	Probability
2	1	$\frac{1}{36}$
3	2	$\frac{2}{36}$
4	3	$\frac{3}{36}$
5	4	$\frac{4}{36}$
6	5	$\frac{5}{36}$
7	6	$\frac{6}{36}$
8	5	$\frac{5}{36}$
9	4	$\frac{4}{36}$
10	3	$\frac{3}{36}$
11	2	$\frac{2}{36}$
12	1	$\frac{1}{36}$
		$Total = \frac{36}{36} = 1$

In the previous example the random variable of interest to us was the number of dots appearing on both dice together. Often, for mathematical convenience, we use special symbols. We use lowercase letters to represent all possible values of the random variable. We denote the probability of the random variable x by the symbol $p(x)$. Thus, for the preceding example the probability that the random variable assumes the value of 2 is $\frac{1}{36}$. Symbolically, $p(2) = \frac{1}{36}$. Similarly, $p(7) = \frac{6}{36}$.

EXAMPLE 7 An airplane has four engines, each of which operates independently of the other. If the random variable is the number of engines that are functioning properly, the random variable has values 0, 1, 2, 3, 4. For a particular airplane we have the following probability function.

Number of Engines Functioning Properly	Probability
0	$\dfrac{1}{19}$
1	$\dfrac{2}{19}$
2	$\dfrac{4}{19}$
3	$\dfrac{5}{19}$
4	$\dfrac{7}{19}$
	$Total = \dfrac{19}{19} = 1$

Notice that in Example 7, as well as in Examples 1 to 6, the sum of all the probabilities is 1. This will be true for any probability function. Also, the probability that the random variable assumes any one particular value is between 0 and 1. Again, this is true in every case. We state this formally as

RULE

a. The sum of all the probabilities of a probability function is always 1, or
$$\sum_{all\ x} p(x) = 1.$$

b. The probability that a random variable assumes any one value in particular is between 0 and 1 inclusive. Zero means that it can never happen and 1 means that it must happen, or $0 \le p(x) \le 1$

COMMENT   Some authors distinguish between **probability functions** and **probability distributions**. In this book we will use these terms interchangeably.

In all examples mentioned thus far some of the random variables assumed many values and some took on only two values. The following random variables take on many values:

1. The number of flat tires that Calvin gets when he drives over nails.
2. The sum obtained when rolling a pair of dice. There are 11 possible values, as indicated on page 337.
3. The number of people in an elevator.

Consider the following events:

1. The possible outcome when one coin is flipped (heads or tails).
2. The outcome of either having a cavity or not having a cavity (yes or no).
3. The sex of a newborn child (male or female).
4. The results of an exam (pass or fail only).

The outcomes "head" and "tail" are not the values of random variables because they are not in numerical form. However, if we let "head" equal 0 and "tail" equal 1, the result is a random variable. Remember, a random variable *must* take on numerical values only. We now have a random variable that can take on only two values.

Binomial variable
Binomial distribution

If a random variable has only two possible values, and if the probability of these values remains the same for each trial regardless of what happened on any previous trial (that is, the trials are independent), the variable is called a **binomial variable** and its probability distribution is called a **binomial distribution**. Actually, this is rather a special case of the binomial distribution. Since the binomial distribution is so important in statistics, we will discuss it in great detail in Section 6.5.

EXERCISES FOR SECTION 6.2

1. For each of the following situations, find the possible values that the random variables may have.

Situation	Random Variable of Interest
a. An airport runway	The number of airplane takeoffs on a particular day
b. A business office	The number of FAXs received per day
c. An emergency room of a hospital	The number of heart attack victims seen per month
d. A ski slope	The number of inches of snow that falls on the slope
e. A ski slope	The outdoor temperature
f. A polluted lake	The number of coliform bacteria per cc of water in the lake
g. A marriage in an Arab country	The number of wives that a man has

2. Identify the following random variables as discrete or continuous.
   a. The number of viruses that can infect a computer system.
   b. The number of hours of sleep per night required by a college student.
   c. The number of people in Los Angeles who use pagers (or beepers).
   d. Your math teacher's age.
3. An IRS agent randomly selects ten individual tax returns for the 1995 calendar year. Let x represent the random variable "the number of tax returns on which the standard exemption was claimed." List the possible values of the random variable x.
4. Can the following be a probability distribution for a random variable? If your answer is no, explain why.

Random Variable, x	Probability
0	0.26
1	0.18
2	0.14
3	0.33
4	0.09
5	0.19

5. Given the following probability distribution for the random variable x. What is the probability that the random variable has a value of 3?

Random Variable, x	Probability
1	0.23
2	0.09
3	?
4	0.26
5	0.04
6	0.18

6. Mr. and Mrs. Jones own three cars. Let x be the number that are imported cars. Find the probability distribution of x. (Assume that a car can be an imported car or a domestic car with equal probability.)

7. An archer shoots four arrows at a bull's eye of one target. Let x be the number of times that the archer hits the bull's eye. Find the probability distribution of x. (*Hint:* Assume that the probability that the archer hits the bull's eye on each try is $\frac{1}{2}$.)

8. According to the *Wall Street Journal* survey of businesses that use personal computers, 60% of the officers also use laptop computers. If two businesses that were surveyed are randomly selected and if x represents the number of these businesses where the officers use laptop computers, find the probability distribution of x.

9. According to an *American Automobile Association* survey, 6% of the drivers purchasing gas at a service station on the New York State Thruway will ask to have the oil checked. Three cars are observed pulling up to a gas pump of a service station on the New York State Thruway. Let x represent the number of drivers who will ask to have the oil checked. Find the probability distribution of x.

10. Bill Giordano is in the bursar's office completing his application for financial aid. Bill recalls that the first five digits of his social security number (needed to process his application) are 147–29. However, he has completely forgotten the last four digits. He decides to guess. Let x be the number of digits that he guesses correctly. Find the probability distribution of x.

11. Thirty percent of American parents allow their children under age 13 to stay home alone after school. (*Source: United States Department of Labor*, Washington, D.C., 1994). Three parents are randomly selected. Let x be the number of these parents

who allow their children under age 13 to stay home alone after school. Find the probability distribution of x.

12. Consider the article "Will Women Soon Outrun Men?" (Whipp and Ward, 1992) that appeared in the British Journal *Nature*. According to this article, the progress made by women in track events over the last half-century has been more substantial than the progress made by men. In one particular marathon, three men and three equally rated women completed the 800-meter run in record time. Three of these winners are randomly selected. Let x be the number of women selected. Find the probability distribution of x.

13. A family is known to have four children. Let x be the number of boys in the family.
   a. Find the probability distribution of x.
   b. Find the probability that x has a value of at least 1.

14. A lab technician knows that two of the eight pints of blood in the blood bank contain type A positive blood. The technician selects two of the pints at random, with the first replaced before the second is selected. Let x be the number of pints of type A blood obtained. Find the probability distribution of x.

15. Determine if the following formula can be the probability function of some random variable. Explain your answer.

$$p(x) = \frac{7 - x}{28} \qquad \text{for } x = 0, 1, 2, 3, 4, 5, 6, 7$$

16. Determine if the following formula can be the probability function of some random variable x. Explain your answer.

$$p(x) = \frac{8 - x}{35} \qquad \text{for } x = 1, 2, 3, 4, 5, 6, 7, 8$$

17. A discrete random variable has the following probability distribution.

x	$p(x)$
2	0.32
3	0.04
4	0.38
5	0.17
6	0.09

   a. Find $p(x = 4)$.
   b. Find $p(x = 7)$.
   c. Find $p(x \leq 3)$.
   d. Find $p(x \leq 6)$.
   e. Find $p(x \leq 4 \quad \text{or} \quad x > 5)$.

18. Francisco Benoit is in charge of maintenance for a large Florida rental agency. The following is the probability distribution for the number of customers who will call the car rental agency daily because of malfunctioning cars.

Number of Customers, x	Probability, $p(x)$
5	0.11
6	0.19
7	0.18
8	0.07
9	0.12
10	0.06
11	0.10
12	0.14
13	0.03

Suppose a day is randomly selected.
a. Find $p(x = 9)$.
b. Find $p(x > 7)$.
c. Find $p(x \leq 11)$.
d. Find $p(8 \leq x \leq 12)$.

19. A counselor at a drug rehabilitation center for young juveniles questioned each participant in the program to determine how many years, x, each had been on drugs before enrolling in the program. The relative frequencies corresponding to x are given in the following probability distribution.

Number of Years, x	Probability, $p(x)$
0	0.06
1	0.17
2	0.19
3	0.20
4	0.23
5	0.12
6	0.03

a. Find $p(x \leq 3)$.
b. Find $p(x > 4)$.
c. Find $p(x < 7)$.
d. Find $p(1 \leq x \leq 6)$.

6.3   THE MEAN OF A RANDOM VARIABLE

Imagine that the traffic department of a city is considering installing traffic signals at the intersection of Main Street and Broadway. The department's statisticians have kept accurate records over the past year on the number of accidents reported per day at this particularly dangerous intersection. They have submitted the report on the number of accidents per day and their respective probabilities shown in Table 6.3.

TABLE 6.3   Report on Accidents at Intersection of Main Street
and Broadway

Number of Accidents x	Probability $p(x)$	Product $x \cdot p(x)$
0	$\dfrac{1}{32}$	$0\left(\dfrac{1}{32}\right) = 0$
1	$\dfrac{1}{32}$	$1\left(\dfrac{1}{32}\right) = \dfrac{1}{32}$
2	$\dfrac{9}{32}$	$2\left(\dfrac{9}{32}\right) = \dfrac{18}{32}$
3	$\dfrac{10}{32}$	$3\left(\dfrac{10}{32}\right) = \dfrac{30}{32}$
4	$\dfrac{8}{32}$	$4\left(\dfrac{8}{32}\right) = \dfrac{32}{32}$
5	$\dfrac{3}{32}$	$5\left(\dfrac{3}{32}\right) = \dfrac{15}{32}$

Let us multiply each of the possible values for the random variable x, which represents the number of accidents given in Table 6.3, by the respective probabilities. The results of these multiplications are shown in the third column of Table 6.3. If we now add the products, we get

$$0 + \left(\frac{1}{32}\right) + \left(\frac{18}{32}\right) + \left(\frac{30}{32}\right) + \left(\frac{32}{32}\right) + \left(\frac{15}{32}\right) = \frac{96}{32} = 3$$

This result is known as the **mean** of the random variable. It tells us that on the average there are three accidents per day at this dangerous intersection.

Recall that in Chapter 3, when we discussed measures of central tendency and measures of variation, we distinguished between sample statistics and population parameters. Thus, we used the symbols $\bar{x}$, s^2, and s as symbols for the sample mean, sample variance, and sample standard deviation, respectively. Also, we used the symbols μ, σ^2, and σ as symbols for the population mean, population variance, and population standard deviation, respectively. In our case the mean number of accidents represents the mean of the entire population of observed data. Therefore, we will use the symbol μ to represent the average number of accidents. Usually, when dealing with probability distributions we work with the entire population. Hence, we use μ as the symbol for mean.

Now consider the following. Suppose we intend to flip a coin four times. What is the average number of heads that we can expect to get? To answer this question we first find the probability distribution. Let x represent the random variable "the number of heads obtained in four flips of the coin." Since the coin is flipped four times, we may

TABLE 6.4 Number of Heads That Can Be Obtained
When a Coin Is Flipped Four Times

Random Variable x	Probability $p(x)$	Product $x \cdot p(x)$
0	$\dfrac{1}{16}$	$0\left(\dfrac{1}{16}\right) = 0$
1	$\dfrac{4}{16}$	$1\left(\dfrac{4}{16}\right) = \dfrac{4}{16}$
2	$\dfrac{6}{16}$	$2\left(\dfrac{6}{16}\right) = \dfrac{12}{16}$
3	$\dfrac{4}{16}$	$3\left(\dfrac{4}{16}\right) = \dfrac{12}{16}$
4	$\dfrac{1}{16}$	$4\left(\dfrac{1}{16}\right) = \dfrac{4}{16}$

get 0, 1, 2, 3, or 4 heads. Thus, the random variable x can have the values 0, 1, 2, 3, or 4. The probabilities associated with each of these values is indicated in Table 6.4. You should verify that the probabilities given in this table are correct.

We now multiply each possible outcome by its probability, the results of which are shown in the third column of Table 6.4. If we now add these products, the result is again called the mean of the probability distribution. In our case the mean is

$$0 + \left(\frac{4}{16}\right) + \left(\frac{12}{16}\right) + \left(\frac{12}{16}\right) + \left(\frac{4}{16}\right) = \frac{32}{16} = 2$$

This tells us that on the average we can expect to get two heads.

We generalize the results of the previous two examples as follows:

Definition 6.2

Mean

The **mean** of a random variable for a given probability distribution is the number obtained by multiplying all possible values of the random variable having this particular distribution by their respective probabilities and adding these products together.

FORMULA 6.1

The mean of a random variable for a given probability distribution is denoted by the Greek letter μ, read as mu. Thus,

$$\mu = \Sigma\, x \cdot p(x)$$

where this summation is taken over all the values that the random variable x can assume and the quantities $p(x)$ are the corresponding probabilities.

Mathematical expectation
Expected value

COMMENT Many books refer to the mean of a random variable for a given probability distribution as its **mathematical expectation** or its **expected value**. We will use the word **mean**. The mean is the mathematical expectation that was discussed in Section 4.6 of Chapter 4.

We illustrate the use of Formula 6.1 with several examples.

EXAMPLE 1 Matthew is a doorman. The following table gives the probabilities that customers will give tips of varying amounts of money.

Amounts of Money (in cents), x	30	35	40	45	50	55	60
Probability, p(x)	0.45	0.25	0.12	0.08	0.05	0.03	0.02

Find the mean for this distribution.

SOLUTION Applying Formula 6.1 we have

$$\mu = \Sigma \, x \cdot p(x) = 30(0.45) + 35(0.25) + 40(0.12) + 45(0.08)$$
$$+ \, 50(0.05) + 55(0.03) + 60(0.02)$$
$$= 13.5 + 8.75 + 4.8 + 3.6 + 2.5 + 1.65 + 1.2$$
$$= 36$$

The mean is 36. Thus, Matthew can expect to receive an average tip of 36 cents.

EXAMPLE 2 Rosemary works for the U.S. Census Bureau in Washington. For a particular midwestern town the number of children per family and their respective probabilities is as follows:

Number of Children, x	0	1	2	3	4	5	6
Probability, p(x)	0.07	0.17	0.31	0.27	0.11	0.06	0.01

Find the mean for this distribution.

SOLUTION Applying Formula 6.1 we have

$$\mu = \Sigma \, x \cdot p(x) = 0(0.07) + 1(0.17) + 2(0.31) + 3(0.27)$$
$$+ \, 4(0.11) + 5(0.06) + 6(0.01)$$
$$= 0 + 0.17 + 0.62 + 0.81 + 0.44 + 0.30 + 0.06$$
$$= 2.4$$

The mean is 2.4. How can the average number of children per family be 2.4? Should it not be a whole number such as 2 or 3, not 2.4?

COMMENT   It may seem to you that Formula 6.1 is a new formula for calculating the mean. Actually, this is not the case. Recall that in Chapter 3 we defined the mean of a frequency distribution as mean value $= \sum \frac{xf}{n}$ or $\sum x \frac{f}{n}$. In Chapter 4 we defined the probability of an event as the relative frequency of the event, that is, $p(x) = \frac{f}{n}$. If we substitute this value for $\frac{f}{n}$ in the preceding formula, we get mean value $= \sum x \cdot p(x)$.

6.4   MEASURING CHANCE VARIATION

Suppose a manufacturer guarantees that a tire will last 20,000 miles under normal driving conditions. If a tire is selected at random and lasts only 12,000 miles, can the difference between what was expected and what actually happened be reasonably attributed to chance, or is there something wrong with the claim?

Similarly, if a coin is flipped 100 times, we would expect to obtain an average number of 50 heads. If a coin was actually flipped 100 times and resulted in only 25 heads, can we conclude that the difference between what was expected and what actually happened is to be attributed to chance, or is it possible that the coin is loaded?

To answer these questions, we need some method of measuring the variations of a random variable that are due to chance. Thus, we will discuss the variance and standard deviation of a probability distribution.

You will recall that in Chapter 3 (page 150) we discussed variation of a set of numbers. We now extend this idea to variation of a probability distribution. We let μ represent the mean, $x - \mu$ represent the difference of any number from the mean, and $(x - \mu)^2$ represent the square of the difference. The difference of a number from the mean is called the **deviation from the mean**. We multiply each of the squared deviations from the mean by their respective probabilities. The sum of these products is called the **variance of a random variable with the given probability distribution**. Formally, we have

Deviation from the mean

Definition 6.3

Variance

The **variance** of a random variable with a given probability distribution is the number obtained by multiplying each of the squared deviations from the mean by their respective probabilities and by adding these products.

FORMULA 6.2

The variance of a random variable with a given probability distribution is denoted by the Greek letter σ^2 (read as sigma squared). Thus,

$$\sigma^2 = \Sigma(x - \mu)^2 \cdot p(x)$$

where this summation is taken over all the values that random variable x can take on. The quantities $p(x)$ are the corresponding probabilities and $(x - \mu)^2$ is the square of the deviations from the mean.

Definition 6.4
Standard deviation

The **standard deviation** of a random variable with a given probability distribution is the square root of the variance of the probability distribution. We denote the standard deviation by the symbol σ (sigma). Thus,

$$\sigma = \sqrt{\text{variance}}$$

EXAMPLE 1 A random variable has the following probability distribution:

x	0	1	2	3	4	5
$p(x)$	$\dfrac{7}{24}$	$\dfrac{5}{24}$	$\dfrac{1}{8}$	$\dfrac{1}{8}$	$\dfrac{1}{12}$	$\dfrac{1}{6}$

Find the mean, variance, and standard deviation for this distribution.

SOLUTION We first find μ by using Formula 6.1 of Section 6.3 and then proceed to use Formula 6.2. We arrange the computations in the form of a chart, as follows:

x	$p(x)$	$x \cdot p(x)$	$x - \mu$	$(x - \mu)^2$	$(x - \mu)^2 \cdot p(x)$
0	$\dfrac{7}{24}$	$0\left(\dfrac{7}{24}\right) = 0$	$0 - 2 = -2$	$(-2)^2 = 4$	$4\left(\dfrac{7}{24}\right) = \dfrac{28}{24}$
1	$\dfrac{5}{24}$	$1\left(\dfrac{5}{24}\right) = \dfrac{5}{24}$	$1 - 2 = -1$	$(-1)^2 = 1$	$1\left(\dfrac{5}{24}\right) = \dfrac{5}{24}$
2	$\dfrac{1}{8}$	$2\left(\dfrac{1}{8}\right) = \dfrac{2}{8}$	$2 - 2 = 0$	$0^2 = 0$	$0\left(\dfrac{1}{8}\right) = 0$
3	$\dfrac{1}{8}$	$3\left(\dfrac{1}{8}\right) = \dfrac{3}{8}$	$3 - 2 = 1$	$1^2 = 1$	$1\left(\dfrac{1}{8}\right) = \dfrac{1}{8}$
4	$\dfrac{1}{12}$	$4\left(\dfrac{1}{12}\right) = \dfrac{4}{12}$	$4 - 2 = 2$	$2^2 = 4$	$4\left(\dfrac{1}{12}\right) = \dfrac{4}{12}$
5	$\dfrac{1}{6}$	$5\left(\dfrac{1}{6}\right) = \dfrac{5}{6}$	$5 - 3 = 3$	$3^2 = 9$	$9\left(\dfrac{1}{6}\right) = \dfrac{9}{6}$

We then have

$$\mu = \Sigma\, x \cdot p(x)$$

$$= 0\left(\frac{7}{24}\right) + 1\left(\frac{5}{24}\right) + 2\left(\frac{1}{8}\right) + 3\left(\frac{1}{8}\right) + 4\left(\frac{1}{12}\right) + 5\left(\frac{1}{6}\right)$$

$$= 0 + \left(\frac{5}{24}\right) + \left(\frac{2}{8}\right) + \left(\frac{3}{8}\right) + \left(\frac{4}{12}\right) + \left(\frac{5}{6}\right)$$

$$= 0 + \left(\frac{5}{24}\right) + \left(\frac{6}{24}\right) + \left(\frac{9}{24}\right) + \left(\frac{8}{24}\right) + \left(\frac{20}{24}\right)$$

$$= \frac{48}{24} = 2$$

Also,

$$\sigma^2 = \Sigma(x - \mu)^2 \cdot p(x)$$

$$= \left(\frac{28}{24}\right) + \left(\frac{5}{24}\right) + 0 + \left(\frac{1}{8}\right) + \left(\frac{4}{12}\right) + \left(\frac{9}{6}\right)$$

$$= \frac{80}{24} \approx 3.3333$$

Thus, the mean is 2, the variance is approximately 3.3333, and the standard deviation is $\sqrt{3.3333}$, or approximately 1.8257.

E X A M P L E   2   A dress manufacturer claims that the probability that a customer will buy a particular size dress is as follows:

Size, x	8	10	12	14	16	18
Probability, $p(x)$	0.11	0.21	0.28	0.17	0.13	0.10

Find the mean, variance, and standard deviation for this distribution.

SOLUTION   We first find μ by using Formula 6.2 and then arrange the data in tabular form as follows:

x	$p(x)$	$x \cdot p(x)$	$x - \mu$	$(x - \mu)^2$	$(x - \mu)^2 \cdot p(x)$
8	0.11	0.88	$8 - 12.6 = -4.6$	21.16	$(21.16)(0.11) = 2.3276$
10	0.21	2.10	$10 - 12.6 = -2.6$	6.76	$(6.76)(0.21) = 1.4196$
12	0.28	3.36	$12 - 12.6 = -0.6$	0.36	$(0.36)(0.28) = 0.1008$
14	0.17	2.38	$14 - 12.6 = 1.4$	1.96	$(1.96)(0.17) = 0.3332$
16	0.13	2.08	$16 - 12.6 = 3.4$	11.56	$(11.56)(0.13) = 1.5028$
18	0.10	1.80	$18 - 12.6 = 5.4$	29.16	$(29.16)(0.10) = 2.9160$

$$\mu = \Sigma x \cdot p(x) = 0.88 + 2.10 + 3.36 + 2.38 + 2.08 + 1.80 = 12.6$$

$$\sigma^2 = 2.3276 + 1.4196 + 0.1008 + 0.3332 + 1.5028 + 2.9160 = 8.6$$

Thus, the mean is 12.6, the variance is 8.6, and the standard deviation is $\sqrt{8.6}$, or approximately 2.9326.

Formula 6.2, like the formula for the variance of a set of numbers (see page 151), requires us first to compute the mean and then to find the square of the deviations from the mean. In many cases we do not wish to do this. For such situations we can use an alternate formula to calculate the variance of a probability distribution.

FORMULA 6.3

The variance of a discrete random variable with a given probability distribution is

$$\sigma^2 = \Sigma x^2 \cdot p(x) - [\Sigma x \cdot p(x)]^2$$

Formula 6.3 may seem strange but it is similar to Formula 3.5 on page 154. Let us see how Formula 6.3 is used.

E X A M P L E   3 Calculate the variance for the probability distribution given in Example 2 by using Formula 6.3.

SOLUTION   We arrange the data as follows:

x	$p(x)$	$x \cdot p(x)$	x^2	$x^2 \cdot p(x)$
8	0.11	0.88	64	$64(0.11) =$ 7.04
10	0.21	2.10	100	$100(0.21) =$ 21.00
12	0.28	3.36	144	$144(0.28) =$ 40.32
14	0.17	2.38	196	$196(0.17) =$ 33.32
16	0.13	2.08	256	$256(0.13) =$ 33.28
18	0.10	1.80	324	$324(0.10) =$ 32.40
		Total = 12.60		Total = 167.36

Using Formula 6.3, we find that the variance is

$$\sigma^2 = \Sigma x^2 \cdot p(x) - [\Sigma x \cdot p(x)]^2$$
$$= 167.36 - (12.6)^2$$
$$= 167.36 - 158.76$$
$$= 8.6$$

The variance is 8.6. This is the same result that we got using Formula 6.2.

EXERCISES FOR SECTION 6.4

1. A hospital administrator intends to improve the level of emergency room service by increasing support personnel. To justify the increase, the administrator has to estimate the mean time that patients must wait before being attended by a physician. This, of course, depends upon the number of patients present in the emergency room. Based upon extensive research, the following is available:

Waiting Time x	Probability $p(x)$
10 min	0.06
20 min	0.11
30 min	0.26
40 min	0.21
50 min	0.19
60 min	0.17

Find the mean, variance, and standard deviation for this distribution.

2. In an effort to determine a fair and equitable reimbursement rate for various dental services performed, most insurance companies keep accurate records on the charges submitted by clients. For the year 1993, the fee charged by many dentists in New York City for period prophyx curettage (dental procedure #4) the Equix Insurance reported that the following charges were submitted:

Charge x	Probability $p(x)$
$ 60	0.16
70	0.18
75	0.27
85	0.19
90	0.11
100	0.09

Find the mean, variance, and standard deviation for this distribution.

3. Child abuse is a serious situation. The following is the probability distribution for the number of reports of child abuse that one social agency in Miami receives daily:

Number of Reports x	Probability $p(x)$
0	0.05
1	0.43
2	0.17
3	0.25
4	0.06
5	0.03
6	0.01

Find the mean, variance, and standard deviation for this distribution.

4. The U.S. Bureau of the Census collects data on family size and publishes its results in *Current Population Reports*. In Danville the number of children that a family has and the corresponding probabilities are as follows:

Size of Family x	Probability p(x)
0	0.09
1	0.31
2	0.24
3	0.18
4	0.08
5	0.04
6	0.03
7	0.02
8	0.01

Find the mean, variance, and standard deviation for this distribution.

5. The Food and Nutrition Board of the *National Academy of Sciences* recommends that adults consume at least 800 milligrams of calcium daily. Nevertheless, many social agencies contend that people with incomes below the official poverty level consume less than the recommended daily allowance (RDA) of 800 mg. Various studies of people with incomes below the poverty level disclosed the following results:

Daily Intake of Calcium (mg) x	Probability p(x)
200	0.04
300	0.06
400	0.11
500	0.16
600	0.18
700	0.21
800	0.24

Find the mean, variance, and standard deviation for this distribution.

6. In *Residential Energy Consumption and Expenditures* published by the U.S. Energy Administration, information on residential energy consumption is presented. For one northeastern city, the following is available:

Annual Residential Energy Consumption (in million BTU) x	Probability $p(x)$
80	0.11
90	0.14
100	0.19
110	0.21
120	0.18
130	0.13
140	0.04

Find the mean, variance, and standard deviation for this distribution.

7. A survey of the number of video cassette recorders (VCRs) that a family in Dover owns produced the following results:

Number of VCRs x	Probability $p(x)$
0	0.07
1	0.36
2	0.29
3	0.13
4	0.09
5	0.06

Find the mean, variance, and standard deviation for this distribution.

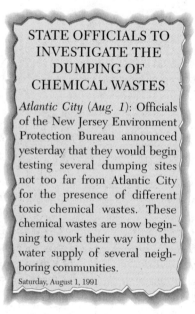

STATE OFFICIALS TO INVESTIGATE THE DUMPING OF CHEMICAL WASTES

Atlantic City (Aug. 1): Officials of the New Jersey Environment Protection Bureau announced yesterday that they would begin testing several dumping sites not too far from Atlantic City for the presence of different toxic chemical wastes. These chemical wastes are now beginning to work their way into the water supply of several neighboring communities.

Saturday, August 1, 1991

8. Refer to the newspaper article on the previous page. The state officials will be testing for the presence of up to eight possible toxic chemical pollutants. The probability that they will find different types of chemical pollutants is as follows.

Number of Toxic Pollutants Found x	Probability $p(x)$
0	0.08
1	0.14
2	0.16
3	0.18
4	0.21
5	0.17
6	0.03
7	0.02
8	0.01

Find the mean, variance, and standard deviation for this distribution.

9. Melinda Torres is fleet operations engineer for Suburban Transit Corp. During each commuting day, busses often break down and interfere with the smooth operation of the company, as the broken busses cannot be used on return runs. The following information on bus breakdowns is available:

Daily Number of Bus Breakdowns x	Probability $p(x)$
2	0.10
3	0.23
4	0.22
5	0.19
6	0.14
7	0.08
8	0.03
9	0.01

Find the mean, variance, and standard deviation.

10. A survey[*] of the families in Belleville found that 12% of the households had no motor vehicles, 36% had one vehicle, 33% had two vehicles, 18% had three vehicles, and 1% had four vehicles. Find the mean, variance, and standard deviation for this distribution.

[*]Rogers and Peters, 1994.

11. The MINITAB RANDOM program was used to randomly generate the digits 0 through 9. The program was repeated 1000 times. The following results were obtained:

Number Generated x	Frequency
0	88
1	96
2	105
3	103
4	111
5	84
6	93
7	97
8	108
9	115

a. Find μ
b. Find σ^2
c. Find σ

12. A die is altered by painting an additional dot on the face that originally had one dot as shown below. Let x be the number of dots that shows when the die is rolled.

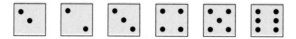

a. Find the probability distribution of x.
b. Find μ, σ^2, and σ for this distribution.

13. A local charity is planning to hold an outdoor fund-raising drive from which it hopes to raise $50,000. The probability of nice weather is 0.49. If it rains, the fund-raising drive will have to be canceled and the charity stands to lose $23,000 (rental fees, advertising, etc.). The probability of it raining is 0.51.
a. Find the charity's expected income from the fund-raising drive.
b. An insurance company is offering rain insurance, that is, for a premium of $1500, it will guarantee the $50,000 in the event of rain. Should the charity buy this type of insurance policy? Explain your answer.

14. A camera store receives a shipment of 20 cameras. It is known that eight of the cameras are defective. A sample of three cameras is selected with replacement. Let x be the number of defective cameras obtained.
a. Find the probability distribution of x.
b. Find the mean and standard deviation for this distribution.

6.5 THE BINOMIAL DISTRIBUTION

Consider the following probability problem:

EXAMPLE 1 Paula is about to take a five-question, true–false quiz. She is not prepared for the exam and decides to guess the answers without reading the questions.

ANSWER SHEET

Directions: For each question darken the appropriate space.

1. [True]   [False]
2. [True]   [False]
3. [True]   [False]
4. [True]   [False]
5. [True]   [False]

Find the probability that she gets

  a. all the answers correct.
  b. all the answers wrong.
  c. three out of the five answers correct.

SOLUTION   Let us denote a correct answer by the letter "c" and a wrong answer by the letter "w." There are two equally likely possible outcomes for question 1, c or w. Similarly, there are two equally likely possible outcomes for question 2 regardless of whether the first question was correct or incorrect. There are two equally likely possible outcomes for each of the remaining questions 3, 4, and 5. Thus, there are 32 equally likely possible outcomes since

$$2 \times 2 \times 2 \times 2 \times 2 = 32$$

These outcomes we list as follows:

ccccc	cwccc	wcccc	wwccc
cccwc	cwcwc	wccwc	wwcwc
ccccw	cwccw	wcccw	wwccw
cccww	cwcww	wccww	wwcww
ccwcc	cwwcc	wcwcc	wwwcc
ccwcw	cwwcw	wcwcw	wwwcw
ccwwc	cwwwc	wcwwc	wwwwc
ccwww	cwwww	wcwww	wwwww

TABLE 6.5   Number of Correct Answers

0	1	2	3	4	5
wwwww	cwwww	ccwww	cccww	wcccc	ccccc
	wcwww	cwcww	ccwcw	cwccc	
	wwcww	cwwcw	ccwwc	ccwcc	
	wwwcw	cwwwc	cwcwc	cccwc	
	wwwwc	wccww	cwccw	ccccw	
		wcwcw	cwwcc		
		wcwwc	wccwc		
		wwcwc	wcccw		
		wwccw	wcwcc		
		wwwcc	wwccc		

Once we have listed all possible outcomes, we can construct a chart similar to Table 6.5.

Now we can calculate the probability associated with each outcome. We have

$$p(0 \text{ correct}) = \frac{1}{32}$$

$$p(1 \text{ correct}) = \frac{5}{32}$$

$$p(2 \text{ correct}) = \frac{10}{32}$$

$$p(3 \text{ correct}) = \frac{10}{32}$$

$$p(4 \text{ correct}) = \frac{5}{32}$$

$$p(5 \text{ correct}) = \frac{1}{32}$$

We can picture these results in the form of a histogram, as shown in Figure 6.2. The relative frequency histogram for Figure 6.2 is shown in Figure 6.3.

We can now answer the question raised at the beginning of the problem. We have

a. $p(\text{all correct answers}) = \dfrac{1}{32}$

b. $p(\text{all wrong answers}) = \dfrac{1}{32}$

c. $p(3 \text{ out of } 5 \text{ corect answers}) = \dfrac{10}{32}$

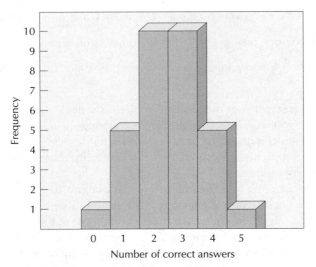

FIGURE 6.2   Histogram of the number of correct
answers obtained by guessing at five true–false questions.

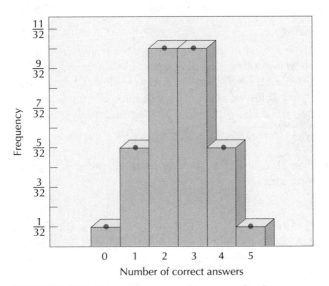

FIGURE 6.3   Relative frequency histogram for the
frequency distribution of Figure 6.2.

COMMENT   These answers assume that since Paula is not prepared, then p(correct answer) $= p$(wrong answer) $= \dfrac{1}{2}$. On the other hand, these answers would be wrong if Paula were better prepared and had a better probability of, say, 0.8, of getting a correct answer.

Many experiments or probability problems result in outcomes that can be grouped into two categories, success or failure. For example, when a coin is flipped there are two possible outcomes, heads or tails; when a hunter shoots at a target there are two possible outcomes, hit or miss; and when a baseball player is at bat the result is get on base or not get on base.

Statisticians apply this idea of success and failure to wide-ranging problems. For instance, if a quality-control engineer is interested in determining the life of a typical light bulb in a large shipment, each time a bulb burns out he has a "success." Similarly, if we were interested in determining the probability of a family having ten boys, assuming they planned to have ten children, then each time a boy is born we have a "success."

Binomial probability experiment

Bernoulli experiment

As we mentioned in Section 6.2 (page 339), experiments that have only two possible outcomes are referred to as **binomial probability experiments** or **Bernoulli experiments** in honor of the mathematician Jacob Bernoulli (1654–1705), who studied them in detail. Binomial probability experiments are characterized by the following.

Definition 6.5 | A **binomial probability experiment** is an experiment that satisfies the following properties:

1. There is a fixed number, n, of repeated trials whose outcomes are independent.
2. Each trial results in one of two possible outcomes. We call one outcome a success and denote it by the letter S and the other outcome a failure, denoted by F.
3. The probability of success on a single trial equals p and remains the same from trial to trial. The probability of a failure equals $1 - p = q$. Symbolically,

$$p(\text{success}) = p \quad \text{and} \quad p(\text{failure}) = q = 1 - p$$

4. We are interested in the number of successes obtained in n trials of the experiment.

COMMENT   Although very few real-life situations satisfy all of the above requirements, many of them can be thought of as satisfying, at least approximately, these requirements. Thus, we can apply the binomial distribution to many different problems.

Since all binomial probability experiments are similar in nature and result in either success or failure for each trial of the experiment, we seek a formula for determining the probability of obtaining x successes out of n trials of the experiment, where the probability of success on any one trial is p and the probability of failure is q. To achieve this goal let us consider the following.

A coin is tossed four times. What is the probability of getting exactly one head? We could list all the possible outcomes and count the number of favorable ones. This is shown below with the favorable outcomes circled.

HHHH	HHTT	THHH	(THTT)
HHHT	HHTH	THHT	THTH
HTHH	(HTTT)	TTHH	TTTT
HTTH	HTHT	(TTTH)	(TTHT)

Thus, the probability of getting exactly one head is $\dfrac{4}{16}$.

However, in many cases it is not possible or advisable to list all the possible outcomes. It is for this reason that we consider an alternate approach.

When a coin is tossed there are two possible outcomes, heads or tails. Thus,

$$p(\text{heads}) = \frac{1}{2} \quad \text{and} \quad p(\text{tails}) = \frac{1}{2}$$

Since we are interested in getting only a head, we classify this outcome as a success and write

$$p(\text{head}) = p(\text{success}) = \frac{1}{2} \quad \text{and} \quad p(\text{tail}) = p(\text{failure}) = \frac{1}{2}$$

Therefore,

$$p = \frac{1}{2}, q = \frac{1}{2} \quad \text{and} \quad p + q = \frac{1}{2} + \frac{1}{2} = 1$$

HISTORICAL NOTE

The Swiss mathematician Jacob Bernoulli (1654–1705) was born in Basel, where he became professor at the University. He was one of the three most celebrated of the eight members of the Bernoulli family. He and his brother Johann were among the first to understand and use the new method of the German mathematician Gottfried Wilhelm von Leibniz. He also wrote an essay on comets, suggested by the comet of 1680. His contributions to probability theory are contained in his book *Ars Conjectandi*, published after his death in 1713. This book also contains a reprint of an earlier treatise of Huygens. In 1657 the great Dutch mathematician Christian Huygens had written the first formal book on probability based on the Pascal-Fermat correspondences discussed in Chapter 1 (page 14). Huygens also introduced the important ideas of mathematical expectation discussed in Chapter 4. All of these ideas are contained in Bernoulli's book.

Each toss is independent of what happened on the preceding toss. We are interested in obtaining one head in four tosses. One possible way in which this can happen, along with the corresponding probabilities, is as follows.

Outcome	head	tail	tail	tail
Success or Failure	Success	Failure	Failure	Failure
Probability	$\frac{1}{2}$	$\frac{1}{2}$	$\frac{1}{2}$	$\frac{1}{2}$
Symbolically	p	q	q	q

Since the probability of success is p and the probability of failure is q, we can summarize this as $p \cdot q^3$. Remember q^3 means $q \cdot q \cdot q$. We would then say that the probability of getting one head is

$$\left(\frac{1}{2}\right) \cdot \left(\frac{1}{2}\right)^3 = \left(\frac{1}{2}\right) \cdot \left(\frac{1}{2}\right) \cdot \left(\frac{1}{2}\right) \cdot \left(\frac{1}{2}\right) = \frac{1}{16}$$

However, we have forgotten one thing. The head may occur on the second, third, or fourth toss.

Different ways in which one head can be obtained when a coin is tossed 4 times	head	tail	tail	tail
	tail	head	tail	tail
	tail	tail	head	tail
	tail	tail	tail	head

There are then four ways in which we can get one head. Thus, the $\frac{1}{16}$ that we calculated before can occur in four different ways. Therefore,

$$p(\text{exactly 1 head}) = 4\left(\frac{1}{16}\right) = \frac{4}{16}$$

Notice that this is exactly the same answer we obtained by listing all the possible outcomes.

Similarly, if we were interested in the probability of getting exactly two heads, we could consider one particular way in which this can happen.

Outcome	head	head	tail	tail
Success or Failure	Success	Success	Failure	Failure
Probability	$\dfrac{1}{2}$	$\dfrac{1}{2}$	$\dfrac{1}{2}$	$\dfrac{1}{2}$
Symbolically	p	p	q	q

The probability is thus

$$p^2 \cdot q^2 = \left(\frac{1}{2}\right)^2 \left(\frac{1}{2}\right)^2 = \left(\frac{1}{2}\right)\left(\frac{1}{2}\right)\left(\frac{1}{2}\right)\left(\frac{1}{2}\right) = \frac{1}{16}$$

Again, we must multiply this result by the number of ways that these two heads can occur in the four trials. This is the number of combinations of four things taken two at a time. We can use Formula 4.4 of Chapter 4. We get

$$_4C_2 = \frac{4!}{2!(4-2)!} = \frac{4!}{2!2!} = 6$$

Thus, the probability of getting two heads in four flips of a coin is

$$6\left(\frac{1}{16}\right) = \frac{6}{16} = \frac{3}{8}$$

More generally, if we are interested in the probability of getting x successes out of n trials of an experiment, then we consider one way in which this can happen. Here we have assumed that all the x successes occur first and all the failures occur on the remaining $n - x$ trials.

Success or Failure	Success · Success · Success · · · Failure · Failure · Failure
Probability	$p \cdot p \cdot p$ · · · $q \cdot q \cdot q$

$\underbrace{}_{x \text{ of them}}$ $\underbrace{}_{n - x \text{ of them}}$

This gives $p^x q^{n-x}$. We then multiply this result by the number of ways that exactly x successes can occur in n trials.

The number of ways that exactly x successes can occur in a set of n trials is given by

$$_nC_x = \frac{n!}{x!(n-x)!}$$

Here we have replaced r by x in the number of possible combinations formula (Formula 4.4). This leads us to the following **binomial distribution formula**.

FORMULA 6.4

Binomial
Distribution
Formula

Consider a binomial experiment that has two possible outcomes, success or failure. Let p(success) $= p$ and p(failure) $= q$. If this experiment is performed n times, then the probability of getting x successes out of the n trials is

$$p(x \text{ successes}) = {}_nC_x p^x q^{n-x} = \frac{n!}{x!(n-x)!} p^x q^{n-x}$$

We illustrate the use of Formula 6.4 with numerous examples.

EXAMPLE   2

Admission to
Medical School

Ninety percent of the graduates of State University who apply to a particular medical school are admitted. This year six graduates from State University have applied for admission to the medical school. Find the probability that only four of them will be accepted.

SOLUTION   Since six students have applied for admission, $n = 6$. We are interested in the probability that four are accepted, so $x = 4$. Also, 90% of the graduates who apply are admitted, so $p = 0.90$, and 10% of the graduates who apply are not admitted, so $q = 0.10$. Now we apply Formula 6.4. We have

$$\text{Prob}(4 \text{ are accepted}) = \text{Prob}(x = 4)$$

$$= \frac{6!}{4!(6-4)!}(0.90)^4(0.10)^{6-4}$$

$$= \frac{6!}{4!2!}(0.9)^4(0.1)^2$$

$$= \frac{6 \cdot 5 \cdot 4 \cdot 3 \cdot 2 \cdot 1}{4 \cdot 3 \cdot 2 \cdot 1 \cdot 2 \cdot 1}(0.9)(0.9)(0.9)(0.9)(0.1)(0.1)$$

$$= 0.0984$$

Thus, the probability that only four of them will be accepted is 0.0984.

EXAMPLE   3

Consider the newspaper article on the next page. Bill and his friends own five cars, which they park in front of his house. Find the probability that none of them will be stolen this year.

SOLUTION   Since Bill and his friends own five cars, $n = 5$. We are interested in the probability that none of the cars will be stolen so that $x = 0$. According to the newspaper article, 1 out of every 120 cars is stolen so that $p = \dfrac{1}{120} = 0.008$ and $q = \dfrac{119}{120} = 0.992$.

Applying Formula 6.4 gives

$$\text{Prob}(0 \text{ stolen cars}) = \text{Prob}(x = 0) = \frac{5!}{0!5!}(0.008)^0(0.992)^5$$

CAR THEFTS ON THE RISE AGAIN

Washington (April 7): Look out your window. Is your car still in your driveway or in front of your house? If it is there, then you're lucky. According to the latest FBI study released yesterday, there are an average of 2300 vehicles stolen per day in the United States. This puts the chances of your car being stolen at about 1 in 120. According to the survey, about 60% of the vehicles are stolen from private residences, apartments, or streets in residential areas between the hours of 6:00 P.M. and 6:00 A.M.

Most of the cars are stolen to be stripped for their parts or to be used for joyriding.

Wednesday—April 7, 1993

Remember that $(0.008)^0 = 1$, so that

$$\text{Prob(0 stolen cars)} = 1(0.992)^5 = 0.9606$$

Thus, the probability that none of these cars will be stolen this year is 0.9606.

E X A M P L E   4   Mario is taking a multiple-choice examination that consists of five questions. Each question has four possible answers. Mario guesses at every answer. What is the probability that he passes the exam if he needs *at least* four correct answers to pass?

SOLUTION   In order to pass, Mario needs to get at least four correct answers. Thus, he passes if he gets four answers correct or five answers correct. Each question has four possible answers so that $p(\text{correct answer}) = \frac{1}{4}$ and $p(\text{wrong answer}) = \frac{3}{4}$. Also, there are five questions, so $n = 5$. Therefore,

$$p(4 \text{ answers correct}) = \frac{5!}{4!1!}\left(\frac{1}{4}\right)^4\left(\frac{3}{4}\right)$$

$$= \frac{5 \cdot 4 \cdot 3 \cdot 2 \cdot 1}{4 \cdot 3 \cdot 2 \cdot 1 \cdot 1}\left(\frac{1}{4}\right)\left(\frac{1}{4}\right)\left(\frac{1}{4}\right)\left(\frac{1}{4}\right)\left(\frac{3}{4}\right)$$

$$= \frac{15}{1024}$$

and

$$p(5 \text{ answers correct}) = \frac{5!}{5!0!}\left(\frac{1}{4}\right)^5\left(\frac{3}{4}\right)^0 \qquad \text{(Any number to the 0 power is 1.)}$$

$$= \frac{5 \cdot 4 \cdot 3 \cdot 2 \cdot 1}{5 \cdot 4 \cdot 3 \cdot 2 \cdot 1 \cdot 1}\left(\frac{1}{4}\right)\left(\frac{1}{4}\right)\left(\frac{1}{4}\right)\left(\frac{1}{4}\right)\left(\frac{1}{4}\right) \cdot 1$$

$$= \frac{1}{1024}$$

Adding the two probabilities, we get

$$p(\text{at least 4 correct answers}) = p(4 \text{ answers correct}) + p(5 \text{ answers correct})$$

$$= \frac{15}{1024} + \frac{1}{1024}$$

$$= \frac{16}{1024} = \frac{1}{64}$$

Hence, the probability that Mario passes is $\frac{1}{64}$.

EXAMPLE 5 A shipment of 100 tires from the Apex Tire Corporation is known to contain 20 defective tires. Five tires are selected at random and each tire is replaced before the next tire is selected. What is the probability of getting *at most* two defective tires?

SOLUTION   We are interested in the probability of getting *at most* two defective tires. This means zero defective tires, one defective tire, or two defective tires. Thus, the probability of at most two defective tires equals

$$p(0 \text{ defective}) + p(1 \text{ defective}) + p(2 \text{ defective})$$

The probability of a defective tire is $\frac{20}{100}$, or $\frac{1}{5}$. Therefore, the probability of getting a nondefective tire is $\frac{4}{5}$. Now,

$$p(0 \text{ defective}) = \frac{5!}{0!5!}\left(\frac{1}{5}\right)^0\left(\frac{4}{5}\right)^5 = \frac{1024}{3125}$$

$$p(1 \text{ defective}) = \frac{5!}{1!4!}\left(\frac{1}{5}\right)^1\left(\frac{4}{5}\right)^4 = \frac{1280}{3125}$$

$$p(2 \text{ defective}) = \frac{5!}{2!3!}\left(\frac{1}{5}\right)^2\left(\frac{4}{5}\right)^3 = \frac{640}{3125}$$

Adding, we get

$$p(\text{at most 2 defectives}) = \frac{1024}{3125} + \frac{1280}{3125} + \frac{640}{3125} = \frac{2944}{3125}$$

Hence, the probability of getting at most two defective tires is $\frac{2944}{3125}$, or approximately 0.9421.

EXAMPLE 6 If the conditions are the same as in the previous problem except that now 15 tires are selected, what is the probability of getting *at least* one defective tire?

SOLUTION   We could proceed as we did in Example 4. Thus,

$$p(\text{at least 1 defective}) = p(1 \text{ defective}) + p(2 \text{ defective}) + \cdots + p(15 \text{ defective})$$

However, this involves a tremendous amount of computation. Recall (see the Rule on page 338) that the sum of all the values of a probability function must be 1. Thus,

$$p(0 \text{ defective}) + p(1 \text{ defective}) + p(2 \text{ defective}) + \cdots + p(15 \text{ defective}) = 1$$

Therefore, if we subtract $p(0 \text{ defective})$ from both sides, we have

$$p(1 \text{ defective}) + p(2 \text{ defective}) + \cdots + p(15 \text{ defective}) = 1 - p(0 \text{ defective})$$

Now

$$p(0 \text{ defective}) = \frac{15!}{0!15!}\left(\frac{1}{5}\right)^0\left(\frac{4}{5}\right)^{15} = 0.0352$$

Consequently, the probability of obtaining at least 1 defective tire is $1 - 0.0352$, or 0.9648.

COMMENT   Calculating binomial probabilities can sometimes be quite time-consuming. To make the job a little easier, we can use Table III of the Appendix, which gives us the binomial probabilities for different values of n, x, and p. No computations are needed. We only need to know the values of n, x, and p.

EXAMPLE 7 Given a binomial distribution with $n = 11$ and $p = 0.4$, use Table III of the Appendix to find the probability of getting

    a. exactly four successes.
    b. at most three successes.
    c. five or more successes.

SOLUTION   We use Table III of the Appendix with $n = 11$ and $p = 0.4$. We first locate $n = 11$ and then move across the top of the table until we reach the $p = 0.4$ column.

a. To find the probability of exactly four successes, we look for the value given in the table for $x = 4$. It is 0.236. Thus, when $n = 11$ and $p = 0.4$, the probability of exactly four successes is 0.236.

b. To find the probability of getting at most three successes, we look in the table for the values given for $x = 0$, $x = 1$, $x = 2$, and $x = 3$. These probabilities are 0.004, 0.027, 0.089, and 0.177, respectively. We add these (Why?) and get

$$0.004 + 0.027 + 0.089 + 0.177 = 0.297$$

Thus, the probability of at most three successes is 0.297.

c. To find the probability of five or more successes, we look in the chart for the values given for $x = 5$, $x = 6$, $x = 7$, $x = 8$, $x = 9$, $x = 10$, and $x = 11$. These probabilities are 0.221, 0.147, 0.070, 0.023, 0.005, 0.001, and 0, respectively. We add these and get

$$0.221 + 0.147 + 0.070 + 0.023 + 0.005 + 0.001 + 0 = 0.467$$

Thus, the probability of five or more successes is 0.467.

COMMENT   In the previous example you will notice that there is no value given in Table III when $x = 11$. It is left blank. Whenever there is a blank in the chart, this means that the probability is approximately 0. This is the reason that we used 0 as the probability in our calculations.

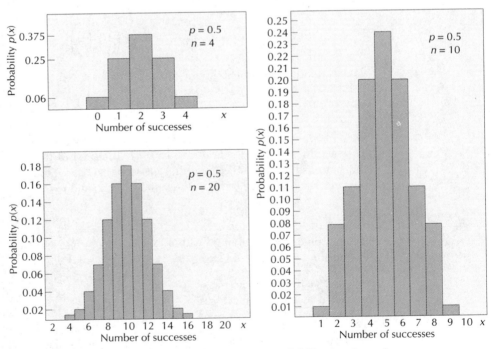

FIGURE 6.4   Binomial distributions with $p = 0.50$ and different values of n.

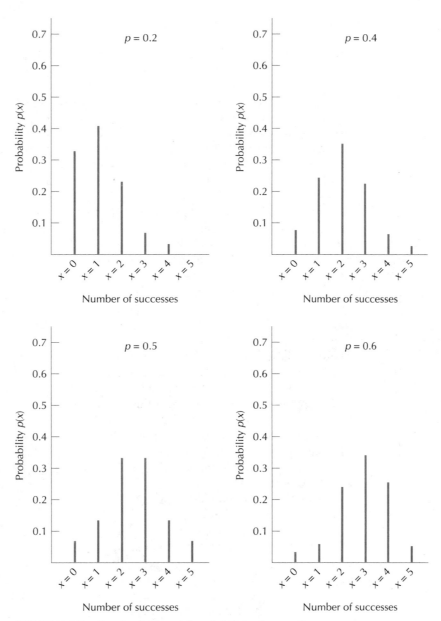

FIGURE 6.5 Graphs of binomial probabilities for $n = 5$.

When $p = 0.5$, then the binomial distribution is symmetrical and begins to resemble the normal distribution as the value of n gets larger. This is shown in Figure 6.4.

As a matter of fact, we can use the values given in Table III and draw the graphs of binomial probabilities. In Figure 6.5 we have drawn several such graphs using different values of p. In each case, however, $n = 5$. What do the graphs look like? We will have more to say about this in the next chapter.

EXERCISES FOR SECTION 6.5

1. According to the New York State Society of Anesthesiologists, 60% of operations performed in the United States today are done on an outpatient basis. (*Source: Journal of the American Medical Association*) If six operations that were performed recently are randomly selected, find the probability that exactly three of them were performed on an outpatient basis.

2. According to a study conducted by the University of Texas Dental School at San Antonio, an estimated 95% of Americans grind their teeth while asleep at some point in their life. (*Source: Vitality*, December 1993, p. 8) If eight Americans are randomly selected, what is the probability that four of them grind their teeth while asleep?

3. Although 49 states prohibit the sale of cigarettes to minors, one study (*Redbook*, 1993) found that 63% of underage children could buy them. Purchases by children account for $221 million in annual profit for the tobacco industry. If seven underage children are attempting to buy cigarettes, what is the probability that *all* seven are successful in buying the cigarettes?

4. According to the *American Orthopaedic Foot and Ankle Society* of Rosemont, 88% of American women wear shoes too small for their feet. If ten women are randomly selected, what is the probability that five of these are wearing shoes too small for their feet?

5. There is a 50% chance that United States youngsters view about three hours of TV commercials each week or 20,000 ads per year. (*Source: Worldwatch*, 1993) If seven U.S. youngsters are randomly selected, find the probability that
   a. exactly three of them view about three hours of TV commercials per week.
   b. at most three of them view about three hours of TV commercials per week.
   c. at least three of them view about three hours of TV commercials per week.

6. The Environmental Protection Agency reports that 19% of Americans recycled garbage in 1992. (*Source: Natural Health*, 1993) If six Americans are randomly selected, find the probability that
   a. exactly three of them recycle garbage.
   b. all of them recycle garbage.
   c. none of them recycle garbage.

7. *Ladies Home Journal* (1993) reports that after a long day, when they come home, 66% of dog owners greet their pet before they greet their spouse or children. The same is true with 57% of cat owners. If ten dog owners are randomly selected, what is the probability that when they come home, at least four of them will greet their pet before they greet their spouse or children?

8. Sixty percent of American victims of health care fraud are senior citizens.° If ten victims of health care fraud are randomly selected, what is the probability that

---

°*Source: Crime Victimization in the United States, 1991, A National Crime Survey Report.* U.S. Department of Justice: Washington, D.C., 1992.

a. all of them are senior citizens?

b. half of them are senior citizens?

c. none of them are senior citizens?

9. According to *Self* (1993), a gynecologist is the only doctor seen by 70% of American women. If nine women are randomly selected, find the probability that a gynecologist is the only doctor seen by

a. all of them.

b. four of them.

10. Refer to the newspaper article given on page 332 at the beginning of this chapter. If eight women over 40 years of age are randomly selected, what is the probability that *at most* three of them will develop breast cancer?

11. Consider the accompanying newspaper clipping. If seven voters are randomly selected, find the probability that

a. exactly four of them support the president.

b. all of them support the president.

c. none of them supports the president.

> ## THREE OUT OF FIVE SUPPORT THE PRESIDENT
>
> Washington, *Jan. 10*: A survey conducted by Hodge and Jones indicates that 3 out of 5 Americans support the President's health plan and think that it is in the country's best interest. The pollsters sampled 10,000 people across the country to obtain a cross section of voter sentiment.
>
> Thursday -January 10, 1994

12. According to the American Association of Retired Persons' magazine *Modern Maturity*, one in ten American families own a recreational vehicle (RV). More than 25 million people vacation each year in RVs. If nine vacationing American families are randomly selected, find the probability that only two of them own an RV.

13. A survey conducted by Louis Harris and Associates, Inc. (NY), disclosed that 63% of all Americans were overweight in 1993, up from 58% in 1983. If eight Americans are randomly selected, what is the probability that six of them are overweight?

14. According to the National Institute of Health,° 32% of all women will suffer a hip fracture because of osteoporosis by the age of 90. If six women aged 90 are randomly selected, find the probability that

°*American Demographics*, October 1985, p. 20.

a. exactly four of them will suffer (or have suffered) a hip fracture because of osteoporosis.

b. at most four of them will suffer (or have suffered) a hip fracture because of osteoporosis.

c. at least four of them will suffer (or have suffered) a hip fracture because of osteoporosis.

15. Consider the accompanying newspaper article. If nine people are randomly selected, find the probability that at most three of them are unemployed.

UNEMPLOYMENT RATE UP AGAIN

Newark, *March 10*: The Bureau of Labor Statistics announced yesterday that the unemployment rate for our region was up again and now stood at a seasonally adjusted 9.6%. This news, coupled with last week's announcement of the rise in the cost of living, forecasts economic hard times in the near future for our region.

March 10, 1990

If there are many vacant toll booths, what is the probability that cars will use some particular toll booths over others?

6.6 THE MEAN AND STANDARD DEVIATION OF THE BINOMIAL DISTRIBUTION

Consider the binomial distribution given on page 356, which is also repeated below. Recall that x represents the number of correct answers that Paula obtained on an exam of five questions, where the probability of a correct answer is $\dfrac{1}{2}$.

x	$p(x)$	$x \cdot p(x)$	x^2	$x^2 \cdot p(x)$
0	$\dfrac{1}{32}$	0	0	0
1	$\dfrac{5}{32}$	$\dfrac{5}{32}$	1	$\dfrac{5}{32}$
2	$\dfrac{10}{32}$	$\dfrac{20}{32}$	4	$\dfrac{40}{32}$
3	$\dfrac{10}{32}$	$\dfrac{30}{32}$	9	$\dfrac{90}{32}$
4	$\dfrac{5}{32}$	$\dfrac{20}{32}$	16	$\dfrac{80}{32}$
5	$\dfrac{1}{32}$	$\dfrac{5}{32}$	25	$\dfrac{25}{32}$
		$Total = \dfrac{80}{32} = \dfrac{5}{2}$		$Total = \dfrac{240}{32} = \dfrac{15}{2}$

Let us calculate the mean and variance for this distribution. Using Formula 6.1 of Section 6.3 we get

$$\mu = \Sigma x \cdot p(x) = 0 + \left(\frac{5}{32}\right) + \left(\frac{20}{32}\right) + \left(\frac{30}{32}\right) + \left(\frac{20}{32}\right) + \left(\frac{5}{32}\right)$$

$$= \frac{80}{32} = \frac{5}{2}$$

To calculate the variance we use Formula 6.3 of Section 6.4. From the preceding table we have

$$\sigma^2 = \Sigma x^2 \cdot p(x) - [\Sigma x \cdot p(x)]^2$$

$$= \frac{15}{2} - \left(\frac{5}{2}\right)^2$$

$$= \frac{30}{4} - \frac{25}{4} = \frac{5}{4}$$

Thus, $\mu = \dfrac{5}{2}$, or 2.5, and $\sigma^2 = \dfrac{5}{4}$, or 1.25.

Notice that if we multiply the total number of exam questions, which is 5, by the probability of a correct answer, which is $\frac{1}{2}$, we get $5\left(\frac{1}{2}\right) = 2.5$. This is exactly the same answer we get for the mean by applying Formula 6.1. We might be tempted to conclude that $\mu = np$. This is indeed the case.

Similarly, if we multiply the total number of questions with the probability of a correct answer and with the probability of a wrong answer, we get $5\left(\frac{1}{2}\right)\left(\frac{1}{2}\right) = 1.25$

Here we might conclude that $\sigma^2 = npq$. Again, this is indeed the case. We can generalize these ideas in the following.

FORMULA 6.5

Mean of a binomial distribution

The **mean of a binomial distribution**, μ, is found by multiplying the total number of trials with the probability of success on each trial. If there are n trials of the experiment and if the probability of success on each trial is p, then

$$\mu = np$$

FORMULA 6.6

Variance of a binomial distribution

The **variance of a binomial distribution** is given by

$$\sigma^2 = npq$$

The **standard deviation** σ is the square root of the variance. Thus,

$$\sigma = \sqrt{\text{variance}} = \sqrt{npq}$$

The proofs of Formulas 6.5 and 6.6 will be given as an exercise at the end of this section.

EXAMPLE   1   A die is rolled 600 times. Find the mean and standard deviation of the number of 1's that show.

SOLUTION   This is a binomial distribution. Since the die is rolled 600 times, $n = 600$. Also, there are six possible outcomes so that $p = \frac{1}{6}$ and $q = \frac{5}{6}$. Thus, using Formulas 6.5 and 6.6, we have

$$\mu = np$$
$$= 600\left(\frac{1}{6}\right) = 100$$

and

$$\sigma^2 = npq$$

$$= 600\left(\frac{1}{6}\right)\left(\frac{5}{6}\right) = 83.3333$$

Therefore, the mean is 100 and the standard deviation, which is the square root of the variance, is $\sqrt{83.3333}$ or approximately 9.1287. This tells us that if this experiment were to be repeated many times, we could expect an average of 100 1's per trial with a standard deviation of $\sqrt{83.3333}$ or 9.1287.

EXAMPLE   2   A large mail-order department store finds that approximately 17% of all purchases are returned for credit. If the store sells 100,000 different items this year, about how many items will be returned? Find the standard deviation.

SOLUTION   This is a binomial distribution. Since 100,000 items were sold, $n = 100,000$. Also, the probability that a customer will return the item is 0.17 so that $p = 0.17$. Therefore, the probability that the customer will not return the item is 0.83. Thus, using Formulas 6.5 and 6.6, we have

$$\mu = np$$

$$= (100,000)(0.17)$$

$$= 17,000$$

and

$$\sigma = \sqrt{npq}$$

$$= \sqrt{(100,000)(0.17)(0.83)}$$

$$= \sqrt{14110}$$

$$\approx 118.7855$$

Thus, the department store can expect about 17,000 items to be returned with a standard deviation of 118.7855.

Formulas 6.5 and 6.6 will be applied in greater detail in Chapter 7.

EXERCISES FOR SECTION 6.6

1. Atlas Travel Co. finds that 9% of all passengers booking travel packages also purchase flight cancellation insurance. If Atlas estimates that it will book 3000 travel packages this year, about how many flight cancellation insurance policies will it sell? Find the standard deviation.

2. The Taubus Electronics Company manufacturers and sells cordless telephones. Records show that approximately 7% of all cordless telephones sold by the company require servicing before the one-year manufacturer guarantee expires. If the company plans to manufacture 80,000 cordless phones this year, about how many can the company expect to service before the one-year manufacturer guarantee expires? Find the standard deviation.

3. Because of rising imports and sluggish domestic sales, the president of the ABCO Textile Corporation announces that 151 workers will have to be laid off during the next fiscal year. The company employs 7623 workers in its various divisions. Past experience indicates that about 2% of the workers die, resign, or retire during a fiscal year. Neglecting any other considerations and assuming that the past attrition rate will continue, can these cutbacks be accomplished through attrition (that is, by deaths, resignations, or retirements)?

4. The Consumer Affairs Commission finds that 44% of all complaints received involving unscrupulous merchants and fraudulent advertisements are valid. If the Commission received 4792 complaints during the year, about how many of them will be valid? Find the standard deviation.

5. Many Americans do not fulfill their civic responsibilities. A study by Blakely and Ribet (*Are You a Civic Minded American?*, 1994) found that 38% of the residents of one community ignore requests for jury duty. If jury-duty requests are mailed to 4730 residents of this community, about how many residents will ignore these requests? Find the standard deviation.

6. Many people invest money in mutual funds. Some are no-loads which do not charge anything for investing as opposed to load funds which charge a certain percentage as a sales charge. One particular load fund has a 7% sales charge. If the fund expects to raise $2,400,000 in new funds this year, about how much will it earn from sales charges? Find the standard deviation.

7. Once a passenger pays for an airline ticket, most airlines charge $35 for any travel plan changes. One particular airline has found that about 11% of passenger reservations are involved in travel plan changes of one type or another during the busy Thanksgiving Day travel period. If 22,000 passengers have confirmed reservations for travel during this period and have already paid for their tickets, about how much money can the company expect to earn from travel plan changes? Find the standard deviation.

1 OUT OF 6 SCHOOLS HAS AN ASBESTOS PROBLEM

New York (Sept. 11): Responding to numerous complaints from concerned parents, safety engineers yesterday visited many of the city's schools to test the air quality and health conditions at these facilities. It was found that unsafe conditions due to falling asbestos existed at 1 out of every 6 schools visited. The Board of Education has not yet responded to the report.

Saturday—September 11, 1993

8. Consider the newspaper article on the previous page. If 600 randomly selected schools are inspected, about how many of them can be expected to have an asbestos problem? Find the standard deviation.

9. Many colleges nationwide find that not all applicants who are accepted for admission to a college will actually attend that college. Past experience at Eastview College shows that about 88% of all the students accepted will actually attend the college. This year the college decided to send out 1800 acceptance letters. Assuming that the students make their decisions independently, find the mean and standard deviation of the number of students who will attend the college.

10. Refer back to the previous question. If the college would like to have an entering freshmen class of 1300 students, how many acceptance letters should it send out?

11. Recent data (1988–1991) from the National Health and Nutrition Examination Survey (Sempos, C.T. et al.: Prevalence of High Blood Cholesterol Among U.S. Adults: An update based on guidelines from the Second Report of the National Cholesterol Education Program Adult Treatment Panel, *JAMA* 269(23):3009–3014, June 16, 1993) shows that 31% of the adult U.S. population had borderline high blood cholesterol levels of between 200 and 239 mg/dl and 20% has high blood cholesterol levels of over 240 mg/dl. One noted cardiologist estimates that 5% of American adults will need drug therapy for hypercholesterolemia. If this doctor sees 5400 patients a year, how many of these patients can be expected to need drug therapy for hypercholesterolemia? Find the standard deviation.

12. One large Chicago mail-order department store finds that approximately 16% of all purchases are returned for credit. If the department store sells 75,000 different items during the year, about how many items will *not* be returned for credit?

*6.7 THE POISSON DISTRIBUTION

There are many practical problems in which we may be interested in finding the probability that x "successes" will occur over a given interval of time or a region of space. This is especially true when we do not expect many successes to occur over the time interval (which may be of any length, such as a minute, a day, a week, a month, or a year). For example, we may be interested in determining the number of days that school will be closed due to snowstorms, or we may be interested in determining the number of times that a baseball game will be postponed in a given season because of rain. For these and similar problems we can use the Poisson probability function formula.

Before giving this formula, we wish to emphasize that certain underlying assumptions must be satisfied for this formula to be applicable. Among these assumptions are the following:

1. Each "success" occurs independently of the others.
2. The probability of "success" in any interval or region is very small (or that most of the probability is concentrated at the low end of the domain of x).
3. The probability of a success in any one small interval is the same as that for any other small interval of the same size.

4. The number of successes in any interval is independent of the number of successes in any other nonoverlapping interval.

When these assumptions are satisfied we can use the following:

. FORMULA 6.7

Poisson
Probability
Distribution

The **Poisson probability distribution** representing the number of successes occurring in a given time interval or specified region is given by

$$p(x \text{ successes}) = \frac{e^{-\mu}\mu^x}{x!} \qquad x = 0, 1, 2, 3, \ldots$$

where μ is the average number of successes occurring in the given time interval or specified region. Thus, for the Poisson distribution, the mean is μ, the variance is μ, and the standard deviation is $\sqrt{\mu}$.

COMMENT   The symbol e that appears in the formula is used often in mathematics to represent an irrational number whose value is approximately equal to 2.71828. Table XII in the Appendix gives the values of $e^{-\mu}$ for different values of μ.

Let us illustrate the use of this formula with several examples.

EXAMPLE   1   An animal trainer finds that the number, x, of animal bites per month that her crew experiences follows an approximate Poisson distribution with a mean of 7.5. Find the variance and standard deviation of x, the number of animal bites per month.

SOLUTION   For a Poisson distribution the mean and variance are both equal to μ. Thus, in our case, we have

$$\mu = 7.5 \qquad \text{and} \qquad \sigma^2 = 7.5$$

The standard deviation is $\sqrt{7.5}$, or approximately 2.7386.

EXAMPLE   2
School Closings

Official records in a particular city show that the average number of school closings in a school year due to snowstorms is four. What is the probability that there will be six school closings this year because of snowstorms?

SOLUTION   Based on the given information, $\mu = 4$. We are interested in the probability that $x = 6$. Using the Poisson distribution formula, we get

$$p(x = 6) = \frac{e^{-4}(4)^6}{6!}$$

From Table XII, $e^{-4} = 0.0183156$ so that

$$p(x = 6) = \frac{e^{-4}(4)^6}{6!}$$

$$= \frac{(0.0183156)(4096)}{720} = 0.1042$$

Thus, the probability that there will be six school closings this year because of snow-storms is 0.1042.

EXAMPLE 3

Rainouts

From past experience a baseball club owner knows that about six games, on average, will have to be postponed during the season because of rain. Find the probability that this season

    a. three games will have to be postponed because of rain.
    b. no games will have to be postponed because of rain.

SOLUTION   Based on the given information, $\mu = 6$.

    a. Here we are interested in the probability that $x = 3$. We have

$$p(x = 3) = \frac{e^{-6}(6)^3}{3!}$$

From Table XII, $e^{-6} = 0.00247875$ so that

$$p(x = 3) = \frac{(0.00247875)(216)}{6} = 0.0892$$

Thus, the probability that three games will have to be postponed during the season because of rain is 0.0892.

    b. Here we are interested in the probability that $x = 0$. We have

$$p(x = 0) = \frac{e^{-6}(6)^0}{0!} \quad \text{(Remember } 6^0 = 1 \text{ and } 0! = 1\text{)}$$

$$= \frac{(0.00247875)1}{1}$$

$$= 0.00247875$$

Thus, the probability that no games will have to be postponed is 0.00247875.

EXERCISES FOR SECTION 6.7

1. The National Fire Protection Association reports that the number of forest fires reported per week in one national park follows a Poisson distribution with $\mu = 4$. What is the probability that the number of forest fires reported in this national park in a given week will be less than or equal to 2?

2. According to the *Statistical Abstract of the U.S. 1989*, p. 493, there were an average of 67.6 bank failures per year among FDIC banks during the period 1978–1987. Assume that for one particular region, the number of bank failures can be approximated by a Poisson distribution whose mean μ is 8. What is the probability that the number of bank failures will be less than or equal to 4 for this region?

3. A helicopter pilot on traffic patrol in Los Angeles claims that the number of car breakdowns occurring per morning drive-to-work period follows a Poisson distribution with an average of five per period. What is the probability that in a particular morning drive-to-work period
   a. there will be at most two car breakdowns?
   b. there will be at least two car breakdowns?
   c. there will be exactly two car breakdowns?

4. One software developer has established a toll-free support hotline for customers to call when they have problems with the software. After several years of operation, the software developer finds that the number of customers using the hotline follows a Poisson process with an average of seven calls per hour. What is the probability that on a particular workday, the hotline will receive
   a. at least three calls per hour?
   b. at most three calls per hour?
   c. exactly three calls per hour?

5. One study finds that the average annual rate of suicides on U.S. college campuses is 1 in 10,000 students, or about 0.0001. If a typical U.S. college with 15,000 students is randomly selected, what is the probability that there will be at least two suicides on this campus? (Assume that the Poisson distribution is applicable with $\mu = 15000$ $(0.0001) = 1.5$.)

6. The Marvo Insurance Company finds that the average number of auto accident claims involving uninsured motorists filed against the company is two per day. Assuming a Poisson process, what is the probability that on any given day
   a. no claim involving an uninsured motorist will be filed against the company?
   b. at most two claims involving an uninsured motorist will be filed against the company?
   c. at least two claims involving an uninsured motorist will be filed against the company?

7. Police department officials indicate that the average number of homicides per day in New York City is 5.4. Assuming a Poisson process, what is the probability that on any given day there will be
   a. at most four homicides?
   b. exactly four homicides?

8. One large fast-food chain receives an average of four complaints per day at its corporate headquarters. Assuming that the number of complaints received may be represented by a Poisson process, what is the probability that no complaints will be received on the next two business days?

9. The probability that a nurse at Brooks Hospital will call in sick on any day is 0.002. What is the probability that at most three of the hospital's 2400 nurses will call in

sick on any particular day? (*Hint:* Assume a Poisson process with $\mu = 2400 \times 0.002 = 4.8$)

10. The average number of coliform bacteria found in one polluted lake follows a Poisson distribution with an average of four bacteria per square centimeter. What is the probability that at most two bacteria per square centimeter will be found?

11. Marcy Ovington is the director of the Social Services agency for one city. She receives an average of five complaints a day about lack of child support by the other parent. Assuming that the number of lack of child support claims received follows a Poisson distribution, what is the probability that on any day, Marcy's agency will receive
   a. at least three lack of child support claims?
   b. at most three lack of child support claims?
   c. exactly three lack of child support claims?
   d. between two and four lack of child support claims?

12. A video store owner finds that the demand for one particular video has been following a Poisson distribution with $\mu = 5$. How large a stock of videos should the owner have on hand so as to be able to supply the customer demand with a probability of at least 0.90?

*6.8   HYPERGEOMETRIC DISTRIBUTION

In our discussion of the binomial distribution, we noted that when we select objects without replacement we cannot use the binomial distribution formula since the probability of success from trial to trial is not constant and the trials are not independent.

Before proceeding with our discussion, let us pause for a moment to review some notation. Recall that when we discussed combinations we used the notation $_nC_r$ to represent the number of combinations of n things taken r at a time. As mentioned, another notation that is often used to represent the same idea is $\binom{n}{r}$. Thus,

$$_5C_3 = \binom{5}{3} = \frac{5!}{3!2!}$$

$$_{10}C_8 = \binom{10}{8} = \frac{10!}{8!2!}$$

and

$$_nC_r = \binom{n}{r} = \frac{n!}{r!(n-r)!}$$

Suppose 5 cards are randomly drawn without replacement from a 52-card deck. Can we find the probability of obtaining 3 red and 2 black cards? The answer is yes if we proceed as follows. The 3 red cards can be selected from the 26 available red cards in $\binom{26}{3}$ possible ways. Similarly, the 2 black cards can be selected from the 26 available black cards in $\binom{26}{2}$ possible ways. Thus, the total number of ways of selecting 3 red cards and 2 black cards in five draws is $\binom{26}{3} \cdot \binom{26}{2}$. Now we determine how many different ways there are for selecting 5 cards from a 52-card deck. This is simply $\binom{52}{5}$. Using the definition of probability, we have

$$\text{Prob}\begin{pmatrix} 3 \text{ red cards and 2 black} \\ \text{cards in 5 draws from a} \\ 52\text{-card deck} \end{pmatrix} = \frac{\binom{26}{3} \cdot \binom{26}{2}}{\binom{52}{5}}$$

$$= \frac{\left(\dfrac{26!}{3!23!}\right) \cdot \left(\dfrac{26!}{2!24!}\right)}{\left(\dfrac{52!}{5!47!}\right)}$$

$$= \frac{(2600)(325)}{2{,}598{,}960} = 0.3251$$

More generally, suppose we are interested in the probability of selecting x successes from k items labeled success and $n - x$ failures from $N - k$ items labeled failure when a random sample of size n is selected from N items. We can then apply the hypergeometric probability function to determine this probability. We have

FORMULA 6.8

Hypergeometric Probability Function

The probability of obtaining x successes when a sample of size n is selected without replacement from N items of which k are labeled success and $N - k$ are labeled failure is given by

$$\frac{\binom{k}{x}\binom{N-k}{n-x}}{\binom{N}{n}} \qquad x = 0, 1, 2, \ldots, n$$

Let us illustrate the use of this formula with several examples.

EXAMPLE 1   A production run of 100 radios is received by the shipping department. It is known that ten of the radios in the production run are defective. The quality-control engineer randomly selects eight of the radios from the production run. Find the probability that six of the radios selected are defective.

SOLUTION   We apply the hypergeometric probability function. Here $N = 100$, $n = 8$, $k = 10$, and $N - k = 100 - 10 = 90$. We are interested in the probability that $x = 6$. Thus,

$$\text{Prob}(x = 6) = \frac{\binom{10}{6}\binom{90}{2}}{\binom{100}{8}}$$

$$= \frac{\left(\dfrac{10!}{6!4!}\right) \cdot \left(\dfrac{90!}{2!88!}\right)}{\left(\dfrac{100!}{8!92!}\right)}$$

$$= \frac{(210)(4005)}{186087894300} = 0.0000045$$

Thus, the probability that the quality-control engineer selects six defective radios is 0.0000045.

EXAMPLE 2   A faculty–student committee is to be selected at random from three students and six faculty members. The committee is to consist of five people. Find the probability that the committee will contain

    a. no students.
    b. one student.
    c. two students.
    d. three students.

SOLUTION   Let x be the number of students selected to be on the committee.

    a. Here $x = 0$, $N = 9$, $n = 5$, and $k = 3$. Thus,

$$\text{Prob}(0 \text{ students}) = \text{Prob}(x = 0) = \frac{\binom{3}{0}\binom{6}{5}}{\binom{9}{5}}$$

$$= \frac{(1)(6)}{126} = \frac{6}{126}$$

b. Here $x = 1$, $N = 9$, $n = 5$, and $k = 3$. Thus,

$$\text{Prob(1 student)} = \text{Prob}(x = 1) = \frac{\binom{3}{1}\binom{6}{4}}{\binom{9}{5}}$$

$$= \frac{(3)(15)}{126} = \frac{45}{126}$$

c. Here $x = 2$, $N = 9$, $n = 5$, and $k = 3$. Thus,

$$\text{Prob(2 students)} = \text{Prob}(x = 2) = \frac{\binom{3}{2}\binom{6}{3}}{\binom{9}{5}}$$

$$= \frac{(3)(20)}{126} = \frac{60}{126}$$

d. Here $x = 3$, $N = 9$, $n = 5$, and $k = 3$. Thus,

$$\text{Prob(3 students)} = \text{Prob}(x = 3) = \frac{\binom{3}{3}\binom{6}{2}}{\binom{9}{5}}$$

$$= \frac{(1)(15)}{126} = \frac{15}{126}$$

EXERCISES FOR SECTION 6.8

For each of the following exercises, assume that the sampling is to be done without replacement.

1. There are 100 college students who have a valid driver's license in an auditorium. It is known that 15 of these people can also drive a motorcycle. If a random sample of 10 of these 100 college students is taken what is the probability that the sample contains
   a. two people who can also drive a motorcycle?
   b. at most two people who can also drive a motorcycle?
2. It is known that 23 of the 68 cars parked in a municipal parking facility are equipped with air bags and ABS (antilock braking system). If a random sample of 11 of these cars is taken (without replacement), what is the probability that the sample contains at most three cars with airbags and ABS?
3. One leading auto transmission repair company on the East coast advertises extensively with the claim "50% of all cars with transmission troubles that we service do

not need a new transmission but rather need minor adjustments." A random sample of five of 12 cars with transmission troubles is taken. Assuming that the repair company's claim is valid, what is the probability that all five of them will not need a new transmission?

4. U.S. Customs officials claim that one fourth of the 100 migrant workers working on the farms in one region of California are illegal aliens. If 12 migrant workers from this region are randomly selected, what is the probability that none of them is an illegal alien?

5. A random sample of five homes is taken from a region in which it is known that 15 of 25 homes have functioning smoke detectors (in the event of fire). What is the probability that two of the homes have functioning smoke detectors?

3 STRIKES AND YOU'RE IN

Trenton (Jan. 30): Governor Christie Whitman announced yesterday her support of a new tough anti-crime proposal whereby any person convicted of committing violent crimes 3 times would face a mandatory life imprisonment. Several other states are considering similar such proposals.

Saturday—January 29, 1994

6. Consider the accompanying newspaper article. It is known that 35 of 70 of the state legislators are in favor of the new anticrime proposal of life imprisonment for people convicted three times of violent crimes. A sample of ten of these legislators is randomly selected. What is the probability that half of the legislators in the sample are in favor of the new anticrime proposal?

7. A task force of four people is to be randomly selected from eight professors and five students to search for a replacement for the retiring president of the college. What is the probability that the task force will consist of
   a. no students?
   b. one student?
   c. two students?
   d. three students?
   e. four students?

8. According to the Automobile Manufacturer's Association, 52% of all cars sold in the United States in 1992 were purchased by females. One dealership sold 100 cars in May to females, 40 of which were equipped with an antilock braking system. If

10 automobiles that were sold by this dealership in May 1992 to females are randomly selected, what is the probability that five are equipped with an ABS?

9. As reported by the U.S. Bureau of Justice statistics in *Profile of Jail Inmates*, 94% of the jail inmates in 1993 were male. One particular correctional facility has 470 male and 30 female inmates. If 30 of these 500 jail inmates are randomly selected, what is the probability that 25 of them are male?

10. In a survey conducted by *Ladies Home Journal* (June 1988), 80% of the married women interviewed indicated that they would marry the same man again, if given the chance. It is known that 16 of the 20 married women of the Briarwood Club would marry the same man again, if given the chance. If 10 of these 20 married women are randomly selected, what is the probability that at least seven of them would marry the same man again, if given the chance?

6.9 USING COMPUTER PACKAGES

The MINITAB statistical package can be used to compute binomial probabilities without having to perform any computations. The MINITAB command **PDF; BINOMIAL** accomplishes this. We illustrate the use of this command with the following example.

E X A M P L E   1   Medical researchers have found that approximately 15% of the population who take a particular antihistamine drug will develop some form of reaction to the drug. If the drug is administered to six people, find the probability that

a. exactly two of these people will develop some form of reaction to the drug.
b. exactly three of these people will develop some form of reaction to the drug.

SOLUTION   In order to get MINITAB to compute these probabilities, we first type the command PDF followd by a semicolon. Then, on the next line, we type the subcommand BINOMIAL followed by the values of n and p. In our case $n = 6$ and $p = 0.15$.

When these commands are applied to our case we obtain the following MINITAB printout:

```
MTB > PDF;
SUBC> BINOMIAL N=6, P=0.15.
   BINOMIAL WITH N =   6   P = 0.150000
      K              P( X = K)
      0               0.3771
      1               0.3993
      2               0.1762
      3               0.0415
      4               0.0055
      5               0.0004
      6               0.0000
```

From this printout we can easily read our answers. The probability that exactly two of these people (X = 2) will develop some form of reaction is 0.1762. Also, the probability that exactly three of these people will develop some form of reaction is 0.0415.

By using the above printout, we can easily answer such questions as "What is the probability that at most five of these people will develop some form of reaction?" We simply add the appropriate probabilities as listed; that is, we add 0.3771 + 0.3993 + 0.1762 + 0.0415 + 0.0055 + 0.0004.

We can also get MINITAB to give us individual probabilities. Thus, in the previous example, if we were interested only in the probability that exactly three of these people will develop some form of reaction, we would simply type a "3" after the PDF command. We would then have the following MINITAB printout:

```
MTB > PDF 3;
SUBC> BINOMIAL N=6, P=0.15.
     K              P( X = K)
   3.00              0.0415
```

The MINITAB statistical package can also be used to compute Poisson probabilities. The MINITAB command **PDF; POISSON** accomplishes this, as can be seen in the following example.

EXAMPLE 2 Records for Camp Morgan indicate that the probability a camper will be involved in an accident requiring sutures (stitches) follows a Poisson distribution with mean $\mu = 12$ campers per summer. Using MINITAB, find the probability that

    a. between four and six campers inclusive will require sutures during the summer.
    b. at most eight campers will require sutures during the summer.

SOLUTION   In this case, the random variable x represents the number of campers requiring sutures during the summer.

    a. Here we must find

$$\text{Prob}(4 \leq x \leq 6)$$

We first place the values 4, 5, and 6 in column C1. We can then apply the PDF command followed by the POISSON subcommand as follows:

```
MTB  > SET THE FOLLOWING IN C1

DATA > 4 5 6

DATA > END

MTB  > PDF C1;

SUBC > POISSON 12.
```

```
        K        P(X = K)
       4.00       0.0053
       5.00       0.0127
       6.00       0.0255

MTB > STOP
```

Thus,

$$p(4 \le x \le 6) = p(x = 4) + p(x = 5) + p(x = 6)$$
$$= 0.0053 + 0.0127 + 0.0255 = 0.0435$$

COMMENT   When using the Poisson subcommand, we must follow this by the mean and then a period for MINITAB to generate the specified output.

b. To find the probability that at most eight campers will require sutures during the summer, we must find $p(x \le 8)$. To accomplish this, we use the CDF command as follows:

```
MTB > CDF 8;

SUBC > POISSON 12.

        K        P( X LESS OR = K)
       8.00            0.1550

MTB > STOP
```

Thus, $p(x \le 8) = 0.1550$.

EXERCISES FOR SECTION 6.9

1. A farmer plants ten trees where the probability that each tree will take root and grow properly is 0.6. Use MINITAB to obtain the probability distribution of x, the number of trees that will actually take root and grow properly.

2. A survey in the *American Journal of Diseases of Children* (May 1987) found that 30% of the population of children 6–11 years old are obese. If eight children 6–11 years old are randomly selected, use MINITAB to obtain the probability distribution of x, the number of these children that are overweight.

3. A large automobile insurance company advertises that 95% of its policyholders renew their policy with the company because of their level of satisfaction. Suppose ten of the company's policyholders are randomly selected. Let x denote the number of policyholders who would renew their policy with the company. MINITAB was used to obtain the following printout:

```
BINOMIAL WITH N = 10 P = 0.950000
   K            P( X = K)
   4              0.0000
   5              0.0001
   6              0.0010
   7              0.0105
   8              0.0746
   9              0.3151
  10              0.5987
```

Using the above printout, find the probability that out of the ten policyholders selected

a. seven will renew their policy with the company.

b. more than seven will renew their policy with the company.

c. between seven and nine, inclusive, will renew their policy with the company.

4. Past experience indicates that the number of fish caught per man hour of fishing effort at a particular location on a stocked lake follows a Poisson distribution with mean μ equal to 5.7. If a person fishes there for one hour, use MINITAB to obtain the probability distribution of x, the number of fish that the person will catch in one hour.

5. Refer back to the previous exercise. If a person fishes there for two hours, use MINITAB to obtain the probability distribution of x, the number of fish that the person will catch in two hours.

6. Refer back to Exercise 4. Use MINITAB to find the probability that the person will catch at most ten fish.

7. Market research indicates that the number of telephone inquiries that a person receives when placing an "auto for sale" ad in one local newspaper follows a Poisson distribution with $\mu = 6.4$ calls for each time the ad is run. Use MINITAB to obtain the probability distribution of x, the number of telephone inquiries that a person will receive when placing an "auto for sale" ad once in the local newspaper.

8. Refer back to the previous exercise. Use MINITAB to find the probability that the person will receive at most 25 calls when the ad is run on seven different occasions in the local newspaper.

6.10 SUMMARY

In this chapter we discussed how the ideas of probability can be combined with frequency distributions. Specifically, we introduced the idea of a random variable and its probability distribution. These enable an experimenter to analyze outcomes of experiments and to speak about the probability of different outcomes.

We then discussed the mean and variance of a probability distribution. These allow us to determine the expected number of favorable outcomes of an experiment.

Although we did not emphasize the point, all the events discussed were mutually exclusive. Thus, we were able to add probabilities by the addition rule for probabilities. Also, the events were independent. This allowed us to multiply probabilities by the multiplication rule for probabilities of independent events.

We discussed one particular distribution in detail, the binomial distribution, since it is one of the most widely used distributions in statistics. In addition to the binomial distribution formula itself, which allows us to calculate the probability of getting a specified number of successes in repeated trials of an experiment, formulas for calculating its mean, variance, and standard deviation were given. These formulas were applied to numerous examples. Because of its importance, we will discuss the binomial distribution further in Chapter 7.

We also discussed the Poisson distribution, which can be used only when certain assumptions are satisfied. These were given on page 375.

Finally we presented the hypergeometric probability function, which is used when the sampling is without replacement. Thus, we cannot use the binomial distribution since the probability of success is not the same from trial to trial.

Study Guide

The following is a chapter outline in capsule form. You should now be able to demonstrate your knowledge of the ideas mentioned by giving definitions, descriptions, or specific examples. Page references are given in parentheses.

If an experiment is performed and some quantitative variable denoted by x is measured or observed, the quantitative variable x is called a **random variable** since the values that x may assume in the given experiment depend on chance. A random variable can take on numerical values only. (page 334)

Whenever all the possible values that a random variable may assume can be listed (or counted), the random variable is said to be **discrete**. (page 334)

Random variables that can assume values corresponding to any of the points contained in one or more intervals on a line are called **continuous random variables**. (page 335)

A **probability function** or **probability distribution** is a correspondence that assigns probabilities to the values of a random variable. The sum of all the probabilities of a probability function is always 1. (page 338)

If a random variable has only two possible values, and if the probability of these values remains the same for each trial regardless of what happened on any previous trial, the variable is called a **binomial variable** and its probability distribution is called a **binomial distribution**. (page 339)

The **mean** of a random variable, denoted by μ, for a given probability distribution is the number obtained by multiplying all the possible values of the random variable having this particular distribution by their respective probabilities and adding these products together. The mean is often called the **mathematical expectation** or the **expected value**. (page 345)

The difference of a number from the mean is called the **deviation from the mean**. (page 346)

The **variance** of a random variable with a given probability distribution is the number obtained by multiplying each of the squared deviations from the mean by their respective probabilities and adding these products. (page 346)

The **standard deviation** of a random variable with a given probability distribution is the square root of its variance. (page 347)

A **Bernoulli experiment** or **binomial probability experiment** is an experiment that satisfies the following properties:

1. There is a fixed number, n, of repeated trials whose outcomes are independent.
2. Each trial results in one of two possible outcomes. We call one outcome a success and denote it by the letter S and the other outcome a failure denoted by F.
3. The probability of success on a single trial equals p and remains the same from trial to trial. The probability of a failure equals $1 - p = q$. Symbolically,

$$p(\text{success}) = p \quad \text{and} \quad p(\text{failure}) = q = 1 - p$$

4. We are interested in the number of successes obtained in n trials of the experiment. (page 358)

The **binomial distribution formula** states the following: Consider a binomial experiment that has two possible outcomes, success or failure. Let $p(\text{success}) = p$ and $p(\text{failure}) = q$. If this experiment is performed n times, then the probability of getting x successes out of the n trials is

$$p(x \text{ successes}) = {}_nC_x p^x q^{n-x} = \frac{n!}{x!(n-x)!} p^x q^{n-x} \qquad \text{(page 362)}$$

The **mean of a binomial distribution**, μ, is found by multiplying the total number of trials with the probability of success on each trial. If there are n trials of the experiment, and if the probability of success on each trial is p, then $\mu = np$. (page 372)

The **variance of a binomial distribution** is given by $\sigma^2 = npq$. (page 372)

The **standard deviation** is the square root of the variance. Thus,

$$\sigma = \sqrt{\text{variance}} = \sqrt{npq} \qquad \text{(page 372)}$$

The **Poisson probability distribution** representing the number of successes occurring in a given time interval or specified region is given by

$$p(x \text{ successes}) = \frac{e^{-\mu}\mu^x}{x!} \qquad x = 0, 1, 2, 3, \ldots$$

where μ is the average number of successes occurring in the given time interval or specified region. Thus, for the Poisson distribution the mean is μ, the variance is μ, and the standard deviation is $\sqrt{\mu}$. (page 376)

Hypergeometric probability function. The probability of obtaining x successes when a sample of size n is selected without replacement from N items of which k are labeled success and $N - k$ are labeled failure is given by

$$\frac{\binom{k}{x}\binom{N-k}{n-x}}{\binom{N}{n}} \qquad x = 0, 1, 2, \ldots, n \qquad \qquad \text{(page 380)}$$

Formulas to Remember

You should be able to identify each symbol in the following formulas, understand the relationships among the symbols expressed in each formula, understand the significance of each formula, and use the formulas in solving problems.

1. Mean of a probability distribution: $\mu = \Sigma x \cdot p(x)$

2. Variance of a probability distribution: $\sigma^2 = \Sigma (x - \mu)^2 \cdot p(x)$

3. Variance of a probability distribution: $\sigma^2 = \Sigma x^2 \cdot p(x) - [\Sigma x \cdot p(x)]^2$

4. Standard deviation of a probability distribution: $\sigma = \sqrt{\text{variance}}$

5. Binomial distribution

$$p(x \text{ successes out of } n \text{ trials}) = \frac{n!}{x!(n-x)!} p^x q^{n-x}$$

6. Mean of the binomial distribution: $\mu = np$

7. Variance of the binomial distribution: $\sigma^2 = npq$

8. Poisson distribution: $p(x \text{ successes}) = \dfrac{e^{-\mu}\mu^x}{x!}$ $x = 0, 1, 2, 3, \ldots$

9. Mean of the Poisson distribution is μ.

10. Variance of a Poisson distribution is μ.

11. $_nC_x = \dbinom{n}{x} = \dfrac{n!}{x!(n-x)!}$

12. Hypergeometric distribution:

$$p(x \text{ successes}) = \frac{\binom{k}{x}\binom{N-k}{n-x}}{\binom{N}{n}} \qquad x = 0, 1, 2, \ldots, n$$

Testing Your Understanding of This Chapter's Concepts

1. A die is altered by painting an additional dot on the face that originally had one dot (see the following diagram). A pair of such altered dice is rolled. Let x be the number of dots showing on both dice together.

    a. Find the probability distribution of x.

    b. Find μ, σ^2, and σ for this distribution.

2. Many large supermarkets use computers that scan the items at the checkout counter. The scanner then prints the correct price of the item on the register. Some stores have such faith in these computers that they make the following claim: "If the computer scanner generates a price that is different from the price marked on the item, we will pay you double the difference." If the computer scanners are alleged to be 95% accurate, that is, they will correctly scan the price 95% of the time, what is the probability that in nine items scanned, there will be *at most* one item scanned incorrectly?

3. Given the probability function

$$p(x) = \frac{9 - x}{45} \qquad \text{for } x = 0, 1, 2, 3, 4, 5, 6, 7, \text{ or } 8$$

Find the mean and standard deviation for this distribution.

4. Suppose x is a random variable that follows the hypergeometric distribution with $N = 10$, $n = 6$, and $k = 7$.

    a. Find the probability distribution of x.

    b. Find the mean and variance of x.

c. What is the probability that x will fall within the interval $\mu - 2\sigma$ and $\mu + 2\sigma$? (*Hint:* Use Chebyshev's Theorem discussed in Chapter 3. We can obtain the exact results by computing $p(x = 3) + p(x = 4) + p(x = 5) = \dfrac{203}{210}$.)

5. If a binomial experiment is repeated n times, then the number of branches in the tree diagram of the outcomes is 2^n. If a binomial experiment is repeated three times, draw a tree diagram showing that the number of branches is 2^3 or 8.

6. In a coin bank there are five pennies, three nickels, four dimes, two quarters, and one half-dollar. Two coins are randomly selected with the first coin replaced before the second coin is drawn. Let x denote the sum of money obtained using both coins.
   a. Find the probability distribution of x.
   b. Find the mean of this distribution.
   c. Find the standard deviation of this distribution.

Chapter Test

Multiple Choice Questions

1. The following represents a probability distribution for a random variable x. Find the probability that $x = 7$.

x	6	7	8	9	10
$p(x)$	0.18	?	0.32	0.25	0.17

   a. 0   b. 0.92   c. 0.8   d. 0.08   e. none of these

2. An auto mechanic installs eight automatic transmissions in eight cars. If the probability that any one of the transmissions will not need any adjustments is $\dfrac{1}{5}$, what is the probability that exactly three of the transmissions will not need any adjustments?

   a. $\dfrac{8!}{3!}\left(\dfrac{1}{5}\right)^3\left(\dfrac{4}{5}\right)^5$ b. $\dfrac{8!}{3!5!}\left(\dfrac{1}{5}\right)^3\left(\dfrac{4}{5}\right)^5$ c. $\dfrac{8!}{5!}\left(\dfrac{1}{5}\right)^3\left(\dfrac{4}{5}\right)^5$

   d. $\dfrac{8!}{3!5!}\left(\dfrac{1}{5}\right)^5\left(\dfrac{4}{5}\right)^3$ e. none of these

3. It is estimated that 20% of the members of a health club have high blood pressure. If 300 members of the club are randomly selected, about how many of them can be expected to have high blood pressure?

   a. 240   b. 60   c. 300   d. 20   e. none of these

4. Refer back to the previous question. Find the standard deviation.

   a. 48   b. 6.9282   c. 0.2   d. 0.16   e. none of these

5. The following table gives the probabilities of the number of police officers who will call in sick (daily) for a large police force.

Number of Police Officers, x	5	6	7	8	9	10	11
Probability, $p(x)$	0.07	0.14	0.17	0.21	0.18	0.19	0.04

Find the mean of this distribution.
    a. 8   b. 7.82   c. 8.13   d. 8.02   e. none of these

6. Refer back to the previous question. Find the standard deviation.
    a. 66.98   b. 64.3204   c. 2.6596   d. 1.6308
    e. none of these

7. Can the following be a probability distribution for a random variable x?

Random Variable, x	1	2	3	4	5
Probability, $p(x)$	$\frac{1}{3}$	$\frac{1}{7}$	$\frac{1}{21}$	$\frac{4}{21}$	$\frac{3}{7}$

    a. yes   b. no   c. not enough information given
    e. none of these

8. A coin is tossed in such a way that the probability of it coming up heads 0.3. The coin is tossed five times. What is the probability of getting exactly three heads?

    a. $\dfrac{5!}{3!}(0.3)^3(0.7)^2$ b. $\dfrac{5!}{2!}(0.3)^3(0.7)^2$ c. $\dfrac{5!}{3!2!}(0.3)^3(0.7)^2$

    d. $\dfrac{5!}{3!2!}(0.3)^2(0.7)^3$ e. none of these

Supplementary Exercises

9. A research organization decides to mail out ten questionnaires to people selected at random. If the probability of any one person answering the questionnaire is $\frac{1}{7}$, find the probability that at least six people will answer the questionnaire.

10. A washing machine manufacturer claims that daily service calls for the company's washing machines follow a Poisson distribution with $\mu = 5$. What is the probability that on any given day there will be three calls for service?

11. An urn contains nine red balls and 15 green balls. Five balls are randomly selected from the urn without replacement. Find the probability that three of the balls are red.

12. In a certain region, the probability that a house contains a dangerous accumulation of radon gas is $\frac{1}{5}$. If six houses in the region are randomly selected, what is the probability that exactly four of these houses contain dangerous accumulations of radon gas?

13. Assume that the daily number of train cancellations on a particular railroad follows a Poisson distribution with $\mu = 7$. What is the probability that on any given day there will be at most five cancellations?

14. *Cancer.* According to medical researchers, about 20% of individuals diagnosed as having one particular form of cancer die within a year. If 12 individuals are diagnosed as having this form of cancer, find the probability that *at most* five of them will die within a year.

ONE FORM OF CANCER UNDER CONTROL

Los Angeles (May 6)—Cancer researchers announced yesterday that they had developed a treatment for one form of deadly cancer. Currently, the rate of success with this new treatment is 80%. With further improvement, this rate of success is expected to rise substantially.

May 6, 1988

15. Doctors estimate that the probability of a pregnant woman's having identical twins is approximately 0.004. If eight births at Washington Hospital are randomly selected, what is the probability that at least one of the births in the sample was an identical twin?

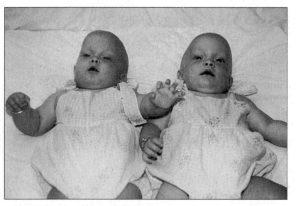

What is the probability of having identical twin babies?
(© *Caroline Brown, Fran Heyl Associates*)

16. According to the U.S. Census Bureau, about 22% of American children under the age of 6 live in households with incomes below the official poverty level. A random sample of 400 children under the age of 6 is taken. What is the mean number of children in the sample who come from households with incomes below the official poverty level? Find the standard deviation.

17. According to the FBI statistics,* only 52% of all rape cases are reported to the police. If eight rape victims are randomly selected, find the probability that at most four of these victims reported the crime to the police.

18. It is an accepted practice in the airline industry to overbook flights. One particular airline has found that 9% of the people who make reservations do not show. If the airline has accepted 220 reservations for a plane that has 210 seats, will the airline have a seat for each passenger who has reserved a seat and who shows up?

19. An auto supplies store owner finds that the daily demand for a particular simonize paste–auto wax follows a Poisson distribution with $\mu = 6$. The store owner wishes to have a sufficient supply of the auto wax so as to satisfy customer demand. How large a stock should the owner have on hand to be able to satisfy the customer demand with a probability of *at least* 0.90?

20. At the end of the exit ramp of the Parker Expressway there is a traffic light that is green 80% of the time for traffic exiting from the expressway. What is the probability that a car exiting the expressway at this location on seven different occasions will miss the green light on each occasion?

21. The U.S. Census Bureau reported that in 1989 one wife in five earned more than her working husband when both worked. If eight couples in which both the husband and wife work are randomly selected, what is the probability that in at most three of these couples the wife earns more than her husband?

22. An electronics store owner finds that the number of two-line answering machines sold follows a Poisson distribution with $\mu = 8$ per week. What is the probability that
    a. between two and four machines will be sold in any week?
    b. at most five machines will be sold in any week?

23. According to the U.S. Bureau of Vital Statistics in Washington, about 53% of all live births in the United States are male; that is, the probability that a newborn child will be a male is about 0.53. If Caledorn Hospital expects 1800 live births this year, what is the expected number of males to be born? Find the standard deviation.

24. The mathematics department of Stapleton University consists of 20 faculty members with ranks as follows:

Position	Rank
Professor	4
Associate Professor	6
Assistant Professor	8
Instructor	2

A five-member appointments committee is to be created. What is the probability that the committee will contain at least one member from each of the ranks specified?

*Source: *Criminal Victimization in the United States, 1991*, *A National Crime Survey Report*. U.S. Department of Justice: Washington, D.C., 1993.

25. A student has not studied at all for the upcoming final examination. The exam consists of 80 multiple-choice questions, with each question having four possible answers. If the student decides to randomly guess the answer for each question, find the mean and standard deviation of the number of questions answered correctly.

26. Consider the accompanying newspaper clipping. If 800 packages of meat are randomly selected, about how many of them can be expected to weigh at least 2 ounces less than its stated weight? Find the standard deviation.

CONSUMER'S AFFAIRS DEPT. TO PROSECUTE APEX

Marborough (Nov. 8)—The Consumer's Affairs Department announced this morning that it would begin prosecuting the Apex Supermarket Chain for short-weighting the packages of meat it sells. In random samples of packages of meat at all of Apex's stores, it was found that approximately 38% of the packages weighed at least 2 ounces less than what was indicated on the package. The unsuspecting customers conceivably were bilked out of millions of dollars.

November 8, 1990

27. When a driver's license is lost or stolen, it must be replaced immediately. The following is the probability distribution for the number of driver's licenses reported lost or stolen daily to the Meadowbrook Motor Vehicle Office.

Number of Licenses Reported Lost, x	Probability $p(x)$
4	0.05
5	0.43
6	0.25
7	0.17
8	0.06
9	0.03
10	0.01

Find μ and σ.

Thinking Critically

1. A swimming team consists of five females and three males. The coach writes the name of each swimmer on a separate piece of paper and places the papers in a hat. The coach then selects two names at random, with the first name replaced before the second is selected. Let x be the number of females selected. Find the probability distribution of x.

2. The locker room of the swimming team mentioned in the previous exercise contains five white towels, seven pink towels, four blue towels, and two brown towels. Three towels are randomly selected with each towel replaced before the next towel is taken. Let x denote the number of pink towels obtained. Find the probability distribution of x.

3. When we square the binomial expression $(p + q)$, that is, when we evaluate $(p + q)^2$, we get $(p + q)^2 = p^2 + 2pq + q^2$. (Verify this.) When a binomial experiment is performed twice, the probability of two successes in the both trials is p^2, the probability of one success in the two trials is $2pq$, and the probability of no successes in both trials is q^2. Note that these are the first, second, and third terms, respectively, in the expansion of $(p + q)^2$. Evaluate $(p + q)^3$ and compare your results with a binomial experiment where $n = 3$. Comment. Can you generalize?

4. Consider a binomial probability distribution for which $n = 8$ and $p = 0.5$.
   a. Draw the graph of this probability distribution.
   b. Locate the interval $\mu \pm 2\sigma$ on this graph.

5. Using the rules for summation given in Chapter 3, show how Formula 6.3 can be derived from Formula 6.2.

6. A parole officer has just interviewed each inmate in the state's correctional facilities to determine the number of previous convictions, x, each had prior to the one for which he or she is now serving time. The following results were obtained:

Number of Previous Convictions x	Probabilities $p(x)$
0	0.03
1	0.62
2	0.19
3	0.09
4	0.07

   a. Find μ and σ.
   b. Find the probability that at least x number of previous convictions will be in the interval $\mu \pm 2\sigma$. (*Hint:* Use Chebyshev's Theorem discussed in Chapter 3.)

7. The standard deviation for the hypergeometric distribution can be obtained from

$$\sigma = \sqrt{np(1 - p)} \cdot \sqrt{\frac{N - n}{N - 1}}$$

where the expression $\sqrt{\frac{N - n}{N - 1}}$ is a *finite population correction factor* that arises because of the process of sampling without replacement from finite populations. (This correction factor will be discussed in later chapters.) Can you see how this formula is derived?

Case Studies

1. A *National Center for Statistics* survey found that one in 15 U.S. workers had one or more job-related injuries this past year. Injuries on the job were twice as common among men than women and more likely among workers aged 18–29 years than workers aged 45–64 years. The most common on-the-job injuries reported were

Injury	Occurrence
Back sprains or strains	26%
Cuts or punctures	21%
Bruises or abrasions	13%

Source: USA TODAY, July 28, 1993.

Insurance company records indicate that there were seven claims for job-related injuries reported by the 106 workers of the All-Boro Trucking Co. in 1994. If 15 workers are randomly selected, what is the probability that at most four of them had job-related injuries this past year?

2. Consider the accompanying newspaper article. Assume that the number of cans thrown away by a resident of Seattle follows a Poisson distribution with $\mu = 1$ per week. Find the probability that a randomly selected Seattle resident will toss away at most two cans a week.

PAY-AS-YOU-THROW WILL MAKE YOU THROW AWAY LESS

New York: In an effort to reduce garbage, over 1000 U.S. communities now charge a fee for every bag or can of garbage that the residents throw away. The trend—which began in Washington and Minnesota and is spreading to Illinois and Pennsylvania—seems to be successful in persuading consumers to recycle and to buy products wrapped in less (but longer lasting) packaging. In Seattle residents threw away an average of only one can a week in 1992 as opposed to 3.5 they threw away in 1981.

Source: The *Wall Street Journal*, © 1993 Dow Jones & Company, Inc. (All Rights Reserved Worldwide)

7

THE NORMAL DISTRIBUTION

Chapter Outline

DID YOU KNOW THAT

a   many measurements associated with large groups of human beings (such as height, weight, etc.) are normally distributed?

b   if the time required to complete a task such as assembling a bike is normally distributed with a mean of 75 minutes, then as many people who need more than 75 minutes to assemble the bike will need less than 75 minutes?

c   the yearly number of major earthquakes, the world over, is a random variable having approximately a normal distribution with mean, $\mu = 20.8$? (*Source: Statistical Abstracts of the United States*, 1993)

d   although the time required by the liver to eliminate alcohol from the bloodstream is approximately normally distributed with a mean of eight hours, medical researchers are experimenting with a new drug that may shorten the process to minutes? (*Source: Time*, December 20, 1993, p. 15)

Would you believe that there are approximately 20.8 major earthquakes a year worldwide? See discussion on page 422. (*Earthquake Engineering Research Institute*)

In this chapter we introduce what is considered to be the most important continuous probability distribution. There are many reasons why great emphasis is placed on the normal distribution. As indicated in Chapter 2, a number of physical characteristics have probability distributions that are very similar to the bell-shaped normal curve. Hence, that accounts for interest in it. Also, as we will see in subsequent chapters, the normal distribution is used extensively in statistical inference. Thus, in this chapter, we will analyze the normal distribution in great detail.

Chapter Objectives

- **To discuss** in detail a probability distribution that is bell shaped or mound shaped, called the normal distribution. (Section 7.2)

- **To indicate** how the normal distribution can be used to calculate probabilities. (Section 7.3)

- **To apply** the normal distribution to many different situations. This will be accomplished by converting to z-scores or standard scores and using the standard normal distribution. (Section 7.4)

- **To show** how the normal distribution can be used to simplify lengthy computations involving the binomial distribution. (Section 7.5)

- **To point out** how statistical quality control charts are used in industry. This represents an additional application of the normal distribution. (Section 7.6)

- **To mention** briefly some historical facts about some mathematicians who used the normal distribution. (Throughout Chapter)

S T A T I S T I C S   I N   A C T I O N

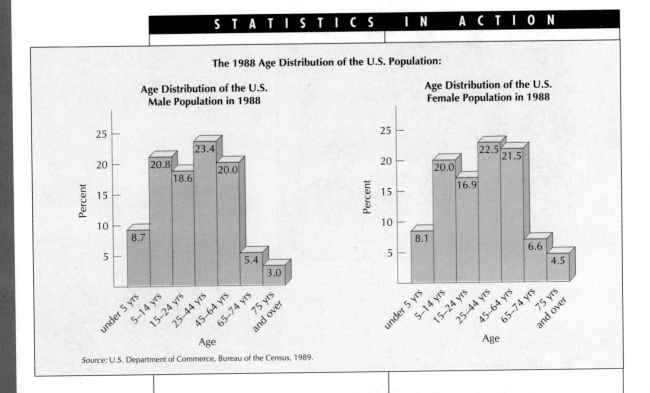

The 1988 Age Distribution of the U.S. Population:

Age Distribution of the U.S. Male Population in 1988

Age Distribution of the U.S. Female Population in 1988

Source: U.S. Department of Commerce, Bureau of the Census, 1989.

The above graph gives the age distribution of the U.S. population for both males and females. Contrary to expectation, the ages for both sexes are not normally distributed. Many people believe that in order to achieve zero population growth, the age distribution should look like the graph on the right, where we have drawn the bars side by side horizontally for ease in comparison.

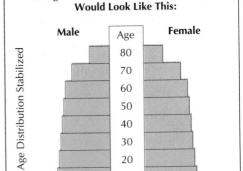

The Age Distribution in a Stable Population Would Look Like This:

Source: ZPG HAS NOT YET BEEN REACHED, Zero Population Growth, Inc., Washington, D.C.

7.1   INTRODUCTION

Discrete random variable

Until now the random variables that we discussed in detail assumed only the limited values of 0, 1, 2, ... Thus, when a coin is flipped many times the number of heads that comes up is 0, 1, 2, 3, ... Also, the number of defective bulbs in a shipment of 100 bulbs is 0, 1, 2, 3, . . . , 100. Since these random variables can assume only the values 0, 1, 2, . . . , they are called **discrete random variables**.

Continuous random variables

As opposed to the preceding examples, consider the following random variables: the length of a page of this book, the height of your pet dog, the temperature at noon on New Year's Day, or the weight of a bag of sugar on your grocer's shelf. Since each of these variables can assume an infinite number of values on a measuring scale, they are called **continuous random variables**. Thus, the weight of a bag of sugar can be 5 pounds, 5.1 pounds, 5.161 pounds, 5.16158 pounds, 5.161581 pounds, and so forth, depending on the accuracy of the scale. Similarly, the temperature at noon on New Year's Day may be 38°, 38.2°, 38.216°, and so forth.

Although there are other distributions of continuous random variables that are important in statistics, the normal distribution is by far the most important. In this chapter we discuss in detail the nature of the normal distribution, its properties, and its uses.

HISTORICAL NOTE

Among the many different kinds of distributions of random variables that are used by statisticians, the normal distribution is by far the most important. This type of distribution was first discovered by the English mathematician Abraham De Moivre (1667–1754). De Moivre spent many hours with London gamblers. In his *Annuities upon Lives*, which played an important role in the history of actuarial mathematics, and his *Doctrine of Chances*, which is a manual for gamblers, he essentially developed the first treatment of the normal probability curve, which is important in the study of statistics. De Moivre also developed a formula, known as Stirling's formula, that is used for approximating factorials of large numbers.

A rather interesting story is told of De Moivre's death. According to the story, De Moivre was ill and each day he noticed that he slept a quarter of an hour longer than on the preceding day. Using progressions, he computed that he would die in his sleep on the day after he slept 23 hours and 45 minutes. On the day following a sleep of 24 hours De Moivre died.

Many years later the French mathematician Pierre-Simon Laplace (1749–1827) applied the normal distribution to astronomy and other practical problems. The normal distribution was also used extensively by the German mathematician Carl Friedrich Gauss (1777–1855) in his studies of physics and astronomy. Gauss is considered by many as the greatest mathematician of the nineteenth century. At the age of 3 he is alleged to have detected an error in his father's bookkeeping records.

Bell-shaped distribution Gaussian distribution

The normal distribution is sometimes known as the **bell-shaped** or **Gaussian distribution** in honor of Gauss, who studied it extensively.

7.2   THE GENERAL NORMAL CURVE

Refer back to the frequency polygon given on page 62 of Chapter 2. It is reproduced below. Experience has taught us that for many frequency distributions drawn from large populations, the frequency polygons approximate what is known as a **normal** or **bell-shaped curve** as shown in Figure 7.2.

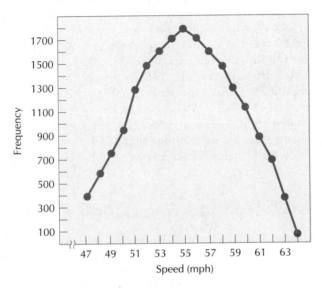

FIGURE 7.1   Distribution of the speeds of cars as they passed through a radar trap.

Heights and weights of people, IQ scores, waist sizes, or even life expectancy of cars, to name but a few, are all examples of distributions whose frequency polygons approach a normal curve when the samples taken are from large populations. When the graph of a frequency distribution resembles the bell-shaped curve shown in Figure 7.2, the graph is called a **normal curve** and its frequency distribution is known as a **normal distribution**. The word normal is simply a name for this particular distribution. It does not indicate that this distribution is more typical than any other.

Normal curve
Normal distribution

FIGURE 7.2   Normal curve.

Since the normal distribution has wide-ranging applications, we need a careful description of a normal curve and some of its properties.

As mentioned before, the graph of a normal distribution is a bell-shaped curve. It extends in both directions. Although the curve gets closer and closer to the horizontal axis, it never really crosses it, no matter how far it is extended. The normal distribution is a probability distribution satisfying the following properties:

1. The mean is at the center of the distribution and the curve is symmetrical about the mean. This tells us that we can fold the curve along the dotted line shown in Figure 7.3 and either portion of the curve will correspond with the other portion.

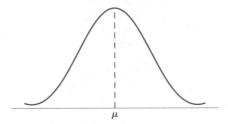

μ FIGURE 7.3

2. Its graph is the bell-shaped curve, most often referred to as the normal curve.
3. The mean equals the median.
4. The scores that make up the normal distribution tend to cluster around the middle, with very few values more than 3 standard deviations away from the mean on either side.

Normal distributions can come in different sizes and shapes. Some are tall and skinny or flat and spread out as shown in Figure 7.4. However, for a given mean and a given standard deviation, there is one and only one normal distribution. The normal distribution is completely specified once we know its mean and standard deviation.

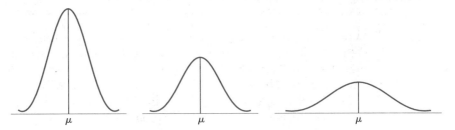

FIGURE 7.4   Different normal distributions.

We mentioned in Chapter 2 (see page 45) that the area under a particular rectangle of a histogram gives us the relative frequency and hence the probability of obtaining values within that rectangle. We can generalize this idea to any distribution. We say that *the area under the curve between any two points a and b gives us the probability that the random variable having this particular continuous distribution will assume values between a and b*. This idea is very important since calculating probabilities for the normal distribution will depend on the areas under the curve. Also, since the sum of the probabilities of a random variable assuming all possible values must be 1 (see page 338), the total area under its probability curve must also be 1.

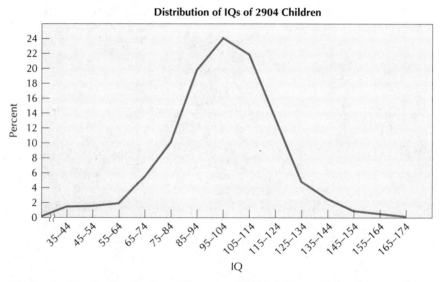

The graph gives the distribution of IQ scores of 2904 children. Notice the type of frequency distribution that is illustrated. This is approximately a normal distribution.

7.3 THE STANDARD NORMAL CURVE

A normal distribution is completely specified by its mean and standard deviation. Thus, although all normal distributions are basically mound shaped, different means and different standard deviations will describe different bell-shaped curves. However, it is possible to convert each of these different normal distributions into one standardized form. You may be wondering, why bother? The answer is rather simple.

Since areas under a normal curve are related to probability, we can use special normal distribution tables for calculating probabilities. Such a table is given in the Appendix at the end of this book. However, because the mean and standard deviation can be any values, it would seem that we need an endless number of tables. Fortunately, this is not the case. We only need one standardized table. Thus, the area under the curve between 40 and 60 of a normal distribution with a mean of 50 and a standard deviation of 10 will be the same as the area between 70 and 80 of another normally distributed random variable with mean 75 and standard deviation 5. They are both within 1 standard deviation unit from the mean. It is for this reason that statisticians use a standard normal distribution.

Definition 7.1
*Standardized
normal distribution*

A **standardized normal distribution** is a normal distribution with a mean of 0 and a standard deviation of 1. The curve of a typical standard distribution is shown in Figure 7.5.

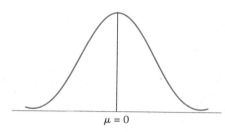

$\mu = 0$

FIGURE 7.5   Curve of typical standard distribution.

Table IV in the Appendix gives us the areas of a standard normal distribution between $z = 0$ and $z = 3.90$. We read this table as follows: The first two digits of the z-score are under the column headed by z, the third digit heads the other columns. Thus, to find the area from $z = 0$ to $z = 2.43$, we first look under z to 2.4 and then read across from $z = 2.4$ to the column headed by 0.03. The area is 0.4925, or 49.25%.

Similarly, to find the area from $z = 0$ to $z = 1.69$, we first look under $z = 1.6$ and then read across from $z = 1.6$ to the column headed by 0.09. The area is 0.4545.

E X A M P L E   1   Find the area between $z = 0$ and $z = 1$ in a standard normal curve.

SOLUTION   We first draw a sketch as shown in Figure 7.6. Then using Table IV for $z = 1.00$, we find that the area between $z = 0$ and $z = 1$ is 0.3413. This means that the probability of a score with this normal distribution falling between $z = 0$ and $z = 1$ is 0.3413.

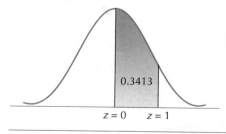

0.3413

$z = 0$ $z = 1$ **FIGURE 7.6**

E X A M P L E   2   Find the area between $z = -1.15$ and $z = 0$ in a standard normal curve.

SOLUTION   We first draw a sketch as shown in Figure 7.7. Then using Table IV, we look up the area between $z = 0$ and $z = 1.15$. The area is 0.3749, not -0.3749. A negative value of z just indicates that the value is to the left of the mean. The area under the curve (and the resulting probability) is *always* a positive number. Thus, the probability of getting a z-score between 0 and -1.15 is 0.3749.

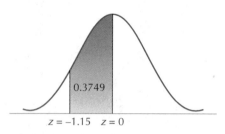

0.3749

$z = -1.15$ $z = 0$ **FIGURE 7.7**

E X A M P L E   3   Find the area between $z = -1.63$ and $z = 2.22$ in a standard normal curve.

SOLUTION   We draw the sketch shown in Figure 7.8. Since Table IV gives the area only from $z = 0$ on, we first look under the normal curve from $z = 0$ to $z = 1.63$. We get 0.4484. Then we look up the area between $z = 0$ and $z = 2.22$. We get 0.4868. Finally, we add these two together and get

$$0.4484 + 0.4868 = 0.9352$$

Thus, the probability that a z-score is between $z = -1.63$ and $z = 2.22$ is 0.9352.

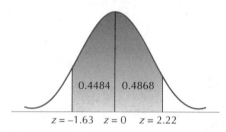

0.4484 | 0.4868

$z = -1.63$ $z = 0$ $z = 2.22$ **FIGURE 7.8**

By following a procedure similar to that used in Example 3, you should verify the following.

PROPERTIES OF STANDARD NORMAL DISTRIBUTION

---

1. The probability that a z-score falls within 1 standard deviation of the mean on either side, that is, between $z = -1$ and $z = 1$, is approximately 68%.
2. The probability that a z-score falls within 2 standard deviations of the mean, that is, between $z = -2$ and $z = 2$, is approximately 95%.
3. The probability that a z-score falls within 3 standard deviations of the mean is approximately 99.7%.

Thus, approximately 99.7% of z-scores fall within $z = -3$ and $z = 3$ (see Figure 7.9).

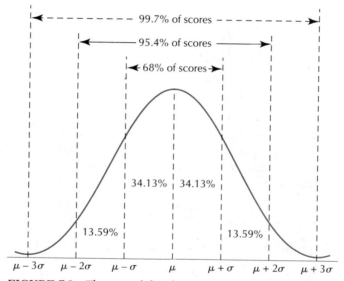

FIGURE 7.9   The normal distribution.

It should be noted that the preceding three statements are true because we are dealing with a normal distribution. A more general result, which is true for any set of measurements—population or sample—and regardless of the shape of the frequency distribution is known as **Chebyshev's Theorem**, which we discussed in Chapter 3. We restate it here.

Chebyshev's Theorem

CHEBYSHEV'S THEOREM

Let k be any number equal to or greater than 1. Then the proportion of any distribution that lies within k standard deviations of the mean is at least $1 - \dfrac{1}{k^2}$

Although Chebyshev's Theorem is more general in scope, if we know that we have a normal distribution, we use the results given above. Since in this chapter our interest is in the normal distribution, we will not use Chebyshev's Theorem.

In many cases we have to find areas between two given values of z or areas to the right or left of some value of z. Finding these areas is an easy task provided we remember that the area under the entire normal distribution is 1. Thus, since the normal distribution is symmetrical about $z = 0$, we conclude that the area to the right of $z = 0$ and the area to the left of $z = 0$ are both equal to 0.5000.

E X A M P L E   4   Find the area between $z = 0.87$ and $z = 2.57$ in a standard normal distribution (see Figure 7.10).

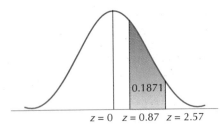

z = 0   z = 0.87   z = 2.57   **FIGURE 7.10**

SOLUTION   We cannot look this up directly since the chart starts at 0, not at 0.87. However, we can look up the area between $z = 0$ and $z = 2.57$ and get 0.4949, then look up the area between $z = 0$ and $z = 0.87$ and get 0.3078. Next we take the difference between the two and get

$$0.4949 - 0.3078 = 0.1871$$

E X A M P L E   5   Find the probability of getting a z-value less than 0.43 in a standard normal distribution.

SOLUTION   The probability of getting a z-value less than 0.43 really means the area under the curve to the left of $z = 0.43$. This represents the shaded portion of Figure 7.11. We look up the area from $z = 0$ to $z = 0.43$ and get 0.1664 and add this to 0.5000 to get

$$0.5000 + 0.1664 = 0.6664$$

Thus, the probability of getting a z-value less than 0.43 is 0.6664.

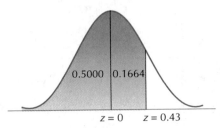

z = 0   z = 0.43   **FIGURE 7.11**

E X A M P L E   6   Find the probability of getting a z-value in a standard normal distribution that is

        a. greater than -2.47
        b. greater than 1.82
        c. less than -1.53

SOLUTION

    a. Using Table IV, we first find the area between $z = 0$ and $z = 2.47$ (see Figure 7.12). We get 0.4932. Then we add this to 0.5000, which is the area to the right of $z = 0$, and get

$$0.4932 + 0.5000 = 0.9932$$

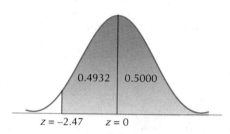

0.4932   0.5000

$z = -2.47$ $z = 0$ **FIGURE 7.12**

    b. Here we are interested in finding the area to the right of $z = 1.82$. See Figure 7.13. We find the area from $z = 0$ to $z = 1.82$. It is 0.4656. Since we are interested in the area to the right of $z = 1.82$, we must *subtract* 0.4656 from 0.5000, which represents the *entire* area to the right of $z = 0$. We get

$$0.5000 - 0.4656 = 0.0344$$

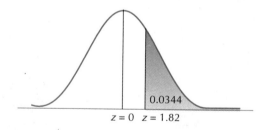

0.0344

$z = 0$ $z = 1.82$ **FIGURE 7.13**

       Thus, the probability that a z-score is greater than $z = 1.82$ is 0.0344.
    c. Here we are interested in the area to the left of $z = -1.53$. See Figure 7.14. We calculate the area between $z = 0$ and $z = -1.53$. It is 0.4370. We subtract this from 0.5000. Our result is

$$0.5000 - 0.4370 = 0.0630$$

Thus, the probability that a z-score is less than $z = -1.53$ is 0.0630.

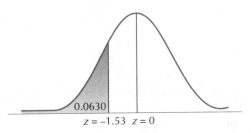

0.0630

$z = -1.53$ $z = 0$ **FIGURE 7.14**

In the preceding examples we interpreted the area under the normal curve as a probability. If we know the probability of an event, we can look at the probability chart and find the z-value that corresponds to this probability.

E X A M P L E   7   If the probability of getting less than a certain z-value is 0.1190, what is the z-value?

SOLUTION   We first draw the sketch shown in Figure 7.15. Since the probability is 0.1190, which is less than 0.5000, we know that the z-value must be to the left of the mean. We subtract 0.1190 from 0.5000 and get

$$0.5000 - 0.1190 = 0.3810$$

See Figure 7.15. This means that the area between $z = 0$ and some z-value is 0.3810. Table IV tells us that the z-value is 1.18. However, this is to the left of the mean. Therefore, our answer is $z = -1.18$.

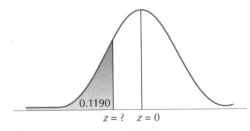

0.1190

$z = ?$ $z = 0$ **FIGURE 7.15**

E X A M P L E   8   If the probability of getting larger than a certain z-value is 0.0129, what is the z-value?

SOLUTION   In this case we are told that the area to the right of some z-value is 0.0129. This z-value must be on the right side. If it were on the left side, the area would have to be at least 0.5000. Why? See Figure 7.16. Thus, we subtract 0.0129 from 0.5000 and

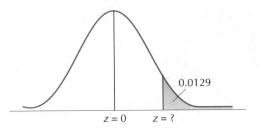

FIGURE 7.16

get 0.4871. Then we look up this probability in Table IV and find the z-value that gives this probability. It is $z = 2.23$. If $z = 2.23$, the probability of getting a z-value greater than 2.23 is 0.0129.

The General Normal Distribution

Converting to standard scores

If we are given a normal distribution with a mean different from 0 and a standard deviation different from 1, we can **convert** this normal distribution into a standardized normal distribution by converting each of its scores into standard scores. To accomplish this we use Formula 3.8 of Chapter 3, which we will now call Formula 7.1:

FORMULA 7.1

$$z = \frac{x - \mu}{\sigma}$$

Expressing the scores of a normal distribution as standard scores allows us to calculate different probabilities, as the following examples show.

EXAMPLE 9   In a normal distribution $\mu = 25$ and $\sigma = 5$. What is the probability of obtaining a value

a. greater than 30?
b. less than 15?

SOLUTION

a. We use Formula 7.1. We have $\mu = 25$, $x = 30$, and $\sigma = 5$ so that

$$z = \frac{x - \mu}{\sigma} = \frac{30 - 25}{5} = \frac{5}{5} = 1$$

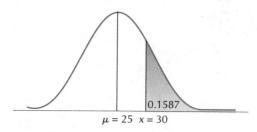

0.1587

$\mu = 25$ $x = 30$ **FIGURE 7.17**

See Figure 7.17. Thus, we are really interested in the area to the right of $z = 1$ of a standardized normal curve. The area from $z = 0$ to $z = 1$ is 0.3413. The area to the right of $z = 1$ is then

$$0.5000 - 0.3413 = 0.1587$$

Therefore, the probability of obtaining a value greater than 30 is 0.1587.

b. We use Formula 7.1. We have $\mu = 25$, $x = 15$, and $\sigma = 5$ so that

$$z = \frac{x - \mu}{\sigma} = \frac{15 - 25}{5} = \frac{-10}{5} = -2$$

See Figure 7.18. Thus, we are interested in the area to the left of $z = -2$. The area from $z = 0$ to $z = -2$ is 0.4772. Thus, the area to the left of $z = -2$ is

$$0.5000 - 0.4772 = 0.0228$$

The probability of obtaining a value less than 15 is therefore 0.0228.

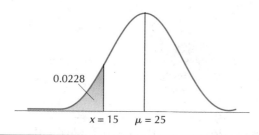

0.0228

$x = 15$ $\mu = 25$ **FIGURE 7.18**

EXAMPLE 10   Find the percentile rank of 20 in a normal distribution with $\mu = 15$ and $\sigma = 2.3$.

SOLUTION   The problem is to find the area to the left of 20 in a normal distribution with $\mu = 15$ and $\sigma = 2.3$. We use Formula 7.1 with $\mu = 15$, $x = 20$, and $\sigma = 2.3$ so that

$$z = \frac{x - \mu}{\sigma} = \frac{20 - 15}{2.3} = \frac{5}{2.3} = 2.17$$

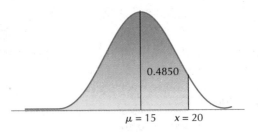

0.4850

$\mu = 15$ $x = 20$ **FIGURE 7.19**

The area between $z = 0$ and $z = 2.17$ is 0.4850. See Figure 7.19. Thus, the area to the left of 20 is

$$0.5000 + 0.4850 = 0.9850$$

The percentile rank of 20 is therefore 98.5.

EXAMPLE 11   In a certain club heights of members are normally distributed with $\mu = 63$ inches and $\sigma = 2$ inches. If Sam is in the 90th percentile, find his height.

SOLUTION   Since Sam is in the 90th percentile, this means that 90% of the club members are shorter than he is. So, the problem here is to find a z-value that has 90% of the area to the left of z. Therefore, we look in the area portion of Table IV to find a z-value that has 0.4000 of the area to its left. See Figure 7.20. We use 0.4000, not 0.9000, since 0.5000 of this is to the left of $z = 0$. The closest entry is 0.3997, which corresponds to $z = 1.28$. Now we convert this score into a raw score by using Formula 3.9, on page 182. We have

$$x = \mu + z\sigma$$
$$= 63 + 1.28(2)$$
$$= 63 + 2.56$$
$$= 65.56$$

Thus, Sam's height is approximately 65.56 inches.

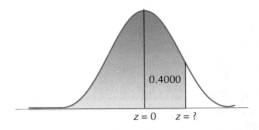

0.4000

$z = 0$ $z = ?$ **FIGURE 7.20**

EXAMPLE   12   Use the same information as Example 11 of this section except Bill's percentile rank is 40. Find his height.

SOLUTION   Since Bill's percentile rank is 40, this means that 40% of the club members are shorter than he is. The problem here is to find a z-value that has 40% of the area to the left of z. See Figure 7.21. Since we are given the area to the left of z, we must subtract 0.4000 from 0.5000:

$$0.5000 - 0.4000 = 0.1000$$

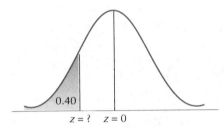

FIGURE 7.21

Then we look in the area portion of Table IV to find a z-value that has 0.1000 as the area between $z = 0$ and that z-value. The closest entry is 0.0987, which corresponds to $z = 0.25$. Since this z-value is to the left of the mean, we have $z = -0.25$. When we convert this score into a raw score, we have

$$x = \mu + z\sigma$$
$$= 63 + (-0.25)2$$
$$= 63 - 0.50 = 62.50$$

Therefore, Bill's height is approximately 62.50 inches.

EXERCISES FOR SECTION 7.3

1. In a standard normal distribution, find the area that lies
   a. between $z = 0$ and $z = 1.82$.
   b. between $z = -0.42$ and $z = 0$.
   c. to the right of $z = 1.86$.
   d. to the left of $z = -0.86$.
   e. to the right of $z = 2.39$.
   f. between $z = -1.64$ and $z = 1.88$.
   g. between $z = -1.78$ and $z = 2.73$.
   h. between $z = -1.84$ and $z = -0.53$.
2. Find the percentage of z-scores in a standard normal distribution that are
   a. above $z = -2.73$.
   b. below $z = 1.74$.

   c. between $z = 1.56$ and $z = 2.28$.
   d. above $z = 2.38$.
   e. above $z = 3.89$.
   f. between $z = -1.54$ and $z = -0.86$.
   g. between $z = -1.49$ and $z = 1.36$.
   h. between $z = -2.84$ and $z = -1.85$.
3. In a normal distribution, find the z-score that cuts off the top
   a. 28%.
   b. 17%.
   c. 35%.
   d. 5%.
4. In a normal distribution, find the z-score(s) that cut(s) off the middle
   a. 18%.
   b. 26%.
   c. 37%.
5. Find z, if the area under a standard normal curve
   a. between $z = 0$ and z is 0.4808.
   b. to the left of z is 0.9854.
   c. to the left of z is 0.2033.
   d. to the right of z is 0.0037.
   e. between $z = 1.14$ and z is 0.1155.
   f. between $z = 2.06$ and z is 0.0180.
6. In a normal distribution with $\mu = 40$ and $\sigma = 6$, find the percentage of scores that are
   a. between 35 and 48.
   b. between 32 and 39.
   c. between 38 and 45.
   d. greater than 39.
   e. less than 44.
7. In a normal distribution with $\mu = 58$ and $\sigma = 7$, find the percentile rank of
   a. a score of 50.
   b. a score of 62.
   c. a score of 45.
8. All candidates for a stock broker job at McGill Securities must take a special marketing exam. It is known that test scores are approximately normally distributed with a mean of 120 and a standard deviation of 11. Find the score of
   a. Felipe, if he is in the 65th percentile.
   b. Heather, if she is in the 27th percentile.
   c. Bob, if he is in the 96th percentile.
9. A normal distribution has mean $\mu = 72$ and unknown standard deviation σ. However, it is known that 30% of the area lies to the right of 80. Find σ.
10. A normal distribution has unknown mean, μ, with a standard deviation of 12.37. However, it is known that the probability that a score is less than 85 is 0.7123. Find μ.

11. The following is a standard normal curve with various z-values marked on it. If this curve also represents a normal distribution with $\mu = 47$ and $\sigma = 6$, replace the z-values with raw scores.

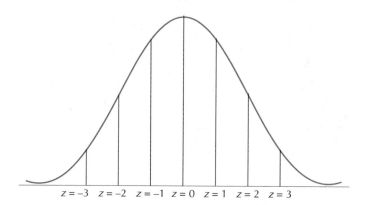

$z = -3$ $z = -2$ $z = -1$ $z = 0$ $z = 1$ $z = 2$ $z = 3$

12. Determine the shaded area under each of the following standard normal curves:

(a)

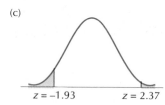

$z = -1.31$ $z = 1.31$

(b)

$z = -2.33$ $z = 2.33$

(c)

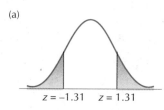

$z = -1.93$ $z = 2.37$

(d)

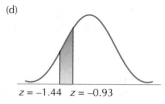

$z = -1.44$ $z = -0.93$

(e)

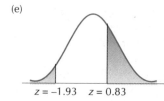

$z = -1.93$ $z = 0.83$

(f)

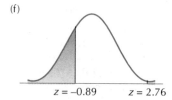

$z = -0.89$ $z = 2.76$

13. The graph at the top of the next page is a sketch of a normal curve whose mean, μ, is 27 and whose standard deviation, σ, is 3. Find the z-scores corresponding to all the x-values given and also find the areas of the indicated regions.

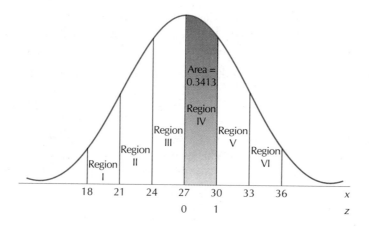

7.4   SOME APPLICATIONS OF THE NORMAL DISTRIBUTION

As mentioned earlier, the importance of the normal distribution lies in its wide-ranging applications. In this section we apply the normal distribution to some concrete examples.

E X A M P L E   1   From past experience it has been found that the weight of a newborn infant at a maternity hospital is normally distributed with a mean $7\frac{1}{2}$ pounds (which equals 120 ounces) and a standard deviation of 21 ounces. If a newborn baby is selected at random, what is the probability that the infant weighs less than 4 pounds 15 ounces (which equals 79 ounces)?

SOLUTION   We use Formula 7.1 of Section 7.3. Here $\mu = 120$, $\sigma = 21$, and $x = 79$ so that

$$z = \frac{x - \mu}{\sigma} = \frac{79 - 120}{21} = \frac{-41}{21}, \text{ or } -1.95$$

Thus, we are interested in the area to the left of $z = -1.95$. The area from $z = 0$ to $z = -1.95$ is 0.4744 so that the area to the left of $z = -1.95$ is $0.5000 - 0.4744$, or 0.0256. See Figure 7.22. Therefore, the probability that a randomly selected baby weighs less than 4 pounds 15 ounces is 0.0256.

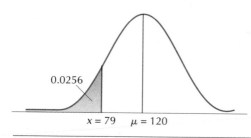

FIGURE 7.22

E X A M P L E   2   The Acme Tire Corporation claims that the useful life of its tires is normally distributed with a mean life of 28,000 miles and with a standard deviation of 4000 miles. What percentage of the tires are expected to last more than 35,000 miles?

SOLUTION   Here $\mu = 28{,}000$, $\sigma = 4000$, and $x = 35{,}000$. Using Formula 7.1, we get

$$z = \frac{x - \mu}{\sigma} = \frac{35{,}000 - 28{,}000}{4000} = 1.75$$

See Figure 7.23. We are interested in the area to the right of $z = 1.75$. The area between $z = 0$ and $z = 1.75$ is 0.4599, so the area to the right of $z = 1.75$ is 0.0401. Thus, approximately 4% of the tires can be expected to last more than 35,000 miles.

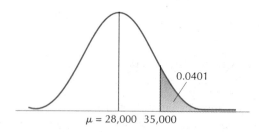

0.0401

$\mu = 28{,}000$ 35,000   **FIGURE 7.23**

E X A M P L E   3   In a recent study it was found that in one town the number of hours that a typical 10-year-old child watches television per week is normally distributed with a mean of 12 hours and a standard deviation of 1.5 hours. If Gary is a typical 10-year-old child in this town, what is the probability that he watches between 9 and 14 hours of television per week?

SOLUTION   We first find the probability that Gary will watch television between 12 and 14 hours per week and add to this the probability that he will watch television between 9 and 12 hours per week. Using Formula 7.1, we have

$$z = \frac{x - \mu}{\sigma} = \frac{14 - 12}{1.5} = 1.33$$

The area between $z = 0$ and $z = 1.33$ is 0.4082. See Figure 7.24. Similarly,

$$z = \frac{x - \mu}{\sigma} = \frac{9 - 12}{1.5} = \frac{-3}{1.5} = -2$$

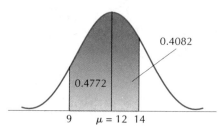

0.4082

0.4772

9     $\mu = 12$ 14   **FIGURE 7.24**

The area between $z = 0$ and $z = -2$ is 0.4772. See Figure 7.24. Adding these two probabilities, we get $0.4082 + 0.4772 = 0.8854$. Thus, the probability that Gary watches between 9 and 14 hours of television per week is 0.8854.

E X A M P L E   4   Daisy discovers that the amount of time it takes her to drive to work is normally distributed with a mean of 35 minutes and a standard deviation of 7 minutes. At what time should Daisy leave her home so that she has a 95% chance of arriving at work by 9 A.M.?

SOLUTION   We first find a z-value that has 95% of the area to the left of z. See Figure 7.25. Thus, we look in the area portion of Table IV to find a z-value that has 0.4500 of the area to its left. (Remember 0.5000 of the area is to the left of $z = 0$.) From Table IV we find that z is midway between 1.64 and 1.65. We will use 1.645. We then find the raw score corresponding to $\mu = 35$, $\sigma = 7$, and $z = 1.645$. We have

$$x = \mu + z\sigma$$
$$= 35 + (1.645)7$$
$$= 35 + 11.515$$
$$= 46.515$$

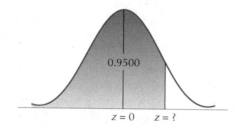

0.9500

$z = 0$ $z = ?$ **FIGURE 7.25**

So, if Daisy leaves her house 46.515 minutes before 9 A.M., she will arrive on time about 95% of the time. She should leave her home at 8:13 A.M.

E X A M P L E   5   In one study a major television manufacturing corporation found that the life of a typical color television tube is normally distributed with a standard deviation of 1.53 years. If 7% of these tubes last more than 6.9944 years, find the mean life of a television tube.

SOLUTION   We are told that 7% of the tubes last more than 6.9944 years. See Figure 7.26. Thus, approximately 43% of the tubes last between μ and 6.9944 years. This means

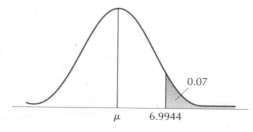

0.07

μ 6.9944   **FIGURE 7.26**

that on a standardized normal distribution the area between $z = 0$ and z is 0.43. Using Table IV, we find that 0.43 of the area is between $z = 0$ and $z = 1.48$. Then

$$x = \mu + z\sigma$$

$$6.9944 = \mu + (1.48)(1.53)$$

$$6.9944 = \mu + 2.2644$$

$$6.9944 - 2.2644 = \mu$$

$$4.73 = \mu$$

Thus, the mean life of a television tube is 4.73 years.

EXERCISES FOR SECTION 7.4

1. Botanists have determined that the length of time it takes one variety of plant seeds to germinate is normally distributed with a mean of 26 days and a standard deviation of 4 days. Find the probability that a randomly selected plant seed of this variety will germinate in less than 20 days.

2. Many supermarkets now use plastic bags instead of paper bags for packing groceries. One nationwide supermarket chain uses plastic bags that are manufactured so that the breaking strength of the bag is normally distributed with a mean of 8 pounds per square inch and a standard deviation of 1.3 pounds per inch. What is the probability that a randomly selected plastic bag used by this supermarket will have a breaking strength between 7 and 10.5 pounds?

3. According to an article in the February 1994 issue of *Consumer Reports* (p. 80), the U.S. Energy Information Administration reports that the life of a typical refrigerator is normally distributed with a mean of 15 years and a standard deviation of 3.7 years. Find the probability that a randomly selected refrigerator will last more than 17 years.

4. The yearly number of major earthquakes, the world over, is approximately normally distributed with a mean of 20.8 and a standard deviation of 4.5 (*Statistical Abstracts of the U.S.*, 1993). What is the probability that in a given year there will be less than 16 major earthquakes the world over?

5. One study (*Time:* December 20, 1993, p. 15) found that the time required by the liver to eliminate alcohol from the bloodstream is approximately normally distributed with a mean of 8 hours and a standard deviation of 1.3 hours. What is the probability that a randomly selected person's liver will require more than 9 hours to eliminate alcohol from the bloodstream?

6. On a recent physical endurance test, the score obtained by the contestants was found to be normally distributed with a mean of 102 and a standard deviation of 5. The judges have decided to award a trophy to the top 10% of the contestants. What is the minimum score that a contestant must obtain in order to receive a trophy?

7. A traffic-control engineer of the State Department of Transportation has found that the waiting time for a commuter in the toll plaza at a certain bridge during rush

hours is approximately normally distributed with a mean of 8.7 minutes and a standard deviation of 1.7 minutes. If a commuter in the toll plaza is randomly selected, what is the probability that a commuter will have to wait
a. less than 5.3 minutes?
b. between 6 and 9 minutes?

Is the amount of pollution emitted by cars as they wait for the signal to change at a busy intersection normally distributed? *(George Semple)*

8. Professor Rogers has found that the grades on the nursing final exam are normally distributed with a mean of 64 and a standard deviation of 11.
   a. If the passing grade is 54, what percent of the class will fail?
   b. If Professor Rogers wants only 85% of the class to pass, what should be the passing grade?
   c. If Professor Rogers wants only 6% of the class to get grades of A, what grade must a student have in order to get an A?
9. The life of a new garbage compactor is approximately normally distributed with a mean of 5 years and a standard deviation of 1.1 years. The manufacturer will repair or replace any defective compactor free of charge (including labor) while the compactor is under warranty. For how many years should the company guarantee its compactors, if the manufacturer does not wish to replace more than 5% of them?
10. A new filling machine has been installed on an assembly line that is supposed to fill each bag with 5 pounds (80 ounces) of sugar. However, the machine is not functioning properly and can be adjusted according to the vendor's specifications. It is known that the filling process is approximately normally distributed with a standard deviation of 2.4 ounces. At what level should the mean be set so that
   a. only 5% of the time will the bags contain less than 80 ounces of sugar?
   b. at most 5% of the time will the bags contain more than 83 ounces of sugar?

11. Refer back to Exercise 10. The machine has been adjusted by mechanics so that the amount of sugar dispensed into each bag is normally distributed with a standard deviation of 1.31 ounces. If 10% of the filled bags contain more than 86 ounces of sugar, what is the new mean amount of sugar per bag filled by this machine?

12. University officials claim that the time required to complete the registration process now that a new computer on-line system has been installed is approximately normally distributed with a mean of 28 minutes and a standard deviation of 6.2 minutes. What is the probability that a randomly selected student will need more than 35 minutes to complete the registration process?

13. A state is administering a qualifying education exam that will enable successful candidates who pass the exam to be licensed teachers within the state. It is known that the time required to complete the exam is approximately normally distributed with a mean of 180 minutes and a standard deviation of 25 minutes. If the state officials wish to assure enough time so that only 85% of the candidates complete the exam, how much time should the officials allow?

14. Heather Carter is the CEO of the Syril Corp. Heather finds that the time it takes her to drive to work each day is approximately normally distributed with a mean of 25 minutes and a standard deviation of 8 minutes. Heather has an important Board of Directors meeting at 9:00 A.M. on Monday morning. At what time should she leave her home so that she has a 95% probability of arriving on time for her 9:00 A.M. meeting?

15. Many airlines have installed toll-free 800 telephone numbers so as to enable passengers to book reservations easily and efficiently. However, these numbers are often busy and a caller is put on "hold." A survey found that for one airline, the holding time is approximately normally distributed with a mean of 4 minutes and a standard deviation of 1.2 minutes. If a call is randomly selected, what is the probability that the caller will have to wait
    a. between 3 to 5 minutes before speaking with a reservation agent?
    b. at least 3.5 minutes before speaking with a reservation agent?
    c. at most 1 minute before speaking with a reservation agent?

16. A medical research company administers a special exam to its employees. The scores of the employees are normally distributed with a mean of 125 and a standard deviation of 13. Furthermore, it is known that a particular chemical research job bores people who score over 135 and demands a minimum score of 116 because of the danger of working with the chemicals. What percentage of the company's employees who took the exam can be used for this particular job (on the basis of the exam results)?

17. The drying time of one particular paint is approximately normally distributed with a mean of 45 minutes. It is also known that 15% of walls painted with this paint need more than 55 minutes to dry completely. Find the standard deviation.

18. A manufacturer of ten-speed bicycles advertises that the time required to assemble its popular F4 model bike is approximately normally distributed with a mean of 75 minutes. Furthermore, only 10% of the people will need more than 90 minutes to assemble the bike. Find the standard deviation.

7.5 THE NORMAL CURVE APPROXIMATION TO THE BINOMIAL DISTRIBUTION

An important application of the normal distribution is the approximation of the binomial distribution. (The binomial distribution, which we discussed in Chapter 6, is often used in sampling without replacement situations if the sample size is small in comparison with the population size.) To see why such an approximation is needed, suppose we wish to determine the probability of getting at least eight heads when a coin is flipped ten times. This is a binomial distribution problem where $n = 10$, $p = \dfrac{1}{2}$, $q = \dfrac{1}{2}$, and $x = 8$, 9, or 10. We can use Formula 6.4 of Chapter 6 (page 362) to determine these probabilities. We get

$$p(8 \text{ heads}) = \frac{10!}{8! \, 2!} \left(\frac{1}{2}\right)^8 \left(\frac{1}{2}\right)^2 = 0.0439$$

$$p(9 \text{ heads}) = \frac{10!}{9! \, 1!} \left(\frac{1}{2}\right)^9 \left(\frac{1}{2}\right)^1 = 0.0098$$

$$p(10 \text{ heads}) = \frac{10!}{10! \, 0!} \left(\frac{1}{2}\right)^{10} \left(\frac{1}{2}\right)^0 = 0.0010$$

Adding these probabilities gives

$$\begin{aligned} p(\text{at least 8 heads}) &= p(8 \text{ heads}) + p(9 \text{ heads}) + p(10 \text{ heads}) \\ &= 0.0439 \quad\;\; + 0.0098 \quad\;\; + 0.0010 \\ &= 0.0547 \end{aligned}$$

Although evaluating the probabilities in this problem is not difficult, the calculations involved are time-consuming. It turns out that the normal distribution can be used as a fairly good approximation to the binomial distribution. To accomplish this, let us first reexamine the previous problem. Since x is a variable that has a binomial distribution, we can use the binomial distribution formula to compute the following probabilities.

Number of Heads x	Probability $p(x)$
0	0.0010
1	0.0098
2	0.0439
3	0.1172
4	0.2051
5	0.2461
6	0.2051
7	0.1172
8	0.0439
9	0.0098
10	0.0010

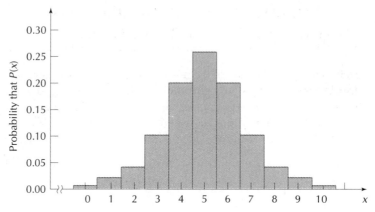

FIGURE 7.27

The histogram for this probability distribution is shown in Figure 7.27 where the height of each rectangle represents the probability that x will assume a particular value.

Recall that for a binomial distribution the mean $\mu = np$ and the standard deviation $\sigma = \sqrt{npq}$. These formulas were given in Section 6.6 of Chapter 6. Applying them to our example gives

$$\mu = np = 10\left(\frac{1}{2}\right) = 5$$

and

$$\sigma = \sqrt{npq} = \sqrt{10\left(\frac{1}{2}\right)\left(\frac{1}{2}\right)} = \sqrt{2.5} \approx 1.5811$$

If we analyze the histogram given in Figure 7.27, we note that it is approximately bell shaped. In Figure 7.28 we have superimposed a normal curve whose mean is 5 and standard deviation is 1.5811.

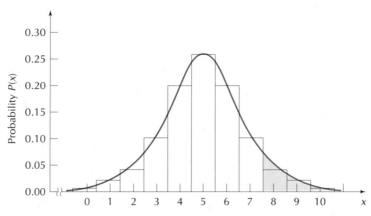

FIGURE 7.28

In the previous example we were interested in the probability that $x = 8$, $x = 9$, or $x = 10$. This represents the cross-hatched area in Figure 7.28. Note that this region can be approximated by the area under the normal curve between $x = 7.5$ and $x = 10.5$. This represents the shaded portion of the diagram. The 0.5 that we subtracted from 8 and the 0.5 that we added to 10 are referred to as **correction factors for continuity**. Thus, we can approximate our binomial distribution by a normal curve with $\mu = 5$ and $\sigma = 1.5811$. We get this approximation by calculating the area under the normal curve between $x = 7.5$ and $x = 10.5$. See Figure 7.29.

Correction factors for continuity

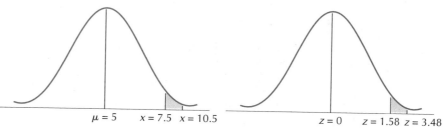

$\mu = 5$ $x = 7.5$ $x = 10.5$ $z = 0$ $z = 1.58$ $z = 3.48$

FIGURE 7.29

When using the normal curve as an approximation for the binomial, we must calculate probabilities using an extra 0.5 either added to or subtracted from the number as a correction factor.

COMMENT   A rigorous mathematical justification as to why we add or subtract 0.5 is beyond the scope of this text. (Suffice it to say that we are representing an integer value in the binomial distribution by those values of a normal distribution that round off to the integer.)

Returning to our example, we have

$$z = \frac{x - \mu}{\sigma} = \frac{7.5 - 5}{1.5811} = 1.58$$

and

$$z = \frac{x - \mu}{\sigma} = \frac{10.5 - 5}{1.5811} = 3.48$$

The area between $z = 0$ and $z = 1.58$ is 0.4429 and the area between $z = 0$ and $z = 3.48$ is 0.4997 so that the area between $z = 1.58$ and $z = 3.48$ is $0.4997 - 0.4429 = 0.0568$.

Using the binomial distribution, we find that the probability of getting at least eight heads is 0.0547. Using the normal curve approximation to the binomial distribution, we find the probability of getting at least eight heads is 0.0568. Although the answers differ slightly, the answer we get by using the normal curve approximation is accurate enough for most applied problems. Furthermore, it is considerably easier to calculate.

More generally, if we were interested in the probability of getting 13 heads in 20 tosses of a coin, we can approximate this by calculating the area between $x = 12.5$ and $x = 13.5$.

COMMENT   Any time you approximate a binomial probability with the normal distribution, depending on the situation make sure to add or subtract 0.5 from the number.

The normal curve approximation to the binomial distribution is especially helpful when we must calculate the probability of many different values. The following examples illustrate its usefulness.

E X A M P L E   1   Melissa is a nurse at Maternity Hospital. From past experience she determines that the probability that a newborn child is a boy is $\frac{1}{2}$. (In the United States today the probability that a newborn child is a boy is approximately 0.53, not 0.50.) What is the probability that among 100 newborn babies there are at least 60 boys?

SOLUTION   We can determine the probability *exactly* by using the binomial distribution or we can get an *approximation* by using the normal curve approximation. To determine the answer exactly, we say that a newborn child is either a boy or a girl with equal probability. Thus, $p = \frac{1}{2}$ and $q = \frac{1}{2}$. The probability that there are at least 60 boys means that we must calculate the probability of having 60 boys, 61 boys, . . . , and, finally, the probability of having 100 boys. Using the binomial distribution formula, we have

$$p(\text{at least 60 boys}) = p(60 \text{ boys}) + p(61 \text{ boys}) + \cdots + p(100 \text{ boys})$$

$$= \frac{100!}{60! \cdot 40!}\left(\frac{1}{2}\right)^{60}\left(\frac{1}{2}\right)^{40} + \frac{100!}{61! \; 39!}\left(\frac{1}{2}\right)^{61}\left(\frac{1}{2}\right)^{39}$$

$$+ \cdots + \frac{100!}{100! \; 0!}\left(\frac{1}{2}\right)^{100}\left(\frac{1}{2}\right)^{0}$$

Calculating these probabilities requires some lengthy computations. However, the same answer can be closely approximated, and more quickly, by using the normal curve approximation. We first determine the mean and standard deviation.

$$\mu = np = 100\left(\frac{1}{2}\right) = 50$$

$$\sigma = \sqrt{npq} = \sqrt{100\left(\frac{1}{2}\right)\left(\frac{1}{2}\right)} = \sqrt{25} = 5$$

Then we find the area to the right of 59.5 as shown in Figure 7.30. We use 59.5 rather than 60.5 in our approximation since we want to include exactly 60 boys in our calculations. If the problem had specified more than 60 boys, we would have used 60.5 rather than 59.5 since more than 60 means do not include 60.

In our case we have

$$z = \frac{x - \mu}{\sigma} = \frac{59.5 - 50}{5} = \frac{9.5}{5} = 1.9$$

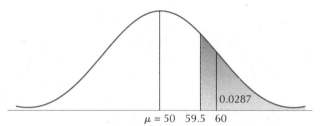

$\mu = 50$ 59.5   60   FIGURE 7.30

From Table IV in the Appendix we find that the area to the right of $z = 1.9$ is 0.5000 − 0.4713, or 0.0287.

Thus, the probability that among 100 newborn children there are at least 60 boys is 0.0287.

EXAMPLE   2   Refer back to Example 1 of this section. Find the probability that there will be between 45 and 60 boys (not including these numbers) among 100 newborn babies at Maternity Hospital.

SOLUTION   The probability is approximated by the area under the normal curve between 45.5 and 59.5. We do not use 44.5 or 60.5 since 45 and 60 are not to be included. To find the area between 45.5 and 50, we have

$$z = \frac{x - \mu}{\sigma} = \frac{45.5 - 50}{5} = -0.90$$

The area is thus 0.3159.

Also, to find the area between 50 and 59.5, we have

$$z = \frac{x - \mu}{\sigma} = \frac{59.5 - 50}{5} = 1.9$$

This area is 0.4713.

Adding these two areas, we get

$$0.3159 + 0.4713 = 0.7872$$

Therefore, the probability that there are between 45 and 60 boys among 100 newborn babies at Maternity Hospital is 0.7872.

EXAMPLE   3   A large television network is considering canceling its weekly 7:30 P.M. comedy show because of a decrease in the show's viewing audience. The network decides to phone 5000 randomly selected viewers and to cancel the show if fewer than 1900 viewers are actually watching the show. What is the probability that the show will be canceled if

a. only 40% of all television viewers actually watch the comedy show?
b. only 39% of all television viewers actually watch the comedy show?

SOLUTION

a. Since a randomly selected television viewer that is phoned either watches the show or does not watch the show, we can consider this as a binomial distribution with $n = 5000$ and $p = 0.40$. We first calculate μ and σ:

$$\mu = 5000(0.40) = 2000$$

$$\sigma = \sqrt{5000(0.40)(0.60)} = \sqrt{1200} \approx 34.6410$$

Since the show will be canceled only if fewer than 1900 people watch it, we are interested in the probability of having 0, 1, 2, . . . , 1899 viewers. Using a normal curve approximation, we calculate the area to the left of 1899.5. We have

$$z = \frac{x - \mu}{\sigma} = \frac{1899.5 - 2000}{34.6410} = -2.90$$

The area to the left of $z = -2.90$ is

$$0.5000 - 0.4981 = 0.0019$$

See Figure 7.31. Thus, the probability that the show is canceled is 0.0019.

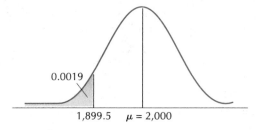

0.0019

1,899.5 $\mu = 2,000$ **FIGURE 7.31**

b. In this case the values of μ and σ are different since the value of p is 0.39. We have

$$\mu = 5000(0.39) = 1950$$

$$\sigma = \sqrt{5000(0.39)(0.61)} = \sqrt{1189.5} \approx 34.4891$$

Since the show will be canceled if fewer than 1900 people watch it, we calculate the area to the left of 1899.5. We have

$$z = \frac{1899.5 - 1950}{34.4891} = -1.46$$

The area to the left of $z = -1.46$ is

$$0.5000 - 0.4279 = 0.0721$$

See Figure 7.32. Thus, the probability that the show is canceled is 0.0721.

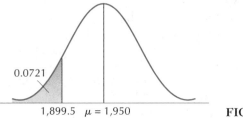

FIGURE 7.32

COMMENT The degree of accuracy of the normal curve approximation to the binomial distribution depends on the values of n and p. Figure 7.33 indicates how fast the histogram for a binomial distribution approaches that of a normal distribution as n gets larger. As a rule, the normal curve approximation can be used with fairly accurate results when *both np* and *nq* are greater than 5.

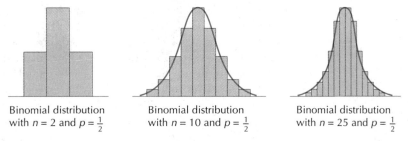

Binomial distribution with $n = 2$ and $p = \frac{1}{2}$

Binomial distribution with $n = 10$ and $p = \frac{1}{2}$

Binomial distribution with $n = 25$ and $p = \frac{1}{2}$

FIGURE 7.33 Histogram for a binomial distribution approaches that of a normal distribution as n gets larger.

COMMENT The normal curve approximation to the binomial distribution actually depends on a very important theorem in statistics known as the Central Limit Theorem. This theorem will be discussed in a later chapter.

EXERCISES FOR SECTION 7.5

In each of the following exercises use the normal curve approximation to the binomial distribution.

1. According to *Health* (May/June 1993, p. 12), 24% of all births in the United States in 1993 were by Caesarean section. If 500 births that occurred in the United States during 1993 are randomly selected, what is the probability that at most 100 of them were by Caesarean section?

2. The U.S. Department of Energy estimates that 75% of all American homes had a microwave oven in 1994. If 400 American homes are randomly selected, what is the probability that at least 275 of them had microwave ovens in 1994?

3. A survey in *USA TODAY* (May 25, 1990) found that 66% of California motorists use their seat belts. If 350 California motorists are randomly selected, what is the probability that at least 200 of them use their seat belts?

4. The U.S. Department of Education (1995) estimates that about 17% of public school teachers moonlight during the school year. If 300 teachers are randomly selected, what is the probability that less than 40 of them moonlight during the school year?

5. *Money* magazine (Vol. 1990, No. 3, March 1990, p. 90) found that only 4% of the tax professionals consulted were able to correctly compute the tax due on a hypothetical 1040 Federal Tax Return. If 200 tax professionals are randomly selected, what is the probability that more than seven of them can correctly compute the tax due on the hypothetical 1040 Federal Tax Return?

6. The U.S. Census Bureau in its *Current Population Reports* (1995) indicates that 25% of U.S. children were not living with both parents in 1993. If 375 children are randomly selected, find the probability that at most 100 of them are not living with their parents.

7. The U.S. Energy Information Administration indicates in its *Residential Energy Consumption Survey: Housing Characteristics* (1994) that 36.7% of U.S. households use an automatic dishwasher. If 40 U.S. households are randomly selected, what is the probability that more than 15 of them use an automatic dishwasher?

8. Radon, the second largest cause of lung cancer after smoking, is a radioactive gas produced by the natural decay of radium in the ground that seeps into homes. Studies by the EPA indicate that 7% of the homes in one particular region contain dangerous levels of radon gas in them. If 100 homes in this region are randomly selected, what is the probability that exactly seven of them will contain dangerous levels of radon gas?

9. If 35% of all households in Bayerville have two cars, what is the probability that a random sample of 80 households in Bayerville will contain exactly 20 households that have two cars?

10. According to the U.S. Census Bureau, both partners work in 57% of the married couples of Dover. If 450 couples in Dover are randomly selected, what is the probability that both partners work in less than 230 of them?

11. Phil is taking a true–false examination consisting of 50 questions. Since he has not studied for the exam, he decides to guess at the answers. If Phil needs at least 25 questions to pass, what is the probability that he passes the exam?

12. *The Wall Street Journal* (February 8, 1990) reported that a poll conducted by the National Geographical Society disclosed that only 55% of adults could locate New York State on a U.S. map. If 80 adults are randomly selected, what is the probability that more than 40 of them can locate New York State on a map of the United States?

13. About two out of every three gas purchases at Pete's Exxon Station are paid for by credit cards. If 480 customers buying gas at this station are randomly selected, what

is the probability that less than 300 of them will pay for their purchases by credit cards?

14. The EPA estimates that 19% of the cars registered in one state cannot meet the state's tough new air pollution exhaust system standards. If 2000 cars in this state are randomly selected, what is the probability that at most 400 of them cannot meet the state's tough new air pollution exhaust system standards?

15. An airline company finds that 12% of all people who make reservations do not actually show up. If the airline has accepted 140 reservations for a particular flight and if there are 130 seats available, find the probability that the airline will have a seat for each person who has reserved one and who shows up.

Government regulations require that airlines that routinely overbook flights offer cash incentives to persuade passengers to reschedule a flight when they are overbooked. (*Joseph Newmark*)

16. It is claimed that two out of every five students at Bisman College are receiving some form of student tuition assistance. If 300 students at this college are randomly selected, what is the probability that at most 100 of them are receiving some form of tuition assistance?

17. Refer back to the previous exercise. If 300 students at this college are randomly selected, what is the probability that more than 120 of them are receiving some form of tuition assistance?

18. A life insurance salesperson knows from past experience that she can expect to sell a life insurance policy to about 17% of her customers after promotional sales presentations. If 500 customer presentations are randomly selected, what is the probability that she will sell at least 60 life insurance policies to the customers?

19. A new drug claims to be 95% effective in relieving some arthritic pain. If the drug is administered to 400 randomly selected people, what is the probability that the drug will be effective in relieving the arthritic pain of 370 of these people?

20. One airline company finds that 18% of all reservations are subsequently changed (different day, different time, etc). If 200 reservations are randomly selected, what is the probability that between 30 and 40 of them will subsequently be changed?

7.6   APPLICATION TO STATISTICAL QUALITY CONTROL CHARTS

In recent years numerous articles and books have been written on how statistical quality controls operate. This is an important branch of applied statistics. What are quality control charts? To answer this question, we must consider the mass production process.

Industrial experience shows that most production processes can be thought of as normally distributed. So, when a manufacturer adjusts the machines to fill a jar with 10 ounces of coffee, although not all the jars will actually weigh 10 ounces, the weight of a typical jar will be very close to 10 ounces. When too many jars weigh more than 10 ounces, the manufacturer will lose money. When too many jars weigh less than 10 ounces, he or she will lose customers. The manufacturer is therefore interested in maintaining the weight of the jars as close as possible to 10 ounces.

If the production process behaves in the manner just described, the weight of a typical jar of coffee is either acceptable or not acceptable. Thus, it can be thought of as a binomial variable. We can then use the normal approximation.

Quality-control charts   Rather than weigh each individual jar of coffee, the manufacturer can use **quality control charts**. This is a simple graphical method that has been found to be highly useful in the solution to problems of this type.

Figure 7.34 is a typical quality control chart. The horizontal line represents the time scale. The vertical line has three markings: μ, $\mu + 3\sigma$, and $\mu - 3\sigma$.

The middle line is thought of as the mean of the production process although in reality it is usually the mean weight of past daily samples. The two other lines serve as control limits for the daily production process. These lines have been spaced 3 standard deviation units above and below the mean. From the normal distribution table we find that approximately 99.7% of the area should be between $\mu - 3\sigma$ and $\mu + 3\sigma$. Thus, the probability that the average weight of many jars of coffee falls outside the control bands is only 0.003. This is a relatively small probability. Therefore, if a sample of sufficient size is taken and the average weight is outside the control bands, the manufacturer can then assume that the production process is not operating properly and that immediate adjustment is necessary to avoid losing money or customers. Each of the

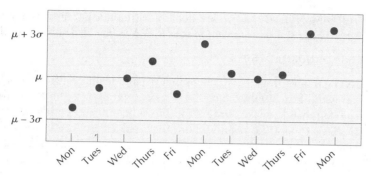

FIGURE 7.34   Typical quality control chart.

dots plotted in Figure 7.34 represents the mean of the weights of n jars of coffee determined on different days.

Figure 7.34 indicates that the production process went out of control on a Friday.

7.7   USING COMPUTER PACKAGES

MINITAB can be used to sketch the graph of any normal distribution once we specify the mean and the standard deviation. MINITAB accomplishes this by plotting a set of (x, y) points. To illustrate the procedure, suppose we wish to sketch the graph of the normal distribution with mean $\mu = 50$ and standard deviation $\sigma = 2$. We enter the following:

```
MTB  > SET C1
DATA > 44:56/0.5
DATA > END
MTB  > PDF C1 C2;
SUBC > NORMAL MU=50 SIGMA=2.
MTB  > PLOT C2 C1
```

Let us analyze the above commands. Since very few values fall more than 3 standard deviations away from the mean on either side, we will start at 44, which represents 3σ below the mean of 50, and end at 56, which represents 3σ above the mean. We will arbitrarily use spacings of 0.5 between each x. These instructions are incorporated in the statement 44:56/0.5.

The PDF (probability distribution function) command in MINITAB calculates the y-value to be plotted for the appropriate probability function. The subcommand instructs MINITAB that we are dealing with a normal distribution whose mean is $\mu = 50$ and whose standard deviation is $\sigma = 2$. We can also abbreviate this subcommand as

`SUBC > NORMAL 50 2.`

MINITAB interprets this to mean that we are dealing with a normal distribution whose mean is the first number and whose second number is the standard deviation. As with all subcommands, make sure to end the subcommand with a period. Then we instruct MINITAB to plot the results where the x-values are in C1 and the y-values are in C2. When using the plot command, the y-values are given first and then the x-values. When we execute the above MTB command, we obtain the graph shown in Figure 7.35.

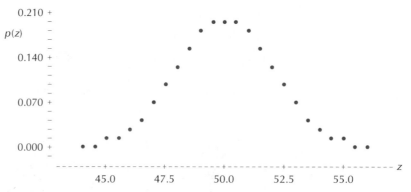

FIGURE 7.35   Graph of the normal probability distribution with $\mu = 50$ and $\sigma = 2$.

We can have MINITAB construct a cumulative probability table for a normal distribution by using the CDF (Cumulative Density Function) command. This is often useful when we deal with percentiles. By subtracting, we can find different probabilities or the probability that x will fall within some interval. This is illustrated in the following commands where for simplicity we have restricted our example to the *standard* normal distribution.

```
MTB  > SET C1
DATA > -4:4
DATA > END
MTB  > CDF C1 C2;
SUBC > NORMAL MU=0 SIGMA=1.
MTB  > NAME C1='z' C2='P(z)'
MTB  > PRINT C1 C2
```

Before we print our results and also to make them easier to analyze, we instruct MINITAB to label the contents of C1 by the name 'z'. We enclose any name within single quotation marks. For the contents of C2 we chose the name '$P(z)$' to represent probability of z. The above commands produce the following printout:

ROW	z	P(z)
1	−4	0.000032
2	−3	0.001350
3	−2	0.022750
4	−1	0.158655
5	0	0.500000
6	1	0.841345
7	2	0.977250
8	3	0.998650
9	4	0.999968

FIGURE 7.36   A standard normal probability distribution.

From the above printout, we find that the probability that z will fall between $z = 0$ and $z = 1$ is $0.841345 - 0.500000$, or 0.341345. Similar results can be obtained for other values of z.

MINITAB easily allows us to use the normal distribution to approximate the binomial distribution. This use will be illustrated in the exercises for this section.

EXERCISES FOR SECTION 7.7

1. Use the information given in Figure 7.36 to find the following:
   a. $p(z < -2)$ c. $p(-3 < z < 1)$
   b. $p(z > 1)$ d. $p(-2 < z < 3)$
2. Using MINITAB, sketch the graph of the normal distribution with a mean of 40 and a standard deviation of 4.
3. Refer back to Example 1 on page 428 of Section 7.5. In that case we were interested in calculating $p(x \geq 59.5)$, where $\mu = 50$ and $\sigma = 5$. This can easily be accomplished by using the following commands:

```
MTB  > SET C1

DATA > 59.5 100

DATA > END

MTB  > CDF C1;

SUBC > NORMAL MU=50 SIGMA=5.
```

MINITAB prints out the following:

x	P(X < = x)
59.5000	0.9713
100.0000	1.0000

Thus $p(x \geq 59.5) = 1.000 - 0.9713 = 0.0287$. This is the same answer that we obtained using the standard normal distribution table (Table IV in the Appendix).

Use MINITAB to obtain the requested probability in Example 2 on page 429.

4. Using MINITAB, verify the probabilities obtained in Example 3 on page 429.

7.8   SUMMARY

In this chapter we discussed the difference between a discrete random variable and a continuous random variable. We studied the normal distribution in detail since it is the most important probability distribution of a continuous random variable. Because of their usefulness, normal curve area charts have been constructed. These charts allow us to calculate the area under the standard normal curve. Thus, we can determine the probability that a random variable will fall within a specified range.

Not only can these charts be used to calculate probabilities for variables that are normally distributed, but they can also be used to obtain a fairly good approximation to binomial probabilities. This is especially helpful when we must calculate the probability that a binomial random variable assumes many different values. Numerous applications of these ideas were given.

The normal distribution was then applied to the construction of quality control charts, which are so important in many industrial processes. Today statistical quality control is an important branch of applied statistics. Without discussing them in detail, we indicated the usefulness of these quality control charts.

Study Guide

The following is a chapter outline in capsule form. You should now be able to demonstrate your knowledge of the ideas mentioned by giving definitions, descriptions, or specific examples. Page references are given in parentheses.

Discrete random variables can assume only the values 0, 1, 2, ... (page 403)

Continuous random variables can assume an infinite number of values on a measuring scale. (page 403)

When the graph of a frequency distribution resembles the bell-shaped curve, the graph is called a **normal curve** and its frequency distribution is known as a **normal distribution** (also called a bell-shaped or Gaussian distribution). (page 404)

The **normal distribution** has the following properties:

1. The mean is at the center of the distribution and the curve is symmetrical about the mean.
2. Its graph is a bell-shaped curve.
3. The mean equals the median.
4. The scores that make up the normal distribution tend to cluster around the middle with very few values more than 3 standard deviations away from the mean on either side. (page 405)

A **standardized normal distribution** is a normal distribution with a mean of 0 and a standard deviation of 1. (page 407)

For a **standard normal distribution**:

1. The probability that a z-score falls within 1 standard deviation of the mean on either side, that is, between $z = -1$ and $z = 1$, is approximately 68%.
2. The probability that a z-score falls within 2 standard deviations of the mean, that is, between $z = -2$ and $z = 2$, is approximately 95%.
3. The probability that a z-score falls within 3 standard deviations of the mean is approximately 99.7%. (page 409)

Chebyshev's Theorem: Let k be any number equal to or greater than 1. Then the proportion of any distribution that lies within k standard deviations of the mean is at least $1 - \dfrac{1}{k^2}$. (page 409)

If we are given a normal distribution with a mean different from 0 and a standard deviation different from 1, we can **convert** this normal distribution into a standardized normal distribution by converting each of its scores into standard scores. We use the formula $z = \dfrac{x - \mu}{\sigma}$. Thus, to find the probability that a normal random variable falls between the values x_1 and x_2, we first find the z-values corresponding to x_1 and x_2. We get $z_1 = \dfrac{x_1 - \mu}{\sigma}$ and $z_2 = \dfrac{x_2 - \mu}{\sigma}$. Then locate the values of x_1 and x_2 on the x-distribution and the values of z_1 and z_2 on a z-distribution as shown below.

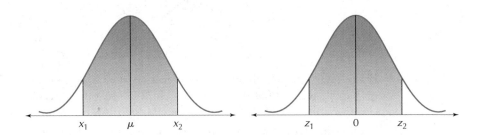

The areas under the standard normal curve between z_1 and z_2 (obtained from Table IV in the Appendix) gives us the probability that x will fall between x_1 and x_2. (page 406)

If we want to determine percentiles when working with a normal random variable, then the Pth **percentile** of the normal random variable x is that value of x that has an area of $P/100$ to its left. Thus, the 15th percentile has an area of 0.15 to its left, the 50th percentile has an area of 0.50 to its left, and so on. (page 415)

The **normal distribution can be used as a fairly good approximation to the binomial distribution** provided that np and nq are *both* greater than 5. First calculate the mean using the formula $\mu = np$ and standard deviation $\sigma = \sqrt{npq}$. Then if the binomial probability to be approximated is of the form "find the probability that $x \leq a$

or $x > a$," the continuity correction factor is $(a + 0.5)$ and the approximating standard normal z-value is $z = \dfrac{(a + 0.5) - \mu}{\sigma}$ as shown below. If the binomial probability to be

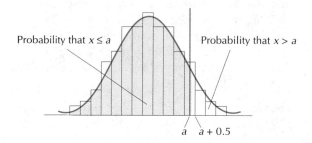

Probability that $x \leq a$ Probability that $x > a$

a $a + 0.5$

approximated is of the form "find the probability that $x \geq a$ or $x < a$," the continuity correction factor is $(a - 0.5)$ and the standard normal z-value is $z = \dfrac{(a - 0.5) - \mu}{\sigma}$ as shown below.

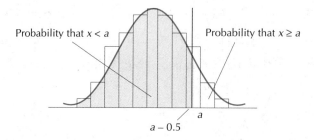

Probability that $x < a$ Probability that $x \geq a$

a

$a - 0.5$

If the binomial probabilities to be approximated is an interval of the form "find the probability that x is between a and b inclusive," treat the ends of the intervals separately. Calculate the two z-values by using one of the procedures given above. (page 427)

Quality-control charts are often used in industry to determine if the production process is operating smoothly. (page 434)

Formulas to Remember

You should be able to identify each symbol in the following formulas, understand the relationships among the symbols expressed in each formula, understand the significance of each formula, and use the formulas in solving problems.

1. $z = \dfrac{x - \mu}{\sigma}$

2. $x = \mu + z\sigma$

3. Chebyshev's Theorem: At least $1 - \dfrac{1}{k^2}$ of a set of measurements will lie within k standard deviations of the mean: $k = 1, 2, \ldots$.

Testing Your Understanding of This Chapter's Concepts

1. Which normal distribution curve has a wider spread, one with mean $\mu = 4$ and standard deviation $\sigma = 2$ or one with mean $\mu = 2$ and standard deviation 4?
2. Find the three x-values that divide the area under the normal curve with parameter $\mu = 12.9$ and $\sigma = 2.9$ into four 0.25 areas.
3. Determine Q_3, the third quartile, for a normal distribution with mean $\mu = 40$ and standard deviation $\sigma = 5$.
4. Compute the probability that the standard normal variable z will have a value that lies within 1.5 standard deviations of the mean. How does your answer compare with the results obtained by using Chebyshev's Theorem?
5. Compute the following probabilities for a standard normal variable.
   a. $p(-1.22 \leq z \leq 2.03)$ b. $p(-1.22 < z \leq 2.03)$
   c. $p(-1.22 \leq z < 2.03)$ d. $p(-1.22 < z < 2.03)$
   Your answers should be the same for all four parts of this exercise. Can you explain why?
6. Draw a normal curve (making sure to label the first three standard deviations on it) that has the following characteristics:
   a. mean = 7 and standard deviation = 3.
   b. mean = 3 and standard deviation = 5.
   c. mean = 7 and standard deviation = 5.
7. Would it be appropriate to use the normal curve approximation to calculate binomial probabilities where $n = 6$ and $p = \dfrac{1}{2}$? Explain your answer.
8. A student claims that when drawing normal curves the value of the mean, μ, has no effect on the shape of the curve. Do you agree? Explain your answer.

Chapter Test

Multiple Choice Questions

1. Find the percentage of z-scores in a standard normal distribution that are between $z = -1.48$ and $z = 2.03$.
   a. 0.4306   b. 0.4788   c. 0.0482   d. 0.9094   e. none of these
2. A gas station attendant claims that 65% of all cars in the city are equipped with studded tires. If 200 cars in the city are randomly selected, what is the probability that at least 130 of them will be equipped with studded tires?
   a. 0.5000   b. 0.5279   c. 0.4721   d. 0.0596
   e. none of these
3. The number of mortgage applications approved by a particular commercial bank is approximately normally distributed with a mean of 15 and a standard deviation of

2 applications per day. If a day is randomly selected, what is the probability that the bank will approve more than 10 applications on that day?

  a. 0.9878   b. 0.0062   c. 0.4938   d. 0.9938
  e. none of these

4. Refer back to the previous question. What is the probability that on any randomly selected day the bank will approve at most 19 applications?

  a. 0.9772   b. 0.4772   c. 0.0228   d. 0.9878
  e. none of these

5. Melissa owns a fish hatchery. She finds that the lengths of the fish in the hatchery are normally distributed with $\mu = 8$ cm and $\sigma = 1.2$ cm. Approximately what percentage of the fish in the hatchery are more than 8.5 cm in length?

  a. 41.67%   b. 16.28%   c. 33.72%   d. 33.16%
  e. none of these

Experience has shown that the lengths of fish in a hatchery are normally distributed. (*Al Ivany, N.J. Division of Fish, Game, and Wildlife*)

6. The life of a certain flashlight battery is normally distributed with a mean of 87 hours and a standard deviation of 5.1 hours. Find the probability that a randomly selected battery from this group will last at most 76 hours.
   a. 0.4846   b. 0.0154   c. 0.5014   d. 0.9846   e. none of these

7. The life of a certain air-conditioner compressor is normally distributed with a mean of 5 years and a standard deviation of 1.1 years. The manufacturer will replace any defective air-conditioner compressor free of charge while under the guarantee. If the manufacturer does not wish to replace more than 6% of the air-conditioner compressors, for how many years should they be guaranteed?
   a. 4.934 yr   b. 4.85 yr   c. 3.2895 yr   d. 4.835 yr   e. none of these

Supplementary Exercises

8. A cookie baking company claims that 5% of each day's production contains boxes with broken cookies. If 300 boxes in a day's production are randomly selected, what is the probability that exactly 18 of the boxes will contain broken cookies?

9. The amount of time required to install a special gas burner is approximately normally distributed with a mean of 95 minutes and a standard deviation of 7 minutes. What is the probability that a randomly selected gas burner of the type described required between 90 and 103 minutes to install?

10. The heights of the players on a certain college basketball team are normally distributed with a mean of 74 inches and a standard deviation of 2.1 inches. What is Dave's percentile rank if he is 76 inches tall?

11. It is claimed that 40% of all women at The Baxt Corporation use roll-on deodorant. A survey of 160 randomly selected women from The Baxt Corp. is taken. What is the probability that exactly 70 of these women will be found to use roll-on deodorant?

12. In its February 1994 issue, *Consumer Reports* presented information on the quality of various brands of paint in terms of hiding power, fading, spatter, sticking, and water resistance. In a study it was found that the drying time of one new brand of latex interior paint is approximately normally distributed with a mean of 4 hours and a standard deviation of 0.23 hour. What percentage of the time will this paint dry in less than 3.6 hours?

13. It is claimed that 35% of all students at Dorx College have a credit card issued in their name. What is the probability that a survey of 100 randomly selected students at this college will contain at least 40 students who have a credit card issued in their own name?

14. The traffic department of a certain city has found that the probability that the lamp posts on the city's highways are functioning properly is 0.93. If 400 lamp posts on the city's highways are randomly selected, what is the probability that at least 375 of them will be functioning properly?

15. Many municipalities rely on the income obtained from various parking tickets as an important source of income. After analyzing the ticketing patterns of the officers of the New York City Police Department, one statistician claims that the number of

parking tickets issued daily by the city's officers is approximately normally distributed with a mean of 27,000 and a standard deviation of 1100. If a day is randomly selected, what is the probability that the number of parking tickets issued will be between 25,000 and 28,000?

16. Many newspapers, such as *The New York Times*, present information as to "where to eat out." In one study, it was found that 70% of the people in Manhattan eat in a restaurant several times during the month. If 300 people in Manhattan are randomly selected, what is the probability that at least 205 of these people eat in a restaurant several times during the month?

17. The U.S. Department of Transportation requires tire manufacturers to provide a wealth of information on the sidewall of every tire. The Government, through the Uniform Tire Quality Grading System, specifies tests in which the tread life of each tire is determined. The tire size, traction ability, temperature grade, load index, speed rating, manufacturing date, and tread-wear index all appear on the sidewall (*Source: Consumer Reports*, February 1994, pp. 75–79). One tire manufacturer claims that the life of the company's steel-belted all-season radial tire is normally distributed with an average of 35,000 miles and a standard deviation of 2200 miles. What is the probability that a randomly selected steel-belted all-season radial tire manufactured by this company will last at most 30,000 miles?

18. The U.S. National Center for Health Statistics collects data on cigarette smokers by sex and age and publishes its results in *Vital and Health Statistics*. In one study it was found that 45% of all students at Bork College smoke. What is the probability that a survey of 700 randomly selected students at this school will contain at most 300 smokers?

19. A fund raiser has found that the amount of money that can be raised through a telephone solicitation is approximately normally distributed with a mean of $36 and a standard deviation of $4. If a telephone solicitation is randomly selected, what is the probability that the amount of money pledged is at least $30?

20. *USA TODAY* (July 24, 1993) indicates that only 19% of the people surveyed trust their local officials. If a random survey of 300 people is taken, what is the probability that more than 60 of them trust their local officials?

21. The U.S. Immigration and Naturalization Service reports in *Statistical Yearbook*, 1992, that 53% of the immigrants that arrived in the United States during 1990 were from Asia. If a sample of 250 immigrants that arrived in the United States during 1990 is randomly selected, what is the probability that at most 130 of them were from Asia?

22. A survey found that 3% of American children live with their father only. (*Source: U.S. News and World Report*, April 26, 1993) If 400 American children are randomly selected, what is the probability that the number of them who live with their father only is between 10 and 15?

23. It is claimed that 65% of all professional football players suffer serious injuries as a result of their career. (*Source: Injuries*, NFL Players Association, Washington, D.C.) If 200 professional football players are randomly selected, what is the probability

that between 125 and 130 of them will suffer serious injuries as a result of their career?

24. According to *Health* (May/June 1993, p. 12), only 7% of all babies born in Japan during 1992 were delivered by Caeserean section. If 250 births recorded in Japan during 1992 are randomly selected, what is the probability that at least 15 of them were delivered by Caeserean section?

25. More than 45 million Americans suffer from recurring headaches. Seventy percent of migraine sufferers are women. (*Source: U.S. News and World Report*, June 28, 1993) If 150 migraine sufferers are randomly selected, what is the probability that at most 100 of them are women?

26. Consider the accompanying newspaper article. If 150 people from the labor force are randomly selected, find the probability that
    a. at least 15 of them are unemployed.
    b. at most 12 of them are unemployed.
    c. exactly 14 of them are unemployed.
    d. the number of unemployed is between 11 and 17.

UNEMPLOYMENT RATE UP AGAIN

Winchester, *March 1*: According to the figures released yesterday, the rate of unemployment for our area increased for a third month in a row and now stands at 9.2%. Such a high rate of unemployment for our area can only mean that more jobs must be created by the state.

May 1, 1989

27. Consider the newspaper article on the next page. One researcher claims that despite medical claims to the contrary, two out of nine potential donors are still reluctant to donate blood because of their fear of contracting AIDS. Assuming that this researcher's claim is true, find the probability that in a random survey of 85 potential donors, at most 17 of them will be reluctant to donate blood because of their fear of contracting AIDS.

**BLOOD DONATIONS
AT A LOW**

December 7: Blood bank officials announced that many volunteers were reluctant to donate blood because of their fear of contracting AIDS. As a result, the blood supply at the city's hospitals is dropping rapidly.

December 7, 1990

Thinking Critically

1. A certain normal distribution has unknown mean μ and unknown standard deviation σ. However, it is known that 15.8% of the scores are less than 29 and 2.28% of the scores are more than 67. Find μ and σ.

2. Assume that the life of a photocopying machine bulb is normally distributed with a mean life of 300 hours. If a company that uses these bulbs requires that at least 90% of the bulbs have lives exceeding 260 hours, what is the largest value that σ can have and still keep the company satisfied?

3. Using Chebyshev's Theorem, how many standard deviations on either side of the mean would have to be considered in order for us to include at least 99.7% of the numbers?

4. The Macon Shirt Company has just installed a new sleeve-cutting machine. It is known that the lengths of the sleeves cut by this machine are normally distributed. Moreover, 8.08% of the shirts have sleeve lengths that are more than 15.948 inches, and 18.67% of the shirts have sleeve lengths that are less than 15.2152 inches. What is the mean sleeve length and standard deviation of the sleeve length of the shirts produced by this machine?

5. We mentioned earlier that the normal curve approximation to the binomial can be used with fairly accurate results when *both* np and nq are greater than 5. To see what happens when these conditions are not satisfied, consider a binomial distribution with $n = 12$ and $p = 0.05$. Here $np = 0.6$. Using the binomial distribution formula, compute the probability of obtaining at most one success. Now use the normal curve approximation. How do the answers compare? Comment.

6. A random variable x is normally distributed with mean $\mu = 51$ and standard deviation $\sigma = 4$. How many standard deviations is $x = 54$ away from the mean?

Case Study

1. As reported in *Statistical Abstract of the United States*, the probability that a newborn baby will be a girl is 0.487. Last year there were 537 births reported by Mount Sinai Hospital. A diaper manufacturer has made arrangements with the maternity ward administrator to distribute as part of a promotions campaign gift packages containing among other things a month's supply of diapers designed for baby girls. The manufacturer has just delivered 230 gift packages to the hospital. What is the probability that the administrator will have a gift package for each mother that delivers a baby girl within the year?

8

SAMPLING

Chapter Outline

DID YOU KNOW THAT

a smokers may have up to twice the risk of developing colon cancer as nonsmokers? (*Source: The Journal of the National Cancer Institute*, 1994)

b elementary and high school teachers in the United States work an average of 50 hours per school week? (*Source: Wall Street Journal*, January 9, 1990)

c the average age at which men in the United States marry for the first time is 24.8 years? (*Source: Statistical Abstract of the United States*, 1994)

d 43.9% of all public elementary schools in the United States have between 250 and 499 students? (*Source:* National Center for Education Statistics in *Digest of Education Statistics*, 1994)

Although a 1040 U.S. tax return can be properly prepared, the IRS randomly selects such returns and carefully scrutinizes them to verify compliance with the tax code even though no mistakes have been detected. See discussion on page 457. *(George Semple)*

In this chapter we discuss several sampling techniques and how the results obtained from random samples, for example, the sample mean, $\bar{x}$, can be used to make inferences about the population mean μ. We also consider an important theorem—the Central Limit Theorem—which gives us some very significant information about the sampling distribution of the mean. These ideas will be used extensively in later chapters when we discuss statistical inferences.

Chapter Objectives

- **To discuss** what a random sample is and how it is obtained. (Section 8.2)

- **To work with** a table of random digits where each number that appears is obtained by a process that gives every digit an equally likely chance of being selected. (Section 8.2)

- **To understand** stratified sampling, which is a sampling procedure that is used when we want to obtain a sample with a specified number of people from different categories. (Section 8.3)

- **To see** that when repeated samples are taken from a population the frequency distribution of the values of the sample means is called the distribution of the sample means. (Section 8.5)

- **To analyze** the standard error of the mean, which represents the standard deviation of the distribution of sample means. (Section 8.5)

- **To apply** the Central Limit Theorem. This tells us that the distribution of the sample means is basically a normal distribution. We discuss how to use this theorem to make predictions about and calculate probabilities for the sample means. (Sections 8.6 and 8.7)

STATISTICS IN ACTION

On April 6, 1976, both ABC and NBC television networks projected that Morris Udall would win the Democratic primary in Wisconsin. Their predictions were based on samples from selected precincts and did not take certain districts into consideration. When all the rural votes were counted, Jimmy Carter came out on top. Many newspapers were so confident of their predictions that they printed, erroneously, the morning editions of their newspapers with the headline "CARTER UPSET BY UDALL."

The second article specifies that the bonds to be recalled will be selected in a random process. How are such random selections made?

(Gary Settle/NYT Pictures)

8.1 INTRODUCTION

Suppose we are interested in determining how many people in the United States believe in the fiscal soundness of the Social Security system. Must we ask each person in the country before making any statement concerning the fiscal soundness of the Social Security system? As there are so many people in the United States, it would require an enormous amount of work to interview each person and gather all the data.

CITY TO CALL $5 MILLION OF THE 1999 SERIES 14 BONDS

New Dorp, June 30: City officials announced yesterday that their financial condition had improved sufficiently to enable it to call $5 million of the 1999 series 14 bonds. These bonds, which were originally issued in 1975, have a 12% annual interest rate and are costing the city $7 million in annual interest charges. The bonds to be recalled will be selected in a random manner and will be made public tomorrow. The holders of all called bonds will no longer earn any interest as of tomorrow.

Friday, June 30, 1995

Do we actually need the complete population data? Can a properly selected sample give us enough information to make predictions about the entire population? In many cases, obtaining the complete population data may be quite costly or even impossible.

Similarly, suppose we are interested in purchasing an electric bulb. Must we use (or test) all electric bulbs produced by a particular company in order to determine the bulb's average life? This is very impractical. Maybe we can estimate the average life of a bulb by testing a sample of only 100 bulbs.

Population sample

As we noticed earlier in Chapter 1, Definition 1.2, this is exactly what inferential statistics involves. Samples are studied to obtain valuable information about a larger group called the **population**. Any part of the population is called a **sample**. The purpose of sampling is to select the part that truly represents the entire population.

Any sample provides only partial information about the population from which it is selected. Thus, it follows that any statement we make (based on a sample) concerning the population may be subject to error. One way of minimizing this error is to make sure that the sample is randomly selected. How is this done?

In this chapter we discuss how to select a random sample and how to interpret different sample results.

8.2 SELECTING A RANDOM SAMPLE

The purpose of most statistical studies is to make generalizations from samples about the entire population. Yet not all samples lend themselves to such generalizations. Thus, we cannot generalize about the average income of a working person in the United States by sampling only lawyers and doctors. Similarly, we cannot make any generalizations about the Social Security system by sampling only people who are receiving Social Security benefits.

Over the years many incorrect predictions have been made on the basis of nonrandom samples. For example, in 1936 the *Literary Digest* was interested in determining who would win the coming presidential election. It decided to poll the voters by mailing ten million ballots. On the basis of the approximately two million ballots returned, it predicted that Alfred E. Landon would be elected. An October 31 headline read

Landon	1,293,669
Roosevelt	972,897

Final returns in the Digest's poll of ten million voters

Source: The Literary Digest, 1936.

Actually, Franklin Roosevelt carried 46 of the 48 states and many of them by a landslide. The ten million people to whom the *Digest* sent ballots were selected from telephone listings and from the list of its own subscribers. The year 1936 was a depression year and many people could not afford telephones or magazine subscriptions. Thus, the *Digest* did not select a random sample of the voters of the United States. The *Literary Digest* soon went out of business. In 1976 Maurice Bryson[*] argued persuasively that the major problem was the *Digest's* reliance on voluntary response.

[*]M. Bryson, "The *Literary Digest* Poll: Making of a Statistical Myth," *American Statistician*, November 1976.

How does a polltaker conduct a random sample? (© *Caroline Brown, Fran Heyl Associates*)

Again, in 1948 the polls predicted that Dewey would win the presidential election. One newspaper even printed the morning edition of its newspaper with the headline "DEWEY WINS BY A LANDSLIDE." Of course, Truman won the election and laughed when presented with a copy of the newspaper predicting his defeat.

In both examples the reason for the incorrect prediction is that it was based on information obtained from poor samples. It is for this reason that statisticians insist that samples be randomly selected.

Definition 8.1
Random sample

A **random sample** of *n* items is a sample selected from a population in such a way that every different sample of size *n* from the population has an equal chance of being selected.

Definition 8.2
Random sampling

Random sampling is the procedure by which a random sample is obtained.

It may seem that the selection of a random sample is an easy task. Unfortunately, this is not the case. You may think that we can get a random sample of voters by opening a telephone book and selecting every tenth name. This will not give a random sample since many voters either do not have phones or else have unlisted numbers. Furthermore, many young voters and most women are not listed. These people, who are members of the voting population, do not have an equal chance of being selected.

To illustrate further the nature of random sampling, suppose that the administration of a large southern college with an enrollment of 30,000 students is considering revising its grading system. The administration is interested in replacing its present grading system with a pass–fail system. Since not all students agree with this proposed change, the administration has decided to poll 1000 students. How is this to be done? Polling a thousand students in the school cafeteria or in the student lounge will not result in a random sample since there may be many students who neither eat in the cafeteria nor go to the student lounge.

One way of obtaining a random sample is to write each student's name on a separate piece of paper and then put all the pieces in a large bowl where they can be thoroughly mixed. A paper is then selected from the bowl. This procedure is repeated until 1000 names are obtained. In this manner a random sample of 1000 names can be obtained. Great care must be exercised to make sure that the bowl is thoroughly mixed after a piece of paper is selected. Otherwise, the papers on the bottom of the bowl do not have an equal chance of being selected and the sample will not be random.

Pictured above is a census taker obtaining valuable information. Can such information be used to form a random sample? See discussion on page 453. *(U.S. Department of Commerce, Bureau of the Census)*

Table of random numbers
Table of random digits

Although using slips of paper in a bowl will result in a random sample if properly done, this fish bowl method becomes unmanageable as the number of people in the population increases. The job of numbering slips of paper can be completely avoided by using a **table of random numbers** or a **table of random digits**.

What are random numbers? They are the digits 0, 1, 2, ... , 9 arranged in a random fashion, that is, in such a way that all the digits appear with approximately the same frequency. Table VI in the Appendix is an example of a table of random numbers. How are such tables constructed? Today most tables of random numbers are constructed with the help of electronic computers. Yet a simple spinner such as the one shown in Figure 8.1 will also generate such a table of random numbers. After each spin we use the digit selected by the arrow. The resulting sequence of random digits could then be used to construct random number tables of five digits each, for example, as that shown in Table VI. Although this method will generate random numbers, it is not practical. After a while some numbers will be favored over others as the spinner begins to wear out.

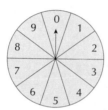

FIGURE 8.1   A spinner.

Thus, the best way to select a random sample is to use a table of random digits obtained with the help of an electronic computer. Table VI in the Appendix is such a table. In this table the various digits are scattered at random and arranged in groups of five for greater legibility. Today even some calculators can generate random digits.

Let us now return to our example. As a first step in using this table each student is assigned a number from 00001 to 30,000. To obtain a sample of 1000 students we merely read down column 1 and select the first 1000 students whose numbers are listed. Thus, the following students would be selected:

$$10,480 \quad 22,368 \quad 24,130 \quad 28,918 \quad 09,429 \ldots$$

Notice that we skip the number 42,167 since no student has this number. The same is true for the numbers 37,570, 77,921, 99,562 .... We disregard any numbers larger than 30,000 that are obtained from this table since no students are associated with these numbers.

To illustrate further the proper use of this table, suppose that a hotel has 150 guests registered for a weekend. The management wishes to select a sample of 15 people at random to rate the quality of its service. It should proceed as follows: Assign each guest a different number from 1 to 150 as 001, 002, . . . , 150. Then select any column in the table of random numbers. Suppose the fourth column is selected. Three-digit numbers are then read off the table by reading down the column. If necessary, the management continues on to another column or another page. Starting on top of column 4, they get

$$020 \quad 853 \quad 972 \quad 616 \quad 166 \quad 427 \quad 699 \ldots$$

Since the guests have numbers between 001 and 150 only, they ignore any number larger than 150. From column 4 they get

$$020 \quad 079 \quad 102 \quad 034 \quad 081 \quad 099 \quad 143 \quad 073 \quad 129$$

From column 5 they get

$$078 \quad 061 \quad 091 \quad 133 \quad 040 \quad 023$$

Thus, the management should interview the guests whose numbers are 20, 79, 102, 34, 81, 99, 143, 73, 129, 78, 61, 91, 133, 40, and 23.

COMMENT   Whenever we speak of a random sample in this text we will assume that it has been selected in the manner just described.

EXERCISES FOR SECTION 8.2

1. The Havesbrow Securities Corporation received 759 requests for more information on a new stock offering during the first week of January. Company officials would like to determine whether the requests were handled courteously and efficiently. The requests are numbered 1, 2, 3, 4, . . . , 759. Twenty of those requesting more information will be randomly selected and contacted. If the officials use column 7 of Table VI in the Appendix, which people requesting more information will be contacted?

2. Each high density–disk drive manufactured by the Yakal Computer Company during year carries a four-digit serial number from 0001 to 9999, in addition to the date of manufacture. During 1995 the company manufactured 6183 high density–disk drives. It is subsequently discovered that some of them may be defective. It is decided to randomly select 20 of the drives and to thoroughly check them. If column 8 of Table VI in the Appendix is used, which disk drives will be inspected?

3. Many bonds issued by various localities have a provision which states that they are subject to call before the stated maturity date. In 1985, Monticello issued 10,000 bonds, each numbered in order 1, 2, 3, . . . , 10,000. In 1998, as a result of falling interest rates, town officials decide to redeem 15 of these bonds ahead of their

maturity date by a random selection process. If they use columns 5, 6, 7, and 8 of Table VI in the Appendix, which bonds will they select to be redeemed?

4. Each member of the Apex Security Force employed by a shopping mall carries a badge. The badges are numbered, in order, from 1 to 228. A random sample of 17 officers is to be selected for training in new anti-shoplifting techniques. By using columns 9 and 10 of Table VI in the Appendix, decide which security force members will be selected for training.

5. One large department store issued 92,437 credit cards during 1994. The sales department has decided to invite 35 of these customers, randomly selected, to a private 40% off Columbus Day Sale. If column 8 of Table VI in the Appendix is used, which credit-card customers will be invited?

6. Each car that is manufactured in the United States has a serial number. After a series of complaints involving rusting engine mounts, one of the auto manufacturers has decided to randomly select and check 40 cars sold on the East Coast. According to company records, cars sold on the East Coast during 1993 had serial numbers whose last five digits were between 20,000 and 61,000. If columns 11 and 12 of Table VI in the Appendix are used, which cars will be checked?

7. The Internal Revenue Service randomly selects properly completed tax returns and subjects them to careful scrutiny to verify compliance with the tax code. One regional office has decided to randomly select 30 individual tax returns for 1994 and thoroughly audit them during the first week of January. Each return is filed according to the taxpayer's Social Security number. If the last four digits of the Social Security number and column 1 of Table VI in the Appendix are used, which tax returns will be audited?

8. There are 379 Senior Citizen Centers in Brookville, each of which has a permit from the Health Department to serve meals to the senior citizens. The centers are numbered 1, 2, 3, . . . , 379. After receiving numerous complaints, officials of the Health Department have decided to check on the sanitary conditions at 19 of these centers. If column 5 of Table VI in the Appendix is used, which centers will be selected to be checked?

9. A new study in the *New England Journal of Medicine* suggests that men 65 and older who simply monitor prostate cancer live nearly as long as those who seek aggressive treatments such as surgery or radiation. The Sloan Association has a list of 3281 volunteers with prostate cancer, each assigned a number 1 to 3281, who have agreed to be candidates for this new monitoring program. It is decided that only 20 of these volunteers (randomly selected) will be allowed to participate in the new monitoring program. By using columns 4 and 5 of Table VI in the Appendix, decide which of the volunteers will be chosen to participate in the new monitoring program.

10. Consider the newspaper article on the next page. One of the reasons cited for the dramatic increase in leasing is that "leasing means less financial commitment and paying only for what you use." A market analyst has the leasing agreements of 281 customers available and would like to interview 20 of these randomly selected customers to determine their reason for leasing. If only columns 13 and 14 of

Table VI in the Appendix are used, which of these customers will be selected to be interviewed?

Leasing Trend
1984 - 1992

Leased vehicles as a percent of total new autos moved

1992
23%

1984
8%

Source: CNW Marketing Research

8.3   STRATIFIED SAMPLING

Stratified samples

Although random sampling, as discussed in Section 8.2, is the most popular way of selecting a sample, there are times when **stratified samples** are preferred. To obtain a stratified sample, we divide the entire population into a number of groups or strata. The purpose of such stratification is to obtain groups of people that are more or less equal in some respect. We select a random sample, as discussed in Section 8.2 from each of these groups or from each stratum. This stratified sampling procedure ensures that no group is missed and improves the precision of our estimates. If we use stratified sampling, then in order to estimate the population mean, we must use *weighted averages* of the strata means, weighted by the population size for that stratum.

Thus, in the example discussed in the beginning of Section 8.2 the administration of the college may first divide the entire student body into four groups: freshmen, sophomores, juniors, and seniors. Then it can select a random sample from each of these groups. The groups are often sampled in proportion to their actual percentages. In this manner the administration can obtain a more accurate poll of student opinion by stratified sampling. However, the cost of obtaining a stratified sample is often higher than that of obtaining a random sample since the administration must spend money to research dividing the student body into four groups.

The method of stratifying samples is especially useful in pre-election polls. Past experience indicates that different subpopulations often demonstrate particularly different voting preferences.

COMMENT   Statistical analyses and tests based on data obtained from stratified samples are somewhat different from what we have discussed in this book. We will not analyze such procedures here.

In addition to the random and stratified sampling techniques discussed until now, there are other sampling techniques that can be used. Among these are the following:

Systematic sampling

Systematic sampling: This is a sampling procedure that is commonly used in business surveys, production processes, and for selections from name files. To use systematic sampling, we first categorize the elements of the population in some way such as alphabetically or numerically. Then we randomly select a starting point. We include in our sample every ith item of the categorized population until we obtain a sample of size n. For example, a systematic sampling procedure might begin with a randomly chosen start (from 0001 to 1200) selected from a pile of invoices. After the first invoice is randomly selected we choose, for example, every 80th invoice thereafter. In this manner we are able to obtain a good cross-sectional representation of the population.

Cluster sampling

Primary subgroups
Clusters

Cluster sampling: This is another sampling technique that is often used. With this method, the target population that we wish to analyze is divided into mutually exclusive subgroups called **primary subgroups** or **clusters**, each of which should be representative of the entire population. Then a random sample of these clusters is selected so as to provide estimates of the population values. The objective in using this procedure is to form clusters (or subgroups) that are small images of the target population. By localizing the sample units to relatively few clusters or regions, we can realize substantial cost reduction.

To illustrate the technique of cluster sampling, suppose a newspaper company cannot decide if it should institute home delivery of newspapers in a particular city. The company plans to conduct a survey to determine the fraction of households in the city that would use the newspaper home delivery service. The sampling procedure that it can use is to choose a city block (cluster) at random and then to survey every household on that block. This cluster sampling procedure is considerably more economical than simple random sampling.

8.4 CHANCE VARIATION AMONG SAMPLES

Imagine that a cigarette manufacturer is interested in knowing the average tar content of a new brand of cigarettes that is about to be sold. The Food and Drug Administration requires such information to be indicated alongside all advertisements that appear in magazines, newspapers, and so on.

The manufacturer decides to send random samples of 100 cigarettes each to 20 different testing laboratories. With the information obtained from these samples, the manufacturer hopes to be able to estimate the mean or average milligram tar content of the cigarette.

Since we will be discussing both samples and populations, let us pause for a moment to indicate the notation that we will use to distinguish between samples and populations. See Table 8.1.

TABLE 8.1	Notation for Sample and Population		
Term		Sample	Population
Mean		$\bar{x}$	μ
Standard deviation		s	σ
Number		n	N

From Chapter 3 we have the following formulas:

FORMULA 8.1

Mean

Sample	Population
$\bar{x} = \dfrac{\Sigma x}{n}$	$\mu = \dfrac{\Sigma x}{N}$

FORMULA 8.2

Standard Deviation

Sample	Population
$s = \sqrt{\dfrac{\Sigma(x - \bar{x})^2}{n - 1}}$	$\sigma = \sqrt{\dfrac{\Sigma(x - \mu)^2}{N}}$

Let us now return to our example. Since each sample sent to a laboratory is randomly selected, it is reasonably safe to assume that there will be differences among the means of each sample. The 20 laboratories report the following average milligram content per cigarette:

14.8	16.2	14.8	15.8	15.3	13.9	16.9	15.9	14.3	15.2
14.9	16.2	15.6	15.5	13.4	15.1	15.7	14.8	14.4	15.3

These figures indicate that the sample means vary considerably from sample to sample. The manufacturer decides to take the average of these 20 sample means and gets

$$\text{Average of 20 sample means} = \frac{\Sigma \bar{x}}{n} = \frac{14.8 + 16.2 + \cdots + 15.3}{20} = \frac{304}{20} = 15.2$$

The manufacturer now uses this overall average of the sample means, 15.2, as an estimate of the true population mean.

How reliable is this estimate? Although we cannot claim for certain that the population mean is 15.2, we can feel reasonably confident that 15.2 is not a bad estimate of the population mean since it is based on 20×100, or 2000, observations.

Thus, we can obtain a fairly good estimate of the population mean by calculating the mean of samples. If we let $\mu_{\bar{x}}$, read as mu sub x bar, represent the mean of the samples, then we say that $\mu_{\bar{x}}$ is a good estimate of μ. Generally speaking, if a random sample of size n is taken from a population with mean μ, then the mean of $\bar{x}$ will always equal the mean of the population (regardless of sample size), that is, $\mu_{\bar{x}} = \mu$.

What about the standard deviation? Let us calculate the standard deviation of the sample means. Recall that the formula for the population standard deviation is

$$\sqrt{\frac{\Sigma(x - \mu)^2}{N}}$$

Since μ is unknown, we have to replace it with an estimate. The most obvious replacement is $\mu_{\bar{x}}$. To account for this replacement, we divide by $N - 1$ instead of by N. Thus, the formula for the standard deviation for the sample means is given by Formula 8.3.

FORMULA 8.3

Standard deviation of the sample means

The **standard deviation of the sample means** is given by

$$\sqrt{\frac{\Sigma(\bar{x} - \mu_{\bar{x}})^2}{n - 1}}$$

where n is the number of sample means.

E X A M P L E   1   Calculate the standard deviation of the sample means for the data of the average tar content of the 20 laboratories.

SOLUTION   We arrange the data as follows:

$\bar{x}$	$\bar{x} - \mu_{\bar{x}}$	$(\bar{x} - \mu_{\bar{x}})^2$
14.8	$14.8 - 15.2 = -0.4$	0.16
16.2	$16.2 - 15.2 = 1$	1.00
14.8	$14.8 - 15.2 = -0.4$	0.16
15.8	$15.8 - 15.2 = 0.6$	0.36
15.3	$15.3 - 15.2 = 0.1$	0.01
13.9	$13.9 - 15.2 = -1.3$	1.69
16.9	$16.9 - 15.2 = 1.7$	2.89
15.9	$15.9 - 15.2 = 0.7$	0.49
14.3	$14.3 - 15.2 = -0.9$	0.81
15.2	$15.2 - 15.2 = 0$	0
14.9	$14.9 - 15.2 = -0.3$	0.09
16.2	$16.2 - 15.2 = 1$	1.00
15.6	$15.6 - 15.2 = 0.4$	0.16
15.5	$15.5 - 15.2 = 0.3$	0.09
13.4	$13.4 - 15.2 = -1.8$	3.24
15.1	$15.1 - 15.2 = -0.1$	0.01
15.7	$15.7 - 15.2 = 0.5$	0.25
14.8	$14.8 - 15.2 = -0.4$	0.16
14.4	$14.4 - 15.2 = -0.8$	0.64
15.3	$15.3 - 15.2 = 0.1$	0.01
304		13.22
$\Sigma\bar{x} = 304$		$\Sigma(\bar{x} - \mu_{\bar{x}})^2 = 13.22$

Using Formula 8.3, we have

$$\text{Standard deviation of sample means} = \sqrt{\frac{\Sigma(\bar{x} - \mu_{\bar{x}})^2}{n - 1}}$$

$$= \sqrt{\frac{13.22}{20 - 1}} = \sqrt{\frac{13.22}{19}}$$

$$= \sqrt{0.6958}$$

$$\approx 0.8341$$

Thus, the standard deviation of the sample means is approximately 0.83.

In practice, the standard deviation is not calculated by using Formula 8.3 since the computations required are time-consuming. Instead, we can use a shortcut formula given as Formula 8.4. The advantage in using Formula 8.4 is that we do not have to calculate $\mu_{\bar{x}}$ and $\bar{x} - \mu_{\bar{x}}$ and square $\bar{x} - \mu_{\bar{x}}$. We only have to calculate $\Sigma \bar{x}$ and $\Sigma \bar{x}^2$. These represent the sum of the $\bar{x}$'s and the sum of the squares of the $\bar{x}$'s, respectively. Then we use Formula 8.4.

FORMULA 8.4

Standard deviation of the sample means

The **standard deviation of the sample means** is given by

$$\sqrt{\frac{n(\Sigma \bar{x}^2) - (\Sigma \bar{x})^2}{n(n-1)}}$$

where n is the number of sample means.

EXAMPLE   2   Using Formula 8.4, find the standard deviation of the sample means for the data of Example 1 in this section.

SOLUTION   We arrange the data as follows:

$\bar{x}$	$\bar{x}^2$
14.8	219.04
16.2	262.44
14.8	219.04
15.8	249.64
15.3	234.09
13.9	193.21
16.9	285.61
15.9	252.81
14.3	204.49
15.2	231.04
14.9	222.01
16.2	262.44
15.6	243.36
15.5	240.25
13.4	179.56
15.1	228.01
15.7	246.49
14.8	219.04
14.4	207.36
15.3	234.09
304	4634.02

$$\Sigma \bar{x} = 304 \qquad \Sigma \bar{x}^2 = 4634.02$$

Using Formula 8.4, we have

$$\text{Standard deviation of sample means} = \sqrt{\frac{n(\Sigma \bar{x}^2) - (\Sigma \bar{x})^2}{n(n-1)}}$$

$$= \sqrt{\frac{20(4634.02) - (304)^2}{20(19)}}$$

$$= \sqrt{\frac{92680.4 - 92416}{380}}$$

$$= \sqrt{\frac{264.4}{380}} = \sqrt{0.6958} \approx 0.8341$$

Thus, the standard deviation of the sample means is approximately 0.83. This is the same result we obtained using Formula 8.3.

EXAMPLE 3   A large office building has six elevators, each with a capacity for ten people. The operator of each elevator has determined the average weight of the people in the elevators when operating at full capacity. The results follow.

Elevator	1	2	3	4	5	6
Average weight (lb.)	125	138	145	137	155	140

Find the overall average of these sample means. Also, find the standard deviation of these sample means by first using Formula 8.3 and then by using Formula 8.4.

SOLUTION   We arrange the data as follows.

$\bar{x}$	$\bar{x} - \mu_{\bar{x}}$	$(\bar{x} - \mu_{\bar{x}})^2$	$\bar{x}^2$
125	$125 - 140 = -15$	225	15,625
138	$138 - 140 = -2$	4	19,044
145	$145 - 140 = 5$	25	21,025
137	$137 - 140 = -3$	9	18,769
155	$155 - 140 = 15$	225	24,025
$\underline{140}$	$140 - 140 = 0$	$\underline{0}$	$\underline{19,600}$
840		488	118,088
$\Sigma \bar{x} = 840$		$\Sigma(\bar{x} - \mu_{\bar{x}})^2 = 488$	$\Sigma \bar{x}^2 = 118{,}088$

Then

$$\mu_{\bar{x}} = \frac{\Sigma\,\bar{x}}{n} = \frac{840}{6} = 140$$

Using Formula 8.3, we get

$$\text{Standard deviation of sample means} = \sqrt{\frac{\Sigma(\bar{x} - \mu_{\bar{x}})^2}{n - 1}} = \sqrt{\frac{488}{5}} = \sqrt{97.6} \approx 9.8793$$

Using Formula 8.4, we get

$$\text{Standard deviation of sample means} = \sqrt{\frac{n(\Sigma\,\bar{x}^2) - (\Sigma\,\bar{x})^2}{n(n - 1)}} = \sqrt{\frac{6(118,088) - (840)^2}{6(5)}}$$

$$= \sqrt{\frac{708528 - 705600}{30}} = \sqrt{97.6} \approx 9.8793$$

Thus, the mean of the samples is 140 and the standard deviation of the sample means (by Formula 8.3 or Formula 8.4) is approximately 9.88.

8.5   DISTRIBUTION OF SAMPLE MEANS

Let us refer back to the example discussed at the beginning of Section 8.4. The manufacturer decides to draw the histogram for the average cigarette tar content that was obtained from the 20 laboratories. This is shown in Figure 8.2. Notice that the value of

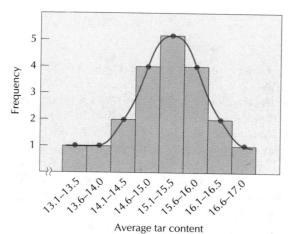

Average tar content   **FIGURE 8.2**

$\bar{x}$ is actually a random variable since its value is different from sample to sample. In repeated samples different values of $\bar{x}$ were obtained. Yet they are all close to the 15.2 we obtained as the average of the sample means. Moreover, exactly 70% of the sample means are between 14.37 and 16.03, which represents 1 standard deviation away from the mean in either direction. Also, 90% of the sample means are between 13.54 and 16.86, which represents 2 standard deviations away from the mean in either direction. Thus, Figure 8.2 actually represents the distribution of $\bar{x}$ since it tells us how the means of the samples vary from sample to sample. We refer to the distribution of $\bar{x}$ as the **distribution of the sample means** or as **the sampling distribution of the mean.** Although the first terminology is much clearer, the second is more commonly used.

Distribution of sample means

Sampling distribution of the mean

Strictly speaking, Figure 8.2 is not a complete distribution of $\bar{x}$ since it is based on only 20 sample means. To obtain the complete distribution of sample means, we would have to take thousands of samples of 100 cigarettes each. Of course, in practice, we do not take thousands of samples from the same population.

COMMENT   Notice that the sample means form an approximate normal distribution. We will have more to say about this in Section 8.6.

What can we say about this distribution? What is its mean? Its standard deviation? How does this distribution compare with the distribution of *all* the cigarettes? To answer this question, the manufacturer decides to draw the frequency polygon for the tar content of all 2000 cigarettes. This is shown in Figure 8.3.

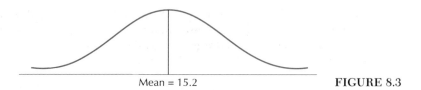

Mean = 15.2   FIGURE 8.3

Let us now compare these two distributions. Notice that both distributions are centered around the same number, 15.2. Thus, it is reasonable to assume that $\mu_{\bar{x}} = \mu$. Also, notice that the distribution of the sample means is not spread out as much as (that is, has a smaller standard deviation than) the distribution of the tar content of all the cigarettes. The reason for this should be obvious. When *all* the cigarettes are considered, several have a very high tar content and several have a very low tar content. These appear on the tail ends of the distribution of Figure 8.3. However, it is unlikely that an entire sample of 100 cigarettes will have a tar content of 18.5. Thus, the distribution of $\bar{x}$ has very little frequency at large distances from the mean.

We use the symbol $\sigma_{\bar{x}}$ to represent the standard deviation of the sampling distribution of the mean. We have the following formula for $\sigma_{\bar{x}}$.

FORMULA 8.5

Standard error of the mean

The standard deviation of the sampling distribution of the mean is referred to as the **standard error of the mean**. If random samples of size n are selected from a population whose mean is μ and whose standard deviation is σ, then the theoretical sampling distribution of $\bar{x}$ has mean $\mu_{\bar{x}} = \mu$ and a standard deviation of

$$\sigma_{\bar{x}} = \frac{\sigma}{\sqrt{n}} \cdot \sqrt{\frac{N-n}{N-1}} \qquad \begin{array}{l}\text{standard error of the mean for}\\ \textit{finite}\text{ populations of size }N\end{array}$$

and

$$\sigma_{\bar{x}} = \frac{\sigma}{\sqrt{n}} \qquad \begin{array}{l}\text{standard error of the mean for}\\ \textit{infinite}\text{ populations}\end{array}$$

Finite population correction factor

COMMENT   The factor $\sqrt{\dfrac{N-n}{N-1}}$ in the first formula for $\sigma_{\bar{x}}$ is referred to as the **finite population correction factor**. It is usually ignored; that is, it has very little effect in the calculation of $\sigma_{\bar{x}}$, unless the sample constitutes at least 5% of the population.

COMMENT   It should be obvious from Formula 8.5 that the larger the sample size is, the smaller the variation of the means will be. Thus, as we take larger and larger samples, we can expect the mean of the samples, $\mu_{\bar{x}}$, to be close to the mean of the population, μ. This is illustrated in Figure 8.4.

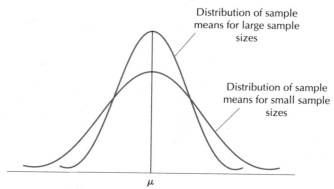

Distribution of sample means for large sample sizes

Distribution of sample means for small sample sizes

μ

FIGURE 8.4

The standard error of the mean, $\sigma_{\bar{x}}$, plays a very important role in statistics as will be illustrated in the remainder of this chapter.

To illustrate the concept of distribution of sample means and to see how Formula 8.5 is used, let us consider Mr. and Mrs. Avery, who have five children. The ages of the children are 2, 5, 8, 11, and 14 years. Suppose we first calculate the mean and standard deviation of these ages. We have

Age, x	$x - \mu$	$(x - \mu)^2$
2	$2 - 8 = -6$	36
5	$5 - 8 = -3$	9
8	$8 - 8 = 0$	0
11	$11 - 8 = 3$	9
14	$14 - 8 = 6$	36
40		90
$\Sigma x = 40$		$\Sigma(x - \mu)^2 = 90$

Then

$$\mu = \frac{\Sigma x}{N} = \frac{40}{5} = 8 \quad \text{and} \quad \sigma = \sqrt{\frac{\Sigma(x - \mu)^2}{N}}$$

$$= \sqrt{\frac{90}{5}}$$

$$= \sqrt{18} \approx 4.2426$$

Thus, the mean age is 8 and the standard deviation is approximately 4.2426 years.

Let us now calculate the mean of each sample of size 2 that can be formed from these ages. We have

If Ages Selected Are	Sample Mean, $\bar{x}$, Is
2 and 5 years	3.5
2 and 8 years	5
2 and 11 years	6.5
2 and 14 years	8
5 and 8 years	6.5
5 and 11 years	8
5 and 14 years	9.5
8 and 11 years	9.5
8 and 14 years	11
11 and 14 years	12.5

There are ten possible samples of size 2 that can be formed from these five ages. What is the average and standard deviation of these sample means? To determine these, we set up the following chart using the sample means as data.

$\bar{x}$	$\bar{x} - \mu_{\bar{x}}$	$(\bar{x} - \mu_{\bar{x}})^2$
3.5	$3.5 - 8 = -4.5$	20.25
5.0	$5.0 - 8 = -3$	9.00
6.5	$6.5 - 8 = -1.5$	2.25
8.0	$8.0 - 8 = 0$	0.00
6.5	$6.5 - 8 = -1.5$	2.25
8.0	$8.0 - 8 = 0$	0.00
9.5	$9.5 - 8 = 1.5$	2.25
9.5	$9.5 - 8 = 1.5$	2.25
11.0	$11.0 - 8 = 3$	9.00
12.5	$12.5 - 8 = 4.5$	20.25
80		67.5
$\Sigma \bar{x} = 80$		$\Sigma(\bar{x} - \mu_{\bar{x}})^2 = 67.5$

Thus, the average of the sample means is $\mu_{\bar{x}} = \dfrac{\Sigma \bar{x}}{\text{No. of samples}} = \dfrac{80}{10} = 8$ and the standard deviation of the sample means is

$$\sigma_{\bar{x}} = \sqrt{\frac{\Sigma(\bar{x} - \mu_{\bar{x}})^2}{\text{No. of samples}}} = \sqrt{\frac{67.5}{10}} = \sqrt{6.75} \approx 2.598$$

In our case the population size, N, is 5 and the sample size n, is 2 so that the sample size is $\dfrac{2}{5}$ or 40% of the population size. Using Formula 8.5, we get

$$\sigma_{\bar{x}} = \frac{\sigma}{\sqrt{n}} \sqrt{\frac{N - n}{N - 1}}$$

$$= \frac{4.2426}{\sqrt{2}} \sqrt{\frac{5 - 2}{5 - 1}} \qquad \text{(Remember } \sigma = 4.2426 \text{, as previously calculated on page 468.)}$$

$$= \frac{4.2426}{1.4142} \sqrt{\frac{3}{4}} \approx \frac{4.2426}{1.4142} (0.8660)$$

$$\approx 2.598$$

This is exactly the value that we obtained previously. Thus, the average of the sample means is exactly the same as the population mean, that is, $\mu_{\bar{x}} = \mu$, and the standard deviation of the sample means is considerably less than the population standard deviation.

COMMENT   In the balance of this chapter we will always assume that our sample size is less than 5% of the population size. Consequently, we will use $\sigma_{\bar{x}} = \dfrac{\sigma}{\sqrt{n}}$ as the standard error of the mean.

Let us summarize our discussion up to this point. Using the distribution of the sample means of the laboratories and the distribution of the tar content of all 2000 cigarettes, or the distribution of the sample means of the ages, we conclude the following:

The mean of the distribution of sample means and the mean of the original population are the same.

The standard deviation of the distribution of the sample means is less than the standard deviation of the original population. The exact relationship is referred to as the standard error of the mean and is found by using Formula 8.5.

The distribution of the sample means is approximately normally distributed.

COMMENT   The last statement is so important that it is referred to as the **Central Limit Theorem**. Since much of the work of statistical inference is based on this theorem, we will discuss its importance, as well as its applications in detail, in the following sections.

EXERCISES FOR SECTION 8.5

1. As reported in 1994 in the *Journal of the National Cancer Institute*, studies have linked smoking to the development of intestinal polyps. A survey of 170,000 people has shown that smokers have up to twice the risk of developing colon cancer as nonsmokers. Dr. Bryan Gupta reports having diagnosed 6, 10, 4, 7, 9, and 8 of his smoking patients as having developed intestinal polyps during the years 1989–1994, respectively.
   a. Make a list of all the possible samples of size 2 that can be drawn from this list of numbers.
   b. Determine the mean of each of these samples and form a sampling distribution of these sample means.
   c. Find the mean, $\mu_{\bar{x}}$, of this sampling distribution.
   d. Find the standard deviation, $\sigma_{\bar{x}}$, of this sampling distribution.
2. Five of our country's largest credit-card issuers were asked to indicate the annual percentage rate that they charged on any unpaid balances. The percent charged by these banks was 18, 21, 14, 19, and 16.
   a. Make a list of all the possible samples of size 2 that can be drawn from this list of numbers.
   b. Determine the mean of each of these samples and form a sampling distribution of these sample means.
   c. Find the mean, $\mu_{\bar{x}}$, of this sampling distribution.
   d. Find the standard deviation, $\sigma_{\bar{x}}$, of this sampling distribution.
3. The number of false fire alarms reported in one region of New York City over a six-day period was as follows:

Mon.	Tues.	Wed.	Thurs.	Fri.	Sat.
8	5	12	9	6	7

a. Make a list of all the possible samples of size 3 that can be drawn from these numbers.
b. Determine the mean of each of these samples.
c. Find the mean, $\mu_{\bar{x}}$.
d. Find the standard deviation, $\sigma_{\bar{x}}$.

4. According to a new study by the U.S. Geological Survey, concentrations of sulfate and nitrate—two components of acid rain—declined significantly between 1980 and 1991. (*Source: Time*, July 19, 1993, p. 20). One study of 20 lakes and rivers found an average concentration of 33, 42, 37, 53, 47, 41, 55, 38, 29, 38, 38, 45, 53, 58, 27, 45, 52, 31, 46, and 32 units in each of these lakes or rivers.
a. Calculate the mean of the 20 sample means.
b. Draw the histogram for these sample means (similar to what was done in Figure 8.2).

5. There are five agents that work for the state's Department of Environmental Protection. The number of cases involving illegal dumping of toxic pollutants that these agents investigated last year was 12, 21, 16, 15, and 26.
a. Determine the mean, μ, and standard deviation, σ, of the population of the number of cases that these agents investigated last year.
b. List all the possible samples of size 2 *and* size 3 that can be selected from these numbers, and determine the mean for each of these samples.
c. Find the mean, $\mu_{\bar{x}}$, and standard deviation, $\sigma_{\bar{x}}$, for the sample means in each case.
d. Show that for both the samples of size 2 and the samples of size 3

$$\sigma_{\bar{x}} = \frac{\sigma}{\sqrt{n}} \sqrt{\frac{N-n}{N-1}}$$

6. State tax officials claim that the average amount of money claimed by all the taxpayers within the state for charitable deductions during 1994 was $964 with a standard deviation of $102. Many samples of size 64 are taken. Find the mean of these samples and the standard error of the mean.

7. Refer to Exercise 6. What would the sample mean *and* standard error of the mean be if the samples are of
a. size 49 each.
b. size 100 each.

8.6 THE CENTRAL LIMIT THEOREM

One of the most important theorems in probability is the **Central Limit Theorem**. This theorem, first established by De Moivre in 1733 (see the discussion on page 473), was named "The Central Limit Theorem of Probability" by G. Polya in 1920. The theorem may be summarized as follows.

THE CENTRAL LIMIT THEOREM

If large random samples of size n (usually samples of size $n > 30$) are taken from a population with mean μ and standard deviation σ, and if a sample mean $\bar{x}$ is computed for each sample, then the following three facts will be true about the distribution of sample means.

1. The distribution of the sample means will be approximately normally distributed.
2. The mean of the sampling distribution will be equal to the mean of the population. Symbolically,

$$\mu_{\bar{x}} = \mu$$

3. The standard deviation of the sampling distribution will be equal to the standard deviation of the population divided by the square root of the number of items in each sample. Symbolically,

$$\sigma_{\bar{x}} = \frac{\sigma}{\sqrt{n}}$$

COMMENT If the sample size is large enough, the sampling distribution will be normal, even if the original distribution is not. "Large enough" usually means larger than 30 items in the sample.

This last comment is extremely important as can be seen in the following graphs where the graph of the sampling distribution of $\bar{x}$ for different populations and different sample sizes approach a normal distribution when $n = 30$.

Original population	Sampling distribution of $\bar{x}$ when $n = 2$	Sampling distribution of $\bar{x}$ when $n = 5$	Sampling distribution of $\bar{x}$ when $n = 30$

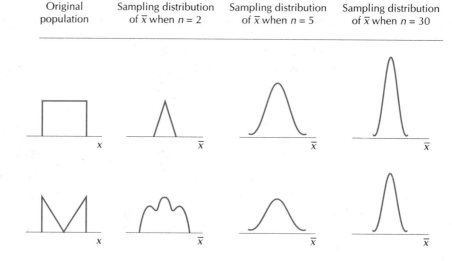

If the original distribution is normal, then the sampling distribution of the mean from 500 samples of size 2, 8, and 32 quickly resembles a normal distribution. Note what happens to $\sigma_{\bar{x}}$ as n gets larger.

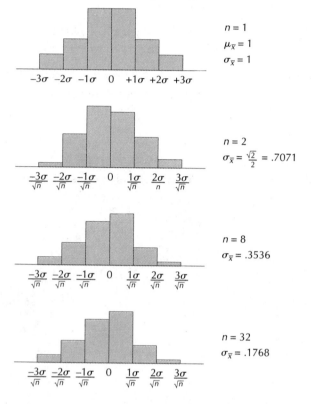

$n = 1$
$\mu_{\bar{x}} = 1$
$\sigma_{\bar{x}} = 1$

$n = 2$
$\sigma_{\bar{x}} = \frac{\sqrt{2}}{2} = .7071$

$n = 8$
$\sigma_{\bar{x}} = .3536$

$n = 32$
$\sigma_{\bar{x}} = .1768$

COMMENT   The n referred to in the Central Limit Theorem refers to the size of each sample, and not to the number of samples.

HISTORICAL NOTE

In the previous chapter we presented some facts about De Moivre and Laplace. The theorem that allows us to approximate the binomial distribution with the normal distribution is known as the De Moivre–Laplace Limit Theorem. It was originally proved for the special case $p = \frac{1}{2}$ by De Moivre in 1733 and then extended to general p by Laplace in 1812. The same is true for the Central Limit Theorem.

The first version of the Central Limit Theorem was proved by De Moivre around 1733 for the special case in which x is a Bernoulli variable with $p = \frac{1}{2}$. Although Laplace extended the theorem for any arbitrary p, a truly rigorous proof of the Central Limit Theorem in full generality was first presented by the Russian mathematician Liapounoff in the period 1901–1902.

Since the Central Limit Theorem is so important, we will discuss its applications in the next section.

8.7 APPLICATIONS OF THE CENTRAL LIMIT THEOREM

In this section we use the Central Limit Theorem to predict the behavior of sample means. To apply the standardized normal distribution discussed in Chapter 7, we have to change Formula 7.1 somewhat. Recall that

$$z = \frac{x - \mu}{\sigma}$$

It can be shown that when dealing with sample means this formula becomes that given as Formula 8.6.

FORMULA 8.6

$$z = \frac{\bar{x} - \mu}{\sigma/\sqrt{n}}$$

The following examples illustrate how the Central Limit Theorem is applied.

EXAMPLE 1 The average height of all the workers in a hospital is known to be 65 inches with a standard deviation of 2.3 inches. If a sample of 36 people is selected at random, what is the probability that the average height of these 36 people will be between 64 and 65.5 inches?

SOLUTION We use Formula 8.6. Here $\mu = 65$, $\sigma = 2.3$, and $n = 36$. Thus, $\bar{x} = 64$ corresponds to

$$z = \frac{64 - 65}{2.3/\sqrt{36}} = \frac{-1}{0.3833} = -2.61$$

and $\bar{x} = 65.5$ corresponds to

$$z = \frac{65.5 - 65}{2.3/\sqrt{36}} = \frac{0.5}{0.3833} = 1.30$$

Thus, we are interested in the area of a standard normal distribution between $z = -2.61$ and $z = 1.30$. See Figure 8.5.

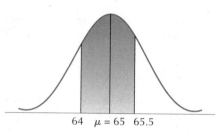
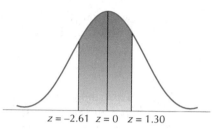

FIGURE 8.5

From Table IV in the Appendix we find that the area between $z = 0$ and $z = -2.61$ is 0.4955 and that the area between $z = 0$ and $z = 1.30$ is 0.4032. Adding, we get

$$0.4955 + 0.4032 = 0.8987$$

Thus, the probability that the average height of the sample of 36 people is between 64 and 65.5 inches is 0.8987.

E X A M P L E   2   The average amount of money that a depositor of the Second National City Bank has in an account is $5000 with a standard deviation of $650. A random sample of 36 accounts is taken. What is the probability that the average amount of money that these 36 depositors have in their accounts is between $4800 and $5300?

SOLUTION   We use Formula 8.6. Here $\mu = 5000$, $\sigma = 650$, and $n = 36$. Thus, $\bar{x} = 4800$ corresponds to

$$z = \frac{4800 - 5000}{650/\sqrt{36}} = \frac{-200}{108.3333} = -1.85$$

and $\bar{x} = 5300$ corresponds to

$$z = \frac{5300 - 5000}{650/\sqrt{36}} = \frac{300}{108.3333} = 2.77$$

Thus, we are interested in the area between $z = -1.85$ and $z = 2.77$. See Figure 8.6.

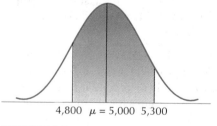
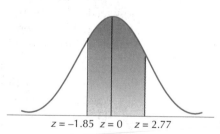

FIGURE 8.6

From Table IV in the Appendix we find that the area between $z = 0$ and $z = -1.85$ is 0.4678 and that the area between $z = 0$ and $z = 2.77$ is 0.4972. Adding these two, we get

$$0.4678 + 0.4972 = 0.9650$$

Thus, the probability is 0.9650 that the average amount of money these depositors have in their accounts is between \$4800 and \$5300.

E X A M P L E   3   The average purchase by a customer in a large novelty store is \$4.00 with a standard deviation of \$0.85. If 49 customers are selected at random, what is the probability that their average purchases will be less than \$3.70?

SOLUTION   We usè Formula 8.6. Here $\mu = 4.00$, $\sigma = 0.85$, and $n = 49$. Thus, $\bar{x} = 3.70$ corresponds to

$$z = \frac{3.70 - 4.00}{0.85/\sqrt{49}} = \frac{-0.30}{0.1214} = -2.47$$

Therefore, we are interested in the area to the left of $z = -2.47$ (Figure 8.7).

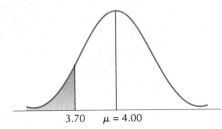

3.70 $\mu = 4.00$ $z = -2.47$ $z = 0$

FIGURE 8.7

From Table IV in the Appendix we find that the area from $z = 0$ to $z = -2.47$ is 0.4932. Thus, the area to the left of $z = -2.47$ is $0.5000 - 0.4932$, or 0.0068. Therefore, the probability that the average purchase will be less than \$3.70 is 0.0068.

E X A M P L E   4   The Smith Trucking Company claims that the average weight of its delivery trucks when fully loaded is 6000 pounds with a standard deviation of 120 pounds. Thirty-six trucks are selected at random and their weights recorded. Within what limits will the average weights of 90% of the 36 trucks lie?

SOLUTION   Here $\mu = 6000$ and $\sigma = 120$. We are looking for two values within which the weights of 90% of the 36 trucks will lie. From Table IV we find that the area between $z = 0$ and $z = 1.645$ is approximately 0.45. Similarly, the area between $z = 0$ and $z = -1.645$ is approximately 0.45. See Figure 8.8. Using Formula 8.6, we have

$$z = \frac{\bar{x} - \mu}{\sigma/\sqrt{n}}$$

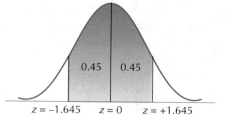

FIGURE 8.8

If $z = 1.645$, then

$$1.645 = \frac{\bar{x} - 6000}{120/\sqrt{36}}$$

$$= \frac{\bar{x} - 6000}{20}$$

$$1.645(20) = \bar{x} - 6000$$

$$32.9 + 6000 = \bar{x}$$

$$6033 = \bar{x}$$

If $z = -1.645$, then

$$-1.645 = \frac{\bar{x} - 6000}{120/\sqrt{36}}$$

$$= \frac{\bar{x} - 6000}{20}$$

$$-1.645(20) = \bar{x} - 6000$$

$$6000 - 32.9 = \bar{x}$$

$$5967 = \bar{x}$$

Thus, 90% of the trucks will weigh between 5967 and 6033 pounds.

EXERCISES FOR SECTION 8.7

1. One study, reported by the Pharmaceutical Manufacturer's Association, found that the average annual per-person spending on prescription drugs in the United States is $210. The standard deviation is $38. A random survey of 64 people is taken. What is the probability that the average annual spending on prescription drugs for these people is between $200 and $225?

2. One tire manufacturer claims that the company's steel-belted radial tires last an average of 35,000 miles with a standard deviation of 7100 miles. A random sample of 81 of this company's tires is taken. What is the probability that these tires will last at least 36,000 miles?

3. The National Center for Education Statistics claims in its *Digest of Education Statistics* that 43.9% of all public elementary schools have between 250 and 499 students. For one particular city, the average number of students per public elementary school is 475 with a standard deviation of 33 students. A random sample of 36 public elementary schools in this city is taken. What is the probability that the average number of students per school in these schools is less than 465?

4. Consider the accompanying newspaper article. If a random sample of 50 taxi trips from Kennedy Airport to midtown Manhattan is taken, what is the probability that the fare charged will be less than $37?

TAXI DRIVER SLAPPED WITH STIFF FINE

New York, November 3: The Taxi and Limousine Commission leveled a stiff fine yesterday against Rauel DeGregorio for overcharging an unsuspecting foreign tourist. It is alleged that the tourist was charged $192 for a trip from Kennedy Airport to midtown Manhattan. The charge for such a trip averages around $35 (with a standard deviation of $4.50).

November 3, 1993

5. The Motor Vehicle Manufacturers Association of the United States publishes information on the ages of cars and trucks in use in *Motor Vehicle Facts and Figures*. For one particular city it is claimed that the average age of a car in use is 6.8 years with a standard deviation of 2.3 years. What is the probability that a random sample of 60 cars will have an average age that is more than 6.1 years?

6. The Gold Baking Company claims that the average weight of all its crumb coffee cakes is 12 ounces with a standard deviation of 1.01 ounces. To check on the accuracy of the stated weights, an agent from the Consumer's Fraud Bureau selects a random sample of 40 of these cakes. Within what limits should the weights of 95% of these cakes lie?

7. Based on data from the *Statistical Abstract of the United States*, the average length of a newborn baby is 20.1 inches with a standard deviation of 1.22 inches. A random sample of 45 newborn babies is taken. Within what limits should the lengths of 95% of those babies be?

8. Each year, on Memorial day, the American Automobile Association (AAA) conducts a nationwide survey to determine the average cost of a gallon of regular unleaded gas. One year, the AAA reported, "Gas will be plentiful with a gallon of regular unleaded gas selling for an average $1.059 per gallon nationwide." Suppose we know that the standard deviation is equal to $0.32. If a random survey of 43 gas stations is taken, what is the probability that the average charge for a gallon of regular unleaded gas will be less than $0.99?

9. A major cereal company packages its large boxes of corn flakes so that they contain an average of 680 grams with a standard deviation of 18 grams. A sample of 45 of these boxes is randomly selected. What is the probability that the average weight of these samples will be at least 675 grams?

10. The U.S. Department of Education reports that elementary and high school teachers work an average of 50 hours per school week (*Source: The Wall Street Journal*, January 9, 1990). If a sample of 40 teachers is randomly selected, what is the probability that these teachers work an average of at least 55 hours per school week? (Assume that the standard deviation is 14 hours.)

11. Consider the accompanying newspaper article. If 81 homes in this city are randomly selected, what is the probability that the average age of these houses will be between 43 and 46 years old?

NEW HOUSING STARTS DOWN DRAMATICALLY

January 25: The economic recession that is currently gripping our nation is affecting the building industry particularly hard. In our region only 17 new housing starts were reported this past month. In a city where the average age of a house is 43.3 years (with a standard deviation of 8.2 years) such devastating facts can wreak havoc on our already staggering economy.

January 25, 1994

12. The average age at which men in the United States marry for the first time is 24.8 years (*Source: Statistical Abstract of the United States*, 1994). If a sample of 55 married men is randomly selected, what is the probability that the average age at which these men married for the first time is at most 25.2 years? (Assume that the standard deviation is 2.7 years.)

8.8   USING COMPUTER PACKAGES

The MINITAB statistical software can be used to generate, for example, a sampling distribution of x based on 25 samples of size $n = 4$ measurements, each drawn from a normal population with $\mu = 10$ and $\sigma = 3$. We first have to use RANDOM and AVERAGE to generate each sample of size $n = 4$ measurements and to compute the sample mean. To accomplish this, we generate one sample of size $n = 4$ as follows:

Sample 1

```
MTB  > RANDOM 4 OBS, INTO C1;
SUBC > NORMAL MU = 10, SIGMA = 3.
```

The following appears in C1:

```
8.9321   11.6106   10.1234   9.7862
MTB > AVERAGE THE OBSERVATIONS IN C1
        MEAN = 10.113
```

To generate a second sample, we repeat the RANDOM and AVERAGE statements again and set the results into C2. We repeatedly use the same procedure until we obtain the 25 samples of size $n = 4$ measurements that we desire. This can be very time-consuming.

After the 25 sample means are obtained, we enter these results into MINITAB to obtain some numerical descriptive measures as was done in Chapter 3. For our example we enter the 25 sample means into MINITAB as follows:

```
MTB  > SET THE FOLLOWING DATA INTO C26
DATA >    10.113    10.101    9.996    10.002   10.123
DATA >     9.468     9.823    9.982    10.176   10.082
           .         .        .        .
           .         .        .        .
           .         .        .        .

DATA > END
```

Then we have MINITAB obtain numerical descriptive measures for the data. We get

```
MTB > DESCRIBE C26
```

The computer output for our example will be

	N	MEAN	MEDIAN	TRMEAN	ST DEV	SEMEAN
C26	25	10.016	10.183	10.213	1.386	0.0812

	MIN	MAX	Q1	Q3
C26	7.938	10.816	8.096	11.547

```
MTB > STOP
```

COMMENT   The mean of the sample means is 10.016. This is fairly close to the theoretical value of $\mu = 10$. Similarly, the standard deviation of the sample means is 1.386, which is close to the theoretical value of $\dfrac{\sigma}{\sqrt{n}} = \dfrac{3}{\sqrt{4}} = 1.5$. These experimental values would be closer to their respective theoretical values had we taken many more samples of size $n = 4$ rather than only the 25 samples that we generated.

COMMENT   An alternate time-saving way to generate all 25 samples at the same time is to change the first MINITAB commands to

```
MTB  > RANDOM 4 OBS, INTO C1 — C25;
SUBC > NORMAL MU = 10, SIGMA = 3.
```

This will generate all 25 samples at the same time.

EXERCISES FOR SECTION 8.8

1. Use MINITAB to generate 50 random samples of size $n = 10$ from a normal distribution with $\mu = 75$ and $\sigma = 10$.
2. Use MINITAB to generate 100 random samples of size $n = 20$ from a normal distribution with $\mu = 40$ and $\sigma = 8$.
3. A medical researcher wishes to test a new drug. One hundred people have volunteered to try the new drug. It is decided that 50 of these volunteers will be randomly selected and given the new drug. The remaining volunteers will be given a placebo. Use MINITAB to select the 50 volunteers who will be given the new medication. Assume that the 100 patients are numbered consecutively 1 through 100.
4. Use MINITAB to generate 75 random samples of size $n = 10$ from a normal distribution with $\mu = 20$ and $\sigma = 3$. For each sample, have MINITAB compute the sample mean and store it in an appropriate column. Then have MINITAB compute its mean and standard deviation. Finally, instruct MINITAB to draw the histogram. How do the results obtained compare with the theoretical distribution of the sample means?
5. Use MINITAB to generate 80 random samples of size $n = 15$ from a normal distribution with $\mu = 18$ and $\sigma = 4$. For each sample, have MINITAB compute the sample mean and store it in an appropriate column. Then have MINITAB compute its mean and standard deviation. Finally instruct MINITAB to draw the histogram. How do the results obtained compare with the theoretical distribution of the sample means?

8.9   SUMMARY

In this chapter we discussed the nature of random sampling and how to go about selecting a random sample. The most convenient way of selecting a random sample is to use a table of random digits.

In some situations, as we pointed out, stratified samples are preferred.

When repeated random samples are taken from the same population, different sample means are obtained. The average of these sample means can be used as an estimate of the population mean. Of course, these sample means form a distribution. If the samples are large enough, the Central Limit Theorem tells us that they will be normally distributed. Furthermore, the mean of the sampling distribution is the same as the population mean. The standard deviation of the sampling distribution is less than the population standard deviation. The Central Limit Theorem led to many useful applications.

Study Guide

The following is a chapter outline in capsule form. You should now be able to demonstrate your knowledge of the ideas mentioned by giving definitions, descriptions, or specific examples. Page references are given in parentheses.

Samples are studied to obtain valuable information about a larger group called the **population**. Any part of the population is called a **sample**. (page 452)

A **random sample** of n items is a sample selected from a population in such a way that every different sample of size n from the population has an equal chance of being selected. (page 453)

Random sampling is the procedure by which a random sample is obtained. (page 453)

A **table of random numbers** or a **table of random digits** is a chart that lists the digits $0, 1, 2, \ldots, 9$ arranged in such a way that all the digits appear with approximately the same frequency. (page 455)

To obtain a **stratified sample**, we first divide the entire population into a number of groups or strata. Then we select a random sample from each of these groups or from each stratum. (page 458)

To use **systematic sampling**, we first categorize the elements of the population in some way, such as alphabetically or numerically. Then we randomly select a starting point. We include in our sample every ith term of the categorized population until we obtain a sample of size n. (page 459)

When using **cluster sampling**, we first divide the target population to be analyzed into mutually exclusive subgroups called **primary subgroups** or **clusters**, each of which should be representative of the entire population. Then a random sample of these clusters is selected so as to provide estimates of the population values. (page 459)

A fairly good estimate of the population mean can be obtained by calculating the mean of samples from the population; that is, $\mu_{\bar{x}} = \mu$. (page 461)

The **standard deviation of the sample means** is given by

$$\sqrt{\frac{\Sigma(\bar{x} - \mu_{\bar{x}})^2}{n - 1}} \quad \text{or} \quad \sqrt{\frac{n(\Sigma \bar{x}) - (\Sigma \bar{x})^2}{n(n - 1)}}$$

n always represents the number of sample means. (pages 461–463)

The **distribution of the sample means**, $\bar{x}$, shows us how the means vary from sample to sample. It is also referred to as the **sampling distribution of the mean**. (page 466)

The standard deviation of the sampling distribution of the mean is referred to as the **standard error of the mean** and is given by

$$\sigma_{\bar{x}} = \frac{\sigma}{\sqrt{n}} \cdot \sqrt{\frac{N-n}{N-1}} \qquad \text{for finite populations of size } N$$

and

$$\sigma_{\bar{x}} = \frac{\sigma}{\sqrt{n}} \qquad \text{for infinite populations} \qquad \text{(page 467)}$$

The factor $\sqrt{\dfrac{N-n}{N-1}}$ used in calculating the standard error of the mean is referred to as the **finite population correction factor**. (page 467)

If large random samples of size n (usually samples of size $n > 30$) are taken from a population with mean μ and standard deviation σ, and if a sample mean $\bar{x}$ is computed for each sample, then the **Central Limit Theorem** says that the following three facts will be true about the distribution of sample means.

1. The distribution of the sample means will be approximately normally distributed.
2. The mean of the sampling distribution will be equal to the mean of the population; that is, $\mu_{\bar{x}} = \mu$.
3. The standard deviation of the sampling distribution will be equal to the standard deviation of the population divided by the square root of the number of items in each sample; that is, $\sigma_{\bar{x}} = \dfrac{\sigma}{\sqrt{n}}$. (page 473)

Formulas to Remember

You should be able to identify each symbol in the following formulas, understand the relationships among the symbols expressed in each formula, understand the significance of each formula, and use the formulas in solving problems.

1. Population mean: $\quad \mu = \dfrac{\Sigma x}{N}$

2. Sample mean: $\quad \bar{x} = \dfrac{\Sigma x}{n}$

3. Population standard deviation: $\sigma = \sqrt{\dfrac{\Sigma(x - \mu)^2}{N}}$

4. Standard deviation of sample means: $\sqrt{\dfrac{\Sigma(\bar{x} - \mu_{\bar{x}})^2}{n - 1}}$

5. Computational formula for standard deviation of sample means:

$$\sqrt{\dfrac{n(\Sigma\, \bar{x}^2) - (\Sigma\, \bar{x})^2}{n(n - 1)}}$$

6. Standard error of the mean when sample size is less than 5% of the population size: $\sigma_{\bar{x}} = \dfrac{\sigma}{\sqrt{n}}$. The finite population correction factor $\sqrt{\dfrac{N - n}{N - 1}}$ adjusts the standard error to describe most accurately the amount of variation. It should be applied whenever the sample size is 5% or more of the population size.

7. When dealing with sample means (Central Limit Theorem): $z = \dfrac{\bar{x} - \mu}{\sigma/\sqrt{n}}$

Testing Your Understanding of This Chapter's Concepts

1. What relationship exists, if any, between the population standard deviation, the standard error of the mean, and the sample size?
2. A student claims that if a sample is selected from an infinite population, then we can assume that the sample is less than 5% of the population, and so we can ignore the finite population correction factor. Do you agree? Explain your answer.
3. False fire alarms can often cost lives. On one summer day in the five boroughs of New York City, there were 100 false alarms reported by the fire department as follows:

Borough	Number of False Alarms Reported
Brooklyn	23
Staten Island	18
Queens	16
Bronx	31
Manhattan	12

a. Determine the mean, μ, and standard deviation, σ, of the population of the number of false fire alarms reported in each of the boroughs.
b. List all possible samples of size 2 *and* size 3 that can be selected from these numbers and determine the mean of each of these samples.

c. Find the mean, $\mu_{\bar{x}}$, and the standard deviation, $\sigma_{\bar{x}}$, for the sample means in each case.

d. Show that for both the samples of size 2 and size 3

$$\sigma_{\bar{x}} = \frac{\sigma}{\sqrt{n}} \cdot \sqrt{\frac{N - n}{N - 1}}$$

e. Draw the histogram for the sample means (similar to what was done in Figure 8.2).

Chapter Test

Multiple Choice Questions

1. The average length of a certain type of book is 488 pages with a standard deviation of 18.3 pages. What is the probability that the average length of 49 randomly selected books of the type described will be greater than 495 pages?
   a. 0.0037   b. 0.9963   c. 0.4963   d. 0.6480   e. none of these

2. Five schools have each been testing 100 bulbs of a particular brand. The average life of the bulbs at each of these schools was 2100, 3200, 2600, 2800, and 2900 hours. Find the overall average of these sample means.
   a. 3400   b. 2720   c. 136   d. 2266.67   e. none of these

3. In the previous exercise, find the standard deviation of these samples.
   a. 365.513   b. 149.22   c. 408.656   d. 20.0215   e. none of these

4. The average life of a washing machine produced by the Bob Corp. is 8.3 years with a standard deviation of 2.13 years. If a random sample of 36 washing machines produced by this company is selected, what is the probability that the average life of these machines will be between 7.3 and 9 years?
   a. 0.9732   b. 0.4756   c. 0.4976   d. 0.3101   e. none of these

5. Refer back to the previous question. If a random survey of 36 washing machines produced by this company is selected, within what limits should the life of 95% of these machines lie?
   a. between 7.7160 and 8.8840 years   b. between 7.7178 and 8.8822 years
   c. between 7.6042 and 8.9958 years   d. between 4.7962 and 11.8039 years
   e. none of these

6. The average weight of an employee at the Bork Corp. is 135 pounds with a standard deviation of 11 pounds. Within what limits should the average weights of 95% of 50 randomly selected employees of this company lie?
   a. between 131.951 and 138.049 pounds
   b. between 132.441 and 137.559 pounds
   c. between 116.905 and 153.095 pounds
   d. between 132.468 and 137.532 pounds
   e. none of these

Supplementary Exercises

7. There are 817 nurses on the staff of a hospital, each of whom has been assigned a number from 1 to 817. A committee of ten nurses is to be randomly selected from among them to discuss changing the shift patterns. By using column 5 of Table VI, which nurses should be selected?

8. The average cost for replacing an automatic transmission of a certain model car is $575 with a standard deviation of $28. A sample of the repair bills for replacing 45 automatic transmissions of the type mentioned is taken. What is the probability that the average replacement cost is between $565 and $580?

9. The average gas purchase at Pete's service station is $15 with a standard deviation of $2.25. If 39 gas purchases are selected at random, what is the probability that the average purchase will be less than $14?

10. The average amount of money that the teachers of the Armington School District have accumulated in the tax-deferred annuity program is $2250 with a standard deviation of $117. If a sample of 45 participating teachers is randomly selected, what is the probability that these teachers will have accumulated an average of at least $2300?

11. The average height of an employee of the Gant Corp. is 65 inches with a standard deviation of 3 inches. Within what limits should the average heights of 95% of 60 randomly selected employees lie?

12. There are 768 police officers in a large city assigned to traffic patrol, each of whom has been assigned a number from 1 to 768. A committee of ten police officers is to be randomly selected from among them to discuss reassignment problems. By using column 6 of Table VI, which police officers should be selected?

13. The average cost of replacing the loading mechanism on a particular VCR is $84 with a standard deviation of $9. A sample of the repair bills for replacing 40 loading mechanisms of the type mentioned is taken. What is the probability that the average replacement cost is between $80 and $87?

14. The average bill at Maggie's Restaurant is $21 with a standard deviation of $2.85. If 43 bills are selected at random, what is the probability that the average of these bills will be less than $20?

15. The pilots of a particular airline have accumulated an average of 5000 flying hours with a standard deviation of 175 hours. If a sample of 47 pilots of this airline is randomly selected, what is the probability that these pilots will have accumulated an average of at least 4975 hours of flying time?

16. Five companies have each been field testing 100 beepers manufactured by the Caldwell Corp. The average life of these beepers for each of these companies was 710, 800, 740, 760, and 780 days. Find the overall average of these sample means.

17. In the previous exercise, find the standard deviation of these sample means.

18. The average life of a microwave oven produced by the West Corp. is 9.7 years with a standard deviation of 1.23 years. If a random sample of 64 microwave ovens produced by this company is selected, what is the probability that the average life of these microwave ovens will be between 9.5 and 10 years?

19. Refer back to the previous question. If a random survey of 64 microwave ovens produced by this company is selected, within what limits should the lives of 95% of these microwave ovens lie?

20. The average age of a doctor at Washington Hospital is 43 years with a standard deviation of 3.2 years. Within what limits should the average age of 95% of 40 randomly selected doctors from this hospital lie?

21. There are 812 doctors at Washington Hospital, each of whom has been assigned a number from 1 to 812. A committee of ten doctors is randomly selected from among them to discuss the rising malpractice insurance costs. By using column 7 of Table VI, which doctors should be selected?

22. The average cost for replacing a disk drive of a particular model computer is $95 with a standard deviation of $10.50. A sample of the repair bills for replacing 37 disk drives of the type mentioned is taken. What is the probability that the average replacement cost is between $90 and $100?

23. The average monthly electric bill in a certain city is $28 with a standard deviation of $4.75. If 52 electric bills are randomly selected, what is the probability that the average of these bills will be less than $26?

24. The average weight of a certain size of grapefruit is 10 ounces with a standard deviation of 1.92 ounces. What is the probability that the average weight of 64 randomly selected grapefruits of the size described will be greater than 10.5 ounces?

25. Five supermarkets have been testing 100 special shopping carts produced by a particular company. The average life of the carts at each of these supermarkets was 410, 510, 480, 500, and 490 days. Find the overall average of these sample means.

26. In the previous exercise, find the standard deviation of these samples.

27. The average life of a refrigerator produced by the West Corp. is 11.5 years with a standard deviation of 1.47 years. If a random sample of 81 refrigerators produced by this company is selected, what is the probability that the average life of these refrigerators will be between 11.2 and 11.9 years?

28. Refer back to the previous question. If a random sample of 81 refrigerators is selected, within what limits should the life of 95% of these refrigerators lie?

29. A population is known to have a mean of 193 and a standard deviation of 28. Many samples of size 100 are taken. Find the mean of these samples and the standard error.

Thinking Critically

1. Explain why it is true that a sample mean from a large sample size gives us a better estimate of the population mean than a sample mean from a smaller sample size.

2. Explain the difference between σ_x and $\sigma_{\bar{x}}$. (*Hint:* Use sketches of normal distributions to illustrate your answers.)

3. Assume that we wish to estimate the mean of an infinite population with standard deviation σ on the basis of a random sample of size n where n is large. It can be shown that if the mean of the sample is used as an estimate, then there is a 50% probability that the magnitude of the error will be less than

$$\frac{(0.06745)\sigma}{\sqrt{n}}$$

This quantity is often called **the probable error of the mean**. Assume that a random sample of size 400 is drawn from a very large population detailing the life of tires used by police cars. If $\sigma = 1281$ miles, what is the probable error of the mean?

4. When samples are selected from small populations *without* replacement, then $\sigma_{\bar{x}}$ does not equal $\dfrac{\sigma}{\sqrt{n}}$. We must introduce a correction factor. Can you explain why?

5. If samples are selected from small populations *with* replacement, will $\sigma_{\bar{x}}$ equal $\dfrac{\sigma}{\sqrt{n}}$? Explain your answer.

6. The drivers of a particular overnight delivery service have been with the company an average of 8.6 years with a standard deviation of 1.46 years.

   a. If a sample of 39 drivers of this company is randomly selected, what is the probability that these drivers will have been with the company an average of at least 8 years?

   b. If the sample of 78 drivers of this company is randomly selected, what is the probability that these drivers will have been with the company an average of at least 8 years?

   c. Compare your answers obtained in parts (a) and (b). What conclusions can be drawn?

Case Study Advertisements

1. Many companies gear their mass media advertisement to particular age-groups. The amount of money allocated for television or radio ads during specified time periods is determined by the number of people and the age of the people who will view or hear the ads. Thus, a 90-second ad during the Super Bowl can cost many thousands of dollars.

   What type of music and for how long teenagers listen to music is an important fact that a music recording company must determine. Urban, rural, and suburban teenagers have different listening habits when it comes to music maintains Kith P. Thompson, Professor of Music Education at Penn State University. Studies show that urban young people listen to music an average of 4.43 hours per day, rural teenagers listen an average of 3.54 hours per day, and suburban teenagers listen an average of 1.98 hours per day. (*Source: USA TODAY*, April 1993, p. 4)

a. Within what limits should the average number of hours spent by 95% of 81 randomly selected urban young people listening to music lie? (Assume that $\sigma = 1.02$ hours.)

b. Within what limits should the average number of hours spent by 95% of 81 randomly selected suburban teenagers listening to music lie? (Assume that $\sigma = 0.66$ hours.)

c. Compare the answers obtained in parts (a) and (b). Comment.

9

ESTIMATION

DID YOU KNOW THAT

a   in 1990, only 13.1 quarts of ice cream were produced per person in America as opposed to 15.7 quarts in 1965? (*Source:* The International Ice Cream Association, 1993)

b   more than two thirds of patients defined as schizophrenic are men? (*Source: American Journal of Psychiatry*, 149:1077–1079, August 1992)

c   the "Wild West" leads the nation in experimentation with drugs? Forty percent of Westerners say they have tried marijuana as opposed to 29% of Southerners? (*Source:* National Institute on Drug Abuse, 1995)

d   51% of all new cars and trucks are purchased by women? When it comes to car leasing, the percent of women who lease cars is considerably higher. (*Source:* Automobile Manufacturers Association, 1994)

e   in 1990 there were only 39 men for every 100 women over 85 years of age as opposed to 1950 when there were 70 men for every 100 women? (*Source:* U.S. Bureau of the Census in *Current Population Trends*, 1995)

Many Americans recycle trash by using special recycling containers. See further discussion on page 507. *(Patrick Farace)*

I n this chapter we indicate how sample statistics can be used to provide estimates for the mean of a population, the proportion (percentage) of a population that has a particular attribute, and the standard deviation of the population. Since such estimates are necessarily subject to sampling error, we can provide information about the accuracy of such estimates by setting up confidence intervals and by selecting a sample of the appropriate size.

Chapter Objectives

- **To discuss** how sample data can often be used to estimate certain unknown quantities. This use of samples is called statistical estimation. (Section 9.1)

- **To point out** that population parameters are statistical descriptions of the population. (Section 9.2)

- **To understand** that sample data can be used to obtain both point and interval estimates. Point estimates give us a single number, whereas interval estimates set up an interval within which the parameter is expected to lie. (Section 9.2)

- **To see** that degrees of confidence give us the probability that the interval will actually contain the quantity that we are trying to estimate. (Section 9.3)

- **To apply** the Central Limit Theorem to set up confidence intervals for the mean and standard deviation. (Section 9.4)

- **To use** the Student's *t*-distribution to set up confidence intervals when the sample size is small. (Sections 9.4 and 9.5)

- **To analyze** how we determine the correct size of a sample for a given allowable error. (Section 9.6)

- **To indicate** how the Central Limit Theorem is also used to set up confidence intervals for population proportions. (Section 9.7)

Consider the information contained in the table below. How do we calculate the average life expectancy for both males and females? Does life expectancy depend on the sex of the person and on the country in which the person lives? Such information has important implications for demographers.

Very often we use currently available information to make predictions about the future. Great care must be exercised in making

LIFE EXPECTATION BY SEX FOR SOME COUNTRIES

AVERAGE LIFETIME (IN YEARS)

Country	Male	Female
North America		
Canada	68.7	75.1
United States	64.7	75.2
Mexico	59.4	63.4
Puerto Rico	68.9	75.2
Europe		
Belgium	67.7	73.5
Denmark	70.7	75.9
England and Wales	68.9	75.1
France	68.5	76.1
Italy	67.9	73.4
Norway	71.1	76.8
Poland	66.8	73.8
Sweden	72.0	77.4
Switzerland	69.2	75.0
U.S.S.R	65.0	74.0
Yugoslavia	65.3	70.1
Asia		
India	41.9	40.6
Israel	70.1	72.8
Japan	70.5	75.9
Jordan	52.6	52.0
Korea	59.7	64.1
Africa		
Egypt	51.6	53.8
Nigeria	37.2	36.7

Source: Information Please Yearbook.

such predictions as there are many factors that can offset the reliability of such estimates. The same is true for the population forecasts presented in the table below.

U.S. POPULATION—NEW FORECAST

The U.S. population will rise from 238.2 million now to 267.5 million by the year 2000—the result of a growth rate slowed from 1.1 percent annually in the '70s to 0.8 for the rest of the century.

That's the latest projection of the National Planning Association, which also predicts that 84 percent of the population expansion will occur in the West and South. Arizona, Nevada, and Florida will have the fastest growth rates, and California, Texas, and Florida will gain the most people.

	1985	2000		1985	2000
California	25.8 mil.	30.4 mil.	Oklahoma	3.2 mil.	3.7 mil.
New York	17.6 mil.	17.5 mil.	Connecticut	3.2 mil.	3.3 mil.
Texas	15.9 mil.	20.0 mil.	Iowa	2.9 mil.	3.0 mil.
Pennsylvania	11.9 mil.	12.1 mil.	Oregon	2.8 mil.	3.4 mil.
Florida	11.6 mil.	15.6 mil.	Mississippi	2.6 mil.	2.9 mil.
Illinois	11.6 mil.	11.9 mil.	Arkansas	2.4 mil.	2.8 mil.
Ohio	10.9 mil.	11.2 mil.	Kansas	2.4 mil.	2.6 mil.
Michigan	9.3 mil.	9.8 mil.	West Virginia	1.9 mil.	1.8 mil.
New Jersey	7.5 mil.	7.8 mil.	Utah	1.6 mil.	2.2 mil.
North Carolina	6.3 mil.	7.6 mil.	Nebraska	1.6 mil.	1.6 mil.
Georgia	5.9 mil.	6.9 mil.	New Mexico	1.4 mil.	1.7 mil.
Massachusetts	5.9 mil.	6.3 mil.	Maine	1.2 mil.	1.3 mil.
Virginia	5.6 mil.	6.4 mil.	New Hampshire	1.0 mil.	1.3 mil.
Indiana	5.6 mil.	5.9 mil.	Idaho	1.0 mil.	1.2 mil.
Missouri	5.0 mil.	5.2 mil.	Hawaii	1.0 mil.	1.1 mil.
Tennessee	4.8 mil.	5.5 mil.	Rhode Island	997,000	1.2 mil.
Wisconsin	4.8 mil.	5.1 mil.	Nevada	944,000	1.3 mil.
Washington	4.5 mil.	5.3 mil.	Montana	801,000	807,000
Louisiana	4.4 mil.	4.7 mil.	South Dakota	692,000	713,000
Maryland	4.3 mil.	4.7 mil.	North Dakota	666,000	672,000
Minnesota	4.2 mil.	4.4 mil.	Delaware	615,000	657,000
Alabama	4.1 mil.	4.6 mil.	Dist. of Columbia	597,000	535,000
Kentucky	3.7 mil.	4.1 mil.	Vermont	541,000	625,000
South Carolina	3.4 mil.	4.0 mil.	Wyoming	510,000	616,000
Arizona	3.2 mil.	4.7 mil.	Alaska	460,000	583,000
Colorado	3.2 mil.	4.1 mil.	**U.S. Total**	**238.2 mil.**	**267.5 mil**

June 17, 1985

Source: U.S. News & World Report, June 17, 1985.

9.1   INTRODUCTION

We have mentioned on several occasions that statistical inference is the process by which statisticians make predictions about a population on the basis of samples. Much information can be gained from a sample. As we mentioned in Chapter 8, the average of the sample means can be used as an estimate of the population mean. Also, we can obtain an estimate of the population standard deviation on the basis of samples. Thus, one use of sample data is to *estimate* certain unknown quantities of the population (Fig. 9.1). This use of samples is referred to as **statistical estimation**.

Statistical estimation

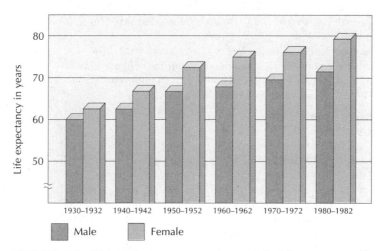

FIGURE 9.1   The graph gives us an estimate of the life expectancy of men and women. Such information is used by insurance companies in determining the premium to charge for life insurance. *(Courtesy of the Metropolitan Life Insurance Company and the Population Reference Bureau, Washington)*

On the other hand, sample data can also be used either to accept or reject specific claims about populations. To illustrate this use of sample data, suppose a manufacturer claims that the average milligram tar content per cigarette of a particular brand is 15 with a standard deviation of 0.5. Repeated samples are taken to test this claim. If these samples show that the average tar content per cigarette is 22, then the manufacturer's claim is incorrect. If these samples show that the average tar content is within "predictable limits," then the claim cannot be rejected. Thus, sample data can also be used either to reject or not reject specific claims about populations. This use of samples is referred to as **hypothesis testing**.

Hypothesis testing

Statistical inference can be divided into two main categories: problems of estimation and tests of hypotheses. In this chapter we discuss statistical estimation. In Chapter 10 we will analyze the nature of hypothesis testing.

9.2   POINT AND INTERVAL ESTIMATES

Population parameters

In most statistical problems we do not know certain population values such as the mean and the standard deviation. Somehow we want to use the information obtained from samples to estimate their values. These values, which really are statistical descriptions of the population, are often referred to as **population parameters**.

Suppose we are interested in determining the average life of an electric refrigerator under normal operating conditions. A sample of 50 refrigerators is taken and their lives are recorded as shown below:

6.9	7.6	5.7	3.6	7.7	6.6	7.2	7.3	10.6	5.9
8.2	5.7	7.6	8.7	7.9	8.8	7.0	8.1	7.3	6.8
5.7	11.1	8.5	8.9	7.6	5.6	9.0	9.2	6.8	8.3
6.1	9.7	9.8	7.4	6.8	7.3	8.3	9.9	7.5	7.8
7.7	7.4	9.1	7.3	5.5	8.1	6.7	8.8	7.6	5.3

Point estimate

The average life of these refrigerators is 7.6 years. Since this is the only information available to us, we would logically say that the mean life of *all* similar refrigerators is 7.6 years. This estimate of 7.6 years for the population mean is called a **point estimate** since this estimate is a single number. Of course, this estimate may be a poor estimate, but it is the best we can get under the circumstances.

Our confidence in this estimate would be improved considerably if the sample size were larger. Thus, we would have much greater confidence in an estimate that is based on 5000 refrigerators or 50,000 refrigerators than in one that is based only on 50 refrigerators.

One major disadvantage with a point estimate is that the estimate does not indicate the extent of the possible error. Furthermore, a point estimate does not specify how confident we can be that the estimate is close in value to the parameter that it is estimating. Yet point estimates are often used to estimate population parameters.

Interval estimation

Another type of estimate that is often used by statisticians, which overcomes the disadvantages mentioned in the previous paragraph, is **interval estimation**. In this method we first find a point estimate. Then we use this estimate to construct an interval within which we can be reasonably sure that the true parameter will lie. Thus, in our example a statistician may say that the mean life of the refrigerators will be between 7.2 and 8.0 years with a 95% degree of confidence. An interval such as this is called a **confidence interval**. The lower and upper boundaries, 7.2 and 8.0, respectively, of the interval are called **confidence limits**. The probability that the procedure used will give a correct interval is called the **degree of confidence**.

Confidence interval
Confidence limits
Degree of confidence

Generally speaking, as we increase the degree of certainty, namely, the degree of confidence, the confidence interval will become wider. Thus, if the length of an interval is very small (with a specific degree of confidence), then a fairly accurate estimate has been obtained.

When estimating the parameters of a population, statisticians use both point and interval estimates. In the next few sections we will indicate how point and interval estimates are obtained.

9.3   ESTIMATING THE POPULATION MEAN
ON THE BASIS OF A LARGE SAMPLE

In Chapter 8 we indicated that the average of the sample means can be used as an estimate of the population mean, μ. Moreover, the larger the sample size is, the better the estimate will be. Yet, as we pointed out in Section 9.2, there are some disadvantages with using point estimates.

The Central Limit Theorem (see page 472) says that the sample means will be normally distributed if the sample sizes are large enough. Generally speaking, statisticians say that a sample size is considered large if it is greater than 30. We can use the Central Limit Theorem to help us construct confidence intervals. This is done as follows.

Since the sample means are approximately normally distributed, we can expect 95% of the $\bar{x}$'s to fall between

$$\mu - 1.96\sigma_{\bar{x}} \quad \text{and} \quad \mu + 1.96\sigma_{\bar{x}}$$

(since from a normal distribution chart we note that 0.95 probability implies that $z = 1.96$ or -1.96)

Recall (Formula 8.5, page 467) that

$$\sigma_{\bar{x}} = \frac{\sigma}{\sqrt{n}}$$

Thus, 95% of the $\bar{x}$'s are expected to fall between

$$\mu - 1.96\,\frac{\sigma}{\sqrt{n}} \quad \text{and} \quad \mu + 1.96\,\frac{\sigma}{\sqrt{n}}$$

This is shown in Figure 9.2. If all possible samples of size n are selected, and the interval $\bar{x} \pm 1.96\,\dfrac{\sigma}{\sqrt{n}}$ is established for each sample, then 95% of all such intervals are expected to contain μ. Thus, a 95% confidence interval for μ is $\bar{x} \pm 1.96\,\dfrac{\sigma}{\sqrt{n}}$.

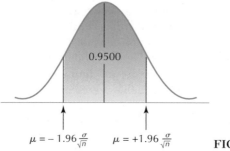

$$\mu = -1.96\,\tfrac{\sigma}{\sqrt{n}} \qquad \mu = +1.96\,\tfrac{\sigma}{\sqrt{n}}$$

FIGURE 9.2

In order to determine the interval estimate of μ, we must first know the value of the population standard deviation, σ. Although this value is usually unknown since the sample size is large, we can use the sample standard deviation as an approximation for σ. We then have the following confidence interval for μ.

FORMULA 9.1

95% confidence interval for μ

Let $\bar{x}$ be a sample mean and s be the sample standard deviation. Then an interval is called a **95% confidence interval for μ** if the lower boundary of the confidence interval is

$$\bar{x} - 1.96 \frac{s}{\sqrt{n}}$$

and if the upper boundary of the confidence interval is

$$\bar{x} + 1.96 \frac{s}{\sqrt{n}}$$

COMMENT   Formula 9.1 tells us how to find a 95% confidence interval for μ. This means that in the long run 95% of such intervals will contain μ. We can be 95% confident that μ lies within the specified interval. We must still realize that 5% of the time the population mean will fall outside this interval. This is true because the sample means are normally distributed and 5% of the values of a random variable in a normal distribution will fall farther away than 2 standard deviations from the mean (see page 409).

COMMENT   Depending on the nature of the problem, some statisticians will often prefer a 99% confidence interval for μ or a 90% confidence interval for μ. The boundaries for these intervals are as follows.

	Lower Boundary	Upper Boundary
99% Confidence Interval	$\bar{x} - 2.58 \frac{s}{\sqrt{n}}$	$\bar{x} + 2.58 \frac{s}{\sqrt{n}}$
90% Confidence Interval	$\bar{x} - 1.645 \frac{s}{\sqrt{n}}$	$\bar{x} + 1.645 \frac{s}{\sqrt{n}}$

Thus, as we reduce the size of the interval, we reduce our confidence that the true mean will fall within that interval.

The following examples illustrate how we establish confidence intervals.

EXAMPLE   1   A coffee vending machine fills 100 cups of coffee before it has to be refilled. On Monday the mean number of ounces in a filled cup of coffee was 7.5. The population standard deviation is known to be 0.25 ounces. Find 95% and 99% confidence intervals for the mean number of ounces of coffee dispensed by this machine.

SOLUTION   We use Formula 9.1. Here $n = 100$, $\sigma = 0.25$, and $\bar{x} = 7.5$. To construct a 95% confidence interval for μ, we have

Lower Boundary	Upper Boundary
$= \bar{x} - 1.96 \dfrac{\sigma}{\sqrt{n}}$	$= \bar{x} + 1.96 \dfrac{\sigma}{\sqrt{n}}$
$= 7.5 - 1.96\left(\dfrac{0.25}{\sqrt{100}}\right)$	$= 7.5 + 1.96\left(\dfrac{0.25}{\sqrt{100}}\right)$
$= 7.5 - 0.05$	$= 7.5 + 0.05$
$= 7.45$	$= 7.55$

Thus, we conclude that the population mean will lie between 7.45 and 7.55 ounces with a confidence of 0.95.

To construct a 99% confidence interval for μ, we have

Lower Boundary	Upper Boundary
$= \bar{x} - 2.58 \dfrac{\sigma}{\sqrt{n}}$	$= \bar{x} + 2.58 \dfrac{\sigma}{\sqrt{n}}$
$= 7.5 - 2.58\left(\dfrac{0.25}{\sqrt{100}}\right)$	$= 7.5 + 2.58\left(\dfrac{0.25}{\sqrt{100}}\right)$
$= 7.5 - 0.06$	$= 7.5 + 0.06$
$= 7.44$	$= 7.56$

Thus, we conclude with a 99% confidence that the population mean will lie between 7.44 and 7.56 ounces. In this example we did not have to use s as an estimate of σ since we were told that the population standard deviation was known to be 0.25.

E X A M P L E   2   A sample survey of 81 movie theaters showed that the average length of the main feature film was 90 minutes with a standard deviation of 20 minutes. Find a

    a. 90% confidence interval for the mean of the population.
    b. 95% confidence interval for the mean of the population.

SOLUTION   We use Formula 9.1. Here $n = 81$, $s = 20$, and $\bar{x} = 90$.

    a. To construct a 90% confidence interval for μ, we have

Lower Boundary	Upper Boundary
$= \bar{x} - 1.645 \dfrac{s}{\sqrt{n}}$	$= \bar{x} + 1.645 \dfrac{s}{\sqrt{n}}$
$= 90 - 1.645\left(\dfrac{20}{\sqrt{81}}\right)$	$= 90 + 1.645\left(\dfrac{20}{\sqrt{81}}\right)$

$$= 90 - 3.66 \qquad\qquad = 90 + 3.66$$
$$= 86.34 \qquad\qquad = 93.66$$

Thus, a 90% confidence interval for μ is 86.34 to 93.66 minutes.

b. To construct a 95% confidence interval for μ, we have

Lower Boundary	Upper Boundary
$= \bar{x} - 1.96 \dfrac{s}{\sqrt{n}}$	$= \bar{x} + 1.96 \dfrac{s}{\sqrt{n}}$
$= 90 - 1.96\left(\dfrac{20}{\sqrt{81}}\right)$	$= 90 + 1.96\left(\dfrac{20}{\sqrt{81}}\right)$
$= 90 - 4.36$	$= 90 + 4.36$
$= 85.64$	$= 94.36$

Thus, a 95% confidence interval for μ is 85.64 to 94.36 minutes.

Notice that as we increase the size of the confidence interval, our confidence that this interval contains μ also increases.

EXAMPLE 3 The management of the Night-All Corporation recently conducted a survey of 196 of its employees to determine the average number of hours that each employee sleeps at night. The company statistician submitted the following information to the management:

$$\Sigma x = 1479.8 \qquad \text{and} \qquad \Sigma(x - \bar{x})^2 = 1755$$

where x is the number of hours slept by each employee. Find a 95% confidence interval estimate for the average number of hours that each employee sleeps at night.

SOLUTION In order to use Formula 9.1, we must first calculate $\bar{x}$ and s. Using the given information, we have

$$\bar{x} = \frac{\Sigma x}{n}$$

$$= \frac{1479.8}{196}$$

$$= 7.55$$

and

$$s = \sqrt{\frac{\Sigma(x - \bar{x})^2}{n - 1}}$$

$$= \sqrt{\frac{1755}{195}}$$

$$= \sqrt{9} = 3$$

Now we can use Formula 9.1, with $\bar{x} = 7.55$, $s = 3$, and $n = 196$. To find the 95% confidence for μ, we have

Lower Boundary	Upper Boundary
$= \bar{x} - 1.96 \dfrac{s}{\sqrt{n}}$	$= \bar{x} + 1.96 \dfrac{s}{\sqrt{n}}$
$= 7.55 - 1.96\left(\dfrac{3}{\sqrt{196}}\right)$	$= 7.55 + 1.96\left(\dfrac{3}{\sqrt{196}}\right)$
$= 7.55 - 0.42$	$= 7.55 + 0.42$
$= 7.13$	$= 7.97$

Thus, the management can conclude with a 95% confidence that the average number of hours that an employee sleeps at night is between 7.13 and 7.97 hours.

EXERCISES FOR SECTION 9.3

RAISING A CHILD WILL BE QUITE EXPENSIVE

Washington, Aug. 1: According to the Family Economic Research Group of the U.S. Department of Agriculture, middle income couples who had babies in 1992 will spend an average of $128,670 by the time the baby is 18 years old.

August 1, 1993

1. Consider the accompanying newspaper article. Assume that the standard deviation of a sample of 100 families was $8473. Find a 90% confidence interval for the average cost to a middle income couple to raise a child that was born in 1992 until the child is 18 years old.

2. According to the U.S. Bureau of the Census in its *Current Population Reports* (Series P60) the average poverty threshold was $8241 for a three-person household in 1993. For a group of 81 three-person households the standard deviation was $972. Find a 99% confidence interval for the average poverty threshold for a three-person household.

3. The average size of a mortgage applied for in 1993 was $116,991 as opposed to $119,999 in 1992. (*Source: US News and World Report*, March 22, 1993, p. 10). A sample of 64 mortgages showed that the standard deviation of the amount applied for was $6019. Find a 95% confidence interval for the average size of a mortgage applied for in 1993.

> ## HEATING MONEY TO BE MADE AVAILABLE
>
> *Washington, Feb. 18:* The Federal Government announced yesterday that $200 million dollars would be made available to needy residents of the Northeast to help offset the effects of the unusually harsh winter that these people were experiencing.
>
> Friday, February 18, 1994

4. Consider the accompanying newspaper article. A sample of 49 homeowners found that the average home heating bill was $2013 with a standard deviation of $276. Find a 90% confidence interval for the average home heating bill.

5. A sample of 49 banks in Middletown showed that the average charge for a returned check was $12 with a standard deviation of $2.35. Find a 95% confidence interval for the average charge for a returned check.

6. Consider the newspaper article on the next page. If the standard deviation was $0.08, find a 90% confidence interval for the average cost of a gallon of number 2 heating oil.

7. The battery packs that accompany portable telephones require a long initial charging period. A random sample of 45 battery packs required an average of 11 hours for charging with a standard deviation of 4.13 hours. Find a 95% confidence interval for the average charging time for a battery pack.

8. Tourism dollars are often the lifeline for many underdeveloped or Third World countries. To stimulate tourism, such countries advertise extensively. A survey of 50 randomly selected tourists returning from one of these countries found that they spent an average of $687 in these countries. The standard deviation was $97. Find a 99% confidence interval for the average amount of money spent by a tourist traveling to one of these countries.

9. A fire marshal surveyed 55 homes in the Boro Hall section of town and found an average of 2.2 smoke detectors in these homes with a standard deviation of 0.57. Find a 90% confidence interval for the average number of smoke detectors in this section of town.

HEATING OIL PRICES TO REMAIN STEADY

Washington, Feb. 15: Energy Department officials predicted yesterday that the current supply of home heating oil would be adequate for this country's winter needs. Furthermore, the price was expected to remain relatively stable despite a sharp increase in demand. A random survey of 50 oil distributors conducted by the Energy Department found that the average current price of a gallon of number 2 heating oil in the Northeastern part of the U.S. was $1.059.

Tuesday, February 15, 1994

10. A computer programming teacher has been keeping records on how long it takes students to execute a particular program. For one group of 40 students, the average time was 28 minutes with a standard deviation of 6 minutes. Find a 99% confidence interval for the average time needed by a student to execute the particular program.

11. A random sample of n measurements was taken from a population with unknown mean μ and standard deviation σ. The following data were obtained:

$$\Sigma x = 875, \qquad \Sigma x^2 = 17,750, \qquad \text{and} \qquad n = 45$$

a. Find a 95% confidence interval for μ.
b. Find a 99% confidence interval for μ.

9.4   ESTIMATING THE POPULATION MEAN ON THE BASIS OF A SMALL SAMPLE

In Section 9.3 we indicated how to determine confidence intervals for μ when the sample size is larger than 30. Unfortunately, this is not always the case. Suppose a sample of 16 bulbs is randomly selected from a large shipment and has a mean life of 100 hours with a standard deviation of 5 hours. Using only the methods of Section 9.3, we cannot

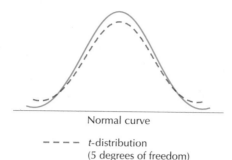

Normal curve

– – – –   t-distribution
(5 degrees of freedom)

FIGURE 9.3   Relationship between the normal distribution and the t-distribution.

*Student's
t-distribution*

determine confidence intervals for the mean life of a bulb since the sample size is less than 30.

Fortunately, in such situations we can base confidence intervals for μ on a distribution that is in many respects similar to the normal distribution. This is the **Student's t-distribution**. Figure 9.3 indicates the relationship between the normal distribution and the t-distribution. Notice that the t-distribution is also symmetrical about zero, which is its mean. However, the shape of the t-distribution depends on a parameter called the **number of degrees of freedom**. In our case the number of degrees of freedom, abbreviated as d.f., is equal to the sample size minus 1. If the population sampled is normally distributed, then $\dfrac{\bar{x} - \mu}{s/\sqrt{n}}$ has a t-distribution. This standardized t-distribution is symmetrical, bell shaped, and has zero as its mean.

*Number of degrees
of freedom*

HISTORICAL NOTE

The t-distribution was first studied by William S. Gosset, who was a statistician for Guinness, an Irish brewing company. Gosset was the first to develop methods for interpreting information obtained from small samples. Yet his company did not allow any of its employees to publish anything. So, Gosset secretly published his findings in 1907 under the name "Student." To this day this distribution is referred to as the Student's t-distribution.

Table VII in the Appendix indicates the value of t for different degrees of freedom. Thus, the 1.96 of Formula 9.1 of Section 9.3 has to be replaced by the $t_{0.025}$ value as listed in this table, depending on the number of degrees of freedom. When using the $t_{0.025}$ value of Table VII, 95% of the area under the curve of the t-distribution will fall between $-t_{0.025}$ and $t_{0.025}$, as shown in Figure 9.4.

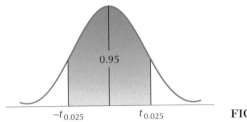

$-t_{0.025}$ $t_{0.025}$ FIGURE 9.4

We then have the following formula:

FORMULA 9.2

*95% small-sample
confidence interval
for μ*

Let $\bar{x}$ be a sample mean and let s be the sample standard deviation. We have the following **95% small-sample confidence interval for μ:**

$$\textbf{Lower boundary} = \bar{x} - t_{0.025}\frac{s}{\sqrt{n}}$$

$$\textbf{Upper boundary} = \bar{x} + t_{0.025}\frac{s}{\sqrt{n}}$$

COMMENT   In addition to the $t_{0.025}$ values, Table VII in the Appendix also lists many other values of t. Thus the $t_{0.005}$ values are used when we want a 99% confidence interval for μ, the $t_{0.050}$ values are used when we want a 90% confidence interval for μ, the $t_{0.10}$ values are used when we want an 80% confidence interval for μ, and so on.

EXAMPLE   1   A survey of 16 taxi drivers found that the average tip they receive is 90 cents with a standard deviation of 8 cents. Find a 95% confidence interval for estimating the average amount of money that a taxi driver receives as a tip.

SOLUTION   We use Formula 9.2. Here we have $n = 16$, $\bar{x} = 90$, and $s = 8$. We first find the number of degrees of freedom, which is $n - 1$, or $16 - 1 = 15$. Then we find the appropriate value of t from Table VII. The $t_{0.025}$ value with 15 degrees of freedom is 2.131. Finally, we establish the confidence interval. We have

Lower Boundary	Upper Boundary
$= \bar{x} - t_{0.025}\dfrac{s}{\sqrt{n}}$	$= \bar{x} + t_{0.025}\dfrac{s}{\sqrt{n}}$
$= 90 - 2.131\left(\dfrac{8}{\sqrt{16}}\right)$	$= 90 + 2.131\left(\dfrac{8}{\sqrt{16}}\right)$

$$= 90 - 4.26 \qquad\qquad = 90 + 4.26$$
$$= 85.74 \qquad\qquad = 94.26$$

Thus, a 95% confidence interval for the average amount of money that a taxi driver will receive as a tip is 85.74 to 94.26 cents.

EXAMPLE 2 A survey of the hospital records of 25 randomly selected patients suffering from a particular disease indicated that the average length of stay in the hospital is 10 days. The standard deviation is estimated to be 2.1 days. Find a 99% confidence interval for estimating the mean length of stay in the hospital.

SOLUTION We use Formula 9.2. Here $n = 25$, $\bar{x} = 10$, and $s = 2.1$. We first find the number of degrees of freedom, which is $n - 1$, or $25 - 1 = 24$. Then we find the appropriate value of t from Table VII in the Appendix. The $t_{0.005}$ value with 24 degrees of freedom is 2.797. Finally, we establish the confidence interval. We have

Lower Boundary	Upper Boundary
$= \bar{x} - t_{0.005}\dfrac{s}{\sqrt{n}}$	$= \bar{x} + t_{0.005}\dfrac{s}{\sqrt{n}}$
$= 10 - 2.797\left(\dfrac{2.1}{\sqrt{25}}\right)$	$= 10 + 2.797\left(\dfrac{2.1}{\sqrt{25}}\right)$
$= 10 - 1.17$	$= 10 + 1.17$
$= 8.83$	$= 11.17$

Thus, a 99% confidence interval for the average length of stay in the hospital is 8.83 to 11.17 days.

EXAMPLE 3 A music teacher asks six randomly selected students how many hours a week each practices playing the electric guitar. The teacher receives the following answers: 10, 12, 8, 9, 16, 5. Find a 90% confidence interval for the average length of time that a student practices playing the electric guitar.

SOLUTION In order to use Formula 9.2, we must first calculate the sample mean and the sample standard deviation. We have

$$\bar{x} = \frac{\Sigma x}{n} = \frac{10 + 12 + 8 + 9 + 16 + 5}{6} = \frac{60}{6} = 10$$

To calculate the sample standard deviation, we arrange the data as follows.

x	$x - \bar{x}$	$(x - \bar{x})^2$
10	$10 - 10 = 0$	0
12	$12 - 10 = 2$	4
8	$8 - 10 = -2$	4

(Table continues)

x	$x - \bar{x}$	$(x - \bar{x})^2$
9	$9 - 10 = -1$	1
16	$16 - 10 = 6$	36
5	$5 - 10 = -5$	25
		Total $= 70$

$$s = \sqrt{\frac{\Sigma(x - \bar{x})^2}{n - 1}} = \sqrt{\frac{70}{6 - 1}} = \sqrt{14} \approx 3.74$$

Now we find the appropriate value of t from Table VII. The $t_{0.05}$ value with $n - 1$, or $6 - 1 = 5$, degrees of freedom is 2.015. Finally, we establish the confidence interval. We have

Lower Boundary	Upper Boundary
$= \bar{x} - t_{0.05} \dfrac{s}{\sqrt{n}}$	$= \bar{x} + t_{0.05} \dfrac{s}{\sqrt{n}}$
$= 10 - 2.015\left(\dfrac{3.74}{\sqrt{6}}\right)$	$= 10 + 2.015\left(\dfrac{3.74}{\sqrt{6}}\right)$
$= 10 - 3.08$	$= 10 + 3.08$
$= 6.92$	$= 13.08$

Thus, a 90% confidence interval for the average length of time that a student practices playing the electric guitar is 6.92 to 13.08 hours.

COMMENT   We wish to emphasize a point made earlier. For small sample inferences to be valid it is assumed that the sample is obtained from some normally distributed population. Often this may not be true. In this case we have to use other statistical procedures.

EXERCISES FOR SECTION 9.4

1. A publisher wishes to determine the list price for a new algebra book. A survey of the list price of eight competing books sold by other companies showed an average price of $39.95 with a standard deviation of $2.85. Construct a 95% confidence interval for the average list price for a new algebra book.
2. Consider the newspaper article on the next page. A survey of 11 randomly selected city dwellers found that they were wasting an average of 57 gallons of water per day. The standard deviation was 6.9 gallons per day. Construct a 90% confidence interval for the average amount of water wasted daily by those people.

CITY STILL IN WATER CRISIS

Los Angeles, Feb. 12: Despite the rains that fell over the weekend, the city is still in the grip of a severe water drought and emergency. The reservoirs are well below their normal levels for this time of the year.

Many city residents are still wasting a lot of water each day in a variety of ways, including letting the water run needlessly, leaky faucets and pipes, unnecessary flushes of the toilet, etc.

The mayor is pleading with the citizens to conserve water.

February 14, 1991

3. The Environmental Protection Agency reports that 17% of Americans recycled garbage in 1990. (*Source: Natural Health*, 1993). One survey of ten randomly selected Americans indicated that they generated 5.7, 6.3, 2.9, 4.2, 5.3, 2.7, 3.6, 3.4, 3.9, and 4.5 pounds of garbage per day. Construct a 90% confidence interval for the average number of pounds of garbage generated per day.

4. According to *Backpacker* (1993), people visit Glacier National Park in Montana for peace and quiet. The noise level (in decibels) at eight different locations in the park produced the following readings:

$$14 \quad 28 \quad 23 \quad 29 \quad 19 \quad 25 \quad 22 \quad 17$$

Construct a 95% confidence interval for the average noise level (in decibels) in the Glacier National Park.

5. Americans eat 50 million hot dogs a day. Two all-beef hot dogs contain more than twice the calories of a comparable portion of skinless chicken breast. One study (*Consumer Reports*, July 1993, p. 417) of nine popular brands of hot dogs, found that each hot dog contained 170, 180, 149, 132, 149, 140, 167, 110, and 126 calories. Construct a 90% confidence interval for the average calorie content per hot dog.

6. Eight volunteers have applied a measured amount of a mosquito repellent and then thrust their arms into a cage of 500 hungry mosquitoes. The number of hours that passed before each of these received one or more bites was 6, 4, 9, 7, 8, 5, 10, and 5. Construct a 95% confidence interval for the average amount of time that elapses before an individual is bitten by a mosquito after having been sprayed with a repellent.

7. Eleven comparison shoppers were asked to buy the same brand of cereal (same size) in 11 different supermarkets. The average price was $3.89 with a standard

deviation of \$0.32. Construct a 99% interval for the average price for this brand of cereal.

A rather common scene these days in many supermarkets is the appearance of consumer comparison shopping and the use of magazine or newspaper coupons. *(Sharon Frankel)*

8. Consider the accompanying newspaper article. Construct a 95% confidence interval for the average price of this prescription drug.

IT PAYS TO SHOP AROUND

By Trudy Hoffman

New York, Oct. 17: Drugstore prices aren't uniform. This reporter shopped at six local pharmacies for the same arthritis drug. These local pharmacies included a drugstore within a chain supermarket, a regional chain, two national chain outlets, and two independent stores. The prices quoted for 60 500-mg tablets of Naprosyn were \$60.49, \$69.95, \$64.87, \$72.00, \$76.00, and \$71.23. If you call around for prices, you may find a place that's offering a particular prescription for less than your usual drugstore.

October 17, 1993

9. Ten members of a health club were randomly selected and asked to indicate the number of hours each spends per week at the club. Their answers were 10, 8, 3, 2, 12, 3, 11, 14, 7, and 5 hours. Construct a 90% confidence interval for the average amount of time spent by a health-club member at the health club.

10. The kidneys of a human body normally remove toxic wastes from the blood. When a person's kidneys deteriorate or malfunction, the person must be treated by dialysis, which is a procedure whereby the toxic removal procedure is performed artificially by a machine. However, it is known that dialysis can result in retention of phosphorus in the blood. This condition must be monitored carefully. The phosphorus levels of nine dialysis patients were measured, and the following results were obtained (in milligrams of phosphorus per deciliter of blood):

$$4.7 \quad 5.3 \quad 5.9 \quad 4.9 \quad 5.1 \quad 6.2 \quad 5.8 \quad 5.5 \quad 6.4$$

Assuming that the phosphorus levels of dialysis patients are normally distributed, construct a 90% confidence interval for the average level of phosphorus in the blood of dialysis patients.

11. Refer to the accompanying newspaper article. Health officials tested the air at seven tunnels to determine the pollution content. The officials found an average of 18 ppm (parts per million) of a certain pollutant in the air at these tunnels. The standard deviation was 3.2 ppm. Find a 95% confidence interval for the average ppm content of this pollutant in the air at these tunnels.

TOLL COLLECTORS PROTEST UNHEALTHY WORKING CONDITIONS

New Dorp, May 17: Toll collectors staged a 3-hour protest yesterday against the unhealthy working conditions occuring at different times of the day at the toll plazas of the city's seven tunnels. A spokesperson for the toll collectors indicated that the level of sulfur oxide pollutants in the air at the plazas was far in excess of the recommended safety level. Between the hours of 10 A.M. and 1 P.M., motorists were able to use the tunnel without paying any toll. In an effort to resolve the dispute, Bill Sigowsky has been appointed as mediator.

May 17, 1990

The exhaust fumes created by the many vehicles using the toll booths make it very unhealthy for the toll collectors. *(Joseph Newmark)*

9.5   THE ESTIMATION OF THE STANDARD DEVIATION

Until now we have been discussing how to obtain point and interval estimates of the population mean μ. In this section we discuss point and interval estimates for the population standard deviation.

Suppose we took a sample of 100 cigarettes of a particular brand and determined that the sample mean tar content was 15.2 with a standard deviation of 1.1. If this procedure were to be repeated many times, each time we would obtain different estimates for the population mean and standard deviation. Nevertheless, the mean of these estimates will approach the true population mean as the sample size gets larger. When this happens we say that the average of the sample means is an **unbiased estimator** of the population mean.

Unbiased estimator

Recall that on page 460, two formulas were given for the variance and standard deviation. Suppose we were to calculate the variance of each sample by using the formula

$$\frac{\Sigma(x - \bar{x})^2}{n}$$

Then we cannot use the average of these values as an unbiased estimate of the population variance. Specifically, the average of such estimates would most likely always be too small, no matter how many samples are included. We can compensate for this by dividing by $n - 1$ instead of n. When this is done, it can be shown that

$$\frac{\Sigma(x - \bar{x})^2}{n - 1}$$

is an unbiased estimate of σ^2. This means that as more and more samples are included, the average of the variances of these samples will approach the true population variance. Thus, a point estimate of σ^2 is

$$\frac{\Sigma(x - \bar{x})^2}{n - 1}$$

We therefore say that $\bar{x}$ is an unbiased estimator of the population mean μ and that s^2 is an unbiased estimator of σ^2. However, this does not mean that s is an unbiased estimator of σ. Nevertheless, when n is large, this bias is small. Thus, we can use s as an estimator of σ.

When dealing with large sample sizes, mathematicians have developed the following 95% confidence interval for σ.

FORMULA 9.3

95% confidence interval for σ

Let s be the sample deviation and let n be the number in the sample. We have the following **95% confidence interval for σ**:

$$\text{Lower boundary} = \frac{s}{1 + \dfrac{1.96}{\sqrt{2n}}} \qquad \text{Upper boundary} = \frac{s}{1 - \dfrac{1.96}{\sqrt{2n}}}$$

EXAMPLE 1 A sample of 50 cigarettes was taken. The sample mean tar content was 15.2 with a standard deviation of 1.1. Find a 95% confidence interval for the population standard deviation.

SOLUTION We use Formula 9.3. Here $n = 50$ and $s = 1.1$. The lower boundary of the confidence interval is

$$\frac{s}{1 + \dfrac{1.96}{\sqrt{2n}}} = \frac{1.1}{1 + \dfrac{1.96}{\sqrt{2(50)}}}$$

$$= \frac{1.1}{1 + 0.196}$$

$$= 0.92$$

The upper boundary of the confidence interval is

$$\frac{s}{1 - \dfrac{1.96}{\sqrt{2n}}} = \frac{1.1}{1 - \dfrac{1.96}{\sqrt{2(50)}}}$$

$$= \frac{1.1}{1 - 0.196}$$

$$= 1.37$$

Thus, a 95% confidence interval for σ is 0.92 to 1.37.

9.6   DETERMINING THE SAMPLE SIZE

Until now we have been discussing how sample data can be used to estimate various parameters of a population. Selecting a sample usually involves an expenditure of money. The larger the sample size, the greater the cost. Therefore, before selecting a sample we must determine how large the sample should be.

Generally speaking, the size of the sample is determined by the desired degree of accuracy. Most problems specify the maximum allowable error. Yet we must realize that no matter what the size of a sample is, any estimate may exceed the maximum allowable error. To be more specific, suppose we are interested in estimating the mean life of a calculator on the basis of a sample. Suppose also that we want our estimate to be within *Maximum* 0.75 years of the true value of the mean. In this case the **maximum allowable error**, *allowable error* denoted as e, is 0.75 years. How large a sample must be taken? Of course, the larger the sample size, the smaller the chance that our estimate will not be within 0.75 years of the value. If we want to be 95% confident that our estimate will be within 0.75 years of the true value, then the sample size must satisfy

$$1.96\sigma_{\bar{x}} = 0.75$$

More generally, if the maximum allowable error is e, then we must have

$$e = 1.96\sigma_{\bar{x}}$$

Since

$$\sigma_{\bar{x}} = \frac{\sigma}{\sqrt{n}}$$

we get

$$e = 1.96\frac{\sigma}{\sqrt{n}}$$

Solving this equation for n gives Formula 9.4.

FORMULA 9.4

Sample size

Let σ be the population standard deviation, e the maximum allowable error, and n the size of the sample that is to be taken from a large population. Then a **sample of size n**

$$n = \left(\frac{1.96\sigma}{e}\right)^2$$

will result in an estimate of μ, which is less than the maximum allowable error 95% of the time.

EXAMPLE   1   Suppose we wish to estimate the average life of a calculator to within 0.75 years of the true value. Past experience indicates that the standard deviation is 2.6 years. How large a sample must be selected if we want our answer to be within 0.75 years 95% of the time?

SOLUTION   We use Formula 9.4. Here $\sigma = 2.6$ and $e = 0.75$ so that

$$n = \left(\frac{1.96\sigma}{e}\right)^2$$

$$= \left(\frac{1.96(2.6)}{0.75}\right)^2 = (6.79)^2 = 46.10$$

Thus, the sample should consist of 47 calculators. (*Note:* In the determination of sample size, any decimal is always rounded to the next highest number.)

EXAMPLE   2   The management of a large company in California desires to estimate the average working experience, measured in years, of its 2000 hourly workers. How large a sample should be taken in order to be 95% confident that the sample mean does not differ from the population mean by more than $\frac{1}{2}$ year? (Past experience indicates that the standard deviation is 2.6 years.)

SOLUTION   We use Formula 9.4. Here $\sigma = 2.6$ and $e = 0.50$. Then,

$$n = \left(\frac{1.96\sigma}{e}\right)^2$$

$$= \left(\frac{1.96(2.6)}{0.50}\right)^2$$

$$= (10.19)^2$$

$$= 103.84$$

Thus, the company officials should select 104 hourly workers to determine the average working experience of an hourly worker.

EXERCISES FOR SECTION 9.6

1. Fifty lawyers were randomly selected and asked to indicate their charge for writing a living will. The standard deviation of the charge was $9.75. Find a 95% confidence interval for the standard deviation of the charge by all lawyers.

2. A random sample of 70 patients given a new pain-relieving drug indicated that the average time elapsed before relief occurred was 17.2 minutes with a standard deviation of 2.3 minutes. Find a 90% confidence interval for the population standard deviation.

3. The American Academy of Pediatrics conducted a survey on the prices of baby formula. A sample of the data for the cost of *Similac* from 75 stores disclosed a standard deviation of $1.03. Find a 95% confidence interval for the standard deviation of these prices.

4. The U.S. Centers for Disease Control claims that there is too much lead in our drinking water. In one study of the drinking water of 75 U.S. cities, the standard deviation of the lead count turned out to be 2 ppb. Find a 99% confidence interval for the standard deviation of the lead count in our drinking water.

For each of Exercises 5–10, assume that the maximum allowable error must not be exceeded with a 95% degree of confidence.

5. A consumer's group wishes to estimate the average life of a steel-belted radial tire. How large a sample should be selected if the group wishes that its estimate be within 100 miles of the true value? (Assume that $\sigma = 425$.)

6. A researcher wishes to determine the average amount of life insurance carried by a married man in Phoenix. How large a sample should be selected if it is desired that the estimate be within $1000 of the true value? (Assume that $\sigma = \$5,000$.)

7. A cosmetics manufacturer wishes to determine how many minutes a woman spends applying facial cosmetics in the morning before departing for work. How large a sample should the manufacturer take so that the estimate be within 2.17 minutes of the true value? (Assume that $\sigma = 5.83$.)

8. A criminologist wishes to estimate the average number of violent crimes per hour in the United States. How large a sample should be selected if it is desired that the estimate be within 6 of the true value? (Assume that $\sigma = 12.8$ crimes.)

9. An economist wishes to estimate the average amount of money spent by a family on entertainment per year. How large a sample should be selected if it is desired that the estimate be within $20 of the true value? (Assume that $\sigma = \$24.75$.)

10. A medical researcher wishes to determine how many hours of sleep a woman 30 to 40 years of age gets per night. How large a sample should be selected if it is desired that the estimate be within 1.35 hours of the true value? (Assume that $\sigma = 5.37$ hours.)

9.7 THE ESTIMATION OF PROPORTIONS

Proportion form
Count form

So far we have discussed the estimation of the population mean and the population standard deviation. Very often statistical problems arise for which the data are available in **proportion** or **count form** rather than in measurement form. For example, suppose a doctor has developed a new technique for predicting the sex of an unborn child. The doctor tests the new method on 1000 pregnant women and correctly predicts the sex of 900 of the children. Is the new technique reliable? The doctor has correctly predicted

Sample proportion the sex of 900 unborn children. Thus, the **sample proportion** is $\dfrac{900}{1000}$, or 0.90. What

True proportion is the **true proportion** of unborn children whose sex the doctor can correctly predict?

Since we will be discussing both sample proportions and true population proportions, we will use the following notation:

p true population proportion

$\hat{p}$ sample proportion   ($\hat{p}$ is pronounced p hat)

In our case the doctor estimates the true population proportion, p, to be 0.90. This estimate is based on the sample proportion $\hat{p}$. How reliable is the estimate of the true population proportion? Suppose the doctor now tests the technique on 1000 different pregnant women. For what proportion of these 1000 unborn children will the doctor be able to correctly predict the sex?

If the true population proportion is p, repeated sample proportions will be normally distributed. The mean of the sample distribution of proportions will be p. The standard deviation of these sample proportions will equal

$$\sqrt{\frac{p(1-p)}{n}}$$

Thus, if the doctor's technique is reliable, then the sample proportion will be normally distributed with a mean of 0.9. The standard deviation will be

$$\sqrt{\frac{(0.9)(1-0.9)}{1000}} = \sqrt{0.00009}, \quad \text{or } 0.0095$$

We can summarize these results as follows.

FORMULA 9.5

Suppose we have a large population, a proportion of which has some particular characteristic. We select random samples of size n and determine the proportion in each sample with this characteristic. Then the sample proportions will be approximately normally distributed with mean p and standard deviation

$$\sigma_{\hat{p}} = \sqrt{\frac{p(1-p)}{n}}$$

Since the sample proportions are approximately normally distributed, we can apply the standardized normal charts as we did for the sample means. The following examples illustrate how this is done.

E X A M P L E 1 From past experience it is known that 70% of all airplane tickets sold by Global Airways are round-trip tickets. A random sample of 100 passengers is taken. What is the probability that at least 75% of these passengers have round-trip tickets?

SOLUTION We use Formula 9.5. Here $n = 100$, $p = 0.70$, and the sample proportion $\hat{p}$ is 0.75. Since the sample proportions are normally distributed, we are interested in the area to the right of $x = 0.75$ in a normal distribution whose mean is 0.70. See Figure 9.5. We first calculate the standard deviation of the sample proportions, denoted as $\sigma_{\hat{p}}$.

$$\sigma_{\hat{p}} = \sqrt{\frac{p(1 - p)}{n}}$$

$$= \sqrt{\frac{0.70(1 - 0.70)}{100}}$$

$$= \sqrt{0.0021}$$

$$= 0.0458$$

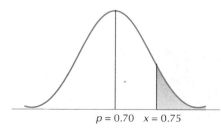

$p = 0.70 \quad x = 0.75$ 　　　 **FIGURE 9.5**

Then

$$z = \frac{\hat{p} - p}{\sigma_{\hat{p}}}$$

$$= \frac{0.75 - 0.70}{0.0458} \qquad (\hat{p} \text{ replaces } \bar{x} \text{ and } p \text{ replaces } \mu \text{ in Formula 8.6 on page 474})$$

$$= 1.09$$

From Table IV the area between $z = 0$ and $z = 1.09$ is 0.3621. Therefore, the area to the right of $z = 1.09$ is $0.5000 - 0.3621$, or 0.1379. Thus, the probability that at least 75% of the passengers have round-trip tickets is 0.1379.

E X A M P L E 2 Fifty-four percent of all nurses in the day shift at Downtown Hospital have type O blood. Thirty-six nurses from the day shift are selected at random. What is the probability that between 51% and 58% of these members have type O blood?

SOLUTION   We use Formula 9.5. Here $p = 0.54$ and $n = 36$. We first calculate $\sigma_{\hat{p}}$:

$$\sigma_{\hat{p}} = \sqrt{\frac{p(1 - p)}{n}}$$

$$= \sqrt{\frac{0.54(1 - 0.54)}{36}}$$

$$= \sqrt{0.0069}$$

$$= 0.0831$$

Now we can use Formula 9.5 with $p = 0.54$, $n = 36$, and $\sigma_{\hat{p}} = 0.0831$:

$$\hat{p} = 0.51 \text{ corresponds to } \quad z = \frac{0.51 - 0.54}{0.0831} = \frac{-0.03}{0.0831} = -0.36$$

$$\hat{p} = 0.58 \text{ corresponds to } \quad z = \frac{0.58 - 0.54}{0.0831} = \frac{0.04}{0.0831} = 0.48$$

Thus, we are interested in the area between $z = -0.36$ and $z = 0.48$. See Figure 9.6.

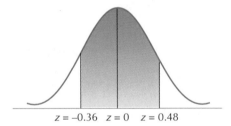

$z = -0.36$ $z = 0$ $z = 0.48$ FIGURE 9.6

From Table IV we find that the area between $z = 0$ and $z = -0.36$ is 0.1406, and the area between $z = 0$ and $z = 0.48$ is 0.1844. Adding, we get

$$0.1406 + 0.1844 = 0.3250$$

Thus, the probability that between 51% and 58% of these nurses have type O blood is 0.3250.

In the same way that the sample mean is used to estimate the population mean we can use the sample proportions, $\hat{p}$, as a point estimate of the population proportion, p. However, this estimate does not indicate the probability of its accuracy. Thus, we set up interval estimates for the true population proportion, p. We have Formula 9.6.

FORMULA 9.6	
95% confidence interval for p	Let $\hat{p}$ be a sample proportion and let p be the true population proportion. Then we have the following **95% confidence interval for p** (assuming we have a large sample): $$\text{Lower boundary} = \hat{p} - 1.96\sqrt{\frac{\hat{p}(1-\hat{p})}{n}}$$ $$\text{Upper boundary} = \hat{p} + 1.96\sqrt{\frac{\hat{p}(1-\hat{p})}{n}}$$

EXAMPLE   3   A union member reported that 80 out of 120 workers interviewed supported some form of work stoppage to further its demands for a shorter workweek. Find a 95% confidence estimate of the true proportion of workers supporting the union's stand on a work stoppage.

SOLUTION   We use Formula 9.6. Here $\hat{p} = \dfrac{80}{120}$, or 0.6667, and $n = 120$. To construct a 95% confidence interval for p, we have

Lower Boundary	Upper Boundary
$= \hat{p} - 1.96\sqrt{\dfrac{\hat{p}(1-\hat{p})}{n}}$	$= \hat{p} + 1.96\sqrt{\dfrac{\hat{p}(1-\hat{p})}{n}}$
$= 0.67 - 1.96\sqrt{\dfrac{0.6667(1-0.6667)}{120}}$	$= 0.67 + 1.96\sqrt{\dfrac{0.6667(1-0.6667)}{120}}$
$= 0.67 - 1.96\sqrt{0.00185}$	$= 0.67 + 1.96\sqrt{0.00185}$
$= 0.67 - 1.96(0.043)$	$= 0.67 + 1.96(0.043)$
$= 0.67 - 0.084$	$= 0.67 + 0.084$
$= 0.586$	$= 0.754$

Thus, the union official concluded with 95% confidence that the true proportion of workers supporting the union claim is between 0.586 and 0.754. This is the 95% confidence interval.

EXAMPLE   4   A new public telephone has been installed in an airport baggage-claim area. A quarter was lost 45 of the first 300 times that it was used. Construct a 95% confidence interval for the true proportion of times that a user will lose a quarter.

SOLUTION   We use Formula 9.6. Here $\hat{p} = \dfrac{45}{300}$, or 0.15, and $n = 300$. To construct a 95% confidence interval for p, we have

Lower Boundary	Upper Boundary
$= \hat{p} - 1.96 \sqrt{\dfrac{\hat{p}(1 - \hat{p})}{n}}$	$= \hat{p} + 1.96 \sqrt{\dfrac{\hat{p}(1 - \hat{p})}{n}}$
$= 0.15 - 1.96 \sqrt{\dfrac{0.15(1 - 0.15)}{300}}$	$= 0.15 + 1.96 \sqrt{\dfrac{0.15(1 - 0.15)}{300}}$
$= 0.15 - 1.96\sqrt{0.0004}$	$= 0.15 + 1.96\sqrt{0.0004}$
$= 0.15 - 1.96(0.02)$	$= 0.15 + 1.96(0.02)$
$= 0.15 - 0.039$	$= 0.15 + 0.039$
$= 0.111$	$= 0.189$

Thus, a 95% confidence interval for the true proportion of times that a user will lose a quarter is 0.111 to 0.189.

EXERCISES FOR SECTION 9.7

1. Find a 95% confidence interval for the proportion of defective items in a large shipment of parts when
   a. $n = 800$ and $\hat{p} = 0.45$
   b. $n = 400$ and $\hat{p} = 0.65$
2. According to the *Wall Street Journal* (June 21, 1995), IRS reports show that approximately seven out of every ten individual income-tax returns filed for 1993 took the standard deduction. (This is true irrespective of the fact that about half of all individual returns are done by professional preparers.) A random survey of 100 taxpayers is taken. Find a 95% confidence interval for the true proportion of taxpayers who take the standard deduction.
3. A 1993 editorial in the *Journal of the American Medical Association* cited a review of 43 recently published studies. In 26 of them, researchers had found that low calcium intake by humans was linked to bone mass, bone loss, or fractures. Find a 95% confidence interval for the true proportion of studies linking low calcium intake by humans to bone mass, bone loss, or fractures.
4. Should you buy mutual funds from your bank? A 1995 survey by the Securities and Exchange Commission found that only 33% of the consumers questioned knew that money-market mutual funds sold by banks are not protected by FDIC insurance. A random survey of 300 consumers is taken. What is the probability that more than 35% of them will know that money-market mutual funds sold by banks are not protected by FDIC insurance?

5. According to *Consumer Reports* (March 1994, p. 167) nearly one out of every ten camcorders bought by consumers since 1988 has needed to be repaired. A random survey of 75 camcorders bought by consumers since 1988 is taken. Find a 90% confidence interval for the true proportion of camcorders bought by consumers since 1988 that have needed to be repaired.

6. The Health Care Financing Administration, or HCFA, is a government agency that certifies some 16,000 nursing homes receiving payment from Medicare or Medicaid. It is claimed that 73% of these nursing homes violate HCFA standards. One recent study found that 37 out of 52 nursing homes surveyed violated federal HCFA standards in some way. What is the probability that the proportion of nursing homes violating federal HCFA standards is more than 73%?

7. According to *Science*, 1994, one of every three U.S. adults is seriously overweight. A random sample of 85 adult Americans is selected. What is the probability that fewer than 30% of these adult Americans is seriously overweight?

8. Administration officials at Dorp University claim that 45% of the student body are part-time students who work during the day. The student government claims that this figure is grossly underestimated. A random sample of 600 part-time students is taken. What is the probability that more than 300 of them work during the day?

9. A report in the *Journal of the American Medical Association* found that patients who spend more than 48 hours in an intensive-care unit of a hospital increase their risk of developing dangerous blood clots by 33%. (The clots result from prolonged inactivity and can lead to heart failure.) To check on this claim, a random survey is taken of 275 people who spent more than 48 hours in an intensive-care unit. What is the probability that the increase in risk of developing dangerous blood clots is less than 30%?

10. Consider the accompanying newspaper article. If a random survey of 525 families across the state is taken, what is the probability that more than 130 of them will be found to be one-person households?

24% OF AMERICAS HOUSEHOLDS ARE ONE-PERSON

Washington, June 26: According to the latest figures released by the U.S. Census Bureau, about 24% of all households in America today are one-person households. This startling revelation has important implications for our concept of "family".

Monday, June 26, 1995

11. Officials of an insurance company that pays reimbursement bills claim that 40% of all druggists in a city will often fill a doctor's prescription with a brand-name drug rather than with a less expensive generic drug even when the doctor does not specify the brand-name drug. A random survey of 525 prescriptions is taken. What is the probability that more than 220 of them will be found to be filled with the more expensive drug rather than with the less expensive generic drug?

9.8 USING COMPUTER PACKAGES

We can use a MINITAB program to determine confidence intervals quite easily. To illustrate, let us use a MINITAB program to determine a 90% confidence interval for the data on the average length of time that a student practices playing the electric guitar given in Example 3 of Section 9.4 (page 505). We have the following:

```
MTB  > SET THE FOLLOWING DATA INTO C1
DATA > 10  12   8    9   16    5
DATA > END
```

After the data are entered, we instruct the computer to determine a 90% confidence interval. We have the following desired results:

```
MTB > TINTERVAL WITH 90 PERCENT CONFIDENCE FOR DATA IN C1

     N    MEAN    STDEV   SEMEAN        90.0 PERCENT C.I.
C1   6   10.00    3.74    1.53       (   6.92,    13.08)
MTB > STOP
```

COMMENT In a similar manner, we can obtain 95% confidence intervals for μ. Minor changes in the code of the program allow us to obtain any desired confidence intervals for μ.

EXERCISES FOR SECTION 9.8

1. According to government records, the total number of businesses that failed and filed for bankruptcy during 1994 in Greenville is as follows:

Month	Number of Business Failures
Jan.	23
Feb.	17
Mar.	16
Apr.	8
May	31
June	14
July	12

Month	Number of Business Failures
Aug.	17
Sept.	15
Oct.	17
Nov.	15
Dec.	13

Using **MINITAB**, construct a 90% confidence interval for the mean number μ of business failures. (Assume that the number of business failures during the year has a normal distribution.)

2. The number of credit card sales reported daily by a department store for the first 20 business days of 1995 was as follows:

52	37	60	41
68	74	62	49
51	64	58	53
62	48	48	67
49	63	61	49

Using **MINITAB**, construct a 95% confidence interval for the mean number μ of credit card sales. (Assume that the number of credit card sales has a normal distribution.)

3. According to official police records, the number of accidents reported per day on the state thruway during the first 70 days of 1995 was as follows:

3	4	13	31	33	16	23	9	17	13
12	7	8	23	16	14	12	16	16	19
5	14	24	17	14	12	14	12	14	11
6	12	21	19	22	11	13	13	10	16
4	19	17	18	19	19	2	10	8	10
10	21	16	23	12	17	1	11	6	11

Using **MINITAB**, construct a 99% confidence interval for the mean number μ of accidents per day.

4. According to the *Wall Street Journal* (June 21, 1995, p. 1), IRS data indicate that individuals whose adjusted gross income was $30,000–$39,000 claimed an average deduction of $1384 for charitable gifts from their 1993 income taxes. A survey of 40 individual returns revealed the following deductions for charitable gifts:

$1253	$1380	$1405	$1350	$1602	$1325	$1300	$1275
1542	1402	1360	1370	1430	1360	1350	1415
1375	1297	1325	1300	1502	1420	1410	1400
1400	1350	1405	1340	1186	1295	1395	1375
1202	1425	1400	1290	1322	1300	1308	1395

Using **MINITAB**, construct a 90% confidence interval for the average amount deducted for charitable gifts.

9.9   SUMMARY

In this chapter we indicated how sample data can be used to estimate the population mean and the population standard deviation. Both point and interval estimates were discussed. In most cases sample data are used to construct confidence intervals within which a given parameter with a specified probability is likely to lie. Sample data can also be used to make estimates about the true population proportion and to construct confidence intervals for the population proportion.

All estimates considered in this chapter were unbiased estimates. This means that the average of these estimates will approach the true population parameter that they are trying to estimate as more and more samples are included. For this reason we divide $\Sigma(x - \bar{x})^2$ by $n - 1$, not by n, in determining the sample variance.

We also indicated how to determine the size of a sample to be used in gathering data. Depending on the maximum allowable error, Formula 9.4 determines the sample size with a 95% degree of confidence.

Study Guide

The following is a chapter outline in capsule form. You should now be able to demonstrate your knowledge of the ideas mentioned by giving definitions, descriptions, or specific examples. Page references are given in parentheses.

Sample data can be used to estimate unknown quantities of the population. This use of sample data is referred to as **statistical estimation**. (page 494)

Sample data can also be used either to accept or reject specific claims about populations. This use of sample data is referred to as **hypothesis testing**. (page 494)

Population values, such as the mean and the standard deviation, are statistical descriptions of the population and are referred to as **population parameters**. (page 495)

An estimate of a population parameter that is a single number is called a **point estimate**. (page 495)

Interval estimation involves constructing an interval, called a **confidence interval**, within which we can be reasonably sure that the true parameter will lie. (page 495)

The lower and upper bounds of a confidence interval are called **confidence limits**. (page 495)

The probability that the procedure used will give a correct interval is called the **degree of confidence**. (page 495)

A **95% confidence interval for μ** means that if all possible samples of size n are selected, and the interval $\bar{x} \pm 1.96 \dfrac{s}{\sqrt{n}}$ is established for each sample, then 95% of all such intervals are expected to contain μ. Similar statements can be made for other confidence intervals. (page 497)

The **Student's t-distribution**, which in many respects is similar to the normal distribution, is used when we deal with information obtained from small samples. (page 503)

The shape of the *t*-distribution depends on a parameter called the **number of degrees of freedom**, abbreviated as d.f., which equals the sample size minus 1. (page 503)

When estimating population parameters such as the mean we can take repeated samples. If the mean of these estimates approaches the true population mean as the sample size gets larger, then the average of the sample means is an **unbiased estimator** of the population mean. Thus, $\bar{x}$ is an unbiased estimator of the population mean μ and s^2 is an unbiased estimator of σ^2. (page 510)

The size for any sample is determined by the desired degree of accuracy. This is usually done by specifying the **maximum allowable error**, denoted as e. (page 512)

Often statistical data are available in **proportion** or **count form**. In such cases we use the **sample proportion** $\hat{p}$ to estimate the **true population** proportion, p. If the true population proportion is p, repeated sample proportions will be normally distributed. The mean of the sample distribution of proportions will be p. The standard deviation of these sample proportions will equal $\sqrt{\dfrac{p(1-p)}{n}}$. (page 515)

Formulas to Remember

You should be able to identify each symbol in the following formulas, understand the relationships among the symbols expressed in each formula, understand the significance of each formula, and use the formulas in solving problems.

1.

Size of Sample	Parameter	Degree of Confidence	Lower Boundary	Upper Boundary
Large	Mean	90%	$\bar{x} - 1.645\dfrac{s}{\sqrt{n}}$	$\bar{x} + 1.645\dfrac{s}{\sqrt{n}}$
Large	Mean	95%	$\bar{x} - 1.96\dfrac{s}{\sqrt{n}}$	$\bar{x} + 1.96\dfrac{s}{\sqrt{n}}$
Large	Mean	99%	$\bar{x} - 2.58\dfrac{s}{\sqrt{n}}$	$\bar{x} + 2.58\dfrac{s}{\sqrt{n}}$
Small	Mean	95%	$\bar{x} - t_{0.025}\dfrac{s}{\sqrt{n}}$	$\bar{x} + t_{0.025}\dfrac{s}{\sqrt{n}}$
Large	Standard deviation	95%	$\dfrac{s}{1+\dfrac{1.96}{\sqrt{2n}}}$	$\dfrac{s}{1-\dfrac{1.96}{\sqrt{2n}}}$
Large	Proportion	95%	$\hat{p} - 1.96\sqrt{\dfrac{\hat{p}(1-\hat{p})}{n}}$	$\hat{p} + 1.96\sqrt{\dfrac{\hat{p}(1-\hat{p})}{n}}$

2. Size of sample where maximum allowable error may not be exceeded with a 95% degree of confidence: $n = \left(\dfrac{1.96\sigma}{e}\right)^2$

3. Mean of sampling proportions: p

4. Standard deviation of sampling proportion: $\sqrt{\dfrac{p(1-p)}{n}}$

Testing Your Understanding of This Chapter's Concepts

1. Which of the following best describes the meaning of a 90% confidence interval for the population mean?
   a. The population will fall within the specified interval with a probability of 0.90.
   b. The probability is 0.90 that the mean is a correct point estimate.
   c. The sample mean is a good estimate of the population mean 90% of the time.
   d. If many samples of the same size were to be taken and 90% confidence intervals constructed in the same manner, we would expect the population mean to be within the specified intervals 90% of the time.
2. A pollster conducted a random survey of 1500 moviegoers and found that 650 of them really enjoyed a particular movie.
   a. Construct a 90% interval for the true proportion of moviegoers that really enjoy the particular movie.
   b. How large a sample must be taken to be sure with a confidence of 95% that the pollster's error is no larger than 0.001? (For this part, assume that $\sigma = 0.0210$.)
3. A random sample of 600 executives found that 75% of the executives carry a beeper and/or a cellular phone with them at all times. Construct a 99.7% confidence interval for the population proportion of executives who carry a beeper and/or a cellular phone with them at all times.
4. The walls of many schools built years ago are beginning to disintegrate thereby releasing asbestos particles into the air. Airborne asbestos particles can be dangerous to your health. An analyst obtained the following fiber counts from 10 air samples from the lunchroom of a school building:

$$9 \quad 5 \quad 6 \quad 8 \quad 10 \quad 12 \quad 3 \quad 9 \quad 7 \quad 4$$

   a. Construct a 95% confidence interval for μ, the average fiber content.
   b. Construct a 95% confidence interval for σ, the population standard deviation. (Assume that it is known from past experience with large sample sizes that $s = 2.8304$.)

Chapter Test

Multiple Choice Questions

1. The average nightly cost for a similar room at 50 motels was $49.95 with a standard deviation of $1.75. Find a 95% confidence interval for the average nightly cost for a motel room.
   a. between $49.46 and $50.44   b. between $49.54 and $50.36
   c. between $49.31 and $50.59   d. none of these

2. A manufacturer wishes to determine the list price for a smoke detector. A survey of the list price of similar smoke detectors sold by nine other competing companies showed an average list price of $10.95 with a standard deviation of $1.49. Find a 95% confidence interval for the average list price for a smoke detector.
   a. between $9.98 and $11.92
   b. between $9.80 and $12.10
   c. between $9.98 and $12.07
   d. none of these

3. Eight plumbers in a city charged an average of $885 for installing baseboard heating. The standard deviation was $45. Find a 90% confidence interval for the average charge for installing baseboard heating in this city.
   a. between $837.30 and $932.70
   b. between $858.91 and $911.09
   c. between $854.85 and $915.15
   d. none of these

4. A sample of 50 elementary school teachers in a city indicated that a teacher spends an average of 15 hours per week preparing lessons. The standard deviation was 1.89 hours. Find a 95% confidence interval for the population standard deviation.
   a. between 1.5803 and 2.3507
   b. between 1.5024 and 2.5472
   c. between 1.6237 and 2.2608
   d. none of these

5. Forty-five orthodontists were asked to indicate their charge for installing braces (no complications). The standard deviation of their charges was computed and found to be $69. Find a 99% confidence interval for the population standard deviation.
   a. between $57.19 and $86.91
   b. between $54.25 and $94.77
   c. between $58.83 and $83.42
   d. none of these

6. An engineer wishes to estimate the tensile strength of a certain material. How large a sample (with a 95% degree of confidence) should the engineer select if the engineer wishes the estimate to be within 2.2 units of the true value? (Assume that $\sigma = 4.79$ units.)
   a. 10
   b. 5
   c. 19
   d. 32
   e. none of these

7. A photographer finds that 7% of all film processed by a certain company will be ruined. Find the probability that a random sample of 81 rolls of film processed by this company will contain at most 5% of the rolls that are ruined.
   a. 0.2580
   b. 0.2611
   c. 0.2389
   d. 0.2420
   e. none of these

Supplementary Exercises

8. A bank official claims that 95% of its employees are courteous and efficient. A random sample of 50 of the bank's employees is taken. What is the probability that fewer than 93% of those employees will be found to be courteous and efficient?

9. A random sample of 250 families in Thomsonville indicated that 140 of them had a pet dog or cat at home. Find a 95% confidence interval for the true proportion of all families in Thomsonville who have a pet dog or cat at home.

10. A consumers' group wishes to estimate the average cost of a steel-belted radial tire. How large a sample (with a 95% degree of confidence) should the group take if it wants the estimate to be within $3.50 of the true value? (Assume that $\sigma = \$6.95$.)

11. The average charge for a home visit by 75 doctors in a particular city was $35 with a standard deviation of $4. Find a 95% confidence interval for the average charge for a home visit by a doctor in this city.

12. A manufacturer wishes to determine the list price for a particular type of silverware. A survey of the list price of similar silverware sold by eight other competing companies showed an average list of $49.95 with a standard deviation of $3.95. Find a 95% confidence interval for the average list price for the silverware.

13. Nine electricians in a city charged an average of $120 for installing comparable electric meter boxes. The standard deviation was $11. Find a 90% confidence interval for the average charge for installing electric meter boxes in this city.

14. A sample of 45 sanitation workers in a city disclosed that a sanitation worker spends an average of $425 per year for uniforms. The standard deviation was $39. Find a 95% confidence interval for the population standard deviation.

15. Forty-five tow truck operators were asked to indicate their charge for towing a special type of vehicle. The standard deviation of their charge was computed and found to be $8. Find a 99% confidence interval for the population standard deviation.

16. A researcher wishes to estimate the average response time by a rat to certain stimuli. How large a sample (with a 95% degree of confidence) should the researcher take, if it is desired that the estimate be within 0.8 seconds of the true value? (Assume that $\sigma = 2.57$.)

17. A social service official claims that 15% of all applications for welfare payments are fraudulent. Find the probability that a random sample of 100 applications will contain at most 12% that are fraudulent.

18. A government official claims that 40% of all workers in the city commute daily to work. A random sample of 75 of the workers in this city is taken. What is the probability that fewer than 36% of these workers will be found to commute daily to work?

19. A random sample of 350 people in Vega Valley indicated that 210 of them had a life insurance policy. Find a 95% confidence interval for the true proportion of all families in Vega Valley who have an insurance policy.

20. A painter wishes to estimate the average drying time of a particular brand of paint under specified conditions. How large a sample (with a 95% degree of confidence) should the painter take, if the painter wants the estimate to be within 12 minutes of the true value? (Assume that $\sigma = 32.37$ minutes.)

21. The average cost for a certain prescription drug at 45 pharmacies was $21.95 with a standard deviation of $2.25. Find a 95% confidence interval for the average cost of this prescription drug.

22. A vendor wishes to determine the price to charge for a shaver. A survey of the prices charged by eight competing vendors for the same shaver showed an average price of $31.95 with a standard deviation of $1.45. Find a 95% confidence interval for the price charged for the shaver.

23. Seven gardeners in a city charged an average of $85 for pruning a small-sized tree. The standard deviation was $7.95. Find a 90% confidence interval for the average charge for pruning a small-sized tree by gardeners in this city.

24. A sample of 50 beauty salons in a city indicated that the beauticians worked an average of 39 hours per week. The standard deviation was 2.84 hours. Find a 95% confidence interval for the population standard deviation.

Thinking Critically

1. For a given confidence level for the mean, if we wish to double the accuracy of the estimate, we must quadruple the sample size. Do you agree with this statement? Explain your answer.
2. Develop a formula for determining the size of a sample n, if the maximum allowable error, e, may not be exceeded with a 90% degree of confidence. (Assume that σ is the population standard deviation.)
3. A random sample of size 400 is selected from a population with unknown mean μ and unknown standard deviation σ. It is known that $\Sigma x = 4000$ and $\Sigma x^2 = 120,000$. Find a 95% confidence interval for μ.
4. Refer to the previous problem. Find a 95% confidence interval for σ.
5. Under what conditions can a sample standard deviation, s, be used in place of σ in the formula $n = \left(\dfrac{1.96\sigma}{e}\right)^2$ when σ is unknown? Explain your answer.
6. The critical t-values as given in Table VII in the Appendix can be approximated by

$$t = \sqrt{df \cdot (e^{\frac{y^2}{df}} - 1)}$$

where $df = n - 1$, $e \approx 2.718$, $y = z\left(\dfrac{8df + 3}{8df + 1}\right)$, and z is the critical z-score. Using this approximation, find the critical t-score for a 90% confidence interval corresponding to $n = 10$. Compare your answer with that given in Table VII. Comment. Some computer programs approximate t-values by using this procedure.

Case Study

According to the *Wall Street Journal* (June 21, 1995, p. 1), only 14.2% of the 1994 individual income-tax returns filed in the year ending September 30 showed a yes vote in the section where you are asked if you want $3 of your tax liability to go to the presidential election campaign. (Checking yes does not increase or decrease a taxpayer's liability or refund.) Nevertheless, the number of tax returns with a yes vote dropped to 16.3 million last year from 19.1 million the prior year. This represents a decrease of 16.7% from the previous year and a record 28.9% decrease from 1978.

    When originally enacted, only $1 could be allocated to the presidential election campaign fund. However, starting with returns for the 1993 tax year, the amount rose to $3. As a result of this increase, the amount of money allocated to this fund has soared, jumping from $27.6 million in 1993 to $69.7 million in 1994 despite the decrease in the number of yes votes.

1. Assume that we select a random sample of 600 of the 1994 income tax returns filed in the year ending September 30. Find a 95% confidence interval for the true proportion of these people who voted yes on the presidential campaign-fund checkoff.

10

HYPOTHESIS TESTING

Chapter Outline

DID YOU KNOW THAT

a　a newborn girl's chance of suffering from breast cancer sometime in her life is 1 in 8? (*Source: Journal of the National Cancer Institute*, 1993)

b　AIDS has muted much of the progress that had been made in treating hemophiliacs? Although the typical hemophiliac could expect to live 57 years in 1979, by 1988 the average life span had dropped to 40 years. (*Source: American Journal of Hematology*, 1993)

c　23% of all U.S. births in 1993 were by Caesarean section as opposed to only 5% in 1965? (*Source: Journal of the American Medical Association*, 1993)

d　taxi drivers have a job-related homicide rate of 27 murders per 100,000? That is 40 times the national average. (*Source: Journal of Occupational Medicine*, 1994)

Several manufacturers of peanut butter claim that a typical three-tablespoon serving of peanut butter contains about 24 grams of fat and 285 calories. How do we verify the accuracy of such claims? See discussion on page 553. (*George Semple*)

In the previous chapter we indicated how sample statistics can be used to help us set up confidence intervals for means, standard deviations, and proportions. In this chapter we indicate how statistics can be used to make decisions about certain assumptions called hypotheses. Statistical inferences of this type are called hypotheses tests. We will analyze procedures for testing hypotheses regarding means, differences between means, and proportions. The appropriate procedure will vary and depend on whether we have a small or large sample size.

Chapter Objectives

- **To analyze** how sample data can be used to reject or accept a claim about some aspect of a probability distribution. The claim to be tested is called the null hypothesis. (Section 10.2)

- **To see** that a test statistic is a number that we compute to determine when to reject the null hypothesis. (Section 10.2)

- **To determine** when critical rejection regions tell us to reject a null hypothesis. This occurs when the test statistic value falls within this region. How these regions are set up depends on the specifications given within the problem. (Section 10.2)

- **To discuss** the two errors that can be made when we use sample data to accept or reject a null hypothesis. We may incorrectly reject a true hypothesis or we may incorrectly accept a false hypothesis. In both cases, an error is made. (Section 10.3)

- **To indicate** what a level of significance is. (Section 10.3)

- **To distinguish** between tests concerning means, differences between means, and proportions. Thus, we discuss the test statistics used when we wish to use sample data to determine whether observed differences between means and proportions are significant. (Sections 10.4 to 10.8)

- **To apply** the hypothesis-testing procedures to wide-ranging problems.

Often we read about various claims, as in the article on the right. How do we determine whether the new test results are accurate? When is any discrepancy between results obtained from two different studies significant?

Now consider the article below. How do we verify the accuracy of the claim made? If we sample 100 Americans, would we expect 55 of them to support this claim? If only 50 of them said that they support the President's health plan, would you say that this is significantly less than the claimed 55%?

VANITY DOES NOT CAUSE CANCER

Washington, Feb. 14: Contrary to earlier reports, a study of the personal habits of 573,369 women as reported in the latest issue of the *Journal of the National Cancer Institute* disclosed that long-term use of hair dyes does not trigger fatal malignancies. These findings have far-reaching implications for those women who rely heavily on hair dyes.

Monday, February 14, 1994

MANY AMERICANS SUPPORT THE PRESIDENT'S HEALTH PLAN

New York, Jan. 3: The results of a nationwide survey by Administration officials indicate that many Americans support President Clinton's proposed health plan. About 55% of those surveyed said they were in favor of revamping our national health system so as to provide universal medical coverage for all Americans under some government plan. One question that remains unanswered is "Who will pay for such a plan?"

Monday, January 3, 1994

10.1 INTRODUCTION

Many television commercials contain unusual performance claims. For example, consider the following TV commercials:

1. Four out of five dentists recommend Brand X sugarless gum for their patients who chew gum.
2. A particular brand of tire will last an average of 40,000 miles before replacement is needed.
3. A certain detergent produces the cleanest wash.
4. Brand X paper towels are stronger and more absorbent.

Hypothesis testing

How much confidence can one have in such claims? Can they be verified statistically? Fortunately, in many cases the answer is yes. Samples are taken and claims are tested. We can then make a decision on whether to accept or reject a claim on the basis of sample information. This process is called **hypothesis testing**. Perhaps this is one of the most important uses of samples.

Hypothesis

As we indicated in Chapter 9, hypothesis testing is an important branch of statistical inference. Sample data provide us with estimates of population parameters. These estimates are in turn used in arriving at a decision either to accept or reject an hypothesis. By an **hypothesis** we mean an assumption about one or more of the population parameters that will either be accepted or rejected on the basis of the information obtained from a sample. In this chapter we discuss methods for determining whether to accept or reject any hypothesis on the basis of sample data.

10.2 TESTING AGAINST AN ALTERNATIVE HYPOTHESIS

Suppose several players are in a gambling casino rolling a die. A bystander notices that in the first 120 rolls of the die, a 6 showed only eight times. Is this reasonable or is the die loaded? The management claims that this unusual occurrence is to be attributed purely to chance and that the die is an honest die. The bystander claims otherwise.

If the die is an honest die, then Formula 6.5 (see page 372) for a binomial distribution tells us that the average number of 6's occurring in 120 rolls of the die is 20, as

$$\mu = np$$
$$= 120\left(\frac{1}{6}\right)$$
$$= 20$$

If the die is loaded, then $\mu \neq 20$. (The symbol $\neq$ means "is not equal to.") Since the die is either an honest die or a loaded die, we must choose between the hypothesis $\mu = 20$ and the hypothesis $\mu \neq 20$. Thus, sample data will be used either to accept or reject the hypothesis $\mu = 20$. Such an hypothesis is called a **null hypothesis** and is denoted by H_0. Any hypothesis that differs from the null hypothesis is called an **alternative hypothesis** and is denoted as $H_1, H_2, \ldots$, and so on. In our example

$$\text{Null hypothesis, } H_0: \quad \mu = 20$$

$$\text{Alternative hypothesis, } H_1: \quad \mu \neq 20$$

Notice that by formulating the alternative hypothesis as $\mu \neq 20$, we are indicating that we wish to perform a **two-sided** or **two-tailed test**. This means that if the die is not honest, then it may be loaded in favor of obtaining 6's more often than is expected or less often than is expected.

In our example the bystander strongly suspects that the die is loaded against obtaining 6's. Thus, his alternative hypothesis would be $\mu < 20$. (The symbol $<$ stands for "is less than.") Similarly, if the bystander suspected that the die was loaded in favor of obtaining 6's more often than is expected, the alternative hypothesis would be $\mu > 20$. (The symbol $>$ stands for "is greater than.") In each of these cases the null hypothesis remains the same, $H_0: \mu = 20$. Such alternative hypotheses indicate that we wish to perform a **one-sided** or **one-tailed test**.

COMMENT   It should be noted that the decision on an alternative hypothesis should be made *before* the results of the sample are known.

A decision as to whether to accept or reject the null hypothesis will be made on the basis of sample data. How is such a decision made? We must realize that even if we know for sure that the die is honest, it is very unlikely that we would get exactly twenty 6's in the 120 rolls of the die. Moreover, if we were to roll the die 120 times on many different occasions, we would find that the number of 6's appearing is around 20, sometimes more and sometimes less. It is therefore obvious that we must set up some interval *Acceptance region* that we call the **acceptance region**. If the number of 6's appearing in 120 rolls of the die is within this acceptance region, then we will accept the null hypothesis. If the number of 6's obtained is outside this region, then we will reject the null hypothesis that the die is an honest die. These possibilities are shown in Figures 10.1, 10.2, and

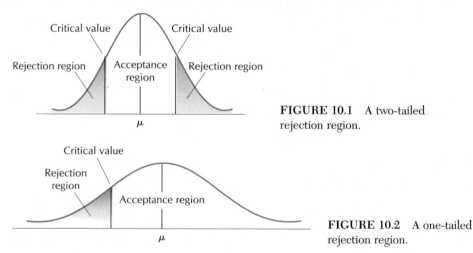

FIGURE 10.1   A two-tailed rejection region.

FIGURE 10.2   A one-tailed rejection region.

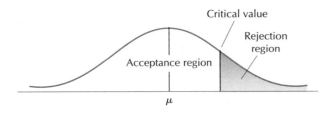

FIGURE 10.3 A one-tailed rejection region.

10.3. The value that separates the rejection region from the acceptance region is called the **critical value**.

The type of symbol in the alternative hypothesis tells us what type of rejection region to use as shown in the following table:

If the Symbol in the Alternative Hypothesis Is	<	≠	>
Then the Rejection Region Consists of	one region on the left side.	two regions, one on each side.	one region on the right side.

Suppose we decide to accept the null hypothesis if the number of 6's obtained is between 15 and 25. Since in our case only eight 6's were obtained, we would reject the null hypothesis. In this case the acceptance region is 15 to 25. The two-tailed rejection region corresponds to the two tails, less than 15 and more than 25, as shown in Figure 10.4. When we reject a null hypothesis, we are claiming that the value of the population parameter, that is, the average number of 6's, is some value other than the one specified in the null hypothesis. Also, when the sample data indicate that we should reject a null hypothesis, we say that the observed difference is **significant**.

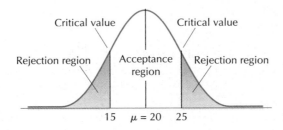

FIGURE 10.4 A two-tailed rejection region.

Our discussions in the last few paragraphs lead us to the following definitions:

Definition 10.1 | The **null hypothesis**, denoted by H_0, is the statistical hypothesis being tested.

Null hypothesis

Definition 10.2 | The **alternative hypothesis** denoted by $H_1, H_2, \ldots$, is the hypothesis that will be

Alternative hypothesis | accepted when the null hypothesis is rejected.

Definition 10.3
One-tailed test
One-sided test

A **one-sided** or **one-tailed test** is a statistical test that has the rejection region located in the left tail or the right tail of the distribution.

Definition 10.4
Two-sided test
Two-tailed test

A **two-sided** or **two-tailed test** is a statistical test that has the rejection region located in both tails of the distribution.

Let us summarize the steps that need to be followed in hypothesis testing:

1. State the null hypothesis, which indicates the value of the population parameter to be tested.
2. State the alternative hypothesis, which indicates the belief that the population parameter has a value other than the one specified in the null hypothesis.

Region of rejection
Critical region
Acceptance region

3. Set up rejection and acceptance regions for the null hypothesis. The **region of rejection** is called the **critical region**. The remaining region is called the **acceptance region**.

Test statistic

4. Compute the value of the test statistic. A **test statistic** is a calculated number that is used to decide whether to reject or accept the null hypothesis. The formula for computing the value of the test statistic depends on the parameter we are testing.
5. Reject the null hypothesis if the test statistic value falls within the rejection region, that is, the critical region. Otherwise, do not reject the null hypothesis.
6. State the conclusion for the particular problem.

Statistical tests
of hypotheses
Statistical tests
of significance

In this chapter we discuss various tests that enable us to decide whether to reject or accept a null hypothesis. Such tests are called **statistical tests of hypotheses** or **statistical tests of significance**.

10.3   TWO TYPES OF ERRORS

Since any decision either to accept or reject a null hypothesis is to be made on the basis of information obtained from sample data, there is a chance that we will make an error. There are two possible errors that we can make. We may reject a null hypothesis when we really should accept it. Thus, returning to the die problem of Section 10.2, we may reject the claim that the die is honest even though it actually is honest. Alternately, we may accept a null hypothesis when we should reject it. Thus, we may say that the die is honest when it really is a loaded die.

These two errors are referred to as a **type-I** and a **type-II error**, respectively. In either case we have made a wrong decision. We define these formally as follows:

Definition 10.5
Type-I error

A **type-I error** is made when a true null hypothesis is rejected; that is, we reject a null hypothesis when we should accept it.

Definition 10.6
Type-II error

A **type-II error** is made when a false null hypothesis is accepted; that is, we accept a null hypothesis when we should reject it.

In the following box we indicate how these two errors are made:

		And We Claim That	
		H_0 Is True	H_0 Is False
If	H_0 Is True	Correct decision (no error)	Type-I error
	H_0 Is False	Type-II error	Correct decision (no error)

When deciding whether to accept or reject a null hypothesis, we always wish to minimize the probability of making a type-I error or a type-II error. Unfortunately, the relationship between the probabilities of the two types of errors is of such a nature that if we reduce the probability of making one type of error, we usually increase the probability of making the other type. In most applied problems, one type of error is more serious than the other. In such situations, careful attention is given to the more serious error.

How much risk should a statistician take in rejecting a true hypothesis, that is, in making a type-I error? Generally speaking, statisticians use the limits of 0.05 and 0.01. Each of these limits is called a **level of significance** or **significance level**. We have the following definition:

Definition 10.7
Significance level | The **significance level** of a test is the probability that the test statistic falls within the rejection region when the null hypothesis is true.

5% level of significance
1% level of significance | The **0.05 level of significance** is used when the statistician wishes that the risk of rejecting a true null hypothesis not exceed 0.05. The **0.01 level of significance** is used when the statistician wishes that the risk of rejecting a true null hypothesis not exceed 0.01.

In this book we will usually assume that we wish to correctly accept the null hypothesis 95% of the time and to incorrectly reject it only 5% of the time. Thus, the maximum probability of a type-I error that we are willing to accept, that is, the significance level, will be 0.05. *The probability of making a type-I error when H_0 is true is denoted by the Greek letter α (pronounced alpha).* Therefore, the probability of making a correct decision is $1 - \alpha$.

As we indicated on pages 534–535, when dealing with one-tailed tests, the critical region lies to the left or to the right of the mean. This is shown in Figure 10.5. When dealing with a two-tailed test, one half of the critical region is to the left of the mean and one half is to the right. The probability of making a type-I error is evenly divided between these two tails, as shown in Figure 10.6.

COMMENT If the test statistic falls within the acceptance region, we do not reject the null hypothesis. When a null hypothesis is not rejected, this does not mean that what

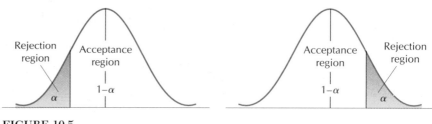

FIGURE 10.5

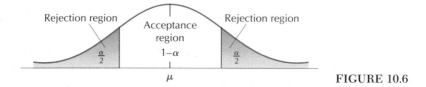

FIGURE 10.6

the null hypothesis claims is guaranteed to be true. It simply means that on the basis of the information obtained from the sample data there is not enough evidence to reject the null hypothesis.

10.4   TESTS CONCERNING MEANS FOR LARGE SAMPLES

In this section we discuss methods for determining whether we should accept or reject a null hypothesis about the mean of a population. We will illustrate the procedure with several examples.

Suppose a manufacturer claims that each family-size bag of pretzels sold weighs 12 ounces, on the average, with a standard deviation of 0.8 ounces. A consumer's group decides to test this claim by accurately weighing 49 randomly selected bags of pretzels. If the mean weight of the sample is considerably different from the population mean, the manufacturer's claim will definitely be rejected. Thus, if the mean weight is 30 ounces or 5 ounces, the manufacturer's claim will be rejected. Only when the sample mean is close to the claimed population mean do we need statistical procedures to determine when to reject or accept a null hypothesis.

Let us assume that the sample mean of the 49 randomly selected bags of pretzels is 11.8 ounces. Since the sample mean, 11.8, is not the same as the population mean, 12, we wish to test the manufacturer's claim at the 5% level of significance.

The population parameter being tested in this case is the mean weight, μ, and the questioned value is 12 ounces. Thus,

$$\text{Null hypothesis, } H_0\text{:} \quad \mu = 12$$

$$\text{Alternative hypothesis, } H_1\text{:} \quad \mu \neq 12$$

The alternative hypothesis of not equal suggests a two-tailed rejection region. Therefore, the α of 0.05 is split equally between the two tails, as shown in Figure 10.7.

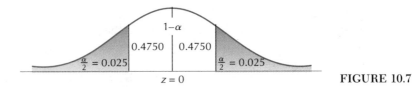

FIGURE 10.7

We now look in Table IV of the Appendix to determine which z-value has 0.4750 of the area between $z = 0$ and this z-value. From Table IV we find that the z-value is 1.96. We label this on the diagram in Figure 10.7 and get the acceptance-rejection diagram shown in Figure 10.8.

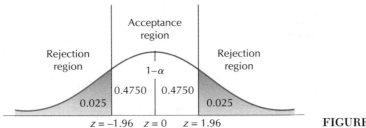

FIGURE 10.8

Since the Central Limit Theorem tells us that the sample means are normally distributed, we use

$$z = \frac{\bar{x} - \mu}{\dfrac{\sigma}{\sqrt{n}}}$$

as the test statistic and reject the null hypothesis if the value of the test statistic falls in the rejection region. In using this test statistic, $\bar{x}$ is the sample mean and μ is the population mean as claimed in the null hypothesis. In our case we have

$$z = \frac{\bar{x} - \mu}{\dfrac{\sigma}{\sqrt{n}}}$$

$$= \frac{11.8 - 12}{\dfrac{0.8}{\sqrt{49}}}$$

$$= \frac{-0.2}{\dfrac{0.8}{7}}$$

$$= -1.75$$

Since this calculated value of z falls within the acceptance region, our decision is that we cannot reject H_0. The difference between the sample mean and the assumed value of the population mean may be due purely to chance. We say that the difference is *not statistically significant*.

If the level of significance had been 0.01, we would split the 0.01 into two equal tails as shown in Figure 10.9. From Table IV we find that the z-value that has 0.4950 of the area between $z = 0$ and this z-value is 2.58. Thus, we reject the null hypothesis if the test statistic falls in the critical region shown in Figure 10.10.

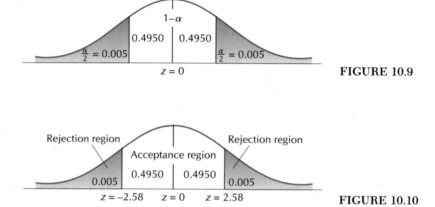

FIGURE 10.9

FIGURE 10.10

Since we obtained a z-value of -1.75, which is in the acceptance region, we do not reject the null hypothesis.

Let us summarize the testing procedure outlined in the previous paragraphs for a two-tailed rejection region.

1. First convert the sample mean into standard units using

$$z = \frac{\bar{x} - \mu}{\dfrac{\sigma}{\sqrt{n}}}$$

2. Then reject the null hypothesis about the population mean if z is less than -1.96 or greater than 1.96 when using a 5% level of significance.
3. If we are using a 1% level of significance, reject the null hypothesis if z is less than -2.58 or greater than 2.58.

We further illustrate the procedure with several examples.

EXAMPLE 1   A light bulb company claims that the 60-watt light bulb it sells has an average life of 1000 hours with a standard deviation of 75 hours. Sixty-four new bulbs were allowed to burn out to test this claim. The average lifetime of these bulbs was found to be 975 hours. Does this indicate that the average life of a bulb is not 1000 hours? (Use a 5% level of significance.)

SOLUTION   In this case the population parameter being tested is μ, the average life of a bulb, and the value questioned is 1000. Since we are testing whether the average life of a bulb is or is not 1000 hours, we have

$$H_0: \quad \mu = 1000$$

$$H_1: \quad \mu \neq 1000$$

We are given that $\bar{x} = 975$, $\sigma = 75$, $\mu = 1000$, and $n = 64$. We first calculate the value of the test statistic, z. We have

$$z = \frac{\bar{x} - \mu}{\dfrac{\sigma}{\sqrt{n}}}$$

$$= \frac{975 - 1000}{\dfrac{75}{\sqrt{64}}}$$

$$= -2.67$$

We use the two-tailed rejection region shown in Figure 10.11. The value of $z = -2.67$ falls in the rejection region. Thus, we reject the null hypothesis that the average life of a bulb is 1000 hours. In this case, the test statistic is sufficiently extreme so that we can reject H_0. The difference is statistically significant.

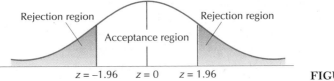

Rejection region   Rejection region
        Acceptance region
$z = -1.96$ $z = 0$ $z = 1.96$ **FIGURE 10.11**

Perhaps you are wondering why we used $\mu \neq 1000$ as the alternative hypothesis and not $\mu < 1000$. After all, who cares if the average life of a bulb is more than 1000 hours? The answer is that the manufacturer cares. When a manufacturer claims that the average life of a bulb is 1000 hours, the manufacturer is concerned when bulbs last more or less than 1000 hours. If the mean life is less than 1000 hours, the manufacturer will lose business and consumer confidence. If the mean life is more than 1000 hours, the company will lose money.

EXAMPLE 2 A bank teller at the Eastern Savings Bank claims that the average amount of money on deposit in a savings account at this bank is $4800 with a standard deviation of $460. A random sample of 36 accounts is taken to test this claim. The average of these accounts is found to be $5000. Does this sample indicate that the average amount of money on deposit is not 4800? (Use a 5% level of significance.)

SOLUTION   In this case the population parameter being tested is μ, the average amount of money on deposit in a savings account. The value questioned is $4800. Since we are testing whether the average amount of money on deposit is $4800 or not, we have

$$H_0: \quad \mu = 4800$$

$$H_1: \quad \mu \neq 4800$$

We are given that $\bar{x} = 5000$, $\mu = 4800$, $\sigma = 460$, and $n = 36$ so that

$$z = \frac{\bar{x} - \mu}{\dfrac{\sigma}{\sqrt{n}}}$$

$$= \frac{5000 - 4800}{\dfrac{460}{\sqrt{36}}}$$

$$= 2.61$$

We use the same two-tailed rejection region as shown in Figure 10.11. The value of $z = 2.61$ falls in the rejection region. Thus, we reject the null hypothesis that the average amount of money on deposit is $4800.

EXAMPLE 3 The average score of all sixth graders in a certain school district on the 1–2–3 math aptitude exam is 75 with a standard deviation of 8.1. A random sample of 100 students in one school was taken. The mean score of these 100 students was 71. Does this indicate that the students of this school are significantly less skilled in their mathematical abilities than the average student in the district? (Use a 5% level of significance.)

SOLUTION   In this case the population parameter being tested is μ, the mean score on the math aptitude exam. The value questioned is $\mu = 75$. We want to determine if the students of this particular school are significantly less skilled in their mathematical abilities. Thus, it is reasonable to set up a one-sided, or a one-tailed, test with the alternative hypothesis being that the population mean for this school is less than 75. We have

$$H_0: \quad \mu = 75$$

$$H_1: \quad \mu < 75$$

When dealing with one-tailed (left-side) alternative hypotheses, we have the rejection regions illustrated in Figure 10.12. These values, like those in the two-tailed tests, are obtained from Table IV in the Appendix. You should verify these results.

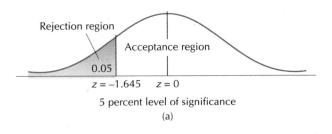

5 percent level of significance
(a)

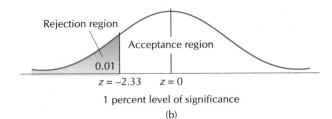

1 percent level of significance
(b)

FIGURE 10.12   One-tailed tests for μ less than some given value.

Now we can calculate the test statistic, z. Here $\bar{x} = 71$, $\mu = 75$, $\sigma = 8.1$, and $n = 100$. We have

$$z = \frac{\bar{x} - \mu}{\frac{\sigma}{\sqrt{n}}}$$

$$= \frac{71 - 75}{\frac{8.1}{\sqrt{100}}}$$

$$= -4.94$$

Since the z-value of -4.94 is in the rejection region, we conclude that the students of this school are significantly less skilled in their mathematical abilities than the average student in the district.

E X A M P L E   4   The We-Haul Trucking Corp. claims that the average hourly salary of its mechanics is $9.25 with a standard deviation of $1.55. A random sample of 81 mechanics showed that the average hourly salary of these mechanics was only $8.95. Does this indicate that the average hourly salary of a mechanic is significantly less than $9.25? (Use a 1% level of significance.)

SOLUTION   In this case the population parameter being tested is μ, the mean hourly salary. The questioned value is $9.25. We wish to know if the hourly salary is less than $9.25. Thus, we use a one-tailed test. We have

$$H_0: \quad \mu = 9.25$$

$$H_1: \quad \mu < 9.25$$

Here we are given that $\bar{x} = 8.95$, $\mu = 9.25$, $\sigma = 1.55$, and $n = 81$ so that

$$z = \frac{\bar{x} - \mu}{\frac{\sigma}{\sqrt{n}}}$$

$$= \frac{8.95 - 9.25}{\frac{1.55}{\sqrt{81}}}$$

$$= -1.74$$

We use the same one-tailed rejection region as shown in Figure 10.12. The value of $z = -1.74$ falls in the acceptance region. Thus, the sample data do not provide us with sufficient justification to reject the null hypothesis. We cannot conclude that the average hourly salary of a mechanic is significantly less than $9.25.

EXAMPLE   5   A trash company claims that the average weight of any of its fully loaded garbage trucks is 11,000 pounds with a standard deviation of 800 pounds. A highway department inspector decides to check on this claim. She randomly checks 36 trucks and finds that the average weight of these trucks is 12,500 pounds. Does this indicate that the average weight of a garbage truck is more than 11,000 pounds? (Use a 5% level of significance.)

Rejection region

Acceptance region

0.05

$z = 0$ $z = 1.645$

5 percent level of significance

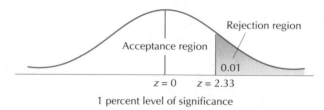

Rejection region

Acceptance region

0.01

$z = 0$ $z = 2.33$

1 percent level of significance

FIGURE 10.13   One-tailed tests for μ greater than some given value.

SOLUTION   In this case the population parameter being tested is μ, the average weight of a garbage truck. The value questioned is 11,000 pounds. The sample data suggest that the mean weight is really more than 11,000 pounds. Thus, the alternative hypothesis will be that the population mean is more than 11,000 pounds. We have

$$H_0: \quad \mu = 11,000$$
$$H_1: \quad \mu > 11,000$$

When dealing with a one-tailed (right-side) alternative hypothesis we have the rejection regions shown in Figure 10.13.

Here we are given that $\bar{x} = 12,500$, $\mu = 11,000$, $\sigma = 800$, and $n = 36$ so that

$$z = \frac{\bar{x} - \mu}{\dfrac{\sigma}{\sqrt{n}}}$$

$$= \frac{12,500 - 11,000}{\dfrac{800}{\sqrt{36}}}$$

$$= 11.25$$

Since the z-value of 11.25 falls within the rejection region, we conclude that the average weight of a garbage truck is not 11,000 pounds. We reject the null hypothesis.

EXAMPLE 6   An insurance company advertises that it takes 21 days on the average to process an auto accident claim. The standard deviation is 8 days. To check on the truth of this advertisement, a group of investigators randomly selects 35 people who recently filed claims. They find that it took the company an average of 24 days to process these claims. Does this indicate that it takes the insurance company more than 21 days on the average to process a claim? (Use a 1% level of significance.)

SOLUTION   In this case the population parameter being tested is μ, the average number of days needed to process a claim. The questioned value is $\mu = 21$. The sample data suggest that $\mu > 21$. Thus, we will use a one-tailed test. We have

$$H_0: \quad \mu = 21$$

$$H_1: \quad \mu > 21$$

Here we are given that $\bar{x} = 24$, $\mu = 21$, $\sigma = 8$, and $n = 35$ so that

$$z = \frac{\bar{x} - \mu}{\dfrac{\sigma}{\sqrt{n}}}$$

$$= \frac{24 - 21}{\dfrac{8}{\sqrt{35}}} = 2.22$$

Since the z-value of 2.22 falls within the acceptance region, we cannot reject the null hypothesis. Thus, we cannot conclude that it takes the insurance company more than 21 days to process a claim.

COMMENT   We wish to emphasize again that the large-sample case tests that we have been discussing work only if the sample size is sufficiently large. This means that n should be at least 30. This is necessary so that the distribution of $\bar{x}$ be approximately normal.

COMMENT   Since σ is unknown in many practical applications, we often have no choice but to use the sample standard deviation, s, as an approximation for σ. Again, s provides a good approximation to σ if the sample size is sufficiently large ($n \geq 30$).

CURE FOUND FOR DEADLY VIRUS

Washington, Jan. 27: A treatment has been found for severe cases of a respiratory virus that infects more than 800,000 U.S. infants a year.

The Food and Drug Administration has approved the drug ribavirin for acute cases of respiratory syncytial virus, or RSV. Administered with an inhalation hood in a hospital for up to seven days, ribavirin stops RSV from reproducing.

Taken orally for 10 days, ribavirin also is the first effective therapy for Lassa fever, a deadly virus found primarily in Africa, say doctors from the Centers for Disease Control.

January 27, 1986

When a new drug is developed, how do we determine whether the average number of people cured by the drug is significant? Is one drug better than another in terms of the average number of people cured?

We summarize our discussion as follows:

LARGE-SAMPLE TEST OF HYPOTHESIS ABOUT A POPULATION MEAN

	One-Tailed Test	Two-Tailed Test	One-Tailed Test
Null Hypothesis Alternative Hypothesis	$H_0: \ \mu = \mu_0$ $H_1: \ \mu < \mu_0$	$H_0: \ \mu = \mu_0$ $H_1: \ \mu \neq \mu_0$	$H_0: \ \mu = \mu_0$ $H_1: \ \mu > \mu_0$
Test Statistic	$z = \dfrac{\bar{x} - \mu_0}{\dfrac{\sigma}{\sqrt{n}}}$ $= \dfrac{\bar{x} - \mu_0}{\dfrac{s}{\sqrt{n}}}$	$z = \dfrac{\bar{x} - \mu_0}{\dfrac{\sigma}{\sqrt{n}}}$ $= \dfrac{\bar{x} - \mu_0}{\dfrac{s}{\sqrt{n}}}$	$z = \dfrac{\bar{x} - \mu_0}{\dfrac{\sigma}{\sqrt{n}}}$ $= \dfrac{\bar{x} - \mu_0}{\dfrac{s}{\sqrt{n}}}$
Rejection Region	$z < -z_\alpha$	$z < -z_{\alpha/2}$ or $z > z_{\alpha/2}$	$z > z_\alpha$

In this chart, z_α is the z-value such that $\text{prob}(z > z_\alpha) = \alpha$.

$z_{\alpha/2}$ is the z-value such that $\text{prob}(z > z_{\alpha/2}) = \dfrac{\alpha}{2}$.

μ_0 represents a particular value for μ as specified in the null hypothesis that we are testing.

EXERCISES FOR SECTION 10.4

1. A manufacturer claims that a particular sensor in a photocopying machine should last an average of 35,000 copies before replacement is needed. To verify the accuracy of this claim, 50 randomly selected sensors are tested. It is found that they lasted an average of 34,000 copies with a standard deviation of 2400 copies. Using a 5% level of significance, should we reject the manufacturer's claim?

2. A farmer's cooperative packages and sells potatoes in bags with the label net weight 5 pounds. The Consumer's Fraud Bureau of one state has received numerous complaints that some of the bags contain less than the specified 5 pounds. The Bureau decides to investigate these complaints by sampling 100 bags of potatoes and accurately measuring their weights. It is found that the average weight of the potatoes in these bags is 4.96 pounds (79.36 ounces) with a standard deviation of 0.2 pounds. Can the farmer's cooperative be accused of "short-changing" the customer? Use a 5% level of significance.

3. Engineers at the Maxwell Product Company claim to have developed a metal bearing rod that is alleged to be extremely durable with a mean useful length of life of 25,400 hours. To check on the accuracy of this claim, 75 of these rods are randomly selected and tested. It is found that these rods have an average life of 25,200 hours

with a standard deviation of 1250 hours. Using a 5% level of significance, can we reject the engineer's claim?

4. According to the Pharmaceutical Manufacturer's Association, the average annual per-person spending on prescription drugs in the United States is $210. If a survey of 85 randomly selected Americans indicated an average per-person spending of $225 with a standard deviation of $45, can we reject the Association's claim? Use a 1% level of significance.

5. As indicated in *Statistical Abstract of the United States*, in 1992 Americans took their dogs to the veterinarian an average of 2.4 times a year at a cost of $82.86. A 1994 survey of 60 randomly selected dog owners found that the owners took their dogs to the veterinarian and were charged an average of $84.93 per visit with a standard deviation of $5.88. Can we conclude that the average cost to a dog owner for veterinarian visits has risen? Use a 5% level of significance.

6. According to the U.S. Department of Justice's *Uniform Crime Statistics*, there were an average of 218 violent crimes per hour in the United States in 1993. A criminologist analyzed 30 communities in one state and found that these communities reported an average of 210 violent crimes per hour with a standard deviation of seven violent crimes per hour. Can we conclude that these communities are below the national average? Use a 5% level of significance.

7. According to the publication *Employment and Earnings* as reported by the Bureau of Labor Statistics, the median U.S. income was $32,073 in 1993. The mayor of Bergenville claims that the average salary in her city was $35,376. A local reporter randomly selects 75 workers and finds that their average salary is $34,876 with a standard deviation of $978. Can we reject the mayor's claim? Use a 5% level of significance.

8. A hospital spokesperson claims that the average daily room charge for a specific procedure is $622. A survey of 50 randomly selected bills for patients who had the

STATE TO INVESTIGATE ABUSIVE HOSPITAL CHARGES

Philadelphia, Jan. 20: The governor's office announced yesterday the establishment of a committee to investigate hospital charges. A Philadelphia woman who had glass removed from her hand was charged $2,236.20 for an hour in the operating room, $52.64 for bandaging, $25.20 for the surgeon's gown and $258.38 for the operating room tray instruments.

January 20, 1984

specific procedure indicated that the average daily room charge was $638 with a standard deviation of $51. Can we reject the hospital spokesperson's claim? Use a 1% level of significance.

9. The Motor Vehicle Manufacturer's Association of the United States publishes information on the ages of cars and trucks in use in *Motor Vehicle Facts and Figures*. The Motor Vehicle Bureau of one state claims that the average age of a vehicle registered within the state is 6.9 years. If a sample of 37 randomly selected vehicles within this state yields an average age of 7.3 years with a standard deviation of 1.3 years, is the state Motor Vehicle Bureau claim accurate? Use a 1% level of significance.

10. In an effort to attract new industry to a certain region, the mayor claims that the average age of a worker in the region is 27 years with a standard deviation of 5.76 years. A prospective company is interested in determining whether the mayor's claim is accurate. A random sample of 60 workers in the region reveals an average age of 30 years. Is there sufficient evidence to conclude that the average age is not 27 years? (Use a 5% level of significance.)

11. Banking industry officials in New York City claim that the average charge for a "bounced" check is $10. To verify this claim, a random sample is taken of the charge imposed by 40 banks for a bounced check. It is found that the average charge is $11.25 with a standard deviation of $0.89. Using a 1% level of significance, should we reject the banking industry official's claim?

10.5 TESTS CONCERNING MEANS FOR SMALL SAMPLES

In the last section we indicated how sample data can be used to reject or accept a null hypothesis about the mean. The sizes of all the samples discussed were large enough to justify the use of the normal distribution. When the sample size is small, we must use the t-distribution discussed in the last chapter instead of the normal distribution. The following examples illustrate how the t-distribution is used in hypothesis testing.

EXAMPLE 1 A manufacturer claims that each can of mixed nuts sold contains an average of 10 cashew nuts. A sample of 15 cans of these mixed nuts has an average of 8 cashew nuts with a standard deviation of 3. Does this indicate that we should reject the manufacturer's claim? (Use a 5% level of significance.)

SOLUTION Since the sample size is only 15, the test statistic becomes

$$t = \frac{\bar{x} - \mu}{\dfrac{s}{\sqrt{n}}}$$

instead of z. Depending upon the number of degrees of freedom, we have the acceptance-rejection regions shown in Figures 10.14 and 10.15. In each case the value

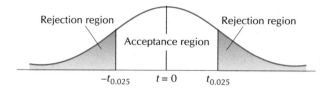

FIGURE 10.14 Two-tailed small-sample rejection region (5% level of significance).

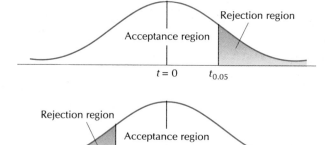

FIGURE 10.15 One-tailed small-sample rejection region (5% level of significance).

of t is obtained from Table VII in the Appendix. It depends on the number of degrees of freedom, which is $n - 1$.

Let us now return to our problem. The population parameter being tested is μ, the average number of cashew nuts in a can of mixed nuts. The questioned value is 10. We have

$$H_0: \quad \mu = 10$$

$$H_1: \quad \mu \neq 10$$

We will use a two-tailed rejection region as shown in Figure 10.14. Here we are given that $\bar{x} = 8$, $\mu = 10$, $s = 3$, and $n = 15$ so that

$$t = \frac{\bar{x} - \mu}{\dfrac{s}{\sqrt{n}}} = \frac{8 - 10}{\dfrac{3}{\sqrt{15}}} = -2.58$$

From Table VII we find that the $t_{0.025}$ value for $15 - 1$, or 14 degrees of freedom is 2.145, which means that we reject the null hypothesis if the test statistic is less than -2.145 or greater than 2.145. Since $t = -2.58$ falls within the critical region, we reject the manufacturer's claim that each can of mixed nuts contains an average of 10 cashew nuts.

EXAMPLE 2 A new weight-reducing pill is being sold in a midwestern city. The manufacturer claims that any overweight person who takes this pill as directed will lose 15 pounds within a month. To test this claim, a doctor gives this pill to six overweight people and finds that they lose an average of only 12 pounds with a standard deviation of 4 pounds. Can we reject the manufacturer's claim? (Use a 5% level of significance.)

SOLUTION   In this case the population parameter being tested is μ, the average number of pounds lost when using this pill. The questioned value is $\mu = 15$. We have

$$H_0: \quad \mu = 15$$

$$H_1: \quad \mu < 15$$

Here we are given that $\bar{x} = 12$, $\mu = 15$, $s = 4$, and $n = 6$ so that

$$t = \frac{\bar{x} - \mu}{\dfrac{s}{\sqrt{n}}} = \frac{12 - 15}{\dfrac{4}{\sqrt{6}}} = -1.84$$

From Table VII we find that the $t_{0.05}$ value for $6 - 1$, or 5 degrees of freedom is 2.015, which means that we reject the null hypothesis if the test statistic is less than -2.015. Since $t = -1.84$ does not fall within the rejection region, we cannot reject the manufacturer's claim that the average number of pounds lost when using this pill is 15.

COMMENT   Again, we wish to emphasize a point made earlier: When testing hypotheses involving small sample sizes, we assume that the relative frequency distribution of the population from which the sample is to be selected is approximately normal.

We summarize the procedure that should be followed when testing hypotheses about a population mean and when we have a small sample size.

SMALL-SAMPLE TEST OF HYPOTHESIS ABOUT A POPULATION MEAN

	One-Tailed Test	Two-Tailed Test	One-Tailed Test
Null Hypothesis *Alternative Hypothesis*	$H_0: \quad \mu = \mu_0$ $H_1: \quad \mu < \mu_0$	$H_0: \quad \mu = \mu_0$ $H_1: \quad \mu \neq \mu_0$	$H_0: \quad \mu = \mu_0$ $H_1: \quad \mu > \mu_0$
Test Statistic	$t = \dfrac{\bar{x} - \mu_0}{\dfrac{s}{\sqrt{n}}}$	$t = \dfrac{\bar{x} - \mu_0}{\dfrac{s}{\sqrt{n}}}$	$t = \dfrac{\bar{x} - \mu_0}{\dfrac{s}{\sqrt{n}}}$
Rejection Region	$t < -t_\alpha$	$t < -t_{\alpha/2}$ or $t > t_{\alpha/2}$	$t > t_\alpha$

In this chart t_α is the *t*-value such that $\text{prob}(t > t_\alpha) = \alpha$.

$t_{\alpha/2}$ is the *t*-value such that $\text{prob}(t > t_{\alpha/2}) = \dfrac{\alpha}{2}$.

The distribution of t has $n - 1$ degrees of freedom.

EXERCISES FOR SECTION 10.5

1. According to the U.S. Department of Energy's Information Administration (*Annual Energy Review*, 1995), the average national utility rates during 1994 were 63 cents for each therm of gas and 8.7 cents for each kilowatt-hours of electricity. A random sample of the rates charged for electricity by ten Northeastern utilities disclosed an average of 9.3 cents per kilowatt-hour with a standard deviation of 0.04 cents. Using a 5% level of significance, does this indicate that the charge for electricity by these ten utilities is significantly above average?

2. According to the American Cancer Society (*Cancer Facts*, 1995), skin cancer is "epidemic" with 700,000 or more new cases per year in the United States. Perhaps 90% of those cancers are linked to overexposure to the sun. Manufacturers have responded by producing sunscreen lotions or creams with sun protection factors (SPF) as high as 30. One particular manufacturer markets a sunscreen lotion with the claim that it has an SPF of 20. A consumer's group decides to challenge this claim. The group randomly selects 15 samples of this product and finds that the average SPF is only 16 with a standard deviation of 1.2. Using a 1% level of significance, should we reject the manufacturer's claim?

3. Refer back to the previous exercise. Government regulations require that when a sunscreen is labeled "waterproof," then it must still work after 80 minutes of immersion or carry claims of continuous protection after 6 to 8 hours in the water. To check on this claim, the consumer's group randomly selects 20 samples and finds that they still work an average of 77 minutes after immersion with a standard deviation of 1.82 minutes. Using a 1% level of significance, should we reject the manufacturer's claim about "waterproof"?

4. A midwestern college claims that its computer science graduates can expect an average starting salary of $24,800 annually. Sixteen 1995 graduates of that college had an average starting salary of $24,000 with a standard deviation of $1850. At the 5% level of significance, can we conclude that the average starting salary of these 1995 computer science graduates is significantly less than $24,800?

5. Consider the newspaper article on the next page. The New York City Taxi and Limousine Commission claims that the average fare for a taxi ride from JFK airport to Manhattan is $26.20. A random sample of 9 trips from the airport to Manhattan had an average fare of $28.15 with a standard deviation of $1.09. At the 5% level of significance, can we conclude that the average fare for a taxi ride from JFK airport to Manhattan is more than $26.20?

6. City officials in Roxville claim that a call to the 911 police emergency number will bring an ambulance in an average of 4.8 minutes. Officials of the EMS claim that ambulance response time has improved since the installation of a new computer system. In a random sample of ten calls for an ambulance, the average response time was 4.6 minutes with a standard deviation of 1.2 minutes. Does this indicate an improvement of service? (Use a 1% level of significance.)

> ## NYC TAXI COMMISSION TO INVESTIGATE $195 AIRPORT FARE
>
> *New York, May 12*: The New York City Taxi and Limousine Commission announced yesterday that it was investigating a complaint lodged by a Japanese couple against a taxi driver. The unsuspecting tourists were charged $195 for a trip from JFK airport to midtown Manhattan.
>
> Friday, May 12, 1995

7. A potato chip manufacturer packs 32-ounce bags of potato chips. The manufacturer wants the bags to contain, on the average, 32 ounces of potato chips. A quality-control engineer randomly selects 13 bags of potato chips and determines that the average weight of the bags is 31.7 ounces with a standard deviation of 0.25 ounce. At the 5% level of significance, should we reject the claim that the average bag contains 32 ounces of potato chips?

8. The three leading manufacturers of peanut butter claim that a typical three-tablespoon serving of peanut butter contains about 24 grams of fat and 285 calories. To check on the accuracy of this claim, a random survey of 18 three-tablespoon servings of peanut butter is taken. The sample results in an average of 288 calories per three-tablespoon serving with a standard deviation of 5 calories. Using a 5% level of significance, test the null hypothesis that the mean calorie content for the sample is the same as the calorie content claimed by the manufacturer.

9. Medicare officials claim that a typical beneficiary nationwide uses an average of 57 days of skilled nursing care in any calendar year. A random survey of twelve skilled-care nursing facilities in the Northeast found that they provided an average of 52 days of skilled nursing care in any calendar year. The standard deviation was 12 days. Using a 5% level of significance, test the null hypothesis that the national average is less than 57 days of skilled nursing care in any calendar year.

10. As reported by the U.S. General Accounting Office, many nursing homes and rehabilitation companies charge exorbitant prices for services rendered. For example, in 1995 nursing homes charged an average of $54 for 15 minutes of speech therapy. The nursing homes deny this. A random survey of 14 nursing homes found that they charged an average of $56 for 15 minutes of speech therapy. The standard deviation was $3. Using a 5% level of significance, test the null hypothesis that the average charge for 15 minutes of speech therapy is $54.

10.6   TESTS CONCERNING DIFFERENCES BETWEEN MEANS FOR LARGE SAMPLES

There are many instances in which we must decide whether the observed difference between two sample means is due purely to chance or whether the population means from which these samples were selected are really different. For example, suppose a teacher gave an IQ test to 50 girls and 50 boys and obtained the following test scores:

	Boys	Girls
Mean	78	81
Standard Deviation	7	9

Is the observed difference between the scores significant? Are the girls smarter?

In problems of this sort the null hypothesis is that there is no difference between the means. Since we will be discussing more than one sample, we use the following notation. Let $\bar{x}_1$, s_1, and n_1 be the mean, standard deviation, and sample size, respectively, of one of the samples, and let $\bar{x}_2$, x_2, and n_2 be the mean, standard deviation, and sample size, respectively, of the second sample. Decisions as to whether to reject or accept the null hypotheses are then based on the test statistic z, where

$$z = \frac{\bar{x}_1 - \bar{x}_2}{\sqrt{\dfrac{s_1^2}{n_1} + \dfrac{s_2^2}{n_2}}}$$

(assuming the samples are independent and both samples are large).

Depending on whether the alternative hypothesis is $\mu_1 \neq \mu_2$, $\mu_1 < \mu_2$, or $\mu_1 > \mu_2$, we have a two-sided test or a one-sided test as indicated in Section 10.4. The following examples illustrate how this test statistic is used.

EXAMPLE   1   Consider the example discussed at the beginning of this section. Is the observed difference between the two IQ scores significant? (Use a 5% level of significance.)

SOLUTION   Let $\bar{x}_1$, s_1, and n_1 represent the boys' mean score, standard deviation, and sample size, and let $\bar{x}_2$, s_2, and n_2 be the corresponding girls' scores. Then the problem is whether the observed difference between the sample means is significant. Thus,

$$H_0: \quad \mu_1 = \mu_2$$

$$H_1: \quad \mu_1 \neq \mu_2$$

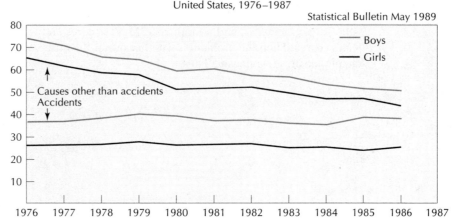

Trend in Death Rates Among Children Aged 1–4, by Sex
United States, 1976–1987

Statistical Bulletin May 1989

FIGURE 10.16 This clipping indicates the trend in death rates, by sex, among children aged 1–4 years. Would you say that the average number of deaths as a result of accidents is significantly higher for boys than for girls? What about child deaths not resulting from accidents? Is the observed difference significant? (Table courtesy of The Metropolitan Life Insurance Company; Source of basic data: Reports of the Division of Vital Statistics, National Center for Health Statistics.)

Here we are given that $\bar{x}_1 = 78$, $s_1 = 7$, $n_1 = 50$, $\bar{x}_2 = 81$, $s_2 = 9$, and $n_2 = 50$ so that

$$z = \frac{\bar{x}_1 - \bar{x}_2}{\sqrt{\dfrac{s_1^2}{n_1} + \dfrac{s_2^2}{n_2}}} = \frac{78 - 81}{\sqrt{\dfrac{7^2}{50} + \dfrac{9^2}{50}}} = \frac{-3}{\sqrt{\dfrac{49}{50} + \dfrac{81}{50}}}$$

$$= \frac{-3}{\sqrt{2.60}} = -1.86$$

We use the two-tailed rejection region of Figure 10.11 (see page 541). The value of $z = -1.86$ falls in the acceptance region. Thus, we cannot conclude that the sample data support the claim that there is a significant difference between the boys' IQ scores and the girls' IQ scores.

EXAMPLE 2 An executive who has two secretaries, Jean and Mark, is interested in knowing whether there is any significant difference in their typing abilities. Mark typed a 40-page report and made an average of 4.6 errors per page. The standard deviation was 0.6. Jean typed a 30-page report and made an average of 2.3 errors per page. The standard deviation was 0.8. Is there any significant difference between their performances? (Use a 5% level of significance.)

SOLUTION   Let $\bar{x}_1$, s_1, and n_1 represent Mark's scores and let $\bar{x}_2$, s_2, and n_2 represent Jean's scores. Then the problem is whether or not the observed difference between the sample means is significant. Thus,

$$H_0: \quad \mu_1 = \mu_2$$
$$H_1: \quad \mu_1 \neq \mu_2$$

Here we are given that $\bar{x}_1 = 4.6$, $s_1 = 0.6$, $n_1 = 40$, $\bar{x}_2 = 2.3$, $s_2 = 0.8$, and $n_2 = 30$ so that

$$z = \frac{\bar{x}_1 - \bar{x}_2}{\sqrt{\dfrac{s_1^2}{n_1} + \dfrac{s_2^2}{n_2}}}$$

$$= \frac{4.6 - 2.3}{\sqrt{\dfrac{(0.6)^2}{40} + \dfrac{(0.8)^2}{30}}}$$

$$= \frac{2.3}{\sqrt{0.009 + 0.0213}}$$

$$= \frac{2.3}{\sqrt{0.0303}} = \frac{2.3}{0.174}$$

$$= 13.22$$

We use the two-tailed rejection region of Figure 10.11 (see page 541). The value of $z = 13.22$ falls in the rejection region. Thus, there is a significant difference between the typing abilities of the two secretaries.

E X A M P L E   3   There are many advertisements on television about toothpastes. One such advertisement claims that children who use Smile toothpaste have fewer cavities than children who use any other brand. To test this claim, a consumer's group selected 100 children and divided them into two groups of 50 each. The children of group I were told to brush daily with only Smile toothpaste. The children of group II were told to brush daily with Vanish toothpaste. The experiment lasted one year. The following number of cavities were identified:

$$\text{Smile:} \quad \bar{x}_1 = 2.31 \qquad s_1 = 0.6$$
$$\text{Vanish:} \quad \bar{x}_2 = 2.68 \qquad s_2 = 0.4$$

Is Smile significantly more effective than Vanish in preventing cavities? (Use a 5% level of significance.)

SOLUTION   In this case the question is whether Smile is better than Vanish. This means that people who use Smile toothpaste will have fewer cavities than those who use Vanish. Thus,

$$H_0: \quad \mu_1 = \mu_2$$

$$H_1: \quad \mu_1 < \mu_2$$

Here we are given that $\bar{x}_1 = 2.31$, $s_1 = 0.6$, $n_1 = 50$, $\bar{x}_2 = 2.68$, $s_2 = 0.4$, and $n_2 = 50$ so that

$$z = \frac{\bar{x}_1 - \bar{x}_2}{\sqrt{\dfrac{s_1^2}{n_1} + \dfrac{s_2^2}{n_2}}}$$

$$= \frac{2.31 - 2.68}{\sqrt{\dfrac{(0.6)^2}{50} + \dfrac{(0.4)^2}{50}}} = \frac{-0.37}{\sqrt{0.0104}} = \frac{-0.37}{0.102}$$

$$= -3.63$$

We use the one-tailed rejection region of Figure 10.12 (see page 543). The value of $z = -3.63$ falls in the rejection region. Thus, we reject the null hypothesis. The sample data would seem to support the manufacturer's claim. Actually, further studies are needed before making any definite decision about the effectiveness of Smile in preventing cavities.

E X A M P L E   4   The local chapter of an affirmative action group claims that a female college graduate earns less than a male college graduate. A survey of 40 men and 30 women indicated the following results.

	Average Starting Salary	Standard Deviation
Women	$29,000	$600
Men	$29,700	$900

Do these figures support the claim that women earn less? (Use a 1% level of significance.)

SOLUTION   Let $\bar{x}_1$, s_1, and n_1 represent the women's scores, and let $\bar{x}_2$, s_2, and n_2 represent the men's scores. The problem is whether or not the observed difference between the sample means is significant. Thus,

$$H_0: \quad \mu_1 = \mu_2$$

$$H_1: \quad \mu_1 < \mu_2$$

Here we are given that $\bar{x}_1 = 29{,}000$, $s_1 = 600$, $n_1 = 30$, $\bar{x}_2 = 29{,}700$, $s_2 = 900$, and $n_2 = 40$ so that

$$z = \frac{\bar{x}_1 - \bar{x}_2}{\sqrt{\dfrac{s_1^2}{n_1} + \dfrac{s_2^2}{n_2}}}$$

$$= \frac{29{,}000 - 29{,}700}{\sqrt{\dfrac{(600)^2}{30} + \dfrac{(900)^2}{40}}} = \frac{-700}{\sqrt{12{,}000 + 20{,}250}}$$

$$= \frac{-700}{\sqrt{32{,}250}} = \frac{-700}{179.58}$$

$$= -3.9$$

We use the one-tailed rejection region of Figure 10.12 (see page 543). The value of $z = -3.9$ falls in the rejection region. Thus, we reject the null hypothesis. There is a significant difference between the starting salary of men and that of women.

We summarize the procedure to be used in the following chart:

LARGE-SAMPLE TEST OF HYPOTHESIS ABOUT THE DIFFERENCE BETWEEN TWO POPULATION MEANS

	One-Tailed Test	Two-Tailed Test	One-Tailed Test
Null Hypothesis Alternative Hypothesis	H_0: $(\mu_1 - \mu_2) = A$ H_1: $(\mu_1 - \mu_2) < A$	H_0: $(\mu_1 - \mu_2) = A$ H_1: $(\mu_1 - \mu_2) \neq A$	H_0: $(\mu_1 - \mu_2) = A$ H_1: $(\mu_1 - \mu_2) > A$
Test Statistic	$z = \dfrac{(\bar{x}_1 - \bar{x}_2) - A}{\sigma_{(\bar{x}_1 - \bar{x}_2)}}$ $\approx \dfrac{(\bar{x}_1 - \bar{x}_2) - A}{\sqrt{\dfrac{s_1^2}{n_1} + \dfrac{s_2^2}{n_2}}}$	$z = \dfrac{(\bar{x}_1 - \bar{x}_2) - A}{\sigma_{(\bar{x}_1 - \bar{x}_2)}}$ $\approx \dfrac{(\bar{x}_1 - \bar{x}_2) - A}{\sqrt{\dfrac{s_1^2}{n_1} + \dfrac{s_2^2}{n_2}}}$	$z = \dfrac{(\bar{x}_1 - \bar{x}_2) - A}{\sigma_{(\bar{x}_1 - \bar{x}_2)}}$ $\approx \dfrac{(\bar{x}_1 - \bar{x}_2) - A}{\sqrt{\dfrac{s_1^2}{n_1} + \dfrac{s_2^2}{n_2}}}$
Rejection Region	$z < -z_\alpha$	$z < -z_{\alpha/2}$ or $z > z_{\alpha/2}$	$z > z_\alpha$

In the preceding chart, A is the numerical value for $(\mu_1 - \mu_2)$ as specified in the null hypothesis. In many applied problems we are interested in testing that there is no difference between the population means. Of course, in such a situation $A = 0$.

COMMENT   For these tests to work we assume that the sample sizes, n_1 and n_2, are sufficiently large and that the two random samples are selected independently from a large population.

EXERCISES FOR SECTION 10.6

1. *Cheating the poor.* A consumer's group claims that many gas stations charge higher prices for a gallon of gas in poorer neighborhoods than in middle-class neighborhoods. To investigate this claim, the group purchased one gallon of unleaded premium gas from each of 41 gas stations located in poorer neighborhoods and one gallon of unleaded premium gas from each of 53 gas stations located in middle-class neighborhoods. The following results were obtained.

	Poorer Neighborhoods	Middle-Class Neighborhoods
Average Price	$1.38	$1.31
Sample Standard Deviation	0.04	0.09
Sample Size	41	53

Is there any significant difference between the average price of a gallon of gas in these neighborhoods? (Use a 1% level of significance.)

Oil companies have been accused by many groups of charging different prices for the same grade of gasoline depending on the area within the city in which it is sold. Is this true, or are the prices determined by competition? *(George Semple)*

2. A reading teacher is experimenting with two different techniques for teaching reading to second grade students. In one class, the 32 students needed an average of 30 minutes to learn to read a paragraph. The standard deviation was 3.6 minutes. In a second class, the 31 students needed an average of 26 minutes to learn to read a paragraph. The standard deviation was 4.9 minutes. Is there a significant difference between these reading techniques? (Use a 5% level of significance.)

3. In a recent high school bowling tournament, the 50 members of the girls team scored an average of 230 points with a standard deviation of 14 points. The 46

members of the boys team scored an average of 235 points with a standard deviation of 28 points. Is there any significant difference between the average scores of the two teams? (Use a 5% level of significance.)

4. According to *Statistical Abstract of the United States*, the average age of men in the United States at the time of their first marriage in 1995 was 24.8 years. A sociologist is interested in knowing if this differs from geographic region to geographic region. A random sample of 40 men in the Northeast found that the average age of these men at the time of their first marriage was 25.1 years with a standard deviation of 2.3 years. A similar survey of 36 men in the South found that the average age of these men at the time of their first marriage was 24.6 years with a standard deviation of 3.1 years. Is there a significant difference between the age of the men at their first marriage in both geographic areas? (Use a 5% level of significance.)

5. A consumer's advocate group, NYPIRG, surveyed 45 banks in New York state and found that the average charge for a "bounced" check was $12.55 with a standard deviation of $2.12. A similar survey of 38 banks in Chicago found that the average charge for a bounced check was $11.12 with a standard deviation of $1.89. Is there a significant difference between the fee charged for a bounced check by banks in New York and banks in Chicago? (Use a 5% level of significance.)

6. According to the U.S. Bureau of Labor Statistics (*Monthly Labor Review*, November 1994) the mean weekly earnings of 100 workers in one U.S. manufacturing industry was $473 with a standard deviation of $18. The mean weekly earnings of 80 workers in another U.S. manufacturing industry was $488 with a standard deviation of $13. Is there a significant difference in the mean weekly earnings of the workers in both industries? (Use a 5% level of significance.)

7. Medical researchers divided a group of 160 volunteers into two groups of equal size. Each individual in group A was given the new X-40 anti-inflammatory medication coated with aspirin. Each individual in group B was given the new medication coated with buffered aspirin. The medical researchers gathered the following information on the average number of hours that the new medication had an effect on the volunteers.

	Drug Coating	
	Aspirin Only	Buffered Aspirin
Average Time of Drug Effectiveness	6 hr	7 hr
Standard Deviation	0.82 hr	1.35 hr

Using a 1% level of significance, test the null hypothesis that there is no significant difference in the lasting effect of the drug whether it is coated with aspirin or with buffered aspirin.

8. The U.S. Department of Agriculture provides statistics on the amount of money that each family spends on food. A random survey of 70 families with four children in one city found that each family spends an average of $62.12 per week on food. The standard deviation was $6.72. A similar survey of 85 families in a second city indicated that each family spends an average of $66.43 per week for food. The standard deviation was $8.84. Is the difference between the average weekly expen-

diture for food by families in these two cities significant? (Use a 5% level of significance.)

9. *Consumer Reports* often tests different paints for their rust-inhibiting ability. In one study, sixty fences that were painted with one brand of rust inhibitor lasted an average of 7 years before repainting was necessary. The standard deviation was 1.86 years. Seventy-five fences painted with a second brand of rust inhibitor lasted an average of 8 years before repainting was necessary. The standard deviation was 2.16 years. Is there any significant difference in the durability of these two brands of rust inhibitor? (Use a 5% level of significance.)

10. *The Wall Street Journal* reports that credit card sales have been steadily increasing. Fifty credit card sales at one department store on July 17, 1995, were made with a Master Card. The average charge was $32.17 with a standard deviation of $4.83. Sixty credit card sales at the same department store and on the same day were made with a Visa Card. The average charge was $29.64 with a standard deviation of $6.97. Is there a significant difference between the average charge using a Master Card or a Visa Card at this department store? (Use a 5% level of significance.)

10.7   TESTS CONCERNING DIFFERENCES BETWEEN MEANS FOR SMALL SAMPLES

In the last section we indicated how we test the difference between sample means. In all the examples the sample sizes were large enough ($n \geq 30$) to justify our use of the normal distribution. If this is not the case, we must use the t-distribution. We assume that the population from which the samples are selected have approximate normal probability distributions and that the random samples are selected independently. We then have the following small-sample-size hypothesis testing procedures.

SMALL-SAMPLE TEST OF HYPOTHESIS ABOUT THE DIFFERENCE BETWEEN MEANS

	One-Tailed Test	Two-Tailed Test	One-Tailed Test
Null Hypothesis *Alternative* *Hypothesis*	H_0: $(\mu_1 - \mu_2) = A$ H_1: $(\mu_1 - \mu_2) < A$	H_0: $(\mu_1 - \mu_2) = A$ H_1: $(\mu_1 - \mu_2) \neq A$	H_0: $(\mu_1 - \mu_2) = A$ H_1: $(\mu_1 - \mu_2) > A$
Test Statistic	$t = \dfrac{(\bar{x}_1 - \bar{x}_2) - A}{s_p \sqrt{\dfrac{1}{n_1} + \dfrac{1}{n_2}}}$	$t = \dfrac{(\bar{x}_1 - \bar{x}_2) - A}{s_p \sqrt{\dfrac{1}{n_1} + \dfrac{1}{n_2}}}$ where $s_p = \sqrt{\dfrac{(n_1 - 1)s_1^2 + (n_2 - 1)s_2^2}{n_1 + n_2 - 2}}$	$t = \dfrac{(\bar{x}_1 - \bar{x}_2) - A}{s_p \sqrt{\dfrac{1}{n_1} + \dfrac{1}{n_2}}}$
Rejection *Region*	$t < -t_\alpha$	$t < -t_{\alpha/2}$ or $t > t_{\alpha/2}$	$t > t_\alpha$

COMMENT   When using the tests outlined in the preceding chart, it is assumed that the variances of the two populations are equal.

COMMENT   When using the tests outlined in the preceding chart, the number of degrees of freedom for the t-distribution is $n_1 + n_2 - 2$.

We illustrate the procedure with an example.

EXAMPLE   1   A chemist at a paint factory claims to have developed a new oil-based paint that will dry very quickly. The manufacturer is interested in comparing this new paint with his currently best-selling paint. In order to accomplish this, he paints each of five different walls with a gallon of his best-selling paint and with a gallon of the new fast-drying paint. The number of minutes needed for each of these paints to dry thoroughly is as follows.

Number of Minutes Needed to Dry

Current Best-Selling Paint	New Fast-Drying Paint
48	42
46	43
44	45
46	43
43	44

Using a 5% level of significance, is the new paint significantly more effective in its drying time than the old paint?

SOLUTION   Using the data in the table, we first compute the sample means and the sample standard deviation. We have

Current Best-Selling Paint	New Fast-Drying Paint
$n_1 = 5$	$n_2 = 5$
$\bar{x}_1 = 45.4$	$\bar{x}_2 = 43.4$
$s_1 = 1.949$	$s_2 = 1.14$

In this case the null hypothesis is $\mu_1 = \mu_2$ and the alternative hypothesis is $\mu_2 < \mu_1$, where μ_1 is the average drying time of the current best-selling paint and μ_2 is the average drying time of the new fast-drying paint.

Based on past experience with other paints, the manufacturer knows that the drying time of paint is approximately normally distributed and that the variances for different paints are about the same. Since the samples were randomly and independently selected, we compute the test statistic. We have

$$t = \frac{\bar{x}_1 - \bar{x}_2}{s_p \sqrt{\dfrac{1}{n_1} + \dfrac{1}{n_2}}}$$

and

$$s_p = \sqrt{\frac{(n_1 - 1)s_1^2 + (n_2 - 1)s_2^2}{n_1 + n_2 - 2}}$$

so that

$$s_p = \sqrt{\frac{(5 - 1)(1.949)^2 + (5 - 1)(1.14)^2}{5 + 5 - 2}} \approx 1.597$$

Therefore,

$$t = \frac{45.4 - 43.4}{1.597 \sqrt{\frac{1}{5} + \frac{1}{5}}} \approx 1.98$$

The number of degrees of freedom is $n_1 + n_2 - 2 = 5 + 5 - 2$ or 8. From Table VII of the Appendix, the $t_{0.05}$ value with 8 degrees of freedom is 1.86.

The test statistic value of $t = 1.98$ falls in the critical rejection region. Hence, we reject the null hypothesis and conclude that based on the data, the newly developed fast-drying paint does indeed dry faster than the current best-selling paint.

COMMENT   If we know for sure that the samples are from nonnormal populations, then we cannot justifiably use the small-sample t test for $(\mu_1 - \mu_2)$. In this situation it is advisable to use the nonparametric Wilcoxon rank-sum test to be discussed in Chapter 14.

Until now we have assumed that the variances of both sampled populations are equal. If this is not the case, then the chart given on page 561 has to be modified somewhat. We have the following:

SMALL-SAMPLE TEST OF HYPOTHESES ABOUT DIFFERENCES BETWEEN MEANS WHEN VARIANCES OF BOTH SAMPLED POPULATIONS ARE NOT EQUAL

Test statistic

$$t = \frac{\bar{x}_1 - \bar{x}_2 - A}{\sqrt{\dfrac{s_1^2}{n_1} + \dfrac{s_2^2}{n_2}}}$$

Degrees of freedom

$$df = \frac{\left(\dfrac{s_1^2}{n_1} + \dfrac{s_2^2}{n_2}\right)^2}{\left[\dfrac{(s_1^2/n_1)^2}{n_1 - 1} + \dfrac{(s_2^2/n_2)^2}{n_2 - 1}\right]}$$

COMMENT   When using the above formula, df may not be an integer. In that case round *down* to the nearest integer.

EXERCISES FOR SECTION 10.7

1. Environmentalists sampled 16 different sites along one polluted lake and found an average coliform bacteria count of 18.8 per unit with a standard deviation of 3.26. Then they sampled 12 different sites along a different polluted lake and found an average coliform bacteria count of 28.6 per unit with a standard deviation of 8.88. Is it true that the average number of coliform bacteria count for one lake is significantly lower than the average number of coliform bacteria count for the other lake? (Use a 5% level of significance.)

2. In *Savings and Home Financing Service* published annually by the Federal Home Loan Board, the median purchase prices for existing single-family or two-family houses for various metropolitan areas are given. In 1994, the average selling price for a two-family house (based on the prices of the last 15 sales) in Brighton was $181,000. The standard deviation was $8700. In neighboring Bathgate, the average selling price for a two-family house (based on the prices of the last 11 sales) was $195,000. The standard deviation was $12,800. Is there any significant difference in the average selling price for a two-family house located in Brighton and one located in Bathgate? (Use a 1% level of significance.)

3. According to the U.S. Bureau of the Census, the mean annual income of U.S. households in 1993 was $39,020. A random survey of 12 families in Bakersfield found a mean annual household income of $39,277. The standard deviation was $4205. A similar survey of 18 families in Plainview disclosed a mean annual household income of $40,116. The standard deviation was $3864. Is there any significant

difference in the mean annual household income in these two communities? (Use a 5% level of significance.)

4. According to a 1994 National Health and Nutrition Examination Survey, the average height of American women aged 18 to 74 years is 63.7 inches. In one survey of 18 women (aged 18 to 74), the average height was 63.1 inches with a standard deviation of 2.33 inches. In a second survey of 20 women (aged 18 to 74), the average height was 64.3 inches with a standard deviation of 1.67 inches. Is the difference between the average height of the women in both surveys significant? (Use a 5% level of significance.)

5. An airline company is about to order some inflatable rubber life rafts for emergency use on its planes. Six samples of life rafts produced by one company needed an average of 7 seconds to be fully inflated. The standard deviation was 1.86 seconds. Five samples of life rafts produced by a second company needed an average of 6 seconds to be fully inflated. The standard deviation was 2.13 seconds. Using a 5% level of significance, is the difference between the mean time required to fully inflate the life rafts produced by the two companies significant?

6. According to *The Wall Street Journal* (June 21, 1995, p. 1), taxpayers whose adjusted gross income (AGI) was between $40,000 to $49,000 deducted an average of $5738 as interest expenses from their AGI on their 1040 U.S. Individual Tax Return in 1993. A newspaper reporter surveyed 14 taxpayers in the specified income range and found that they had deducted an average of $5376 as interest expenses from their AGI in 1993. The standard deviation was $29. A second reporter surveyed 12 families and found that they had deducted an average of $5518 as interest expenses on their 1993 federal income tax return. The standard deviation was $38. Is there any significant difference between the average amount deducted by a taxpayer for interest expenses as presented by both newspaper reporters? (Use a 1% level of significance.)

7. Ten patients at Brooks Hospital required an average hospital stay of 5 days after a particular surgical procedure. The standard deviation was 1.35 days. Eight patients at Mercy Hospital required an average of 6.5 days after the same surgical procedure. The standard deviation was 1.98 days. Using a 5% level of significance, is there any significant difference between the average length of stay at both hospitals after the surgical procedure?

8. According to Bankcard Holders of America, BHA, as reported in *Good Housekeeping* (September 1991, p. 138), the typical credit card holder charges $2750 a year and pays, on average, $315 in interest charges. One research group randomly selected 12 credit card holders and found that they charged an average of $3270 per year. The standard deviation was $476. A second research group surveyed 14 credit card holders and found that these credit card holders charged an average of $3064 per year. The standard deviation was $716. Using a 5% level of significance, is there any significant difference between the average charge as reported by these two research groups?

9. The National Association for the Education of Young Children (NAEYC) reports that child-care costs vary considerably across the country, with the national average (in 1995) around $4000 to $5000 a year. A survey of ten families in Boston found

that they were paying an average of $6375 a year for child care. The standard deviation was $812. A survey of 16 families in Chicago found that they were paying an average of $5628 a year for child-care costs with a standard deviation of $675. Using a 1% level of significance, is the difference between the average child-care costs in Boston significantly different than the costs in Chicago?

10. A large Wall Street brokerage house uses high-volume photocopy machines manufactured by either company A or company B. As a cost-cutting measure, management has decided to use only machines manufactured by company A *or* by company B. The following annual operating cost sample data are available:

Photocopy Machine Manufacturer

Company A			Company B			
$14,698	$13,986	$13,576	$ 8,623	$10,286	$14,053	$13,928
10,961	12,683	14,052	9,845	12,757	15,058	9,256
13,002	12,465		13,517	13,684	12,977	9,998
			12,654	11,568	10,896	12,892

a. Compute the mean and standard deviation for each sample.
b. Management has decided to use only the photocopy machines manufactured by company B if it is found to be significantly cheaper. Using a 5% level of significance, test the null hypothesis that the machines manufactured by company B are at least as expensive to operate as those manufactured by company A.
c. What action should management take? Explain your answer.

10.8   TESTS CONCERNING PROPORTIONS

Suppose a congressman claims that 60% of the voters in his district are in favor of lowering the drinking age to 16 years. If a random sample of 400 voters showed that 221 of them favored the proposal, can we reject the congressman's claim? Questions of this type occur quite often and are usually answered on the basis of observed proportions. We assume that we can use the binomial distribution and that the probability of success is the same from trial to trial. We can therefore apply Formulas 6.5 and 6.6 (see page 372), which give us the mean and standard deviation of a binomial distribution. Thus,

Mean: $\mu = np$

Standard deviation: $\sigma = \sqrt{np(1 - p)}$

The null hypothesis in such tests assumes that the observed proportion, $\hat{p}$, is the same as the population proportion, p. Depending on the situation, we have the following alternative hypotheses:

Null hypothesis: H_0: $\hat{p} = p$

Alternative

hypothesis: H_1: $\hat{p} \neq p$ [which means a two-tailed test]

H_2: $\hat{p} > p$ [which means a one-tailed (right-side) test]

H_3: $\hat{p} < p$ [which means a one-tailed (left-side) test]

The test statistic is z, where

$$z = \frac{\hat{p} - p}{\sqrt{\dfrac{p(1-p)}{n}}}$$

We are assuming that the sample size is large. We reject the null hypothesis if the test statistic falls in the critical, or rejection region.

The following examples illustrate how we test proportions.

EXAMPLE 1 The Dean of Students at a college claims that only 12% of the students commute to school by bike. To test this claim, a students' group takes a sample of 80 students. They find that 14 of these students commute by bike. Is the Dean's claim acceptable? (Use a 5% level of significance.)

SOLUTION In this case the population parameter being tested is p, the true proportion of students who commute by bike. The questioned value is 0.12. Since we are testing whether or not the true proportion is 0.12, we have

$$H_0: \quad p = 0.12$$

$$H_1: \quad p \neq 0.12$$

We are told that 14 of the 80 sampled students commute to school by bike so that $\hat{p} = \dfrac{14}{80} = 0.175$. Thus,

$$z = \frac{\hat{p} - p}{\sqrt{\dfrac{p(1-p)}{n}}} = \frac{0.175 - 0.12}{\sqrt{\dfrac{(0.12)(1-0.12)}{80}}}$$

$$= \frac{0.055}{\sqrt{0.00132}} = \frac{0.055}{0.036}$$

$$= 1.53$$

We use the two-tailed rejection region of Figure 10.11 (see page 541). The value of z = 1.53 falls in the acceptance region. Thus we cannot reject the null hypothesis and the Dean's claim that the true proportion of students who commute to school by bike is 12%.

E X A M P L E   2   In a recent conference a senator claimed that 55% of the American people supported the president's foreign policy. To test this claim, a newspaper editor selected a random sample of 1000 people and 490 of them said that they supported the president. Is the senator's claim justified? (Use a 1% level of significance.)

SOLUTION   In this case the population parameter being tested is p, the true proportion of Americans who support the president. The questioned value is 0.55. Since we are testing whether or not the true proportion is 0.55, we have

$$H_0: \quad p = 0.55$$

$$H_1: \quad p \neq 0.55$$

We are told that 490 of the 1000 people interviewed supported the president so that

$$\hat{p} = \frac{490}{1000} = 0.49$$

Thus,

$$z = \frac{\hat{p} - p}{\sqrt{\dfrac{p(1 - p)}{n}}} = \frac{0.49 - 0.55}{\sqrt{\dfrac{(0.55)(1 - 0.55)}{1000}}}$$

$$= \frac{-0.06}{\sqrt{0.0002475}} = \frac{-0.06}{0.016}$$

$$= -3.75$$

Since the level of significance is 1%, we use the two-tailed rejection of Figure 10.10 (see page 540). The value of $z = -3.75$ falls in the rejection region. Thus, we reject the null hypothesis and the senator's claim that 55% of the American people support the president's foreign policy.

E X A M P L E   3   A latest government survey indicates that 22% of the people in Camelot are illegally receiving some form of public assistance. The mayor of Camelot believes that the figures are exaggerated. To test this claim, she carefully examines 75 cases and finds that 11 of these people are illegally receiving aid. Does this sample support the government's claim? (Use a 5% level of significance.)

SOLUTION   In this case the population parameter being tested is p, the true proportion of people illegally receiving financial aid. The questioned value is 0.22. Since we are testing whether the true proportion is 0.22 or lower, we have

$$H_0: \quad p = 0.22$$

$$H_1: \quad p < 0.22$$

We are told that 11 of the 75 cases examined are illegally receiving aid so that

$$\hat{p} = \frac{11}{75} = 0.15$$

Thus,

$$z = \frac{\hat{p} - p}{\sqrt{\dfrac{p(1 - p)}{n}}} = \frac{0.15 - 0.22}{\sqrt{\dfrac{(0.22)(1 - 0.22)}{75}}}$$

$$= -1.46$$

We use the one-tail rejection region of Figure 10.12 (see page 543). The value of $z = -1.46$ falls within the acceptance region. Thus, we cannot reject the null hypothesis that 22% of the people in Camelot are illegally receiving financial aid.

EXAMPLE 4   Government officials claim that approximately 29% of the residents of a state are opposed to building a nuclear plant to generate electricity. Local conservation groups claim that the true percentage is much higher. To test the government's claim, an independent testing group selects a random sample of 81 state residents and finds that 38 of the people are opposed to the nuclear plant. Can we conclude that the government's claim is inaccurate? (Use a 5% level of significance.)

SOLUTION   In this case the population parameter being tested is p, the true proportion of state residents who are opposed to building the nuclear plant. The questioned value is 0.29. Since we are testing whether the true proportion is 0.29 or higher, we have

$$H_0: \quad p = 0.29$$

$$H_1: \quad p > 0.29$$

We are told that 38 of the 81 residents are opposed to the nuclear plant so that

$$\hat{p} = \frac{38}{81} = 0.47$$

Thus,

$$z = \frac{\hat{p} - p}{\sqrt{\dfrac{p(1 - p)}{n}}} = \frac{0.47 - 0.29}{\sqrt{\dfrac{(0.29)(1 - 0.29)}{81}}}$$

$$= \frac{0.18}{\sqrt{0.002542}} = \frac{0.18}{0.0504} = 3.57$$

We use the one-tail rejection region of Figure 10.13 (see page 544). The value of $z = 3.57$ falls in the rejection region. Thus, we reject the null hypothesis that the true proportion of state residents opposed to the nuclear plant is 0.29.

EXERCISES FOR SECTION 10.8

1. The U.S. Food and Drug Administration (FDA) bans the use of hormones in poultry production. Nevertheless, a recent University of California study (*Good Housekeeping*, September 1995, p. 96) found that 10% of the consumers surveyed said that they ate less poultry because of concern over hormones. To check on this claim, 84 consumers are randomly selected, and it is found that 10 of them eat less poultry because of their concern over hormones. Should we reject the results of the University of California study? (Use a 5% level of significance.)

2. According to a study appearing in *Self* magazine (April 1994), 39% of the adult readers of this magazine believe in ghosts. A recent survey of 80 adult readers of *Self* magazine found that 27 of them believe in ghosts. Using a 5% level of significance, do the results of this recent survey disagree with the earlier study?

3. Is cigarette smoking dangerous to your health? According to a *USA TODAY/CNN* Gallup Poll (*USA TODAY*, March 16, 1994), 45% of U.S. adults believe that cigarette advertising should be banned completely. A recent survey by Hodges and Bradley of 65 U.S. adults found that 35 of them believe that cigarette smoking should be banned. Using a 1% level of significance, do the results of this survey disagree with the earlier *USA TODAY/CNN* Gallup Poll?

4. *The Wall Street Journal* (February 9, 1990) estimated that 25% of 13-year-old U.S. students use calculators in their math class. A 1995 survey of thirty-nine 13-year-old U.S. students found that 26 of them use calculators in their math class. At the 5% level of significance, does the new survey indicate that the percentage of 13-year-old U.S. students using calculators in their math classes has changed significantly when compared with the earlier 1990 estimate?

5. The U.S. Bureau of the Census (1995) estimates that 25% of all American households are one-parent homes. A sociologist believes that the true percentage is much higher. In a random sample of 80 households in one city, it is found that 24 of them are one-parent homes. Should we reject the claim that 25% of all American households are one-parent homes? (Use a 5% level of significance.)

6. The City Transportation Company claims that 96% of its commuter trains or busses arrive on time. In a random sample of 629 arriving trains, it was found that 593 of the trains or busses arrived on time. Should we reject the transportation company's claim? (Use a 5% level of significance.)

7. According to the National Highway Traffic and Safety Administration (NHTSA), about 15% of teen drivers are involved in an auto crash during their first year of driving. A survey of 258 teens found that 46 of them were involved in an auto crash during their first year of driving. Does this indicate that the percentage of the teens

who were involved in an auto crash during their first year of driving is significantly higher than the 15% NHTSA estimate? (Use a 5% level of significance.)

8. According to a 1995 study conducted by Hyatt Hotels and Resorts, 97% of women business executives are convinced that vacations help avoid "burnout." A new survey of 43 women business executives found that 38 of them are convinced that vacations help avoid burnout. Does this new survey indicate that the percentage of women business executives who are convinced that vacations help avoid burnouts is significantly different than the 1995 Hyatt Hotel study? (Use a 5% level of significance.)

9. According to a 1992 Insurance Institute for Highway Safety (IIHS) survey, about 66% of Americans nationwide use auto seat belts. According to a recent survey for *Prevention* magazine that appeared in *The Wall Street Journal* (September 13, 1993), 71% of American adults use auto seat belts. A recent survey of 50 drivers found that 35 of them were using seat belts. Do these results differ significantly from the IIHS claim or the *Prevention* claim about seat-belt usage by Americans? (Use a 5% level of significance.)

10. Consider the accompanying newspaper article. If 85 people are randomly selected, and if 65 of them are in favor of tighter gun-control legislation, can we reject the claim that 70% of all people are in favor of gun-control legislation? (Use a 5% level of significance.)

> ## 70% QF ALL PEOPLE ARE IN FAVOR OF GUN CONTROL LEGISLATION
>
> *Washington, Dec. 26:* A nationwide survey conducted by Wallinsky and Rogers reveals that 70% of the people interviewed are in favor of tighter gun control legislation. This is a direct outgrowth of the rising incidence of crime that is currently sweeping our country in which guns are used.
>
> Monday, December 26, 1994

11. *Affirmative Action.* A construction company claims that 55% of its workers are from minority groups or are women. In a random sample of 80 of the company's workers it is found that 41 of the workers are from minority groups or are women. Can we accept the claim that 55% of the company's workers are from minority groups or are women? (Use a 5% level of significance.)

12. Refer back to the newspaper article at the beginning of this chapter (page 532), which indicates that 55% of those Americans questioned were in favor of revamping our national health system. If a random survey of 60 Americans indicates that only 25 of them support the President's proposed health plan, should we reject the newspaper claim about 55% support? (Use a 5% level of significance.)

10.9   USING COMPUTER PACKAGES

We can use MINITAB to perform hypothesis tests involving a population mean when the sample size is large ($n \geq 30$) and also when the sample size is small ($n \leq 30$). We illustrate the procedure with the following examples.

E X A M P L E   1   Industry representatives claim that the average price of an introductory college-level math textbook is $34.95. Numerous student groups claim that the price is considerably higher. A survey of 36 such books disclosed the following prices:

$32.95	$33.49	$36.45	$34.28	$33.68	$34.95	$32.78	$36.21	$33.89
35.79	31.87	33.51	34.76	35.10	32.69	32.95	33.29	33.78
34.78	35.65	35.29	36.19	31.79	37.25	34.56	35.49	34.95
34.78	35.87	36.85	33.17	35.67	34.99	37.10	36.08	33.09

Using a 5% level of significance, should we reject the industry representative's claim?

SOLUTION   We must determine if H_0: $\mu = 34.95$ or H_1: $\mu > 34.95$ with $\alpha = 0.05$. In order to use MINITAB for hypothesis testing, we must know the value of the population standard deviation, σ. Unfortunately, this information is not supplied. However, since the sample size is sufficiently large ($n = 36$), we can use MINITAB's DESCRIBE command discussed in Chapter 3 to accomplish this. First we enter the data into the computer. Then we use the MINITAB command **ZTEST** followed by the null hypothesis (MU = 34.95), the estimated value of σ (SIGMA = 1.461), and the storage location of the sample data (C1). When these are executed the machine responds SUBC. We must now inform the machine as to what the alternative hypothesis is. If we want to perform a left-tailed test, we type ALTERNATIVE = -1. If we want to perform a right-tailed test we type ALTERNATIVE = $+1$. For a two-tailed test we type nothing. MINITAB is already programmed for a two-tailed test. When applied to our example we have the following MINITAB printout:

ZTEST

```
MTB > SET THE FOLLOWING DATA INTO C1
DATA> 32.95 33.49 36.45 34.28 33.68 34.95 32.78 36.21 33.89
DATA> 35.79 31.87 33.51 34.76 35.10 32.69 32.95 33.29 33.78
DATA> 34.78 35.65 35.29 36.19 31.79 37.25 34.56 35.49 34.95
DATA> 34.78 35.87 36.85 33.17 35.67 34.99 37.10 36.08 33.09
DATA> END
```

```
MTB > DESCRIBE C1
VARIABLE          N      MEAN    MEDIAN    TRMEAN    STDEV    SEMEAN
C1               36    34.610    34.780    34.624    1.461     0.243

VARIABLE        MIN       MAX        Q1        Q3
C1           31.790    37.250    33.340    35.760
MTB > ZTEST, MU = 34.95, SIGMA = 1.461, C1;
SUBC> ALTERNATIVE = +1.

TEST OF MU = 34.950 VS MU G.T. 34.950
THE ASSUMED SIGMA = 1.46

                N      MEAN    STDEV    SE MEAN         Z    P VALUE
C1             36    34.610    1.461      0.243     -1.40       0.92

MTB > STOP
```

In this printout we have some very valuable information. It informs us that MINITAB is testing the null hypothesis of $\mu = 34.950$ versus the alternative hypothesis that μ is G.T. (greater than) 34.950. MINITAB then informs us of the assumed value of σ, the column number that it worked with, the sample size, sample mean, sample standard deviation, and standard error of the mean. The z-value of -1.40 represents the value of the test statistic obtained by computing

$$z = \frac{\bar{x} - \mu}{\frac{s}{\sqrt{n}}} = \frac{34.610 - 34.95}{\frac{1.461}{\sqrt{36}}} = -1.40$$

The P-value is really the most important thing in hypothesis testing. We have

Definition 10.8
P-value

The **P-value** for a hypothesis test is the smallest significance level at which the null hypothesis can be rejected based on the sample data. If the P-value is less than the significance level, α, then we reject H_0; otherwise we do not reject H_0.

In our case the P-value is 0.92. Since this value is more than 0.05, we do not reject H_0. We would then say that based on the sample data, we cannot reject the industry representative's claim.

EXAMPLE 2   A potato chip manufacturer packs 32-ounce bags of potato chips. The manufacturer wants the bags to contain, on the average, 32 ounces of potato chips. A quality-control engineer selects 13 bags of potato chips and records their weights as follows:

32.5, 32.1, 31.9, 31.8, 32.3, 32.2, 31.8, 31.9, 32.3, 32.4, 32.2, 32.1, 32.3

At the 5% level of significance, can we accept the manufacturer's claim that the bags contain more than 32 ounces of potato chips?

TTEST

SOLUTION We must determine if H_0: $\mu = 32$ or H_1: $\mu > 32$ with $\alpha = 0.05$. First we enter the data into the computer. Then we use the MINITAB command **TTEST** followed by the null hypothesis (MU = 32.0) and the storage location of the data (C1). When these are executed the machine responds SUBC>. Since we are performing a right-tailed test, we type ALTERNATIVE = +1. When applied to our example we have the following MINITAB printout:

```
MTB > SET THE FOLLOWING DATA IN C1
DATA> 32.5, 32.1, 31.9, 31.8, 32.3, 32.2, 31.8, 31.9, 32.3,
DATA> 32.4, 32.2, 32.1, 32.3
DATA> END

MTB > TTEST, MU = 32.0, C1;
SUBC> ALTERNATIVE = +1.

TEST OF MU = 32.0000 VS MU G.T. 32.0000

            N      MEAN    STDEV   SE MEAN      T   P VALUE
C1         13   32.1385   0.2293   0.0636    2.18    0.025

MTB > STOP
```

Again, the *P*-value is really the most important thing. In this case the *P*-value of 0.025 is less than $\alpha = 0.05$, so we reject the null hypothesis. We conclude that the data indicate that the average 32-ounce bag of potato chips contains more than 32 ounces of potato chips.

EXERCISES FOR SECTION 10.9

In each of the following, use MINITAB to determine if we should reject the null hypothesis or not. In each case, analyze the problem by means of the *P*-value.

1. In an effort to protect the toll collectors at the city tunnels, health officials frequently monitor the level of the pollutants in the air. On eight randomly selected days, the air at one of these tunnels contained 36, 34, 39, 43, 29, 32, 35, and 36 parts per million (ppm) of a certain pollutant. Union officials claim that the air at these tunnels contains an average of 44 ppm of the same pollutant. Is there any significant difference between the average amount of the pollutant in the air found by the health officials and the union claim? (Use a 1% level of significance.)
2. Medical research indicates that the average reaction time to a certain stimulus is 1.83 seconds. Seven adults have been given a new antibiotic drug. Their average reaction time to the stimulus is found to be 1.93, 1.66, 1.22, 2.83, 3.02, 2.69, and 3.09 seconds. Using a 5% level of significance, does the new antibiotic drug increase the average reaction time to the stimulus?
3. Health officials claim that patients in the Emergency Room at St. Vincent's Hospital wait an average of 31 minutes before being examined by a doctor. Thirty-seven patients in the Emergency Room at St. Vincent's Hospital waited 42, 38, 59, 46, 31, 39, 42, 55, 69, 51, 50, 22, 36, 39, 21, 19, 26, 22, 29, 34, 36, 52, 40, 21, 29, 37, 62, 50,

46, 54, 71, 23, 39, 26, 34, 23, and 38 minutes before being examined by a doctor. Using a 5% level of significance, can we accept the hospital officials' claim that the average amount of time that one must wait in the Emergency Room at St. Vincent's Hospital before being examined by a doctor is 31 minutes?

4. Data gathered from many purchasers of a certain type of personal computer suggest that the time it takes a purchaser to fully connect the computer to the monitor and printer is a random variable having a mean of 85 minutes (min). A group of 10 different potential purchasers of the same computer is randomly selected and shown a film on how to connect the computer to the monitor and printer easily. After viewing the film, each purchaser is asked to fully connect the computer. It is found that they now need 68, 61, 63, 74, 69, 72, 70, 73, 65, and 68 minutes. Using a 5% level of significance, does viewing the film reduce the time required to connect the computer?

5. Welfare department officials of a certain city claim that the average number of cases of child abuse handled daily by its various agencies is 8.62. A newspaper reporter decides to test this claim. The reporter randomly selects 12 days and determines that the average number of child-abuse claims handled on those days was 6, 8, 4, 2, 3, 7, 8, 1, 3, 2, 7, and 9. Using a 5% level of significance, should we reject the claim of the welfare department officials?

6. A large oil company claims that the average cost of a gallon of regular unleaded gasoline to a motorist is $1.33. To test this claim, 16 service stations are randomly selected. It is found that the average price of a gallon of regular unleaded gasoline at these stations is $1.39, $1.42, $1.31, $1.53, $1.36, $1.39, $1.37, $1.29, $1.29, $1.35, $1.35, $1.37, $1.32, $1.36, $1.35, and $1.37. Should we reject the oil company's claim? (Use a 5% level of significance.)

7. Consider the accompanying newspaper article. The union claims that the average number of sick days taken by an officer is 8.4 days per year. A survey of 35 highway patrol officers found that these officers were absent 10, 12, 5, 17, 21, 18, 19, 10, 11,

STATE TO INVESTIGATE ABUSE OF SICK LEAVE PRIVILEGE

New Haven, Aug. 2: The governor's office announced yesterday the establishment of a commission to investigate abuses of the sick leave policy privilege afforded to highway patrol officers. Under this privilege, an officer is allowed an unlimited number of sick days per year.

The officials claim that this privilege is being abused by some officers.

August 2, 1994

6, 7, 9, 10, 3, 18, 16, 17, 12, 5, 7, 6, 8, 9, 3, 12, 15, 17, 16, 10, 12, 4, 11, 17, 15, and 13 days per year. Using a 5% level of significance, should we reject the union's claim?

For Exercises 8 and 9 use the following information.

Although MINITAB does not have a direct set of commands for testing an hypothesis about the difference between population means for *large* and independent samples (even if the variances are normally distributed), if we take *small* and independent samples, we can use the t-distribution to test inferences about the difference between two population means. If we have reason to believe that the population standard deviations are equal, then after we enter the data in columns C1 and C2, we use the following MINITAB commands:

```
MTB > TWOSAMPLE T C1 C2;
SUBC> ALTERNATIVE = 0;
SUBC> POOLED.
```

If we are unsure about the validity of assuming that $\sigma_1^2 = \sigma_2^2$, then we use the following MINITAB commands:

```
MTB > TWOSAMPLE T C1 C2;
SUBC> ALTERNATIVE = +1.
```

°8. To determine if a significant difference exists in the mean length of fish from two hatcheries, 11 fish were randomly selected from hatchery A, and 10 fish were randomly selected from hatchery B. Their lengths, in centimeters, were as follows:

Hatchery A:	16.3	14.8	12.4	14.8	15.3	11.6	12.7	13.3
	12.9	15.3	14.2					
Hatchery B:	12.8	13.9	14.6	15.6	16.8	14.3	18.1	14.1
	12.2	11.7						

Use MINITAB to determine if the difference in the sample means is statistically significant at the 5% level. (Assume that the standard deviations of the sampled populations are equal.)

°9. Researchers at the Arlington Corp. claim that they have developed a new chemical additive that increases gasoline mileage. To determine if this additive does indeed increase gasoline mileage, eight test runs using this additive and nine test runs without using the additive were made. The following miles per gallon were obtained:

Without Additive:	24.6	25.3	28.6	26.1	23.7	26.4	25.9
	25.1	24.3					
With Additive:	26.4	27.3	25.4	26.1	24.2	23.8	24.0
	22.1						

Using MINITAB, test at the 5% level of significance if the use of the new chemical additive increases gasoline mileage. (Do not assume that the standard deviations of the sample populations are equal.)

10.10   SUMMARY

In this chapter we discussed hypothesis testing, which is a very important branch of statistical inference. Hypothesis testing is the process by which a decision is made either to reject or accept a null hypothesis about one of the parameters of the distribution. The decision to accept or reject a null hypothesis is based on information obtained from sample data. Since any decision is subject to error, we discussed type-I and type-II errors.

We reject a null hypothesis when the test statistic falls in the critical, or rejection, region. The critical region is determined by two things:

1. whether we wish to perform a one-tailed or two-tailed test
2. the level of significance

If a null hypothesis is not rejected, we cannot say that the sample data prove that what the null hypothesis says is necessarily true. It merely does not reject it. Some statisticians prefer to say that in this situation they *reserve judgment* rather than accept the null hypothesis.

Study Guide

The following is a chapter outline in capsule form. You should now be able to demonstrate your knowledge of the ideas mentioned by giving definitions, descriptions, or specific examples. Page references are given in parentheses.

Hypothesis testing is the procedure whereby we make a decision to accept or reject a claim on the basis of sample information. (page 533)

By an **hypothesis** we mean an assumption about one or more of the population parameters that will be either accepted or rejected on the basis of sample information. (page 533)

When testing any hypothesis, if the sample value falls within an **acceptance region**, then we will not reject the null hypothesis. (page 534)

The value that separates the rejection region from the acceptance region is called the **critical value**. (page 535)

The **null hypothesis**, denoted by H_0, is the statistical hypothesis being tested. (page 535)

The **alternative hypothesis**, denoted by H_1, H_2, ... is the hypothesis that will be accepted when the null hypothesis is rejected. (page 535)

A **one-sided** or **one-tailed test** is a statistical test that has the rejection region located in the left tail or the right tail of the distribution. (page 536)

A **two-sided** or **two-tailed test** is a statistical test that has the rejection region located in both tails of the distribution. (page 536)

The **region of rejection** is called the **critical region**. The remaining region is called the **acceptance region**. (page 536)

A **test statistic** is a calculated number that is used to decide whether to reject or accept the null hypothesis. (page 536)

Tests that enable us to decide whether to reject or accept a null hypothesis are called **statistical tests of hypotheses** or **statistical tests of significance**. (page 536)

A **type-I error** is made when a true null hypothesis is rejected; that is, we reject it when we should accept it. The probability of making a type I error is denoted by α. (page 536)

A **type-II error** is made when a false null hypothesis is accepted; that is, we accept a null hypothesis when we should reject it. (page 536)

The **significance level** of a test is the probability that the test statistic falls within the rejection region when the null hypothesis is true. (page 537)

The **5% level of significance** is used when the statistician wishes that the risk of rejecting a true null hypothesis not exceed 5%. (page 537)

The **1% level of significance** is used when the statistician wishes that the risk of rejecting a true null hypothesis not exceed 1%. (page 537)

The MINITAB command **ZTEST** followed by the null hypothesis, the estimated value of σ, and the storage location of the sample data allows us to perform a hypothesis test involving a population mean when the sample size is large $n \geq 30$. If the sample size is small, we use the MINITAB command **TTEST** to accomplish the same thing. (pages 572 and 574)

The **P-value** for a hypothesis test is the smallest significance level at which the null hypothesis can be rejected on the basis of sample data. If the P-value is less than the significance level, α, then we reject H_0; otherwise we do not reject H_0. (page 573)

Formulas to Remember

We discussed methods for testing means, proportions, and differences between means. All the tests studied are summarized in Table 10.1. You should be able to identify each symbol in the formulas, understand the relationships among the symbols expressed in each formula, understand the significance of each formula, and use the formulas in solving problems.

TABLE 10.1 Various Tests for Accepting or Rejecting a Null Hypothesis

Population Parameter Tested	Sample Size	Type of Test	Significance			Test Statistic	Reject Null Hypothesis if
Mean	Large	Two-tailed	0.05				$z < -1.96$ or $z > 1.96$
Mean	Large	Two-tailed	0.01			$z = \dfrac{\bar{x} - \mu}{\sigma/\sqrt{n}}$	$z < -2.58$ or $z > 2.58$
Mean	Large	One-tailed	0.05				$z < -1.645$ or $z > 1.645$
Mean	Large	One-tailed	0.01				$z < -2.33$ or $z > 2.33$
Mean	Small	Two-tailed	0.05			$t = \dfrac{\bar{x} - \mu}{s/\sqrt{n}}$	$t < -t_{0.025}$ or $t > t_{0.025}$
Mean	Small	One-tailed	0.05				$t < -t_{0.05}$ or $t > t_{0.05}$
Difference between sample means	Large	Two-tailed	0.05	or	0.01	$z = \dfrac{\bar{x}_1 - \bar{x}_2}{\sqrt{\dfrac{s_1^2}{n_1} + \dfrac{s_2^2}{n_2}}}$	Same as above
	Large	One-tailed	0.05	or	0.01		
Difference between sample means	Small	Two-tailed	0.05	or	0.01	$t = \dfrac{\bar{x}_1 - \bar{x}_2}{s_p \sqrt{\dfrac{1}{n_1} + \dfrac{1}{n_2}}}$ where $s_p = \sqrt{\dfrac{(n_1 - 1)s_1^2 + (n_2 - 1)s_2^2}{n_1 + n_2 - 2}}$ $df = n_1 + n_2 - 2$	Same as above
	Small	One-tailed	0.05	or	0.01		
Proportion	Large	Two-tailed	0.05	or	0.01	$z = \dfrac{\hat{p} - p}{\sqrt{\dfrac{p(1 - p)}{n}}}$	Same as above
	Large	One-tailed	0.05	or	0.01		

Testing Your Understanding of This Chapter's Concepts

1. Due to technological advances, customers who have checking accounts at almost any bank can withdraw funds from their accounts either by going personally to their bank and dealing with a human teller or by using money machines that are available nationwide at numerous locations. The First Maritime Bank is interested in estimating the difference in costs per transaction between a human teller and a cash machine teller. It has gathered the following sample data from its numerous branches.

	Withdrawal Using	
	Human Teller	Cash Machine
Average Cost per Transaction	25 cents	11 cents
Standard Deviation	3.1 cents	2.98 cents
Number Involved in Survey	12 branch offices	14 cash machine sites

Set up a 90% confidence interval for the mean cost savings per transaction by using the cash machines.

2. To qualify for federal funding, many programs stipulate that the recipients must be economically disadvantaged and have an annual family income below a specified level. The following are the family incomes of ten families residing in the Jonathan Williams Houses and 12 families residing in the Martin Luther King Houses.

Income of Families Residing in

Jonathan Williams Houses		Martin Luther King Houses	
$18,363	$12,807	$19,382	$16,862
$16,802	$16,123	$18,043	$19,098
$18,291	$15,492	$15,196	$16,051
$16,080	$19,032	$12,769	$14,954
$16,923	$14,994	$13,983	$16,083
		$12,888	$15,995

a. Compute the mean and standard deviation for each sample.
b. Construct a 95% confidence interval estimate of the difference in average family income using the families residing in the Williams or King houses.
c. At the 5% level of significance, should we reject or accept the null hypothesis of identical means?

Chapter Test

Multiple Choice Questions

1. The average weekly salary of a technician in a certain city is $495 with a standard deviation of $47. One large company, which employs 86 technicians, pays its technicians an average weekly salary of $475. Can this company be accused of paying lower wages than the average rate? (Use a 5% level of significance.)
   a. yes   b. not necessarily   c. not enough information given

2. From past experience, the traffic department of a certain city has found that the yellow traffic lane markings on the highways remain visible for an average of 185 days with a standard deviation of 12.8 days. This year, highway markings at 46 different locations remained visible for an average of 182 days. Does this indicate that we should reject the traffic department's past average of 185 days? (Use a 5% level of significance.)
   a. yes   b. not necessarily   c. not enough information given

3. Educators believe that the average time needed to complete a certain reading test by youngsters is 78 minutes. Eight students have received some special tutoring for the exam. Their average time needed to complete the exam (after the tutoring) is

found to be 71 minutes with a standard deviation of 6.86 minutes. Using a 1% level of significance, does the tutoring significantly affect the average time needed to complete the reading test?

a. yes   b. not necessarily   c. not enough information given

4. The average weight of a deer in a certain national park is 125 pounds with a standard deviation of 6.8 pounds. A game warden caught ten deer in the park and determined that their average weight was 130 pounds. At the 5% level of significance, can we reject the game warden's claim that the average weight of a deer in the park is 125 pounds?

a. yes   b. not necessarily   c. not enough information given

5. Two techniques are being compared. Using technique A, 80 students needed an average of 25 minutes to complete a task. The standard deviation was 2.7 minutes. Using technique B, 111 students needed an average of 36 minutes to complete the task. The standard deviation was 3.9 minutes. At the 5% level of significance, is the difference between the average completion time using both techniques significant?

a. yes   b. not necessarily   c. not enough information given.

Supplementary Exercises

6. A random survey of 50 patients in the emergency room at Harden Hospital indicated that they waited an average of 33 minutes before receiving medical attention. The standard deviation was 3.76 minutes. A similar survey of 45 patients at Lincoln Hospital indicated that they waited an average of 39 minutes before receiving medical attention. The standard deviation was 5.34 minutes. Is the difference between the average waiting time in the emergency rooms at these two hospitals significant? (Use a 5% level of significance.)

7. A college has two campuses at which students can register for next semester's courses. Eight randomly selected students at campus I needed an average of 84 minutes to complete the registration process. The standard deviation was 5.83 minutes. Eleven randomly selected students at campus II needed an average of 78 minutes to complete the registration process. The standard deviation was 6.93 minutes. Using a 5% level of significance, is there any significant difference between the average amount of time needed by students to complete the registration process at both campuses?

8. A mayor claims that 88% of all people in the city are in favor of increasing real estate taxes so as to balance the budget. In a random survey of 70 people in the city, it was found that 50 of them were in favor of increasing the real estate taxes. At the 5% level of significance, should we reject the mayor's claim?

9. An agricultural chemist claims to have developed a new pesticide that is 85% effective in eliminating a certain insect. To test this claim, 58 randomly selected parcels of land are sprayed with this chemical and the chemical successfully eliminates the insect from only 48 of the parcels. At the 5% level of significance, should we reject the chemist's claim?

10. The 12 workers in the production department of a publishing company spend an average of 18 minutes on their morning coffee break. The standard deviation is 2.37 minutes. The 11 workers in the advertising department of the company spend an average of 19.7 minutes on their coffee break. The standard deviation is 1.88 minutes. Is it true that the average number of minutes spent by the workers in the advertising department is significantly different from the average number of minutes spent by the workers in the production department? (Use a 5% level of significance.)

11. The average daily room charge for a semiprivate hospital room in a certain state during 1995 was $627. The standard deviation was $13. In the same year, a random sample of nine hospitals in a city within the state showed an average charge of $632. At the 5% level of significance, can we conclude that the average daily charge at these hospitals significantly exceeds the state average?

12. The average age of a prisoner in a state's jail is 28.76 years with a standard deviation of 2.17 years. One jail in the state has a prisoner average age of 26.89 years for its 117 inmates. At the 5% level of significance, can we conclude that the average age of the prisoners at this jail is significantly lower than the average age of the prisoners at all of the state's jails?

13. Research indicates that the average life of a transmission in a city police car is 34,000 miles. Eight police cars have a special mechanism added to the transmission. It is now found that the average life of the transmission of these eight police cars using this special mechanism is 36,000 miles with a standard deviation of 2200 miles. Using a 1% level of significance, does this special mechanism affect the average life of a transmission of a police car?

14. Police department officials claim that the average height of a police officer in Buscane is 69 inches with a standard deviation of 1.89 inches. A newspaper reporter randomly selected ten police officers in Buscane and found that their average height was 67.1 inches. At the 1% level of significance, can we conclude that the average height of a police officer in Buscane is less than 69 inches?

15. Two groups of children are being compared. Group I, which consisted of 40 children, required an average of 57 minutes to complete a task. The standard deviation was 4.62 minutes. Group II, which consisted of 55 children, required an average of 53 minutes to complete the task. The standard deviation was 5.13 minutes. At the 5% level of significance, is the difference between the average time for both groups significant?

16. A group of 40 baseball fans waited an average of 96 minutes before being admitted to the stadium for a particular baseball game. The standard deviation was 7.1 minutes. A second group of 75 baseball fans waited an average of 111 minutes before being admitted to the same game. The standard deviation was 9.12 minutes. Is the difference between the average waiting times by these two groups before being admitted to the stadium significant? (Use a 5% level of significance.)

17. Seven lakes in a state had an average of 58 units of coliform bacteria. The standard deviation was 4.69 units. Nine lakes in an adjoining state had an average of 63 units of coliform bacteria. The standard deviation was 6.13 units. Using a 5% level of

significance, is there any significant difference between the average amount of coliform bacteria in both states?

18. The nine dolphins in a particular park needed an average of 28 practice days before mastering a dangerous act. The standard deviation was 3.76 days. The 12 dolphins in a different park needed an average of 32 practice days before mastering the same act. The standard deviation was 5.38 days. Is it true that the average number of practice days required by the dolphins in the first park is significantly less than the average number of days required by the dolphins in the second park? (Use a 5% level of significance.)

19. A researcher claims that 88% of all the residents of a city are opposed to any further tax abatements. A random survey of 80 residents of this city disclosed that 69 of them were opposed to any further tax abatements. Should we reject the researcher's claim? (Use a 5% level of significance.)

20. A researcher claims that as a result of many antismoking campaigns, only about 53% of the staff at a school still smokes. To test this, a random survey of 168 staff members at this school is taken. It is found that only 80 of them still smoke. At the 5% level of significance, should we reject the researcher's claim?

21. Fifty-five traffic lights painted with brand A rust-inhibitor paint lasted an average of 7.1 years before repainting was necessary. The standard deviation was 1.48 years. Sixty-five traffic lights painted with brand B rust-inhibitor paint lasted an average of 6.8 years before repainting was necessary. The standard deviation was 1.96 years. At the 5% level of significance, is the difference between the average lasting time of these two brands of rust-inhibitor paints statistically significant?

22. Seven randomly selected assistant professors at Sierra University had an average annual salary of $44,500 during 1995. The standard deviation was $689. Ten randomly selected assistant professors at Gorg College had an annual salary of $46,100. The standard deviation was $941. Using 5% level of significance, is there any significant difference between the average annual salaries of the assistant professors at these two schools?

23. A consumer's group purchased the same dose of a particular prescription drug at 10 different pharmacies in Baltimore and 12 different pharmacies in Chicago. The following results were obtained:

	Baltimore Pharmacies	Chicago Pharmacies
Average Price	$5.43	$5.86
Standard Deviation	0.39	0.62

Is it true that the average price charged by pharmacies in Baltimore is significantly less than the average price charged by pharmacies in Chicago? (Use a 5% level of significance.)

24. A fast-food store owner claims that 90% of her customers order a soft drink when ordering French-fried potatoes. In a random survey of 80 customers ordering French-fried potatoes, it was found that 68 of them ordered a soft drink also. At the 5% level of significance, should we reject the store owner's claim?

25. A politician claims that 70% of all voters in her district are in favor of building the proposed superhighway through the city. If a random survey of 300 voters shows that 200 of them are in favor of the proposed new superhighway, should we reject the politician's claim? (Use a 5% level of significance.)

Thinking Critically

1. In any test of hypotheses, is it ever possible to make a type-I error *and also* a type-II error at the same time? Explain your answer.

2. In one particular experiment, a researcher tested the null hypothesis $\mu \le 45$ against the alternative hypothesis $\mu > 45$. Assuming that the sample mean x has a value of 40,
   a. could the null hypothesis ever be rejected? Explain your answer.
   b. if we accept the null hypothesis, what type of error could we be making? Explain your answer.

3. The formula given on page 561 for the pooled standard deviation is

$$s_p = \sqrt{\frac{(n_1 - 1)s_1^2 + (n_2 - 1)s_2^2}{n_1 + n_2 - 2}}$$

Show that if the sample sizes n_1 and n_2 are equal, then s_p^2 is just the mean of s_1^2 and s_2^2.

4. If the level of significance of an hypothesis test is chosen to be $\alpha = 0$, what is the probability of a type-I or a type-II error? Explain your answer.

Case Study

Many people have long suspected that fear of malpractice suits has prompted many doctors to opt for Caesarean sections rather than let nature take its course.

A new report in the *Journal of the American Medical Association* (1983) analyzed data on 60,490 births in the New York hospitals and insurance premiums paid by doctors in various regions. The report states that there is significant evidence to conclude that the odds of having a Caesarean section were three times as great in areas where premiums and the frequency of malpractice suits were high as opposed to regions where premiums and the frequency of malpractice suits were low. In 1965, only 5% of all U.S.

births were Caesarean sections. In 1993, the rate had risen to the point where 23% of all U.S. births were by Caesarean section. (*Source:* Bureau of Vital Statistics, 1995)

1. An insurance company representative randomly selects 98 maternity claims and finds that 33 of the births were by Caesarean section. Does this indicate that the number of births by Caesarean section is significantly more than the claimed 23%? (Use a 5% level of significance.)

11

LINEAR CORRELATION AND REGRESSION

Chapter Outline

DID YOU KNOW THAT

a   a public health campaign urging parents to put newborns to sleep on their backs more and more has resulted in a nationwide decrease in sudden infant death syndrome (SIDS) by 12% in just six months? (*Source: Archives of Pediatrics and Adolescent Medicine*, 1994)

b   increased use of cocaine by smokers puts an increased strain on the heart and that cocaine use by chronic smokers can be doubly deadly? (*Source: New England Journal of Medicine*, 1994)

c   attendance at air shows has been increasing over the years? (*Source: International Council of Air Shows*, 1995)

d   the time required by women to complete the 800-meter run has been decreasing over the years and that women may soon outrun men? (*Source: Nature* 355:25, January 1992)

The resale value of a car depends, among other things, on the car's age and the number of miles indicated on the car's odometer. Is the geographic area in which the car is being sold a factor in determining its resale value? See discussion on page 650. (*George Semple*)

Often we are given two or more variables and wish to know whether they are related. If they are, can we find some sort of equation for predicting the value of one variable when the values of the other are known? How reliable are such predictions? In this chapter we analyze procedures that can be used to determine when a relationship exists between variables and also to find an equation expressing such a relationship. Thus, we will study correlation and regression.

Chapter Objectives

- **To briefly discuss** some of the pioneers in the field of correlation and regression. (Section 11.1)
- **To decide** whether two variables are related by drawing scatter diagrams. (Section 11.2)
- **To discuss** linear correlation, which tells us whether there is a relationship that will cause an increase (or decrease) in the value of one variable when the other is increased (or decreased). (Section 11.2)
- **To measure** the strength of the linear relationship that exists between two variables. This is the coefficient of linear correlation. (Section 11.3)
- **To determine** the reliability of r. (Section 11.4)
- **To calculate** a regression line that allows us to predict the value of one of the variables if the value of the other variable is known. (Section 11.5)
- **To analyze** the method of least squares that we use to determine the estimated regression line used in prediction. (Section 11.6)
- **To indicate** how the standard error of the estimate tells us how well the least-squares prediction equation describes the relationship between two variables. (Section 11.7)
- **To set up** confidence intervals for our regression estimates. (Sections 11.8 and 11.9)
- **To introduce** the concept of multiple regression where one variable depends on many other variables. (Section 11.10)

Consider the accompanying newspaper clipping. How are such predictions made? How reliable are such predictions?

PAST, PRESENT AND FUTURE OF SPENDING

	1960°	1995	2030°°
Average income	$5,260	$30,000°°°	$117,797
Chrysler Luxury Sedan	$2,964	$35,000	$207,797
Potatoes per lb.	5¢	59¢	$8.25
Child's blouse	$4	$15	$89
Postage stamp	5¢	32¢	$1.90
Haircut	$2.50	$20	$119
Movie ticket	75¢	$7	$42
Life Magazine	20¢	$3	$18

°*Source: The Value of a Dollar, Gale Research Inc. (1994).*
°°*Projections based on an estimated average annual inflation rate of 5.1%.*
°°°*Estimate based on latest figure available (1993) from U.S. Bureau of Economic Analysis: average wage and salary for full-time equivalent employee in a domestic industry 1993: $29,367.*

11.1   INTRODUCTION

Up to this point we have been discussing the many statistical procedures for analyzing a single variable. However, when dealing with the problems of applied statistics in education, psychology, sociology, and so on, we may be interested in determining whether a relationship exists between two or more variables. For example, if college officials have just administered a vocational aptitude test to 1000 entering freshmen, they may be interested in knowing whether there is any relationship between the math aptitude scores and the business aptitude scores. Do students who score well on the math section of the aptitude exam also do well on the business part? On the other hand, is it true that a student who scored poorly on the math part will necessarily score poorly on the business part? Similarly, the college officials may be interested in determining if there is a relationship between high school averages and college performance.

Questions of this nature frequently arise when we have many variables and are interested in determining relationships between these sets of scores. In this chapter we learn how to compute a number that measures the relationship between two sets of scores. This number is called the **correlation coefficient**. The English mathematician Karl Pearson (1857–1936) studied it in great detail and to some extent so did another English mathematician, Sir Francis Galton (1822–1911).

Correlation
coefficient

HISTORICAL NOTE

Sir Francis Galton, a cousin of Charles Darwin, undertook a detailed study of human characteristics. He was interested in determining whether a relationship exists between the heights of fathers and the heights of their sons. Do tall parents have tall children? Do intelligent parents or successful parents have intelligent or successful children? In *Natural Inheritance* Galton introduced the idea referred to today as correlation. This mathematical idea allows us to measure the closeness of the relationship between two variables. Galton found that a very close relationship exists between the heights of fathers and the heights of their sons. On the question of whether intelligent parents have intelligent children, it has been found that the correlation is 0.55. As we will see, this means that it is not necessarily true that intelligent parents have intelligent children. In many cases children will score higher or lower than their parents. Figure 11.1 shows how the correlations for IQ range from 0.28 between parents and adopted children to 0.97 for identical twins reared together.

The precise mathematical measure of correlation as we use it today was actually formulated by Karl Pearson.

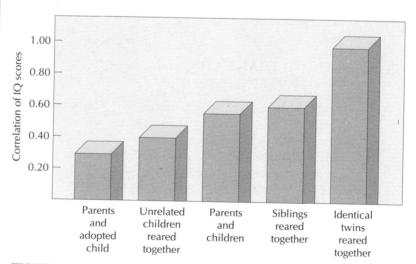

FIGURE 11.1   Correlations between the intelligence of parents and that of their children.

Method of least squares

If there is a high correlation between two variables, we may be interested in representing this correspondence by some form of an equation. So we will discuss the **method of least squares**. The statistical method of least squares was developed by Adrien-Marie Legendre (1752–1833). Although Legendre is best known for his work in geometry, he also did important work in statistics. He developed the method of least squares. This method is used when we want to find the regression equation.

Finally, we will discuss how this equation can be used to make predictions and we will examine the reliability of these predictions.

LIKE PARENTS, LIKE CHILDREN

Jan. 5: Children whose parents are hooked on drugs and alcohol are more likely to fall victim to these substances than other kids. U.S. News & World Report says: "Studies show 65 percent of those youths dependent on drugs or alcohol are from homes where at least one parent is also hooked."

Saturday, January 5, 1991

11.2   SCATTER DIAGRAMS

To help us understand what is meant by a correlation coefficient, let us consider a guidance counselor who has just received the scores of an aptitude test administered to ten students. See Table 11.1.

The counselor may be interested in determining if there is any correlation among these sets of scores. For example, the counselor may wish to know whether a student who scores well on the math aptitude part of the exam will also score well on the business aptitude part. She can analyze the situation pictorially by means of a **scatter diagram**.

Scatter diagram

To make a scatter diagram, we draw two lines, one vertical and one horizontal. On the horizontal line we indicate the math scores, and on the vertical line we indicate the business scores. Although we can put the math scores on the vertical line, we have purposely labeled the math scores on the horizontal line. This is done because we wish to predict the scores on the business aptitude part on the basis of the math scores. After both axes, that is, both lines, are labeled, we use one dot to represent each person's score. The dot is placed directly above the person's math score and directly to the right

TABLE 11.1 Different Aptitude Scores Received by Ten Students

Student	Math Aptitude	Business Aptitude	Language Aptitude	Music Aptitude
A	52	48	26	22
B	49	49	53	23
C	26	27	48	57
D	28	24	31	54
E	63	59	67	13
F	44	40	75	20
G	70	72	31	9
H	32	31	22	50
I	49	50	11	17
J	51	49	19	24

of the business score. Thus, the dot for Student A's score is placed directly above the 52 score on the math axis and to the right of 48 on the business axis. Similarly, the dot for Student B's score is placed directly above the 49 score on the math axis and to the right of 49 on the business axis. The same procedure is used to locate all the dots of Figure 11.2.

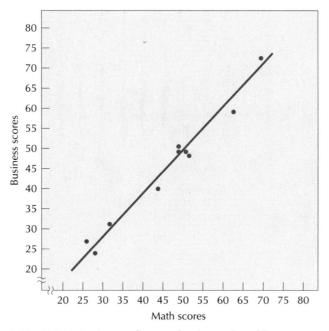

FIGURE 11.2 Scatter diagram for the math and business scores.

Linear correlation

Notice that these dots form an approximate straight line. When this happens we say that there is a **linear correlation** between the two variables. Notice also that the higher the math score is, the higher the business score will be. The line moves in a direction that is from lower left to upper right. When this happens, we say that there is

Positive correlation

a **positive correlation** between the math scores and the business scores. This means that a student with a higher math score will tend to have a higher business score.

Now let us draw the scatter diagram for the business aptitude scores and the music aptitude scores. It is shown in Figure 11.3. In this case we notice that the higher the business score, the lower the music score. Again, the dots arrange themselves in the form of a line, but this time the line moves in a direction that is from upper left to lower

Negative correlation

right. When this happens we say that there is a **negative correlation** between the business scores and the music scores. This means that a student with a high business score will tend to have a low music score.

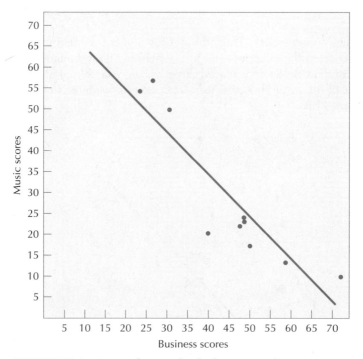

FIGURE 11.3   Scatter diagram for the business and music scores.

Now let us draw the scatter diagram for the language scores and music scores. It is shown in Figure 11.4. In this case the dots do not form a straight line. When this happens we say that there is little or no correlation between the language scores and the music scores.

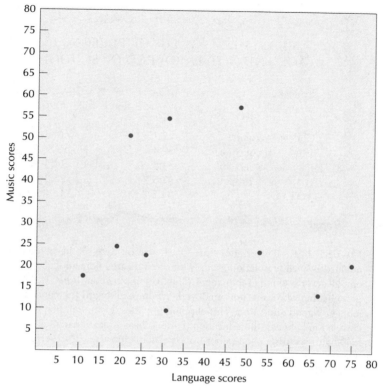

FIGURE 11.4 Scatter diagram for the language and music scores.

COMMENT Although we will concern ourselves with only linear, that is, a straight line, correlation, the dots may suggest different types of curves. These are studied in detail by statisticians. Several examples of such scatter diagrams are given in Figure 11.5. In this text we will analyze only linear correlation.

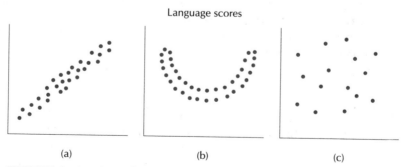

FIGURE 11.5 Scatter diagrams that suggest (a) a linear relationship, (b) a curvilinear relationship, and (c) no relationship.

EMPLOYMENT STATUS OF PERSONS 16 TO 24 YEARS OF AGE AND NOT ENROLLED IN SCHOOL OCTOBER 1990

Educational Attainment	Labor Force Participation Rate	Unemployment Rate	Percent Employed
College graduates	95.2%	5.9%	89.7%
1 to 3 years of college	89.4	8.8	81.5
High school, no college	84.3	12.5	73.7
High school dropouts	67.5	25.3	50.4
TOTAL	**81.8%**	**14.0%**	**70.2%**

FIGURE 11.6   The "Labor Force Participation Rate" is the percent of the total civilian, noninstitutional population group with the indicated educational characteristic who were employed or seeking employment. The "Unemployment Rate" is the percent of those participating who were not employed. We have all heard the concept that if you want a good job, get a good education. This clipping provides documentation that there is a positive correlation between the education that one has (educational attainment) and the chances of holding or securing a job (labor force participation rate). Furthermore, the clipping indicates that there is a negative correlation between educational attainment and unemployment. (*Source:* Current Population Survey *conducted by the U.S. Bureau of the Census.*)

11.3   THE COEFFICIENT OF CORRELATION

Coefficient of linear correlation

Once we have determined that there is a linear correlation between two variables, we may wish to determine the strength of the linear relationship. Karl Pearson developed a **coefficient of linear correlation**, which measures the strength of a relationship between two variables. The value of the coefficient of linear correlation is calculated by means of Formula 11.1.

FORMULA 11.1

The **coefficient of linear correlation** is given by

$$r = \frac{n(\Sigma\ xy) - (\Sigma\ x)(\Sigma\ y)}{\sqrt{n(\Sigma\ x^2) - (\Sigma\ x)^2}\sqrt{n(\Sigma\ y^2) - (\Sigma\ y)^2}}$$

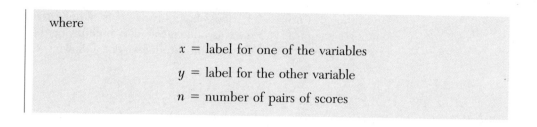

where

$$x = \text{label for one of the variables}$$
$$y = \text{label for the other variable}$$
$$n = \text{number of pairs of scores}$$

When using Formula 11.1 the coefficient of correlation will always have a value between -1 and $+1$. A value of $+1$ means perfect positive correlation and corresponds to the situation where all the dots lie exactly on a straight line. A value of -1 means perfect negative correlation and again corresponds to the situation where all the points lie exactly on a straight line. Correlation is considered high when it is close to $+1$ or -1 and low when it is close to 0. If the coefficient of linear correlation is zero, we say that there is no linear correlation. These possibilities are indicated in Figures 11.7 and 11.8.

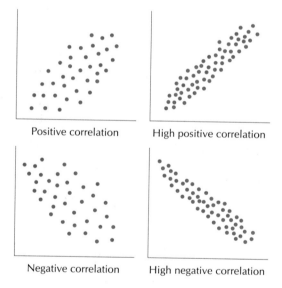

Positive correlation   High positive correlation

Negative correlation   High negative correlation   **FIGURE 11.7**   Possible correlations.

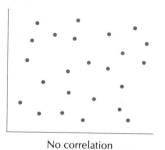

No correlation   **FIGURE 11.8**   No correlation.

Although Formula 11.1 looks complicated, it is rather easy to use. The only new symbol that appears is $\Sigma\, xy$. This value is found by multiplying the corresponding values of x and y and then adding all the products. The following examples illustrate the procedure.

E X A M P L E   1   Find the correlation coefficient for the data of Figure 11.2 (see page 591). It is repeated in the following chart:

Math Score	52	49	26	28	63	44	70	32	49	51
Business Score	48	49	27	24	59	40	72	31	50	49

SOLUTION   We first let x represent the math score and y represent the business score. Then we arrange the data in tabular form as follows and apply Formula 11.1.

x (Math)	y (Business)	x^2	y^2	xy
52	48	2704	2304	2496
49	49	2401	2401	2401
26	27	676	729	702
28	24	784	576	672
63	59	3969	3481	3717
44	40	1936	1600	1760
70	72	4900	5184	5040
32	31	1024	961	992
49	50	2401	2500	2450
51	49	2601	2401	2499
464	449	23,396	22,137	22,729

$\Sigma x = 464 \qquad \Sigma y = 449 \qquad \Sigma x^2 = 23{,}396 \qquad \Sigma y^2 = 22{,}137 \qquad \Sigma xy = 22{,}729$

Then,

$$r = \frac{n(\Sigma\, xy) - (\Sigma\, x)(\Sigma\, y)}{\sqrt{n(\Sigma\, x^2) - (\Sigma\, x)^2}\sqrt{n(\Sigma\, y^2) - (\Sigma\, y)^2}}$$

$$= \frac{10(22{,}729) - (464)(449)}{\sqrt{10(23{,}396) - (464)^2}\sqrt{10(22{,}137) - (449)^2}}$$

$$= \frac{18{,}954}{\sqrt{18{,}664}\sqrt{19{,}769}}$$

$$= \frac{18{,}954}{(136.62)(140.60)}$$

$$= \frac{18{,}954}{19{,}208.77} = 0.9867$$

Thus, the coefficient of correlation is 0.9867. Since this value is close to $+1$, we say that there is a high degree of positive correlation. Figure 11.2 also indicated the same result.

E X A M P L E   2   Find the coefficient of correlation for the data of Figure 11.4 (see page 593). It is repeated in the following chart.

Language Score	26	53	48	31	67	75	31	22	11	19
Music Score	22	23	57	54	13	20	9	50	17	24

SOLUTION   We first let x represent the language score and y represent the music score. Then we arrange the data in tabular form as follows and apply Formula 11.1.

x (Language)	y (Music)	x^2	y^2	xy
26	22	676	484	572
53	23	2809	529	1219
48	57	2304	3249	2736
31	54	961	2916	1674
67	13	4489	169	871
75	20	5625	400	1500
31	9	961	81	279
22	50	484	2500	1100
11	17	121	289	187
19	24	361	576	456
383	289	18,791	11,193	10,594

$$\Sigma x = 383 \qquad \Sigma y = 289 \qquad \Sigma x^2 = 18,791 \qquad y^2 = 11,193 \qquad \Sigma xy = 10,594$$

Now we apply Formula 11.1. We have

$$r = \frac{n(\Sigma\ xy) - (\Sigma\ x)(\Sigma\ y)}{\sqrt{n(\Sigma\ x^2) - (\Sigma\ x)^2}\sqrt{n(\Sigma\ y^2) - (\Sigma\ y)^2}}$$

$$= \frac{10(10,594) - (383)(289)}{\sqrt{10(18,791) - (383)^2}\sqrt{10(11,193) - (289)^2}}$$

$$= \frac{-4747}{\sqrt{41,221}\sqrt{28,409}} = \frac{-4747}{(203.03)(168.55)}$$

$$= \frac{-4747}{34,220.71} = -0.1387$$

Thus, the coefficient of correlation is -0.1387. Since this value is close to zero, there is little correlation. Figure 11.4 indicated the same result.

How will the correlation between x and y be affected if x is coded by adding the same number to (or subtracting the same number from) each score? How is y affected?

Fortunately, it turns out that the correlation coefficient is unaffected by adding or subtracting a number to either x or y or both. Thus, x can be coded in one way—perhaps by adding or subtracting a number—and y can be coded in another way—say, by multiplying by a number. In either case the value of the correlation coefficient is unaffected. Of greater importance is the fact that if we code before calculating the value of r, we do not have to uncode our results.

Let us code the data of Example 2 of this section and see how coding simplifies the computations involved.

EXAMPLE   3   By coding the data, find the coefficient of correlation for the data of Example 2 of this section.

SOLUTION   We will code the data by subtracting 38 from each x-value and 29 from each y-value. Our new distribution of test scores then becomes

Language Score	-12	15	10	-7	29	37	-7	-16	-27	-19
Music Score	-7	-6	28	25	-16	-9	-20	21	-12	-5

Now we calculate r for the coded data. We have the following.

x (Language)	y (Music)	x^2	y^2	xy
-12	-7	144	49	84
15	-6	225	36	-90
10	28	100	784	280
-7	25	49	625	-175
29	-16	841	256	-464
37	-9	1369	81	-333
-7	-20	49	400	140
-16	21	256	441	-336
-27	-12	729	144	324
-19	-5	361	25	95
3	-1	4123	2841	-475
$\Sigma x = 3$	$\Sigma y = -1$	$\Sigma x^2 = 4123$	$\Sigma y^2 = 2841$	$\Sigma xy = -475$

Then,

$$r = \frac{n(\Sigma\, xy) - (\Sigma\, x)(\Sigma\, y)}{\sqrt{n(\Sigma\, x^2) - (\Sigma\, x)^2}\sqrt{n(\Sigma\, y^2) - (\Sigma\, y)^2}}$$

$$= \frac{10(-475) - (3)(-1)}{\sqrt{10(4123) - (3)^2}\sqrt{10(2841) - (-1)^2}}$$

$$= \frac{-4750 + 3}{\sqrt{41{,}230 - 9}\sqrt{28{,}410 - 1}}$$

$$= \frac{-4747}{\sqrt{41{,}221}\sqrt{28{,}409}} = \frac{-4747}{(203.03)(168.55)} = -0.1387$$

Notice that the value of r obtained by coding and the value of r obtained by working with the original uncoded data is exactly the same. Coding simplifies computations if the values with which we code are chosen carefully.

EXERCISES FOR SECTION 11.3

1. For each of the following, indicate whether you would expect a positive correlation, a negative correlation, or zero correlation.
   a. The number of female suicides in the United States over the past 20 years and the number of male suicides in the same period.
   b. The height of an individual and the number of children that the individual has.
   c. The number of violent crime scenes on TV and the reported incidence of crime.
   d. The height of an individual and the shoe size of that individual.
   e. The age of an individual and the individual's blood pressure.
   f. A car's engine size and the number of miles that can be driven per gallon of gas.
   g. A company's advertising expenditure and its sales revenue.
   h. A person's income and the number of credit cards that the person has.
2. The U.S. Energy Information Administration reports data on residential energy consumption and expenditures in *Residential Energy Survey: Consumption and Expenditures*. The following information is available for seven randomly selected households whose primary energy source is electricity.

Annual Energy Consumption (kilowatts), x	Annual Expenditure, y
6000	$1984
4975	$1480
4800	$1375
4521	$1291
4129	$1147
4026	$1088
3479	$ 992

a. Draw a scatter diagram for the data, and then compute the coefficient of correlation.

b. Does an increase in energy consumption result in a higher annual expenditure?

3. The U.S. Department of Transportation through its *National Highway Traffic Safety Administration* performs numerous tests on automobiles. The following table gives some data on the number of highway miles obtained per gallon of gas and engine size for several cars equipped with automatic transmission, as determined by an independent testing agency.

Engine Displacement Size x	Miles per Gallon y
240	29
250	26
260	23
305	21
350	18
480	16

a. Draw a scatter diagram for the data, and then compute the coefficient of correlation.

b. Does engine size affect the number of highway miles per gallon obtained?

4. The owner of Rochelle's Department Store is interested in determining whether a significant correlation exists between the amount of money spent on television advertising and total sales. The following data are available:

Television Advertising Expenditure (in thousands of dollars), x	Sales (in thousands of dollars), y
85	820
95	830
98	840
107	860
111	880

Draw a scatter diagram and then compute the coefficient of correlation.

5. Traffic department records indicate that as the posted speed limit has gradually been changed on a particular stretch of the Clearbrook Expressway, so have the number of reported accidents, as shown in the accompanying chart:

Posted Speed Limit (mph), x	Average Number of Reported Accidents (weekly), y
55	28
50	25
45	21

Posted Speed Limit (mph), x	Average Number of Reported Accidents (weekly), y
40	17
30	11
25	6

Draw a scatter diagram for the data, and then compute the coefficient of correlation.

6. Consider the accompanying newspaper article. Based upon past experience the following data are available:

Average Snowfall (inches), x	Amount of Salt Sold (tonnage), y
20	80
24	95
29	99
35	112
40	119
48	126

MORE SNOW ON THE WAY

New York, Feb. 20: This year's record snowfall has been a boon to the distributors of salt used for melting snow and ice. Bob Vicaro, a spokesperson for the ABC Salt Corp., a company that specializes in the manufacture and distribution of ice melters, claims that his company has been unable to keep up with the demand despite a 24-hour round-the-clock work schedule.

Sunday, February 20, 1994

Compute the coefficient correlation for the data.

7. The owner of a video rental store claims that the average number of video tapes rented daily is related to the rental price charged per tape per day. The following data are available:

Price Charged per Video x	Number of Tapes Rented Daily y
$2.00	227
1.75	240
1.50	270
1.25	290
1.00	320
0.75	380

Compute the coefficient of correlation for the data.

8. An auto club official believes that there is a relationship between the outdoor temperature and the average number of calls for assistance in starting a car with a dead battery as shown below:

Outdoor Temperature, x	Average Number of Calls for Assistance, y
30°F	121
25°F	142
20°F	155
15°F	167
10°F	181
5°F	192
0°F	209

Compute the coefficient of correlation for the data.

9. An analysis of the lading bills of the Acme Trucking Corp. for the freight charge for a standard-sized crate reveals the following:

Destination Distance (hundreds of miles), x	Charge (to the nearest dollar), y
8	50
11	62
13	65
16	70
18	75
19	80

(*Source:* Acme Trucking Corp.)

Compute the coefficient of correlation for the data.

10. As a result of a new state bottling law, which requires that all soda bottles sold within the state be recycled, industry officials are uncertain as to how much deposit to charge for each bottle. Past experience in other states indicates that the percentage of bottles returned for recycling depends upon the amount of money required as a deposit as shown below:

Bottle Deposit Charge (cents per bottle), x	Percentage of Bottles Returned, y
2	50
5	55
7	65
10	70
15	85
25	96

Compute the coefficient of correlation for the data.

11. The Health Insurance Association of America reports on the average length of stay and the average daily charge for a semiprivate room in U.S. hospitals in *Survey of Hospital Semiprivate Room Charges*. Researchers have found that the number of days required for postoperative convalescence in the hospital after undergoing a particular operation is directly related to the age of the patient as shown below:

Age of Patient (in years), x	Number of Days, y
25	2
28	3
31	4
35	5
40	6
45	8
53	9
60	11
65	12

Compute the coefficient of correlation for the data.

12. Refer back to Figure 11.6 on page 594. The following is a list of the number of years of schooling beyond high school and the salaries of eight randomly selected people in Belleville.

Number of Years of Schooling Beyond High School, x	Salary (in thousands of dollars), y
2	25
3	27
5	30
1	24
0	22
7	38
6	33
4	29

   a. Draw a scatter diagram for the data and compute the coefficient of correlation.
   b. Is salary related to the number of years of schooling?

11.4   THE RELIABILITY OF r

Although the coefficient of correlation is usually the first number that is calculated when we are given several sets of scores, great care must be used in interpreting the results. It can undoubtedly be said that among all the statistical measures discussed in this book the correlation coefficient is the one that is most misused. One reason for this misuse is the assumption that because the two variables are related, a change in one will result in a change in the other.

Many people have applied a positive correlation coefficient to prove a cause-and-effect relationship that may not even exist. To illustrate the point, it has been shown that there is a high positive correlation between teacher's salaries and the use of drugs on campus. Does this mean that reducing the teachers' salaries would reduce the use of drugs on campus or does it simply mean that the students at wealthier schools (which pay higher salaries) are more apt to use drugs?

Frequently, two variables may appear to have a high positive correlation even though they are not directly associated with each other. There may be a third variable that is highly correlated to these two variables.

There is another important consideration that is often overlooked. When r is computed on the basis of sample data, we may get a strong correlation, positive or negative, which is due purely to chance, not to some relationship that exists between x and y. For example, if x represents the amount of snowfall and y represents the number of hours that Joe studied on five consecutive days, we may have the following results:

Amount of Snow (in inches), x	1	4	2	6	3
Number of Hours Studied, y	2	6	3	4	4

The value of r in this case is 0.63. Can we conclude that if it snows more, then Joe studies more?

Fortunately, a chart has been constructed that allows us to interpret the value of the correlation coefficient correctly. This is Table V in the Appendix. This chart allows us to determine the significance of a particular value of r. We use this table as follows:

1. First compute the value of r using Formula 11.1.
2. Then look in the chart for the appropriate r-value corresponding to some given n, where n is the number of pairs of scores.
3. The value of r is *not* statistically significant if it is between $-r_{0.025}$ and $r_{0.025}$ for a particular value of n.

Level of significance

The subscript, that is, the little numbers, attached to r is called the **level of significance**. If we use this chart and the $r_{0.025}$ values of the chart as our guideline, we will be correct in saying that when there is no significant statistical correlation between x and y we will not reject H_0 95% of the time.

Table V also gives us the values of $r_{0.005}$. We use these chart values when we want to be correct 99% of the time. In this book we use the $r_{0.025}$ values only. Table V is constructed so that r can be expected to fall between $-r_{0.025}$ and $+r_{0.025}$ approximately 95% of the time and between $-r_{0.005}$ and $+r_{0.005}$ approximately 99% of the time when the true correlation between x and y is zero.

Returning to our example, we have $n = 5$ and $r = 0.63$. From Table V we have $r_{0.025} = 0.878$. Thus, r will *not* be statistically significant if it is between -0.878 and $+0.878$. Since $r = 0.63$ is between -0.878 and $+0.878$, we conclude that the correlation may be due purely to chance. We cannot say that if it snows more, then Joe will necessarily study more.

Similarly, in Example 1 of Section 11.3 (see page 596) we found that $r = 0.9867$. There were ten scores, so $n = 10$. The chart values tell us that if r is between -0.632 and $+0.632$, there is *no* significant statistical correlation. Since the value of r that we obtained is greater than $+0.632$, we conclude that there is a *definite* positive correlation between x and y. Thus, we are justified in claiming that there is a relationship between the math aptitude scores and the business aptitude scores.

EXERCISES FOR SECTION 11.4

1. Refer back to Exercise 2, page 599. Determine if r is significant.
2. Refer back to Exercise 3, page 600. Determine if r is significant.
3. Refer back to Exercise 4, page 600. Determine if r is significant.
4. Refer back to Exercise 5, page 600. Determine if r is significant.
5. Refer back to Exercise 6, page 601. Determine if r is significant.
6. Refer back to Exercise 7, page 602. Determine if r is significant.
7. Refer back to Exercise 8, page 602. Determine if r is significant.
8. Refer back to Exercise 9, page 602. Determine if r is significant.
9. Refer back to Exercise 10, page 603. Determine if r is significant.
10. Refer back to Exercise 11, page 603. Determine if r is significant.
11. Refer back to Exercise 12, page 603. Determine if r is significant.

11.5   LINEAR REGRESSION

Let us return to the example discussed in Section 11.2. In that example a guidance counselor was interested in determining whether a relationship existed between the different aptitudes tested. Once a relationship, in the form of an equation, can be found between two variables the counselor can use this relationship to *predict* the value of one of the variables if the value of the other variable is known. Thus, the counselor may wish to predict how well a student will do on the business portion of the test if she knows the student's score on the math part.

Also, the counselor might be analyzing whether any correlation exists between high school averages and college grade-point averages. Her intention would be to try to find some relationship that will *predict* a college student's academic success from a knowledge of the high school average alone.

COMMENT   It should be noted that the correlation coefficient merely determines whether two variables are related, but it does not specify how. Thus, the correlation coefficient cannot be used to solve prediction problems.

Estimated regression line

When given a prediction problem, we first locate all the dots on a scatter diagram as we did in Section 11.2. Then we try to fit a straight line to the data in such a way that it best represents the relationship between the two variables. Such a fitted line is called an **estimated regression line**. Once we have such a line we try to find an equation that will determine this line. We can then use this equation to predict the value of y corresponding to a given value of x.

Regression methods
Linear regression

Fitting a line to a set of numbers is by no means an easy task. Nevertheless, methods have been designed to handle such prediction problems. These methods are known as **regression methods**. In this book we discuss **linear regression** only. This means that we will try to fit a straight line to a set of numbers.

Curvilinear regression

COMMENT   Occasionally, the dots are so scattered that a straight line cannot be fitted to the set of numbers. The statistician may then try to fit a **curve** to the set of numbers. This is shown in Figure 11.9. We would then have **curvilinear regression**.

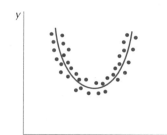

x   **FIGURE 11.9**   Curvilinear regression.

The following examples illustrate how a knowledge of the regression line enables us to predict the value of y for any given value of x. It is standard notation to call the variable to be predicted the **dependent variable** and to denote it by y. The known variable is called the **independent variable** and is denoted by x.

Dependent variable
Independent variable

E X A M P L E 1 A guidance counselor notices that there is a strong positive correlation between the math scores and business scores. Based on a random sample of many students, she draws the scatter diagram shown in Figure 11.10. To this scatter diagram we have drawn a straight line that best represents the relationship between the two variables. This line enables us to predict the value of y for any given value of x. For example, if a student scored 35 on the math portion of the exam, then this line predicts that the student will score about 59 on the business part of the exam. This may be seen by first finding 35 on the horizontal axis, x, then moving straight up until we hit the estimated regression line. Finally, we move directly to the left to see where we cross the vertical axis, y. This is indicated by the dotted line of Figure 11.11. Similarly, we can predict that a student whose math score is 27 will score 51 on the business portion of the exam.

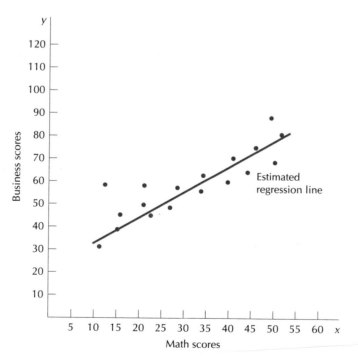

FIGURE 11.10

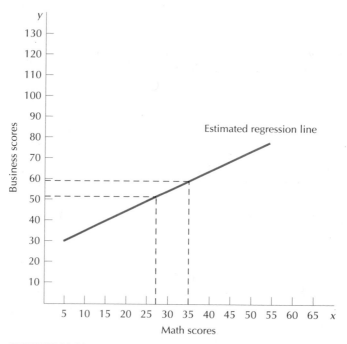

FIGURE 11.11

E X A M P L E   2   A certain organization claims to be able to predict a person's height if given the person's weight. It has collected the following data for ten people.

Weight (in pounds), x	140	135	146	160	142	157	138	164	159	150
Height (in inches), y	63	61	68	72	66	65	64	73	70	69

If a person weighs 155 pounds, what is his predicted height?

SOLUTION   We first draw the scatter diagram as shown in Figure 11.12. Then we draw a straight line that best represents the relationship between the two variables. This line now enables us to predict a person's height if we know the person's weight. The estimated regression line predicts that a person who weighs 155 pounds will be about 70 inches tall.

How does one draw an estimated regression line? Since there is no set procedure, different people are likely to draw different regression lines. So, although we may speak

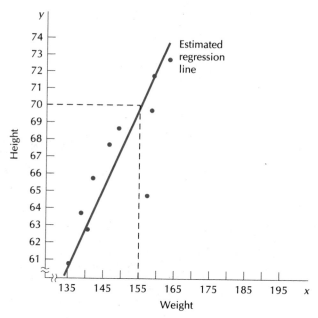

FIGURE 11.12

of finding a straight line that best fits the data, how is one to know when the best fit has been achieved? There are, in fact, several reasonable ways in which best fit can be defined. For this reason statisticians use a mathematical method for determining an equation that best describes the linear relationship between two variables. The method

Least-squares method

is known as the **least-squares method** and is discussed in detail in the next section.

11.6 THE METHOD OF LEAST SQUARES

Deviation

Least-squares method

Whenever we draw an estimated regression line, not all points will lie on the regression line. Some will be above it and some will be below. The difference between any point and the corresponding point on the regression line is called the (vertical) **deviation** from the line. It represents the difference in value between what we predicted and what actually happened. See Figure 11.13. The **least-squares method** determines the estimated regression line in such a way that the sum of the squares of these vertical deviations is as small as possible. Although a background in calculus is needed to understand how we obtain the formula for the least-squares regression line, it is very easy to use the formula.

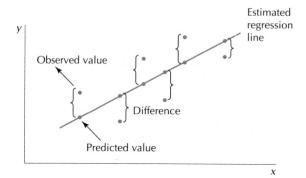

FIGURE 11.13   Difference between y-values and the estimated regression line.

FORMULA 11.2

Regression Equation
Estimated regression line

The equation of the **estimated regression line** is

$$\hat{y} = b_0 + b_1 x$$

where

$$b_1 = \frac{n(\Sigma\, xy) - (\Sigma\, x)(\Sigma\, y)}{n(\Sigma\, x^2) - (\Sigma\, x)^2} \qquad b_0 = \frac{1}{n}(\Sigma\, y - b_1 \cdot \Sigma\, x)$$

and n is the number of pairs of scores.

Regression line
Regression equation

COMMENT   The straight line that best fits a set of data points according to the least-squares criterion is called the **regression line**, whereas the equation of the regression line is called the **regression equation**.

Let us use Formula 11.2 to find the regression equation connecting two variables.

EXAMPLE 1   Fifteen students were asked to indicate how many hours they studied before taking their statistics examination. Their responses were then matched with their grades on the exam, which had a maximum score of 100.

Hours x	0.50	0.75	1.00	1.25	1.50	1.75	2.00	2.25	2.50	2.75	3.00	3.25	3.50	3.75	4.00
Scores y	57	64	59	68	74	76	79	83	85	86	88	89	90	94	96

a. Find the regression equation that will predict a student's score if we know how many hours the student studied.

b. If a student studied 0.85 hour, what is the student's predicted grade?

SOLUTION

a. To enable us to perform the computations, we arrange the data in the form of a chart.

x	y	x^2	xy
0.50	57	0.2500	28.5
0.75	64	0.5625	48
1.00	59	1.0000	59
1.25	68	1.5625	85
1.50	74	2.2500	111
1.75	76	3.0625	133
2.00	79	4.0000	158
2.25	83	5.0625	186.75
2.50	85	6.2500	212.5
2.75	86	7.5625	236.5
3.00	88	9.0000	264
3.25	89	10.5625	289.25
3.50	90	12.2500	315
3.75	94	14.0625	352.5
4.00	96	16.0000	384
33.75	1188	93.4375	2863

$$\Sigma x = 33.75 \qquad \Sigma y = 1188 \qquad \Sigma x^2 = 93.4375 \qquad \Sigma xy = 2863$$

From the chart we have $n = 15$ so that

$$b_1 = \frac{n(\Sigma \, xy) - (\Sigma \, x)(\Sigma \, y)}{n(\Sigma \, x^2) - (\Sigma \, x)^2}$$

$$= \frac{15(2863) - (33.75)(1188)}{15(93.4375) - (33.75)^2}$$

$$= \frac{42{,}945 - 40{,}095}{1401.5625 - 1139.0625}$$

$$= \frac{2850}{262.5}$$

$$= 10.857$$

and

$$b_0 = \frac{1}{n} (\Sigma\, y - b_1 \cdot \Sigma\, x)$$

$$= \frac{1}{15} (1188 - (10.857) \cdot (33.75))$$

$$= \frac{1}{15} (1188 - 366.424)$$

$$= \frac{1}{15} (821.576)$$

$$= 54.772$$

Thus,°

$$\hat{y} = b_0 + b_1 x$$
$$= 54.772 + 10.857x$$

The equation of the predicted regression line then is

$$\hat{y} = 54.772 + 10.857x$$

b. For $x = 0.85$, we get

$$\hat{y} = 54.772 + 10.857(0.85)$$
$$= 54.772 + 9.22845$$
$$= 64.00045$$

Thus, the predicted grade of a student who studies 0.85 hours is approximately 64.

E X A M P L E   2   A West Coast publishing company keeps accurate records of its monthly expenditure for advertising and its total monthly sales. For the first 10 months of 1995, the records showed the following.

Advertising (in thousands), x	43	44	36	38	47	40	41	54	37	46
Sales (in millions), y	74	76	60	68	79	70	71	94	65	78

(Note that units are in dollars.)

a. Find the least-squares prediction equation appropriate for the data.
b. If the company plans to spend $50,000 for advertising next month, what is its predicted sales? Assume that all other factors can be neglected.

°Values for the variables throughout this chapter are calculated using computer accuracy, even though answers are often rounded to two or three decimal places.

SOLUTION

a. We arrange the data in the form of a chart.

x	y	x^2	xy
43	74	1849	3182
44	76	1936	3344
36	60	1296	2160
38	68	1444	2584
47	79	2209	3713
40	70	1600	2800
41	71	1681	2911
54	94	2916	5076
37	65	1369	2405
46	78	2116	3588
426	735	18,416	31,763

$$\Sigma x = 426 \qquad \Sigma y = 735 \qquad \Sigma x^2 = 18,416 \qquad \Sigma xy = 31,763$$

From the chart we have $n = 10$ so that

$$b_1 = \frac{n(\Sigma\,xy) - (\Sigma\,x)(\Sigma\,y)}{n(\Sigma\,x^2) - (\Sigma\,x)^2}$$

$$= \frac{10(31,763) - (426)(735)}{10(18,416) - (426)^2}$$

$$= \frac{317,630 - 313,110}{184,160 - 181,476}$$

$$= \frac{4520}{2684} = 1.684$$

and

$$b_0 = \frac{1}{n}\,(\Sigma\,y - b_1 \cdot \Sigma\,x)$$

$$= \frac{1}{10}\,(735 - (1.684) \cdot (426))$$

$$= \frac{1}{10}\,(735 - 717.384)$$

$$= \frac{1}{10}\,(17.616)$$

$$= 1.762$$

Thus,

$$\hat{y} = b_0 + b_1 x$$
$$= 1.762 + 1.684x$$

The equation of the predicted regression line is

$$\hat{y} = 1.762 + 1.684x$$

b. For $x = 50$, not 50,000, since x is in thousands of dollars, we get

$$\hat{y} = 1.762 + 1.684(50)$$
$$= 1.762 + 84.2$$
$$= 85.962$$

Thus, if the company spends \$50,000 next month for advertising, its predicted sales are \$85.962 million assuming all other factors can be neglected.

There is an alternate way to compute the equation of the regression line. This involves the sample covariance. We first compute the average of the x-values, denoted as $\bar{x}$, and the average of the y-values denoted as $\bar{y}$. Then we compute the sample standard deviation of the x-values. This is denoted as s_x. Finally, we determine the **sample covariance** of the n data points, which is defined by

Sample covariance

$$\text{Sample covariance} = s_{xy} = \frac{\Sigma(x - \bar{x})(y - \bar{y})}{n - 1}$$

We then have the following alternate formula.

FORMULA 11.3

Equation of Regression Line (Alternate Version)

The equation of the regression line for a set of n data points is given by

$$\hat{y} = b_0 + b_1 x$$

where

$$b_1 = \frac{s_{xy}}{s_x^2}, \qquad b_0 = \bar{y} - b_1\bar{x}$$

and s_x is the sample standard deviation of the x-values

COMMENT In a later exercise the reader will be asked to demonstrate that Formula 11.3 is indeed equivalent to Formula 11.2, which was used earlier to calculate the equation of the regression line.

We will illustrate the use of Formula 11.3 for the data presented in Example 2 of this section.

E X A M P L E   3   Find the equation of the regression line for the data of Example 2.

SOLUTION   We arrange the data in the form of a chart.

x	y	$x - \bar{x}$	$y - \bar{y}$	$(x - \bar{x})(y - \bar{y})$	$(x - \bar{x})^2$
43	74	0.4	0.5	0.2	0.16
44	76	1.4	2.5	3.5	1.96
36	60	−6.6	−13.5	89.1	43.56
38	68	−4.6	−5.5	25.3	21.16
47	79	4.4	5.5	24.2	19.36
40	70	−2.6	−3.5	9.1	6.76
41	71	−1.6	−2.5	4.0	2.56
54	94	11.4	20.5	233.7	129.96
37	65	−5.6	−8.5	47.6	31.36
46	78	3.4	4.5	15.3	11.56
426	735	0	0	452	268.4

$\Sigma x = 426$ $\Sigma y = 735$ $\Sigma(x - \bar{x})(y - \bar{y}) = 452$ $\Sigma(x - \bar{x})^2 = 268.4$

Thus,

$$\bar{x} = \frac{\Sigma x}{n} = \frac{426}{10} = 42.6$$

$$\bar{y} = \frac{\Sigma y}{n} = \frac{735}{10} = 73.5$$

$$s_x^2 = \frac{\Sigma(x - \bar{x})^2}{n - 1} = \frac{268.4}{9} = 29.8222$$

$$s_{xy} = \frac{\Sigma(x - \bar{x})(y - \bar{y})}{n - 1} = \frac{452}{9} = 50.2222$$

so that

$$b_1 = \frac{s_{xy}}{s_x^2} = \frac{50.2222}{29.8222} = 1.684$$

and

$$b_0 = \bar{y} - b_1\bar{x} = 73.5 - (1.684)(42.6) = 1.762$$

The equation of the regression line is

$$\hat{y} = b_0 + b_1 x$$
$$= 1.762 + 1.684x$$

This is indeed the same answer that we obtained earlier.

EXERCISES FOR SECTION 11.6

In each of the following exercises, assume that the correlation is high enough to allow for reasonable predictions.

1. As we indicated at the beginning of this chapter, Galton believed that a very close relationship exists between the heights of fathers and the heights of their sons. To test this claim, a scientist selects ten men at random and records their heights and the heights of their sons.

Father's Height (in inches), x	64	65	68	73	72	66	71	75	70	69
Son's Height (in inches), y	66	67	69	74	73	68	72	75	71	69

   a. Determine the least-squares prediction equation.
   b. If a father is 74 inches tall, what is the predicted height of the son?

2. In *Vital and Health Statistics*, the National Center for Health Statistics publishes data on the heights and weights of individuals. In one study of ten male college students aged 18 to 24 years, the following information was obtained:

Height (in inches), x	65	67	62	66	74	71	70	69	73	72
Weight (in pounds), y	143	157	137	153	192	180	175	168	187	182

   a. Determine the least-squares prediction equation.
   b. If a male college student (aged 18–24 years) is randomly selected and if he is 68 inches tall, what is his predicted weight?

3. After analyzing the salaries and batting averages of many baseball players in both leagues, a sports commentator concluded that higher-paid players will have higher batting averages. To determine if there is a relationship, ten players are randomly selected and the following statistics are obtained:

Annual Salary, x	Batting Average (in prior year), y
$ 190,000	0.290
162,000	0.281
298,000	0.299
325,000	0.302
700,000	0.309
1,225,000	0.335
375,000	0.304
600,000	0.306
290,000	0.295
375,000	0.320

    a. Determine the least-squares prediction equation.
    b. Assuming that the sports commentator's belief is accurate, what is the predicted batting average of a baseball player who is earning $875,000?
    (Hint: Coding the salaries may be very helpful.)

4. City officials believe that the number of complaints to the city's heat complaint control board (for lack of heat) is directly related to the outdoor temperature. The following data have been collected:

Outdoor Temperature, x	Number of Complaints to Heat Control Board (24-hour period), y
30°F	58
25°F	67
20°F	84
15°F	90
10°F	102
5°F	113
0°F	127

    a. Determine the least-squares prediction equation.
    b. What is the predicted number of complaints to the city's heat complaint control board when the outside temperature drops to 8°F?

5. A large trucking company that delivers fresh fruit wishes that its truck drivers be forced to work overtime. The union claims that the more hours that a truck driver works, the greater the risk of an accident (due to fatigue). To support its claim, the union has gathered the following statistics on the average number of hours worked by a truck driver (per week) and the average number of accidents (per week).

Number of Hours Worked x	Number of Accidents y
35	1.6
37	2.2
39	3.8
42	4.3
44	5.6
46	6.1
50	7.3

    a. Determine the least-squares prediction equation.
    b. What is the predicted number of accidents when a truck driver is forced to work 48 hours a week?

6. A chain of health-food stores is determining the relationship between the number of times its commercial is broadcast on radio or television weekly and the weekly sales volume. It randomly selects nine weeks and determines the number of times that the commercial was broadcast and the weekly volume of sales as shown below:

Number of Times Commercial Is Broadcast x	Weekly Sales Volume (in thousands) y
3	42
4	47
5	52
7	72
8	85
9	100
10	115
12	185
20	225

a. Determine the least-squares prediction equation.
b. What is the predicted sales volume when the commercial is broadcast 15 times weekly on radio or television?

7. Franklyn Pierce, owner of the Meadowlands Amusement Park, believes that the average daily attendance at the amusement park is related to the admission price charged, as shown below:

Admission Price (per adult), x	Average Daily Attendance, y
$ 8.00	200,000
9.00	175,000
10.00	160,000
12.00	145,000
14.00	130,000
16.00	120,000

a. Determine the least-squares prediction equation.
b. What is the predicted average daily attendance at the amusement park when the admission price is $13.00?
(Hint: Coding the data may be very helpful.)

8. A medical researcher is analyzing the relationship between the amount of alcohol in a person's bloodstream and the number of times that the person can safely perform a certain task. The following data are available:

Amount of Alcohol in Bloodstream (units), x	Number of Times Person Can Safely Perform Task, y
6	36
9	32
10	30
14	24
18	20
23	10
27	4

   a. Determine the least-squares prediction equation.

   b. If a person drinks 20 units of alcohol, what is the predicted number of times that the person can safely perform the task?

9. The price charged by OPEC (Organization of Petroleum Exporting Countries) for a barrel of crude oil and the cost of a gallon of gasoline both rose dramatically during the 1970s to the mid 1980s as shown below:

Year	Average Price per Barrel of Crude Oil, x	Average Cost per Gallon of Regular Leaded Gasoline, y
1973	$ 3.89	$0.39
1974	7.67	0.57
1975	8.19	0.59
1976	8.57	0.62
1977	9.00	0.63
1978	12.59	0.86
1979	21.59	1.19
1980	31.77	1.33
1981	28.52	1.22
1982	26.19	1.16
1983	25.88	1.13
1984	42.09	1.12
1985	12.51	0.86
1986	15.41	0.90

Source: Statistical Abstract of the United States 1988, pp. 476–480.

   a. Determine the least-squares prediction equation.

   b. If crude oil sells for $16.50 a barrel, what is the predicted price per gallon of gasoline?

10. Can physical exercise affect a person's longevity? A study of 16,936 Harvard alumni suggests that men who expend at least 2000 kcal per week in exercise can add years to their lives as shown below:

Age at the Start of Exercises, x	Estimated Additional Life (in years), y
37	2.51
42	2.34
47	2.10
52	2.11
62	1.75
67	1.35
72	0.72
77	0.42

Source: Adapted from R. S. Paffenbarger, Jr., R. T. Hyde, and A. L. Wing, *New England Journal of Medicine,* 1986, Volume 314, p. 605–13. Copyright 1986. Massachusetts Medical Society. All rights reserved.

    a. Determine the least-squares prediction equation.
    b. If the age at the start of exercises is 50, what is the predicted estimate of added years of life?

11. Are the prices of gold and silver related? According to *Minerals Yearbook* published by the U.S. Bureau of Mines, the average prices for one ounce of gold and one ounce of silver over the years was as follows:

Year	Price per Ounce of Gold, x	Price per Ounce of Silver, y
1980	$613	$20.63
1981	460	10.52
1982	376	7.95
1983	424	11.44
1984	361	8.14
1985	318	6.14
1986	368	5.47
1987	448	7.01

    a. Determine the least-squares prediction equation.
    b. If gold sells for $512 an ounce, what is the predicted price of an ounce of silver?

11.7 STANDARD ERROR OF THE ESTIMATE

In Section 11.6 we discussed the least-squares regression line, which predicts a value of y when x has a particular value. Quite often it turns out that the predicted value of y and the observed value of y are different. If the correlation is low, these differences will be large. Only when the correlation is high can we expect the predicted values to be close to the observed values.

In general, the true population regression line is not known, so we use the data to estimate the equation of this true population regression line. For example, consider the data of Example 2 given in Section 11.6 (page 612), which is reproduced here.

x	y
43	74
44	76
36	60
38	68
47	79
40	70
41	71
54	94
37	65
46	78

The equation of the regression line is $\hat{y} = 1.762 + 1.684x$. When $x = 50$ the predicted value of y is 85.962. We cannot expect such predictions to be completely accurate. At different times, when $x = 50$, we may get different y-values. Thus, for each x there is a corresponding population of y-values. The mean of the corresponding y-values lies on some straight line whose equation we do not know but which is of the form $y = \alpha + \beta x$. For each x-value, the distribution of the corresponding population of y-values is normally distributed. Moreover, for each x-value the mean of the corresponding population of *Population* y-values lies on a straight line called the **population regression line**, whose equation *regression line* is of the form $y = \alpha + \beta x$. The population standard deviation, σ, of the population of y-values corresponding to a given x-value is the *same*, regardless of the x-value.

Thus, we assume that the kind of normal distribution occurring when $x = 50$ will also appear at any other value of x. This means that for any x-value the distribution of the population of y-values is a normal distribution and that the variance of that normal distribution is the same for every x. This can be seen in Figure 11.14.

Error terms   We can then conclude that the **error terms** (the vertical distance between the predicted y-values and the true population values) are normally distributed with mean 0 and the same standard deviation σ. How do we estimate σ, the common standard deviation of the normal distributions given in Figure 11.14? Statisticians have devised a *Standard error* method for measuring σ. This is the **standard error of the estimate**. However, before *of the estimate* doing this, let us analyze our least-squares prediction equation, $\hat{y} = 1.762 + 1.684x$. Why bother computing this equation? Why not use the given data to make predictions about the sales by simply ignoring the values of x (the amount of money spent for advertising) and using only the mean value of the sampled y's (the average sales) in making predictions? Thus, we can use

$$\bar{y} = \frac{\Sigma y}{n} = \frac{735}{10} = 73.5$$

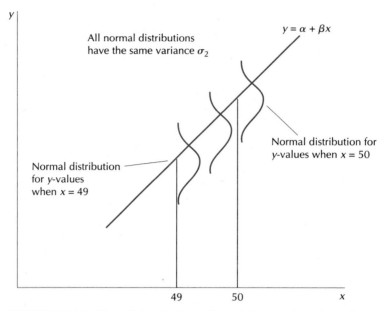

FIGURE 11.14 Normal distributions of population y-values about the regression line.

as our predicted sales (in all cases). How large an error have we made? In the following computations we indicate the (squared) error that is made when we predict a value of $\bar{y} = 73.5$ for the observed y-values.

y	$y - \bar{y}$	$(y - \bar{y})^2$
74	$74 - 73.5 = 0.5$	0.25
76	$76 - 73.5 = 2.5$	6.25
60	$60 - 73.5 = -13.5$	182.25
68	$68 - 73.5 = -5.5$	30.25
79	$79 - 73.5 = 5.5$	30.25
70	$70 - 73.5 = -3.5$	12.25
71	$71 - 73.5 = -2.5$	6.25
94	$94 - 73.5 = 20.5$	420.25
65	$65 - 73.5 = -8.5$	72.25
78	$78 - 73.5 = 4.5$	20.25
		780.50

$$\Sigma(y - \bar{y})^2 = 780.50$$

Thus, the total squared error that is made when we predict $\bar{y} = 73.5$ for the observed y-values is

$$\Sigma(y - \bar{y})^2 = 780.50$$

Total sum of squares, SST   This is referred to as the **total sum of squares, SST**. Therefore,

$$SST = \Sigma(y - \bar{y})^2 = 780.50$$

Instead of using $\bar{y}$ as our predicted value (in all cases), we can use the equation of the regression line values, $\hat{y}$, for our sales prediction. If we believe that our regression equation can be used to predict sales, then the squared error should be less when we use these values. The actual error obtained when we use these values is as follows:

x	y	$\hat{y}$	$y - \hat{y}$	$(y - \hat{y})^2$
43	74	74.17	−0.17	0.03
44	76	75.85	0.15	0.02
36	60	62.41	−2.41	5.81
38	68	65.77	2.23	4.97
47	79	80.89	−1.89	3.57
40	70	69.13	0.87	0.76
41	71	70.81	0.19	0.04
54	94	92.65	1.35	1.82
37	65	64.09	0.91	0.83
46	78	79.21	−1.21	1.46
				19.31

$$\Sigma(y - \hat{y})^2 = 19.31$$

In this case the total squared error when using the equation of the regression line values $\hat{y}$ for predictions is

$$\Sigma(y - \hat{y})^2 = 19.31$$

Error sum of squares, SSE   This is called the **error sum of squares, SSE**. Thus,

$$SSE = \Sigma(y - \hat{y})^2 = 19.31$$

Percentage reduction   It should be obvious that using the regression equation for prediction reduced the total squared error considerably. The **percentage reduction** is

$$\frac{SST - SSE}{SST} = 1 - \frac{SSE}{SST} = 1 - \frac{19.31}{780.50} = 0.9753$$

Coefficient of determination   or 97.53%. The percentage reduction in the total squared error obtained by using the regression equation instead of $\bar{y}$ is denoted by r^2 and is called the **coefficient of determination**. Thus, we have Formula 11.4.

FORMULA 11.4

$$\text{Coefficient of determination} = r^2 = 1 - \frac{SSE}{SST}$$

COMMENT   The coefficient of determination, namely, r^2, can be interpreted as representing the percentage of variation in the observed y-values that is explainable by the regression line.

As mentioned earlier, the population of y-values corresponding to the various x-values all have the same (usually unknown) standard deviation, σ. The value of σ can be estimated from the sample data by computing the **standard error of the estimate** also called the **residual standard deviation**.

Residual standard deviation

To determine the standard error of the estimate, we first calculate the predicted value of y for each x and then compute the difference between the observed value and the predicted value. We then square these differences and divide the sum of these squares by $n - 2$. The square root of the result is called the standard error of the estimate.

FORMULA 11.5

Standard error of the estimate

The **standard error of the estimate** is denoted by s_e and is defined as

$$s_e = \sqrt{\frac{\Sigma(y - \hat{y})^2}{n - 2}} = \sqrt{\frac{SSE}{n - 2}}$$

where $\hat{y}$ is the predicted value, y is the observed value, and n is the number of pairs of scores.

EXAMPLE   1   Find the standard error of the estimate for the least-squares regression equation of Example 1 of Section 11.6 on page 610.

SOLUTION   The least-squares regression equation was

$$\hat{y} = 54.772 + 10.857x$$

Using this equation, we find the predicted value of y corresponding to each value of x. We arrange our computations in the form of a chart.

x	y	$\hat{y}$	$y - \hat{y}$	$(y - \hat{y})^2$
0.50	57	60.2	−3.2	10.24
0.75	64	62.91	1.09	1.19
1.00	59	65.63	−6.63	43.96

x	y	$\hat{y}$	$y - \hat{y}$	$(y - \hat{y})^2$
1.25	68	68.34	−0.34	0.12
1.50	74	71.06	2.94	8.64
1.75	76	73.77	2.23	4.97
2.00	79	76.49	2.51	6.30
2.25	83	79.2	3.8	14.44
2.50	85	81.92	3.08	9.49
2.75	86	84.63	1.37	1.88
3.00	88	87.35	0.65	0.42
3.25	89	90.06	−1.06	1.12
3.50	90	92.78	−2.78	7.73
3.75	94	95.49	−1.49	2.22
4.00	96	98.21	−2.21	4.88
				117.60

$$\Sigma(y - \hat{y})^2 = 117.60$$

Applying Formula 11.5, we get

$$s_e = \sqrt{\frac{\Sigma(y - \hat{y})^2}{n - 2}} = \sqrt{\frac{117.60}{15 - 2}} = \sqrt{\frac{117.60}{13}} \approx \sqrt{9.05} \approx 3.01$$

Thus, the standard error of the estimate is approximately 3.01.

EXAMPLE   2   Find the standard error of the estimate for the least-squares regression equation of Example 2 of Section 11.6 (page 612).

SOLUTION   The least-squares regression equation was

$$\hat{y} = 1.762 + 1.684x$$

Using this equation, we find the predicted value of y corresponding to each value of x. We arrange our computations in the form of a chart.

x	y	$\hat{y}$	$y - \hat{y}$	$(y - \hat{y})^2$
43	74	74.17	−0.17	0.03
44	76	75.85	0.15	0.02
36	60	62.41	−2.41	5.81
38	68	65.77	2.23	4.97
47	79	80.89	−1.89	3.57
40	70	69.13	0.87	0.76
41	71	70.81	0.19	0.04
54	94	92.65	1.35	1.82
37	65	64.09	0.91	0.83
46	78	79.21	−1.21	1.46
				19.31

$$\Sigma(y - \hat{y})^2 = 19.31$$

Applying Formula 11.5, we get

$$s_e = \sqrt{\frac{\Sigma(y - \hat{y})^2}{n - 2}}$$

$$= \sqrt{\frac{19.31}{8}} \approx \sqrt{2.41} \approx 1.55$$

Thus, the standard error of the estimate is approximately 1.55.

The goodness of fit of the least-squares regression line is determined by the value of the standard error of the estimate. A relatively small value of s_e indicates that the predicted and observed values of y are fairly close. This means that the regression equation is a good description of the relationship between the two variables. On the other hand, a relatively large value of s_e indicates that there is a large difference between the predicted and observed values of y. When this happens the relationship between x and y as given by the least-squares equation is not a good indication of the relationship between the two variables. Only when the standard error of the estimate is zero can we say for sure that the least-squares regression equation is a perfect description of the relationship between x and y.

COMMENT   In computing s_e statisticians often use the following equivalent formula.

$$s_e = \sqrt{\frac{\Sigma y^2 - b_0(\Sigma y) - b_1(\Sigma xy)}{n - 2}}$$

where the values of b_0 and b_1 are the same as those obtained earlier in computing the equation of the regression line. The values for the various summations should have already been obtained, thereby reducing the amount of computation needed.

EXERCISES FOR SECTION 11.7

For each of the following, refer back to the exercise indicated and calculate the standard error of the estimate.

1. Exercise 1, page 616.
2. Exercise 2, page 616.
3. Exercise 3, page 616.
4. Exercise 4, page 617.
5. Exercise 5, page 617.
6. Exercise 6, page 618.
7. Exercise 7, page 618.
8. Exercise 8, page 618.
9. Exercise 9, page 619.
10. Exercise 10, page 619.
11. Exercise 11, page 620

*11.8   USING STATISTICAL INFERENCE FOR REGRESSION

We can use the standard statistical inferential procedures discussed in earlier chapters for regression. Specifically, we mentioned earlier that for each x-value, the corresponding population of y-values is normally distributed with mean $y = \beta_0 + \beta_1 x$ and standard deviation σ. However, if β_1 has a value of 0, then x will be totally worthless in predicting y-values since in that case the regression equation would be

$$\hat{y} = \beta_0 + \beta_1 x$$
$$= \beta_0 + 0x$$
$$= \beta_0$$

Thus, the value of x would have absolutely nothing to do with the distribution of y-values. Hence, it is important for us to determine in advance whether x can be used as a predictor of y, that is, if x and y are linearly related. We can decide this by performing the following hypothesis test.

$$H_0: \quad \beta_1 = 0$$

$$H_1: \quad \beta_1 \neq 0$$

If we conclude that the null hypothesis has to be rejected, then this indicates that x and y are linearly related and that we can proceed to use the equation of the regression line for making predictions.

How do we test the null hypothesis that $\beta_1 = 0$? This can be done by using the value of b_1, which actually represents the slope of the sample regression line. We proceed as follows:

STATISTICAL INFERENCE CONCERNING β_1

To test the null hypothesis H_0 that the slope β_1 of the population regression line is zero or not (that is, whether x can be used as a predictor of y or not) do the following:

1. State the null hypothesis and the alternative hypothesis as well as the significance level α.
2. Find $n - 2$. This gives us the number of degrees of freedom for a t-distribution.
3. Find the appropriate critical values $\pm t_{\alpha/2}$ by using Table VII in the Appendix.
4. Compute the value of the test statistic.

$$t = \frac{b_1}{s_e \Big/ \sqrt{\sum x^2 - \dfrac{(\sum x)^2}{n}}}$$

5. If the value of the test statistic falls in the rejection region, reject H_0. Otherwise do not reject H_0.
6. State the conclusion.

Let us illustrate the above procedure with a few examples.

E X A M P L E   1   Refer back to the data of Example 1 on page 610. Does the data indicate that the value of β_1, that is, the slope of the population regression line, is not zero, which would mean that x (the number of hours studied) can be used as a predictor of y (the score received)? Assume that the level of significance is 5%.

SOLUTION   In this case the null hypothesis is $\beta_1 = 0$, and the alternative hypothesis is $\beta_1 \neq 0$. Since $n = 15$, we will have a t-distribution with $15 - 2$ or 13 degrees of freedom. From Table VII in the Appendix, the appropriate $t_{\alpha/2} = t_{0.05/2} = t_{0.025}$ value is 2.160. Now we compute the value of the test statistic. We have

$$t = \frac{b_1}{s_e \Big/ \sqrt{\sum x^2 - \dfrac{(\sum x)^2}{n}}}$$

All the values of the variables needed to use this formula have been found earlier. Thus, we know that $b_1 = 10.857$ (computed on page 611), $\sum x^2 = 93.4375$ (computed on page 611), $\sum x = 33.75$ (computed on page 611), and $s_e = 3.01$ (computed on page 625). Also, $n = 15$ so that

$$t = \frac{10.857}{3.01 \Big/ \sqrt{93.4375 - \dfrac{(33.75)^2}{15}}} = 15.089$$

Since the value of the test statistic 15.089 falls in the rejection region of Figure 11.15, we reject H_0 and conclude that the slope of the population regression line is not 0. Hence, x (number of hours studied) can be used as a predictor of y.

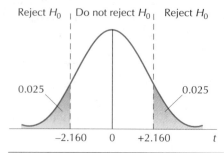

Reject H_0 | Do not reject H_0 | Reject H_0

0.025   0.025

−2.160   0   +2.160 t **FIGURE 11.15**

EXAMPLE   2   Refer back to the data of Example 2 on page 612. Does the data indicate that the value of β_1, that is, the slope of the population regression line, is not zero, which would mean that x (the amount spent on advertising) can be used as a predictor of y (sales)? Use a 5% level of significance.

SOLUTION   The null hypothesis is $\beta_1 = 0$, and the alternative hypothesis is $\beta_1 \neq 0$. Since $n = 10$, we have a t-distribution with $10 - 2$ or 8 degrees of freedom. From Table VII, the $t_{0.025}$ value is 2.306. We then have the acceptance-rejection region shown in Figure 11.16. The value of the test statistic is

$$t = \frac{b_1}{s_e \Big/ \sqrt{\Sigma x^2 - \frac{(\Sigma x)^2}{n}}} = \frac{1.684}{1.55 \Big/ \sqrt{18{,}416 - \frac{(426)^2}{10}}} = 17.799$$

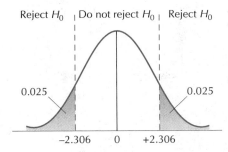

Reject H_0 | Do not reject H_0 | Reject H_0

0.025   0.025

−2.306   0   +2.306   **FIGURE 11.16**

where the value of $b_1 = 1.684$ was computed on page 613, the value of $\Sigma x^2 = 18{,}416$ was computed on page 613, the value of $\Sigma x = 426$ was computed on page 613, and the value of $s_e = 1.55$ was computed on page 626. Since the value of the test statistic 17.799 falls in the rejection region of Figure 11.16, we reject H_0 and conclude that the slope of the population regression line is not 0. Hence, x (the amount spent on advertising) can be used as a predictor of y (sales).

COMMENT   Since all the computations have been done already, we can also determine a confidence interval for the slope β_1 of the population regression line. The end points of the confidence interval are

$$b_1 \pm t_{\alpha/2} \cdot \frac{s_e}{\sqrt{\Sigma x^2 - \frac{(\Sigma x)^2}{n}}}$$

The interested reader should actually verify that a 95% confidence interval for β_1 of Example 1 is $10.857 \pm 2.160\left(\dfrac{3.01}{\sqrt{17.5}}\right)$ or from 9.303 to 12.411. Thus, we can be 95% confident that β_1 is somewhere between 9.303 and 12.411.

Prediction Intervals

After we have determined the least-squares prediction equation for some data, we may wish to set up a prediction interval for a population value of y that we are predicting, corresponding to some particular value of x. Under such circumstances we proceed as follows:

1. Find $n - 2$. This gives us the number of degrees of freedom for a t-distribution.
2. Find the appropriate $t_{\alpha/2}$ values by using Table VII in the Appendix.
3. Find the least-squares prediction equation, $\hat{y}$, by using Formula 11.2. Use it to compute the predicted y-value, $\hat{y}$, corresponding to some particular x_p.
4. Find the standard error of the estimate, s_e, by using Formula 11.5.
5. Set up the appropriate prediction interval by using Formula 11.6, as follows.

FORMULA 11.6

A prediction interval for a particular y corresponding to some particular value of $x = x_p$ is

$$\text{Lower boundary: } \hat{y}_p - t_{\alpha/2} \cdot s_e \cdot \sqrt{1 + \frac{1}{n} + \frac{n(x_p - \bar{x})^2}{n(\Sigma x^2) - (\Sigma x)^2}}$$

$$\text{Upper boundary: } \hat{y}_p + t_{\alpha/2} \cdot s_e \cdot \sqrt{1 + \frac{1}{n} + \frac{n(x_p - \bar{x})^2}{n(\Sigma x^2) - (\Sigma x)^2}}$$

where $t_{\alpha/2}$ represents the t-distribution value obtained from Table VII using $n - 2$ degrees of freedom, s_e is the standard error of the estimate, and $\hat{y}_p$ is the predicted value of y corresponding to $x = x_p$.

Prediction intervals   **COMMENT**   The preceding intervals are frequently called **prediction intervals** because they give intervals for future values of y at a specified x.

We illustrate the use of Formula 11.6 with two examples.

EXAMPLE   3   Using the data of Example 1 on page 610, find a 95% prediction interval for the score of a student who studies for 0.85 hour.

SOLUTION   The least-squares prediction equation was already calculated. It is

$$\hat{y} = 54.772 + 10.857x$$

Also, when $x_p = 0.85$, then the predicted value of y is $\hat{y}_p = 64$. This was calculated earlier. The standard error of the estimate was calculated on page 625. It is 3.01. In this case $n = 15$, so we have a t-distribution with $15 - 2$, or 13 degrees of freedom. From Table VII, the appropriate $t_{\alpha/2} = t_{0.05/2} = t_{0.025}$ value is 2.160. Now we apply Formula 11.6. We get

$$\text{Lower boundary} = \hat{y}_p - t_{\alpha/2} \cdot s_e \cdot \sqrt{1 + \frac{1}{n} + \frac{n(x_p - \bar{x})^2}{n(\Sigma\, x^2) - (\Sigma\, x)^2}}$$

$$= 64 - (2.160)(3.01) \cdot \sqrt{1 + \frac{1}{15} + \frac{15(0.85 - 2.25)^2}{15(93.4375) - (33.75)^2}}$$

$$= 64 - (2.160)(3.01)\sqrt{1 + 0.067 + 0.112}$$

$$= 64 - (2.16)(3.01)(1.086)$$

$$= 64 - 7.061 = 56.939$$

$$\text{Upper boundary} = y_p + t_{\alpha/2} \cdot s_e \cdot \sqrt{1 + \frac{1}{n} + \frac{n(x_p - \bar{x})^2}{n(\Sigma\, x^2) - (\Sigma\, x)^2}}$$

$$= 64 + (2.160)(3.01)\sqrt{1 + \frac{1}{15} + \frac{15(0.85 - 2.25)^2}{15(93.4375) - (33.75)^2}}$$

$$= 64 + (2.160)(3.01)(1.086)$$

$$= 64 + 7.061 = 71.061$$

Thus, a 95% prediction interval for the score of a student who studies for 0.85 hour is 56.939 to 71.061.

E X A M P L E   4   Using the data of Example 2 on page 612 find a 95% prediction interval of the predicted sales of the publishing company if it spends $50,000 for advertisement next month.

SOLUTION   The least-squares prediction equation was already calculated. It is

$$\hat{y} = 1.762 + 1.684x$$

When $x_p = 50$, $\hat{y}_p = 85.962$. The standard error of the estimate, s_e, was calculated on page 626 and is 1.55. In this case $n = 10$ so that we have a t-distribution with $10 - 2$, or 8 degrees of freedom. From Table VII the $t_{0.025}$ value is 2.306. Applying Formula 11.6 gives

$$\text{Lower boundary} = \hat{y}_p - t_{\alpha/2} \cdot s_e \cdot \sqrt{1 + \frac{1}{n} + \frac{n(x_p - \bar{x})^2}{n(\Sigma\, x^2) - (\Sigma\, x)^2}}$$

$$= 85.962 - (2.306)(1.55)\sqrt{1 + \frac{1}{10} + \frac{10(50 - 42.6)^2}{10(18,416) - (426)^2}}$$

$$= 85.962 - (2.306)(1.55)\sqrt{1 + 0.1 + 0.204}$$

$$= 85.962 - (2.306)(1.55)(1.142)$$

$$= 85.962 - 4.082 = 81.88$$

$$\text{Upper boundary} = \hat{y}_p + t_{\alpha/2} \cdot s_e \cdot \sqrt{1 + \frac{1}{n} + \frac{n(\Sigma\, x_p - \bar{x})^2}{n(\Sigma\, x^2) - (\Sigma\, x)^2}}$$

$$= 85.962 + (2.306)(1.55)\sqrt{1 + \frac{1}{10} + \frac{10(50 - 42.6)^2}{10(18{,}416) - (426)^2}}$$

$$= 85.962 + (2.306)(1.55)(1.142)$$

$$= 85.962 + 4.082 = 90.044$$

Thus, a 95% prediction interval for the predicted sales if the company spends $50,000 for advertisement next month is $81.88 million to $90.044 million (assuming that all other factors can be neglected).

EXERCISES FOR SECTION 11.8

For each of the following, test the null hypothesis that $\beta_1 = 0$ and then find a 95% prediction interval for the indicated value.

1. The predicted height of a son when his father is 74 inches tall in Exercise 1 on page 616.
2. The predicted weight of a male college student (aged 18–24 years) who is 68 inches tall in Exercise 2 on page 616.
3. The predicted batting average of a baseball player who is earning $875,000 annually in Exercise 3 on page 616.
4. The predicted number of complaints to the city's heat control board when the outside temperature drops to 8° in Exercise 4 on page 617.
5. The predicted number of accidents when a truck driver is forced to work 48 hours a week in Exercise 5 on page 617.
6. The predicted weekly sales volume when the commercial is broadcast 15 times in Exercise 6 on page 618.
7. The predicted average daily attendance at the amusement park when the admission price is $13.00 in Exercise 7 on page 618.
8. The predicted number of times that a task can be safely performed by a person who drinks 20 units of alcohol in Exercise 8 on page 618.
9. The predicted price per gallon of gasoline when crude oil sells for $16.50 a barrel in Exercise 9 on page 619.
10. The predicted estimate of added years of life when the age at the start of exercises is 50 in Exercise 10 on page 619.

11. The predicted price of an ounce of silver when gold sells for $512 an ounce in Exercise 11 on page 620.

11.9 THE RELATIONSHIP BETWEEN CORRELATION AND REGRESSION

Although it may seem to you that linear correlation and regression are very similar since many of the computations performed in both are the same, the two ideas are quite different. One uses the correlation coefficient (Formula 11.1) to determine whether two variables are linearly related. The correlation coefficient measures the strength of the linear relationship. Regression analysis, on the other hand, is used when we want to answer questions about the relationship between two variables. Just exactly how the two variables are related (that is, can we find an equation connecting the variables) requires regression analysis. The equation connecting the variables may not necessarily be linear.

Extrapolation   Another factor that has to be considered involves **extrapolation**. Two variables may have a high positive correlation. Yet regression analysis *cannot* be used to make predictions for new x-values that are far removed from the original data. Such uses of regression are called extrapolation. To illustrate, in Example 1 on page 610 we determined that the regression equation or the least-squares prediction equation was $\hat{y} = 54.772 + 10.857x$. This equation *cannot* be used to make predictions for values of x that are far removed from the values of x between 0.50 and 4.00.

*11.10 MULTIPLE REGRESSION

Until now we have been interested in finding a linear equation connecting two variables. However, there are many practical situations where several variables may simultaneously affect a given variable. For example, the yield from an acre of land may depend on, among other things, such variables as the amount of fertilizer used, the amount of rainfall, the amount of sunshine, and so on.

There are many formulas that can be used to express relationships between more than two variables. The most commonly used formulas are linear equations of the form

$$\hat{y} = b_0 + b_1 x_1 + b_2 x_2 + \cdots + b_m x_m$$

The main difficulty in deriving a linear equation in more than two variables that best describes a given set of data is that of determining the values of $b_0, b_1, b_2, \ldots, b_m$. When there are two independent variables x_1 and x_2 and the linear equation connecting them is of the form $y = b_0 + b_1 x_1 + b_2 x_2$, then we can apply the method of least squares. This means that we must solve the following equations simultaneously.

FORMULA 11.7

Multiple Regression Formulas

$$\Sigma y = n \cdot b_0 + b_1(\Sigma x_1) + b_2(\Sigma x_2)$$

$$\Sigma x_1 y = b_0(\Sigma x_1) + b_1(\Sigma x_1^2) + b_2(\Sigma x_1 x_2)$$

$$\Sigma x_2 y = b_0(\Sigma x_2) + b_1(\Sigma x_1 x_2) + b_2(\Sigma x_2^2)$$

Solving these equations usually involves a lot of computation. Nevertheless, we illustrate the procedure with an example.

EXAMPLE 1 The following data give the yield, y, per plot of land depending on the quantity of fertilizer used, x_1, and the number of inches of rainfall, x_2. Compute a linear equation that will enable us to predict the average yield, y, per plot in terms of the quantity of fertilizer used, x_1, and the number of inches, x_2, of rainfall.

Yield per Plot (hundreds of bushels), y	Quantity of Fertilizer Used (units), x_1	Rainfall (inches), x_2
20	2	5
25	3	9
28	5	14
30	7	15
32	11	23

SOLUTION We arrange the data in the following tabular format.

x_1	x_2	y	$x_1 y$	$x_2 y$	x_1^2	$x_1 x_2$	x_2^2
2	5	20	40	100	4	10	25
3	9	25	75	225	9	27	81
5	14	28	140	392	25	70	196
7	15	30	210	450	49	105	225
11	23	32	352	736	121	253	529
28	66	135	817	1903	208	465	1056
$\Sigma x_1 =$ 28	$\Sigma x_2 =$ 66	$\Sigma y =$ 135	$\Sigma x_1 y =$ 817	$\Sigma x_2 y =$ 1903	$\Sigma x_1^2 =$ 208	$\Sigma x_1 x_2 =$ 465	$\Sigma x_2^2 =$ 1056

We now substitute these values into the equations given in Formula 11.7. Here $n = 5$. We get

$$135 = 5b_0 + 28b_1 + 66b_2$$

$$817 = 28b_0 + 208b_1 + 465b_2$$

$$1903 = 66b_0 + 465b_1 + 1056b_2$$

This represents a system of three equations in three unknowns. If we solve these equations simultaneously, we get $b_0 = 17.4459$, $b_1 = -0.7504$, and $b_2 = 1.0422$. Thus, the least-squares prediction equation is

$$\hat{y} = 17.4459 - 0.7504x_1 + 1.0422x_2$$

When the farmer uses four units of fertilizer and there are 9 inches of rain, then the predicted yield per plot is

$$y = 17.4459 - 0.7504(4) + 1.0422(9)$$

or approximately 23.8241 hundreds of bushels.

EXERCISES FOR SECTION 11.10

1. Medical research indicates that the systolic blood pressure of an individual is related to the person's age and weight. The following data are available for eight randomly selected males:

Age (years), x_1	Weight (pounds), x_2	Systolic Blood Pressure y
52	170	130
53	175	135
56	180	140
58	186	145
60	195	150
63	200	155
65	208	160
70	215	165

  a. Determine the least-squares prediction equation.
  b. What is the predicted systolic blood pressure of an individual who is 67 years old and who weighs 205 pounds?

2. A shoe manufacturer claims that a person's shoe size is related to his or her height and weight. The following data are available for seven randomly selected people:

Height (inches), x_1	Weight (pounds), x_2	Shoe Size, y
66	150	$8\frac{1}{2}$
68	160	9

Height (inches), x_1	Weight (pounds), x_2	Shoe Size, y
70	180	$9\frac{1}{2}$
72	185	10
74	190	$10\frac{1}{2}$
75	195	11
77	205	$11\frac{1}{2}$

    a. Determine the least-squares prediction equation.
    b. What is the predicted shoe size of a man who is 69 inches tall and who weighs 200 pounds?
3. Many colleges require students to take a mathematics placement exam before allowing them to enroll in any credit-bearing math course. The following is a list of the IQ scores of seven randomly selected students, the number of hours that each prepared for the exam, and the grades that these students achieved on the mathematics placement exam:

IQ Scores, x_1	Preparation for Exam (hours), x_2	Grades on Exam, y
120	18	84
130	17	88
118	25	81
144	8	92
125	19	79
116	20	78
135	10	90

    a. Determine the least-squares prediction equation.
    b. What is the predicted grade on the mathematics placement exam for a student who has an IQ of 123 and who studies 12 hours preparing for the exam?

11.11   USING COMPUTER PACKAGES

Computer packages are ideally suited for drawing scatter diagrams, computing correlation coefficients, deriving regression equations, and for determining the standard error of the estimate. The computer output for such data also provides an analysis of variance, which is a topic that we will discuss in a later chapter. (Again we are assuming that the DOS version of MINITAB is being used. If you are using the Windows version of MINITAB, you must first type GSTD at the MTB> prompt to obtain the results shown here.)

The data from Example 2 of Section 11.6 (page 612) are used to illustrate how MINITAB handles correlation and regression problems.

```
MTB > READ ADVERTISING IN C1, SALES IN C2
DATA > 43 74
DATA > 44 76
DATA > 36 60
DATA > 38 68
DATA > 47 79
DATA > 40 70
DATA > 41 71
DATA > 54 94
DATA > 37 65
DATA > 46 78
DATA > END


     10 ROWS READ

MTB > PLOT SALES IN C2 VS ADVERTISING IN C1
```

This MINITAB program produces the following scatter diagram.

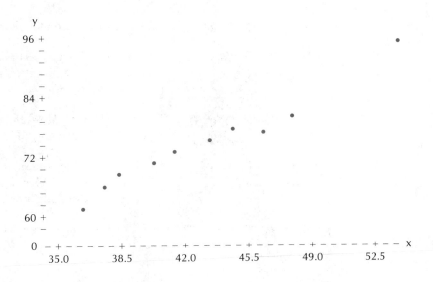

```
MTB > CORRELATION COEFFICIENT BETWEEN ADVERTISING IN C1 AND SALES IN C2
      CORRELATION OF C1 AND C2 = 0.988
MTB > REGRESS SALES IN C2 ON 1 PREDICTOR ADVERTISING IN C1
```

```
        The regression equation is
        C2 = 1.76 + 1.68 C1
        Predictor      Coef      Stdev     t-ratio        p
        Constant      1.759      4.069        0.43     0.677
        C1          1.68405    0.09483       17.76     0.000

        s = 1.554   R-sq = 97.5%   R-sq(adj) = 97.2%

        Analysis of Variance

        SOURCE     DF       SS        MS        F        p
        Regression 1    761.19    761.19   315.39    0.000
        Error       8     19.31      2.41
        Total       9    780.50
```

```
MTB > STOP
```

MINITAB labels the variables as C1 and C2. It also plots these accordingly. If we wish to use x and y instead of C1 and C2, proceed as follows:

```
MTB > NAME C1 'x' C2 'y'
```
This command names C1 as 'x' and C2 as 'y'. Note that x and y must be enclosed within single quotes when naming columns as x and y.

If we rename the columns as x and y, we can then use the following commands:

```
MTB > PLOT 'y' VS 'x'
```
This command plots a scatter diagram with y on the vertical axis and x on the horizontal axis.

```
MTB > CORRELATION 'y' 'x'
```
This command calculates the correlation coefficient between y and x.

```
MTB > REGRESS 'y' 1 'x'
```
This command finds the regression equation. The 1 tells MINITAB that there is only one independent variable in the equation.

When the above commands are applied to our example, we have the following:

```
MTB > PLOT 'y' VS 'x'
```

Character Plot

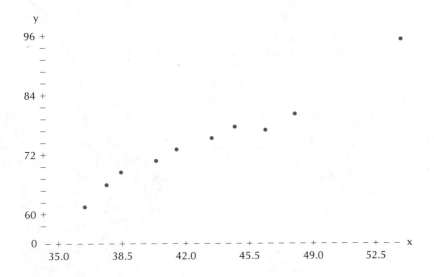

```
MTB > CORRELATION 'y' 'x'
```

Correlations (Pearson)

Correlation of y and x = 0.988

```
MTB > REGRESS 'y' 1 'x'
```

Regression Analysis

The regression equation is

$y = 1.76 + 1.68\ x$

```
Predictor      Coef      Stdev     t-ratio       p
Constant      1.759      4.069        0.43   0.677
x           1.68405    0.09483       17.76   0.000

s = 1.554   R-sq = 97.5%   R-sq(adj) = 97.2%
```

Analysis of Variance

```
SOURCE          DF         SS         MS         F        p
Regression       1     761.19     761.19    315.39    0.000
Error            8      19.31       2.41
Total            9     780.50
```

Let us analyze this printout. The first part gives us the regression equation, which is

$$y = 1.76 + 1.68x.$$

Thus $b_0 = 1.76$ and $b_1 = 1.68$.

The second part of the printout:

```
Predictor        Coef      Stdev     t-ratio        p
Constant        1.759      4.069        0.43    0.677
x             1.68405    0.09483       17.76    0.000
```

gives us the following information. In the column marked Coef we have the values of b_0 and b_1, respectively. In the column marked Stdev we have the standard deviation for b_0 and b_1. In the column t-ratio we have the value of the test statistic t for b_0 and b_1. Finally in the p column we have the p-values for hypothesis tests about p_0 and p_1. We will not analyze all of this information. We will use only the values for b_0 and b_1.

The third part of the printout:

```
s = 1.554    R-sq = 97.5%    R-sq(adj) = 97.2%
```

gives us the values of the standard error of the estimate s_e, the coefficient of determination r^2 and the adjusted r^2 (the value of r^2 adjusted for the degrees of freedom). The last concept is not discussed in this chapter.

The last part of the printout:

```
Analysis of Variance

SOURCE        DF      SS       MS        F       p
Regression     1   761.19   761.19   315.39   0.000
Error          8    19.31     2.41
Total          9   780.50
```

gives us an analysis of variance table. This will be discussed in a later chapter.

If we wish to predict values of y corresponding to some specific x or to set up confidence and/or prediction intervals, we use a subcommand. Returning to the problem we are discussing (Example 2 of Section 11.6), if we wish to predict sales when $50,000 will be spent for advertising, we use the PREDICT subcommand and the REGRESS command as follows:

```
MTB  > REGRESS 'y' 1 'x';
SUBC > PREDICT 50.
```

When the above commands are used, MINITAB will output all of the results that were given earlier and then the following:

```
  Fit   Stdev.Fit         95% C.I.              95% P.I.
85.962      0.857      (83.986, 87.938)     (81.870, 90.054)
```

    ↑   ↑   ↑

This is the value of $\hat{y}$ This is the 95%   This is the 95% prediction

when $x = 50$ confidence interval for   interval for $\hat{y}_p$ for

                   the mean value of y for $x = 50$.

                      $x = 50$.

Thus, if the company plans to spend $50,000 next month for advertising, its predicted sales are $85.962 million.

EXERCISES FOR SECTION 11.11

For each of the following, use MINITAB to
a. draw a scatter diagram for the data.
b. compute the correlation coefficient.
c. determine the regression equation.
d. compute the standard error of the estimate.
e. predict the value of y corresponding to some specific x.
f. set up prediction intervals for the predicted value of y.

1. The U.S. Department of Labor requires all pension plans to be fiscally sound. One particular pension specifies certain penalties for an employee who retires early. The following is available:

Retirement Age x	Pension Reduction y
52	60%
53	54%
55	46%
56	40%
57	35%
58	23%
59	18%
60	12%
61	4%

If an employee plans to retire at age 54, by what percentage will the employee's pension be reduced?

2. Seven randomly selected families had the following annual income and accumulated savings in a bank account:

Family	Income x	Accumulated Savings y
A	$21,000	$15,000
B	28,000	20,000
C	39,000	25,000
D	71,000	37,000
E	53,000	29,000
F	62,000	28,000
G	40,000	30,000

If a family has an annual income of $45,000, what is the predicted amount of money that the family has accumulated as savings in a bank account?

3. A professor in a business school suspects that there is a linear relationship between the height of a college graduate and the starting salary offered to the college graduate. The professor believes that taller people tend to be offered higher starting salaries than shorter people. The following data have been collected for seven recent graduates:

Height (inches), x	Starting Salary (per year), y
64	$21,000
67	24,000
66	23,000
68	26,000
69	26,000
71	27,000
65	25,000

Assuming that the professor's beliefs are valid, what is the predicted starting salary for a college graduate who is 70 inches tall?

4. All visitors to a chocolate factory on the East Coast receive a free sample of the company's products. The company believes that sales at its on-premises gift shop are related to the number of free samples distributed as shown below:

Number of Free Samples, Distributed, x	Sales at On-Site Gift Shop, y
100,000	$600,000
200,000	800,000
500,000	1,100,000

Number of Free Samples, Distributed, x	Sales at On-Site Gift Shop, y
700,000	1,400,000
900,000	1,800,000
1,100,000	2,200,000
1,500,000	3,000,000

If 600,000 free samples are distributed, what is the predicted sales at the on-premises gift shop?

11.12 SUMMARY

In this chapter we analyzed the relationship between two variables. Scatter diagrams were drawn that help us to understand this relationship. We discussed the concept of correlation coefficients, which tell us the extent to which two variables are related. Correlation coefficients vary between the values of -1 and $+1$. A value of $+1$ or -1 represents a perfect linear relationship between the two variables. A correlation coefficient of 0 means that there is no linear relationship between the two variables.

We indicated how to test whether or not a value of r is significant. Furthermore, we mentioned that even when there is an indication of positive correlation between two variables, great care must be shown in how we interpret this relationship.

Once we determine that there is a significant linear correlation between two variables we find the least-squares equation, which expresses this relation mathematically.

We discussed the standard error of the estimate. This is a way of measuring how well the estimated least-squares regression line really fits the data. The smaller the value of s_e is, the better the estimate will be. We also indicated how to set up prediction intervals for values of y obtained through regression analysis. Finally, we discussed multiple regression where the value of one variable depends upon several other variables.

Study Guide

The following is a chapter summary in capsule form. You should now be able to demonstrate your knowledge of ideas mentioned by giving definitions, descriptions, or specific examples. Page references are given in parentheses.

The **correlation coefficient** is a number that measures the relationship between two sets of scores. (page 589)

To analyze the type of relationship that exists between two variables, we draw a **scatter diagram** where one of the variables is pictured on the horizontal axis and the other is pictured on the vertical axis. (page 590)

If the dots form an approximate straight line, then we say that there is a **linear correlation** between the variables. (page 592)

If the line moves in a direction that is from lower left to upper right, then we have **positive correlation**. If the line moves in a direction that is from upper left to lower right, then we have **negative correlation**. (page 592)

The **coefficient of linear correlation** measures the strength of a relationship between two variables. (page 594)

The **reliability of a particular value of r** can be determined by using Table V in the Appendix. (page 604)

The **level of significance** of the reliability of r is indicated by writing r with an appropriate subscript. (page 605)

An **estimated regression line** is a line that best represents the relationship between the two variables. (page 606)

Regression methods involve fitting a straight line or a curve to a set of numbers. (page 606)

Linear regression involves fitting a straight line to a set of numbers. (page 606)

Curvilinear regression involves fitting a curve to a set of numbers. The variable to be predicted is called the **dependent variable** (y) and the known variable is called the **independent variable** (x). (page 607)

The difference between any point and the corresponding point on the regression line is called the (vertical) **deviation** from the line. (page 609)

The **least-squares method** determines the estimated regression line in such a way that the sum of the squares of these vertical deviations is as small as possible. (page 609)

The straight line that best fits a set of data points according to the least-squares criterion is called the **regression line**, whereas the equation of the regression line is called the **regression equation**. (page 610)

The equation of the regression line can also be determined by using a formula to calculate the **sample covariance**. (page 614)

For each x-value the mean of the corresponding population of y-values lies on a straight line called the **population regression line**, whose equation is $y = \alpha + \beta x$. (page 621)

The **error terms** (the vertical distance between the predicted y-values and the true population values) are normally distributed with mean 0 and standard deviation σ. The estimated value of σ is called the **standard error of the estimate** or the **residual standard deviation**. (pages 621 and 624)

The **total sum of squares, SST**, representing the total squared error that is made when we predict y for the observed y-values is given by $SST = \Sigma(y - \bar{y})^2$. (page 623)

The **error sum of squares, SSE**, representing the total squared error that is made when using the equation of the regression line values for prediction is given by $SSE = \Sigma(y - \hat{y})^2$. (page 623)

The **percentage reduction** in the total squared error obtained by using the regression equation instead of y is denoted by r^2 and is called the **coefficient of determination**. (page 623)

After we have determined the least-squares prediction equation for some data we can set up a **prediction interval** for a population value of y that we are predicting, corresponding to some particular value of x. (page 630)

Two variables may have a high positive correlation. Yet regression analysis cannot be used to make predictions using x-values that are far removed from the original data. Such uses of regression are called **extrapolation**. (page 633)

When there are two or more independent variables $x_1, x_2, \ldots$, and the linear equation connecting them is of the form $y = b_0 + b_1x_1 + b_2x_2 + \ldots$, we must use **multiple regression formulas**. (page 634)

Formulas to Remember

You should be able to identify each symbol in the following formulas, understand the relationships among the symbols expressed in each formula, understand the significance of each formula, and use the formulas in solving problems.

1. Coefficient of linear correlation:

$$r = \frac{n(\Sigma\ xy) - (\Sigma\ x)(\Sigma\ y)}{\sqrt{n(\Sigma\ x^2) - (\Sigma\ x)^2}\sqrt{n(\Sigma\ y^2) - (\Sigma\ y)^2}}$$

2. Estimated regression line:

$$\hat{y} = b_0 + b_1x$$

where

$$b_1 = \frac{n(\Sigma\ xy) - (\Sigma\ x)(\Sigma\ y)}{n(\Sigma\ x^2) - (\Sigma\ x)^2} \quad \text{and} \quad b_0 = \frac{1}{n}(\Sigma\ y - b_1 \cdot \Sigma\ x)$$

3. Sample covariance $= s_{xy} = \dfrac{\Sigma(x - \bar{x})(y - \bar{y})}{n - 1}$

4. Equation of regression line (alternate version):

$$\hat{y} = b_0 + b_1x$$

where

$$b_1 = \frac{s_{xy}}{s_x^2}, \quad b_0 = \bar{y} - b_1\bar{x},$$

and s_x is the sample standard deviation of x-values

5. Coefficient of linear correlation (alternate version):

$$r = \frac{s_{xy}}{s_x s_y}$$

6. Total sum of squares, $SST = \Sigma(y - \bar{y})^2$

7. Error of sum of squares, $SSE = \Sigma(y - \hat{y})^2$

8. Coefficient of determination or percentage reduction:

$$r^2 = \frac{SST - SSE}{SST} = 1 - \frac{SSE}{SST}$$

9. Estimate of the common standard deviation, σ, or standard error of the estimate:

$$s_e = \sqrt{\frac{\Sigma(y - \hat{y})^2}{n - 2}} = \sqrt{\frac{SSE}{n - 2}} \quad \text{or} \quad \sqrt{\frac{\Sigma y^2 - b_0(\Sigma y) - b_1(\Sigma xy)}{n - 2}}$$

10. To test whether or not $\beta_1 = 0$, the test statistic is

$$t = \frac{b_1}{s_e \Big/ \sqrt{\Sigma x^2 - \frac{(\Sigma x)^2}{n}}}$$

11. End points of prediction interval for β_1:

$$b_1 \pm t_{\alpha/2} \cdot \frac{s_e}{\sqrt{\Sigma x^2 - \frac{(\Sigma x)^2}{n}}}$$

12. Prediction interval for y corresponding to some given value of $x = x_p$:

Lower boundary: $\hat{y}_p - t_{\alpha/2} \cdot s_e \sqrt{1 + \dfrac{1}{n} + \dfrac{n(x_p - \bar{x})^2}{n(\Sigma x^2) - (\Sigma x)^2}}$

Upper boundary: $\hat{y}_p + t_{\alpha/2} \cdot s_e \sqrt{1 + \dfrac{1}{n} + \dfrac{n(x_p - \bar{x})^2}{n(\Sigma x^2) - (\Sigma x)^2}}$

13. Multiple regression: $\hat{y} = b_0 + b_1 x_1 + b_2 x_2$, where

$$\Sigma y = n \cdot b_0 + b_1(\Sigma x_1) + b_2(\Sigma x_2)$$
$$\Sigma x_1 y = b_0(\Sigma x_1) + b_1(\Sigma x_1^2) + b_2(\Sigma x_1 x_2)$$
$$\Sigma x_2 y = b_0(\Sigma x_2) + b_1(\Sigma x_1 x_2) + b_2(\Sigma x_2^2)$$

Testing Your Understanding of This Chapter's Concepts

1. Where $r = 0$, what is the value of $\sqrt{\dfrac{\Sigma(y - \hat{y})^2}{n - 2}}$?

   a. 1   b. -1 .c. 0   d. a very large number   e. none of these

2. A correlation coefficient of -0.97
   a. indicates a strong negative correlation.
   b. indicates a weak negative correlation.
   c. is insignificant.
   d. is impossible.
   e. none of these.

3. In many respects the standard error of the estimate very closely resembles the standard deviation. Do you agree with this statement? Explain.

4. When $r = \pm1.00$, what is the value of

$$\sqrt{\frac{\Sigma(y - \hat{y})^2}{n - 2}}$$

5. A patient in a medical laboratory was given various doses of a certain drug to determine its effect on the patient's heart rate. The following data summarize the results of this experiment.

Units of Drug Administered x	Heart Beats per Minute y
8	70
6	60
3	50
9	82
15	99
12	90

   a. Determine the least-squares prediction equation for the data.
   b. If it is desired to stabilize the patient's heart rate at 75 beats per minute, how many units of the drug should be administered?

Chapter Test

Multiple Choice Questions

For Questions 1–7, use the following: A used car dealer has been analyzing the prices of used cars and has determined the following information concerning the age of a particular model used car and its price.

Age of Car in Years, x	0	1	2	4	5	6
Price of Used Car, y	$16,000	$11,000	$9000	$5000	$4000	$2500

1. Compute the coefficient of correlation for the data.
   a. −0.973   b. −0.895   c. −0.991   d. −0.937   e. none of these

2. Determine whether r is significant.
   a. yes   b. no   c. not enough information   d. none of these

3. Find the least-squares prediction equation.
   a. $y = -14184.52 + 2089.286x$ b. $y = 14184.52 - 2089.286x$
   c. $y = -2089.286 + 14184.52x$ d. $y = 14184.52 + 2089.286x$
   e. none of these

4. If a car is 3 years old, what is its predicted price?
   a. $8147.79   b. $7844.78   c. $7916.66   d. $8217.69   e. none of these

5. What is the value of the standard error of the estimate?
   a. 1179.45   b. 1397.576   c. 1278.628   d. 1321.469   e. none of these

6. What is the value of the coefficient of determination for the data?
   a. 0.917   b. 0.927   c. 0.946   d. 0.964   e. none of these

7. At the 5% level of significance, does the data indicate that the age of a car can be used as a predictor of its price as a used car?
   a. yes   b. no   c. not enough information given   d. none of these

8. What type of correlation would you think exists between high blood pressure and the incidence of obesity (overweight)?
   a. positive correlation   b. negative correlation
   c. zero correlation   d. not enough information given
   e. none of these

9. A correlation coefficient of +0.96
   a. is impossible.
   b. indicates a weak positive correlation.
   c. indicates a strong positive correlation.
   d. is insignificant.
   e. none of these.

10. When determining the values of b_0, b_1, and b_2, which are used in the multiple regression formula, what does $b_0(\Sigma x_1) + b_1(\Sigma x_1^2) + b_2(\Sigma x_1 x_2)$ equal?
   a. Σy b. $\Sigma x_1 y$ c. $\Sigma x_2 y$ d. $\Sigma(x_1 x_2)$ e. none of these

Supplementary Exercises

For Questions 11–17, use the following: A magazine publisher has determined that the monthly circulation of the magazine depends upon the price charged per issue as shown on the next page:

Price per Issue (in dollars), x	1.00	1.25	1.50	1.75	2.00	2.50
Circulation (in thousands), y	55	49	43	39	37	30

11. Compute the coefficient of correlation for the data.
12. Find the least-squares prediction equation.
13. If the magazine is priced to sell at $2.25, what is its predicted circulation?
14. Find a 95% prediction interval for the predicted circulation when the magazine is priced to sell at $2.25.
15. What is the value of the standard error of the estimate?
16. What is the value of the coefficient of determination for the data?
17. At the 5% level of significance, does the data indicate that the price charged per issue can be used as a predictor of its circulation?
18. The final exam scores for eight students in a statistics course and their final exam scores in a calculus course are shown below:

Statistics Course x	Calculus Course y
64	76
68	84
89	90
39	34
88	85
96	99
81	75
76	83

Draw a scatter diagram for the data. What type of correlation is suggested? Explain your answer.

19. A state judicial system judge believes that the number of cases adjudicated daily is dependent upon the number of judges working on that particular day as shown below:

Number of Judges Working x	Number of Cases Adjudicated y
4	27
6	34
8	41

Number of Judges Working x	Number of Cases Adjudicated y
9	53
10	76
12	98

a. Compute the coefficient of correlation for the data.
b. Find the least-squares prediction equation.
c. Test the null hypothesis that $\beta_1 = 0$.
d. Find the 95% prediction interval for the predicted number of cases to be adjudicated when 11 judges are working.

20. The resale value of a car depends, among other things, on the car's age and the number of miles indicated on the car's odometer. For one particular model car and geographic area, the following selling prices were reported:

Cars Age (in years), x_1	Number of Miles on Odometer, x_2	Selling Price, y
1	14,000	$9800
2	22,000	8700
3	32,000	7600
4	43,000	6000
5	58,000	4900
6	69,000	3800

a. Determine the least-squares prediction equation.
b. What is the predicted selling price of a car that is $3\frac{1}{2}$ years old and whose odometer reading is 41,000 miles?

21. Refer to the information presented in the graph on the next page. A survey was conducted to determine if the sex of a respondent affects the person's response. In each case the person was asked the same question.
a. Draw a scatter diagram for the data. (Let x = male percentage and y = female percentage.)
b. Calculate the coefficient of correlation.
c. Determine the least-squares prediction equation.
d. If 62% of the males are in favor of the death penalty, what is the predicted percent of females who are in favor of the death penalty?
e. Find a 95% prediction interval for the percentage of females who are in favor of the death penalty.

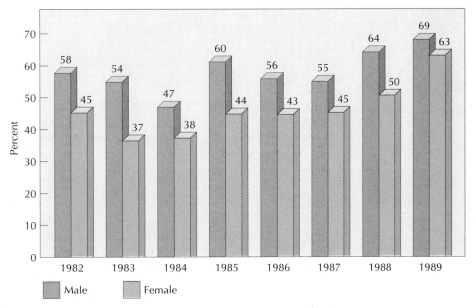

Question: **"Are you in favor of the death penalty for persons convicted of murder?"**
(*Source:* Sourcebook of Criminal Justice Statistics, *Washington, D.C., 1990*)

⫸⫸ Thinking Critically

1. Verify that Formulas 11.2 and 11.3 are equivalent. Hence, either can be used to calculate the equation of the regression line.
2. Show that the formula for calculating the linear correlation coefficient (Formula 11.1) and the following computational formula are equivalent.

$$r = \frac{s_{xy}}{s_x x_y}$$

where

$$s_{xy} = \text{the sample covariance}$$

$$s_x = \text{the sample standard deviation of the } x\text{-values}$$

$$s_y = \text{the sample standard deviation of the } y\text{-values}$$

(*Hint:* Verify the following identity:

$$\frac{s_{xy}}{s_x s_y} = \frac{n(\Sigma\, xy) - (\Sigma\, x)(\Sigma\, y)}{\sqrt{n(\Sigma\, x^2) - (\Sigma\, x)^2}\sqrt{n(\Sigma\, y^2) - (\Sigma\, y)^2}}$$

3. Why is correlation analysis included in the study of linear regression? Can we study these two subjects independently?

Case Studies

1. Attendance at air shows has been increasing over the past years. The following attendance figures, as reported by the *International Council of Air Shows*, are available.

Year x	Attendance y
1989	22.0 million
1990	23.1 million
1991	23.9 million
1992	24.4 million
1993	25.9 million
1994	26.3 million
1995	27.4 million

Source: International Council of Air Shows.

a. Draw a scatter diagram for the data.
b. Compute the least-squares prediction equation.
c. One official would like to use the prediction equation obtained in part (b) to predict attendance for future years at air shows. Can this be done from a statistical point of view? Explain your answer.

2. Progress made by women in track events over the last half-century has been considerably more substantial than the progress made by men. The data below represent world record times for the 800-meter run. (*Source:* Whipp, Brian J. and Susan A. Ward. "Will Women Soon Outrun Men?" *Nature*, 355:25, January 1992)

Year, x	Men, y_1	Women, y_2
1925	111.9	144.0
1935	109.7	135.6
1945	106.6	132.0
1955	105.7	125.0
1965	104.3	118.0
1975	104.1	113.3

Reprinted with permission from *Nature* 1992, Macmillan Magazines, Ltd.

a. Compute the regression equation for men.
b. Compute the regression equation for women.
c. Plot both equations on the same set of axes.
d. What conclusions, if any, can be drawn?
e. Read also the article by Natalie Angier "2 Experts Say Women Who Run May Overtake Men," *The New York Times*, January 7, 1992, C3.

12

ANALYZING COUNT DATA:
THE CHI-SQUARE DISTRIBUTION

DID YOU KNOW THAT

a   there has been a steady rise in the cancer
mortality rate in the United States? It rose from
157 per 100,000 population in 1950 to 174 per
100,000 in 1990, and that if lung cancer deaths
were excluded, cancer mortality would have
declined 14% over this period. (*Source: Cancer
Facts and Figures, 1995*)

b.   about 80% of adults now use seat belts in cars
but the usage rate for child safety seats is only
65%? (*Source:* National Highway Traffic Safety
Administration, 1995)

c.   older adults lose perhaps 1% of their bone mass
per year, but that women are at a disadvantage
since they start out with less? (*Source: American
Journal of Clinical Nutrition*, 1994)

Lawyers or prosecutors have preemptive rights to disqualify a potential juror, possibly on the basis of the juror's sex alone. Is the sex of a juror independent of his or her tendency to be liberal, fair, or vindictive? See discussion on page 686. (Copyright 1995, Comstock, Inc.)

Until now we have been using statistical inference procedures for hypothesis tests and confidence intervals involving means and proportions. In this chapter we apply hypothesis-testing procedures to determine whether two characteristics of a population as given in a contingency table are statistically independent or whether observed data follow some pattern. This is referred to as goodness of fit. The chi-square distribution plays a key role in such analyses.

Chapter Objectives

- **To apply** chi-square tests that provide the basis for testing whether more than two population proportions can be considered as equal. (Section 12.2)

- **To discuss** a chi-square test statistic that can be applied to determine whether an observed frequency is in agreement with the expected mathematical distribution. This is known as the goodness of fit. (Section 12.3)

- **To analyze** contingency tables. These are tabular arrangements of data into a two-way classification. The chi-square test statistic tells us whether the two ways of classifying the data are independent. (Section 12.4)

- **To compute** expected frequencies. These are the numbers, that is, the frequencies, that should appear in each of the boxes of a contingency table. (Section 12.4)

STATISTICS IN ACTION

The first newspaper article gives the Army's new physical fitness standards for both men and women in different age-groups. By looking at the data, we can conclude that people in different age-

THE ARMY'S NEW PHYSICAL FITNESS STANDARDS

Age Groups	Push-ups		Sit-ups		Two-mile run	
	Min.	Max.	Min.	Max.	Min.	Max.
17–21	42	82	52	92	15:54	11:45
	18	58	50	90	18:45	14:45
22–26	40	80	47	87	16:36	12:36
	16	56	45	85	19:36	15:36
27–31	38	78	42	82	17:18	13:18
	15	54	40	80	21:00	17:00
32–36	33	73	38	78	18:00	14:00
	14	52	35	75	22:36	18:36
37–41	32	72	33	73	18:42	14:42
	13	48	30	70	22:36	19:36
42–46	26	66	29	69	19:06	15:12
	12	45	27	67	24:00	20:00
47–51	22	62	27	67	19:36	15:36
	10	41	24	64	24:30	20:30
52 & over	16	56	26	66	20:00	16:00
	09	40	22	62	25:00	21:00

■ MALE   ■ FEMALE

Source: Office of Deputy Chief of Staff for Operations, Individual Training Branch of the U.S. Army, 1990

groups are capable of performing different numbers of push-ups, sit-ups, and so on, or is the age of the person independent of these standards? Such information is very valuable since it enables an individual to measure his or her performance against that of others.

Now consider the second article. Is the geographical area in which a person lives a factor in determining a person's opinion as to whether the new health plan should be enacted? Are there ways of comparing the proportions of responses?

MANY AMERICANS OPPOSE NEW HEALTH PLAN

Washington, March 1: According to the latest PBC poll, more and more Americans are now opposed to the President's new health plan than favor it. This represents a dramatic change from a similar poll taken just 2 months ago. In a survey of 1125 randomly selected people from four geographic areas of the United States, about 51% of those surveyed said that in light of the many uncertainties and the costs associated with the new health plan, they were opposed to it. Each person surveyed was asked the question: "Are you in favor of the President's health plan?" Their responses are summarized below:

Geographical Area

	East	Midwest	West	South
Yes	89	81	263	142
No	102	123	151	174

Tuesday, March 1, 1994

12.1 INTRODUCTION

In Chapter 10 we discussed methods for testing whether the observed difference between two sample means is significant. In this chapter we analyze whether differences among two or more sample proportions are significant or whether they are due purely to chance. For example, suppose a college professor distributes a faculty-evaluation form to the 150 students of his Psychology 12 classes. The following are two examples of the multiple-choice questions appearing on the form:

1. What is your grade-point average? (Assume A = 4 points, B = 3 points, C = 2 points, D = 1 point, and F = 0 points.)
   a. 3.0 to 4.0   b. 2.0 to 2.99   c. below 2.0
2. Would you be willing to take another course with this teacher?
   a. Yes   b. No

PERSONAL CRIMES: VICTIMIZATION RATES FOR PERSONS AGE 12 AND OVER, BY TYPE OF CRIME AND AGE OF VICTIMS, 1980
(Rate per 1,000 population in each age group)

Type of crime	12–15 (16,527,000)	16–19 (15,792,000)	20–24 (17,609,000)	25–34 (29,211,000)	35–49 (33,783,000)	50–64 (30,847,000)	65 and over (20,792,000)
Crimes of violence	52.6	67.9	61.1	38.6	20.8	11.8	9.0
Rape	1.5	2.5	2.1	1.4	0.2^1	0.3	0.2^1
Robbery	12.7	11.3	10.7	7.0	5.5	4.1	3.9
Robbery with injury	3.3	3.5	3.3	2.1	2.1	1.5	1.9
From serious assault	1.4	1.9	1.7	1.4	1.3	0.9	1.0
From minor assault	2.0	1.6	1.6	0.6	0.8	0.6	0.9
Robbery without injury	9.4	7.8	7.4	5.0	3.4	2.6	2.0
Assault	38.5	54.1	48.3	30.2	15.2	7.3	4.9
Aggravated assault	12.9	23.7	22.0	12.6	7.0	2.7	1.6
With injury	5.7	7.7	6.9	3.7	2.1	0.7	0.4^1
Attempted assault with weapon	7.2	16.0	15.2	8.9	4.9	2.0	1.2
Simple assault	25.6	30.4	26.3	17.7	8.2	4.6	3.4
With injury	7.9	8.3	6.5	3.7	1.7	1.0	0.5
Attempted assault without weapon	17.8	22.1	19.7	13.9	6.5	3.6	2.9
Crimes of theft	166.7	159.8	146.3	106.2	79.2	49.4	21.9
Personal larceny with contact	3.1	3.7	3.4	2.6	2.6	3.5	3.4
Purse snatching	0.4^1	0.6^1	0.9	0.6	0.7	1.5	1.4
Pocket picking	2.7	3.1	2.4	2.0	1.9	1.9	2.0
Personal larceny without contact	163.6	156.1	143.0	103.5	76.7	45.9	18.5

NOTE: Detail may not add to total shown because of rounding. Numbers in parentheses refer to population in the group.
^{1}Estimate, based on about 10 or fewer sample cases, is statistically unreliable.

FIGURE 12.1   (*Source: Criminal Victimization in the United States, U.S. Department of Justice—Law Enforcement Assistance Administration, Washington, D.C., 1989*)

The results for these two questions are summarized in the following chart:

		Grade Point Average		
		3.0–4.0	2.0–2.99	Below 2.0
Would You Take This Teacher Again?	*Yes*	28	36	11
	No	22	44	9
	Total	50	80	20

The teacher may be interested in knowing whether these ratings are influenced by the student's grade-point average. In the 3.0–4.0 category, the proportion of students who said that they would take this teacher for another course is $\frac{28}{50}$. In the 2.0–2.99 category the proportion is $\frac{36}{80}$, and in the below 2.0 category the proportion is $\frac{11}{20}$. Is it true that students with a higher grade-point average tend to rate the teacher differently than students with a lower grade-point average?

Now consider the magazine clipping (Figure 12.1), which gives the victimization rates by the type of crime and the age of the victim. Is the type of crime committed related to the age of the intended victim? Such information is vital to law-enforcement officials. Questions of this type occur quite often.

In this chapter we study the chi-square distribution, which is of great help in studying differences between proportions.

12.2   THE CHI-SQUARE DISTRIBUTION

To illustrate the method that is used when analyzing several sample proportions, let us return to the first example discussed in the introduction. Let p_1 be the true proportion of students in the 3.0–4.0 category who will take another course with this teacher. Similarly, let p_2 and p_3 represent the proportion of students in the 2.0–2.99 and below 2.0 categories who will take another course with this teacher. The null hypothesis that we wish to test is

$$H_0: \quad p_1 = p_2 = p_3$$

This means that a student's grade-point average does not affect the student's decision to take this teacher again. The alternative hypothesis is that at least one of p_1, p_2, and p_3 is different. This means that a student's grade-point average does affect the student's decision.

If the null hypothesis is true, the observed difference between the proportions in each of the grade-point average categories is due purely to chance. Under this assumption we combine all the samples into one and consider it as one large sample. We then obtain the following estimate of the true proportion of *all* students in the school who are willing to take another course with the teacher. We get

$$\frac{28 + 36 + 11}{50 + 80 + 20} = \frac{75}{150} = 0.5$$

Thus, our estimate of the true proportion of students who are willing to take another course with this teacher is 0.5.

There are 50 students in the 3.0–4.0 category. We would therefore expect 50(0.5), or 25, of these students to indicate yes they would take another course with this teacher, and we would expect 50(0.5), or 25, to indicate no. Similarly, in the 2.0–2.99 category there are 80 students so that we would expect 80(0.5), or 40, yes answers and 40 no answers. Also, in the below 2.0 category there are 20 students so that we would expect 20(0.5), or 10, yes answers and 10 no answers. The numbers that should appear are called **expected frequencies**. In the following table we have indicated these numbers in parentheses below the ones that were actually observed. We call the numbers that were actually observed **observed frequencies**.

Expected frequencies

Observed frequencies

	Grade Point Average		
	3.0–4.0	2.0–2.99	Below 2.0
Yes	28 (25)	36 (40)	11 (10)
No	22 (25)	44 (40)	9 (10)
Total	50	80	20

Notice that the expected frequencies and the observed frequencies are not the same. If the null hypothesis that $p_1 = p_2 = p_3$ is true, the observed frequencies should be fairly close to the expected frequencies. Since this rarely will happen, we need some way of determining when these differences are significant.

When is the difference between the observed frequencies and the expected frequencies significant? To answer this question, we calculate a test statistic called the **chi-square statistic**.

Chi-square statistic

FORMULA 12.1

Let E represent the expected frequency and let O represent the observed frequency. Then the **chi-square test statistic**, denoted as χ^2 is defined as

$$\chi^2 = \Sigma \frac{(O - E)^2}{E}$$

COMMENT   In using Formula 12.1 we must calculate the square of the difference for each box, that is, cell, of the table. Then we divide the squares of the difference for each cell by the expected frequency for that box. Finally, we add these results together.

Returning to our example, we have

$$\chi^2 = \frac{(28-25)^2}{25} + \frac{(36-40)^2}{40} + \frac{(11-10)^2}{10} + \frac{(22-25)^2}{25} + \frac{(44-40)^2}{40} + \frac{(9-10)^2}{10}$$

$$= \quad 0.36 \quad + \quad 0.40 \quad + \quad 0.10 \quad + \quad 0.36 \quad + \quad 0.40 \quad + \quad 0.10$$

$$= 1.72$$

The value of the χ^2 statistic is 1.72.

It should be obvious from Formula 12.1 that the value of χ^2 will be 0 when there is perfect agreement between the observed frequencies and the expected frequencies since in this case $O - E = 0$. Generally speaking, if the value of χ^2 is small, the observed frequencies and the expected frequencies will be pretty close to each other. On the other hand, if the value of χ^2 is large, this indicates that there is considerable difference between the observed frequency and the expected frequency.

Chi-square distribution   To determine when the value of the χ^2 statistic is significant, we use the **chi-square distribution**. This is pictured in Figure 12.2. We reject the null hypothesis when the value of the chi-square statistic falls in the rejection region of Figure 12.2. Table VIII in the Appendix gives us the critical values, that is, the dividing line, depending on the number of degrees of freedom. Thus, $\chi^2_{0.05}$ represents the dividing line that cuts off 5% of the right tail of the distribution. *The number of degrees of freedom is always 1 less than the number of sample proportions that we are testing.*

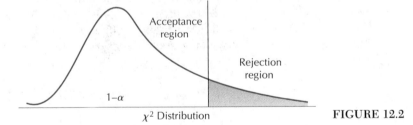

χ^2 Distribution   **FIGURE 12.2**

In our example we are comparing three proportions so that the number of degrees of freedom is $3 - 1$, or 2. Now we look at Table VIII to find the χ^2 value that corresponds to 2 degrees of freedom. We have $\chi^2_{0.05} = 5.991$. Since the test statistic value that we obtained, $\chi^2 = 1.72$, is much less than the table value of 5.991, we do not reject the null hypothesis. The difference between what was expected and what actually happened can be attributed to chance.

Although we will usually use the 5% level of significance, Table VIII in the Appendix also gives us the χ^2 values for the 1% level of significance. We use these values when we wish to find the dividing line that cuts off 1% of the right tail of the distribution.

Let us further illustrate the χ^2 test with several examples.

EXAMPLE 1   There are 10,000 students at a college: 2700 are freshmen, 2300 are sophomores, 3000 are juniors, and 2000 are seniors. Recently a new president was appointed. Two thousand students attended the reception party for the president. The attendance breakdown is shown in the following table.

		Freshmen	Sophomores	Juniors	Seniors
Attended Reception?	Yes	300	700	650	350
	No	2400	1600	2350	1650
	Total	2700	2300	3000	2000

Test the null hypothesis that the proportion of freshmen, sophomores, juniors, and seniors that attended the reception is the same. (Use a 5% level of significance.)

SOLUTION   In order to compute the χ^2 test statistic, we must first compute the expected frequency for each box, or cell. To do this we obtain an estimate of the true proportion of students who attend the reception. We have

$$\frac{300 + 700 + 650 + 350}{2700 + 2300 + 3000 + 2000} = \frac{2000}{10,000} = 0.20$$

Out of 2700 freshmen we would expect 2700(0.20) or 540 to attend and $2700 - 540$, or 2160 not to attend. Similarly, out of 2300 sophomores we would expect 2300(0.20), or 460 to attend and $2300 - 460$, or 1840 not to attend. Also, out of 3000 juniors we would expect 3000(0.20), or 600 to attend and $3000 - 600$, or 2400 not to attend. Finally, out of 2000 seniors, we would expect 2000(0.20), or 400 to attend and $2000 - 400$, or 1600 not to attend. We have indicated these expected frequencies just below the observed values in the following chart:

		Freshmen	Sophomores	Juniors	Seniors
Attended Reception?	Yes	300 (540)	700 (460)	650 (600)	350 (400)
	No	2400 (2160)	1600 (1840)	2350 (2400)	1650 (1600)
	Total	2700	2300	3000	2000

Now we calculate the value of the χ^2 statistic. We have

$$\chi^2 = \Sigma \frac{(O - E)^2}{E}$$

$$= \frac{(300 - 540)^2}{540} + \frac{(700 - 460)^2}{460} + \frac{(650 - 600)^2}{600}$$

$$+ \frac{(350 - 400)^2}{400} + \frac{(2400 - 2160)^2}{2160} + \frac{(1600 - 1840)^2}{1840}$$

$$+ \frac{(2350 - 2400)^2}{2400} + \frac{(1650 - 1600)^2}{1600}$$

$$= 106.67 + 125.22 + 4.17 + 6.25 + 26.67 + 31.30 + 1.04 + 1.56$$

$$= 302.88$$

There are four proportions that we are testing so that there are $4 - 1$, or 3 degrees of freedom. From Table VIII in the Appendix we find that the $\chi^2_{0.05}$ value with 3 degrees of freedom is 7.815. The value of the χ^2 test statistic ($\chi^2 = 302.88$) is definitely greater than 7.815. Hence, we reject the null hypothesis. The proportions of freshmen, soph-omores, juniors, and seniors that attended the reception are not the same.

EXAMPLE 2 A survey of the marital status of the members of three health clubs was taken. The following table indicates the results of the survey.

		Club 1	Club 2	Club 3
	Yes	11	17	8
Married?	No	29	33	22
	Total	40	50	30

Test the null hypothesis that the proportion of members that are married in each of these health clubs is the same. (Use a 5% level of significance.)

SOLUTION We must first compute the expected frequency for each cell. To do this we obtain an estimate of the true proportion of members who are married. We have

$$\frac{11 + 17 + 8}{40 + 50 + 30} = \frac{36}{120} = 0.3$$

Thus, the estimate of the true proportion is 0.3. Out of the 40 members in Club 1 we would expect 40(0.3), or 12, members to be married and $40 - 12$, or 28, not to be married. In Club 2 we would expect 50(0.3), or 15, members to be married and $50 - 15$, or 35, members not to be married. In Club 3 we would expect 30(0.3), or 9, members to be married and $30 - 9$, or 21, not to be married. We have indicated these expected frequencies in parentheses in the following table.

		Club 1	Club 2	Club 3
	Yes	11 (12)	17 (15)	8 (9)
Married?	No	29 (28)	33 (35)	22 (21)
	Total	40	50	30

Now we calculate the value of the χ^2 statistic. We have

$$\chi^2 = \Sigma \frac{(O - E)^2}{E}$$

$$= \frac{(11 - 12)^2}{12} + \frac{(17 - 15)^2}{15} + \frac{(8 - 9)^2}{9} + \frac{(29 - 28)^2}{28}$$

$$+ \frac{(33 - 35)^2}{35} + \frac{(22 - 21)^2}{21}$$

$$= 0.08 + 0.27 + 0.11 + 0.04 + 0.11 + 0.05$$

$$= 0.66$$

There are three proportions that we are testing so that there are $3 - 1$, or 2 degrees of freedom. From Table VIII in the Appendix we find that the $\chi^2_{0.05}$ value with 2 degrees of freedom is 5.991. Since the value of the test statistic, 0.66, is less than 5.991, we do not reject the null hypothesis.

COMMENT   Experience has shown us that the χ^2 test can only be used when the expected frequency in each cell is at least 5. If the expected frequency of a cell is not larger than 5, this cell should be combined with other cells until the expected frequency is at least 5. We will not, however, concern ourselves with this situation.

EXERCISES FOR SECTION 12.2

1. A dermatologist is interested in determining if a woman's age is or is not a factor in whether she applies sunscreen as a protection against the sun's ultraviolet rays. A random survey of 1200 women in Hawaii produced the following results:

		Age-Group (in years)			
		20–29	30–39	40–49	Over 50
Applies	Yes	167	131	49	25
Sunscreen?	No	320	278	148	82

Using a 5% level of significance, test the null hypothesis that there is no significant difference between the corresponding proportion of women in the various age-groups who apply sunscreen.

2. A study was conducted to determine whether an individual's social class has any effect on the individual's belief that the quality of life has changed since 1980. Each of 972 randomly selected individuals was asked the same question: "Do you believe

that the quality of your life has changed since 1980?" A summary of their responses is shown in the accompanying table.

		Social Class			
		Upper	Upper-Middle	Middle-Lower	Lower
Has the Quality of Your Life Changed?	Yes	89	104	101	89
	No	131	158	151	149

Using a 5% level of significance, test the null hypothesis that the percentage of people who believe that the quality of their life has changed is the same for all social classes.

3. According to the U.S. Census Bureau, 25% of U.S. households are one-parent homes (U.S. Census Bureau, 1995). Furthermore, it is claimed that the percentage of children from one-parent homes that attend the city's public schools is the same for all schools in the city. A recent study by Bennet and Rogers (Chicago, 1995) revealed the following information about students at five of the city's schools.

		School				
		A	B	C	D	E
Student from One-Parent Home?	Yes	208	218	143	424	527
	No	661	731	532	1477	799

Using a 1% level of significance, test the null hypothesis that the percentage of students from one-parent homes is the same for these schools.

4. A newspaper reporter conducted a random survey of 3257 people nationwide to determine their views on the fiscal soundness of the Social Security system. The results are summarized below:

		Geographic Location of Respondent				
		Far West	West	South	Northeast	Southeast
Do You Believe in the Fiscal Soundness of the Social Security System?	Yes	301	403	346	275	233
	No	325	427	388	293	266

Using a 1% level of significance, test the null hypothesis that the percentage of people who believe in the fiscal soundness of the Social Security system is the same for all geographic regions.

5. Consider the accompanying newspaper article. A legislator polled 450 of her constituents to determine their views regarding the new ordinance requiring restaurant owners to set up special sections for smokers and nonsmokers. A summary of their responses is shown below:

		Almost Every Day	*Once a Week*	*Once a Month*	*Almost Never*
				Frequency of Restaurant Visits	
In Favor of New Legislation?	*Yes*	98	84	69	27
	No	76	53	32	11

SMOKERS TAKE NOTE

New York, March 1: Following the lead of many other communities, New York City officials announced yesterday new rules and regulations which, in effect, severely restrict smoking in restaurants and other public areas.

Thursday, March 1, 1995

Using a 5% level of significance, test the null hypothesis that the percentage of constituents in favor of the new legislation is the same for *all* the patrons of the restaurants, regardless of frequency of visits.

6. Consider the newspaper article on the next page. The commission surveyed homes on seven streets and obtained the following data:

		Bay Street	Colby Square	Morris Ave.	Francis Blvd.	Clarence Drive	Hylan Avenue	Woodland Way
					Location of Homes			
Dangerous Level of Radon Gas Present?	*Yes*	7	9	3	8	6	10	6
	No	31	29	37	26	39	40	22

Using a 5% level of significance, test the hypothesis that there is no significant difference between the percentage of homes on all the streets containing excessive levels of radon gas.

MORE HOMES CONTAMINATED WITH RADON

Bergen, Sept. 12: A new study released today by the state's environmental commission reveals that the number of homes containing excessive amounts of deadly radon gas is larger than expected. The radon gas is entering many homes on the East Coast (particularly those in the states of New Jersey and Pennsylvania) through holes or cracks in the foundation. It is believed that the gas is coming from deep within the earth.

The Commission urged the governor to appropriate additional funds to enable it to expand its monitoring activities.

September 12, 1994

7. The National Highway Traffic and Safety Administration estimates that about 70% of the U.S. drivers use seat belts. An adjuster for the Marvel Insurance Company polled 500 of the company's policyholders in one Northeast state to determine whether seat-belt usage is dependent upon the geographic location of driver. The following results were obtained:

		Geographic Location of Policyholder Within State			
		West	North	South	East
Do You Always Use Your Seat Belt?	*Yes*	98	84	82	77
	No	40	39	44	36

Using a 5% level of significance, test the null hypothesis that the percentage of policyholders who use seat belts is the same for all geographical locations within the state.

8. A 1993 article in the *Journal of the American Medical Association* found that among men and women aged 50 to 80 those who consumed the most calcium had a 60% lower risk of hip fracture than the others. Even with a good diet, women typically fall 500 to 1000 milligrams short of their recommended daily allowance and must consider supplements. A survey of 356 people in different age-groups produced the following results:

		Age-Group				
		31–40	41–50	51–60	61–70	71–80
Do You Take Calcium	*Yes*	27	38	49	58	53
Supplements?	*No*	12	19	23	41	36

Using a 1% level of significance, test the null hypothesis that the proportion of people who take calcium supplements is not significantly different for the different age-groups.

9. *More jails?* A random survey of 1816 residents located throughout the state was taken to see how many of them were in favor of building new jails and/or drug rehabilitation centers in their neighborhoods. The following table indicates the results of the survey according to the section of the city in which the person interviewed lives:

		Brighton	Beachgate	Bayview	Shorefront
Would You Allow New	*Yes*	129	137	164	158
Jails to be Built in	*No*	317	288	302	321
Your Neighborhood?					

Using a 5% level of significance, test the null hypothesis that there is no significant difference in the proportion of residents who would allow new jails and/or drug rehabilitation centers to be built in their neighborhoods.

10. Refer back to the newspaper article given at the beginning of this chapter on page 657. Consider the information presented in it. Using a 1% level of significance, test the null hypothesis that the percentage of Americans in favor of the President's proposed health plan is not significantly different for the various geographic areas.

11. According to *Information Technology Journal* (1995), more and more business executives are using laptop computers for a variety of functions. A nationwide survey of 294 business executives was conducted to determine what these laptop computers were used for. The following results were obtained:

		Principal Computer Use		
		Word Processing	Billing and Shipping Records	Inventory Control	Sales Analysis
Use of Laptop Computer	Yes	31	52	51	46
	No	29	38	28	19

Using a 5% level of significance, test the null hypothesis that there is no significant difference in the principal computer usage among the uses listed by these business executives.

12. The U.S. Department of Education reports that many high school students who attend "inner" schools in large urban areas have no intention of attending college. A survey of 1106 students in five high schools of one large urban area resulted in the following:

		High School				
		A	B	C	D	E
Do You Plan To Attend College?	Yes	81	76	63	96	62
	No	118	134	106	168	202

Using a 5% level of significance, test the null hypothesis that there is no significant difference in the proportion of students from the "inner" schools who plan to attend college.

13. The American Medical Association estimates that about 12% of the U.S. population is left-handed. A survey of 240 randomly selected visitors to a museum found that 21 of the 158 male visitors were left-handed and 13 of the 82 female visitors were left-handed. Using a 1% level of significance, test the null hypothesis that there is no significant difference between the corresponding proportion of male or female visitors who are left-handed.

12.3 GOODNESS OF FIT

In addition to the applications mentioned in the previous section, the chi-square test statistic can also be used to determine whether an observed frequency distribution is in agreement with the expected mathematical distribution. For example, when a die is rolled, we assume that the probability of any one face coming up is $\frac{1}{6}$. Thus, if a die is rolled 120 times, we would expect each face to come up approximately 20 times since

$$\mu = np = 120\left(\frac{1}{6}\right) = 20.$$

Suppose we actually rolled a die 120 times and obtained the results shown in Table 12.1. In this table we have also indicated the expected frequencies. Are these observed frequencies reasonable? Do we actually have an honest die?

TABLE 12.1   Expected and Observed Frequencies
When a Die Was Tossed 120 Times

Die Shows	Expected Frequency	Observed Frequency
1	20	18
2	20	21
3	20	17
4	20	21
5	20	19
6	20	24
	Total = 120	Total = 120

To check whether the differences between the observed frequencies and the expected frequencies are due purely to chance or are significant, we use the chi-square test statistic of Formula 12.1. We reject the null hypothesis that the observed differences are not significant only when the test statistic falls in the rejection region.

In our case the value of the test statistic is

$$\chi^2 = \Sigma \frac{(O - E)^2}{E}$$

$$= \frac{(18 - 20)^2}{20} + \frac{(21 - 20)^2}{20} + \frac{(17 - 20)^2}{20} + \frac{(21 - 20)^2}{20}$$

$$+ \frac{(19 - 20)^2}{20} + \frac{(24 - 20)^2}{20}$$

$$= 1.60$$

There are $6 - 1$, or 5, degrees of freedom. From Table VIII in the Appendix we find that the $\chi^2_{0.05}$ value with 5 degrees of freedom is 11.070. Since the test statistic has a value of only 1.60, which is considerably less than 11.070, we do not reject the null hypothesis. Any differences between the observed frequencies and the expected frequencies are due purely to chance.

Goodness of fit   The following examples will further illustrate how the chi-square test statistic can be used to test **goodness of fit**, that is, to determine whether the observed frequencies fit with what was expected.

E X A M P L E   1   The number of phone calls received per day by a local chapter of Alcoholics Anonymous is as follows:

	M	T	W	T	F
Number of Calls Received	173	153	146	182	193

Using a 5% level of significance, test the null hypothesis that the number of calls received is independent of the day of the week.

SOLUTION   We first calculate the number of expected calls per day. If the number of calls received is independent of the day of the week, we would expect to receive

$$\frac{173 + 153 + 146 + 182 + 193}{5} = 169.4$$

calls per day. We can then set up the following table:

	M	T	W	T	F
Observed Number of Calls	173	153	146	182	193
Expected Number of Calls	169.4	169.4	169.4	169.4	169.4

Now we calculate the value of the chi-square test statistic. We have

$$\chi^2 = \Sigma \frac{(O - E)^2}{E}$$

$$= \frac{(173 - 169.4)^2}{169.4} + \frac{(153 - 169.4)^2}{169.4} + \frac{(146 - 169.4)^2}{169.4}$$

$$+ \frac{(182 - 169.4)^2}{169.4} + \frac{(193 - 169.4)^2}{169.4}$$

$$= 0.0765 + 1.5877 + 3.2323 + 0.9372 + 3.2878$$

$$= 9.1215$$

There are $5 - 1$, or 4 degrees of freedom. From Table VIII in the Appendix we find that the $\chi^2_{0.05}$ value with 4 degrees of freedom is 9.488. Since the test statistic has a value of 9.1215, which is less than 9.488, we do not reject the null hypothesis and the claim that the number of calls received is independent of the day of the week.

HISTORICAL NOTE

The most widely used measure of correlation was developed by Karl Pearson. He was born on March 27, 1857 in London. Although he earned his B.A. (with honors) in mathematics in 1879, he actually began his pioneering work in statistics in 1893. His association with Walter Weldon (a zoology professor at University College) and Sir Francis Galton (see the discussion in the Historical Note of Chapter 11, page 589) as well as his analysis of published material led him to discover the chi-square goodness-of-fit test and then the correlation coefficient. He retired from University College in 1933 and died in London on April 27, 1936.

EXAMPLE 2 A scientist has been experimenting with rats. As a result of certain injections, the scientist claims that when two black rats are mated, the offspring will be black, white, and gray in the proportion $5:4:3$. $\Big($ This means that the probability of a black offspring is $\dfrac{5}{12}$, the probability of a white offspring is $\dfrac{4}{12}$, and the probability of a gray rat is $\dfrac{3}{12}$. $\Big)$ Many rats were mated after being injected with the chemical. Of 180 newborn rats 71 were black, 69 were white, and 40 were gray. Can we accept the scientist's claim that the true proportion is $5:4:3$? Use a 5% level of significance.

SOLUTION We first calculate the expected frequencies. Out of 180 rats we would expect $180\Big(\dfrac{5}{12}\Big)$, or 75, of them to be black. Similarly, out of 180 rats we would expect $180\Big(\dfrac{4}{12}\Big)$, or 60, of them to be white, and we would expect $180\Big(\dfrac{3}{12}\Big)$, or 45, of them to be gray. We now set up the following table:

Color of Rat	Expected Frequency	Observed Frequency
Black	75	71
White	60	69
Gray	45	40

The value of the χ^2 test statistic is

$$\chi^2 = \Sigma \, \frac{(O - E)^2}{E}$$

$$= \frac{(71 - 75)^2}{75} + \frac{(69 - 60)^2}{60} + \frac{(40 - 45)^2}{45}$$

$$= 2.12$$

There are $3 - 1$, or 2 degrees of freedom. From Table VIII in the Appendix we find that the $\chi^2_{0.05}$ value with 2 degrees of freedom is 5.991. Since the test statistic has a value of 2.12, which is less than 5.991, we do not reject the null hypothesis and the scientist's claim.

EXERCISES FOR SECTION 12.3

1. An obstetrician delivered 58 babies in the fall, 89 babies in the winter, 61 babies in the spring, and 37 babies in the summer. Using a 5% level of significance, test the null hypothesis that the proportion of babies delivered by the obstetrician during the various seasons of the year is independent of the season of the year.

2. Professor Gertrude Brier is a college guidance counselor. She is analyzing the majors of this year's 1464 graduates. She has compiled the following data.

Major	Number of Students with This Major
Computer Science	292
Engineering	268
Nursing	286
Business	319
Other	299

Using a 5% level of significance, test the null hypothesis that the proportions of students majoring in the various categories mentioned are not significantly different.

3. The Acme Baking Company conducted a taste test at the Riverside Shopping Mall. Each of the 290 randomly selected volunteers was given samples of five different new low-calorie cakes to be introduced into the marketplace. The volunteers were then asked to select the one with the greatest taste appeal. The results of the survey were then tallied and presented to the manufacturer. After analyzing the accompanying data, the manufacturer concludes that the volunteers expressed no significant difference in their opinions of the taste appeal of the new cakes and that any observed difference was due merely to random selection.

Cakes with Greatest Appeal	Frequency of Response
A	58
B	49
C	63
D	70
E	50

Using a 1% level of significance, test the null hypothesis that the manufacturer's assumptions are correct, and that customers prefer each of these cakes equally.

4. There are five entrance ramps to the main parking lot at Meadows Stadium. The number of cars entering the parking lot through these ramps for one particular football game was as follows:

Entrance Ramp	Number of Cars Entering Through This Ramp
Northeast	258
Northwest	236
South	322
West	314
Southeast	219

Using a 5% level of significance, test the null hypothesis that all entrance ramps to the stadium parking lot are used by cars with the same frequency.

5. "Fear of heights" is a feeling often demonstrated by many people. A psychologist is interested in knowing whether frequency of expression of such fear varies by age-group. The following data have been collected for 329 people who are afraid of heights.

Age of Person (in years)	10–19	20–29	30–39	40–49	50–59
Number of People Expressing Fear of Heights	68	52	63	71	75

Using a 5% level of significance, does the data indicate that people in some age-groups are more likely to be afraid of heights than people in other age-groups?

6. The Austrian monk Gregor Mendel performed many experiments with garden peas. In one such experiment, the following results were obtained:

387 round and yellow peas
180 wrinkled and yellow peas
121 round and green peas
26 wrinkled and green peas

Using a 5% level of significance, test the null hypothesis that the frequencies of these types of peas are in the ratio $9:4:3:1$.

7. Using gene-splicing techniques, a researcher claims to have developed a new strain of cattle with many desirable traits which were named characteristics A, B, C, and D. The following results were obtained in numerous gene-splicing experiments:

395 had characteristic A
257 had characteristic B
158 had characteristic C
140 had characteristic D

Using a 5% level of significance, test the null hypothesis that the frequencies of these types of characteristics obtained by gene splicing are in the ratio $8:6:3:2$.

8. A botanist recently crossbred several plants. Of 220 new plants, 92 were orange, 68 were yellow, 32 were pink, and 28 were red. Using a 5% level of significance, test the null hypothesis that the true proportion of plants with these colors is $5:3:2:1$.

9. The vending machines in the student cafeteria dispense different types of soft drink. The food manager notices that these machines dispensed the following items in 1000 sales:

Item	Number of Cans Dispensed
Pepsi-Cola	275
Coca-Cola	261
Sunkist Orange	184
Sprite	168
Fresca	112

Using a 5% level of significance, test the null hypothesis that the proportions of these drinks dispensed are in the ratio $7:5:4:3:1$.

10. There were 205 babies born at Wellington Hospital during the first three months of 1996. The ages of the mothers at the time of birth is as follows:

Age of Mothers	Number of Mothers in This Age-Group
17–21	43
22–26	47
27–31	45
32–36	36
37–41	34

Using a 5% level of significance, test the null hypothesis that the proportion of women delivering babies is not significantly different for the different age-groups.

11. Many tourists from foreign countries come to the United States and spend substantial sums of money on a daily basis as shown in the following chart:

	Number of Tourists	Amount Spent per Day per Person
Britain	2.921 million	$68
Germany	1.705 million	76
France	0.863 million	71
South Korea	0.504 million	83
Venezuela	0.420 million	91

Source: U.S. Travel and Tourism Administration, 1995.

Using a 5% level of significance, test the null hypothesis that the amount of money spent per day per tourist is independent of the tourist's home country.

12.4 CONTINGENCY TABLES

A very useful application of the χ^2 test discussed in Section 12.2 occurs in connection with contingency tables. Contingency tables are used when we wish to determine whether two variables of classification are related or dependent one on the other. For example, consider the following chart, which indicates the eye color and hair color of 100 randomly selected girls.

	Brown Eyes	**Blue Eyes**
Light Hair	10	33
Dark Hair	44	13

Is eye color independent of hair color or is there a significant relationship between hair color and eye color?

Contingency tables are especially useful in the social sciences where data are collected and often classified into two main groups. We might be interested in determining whether a relationship exists between these two ways of classifying the data or whether they are independent. We have the following definition:

Definition 12.1
Contingency table

A **contingency table** is an arrangement of data into a two-way classification. One of the classifications is entered in rows and the other in columns.

When dealing with contingency tables, remember that the null hypothesis always assumes that the two ways of classifying the data are independent. We use the χ^2 test statistic as discussed in Section 12.2. The only difference is that we compute the expected frequency for each cell by using the following formula.

FORMULA 12.2

Expected frequency

The **expected frequency** of any cell in a contingency table is found by multiplying the total of the row with the total of the column to which the cell belongs. The product is then divided by the total sample size.

The following examples will illustrate how we apply the χ^2 test to contingency tables.

EXAMPLE 1 Let us consider the contingency table given at the beginning of this section. Is eye color independent of hair color?

SOLUTION In order to compute the χ^2 test statistic, we must first compute the row total, column total, and the total sample size. We have

$$Row\ Totals: \quad 10 + 33 = 43$$
$$44 + 13 = 57$$
$$Column\ Totals: \quad 10 + 44 = 54$$
$$33 + 13 = 46$$
$$Total\ Sample\ Size: \quad 10 + 33 + 44 + 13 = 100$$

We indicate these values in the following table:

	Brown Eyes	Blue Eyes	Row Total
Light Hair	10 (23.22)	33 (19.78)	43
Dark Hair	44 (30.78)	13 (26.22)	57
Column Total	54	46	

The expected value for the cell in the first row first column is obtained by multiplying the first row total with the first column total and then dividing the product by the total sample size. We get

$$\frac{54 \times 43}{100} = 23.22$$

For the first row second column we have

$$\frac{46 \times 43}{100} = 19.78$$

For the second row first column we have

$$\frac{54 \times 57}{100} = 30.78$$

For the second row second column we have

$$\frac{46 \times 57}{100} = 26.22$$

These values are entered in parentheses in the appropriate cell. We now use Formula 12.1 of Section 12.2 and calculate the χ^2 statistic. We have

$$\chi^2 = \Sigma \frac{(O - E)^2}{E}$$

$$= \frac{(10 - 23.22)^2}{23.22} + \frac{(33 - 19.78)^2}{19.78} + \frac{(44 - 30.78)^2}{30.78} + \frac{(13 - 26.22)^2}{26.22}$$

$$= 7.53 + 8.84 + 5.68 + 6.67$$

$$= 28.72$$

The χ^2 test statistic has a value of 28.72. *If the contingency table has r rows and c columns, then the number of degrees of freedom is $(r - 1)(c - 1)$.* In this example there are two rows and two columns, so there are $(2 - 1) \cdot (2 - 1)$, or $1 \cdot 1$, which is 1 degree of freedom. From Table VIII in the Appendix we find that the $\chi^2_{0.05}$ value with 1 degree of freedom is 3.841. Since we obtained a value of 28.72, we reject the null hypothesis and conclude that hair color and eye color are *not* independent.

EXAMPLE 2

Criminal Analysis

A sociologist is interested in determining whether the occurrence of different types of crimes varies from city to city. An analysis of 1100 reported crimes produced the following results.

	Type of Crime				
	Rape	Auto Theft	Robbery and Burglary	Other	Total
City A	76 (61.35)	112 (146.34)	87 (72.66)	102 (96.65)	377
City B	64 (68.83)	184 (164.20)	77 (81.52)	98 (108.44)	423
City C	39 (48.82)	131 (116.45)	48 (57.82)	82 (76.91)	300
Total	179	427	212	282	1100

Do these data indicate that the occurrence of a type of crime is dependent on the location of the city? (Use a 5% level of significance.)

SOLUTION   We first calculate the expected frequency for each cell. We have

$$\text{First Row:} \quad \frac{(179)(377)}{1100} = 61.35 \quad \frac{(427)(377)}{1100} = 146.34$$

$$\frac{(212)(377)}{1100} = 72.66 \quad \frac{(282)(377)}{1100} = 96.65$$

$$\text{Second Row:} \quad \frac{(179)(423)}{1100} = 68.83 \quad \frac{(427)(423)}{1100} = 164.20$$

$$\frac{(212)(423)}{1100} = 81.52 \quad \frac{(282)(423)}{1100} = 108.44$$

$$\text{Third Row:} \quad \frac{(179)(300)}{1100} = 48.82 \quad \frac{(427)(300)}{1100} = 116.45$$

$$\frac{(212)(300)}{1100} = 57.82 \quad \frac{(282)(300)}{1100} = 76.91$$

These numbers appear in parentheses in the preceding chart. Now we calculate the χ^2 test statistic, getting

$$\chi^2 = \Sigma \frac{(O - E)^2}{E}$$

$$= \frac{(76 - 61.35)^2}{61.35} + \frac{(112 - 146.34)^2}{146.34} + \dots + \frac{(82 - 76.91)^2}{76.91}$$

$$= 24.46$$

There are three rows and four columns so that there are 6 degrees of freedom since

$$(3 - 1)(4 - 1) = 2 \cdot 3 = 6$$

From Table VIII in the Appendix we find that the $\chi^2_{0.05}$ value with 6 degrees of freedom is 12.592. Since we obtained a χ^2 value of 24.46, we reject the null hypothesis and conclude that the type of crime and the location of the city are not independent.

COMMENT   It should be noted that the chi-square independence test is appropriate only when applied to sample data and not to data for the entire population.

COMMENT   For the chi-square independence test to be valid, it is assumed that all expected frequencies are at least 1 and that at most 20% of the expected frequencies are less than 5. Otherwise this test procedure should not be used.

   We can summarize the procedures for using the chi-square test for independence as follows:

To perform a chi-square independence test:
1. State the null and alternate hypotheses.
2. Calculate the expected frequency for each cell by using Formula 12.2 so that the entry placed in each cell below the observed frequency is given by

$$\text{Entry} = \frac{(\text{row total}) \cdot (\text{column total})}{\text{total sample size}}$$

3. Assuming the expected frequencies satisfy the necessary preconditions for the chi-square test to be applicable (as given above), determine the significance level.

4. Compute the value of the test statistic $\chi^2 = \dfrac{\Sigma(O - E)^2}{E}$ and compare the results with the critical value χ^2 as given in Table VIII. If the contingency table has r rows and c columns, then the number of degrees of freedom is $(r - 1)(c - 1)$.

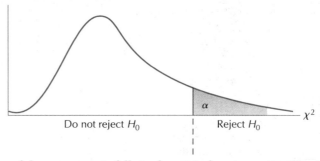

Do not reject H_0 — α —   Reject H_0 — χ^2

5. If the value of the test statistic falls in the critical region, reject H_0. Otherwise, do not reject H_0. Specify your conclusion in words.

EXERCISES FOR SECTION 12.4

1. In 1995, a newspaper reporter (David Hamel, Chicago) conducted a nationwide survey of 1758 people to find out their opinion on legalizing mercy killings. The results of the survey are shown in the following table:

	Number of People	
Region of Residency	For Proposal	Against Proposal
East	184	214
Midwest	130	368
South	148	286
Far West	166	262

Using a 5% level of significance, test the null hypothesis that the region in which the respondent lives is independent of their opinion on legalizing mercy killings.

2. An analysis of the medical records of the Bakst Corporation reveals the following information about 732 of its employees who applied for coverage under the company's extended health-coverage plans:

	Heavy Smoker and Drinker	Heavy Smoker and Nondrinker	Nonsmoker But Heavy Drinker	Nonsmoker and Nondrinker
Male	172	108	124	16
Female	84	146	62	20

Using a 5% level of significance, test the null hypothesis that the sex of the employee is independent of whether or not the employee is a heavy smoker or a drinker.

3. In an effort to generate sales and influence the shopper, many businesses advertise heavily in magazines or media directed primarily toward some particular segment of the population. An advertising agency recently conducted a random survey of shoppers at five different sales outlets to determine who usually buys shirts for a man—the man himself or a woman. The results of the survey are presented in the accompanying chart:

		Sales Location				
		I	II	III	IV	V
Who Bought the Shirt?	*Male*	42	49	51	68	81
	Female	48	57	50	83	92

Using a 1% level of significance, test the null hypothesis that the number of males or females who will buy a shirt for a man is not significantly different at each of these locations.

4. In a recent survey conducted by the Acme Insurance Company, the following information was obtained about 400 cars equipped with some type of anti-theft device:

		Type of Anti-Theft Device		
		Ignition Shut-Off	Steering Wheel Lock	Burglar Alarm
Size of Car	*Compact*	68	37	63
	Intermediate	42	29	56
	Large	27	32	46

Using a 5% level of significance, test the null hypothesis that the type of anti-theft device used is independent of the size of the car.

5. A quality-control engineer at General Electronics Company samples parts from each of the company's production lines on a daily basis. The following data are available for one week's production. (The production lines are closed on Friday.)

		Number of Defective Items Produced on Production Lines				
		I	II	III	IV	V
Day of the Week	*Mon.*	28	22	26	22	21
	Tues.	25	27	29	31	23
	Wed.	24	28	25	26	20
	Thurs.	21	23	25	26	24

Using a 5% level of significance, test the null hypothesis that the number of defective items produced on the production line is independent of the day of the week.

6. Recently, a nationwide survey by the Brown Associates (San Francisco, 1995) of used-car buyers in different age-groups was conducted to determine which feature in a used car was of utmost concern to the buyer. The following results were obtained:

		Car Feature			
		Economy of Operation	Styling	Size	Color
Age (in years) of Respondent	*Under 30*	362	492	401	461
	30–50	527	409	396	303
	Over 50	622	383	471	252

Using a 5% level of significance, test the null hypothesis that the age of the respondent is independent of the car feature selected.

7. The Atlas Travel Agency arranges many tour packages. It recently surveyed 676 of its customers to obtain some information needed to arrange such tour packages.

Each customer was asked the same question: "How should the tour packages be arranged?" The answers of the customers are presented in the accompanying table:

		How Should Tour Be Arranged?		
		Every Day Planned Out	Several Days Planned—The Remainder Free Time	All Free Time—Only Arrange Air Transportation
	Under 50	41	92	112
Age of Customer (in years)	Between 50 and 65	56	81	75
	Over 65	67	103	49

Using a 5% level of significance, test the null hypothesis that the age of the customer is independent of the customer's views as to how the tours should be arranged.

8. An analysis of the cause of death of 867 patients at the Morgantown Hospital and Convalescent Home reveals the following information:

		Cause of Death				
		Disease of Heart and Blood Vessels	Cancer	Pneumonia and Influenza	Accidents	All Other Causes
	Under 50	52	39	22	68	84
Age of Person (in years)	Between 50 and 65	69	44	53	29	78
	Over 65	77	52	91	17	92

Using a 5% level of significance, test the null hypothesis that the cause of death is not related to a person's age.

9. According to the U.S. Census Bureau, the number of marriages ending in divorce each year has been steadily increasing. A sociologist conducted a survey to determine if the highest level of education attained by at least one of the partners is independent of or affects the number of years that a marriage will last (before ending in divorce). A summary of that survey is given below:

		Number of Years Marriage Lasted Before Ending in Divorce			
		0–1	2–5	6–15	16–20
Highest Education Level Attained by at Least One of the Partners	Elementary School	91	82	74	40
	High School	109	91	79	87
	College	133	111	96	68

Using a 5% level of significance, test the null hypothesis that the highest educational level attained by at least one of the partners is independent of the number of years that a marriage will last.

10. *When is an auto accident more likely to occur?* An insurance company statistician is analyzing claims involving car accidents that occurred over a period of many months at different times of the day, as shown in the accompanying chart.

		Frequency for the Number of Reported Accidents		
		0–10 Accidents	11–20 Accidents	Over 20 Accidents
	Between Midnight and 6 A.M.	18	10	5
	Between 6 A.M. and 9 A.M.	33	24	21
Time of Day Accident Occurred	Between 9 A.M. and 3 P.M.	27	19	16
	Between 3 P.M. and 7 P.M.	36	24	20
	Between 7 P.M. and Midnight	19	16	13

Using a 5% level of significance, test the null hypothesis that the frequency with which the different number of reported accidents occur is independent of the time of the day when the accident occurs.

11. Four hundred computer science majors were asked to indicate what they used their personal computers for. The results of the survey are shown below:

		Principal Use of Computer			
		Word Processing	Games	Spreadsheet Analysis	Data Base
	Freshman	24	31	18	25
Class Standing	Sophomore	27	29	16	27
	Junior	25	32	15	28
	Senior	26	34	17	26

Using a 5% level of significance, test the null hypothesis that the principal use of a computer by a computer science major is independent of the class standing of the individual.

12.5   USING COMPUTER PACKAGES

MINITAB is well-suited to perform all calculations needed to test for independence. One only need to enter the cell frequencies to obtain the results. MINITAB is especially useful when the number of rows and columns of a contingency table is large. Let us apply the MINITAB program to the data given in Example 2 of Section 12.4 (page 678).

```
MTB  > READ THE TABLE INTO C1, C2, C3, C4
DATA >     76   112   87   102
DATA >     64   184   77    98
DATA >     39   131   48    82
DATA >     END
```

```
MTB  > CHISQUARE ANALYSIS ON TABLE IN C1, C2, C3, C4
Expected counts are printed below observed counts:
```

	C1	C2	C3	C4	Total
1	76	112	87	102	377
	61.35	146.34	72.66	96.65	
2	64	184	77	98	423
	68.83	164.20	81.52	108.44	
3	39	131	48	82	300
	48.82	116.45	57.82	76.91	
Total	179	427	212	282	1100

```
ChiSq = 3.499 + 8.060 + 2.831 + 0.296 +
        0.339 + 2.387 + 0.251 + 1.005 +
        1.975 + 1.817 + 1.667 + 0.337 = 24.465
```

```
df = 6
```

```
MTB > STOP
```

Although MINITAB does not have a direct command that would allow a goodness-of-fit test, we can still perform such a test by combining several commands. To illustrate the procedure, let us refer back to the data given in Example 2 of Section 12.3 (page 672). We first enter the observed frequencies in column C1 and the expected frequencies in column C2. Then we calculate $\frac{(O - E)^2}{E}$. In MINITAB, this is accomplished by using the command:

```
LET C3 = (C1-C2)**2/C2.
```

Finally, we sum the values listed in column C3. This gives us the value of the χ^2 test statistic. When applied to our example, we have the following:

MTB > READ THE FOLLOWING INTO C1, C2 ← This command enters the data on the observed frequencies and expected frequencies in columns C1 and C2.

DATA > 71 75

DATA > 69 60

DATA > 40 45

DATA > END

MTB > LET C3 = (C1-C2)**2/C2 ← This command calculates $\dfrac{(O - E)^2}{E}$ for each category.

MTB > PRINT C1 C2 C3

Row	C1	C2	C3
1	71	75	0.2133
2	69	60	1.35000
3	40	45	0.55556

MTB > SUM C3 ← This command prints the sum of C3, which is the value of χ^2 test statistic.

SUM = 2.1189

Thus, the value of the chi-square test statistic is 2.1189 or 2.12. From Table VIII in the Appendix, we find that the $\chi^2_{0.05}$ value with $3 - 1$ or 2, degrees of freedom is 5.991. Since the test statistic has a value of 2.12, which is less than 5.991, we do not reject the null hypothesis and the scientist's claim.

EXERCISES FOR SECTION 12.5

Use MINITAB to solve each of the following exercises.

1. According to official records, the number of motorists calling the local AAA for roadside assistance during one week was as follows:

Day of the Week	Mon.	Tues.	Wed.	Thurs.	Fri.
Number of Calls Received	212	276	198	246	253

Using a 5% level of significance, test the null hypothesis that the number of calls received is independent of the day of the week.

2. Are female jurors more liberal than male jurors? A recent study of 284 jurors involved in similar cases produced the following results:

Sex of Juror		Tendency of Juror		
		Liberal	Fair	Vindictive
	Male	41	47	55
	Female	52	48	41

Using a 5% level of significance, test the null hypothesis that the sex of the juror is independent of his or her tendency to be liberal, fair, or vindictive.

3. A sociologist interviewed 224 male viewers and 231 female viewers to determine their opinion as to whether there is too much violence being shown on TV during prime time. The following results were obtained:

Sex of Viewer		Is There Too Much Violence on Prime-Time TV?		
		Definitely	Sometimes	Not Really
	Male	71	101	52
	Female	82	92	57

Using a 5% level of significance, test the null hypothesis that the sex of the viewer is independent of the viewer's opinion.

12.6   SUMMARY

In this chapter we discussed the chi-square distribution and how it can be used to test hypotheses that differences between expected frequencies and observed frequencies are due purely to chance.

We applied the chi-square test statistic to test whether observed frequency distributions are in agreement with expected mathematical frequencies.

The chi-square test statistic can also be used to analyze whether the two factors of a contingency table are independent. This is very useful, especially in the social sciences, where the data are often grouped according to two factors.

When using the χ^2 test statistic we must take great care in determining the number of degrees of freedom. Also, as we pointed out in a comment on page 664, each expected cell frequency must be at least 5 for the χ^2 test statistic to be applied.

Study Guide

The following is a chapter outline in capsule form. You should be able to demonstrate your knowledge of the ideas mentioned by giving definitions, descriptions, or specific examples. Page references are given in parentheses.

When data have been categorized and arranged in tabular format, the numbers that should appear in each cell (box) are called **expected frequencies** and the numbers that were actually observed are called **observed frequencies**. (page 660)

To determine whether the difference between the observed frequencies and the expected frequencies is significant, we calculate the **chi-square test statistic**. (page 660)

To determine when the value of the χ^2 test statistic is significant, we use the **chi-square distribution**. The number of degrees of freedom is always one less than the number of sample proportions that we are testing. (page 661)

The **goodness-of-fit test** is used to determine whether the observed frequencies fit with what was expected. (page 671)

A **contingency table** is an arrangement of data into a two-way classification. One of the classifications is entered in rows and the other in columns. (page 676)

Contingency tables are used to summarize observed and expected frequencies for a **test of independence of two variables** associated with a population. (page 676)

The **expected frequency** of any cell in a contingency table is found by multiplying the total of the row with the total of the column to which the cell belongs. The product is then divided by the total sample size. (page 676)

If a contingency table has r rows and c columns, then the number of degrees of freedom is $(r - 1) \cdot (c - 1)$. (page 677)

Formulas to Remember

You should be able to identify each symbol in the following formulas, understand the relationships among the symbols expressed in each formula, understand the significance of each formula, and use the formulas in solving problems.

1. $\chi^2 = \Sigma \dfrac{(O - E)^2}{E}$

   where O = observed frequency and E = expected frequency.
2. The expected frequency for any cell of a contingency table:

   $$\frac{\text{(total of row to which cell belongs)} \cdot \text{(total of column to which cell belongs)}}{\text{total sample size}}$$
3. The number of degrees of freedom for a contingency table:

   $$(r - 1)(c - 1)$$

   where c = number of columns and r = number of rows

Testing Your Understanding of This Chapter's Concepts

1. When using the chi-square independence test, the null hypothesis is that the two characteristics under consideration are independent whereas the alternative hypothesis is that they are dependent. Why is it true that this test is always right-tailed?

2. When applying the chi-square test of independence, if the null hypothesis is rejected, we can conclude that the two characteristics are dependent. Does this imply a causal relationship between the two characteristics? Explain your answer. Give an example to support your answer.

3. A school principal is interested in knowing whether the grade level of a child determines which parent will come (assuming only one parent comes) on open school day to discuss the child's academic progress with the teacher. The following randomly selected data are available.

		Grade Level of Child in School		
		Elementary School	Junior High School	Senior High School
Which Parent Came	*Father*	32	47	96
on Open School Day?	*Mother*	78	53	29

The following is a **MINITAB** printout of this information, showing the expected frequencies as well as the calculated values for chi-square.

a. Verify the values given for the expected frequencies in each cell and also the chi-square values.

b. Find the value of p.

```
MTB > READ TABLE IN C1, C2, C3
DATA> 32   47   96
DATA> 78   53   29
DATA> END
        2 ROWS READ

MTB > CHISQUARE ANALYSIS ON TABLE IN C1, C2, C3
Expected counts are printed below observed counts:
          C1        C2        C3     Total
    1     32        47        96       175
          57.46     52.24     65.30

    2     78        53        29       160
          52.54     47.76     59.70

Total     110       100       125      335

ChiSq = 11.283 + 0.525 + 14.435 +
        12.341 + 0.575 + 15.788 = 54.947
df = 2

MTB > STOP
```

Chapter Test

Multiple Choice Questions

For Questions 1–5, use the following information: 1158 patients in Walker Hospital were asked to indicate whether or not they were satisfied with the nursing service at the hospital. The following results were obtained:

		Satisfied	Not Satisfied	No Opinion
	Under 30 Years	226	182	123
Age of Patient	*30–50 Years*	117	153	89
	Over 50 Years	88	112	68

1. What is the expected number of patients over 50 years of age who had no opinion on the nursing service at the hospital?
   a. 99.7   b. 103.5   c. 64.8   d. 86.8   e. none of these
2. What is the expected number of patients under 30 years of age who were satisfied with the nursing service at the hospital?
   a. 197.6   b. 205   c. 128.4   d. 133.6   e. none of these
3. The number of degrees of freedom for this problem is
   a. 3   b. 9   c. 4   d. 8   e. none of these
4. The value of χ^2 for this problem is
   a. 11.183   b. 12.743   c. 13.486   d. 12.147   e. none of these
5. Using a 5% level of significance, test the null hypothesis that the age of the patient is independent of the view expressed by the patient.
   a. do not reject null hypothesis   b. reject null hypothesis
   c. not enough information given
6. When a certain plant is crossbred with another plant, the offspring should be red, pink, and yellow in the ratio 2:5:7. Of 280 new plants, 22 were red, 110 were pink and 148 were yellow. At the 5% level of significance, should we reject the null hypothesis that the true proportion is 2:5:7?
   a. do not reject null hypothesis   b. reject null hypothesis
   c. not enough information given
7. The number of express delivery packages received by a publishing company is shown below:

Mon.	Tues.	Wed.	Thurs.	Fri.
12	22	18	19	14

Using a 5% level of significance, test the null hypothesis that the number of express delivery packages received is independent of the day of the week.
a. do not reject null hypothesis   b. reject null hypothesis
c. not enough information given

For Questions 8–10, use the following information: A survey of 400 residents of a senior citizen home was taken to determine their view on the social activities provided by the senior citizen home. The results of the survey according to the sex of the resident are shown below.

		Male	Female
Do You Like the	*Yes*	88	127
Social Activities?	*No*	87	98

8. Find the expected number of males who like the social activities?
   a. 80.9   b. 104.1   c. 94.1   d. 120.9   e. none of these
9. The number of degrees of freedom for the problem is
   a. 1   b. 2   c. 3   d. 4   e. none of these
10. The value of χ^2 for this problem is
    a. 1.493   b. 1.397   c. 1.468   d. 1.502   e. none of these

Supplementary Exercises

11. *Do obese parents have obese children?*   A panel of researchers conducted a survey of 610 families. The results of the survey are shown below:

		Obese Parent(s)	
		Yes	No
Obese Child	*Yes*	198	152
	No	137	123

Using a 1% level of significance, test the null hypothesis that the obesity of a parent is not a factor in the obesity of the child.

12. Teenage drug use is a growing problem facing our society. A psychologist is interested in knowing whether there is a relationship between the use of drugs by teenagers and the financial status of the parents. The following data have been collected.

		Financial Status of Parents		
		Low Income	Middle Income	Upper Income
	Never	20	21	23
Frequency of Drug Use by Teenagers	*Occasionally*	71	57	63
	Frequently	47	39	52

Using a 5% level of significance, test the null hypothesis that the financial status of parents is independent of the frequency of the use of drugs by teenagers.

13. A politician surveyed 680 people to determine whether they were in favor of legalizing gambling and the setting up of gambling casinos. The following table indicates their responses according to their socioeconomic standing:

		Socioeconomic Standing		
		Lower Class	Middle Class	Upper Class
In Favor of Legalizing Gambling Casinos?	Yes	184	167	132
	No	66	63	68

Using a 5% level of significance, test the null hypothesis that the proportion of people in favor of legalizing gambling casinos is the same for all socioeconomic classes.

14. A vending-machine operator believes that the number of cans of Coca-Cola, Pepsi-Cola, Seven-Up, Orange Soda, and Sprite sold per day from his vending machines is in the ratio $7:5:4:2:3$. In a random sample of 630 purchases of cans of soda, the following sodas were purchased:

Soda Purchased	Frequency
Coca-Cola	299
Pepsi-Cola	126
Seven-Up	118
Orange Soda	70
Sprite	17

Using a 5% level of significance, test the null hypothesis that the vending-machine operator's claim is correct.

15. A sociologist is interested in knowing whether women who come from large families also have large families themselves. To investigate this question, a random survey is taken of 1000 women from different-sized families. The data are presented below:

		Number of Children in Family from Which Mother Came			
		1	2–3	4–5	6 and Over
	1	90	88	79	36
Number of Children Mother Now Has	2–3	43	160	85	20
	4–5	28	38	56	37
	6 and Over	39	52	63	86

Using a 5% level of significance, test the null hypothesis that the number of children that a woman has is independent of the size of the family from which she comes.

16. A school principal is interested in knowing whether the grade level of a child determines which parent will come (assuming only one parent comes) on open school days to discuss the child's academic progress with the teacher. The following randomly selected data are available.

		Grade Level of Child in School		
		Elementary School	Junior High School	Senior High School
Which Parent Came on Open School Day?	*Father*	36	48	97
	Mother	79	54	36

At the 5% level of significance, is grade level a factor in determining which parent will come to school to discuss the child's academic progress with the teacher on open school day?

17. A housing sales agency interviewed 1030 potential buyers of apartments in a new housing complex to determine which factor was important in deciding whether or not to buy. The results are presented in the accompanying table.

		Factor				
		Crime-Free Neighborhood	Many Playgrounds	Schools Nearby	Shopping Nearby	Parking
Gender of Respondent	*Male*	178	87	78	57	112
	Female	196	101	71	68	82

Using a 5% level of significance, test the null hypothesis that the gender of the respondent is independent of the factor considered important by a potential buyer.

18. A recent survey° of 1500 smokers and nonsmokers contained the following question: "Do you believe that smoking is harmful to your health?" The responses are shown below:

	Smokers	Nonsmokers
Harmful	208	722
Not Harmful	462	108

What is the expected number of smokers who believe that smoking is harmful?
a. 402   b. 242.1   c. 398.3   d. 424.72   e. none of these

°*Source:* Giligman, "The Effects of Smoking," 1995.

19. The Florida State Attorney General's office receives a number of telephone complaints daily from senior citizens dealing with deceptive business practices. During one week, the number of calls received was as follows:

Day of Week	Number of Calls Received
Mon.	34
Tue.	33
Wed.	38
Thurs.	29
Fri.	31

Using a 5% level of significance, test the null hypothesis that the number of calls received is independent of the day of the week.

20. A hospital administrator wishes to know if there is a relationship between a patient's length of stay in the hospital for a specified illness and the extent of the person's hospitalization insurance. Of people who were patients in the hospital over the past six months for the specified illness, 1440 are randomly selected. The extent of hospital insurance reimbursement is determined and is shown below.

		Extent of Insurance Reimbursement		
		50% to 59%	60% to 69%	70% to 79%	90% or above
	1–5	39	55	69	84
Length of Hospital Stay (in days)	*6–8*	60	71	88	103
	9–12	79	92	104	127
	More Than 12	81	101	132	155

Using a 5% level of significance, test the null hypothesis that the length of hospital stay is independent of the extent of hospital insurance reimbursement.

21. A researcher claims that among people who eat in restaurants, the following preferences (given in percentage instead of ratio form) apply.

Day of Week Preferred	Number of People Indicating Preference
Sunday	25%
Monday	11%
Tuesday	9%
Wednesday	3%
Thursday	5%
Friday	19%
Saturday	28%

Sample results from 429 patrons of the restaurants are given here.

Day	Sun.	Mon.	Tues.	Wed.	Thurs.	Fri.	Sat.
Number	89	38	42	11	23	75	151

Using a 5% level of significance, test the null hypothesis that the given percentages are all correct.

22. Many of our nation's lakes and rivers are polluted with mercury, PCVs, or other toxic chemicals. Data gathered at numerous locations along several rivers by the Charleston Environmental Group are shown in the accompanying table.

		River			
		A	B	C	D
	Low	4	11	10	6
Level of Pollution	Moderate	5	13	12	4
	High	9	12	14	8

Using a 5% level of significance, test the null hypothesis that the level of pollution is independent of the river involved.

23. Is our government doing enough for the homeless? Bates and Young[*] interviewed 625 randomly selected Americans. The following results were obtained.

		City in Which Respondent Lives				
		Los Angeles	New York	Chicago	Detroit	Miami
Is Our	Yes	57	51	65	58	82
Government Doing Enough for the Homeless?	No	67	59	72	63	51

Using a 5% level of significance, test the null hypothesis that the proportion of Americans who think that our government is doing enough for the homeless is the same for the cities mentioned.

24. Safety engineers often analyze industrial accidents to determine whether the day of the week when an accident occurs is independent of the time of the day that it occurs. For one very large company employing many workers the following data on industrial accidents are available.

[*]Bates and Young. *What is Our Government Doing for the Homeless?* San Francisco, 1993.

		Day of the Week				
		Mon.	Tues.	Wed.	Thurs.	Fri.
Time of Day When Accident Occurred	*Morning Shift*	17	16	14	13	19
	Afternoon Shift	15	18	13	12	17
	Evening Shift	20	12	18	14	15

Using a 1% level of significance, test the null hypothesis that the time of day when the accident occurs is independent of the day of the week that the accident occurs.

25. According to a certain theory, the number of flowers of each of four types should be as shown in the accompanying table. The actual numbers obtained as a result of genetic manipulation are also given:

Type of Flower	Expected	Actual
Smooth pink	318	323
Wrinkled pink	109	99
Smooth red	89	95
Wrinkled red	74	73
Total	590	590

At the 5% level of significance, are the data consistent with the theory?

Thinking Critically

1. In a recent survey conducted by the Acme Insurance Company of 66 cars equipped with some anti-theft device, the following information was obtained.

		Type of Anti-Theft Device		
		Ignition Shutoff	Steering Wheel Lock	Burglar Alarm
Size of Car	*Compact*	8	7	15
	Intermediate	3	8	5
	Large	9	3	8

Using a 5% level of significance, test the null hypothesis that the type of anti-theft device used is independent of the size of the car.

2. In the previous exercise it was not really possible to perform the chi-square test of independence since the assumptions regarding the expected frequencies were not

met for each cell. To overcome this difficulty we can combine rows or columns, eliminate rows or columns, or increase the sample size. Combine the last two rows to form a new contingency table and perform the hypothesis test indicated earlier. Compare the results. What can we conclude?

3. Many people have never flown in an airplane because they are afraid of heights. The following data were obtained by asking 300 people in each of five different age-groups whether they had serious fears of heights.

Age-Group (in years)	Number of People Who Fear Heights
11–20	141
21–30	113
31–40	108
41–50	96
51–60	89

Using a 5% level of significance, test the null hypothesis that the proportion of people in each age-group who fear heights is the same. (*Hint:* The contingency table must account for all 1500 people.)

4. For the following contingency table,

		Column	
		1	2
Row	1	a	b
	2	c	d

verify that the chi-square test statistic is: $\chi^2 = \dfrac{(a + b + c + d)(ad - bc)^2}{(a + b)(c + d)(b + d)(a + c)}$.

5. Jeremy is informed by the registrar that in order to graduate, he must successfully complete "MTH 102—Math for Liberal Arts Students." Next semester, two sections of the course will be offered, one taught by Prof. Gertrude Brier and the other taught by Prof. Fran Silvernail. As math has always been one of Jeremy's most difficult subjects, Jeremy would like to register for the section taught by the "easier" teacher. From the math department bulletin board where grades are posted, Jeremy observes that in the previous semester Prof. Brier gave 32 passing grades in a class of 39 students and Prof. Silvernail gave 38 passing grades in a class of 48 students.

a. Construct a 2 × 2 contingency table for the above data and then compute the χ^2 test statistic for independence and use it to find the p-value.

b. Using the hypothesis testing procedure for the difference between proportions discussed in Chapter 10, compute the value of the test statistic, z, and use it to find the p-value.

c. Compare the answers obtained in parts (a) and (b). Comment.

Case Study

The American Cancer Society in its *Cancer Facts and Figures—1995* reports that over 800,000 new cases of highly curable basal cell or squamous skin cancer are reported each year. The most serious skin cancer is melanoma, which was diagnosed in about 35,000 cases in 1995. An estimated 9300 deaths in 1995 were attributed to skin cancer with 7200 from malignant melanoma. Almost all of these cases in the United States are sun related. The sun's ultraviolet rays are strongest between 10 A.M. and 3 P.M. Exposure at these times should be avoided or sunscreen should be used. Early detection is critical. About 82% of melanomas are diagnosed in a local stage. The five-year survival rate for localized malignant melanoma is 93%.

A medical researcher is interested in determining if a person's age is a factor in whether they use sunscreens or not when sunbathing. A survey of 1200 returning vacationers produced the following results:

		Age-Group (in years)			
		Between 20 and 29	Between 30 and 39	Between 40 and 49	Over 50
Use Sunscreen?	Yes	320	278	148	82
	No	167	131	49	25

Using a 5% level of significance, test the null hypothesis that there is no significant difference between the corresponding proportion of vacationers who use sunscreen.

13

ANALYSIS OF VARIANCE

Chapter Outline

DID YOU KNOW THAT

a   about 31% of the U.S. adult population has borderline high cholesterol levels between 200 and 239 mg/dl while another 20% has high cholesterol levels above 240 mg/dl? (*Source: JAMA* 269(23):3009–14, June 16, 1993)

b   fourth graders who had changed schools three or more times in the previous two years achieved an average math proficiency score of 199 on a 500-point scale as opposed to an average score of 224 for those who had not changed schools? (*Source: Nat'l Assessment of Educ. Progress*, U.S. Accounting Office, 1995)

c   one in six of our nation's third graders—more than half a million—have attended at least three different schools according to a 1995 U.S. General Accounting Office study? Furthermore, these "on the move" students are not necessarily from inner city, suburban, or rural schools, as the following chart indicates.

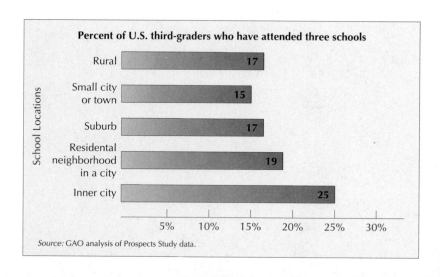

Percent of U.S. third-graders who have attended three schools

School Locations

Rural	17
Small city or town	15
Suburb	17
Residental neighborhood in a city	19
Inner city	25

5%   10%   15%   20%   25%   30%

Source: GAO analysis of Prospects Study data.

Many companies produce nickel-cadmium rechargeable batteries. However, these batteries can be recharged only a specific number of times before becoming useless. See discussion on page 735. (*George Semple*)

I n previous chapters we presented hypothesis-testing procedures for determining whether the difference between two sample means is significant. In this chapter we test hypotheses about several means. The ANOVA techniques to be discussed allow us to test the null hypothesis that *all* sample means are equal against the alternate hypothesis that at least one mean value is different.

Chapter Objectives

- **To analyze** a technique that is used when comparing several sample means. This technique is called single-factor analysis of variance. (Section 13.2)

- **To discuss** how to set up Analysis of Variance (ANOVA) charts and develop formulas to use with ANOVA charts. (Section 13.3)

- **To study** a distribution that we use when comparing variances. This is the *F*-distribution. (Section 13.3)

- **To apply** the *F*-distribution to help us determine whether differences in sample means are significant. (Section 13.3)

- **To learn** how to analyze two-way ANOVA tables where two factors may affect the sample means. (Section 13.4)

S T A T I S T I C S   I N   A C T I O N

The most widely used international yardstick of a country's wealth is Gross National Product (GNP) per capita. Basically, GNP measures the total of *all* goods and services produced by a country's economy in a given year. Some economists use another internationally comparable measure of living standards. This is a country's Purchasing Power Parity (PPP), which compares the cost of goods in the purchasing power of each country's currency in that country with what the same things might cost in the United States. Al-though neither GNP nor PPP provides a perfect picture of a country's standard of living, the two measures provide some surprisingly different conclusions for certain countries.

An analysis of the table below indicates that for many countries the 1993 GNP per capita and the 1993 PPP are considerably different. How do we determine if there is any significant difference between the average GNP and the average PPP? A country-by-country comparison is very time consuming.

Comparison of GNP Per Capita and Purchasing Power Parity

Country	1993 GNP per capita	1993 PPP
Switzerland	$36,410	$23,620
Japan	31,450	21,090
United States°	24,750	24,750
United Kingdom	17,970	17,750
Greece	7,390	8,360
Mexico	3,750	7,100
Brazil	3,020	5,470
Russia	2,350	5,240
Indonesia	730	3,140
Egypt	660	3,530
China	490	2,120
Nigeria	310	1,480
India	290	1,250
Mozambique	80	380

°U.S. dollar is used as the standard.

Source: The World Bank Atlas, 1995.

13.1   INTRODUCTION

Suppose a company is interested in determining whether changing the lighting conditions of its factory will have any effect on the number of items produced by a worker. It can arrange the lighting conditions in four different ways: A, B, C, and D. To determine how lighting affects production, the factory supervisor decides to randomly arrange the lights under each of the four possible conditions for an equal number of days. She will then measure worker productivity under each of these conditions.

After all the data are collected, the manager calculates the number of items produced under each of these lighting conditions. She now wishes to test whether there is any significant difference between these sample means. How does she proceed?

The null hypothesis is

$$H_0: \quad \mu_A = \mu_B = \mu_C = \mu_D$$

This tells us that the mean number of items produced is the same for each lighting condition. The alternative hypothesis is that not all the means are the same. Thus, the lighting condition does affect production. The manager will reject the null hypothesis if one (or more) of the means is different from the others.

To determine if lighting affects production, she could use the techniques of Section 10.5 for testing differences between means. However, she would have to apply those techniques many times since each time she would be able to test only two means. Thus she would have to test the following null hypotheses:

$$H_0: \quad \mu_A = \mu_B \qquad H_0: \quad \mu_A = \mu_D \qquad H_0: \quad \mu_B = \mu_D$$

$$H_0: \quad \mu_A = \mu_C \qquad H_0: \quad \mu_B = \mu_C \qquad H_0: \quad \mu_C = \mu_D$$

In order to conclude that there is no significant difference between the sample means, she would have to accept each of the six separate null hypotheses previously listed. Performing these tests involves a tremendous amount of computation. Furthermore, in doing it this way, the probability of making a Type-I error is quite large.

Since many problems that occur in applied statistics involve testing whether there is any significant difference between several means, statisticians use a special analysis of *ANOVA* variance technique. This is abbreviated as **ANOVA**. These types of problems occur so frequently because many companies often hire engineers to design new techniques or processes for producing products. The company must then compare the sample means of these new processes with the sample mean of the old process. Using the ANOVA technique, we have to test only one hypothesis in order to determine when to reject or not reject a null hypothesis.

ANOVA techniques can be applied to many different types of problems. In this chapter we first discuss one simple application of ANOVA techniques. This is the case in which the data are classified into groups on the basis of one single property. Then we discuss two-way ANOVA techniques, where the data are classified into groups on the basis of two properties.

13.2   DIFFERENCES AMONG r SAMPLE MEANS

Suppose that a chemical researcher is interested in comparing the average number of months that four different paints will last on an exterior wall before beginning to blister. The researcher has available the following data on the lasting time of three paint samples for each of the four different brands of paint.

	Lasting Time (in months)		
Paint A	Paint B	Paint C	Paint D
13.4	14.8	12.9	13.7
11.8	13.9	14.3	14.1
14.4	12.4	13.3	12.1
Average: 13.2	13.7	13.5	13.3

The average lasting time for these paints is 13.2, 13.7, 13.5, and 13.3 months, respectively. Since not all the paints had the same average lasting time, the chemical researcher is interested in knowing whether the observed differences between the sample means is significant or whether they can be attributed purely to chance. Is there a rule available for determining when observed differences between sample means are significant?

Before proceeding further, we wish to emphasize the fact that our discussion will be based on certain assumptions. These are the following:

1. We will always assume that the populations from which the samples are obtained are normally distributed and that the samples are obtained independently of one another.
2. We will also assume that the populations from which the samples are obtained all have the same (often unknown) variance, σ^2.

HISTORICAL NOTE

One of the major contributors to ANOVA techniques was Ronald Fisher. Fisher was born on February 17, 1880, in London, England. After graduating from Cambridge in 1912, he worked at a variety of jobs. In 1919 he was hired by Rothamsted Experimental Station in Harpenden, West Hertford, England, as a statistician to analyze data that had accumulated over a period of many years. It was here that he developed the field of analysis of variance.

In 1925, he published *Statistics for Research Workers*, a book that remained in print for well over 40 years. Fisher's ANOVA techniques and experimental designs have been applied to a variety of biological and agricultural studies.

In 1959, Fisher "retired" and moved to Australia where he got a job in the Division of Mathematical Statistics of the Commonwealth Scientific and Industrial Research Organization. He worked there until his death in 1962.

When these assumptions are satisfied, we can obtain an estimate of σ^2 in two different ways. First we note that if the paints have the same mean, then the average lasting time of the different paints (in our example) is

$$\bar{x} = \frac{13.2 + 13.7 + 13.5 + 13.3}{4} = 13.425$$

The averages will be normally distributed with a variance of σ^2. This can be estimated by the following procedure:

$$s_{\bar{x}}^2 = \frac{(13.2 - 13.425)^2 + (13.7 - 13.425)^2 + (13.5 - 13.425)^2 + (13.3 - 13.425)^2}{4 - 1}$$

$$= \frac{0.1475}{3} = 0.0491667$$

From our earlier work we already know that

$$s_{\bar{x}}^2 \approx \left(\frac{\sigma}{\sqrt{n}}\right)^2 = \frac{\sigma^2}{n}$$

Here n = the number of samples of each brand of paint. Thus, $n \cdot s_{\bar{x}}^2$ can be used as an estimate of σ^2. In our case we get $3(0.0491667) = 0.1475$. Our first estimate of σ^2, which is based on the variation among the sample means (assuming that all the groups have the same mean), has a value of 0.1475.

Since under our second assumption the populations from which the samples are obtained all have the same (often unknown) variance, σ^2, we can obtain another estimate of σ^2 by selecting any one of the sample variances. For each brand of paint we can obtain an estimate of σ^2 based on 2 degrees of freedom. We can then pool the estimates of the σ from each of the brands of paint to obtain an estimate of σ. This estimate of σ is not affected by variation of the population means among the paints. We have the following computations.

Paint Brand	Variance		
A	$s_1^2 =$	$\dfrac{(13.4 - 13.2)^2 + (11.8 - 13.2)^2 + (14.4 - 13.2)^2}{3 - 1}$	$= \dfrac{3.44}{2} = 1.72$
B	$s_2^2 =$	$\dfrac{(14.8 - 13.7)^2 + (13.9 - 13.7)^2 + (12.4 - 13.7)^2}{3 - 1}$	$= \dfrac{2.94}{2} = 1.47$
C	$s_3^2 =$	$\dfrac{(12.9 - 13.5)^2 + (14.3 - 13.5)^2 + (13.3 - 13.5)^2}{3 - 1}$	$= \dfrac{1.04}{2} = 0.52$
D	$s_4^2 =$	$\dfrac{(13.7 - 13.3)^2 + (14.1 - 13.3)^2 + (12.1 - 13.3)^2}{3 - 1}$	$= \dfrac{2.24}{2} = 1.12$

Taking the average of these gives

$$\frac{s_1^2 + s_2^2 + s_3^2 + s_4^2}{4} = \frac{1.72 + 1.47 + 0.52 + 1.12}{4}$$

$$= \frac{4.83}{4} = 1.2075$$

Thus, our second estimate of σ^2, which is based on the variation within the samples, is 1.2075.

How do we compare these two estimates of σ^2? Our first estimate of σ^2, which is based on the variation among the sample means, had a value of 0.1475, whereas our second estimate, which is based on the variation within the samples (or on the fact that the variation is due purely to chance), had a value of 1.2075. It seems reasonable that the variation among the sample means should be larger than that which is due purely to chance.

Fortunately, statisticians use a special F-distribution that facilitates such comparisons. It tells us under what conditions to reject the null hypothesis assuming the variation of the sample means is simply too great to be attributed to chance. This, of course, implies that the differences among the sample means is significant. The exact technique will be discussed in the next section after we introduce a convenient tabular arrangement for displaying our computations.

COMMENT   The straight averaging technique discussed until now works since the example given is a case with equal sample sizes.

13.3   ONE-WAY OR SINGLE-FACTOR ANOVA

Let us return to the factory supervisor problem discussed in the introduction. The manager repeated each of the lighting conditions on five different days. Table 13.1 indicates the number of items produced under each of the conditions.

TABLE 13.1   **Number of Items Produced Under Different Lighting Conditions**

		Day					
		1	2	3	4	5	Average
	A	12	10	15	12	13	12.4
	B	16	14	9	10	15	12.8
Lighting Conditions	C	11	15	8	12	10	11.2
	D	15	14	12	11	13	13

The means for each of these lighting conditions are 12.4, 12.8, 11.2, and 13. Since the average number of items produced is not the same for all the lighting conditions, the question becomes: "Is the variation among individual sample means due purely to chance or are these differences due to the different lighting conditions?" The null hypothesis is H_0: $\mu_A = \mu_B = \mu_C = \mu_D$. If the null hypothesis is true, then the lighting condition does not affect production. We can then consider our results as a listing of the number of items produced on 20 randomly selected days under one of the lighting conditions.

As mentioned earlier, when working with such problems we will assume that the number of items produced is normally distributed and that the variance is the same for each of the lighting conditions. We will also assume that the experiments with the different lighting conditions are independent of each other. (Usually the experiments are conducted in random order so that we have independence.)

Let us now apply the ANOVA techniques to this problem. Notice that we have included the row totals in Table 13.2. In applying the ANOVA technique, we analyze the reasons for the difference among the means. What is the source of the variance?

TABLE 13.2   Number of Items Produced Under Different Lighting Conditions

		Day					
		1	2	3	4	5	Row Total
	A	12	10	15	12	13	62
	B	16	14	9	10	15	64
Lighting Conditions	C	11	15	8	12	10	56
	D	15	14	12	11	13	65
						247	*Grand Total*

Total sum of squares   The first number that we calculate is called the **Total Sum of Squares** and is abbreviated as SS(total). To obtain the SS(total), we first square each of the numbers in the table and add these squares together. Then we divide the square of the grand total (total of all the rows or the total of all the columns) by the number in the sample. (In our case there are 20 in the sample.) Subtracting this result from the sum of the squares, we get in our case

$$[12^2 + 10^2 + 15^2 + 12^2 + 13^2 + 16^2 + 14^2 + 9^2 + 10^2$$

$$+ 15^2 + 11^2 + 15^2 + 8^2 + 12^2 + 10^2 + 15^2 + 14^2 + 12^2 + 11^2 + 13^2] - \left[\frac{(247)^2}{20}\right]$$

$$= 3149 - 3050.45$$

$$= 98.55$$

†Thus, SS(total) = 98.55. (An † will precede important results.)

To find the next important result, we first square each of the row totals and divide the sum of these squares by the number in each row, which is 5. Then we divide the square of the grand total by the number in the sample. We subtract this result from the sum of the squares and get

$$\left(\frac{62^2 + 64^2 + 56^2 + 65^2}{5}\right) - \frac{(247)^2}{20} = 3060.2 - 3050.45 = 9.75$$

Sum of Squares Due to the Factor   This result is called the **Sum of Squares Due to the Factor**. The factor in our example is the lighting conditions. Thus,

$$^\dagger SS(\text{lighting}) = 9.75$$

Sum of Squares of the Error   Now we calculate a number that is called the **Sum of Squares of the Error** abbreviated as SS(error). We first square each entry of the table and add these squares together. Then we square each row total and divide the sum of these squares by the number in each row. Subtracting this result from the sum of the squares, we get

$$(12^2 + 10^2 + 15^2 + 12^2 + \cdots + 11^2 + 13^2) - \frac{62^2 + 64^2 + 56^2 + 65^2}{5}$$

$$= 3149 - 3060.2$$

$$= 88.8$$

Thus,

$$^\dagger SS(\text{error}) = 88.8$$

COMMENT   Actually, SS(error) = SS(total) − SS(factor).

ANOVA table   We enter these results along with the appropriate number of degrees of freedom, which we will determine shortly, in a table known as an **ANOVA table**. The general form of an ANOVA table is shown in Table 13.3.

TABLE 13.3   ANOVA Table

Source of Variation	Sum of Squares (SS)	Degrees of Freedom (df)	Mean Square (MS)	F-Ratio
Factors of experiment				
Error				
Total				

In our case we have the ANOVA table shown in Table 13.4. We have entered the important results that we obtained in the preceding paragraphs in the appropriate space of the table. The degrees of freedom are obtained according to the following rules.

TABLE 13.4   **ANOVA Table**

Source of Variation	Sum of Squares	Degrees of Freedom	Mean Square	F-Ratio
Factors (lighting)	9.75	3	3.25	$\dfrac{3.25}{5.55} = 0.59$
Error	88.8	16	5.55	
Total	98.55	19		

1. The number of degrees of freedom for the factor tested is one less than the number of possible levels at which the factor is tested. If there are r levels of the factor, the number of degrees of freedom is $r - 1$.
2. The number of degrees of freedom for the error is one less than the number of repetitions of each condition multiplied by the number of possible levels of the factor. If each condition is repeated c times, then the number of degrees of freedom is $r(c - 1)$.
3. Finally, the number of degrees of freedom for the total is one less than the total number in the sample. If the total sample consists of n things, the number of degrees of freedom is $n - 1$.

In our case there are four experimental conditions and each condition is repeated five times so that $r = 4$, $c = 5$, and $n = 20$.
Thus,

$$df(\text{factor}) = r - 1 = 4 - 1 = 3$$
$$df(\text{error}) = r(c - 1) = 4(5 - 1) = 4 \cdot 4 = 16$$
$$df(\text{total}) = n - 1 = 20 - 1 = 19$$

We enter these values in an ANOVA table as shown in Table 13.4.

Mean square   Now we calculate the values that belong in the **mean square** column. We divide the sum of squares for the row by the number of degrees of freedom for that row. Thus,

$$MS(\text{lighting}) = \frac{SS(\text{lighting})}{df(\text{lighting})} = \frac{9.75}{3} = 3.25$$

$$MS(\text{error}) = \frac{SS(\text{error})}{df(\text{error})} = \frac{88.8}{16} = 5.55$$

COMMENT: $MS(\text{Error})$ is often abbreviated as MSE

Although we are comparing the sample means for the different lighting conditions, the ANOVA technique compares variances under the assumption that the variance among all the levels is 0. If the variance is 0, then all the means are the same.

Suppose we are given two samples of size n_1 and n_2 for two populations that have roughly the shape of a normal distribution. Then we base tests on the equality of the two population standard deviations (or variances) on the ratio $\frac{s_1^2}{s_2^2}$ or $\frac{s_2^2}{s_1^2}$, where s_1 and s_2

Means are not significantly different

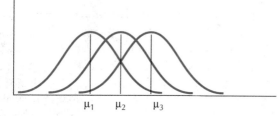

μ_1 μ_2 μ_3

All means are different

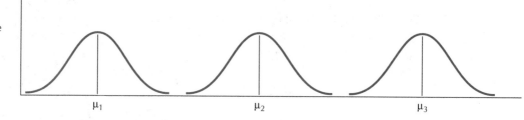

μ_1 μ_2 μ_3

At least one mean is significantly different

μ_1 μ_2 μ_3

FIGURE 13.1

F-distribution are the standard deviations of both samples. The distribution of such a ratio is a continuous distribution called the **F-distribution**. This distribution depends on the number of degrees of freedom of both sample estimates, $n_1 - 1$ and $n_2 - 1$. When using this distribution **we reject the null hypotheses** of equal variances ($\sigma_1 = \sigma_2$) when the test statistic value of F exceeds $F_{\alpha/2}$, where α is the level of significance.

When comparing variances we use the F-distribution. The test statistic is

$$F = \frac{MS(\text{lighting})}{MS(\text{error})}$$

COMMENT   When using the F-distribution we are testing whether or not the variances among the factors are zero (in which case the means are equal). These situations are pictured in Figure 13.1 for the case involving three means.

As mentioned earlier, the values of the F-distribution depend on the number of degrees of freedom of the numerator, that of the denominator, and on the level of significance. Table IX in the Appendix gives us the different values corresponding to different degrees of freedom. From Table IX we find that the F-value with 3 degrees of freedom for the numerator and 16 degrees of freedom for the denominator at the 5% level of significance is 3.24. We reject the null hypothesis if the F-value is greater than 3.24.

In our case the value of the F statistic is

$$\frac{MS(\text{lighting})}{MS(\text{error})} = \frac{3.25}{5.55} = 0.59$$

Since 0.59 is less than 3.24, we do not reject the null hypothesis. Thus, the data do not indicate that the lighting condition affects production.

ANOVA table Let us summarize the procedure to be used when testing several sample means by the ANOVA technique. First draw an ANOVA table such as shown in Table 13.5. In

TABLE 13.5 **Sample ANOVA Table**

Source of Variation	Sum of Squares	Degrees of Freedom	Mean Square	F-Ratio
Factor of the experiment	(1)	(4)	(7)	(9)
Error	(2)	(5)	(8)	
Total	(3)	(6)		

this table we have placed numbers in parentheses in the various cells. The values that belong in each of these cells are obtained by using the following formulas.

Cell (1): $\dfrac{\Sigma(\text{each row total})^2}{\text{number in each row}} - \dfrac{(\text{grand total})^2}{\text{total sample size}}$

Cell (2): $\Sigma(\text{each original number})^2 - \dfrac{\Sigma(\text{each row total})^2}{\text{number in each row}}$

Cell (3): $\Sigma(\text{each original number})^2 - \dfrac{(\text{grand total})^2}{\text{total sample size}}$

Cell (4): There are r levels of the factor, $r - 1$

Cell (5): If there are c repetitions of each of the r levels, $r(c - 1)$

Cell (6): If the total sample consists of n things, $n - 1$

Cell (7): $\dfrac{\text{cell}(1)\text{ value}}{\text{cell}(4)\text{ value}}$

Cell (8): $\dfrac{\text{cell}(2)\text{ value}}{\text{cell}(5)\text{ value}}$

Cell (9): $\dfrac{\text{cell}(7)\text{ value}}{\text{cell}(8)\text{ value}}$

Although these formulas seem complicated, they are easy to use as the following examples illustrate.

EXAMPLE 1
Testing the Tar Content of Cigarettes

Three brands of cigarettes, six cigarettes from each brand, were tested for tar content. Do the following data indicate that there is a significant difference in the average tar content for these three brands of cigarettes? (Use a 5% level of significance.)

								Row Total
	X	14	16	12	18	11	13	84
Brand	Y	10	11	22	19	9	18	89
	Z	24	22	19	18	20	19	122
								295 *Grand Total*

SOLUTION   The null hypothesis is that the sample mean tar content is the same for the three brands of cigarettes. To solve this problem, we set up the following ANOVA table.

Source of Variation	Sum of Squares	Degrees of Freedom	Mean Square	F-Ratio
Cigarettes	(1)	(4)	(7)	(9)
Error	(2)	(5)	(8)	
Total	(3)	(6)		

Now we calculate the values that belong in each of the cells by using the formulas given on page 710. We have

Cell (1): $\dfrac{\Sigma(\text{each row total})^2}{\text{number in each row}} - \dfrac{(\text{grand total})^2}{\text{total sample size}}$

$$= \frac{84^2 + 89^2 + 122^2}{6} - \frac{(295)^2}{18}$$

$$= 4976.83 - 4834.72$$

$$= 142.11$$

Cell (2): $\Sigma(\text{each original number})^2 - \dfrac{\Sigma(\text{each row total})^2}{\text{number in each row}}$

$$= (14^2 + 16^2 + 12^2 + \cdots + 19^2) - \frac{(84^2 + 89^2 + 122^2)}{6}$$

$$= 5187 - 4976.83$$

$$= 210.17$$

Cell (3): $\Sigma(\text{each original number})^2 - \dfrac{(\text{grand total})^2}{\text{total sample size}}$

$$= (14^2 + 16^2 + 12^2 + \cdots + 19^2) - \frac{(295)^2}{18}$$

$$= 5187 - 4834.72$$

$$= 352.28$$

Cell (4): $df = r - 1$
$$= 3 - 1 = 2$$

Cell (5): $df = r(c - 1)$
$$= 3(6 - 1) = 3(5) = 15$$

Cell (6): $df = n - 1$
$$= 18 - 1 = 17$$

Cell (7): $\dfrac{\text{cell(1) value}}{\text{cell(4) value}} = \dfrac{142.11}{2} = 71.06$

$$\text{Cell (8):} \quad \frac{\text{cell(2) value}}{\text{cell(5) value}} = \frac{210.17}{15} = 14.01$$

$$\text{Cell (9):} \quad \frac{\text{cell(7) value}}{\text{cell(8) value}} = \frac{71.06}{14.01} = 5.07$$

Now we enter these numbers on the ANOVA table.

Source of Variation	Sum of Squares	Degrees of Freedom	Mean Square	F-Ratio
Cigarettes	142.11 (1)	2 (4)	71.06 (7)	5.07 (9)
Error	210.17 (2)	15 (5)	14.01 (8)	
Total	352.28 (3)	17 (6)		

From Table IX in the Appendix we find that the F-value with 2 degrees of freedom for the numerator and 15 degrees of freedom for the denominator at the 5% level of significance is 3.68. Since we obtained an F-value of 5.07, which is larger than 3.68, we reject the null hypothesis that the sample mean tar content is the same for the three brands of cigarettes.

COMMENT   Rejection of the null hypothesis H_0: $\mu_1 = \mu_2 = \mu_3$ does not tell us where the difference lies.

EXAMPLE   2   Four groups of five students each were taught a skill by four different teaching techniques. At the end of a specified time the students were tested and their scores were recorded. Do the following data indicate that there is a significant difference in the mean achievement for the four teaching techniques? (Use a 1% level of significance.)

							Row Total
	A	64	73	69	75	78	359
Teaching	B	73	82	71	69	74	369
Technique	C	61	79	71	73	66	350
	D	63	69	68	74	75	349
							1427 *Grand Total*

SOLUTION   The null hypothesis is that the mean achievement for the four teaching techniques is the same. To solve this problem, we set up the following ANOVA table.

Source of Variation	Sum of Squares	Degrees of Freedom	Mean Square	F-Ratio
Teaching method	52.15 [(1)]	3 [(4)]	17.38 [(7)]	0.56 [(9)]
Error	500.4 [(2)]	16 [(5)]	31.28 [(8)]	
Total	552.55 [(3)]	19 [(6)]		

Now we calculate the values that belong in each of the cells by using the formulas given on page 710. We have

Cell (1): $\dfrac{\Sigma(\text{each row total})^2}{\text{number in each row}} - \dfrac{(\text{grand total})^2}{\text{total sample size}}$

$$= \frac{359^2 + 369^2 + 350^2 + 349^2}{5} - \frac{(1427)^2}{20}$$

$$= 101{,}868.6 - 101{,}816.45$$

$$= 52.15$$

Cell (2): $\Sigma(\text{each original number})^2 - \dfrac{\Sigma(\text{each row total})^2}{\text{number in each row}}$

$$= (64^2 + 73^2 + 69^2 + 75^2 + \cdots + 74^2 + 75^2)$$
$$- \left(\frac{359^2 + 369^2 + 350^2 + 349^2}{5}\right)$$

$$= 102{,}369 - 101{,}868.6$$

$$= 500.4$$

Cell (3): $\Sigma(\text{each original number})^2 - \dfrac{(\text{grand total})^2}{\text{total sample size}}$

$$= (64^2 + 73^2 + 69^2 + 75^2 + \cdots + 74^2 + 75^2) - \frac{(1427)^2}{20}$$

$$= 102{,}369 - 101{,}816.45$$

$$= 552.55$$

Cell (4): $df = r - 1 = 4 - 1 = 3$

Cell (5): $df = r(c - 1) = 4(5 - 1) = 4(4) = 16$

Cell (6): $df = n - 1 = 20 - 1 = 19$

$$\text{Cell (7):} \quad \frac{\text{cell(1) value}}{\text{cell(4) value}} = \frac{52.15}{3} = 17.38$$

$$\text{Cell (8):} \quad \frac{\text{cell(2) value}}{\text{cell(5) value}} = \frac{500.4}{16} = 31.28$$

$$\text{Cell (9):} \quad \frac{\text{cell(7) value}}{\text{cell(8) value}} = \frac{17.38}{31.28} = 0.56$$

We enter all these values on the ANOVA table as shown.

From Table IX in the Appendix we find that the F-value with 3 degrees of freedom for the numerator and 16 degrees of freedom for the denominator at the 1% level of significance is 5.29. Since we obtained an F-value of 0.56, we do not reject the null hypothesis that the mean achievement for the four teaching techniques is the same.

*13.4   TWO-WAY ANOVA

Let us return to the example given in Section 13.2 and construct an analysis-of-variance table for the data that are reproduced here except that we have now rewritten the chart in a slightly different manner. (The reason for this will be apparent shortly.)

	Lasting Time (in months)			Row Total	
Paint A	13.4	11.8	14.4	39.6	
Paint B	14.8	13.9	12.4	41.1	
Paint C	12.9	14.3	13.3	40.5	
Paint D	13.7	14.1	12.1	39.9	
Column Total	54.8	54.1	52.2	161.1	Grand Total

We set up the following ANOVA table to which we have added the values that belong in each of the cells by using the formulas given on page 710.

Source of Variation	Sum of Squares	Degrees of Freedom	Mean Square	F-Ratio
Different Paints	0.4425 [1]	3 [4]	0.1475 [7]	0.1222 [9]
Error	9.66 [2]	8 [5]	1.2075 [8]	
Total	10.1025 [3]	11 [6]		

Thus, we have

Cell (1): $\left(\dfrac{(39.6)^2 + (41.1)^2 + (40.5)^2 + (39.9)^2}{3}\right) - \dfrac{(161.1)^2}{12}$

$= 2163.21 - 2162.7675 = 0.4425$

Cell (2): $[(13.4)^2 + (11.8)^2 + \cdots + (12.1)^2]$

$- \left[\dfrac{(39.6)^2 + (41.1)^2 + (40.5)^2 + (39.9)^2}{3}\right]$

$= 2172.87 - 2163.21 = 9.66$

Cell (3): $[(13.4)^2 + (11.8)^2 + \cdots + (12.1)^2] - \dfrac{(161.1)^2}{4.3}$

$= 2172.87 - 2162.7675 = 10.1025$

Cell (4): $4 - 1 = 3$

Cell (5): $4(3 - 1) = 8$

Cell (6): $12 - 1 = 11$

Cell (7): $\dfrac{0.4425}{3} = 0.1475$

Cell (8): $\dfrac{9.66}{8} = 1.2075$

Cell (9): $\dfrac{0.1475}{1.2075} = 0.1222$

From Table IX in the Appendix we find that the F-value with 3 degrees of freedom for the numerator and 8 degrees of freedom for the denominator at the 5% level of significance is 4.07. Therefore, since we obtained an F-value of 0.1222, we do not reject the null hypothesis that there is a significant difference in the average lasting times of the four brands of paint.

After analyzing the data carefully the chemical engineer discovers that the test conditions for each of the brands of paint were not the same and that the samples for each brand of paint were subject to different temperature and humidity conditions. This factor, which was not considered earlier, definitely has to be considered. Thus, instead of just having to determine if the average lasting time for the four different brands of paint is the same, we also have to consider the various test conditions as a possible cause of the variation as well as pure chance. Under these new circumstances the data must be arranged as follows:

	Test Conditions			
	Test Condition I	Test Condition II	Test Condition III	Row Total
Paint A	13.4	11.8	14.4	39.6
Paint B	14.8	13.9	12.4	41.1
Paint C	12.9	14.3	13.3	40.5
Paint D	13.7	14.1	12.1	39.9
Column Total	54.8	54.1	52.2	161.1 *Grand Total*

Blocks
Treatments
Two-way analysis of variance

When discussing this second factor it is customary to refer to it as **blocks** as opposed to the original factor, which is referred to as the **treatments**. We then have the following ANOVA table for such a **two-way analysis of variance**.

Source of Variation	Sum of Squares	Degrees of Freedom	Mean Square	F-Ratio
Treatments	(1)	(5)	(9)	(12)
Blocks	(2)	(6)	(10)	(13)
Error	(3)	(7)	(11)	
Total	(4)	(8)		

The values that belong in each of these cells for a two-way analysis of variance table are obtained by using the following formulas.

$$\text{Cell (1):} \quad \frac{\Sigma(\text{each row total})^2}{\text{number in each row}} - \frac{(\text{grand total})^2}{r \cdot c}$$

where r = number of levels of one factor (row)

c = number of levels of second factor (column)

$$\text{Cell (2):} \quad \frac{\Sigma(\text{each column total})^2}{\text{number in each column}} - \frac{(\text{grand total})^2}{r \cdot c}$$

$$\text{Cell (3):} \quad \Sigma(\text{each original number})^2 - \frac{\Sigma(\text{each row total})^2}{\text{number in each row}}$$
$$- \frac{\Sigma(\text{each column total})^2}{\text{number in each column}} + \frac{(\text{grand total})^2}{r \cdot c}$$

$$\text{Cell (4):} \quad \Sigma(\text{each original number})^2 - \frac{(\text{grand total})^2}{r \cdot c}$$

Cell (5): $r - 1$

Cell (6): $c - 1$

Cell (7): $(r - 1)(c - 1)$

Cell (8): $rc - 1$

Cell (9): $\dfrac{\text{cell}(1) \text{ value}}{\text{cell}(5) \text{ value}}$

Cell (10): $\dfrac{\text{cell}(2) \text{ value}}{\text{cell}(6) \text{ value}}$

Cell (11): $\dfrac{\text{cell}(3) \text{ value}}{\text{cell}(7) \text{ value}}$

Cell (12): $\dfrac{\text{cell}(9) \text{ value}}{\text{cell}(11) \text{ value}}$

Cell (13): $\dfrac{\text{cell}(10) \text{ value}}{\text{cell}(11) \text{ value}}$

Let us apply these formulas to our example. We have the following two-way ANOVA table to which we have added the values that belong in each of the cells by using the formulas given previously.

Source of Variation	Sum of Squares	Degrees of Freedom	Mean Square	F-Ratio
Paints	0.4425 (1)	3 (5)	0.1475 (9)	0.1011 (12)
Condition	0.905 (2)	2 (6)	0.4525 (10)	0.3101 (13)
Error	8.755 (3)	6 (7)	1.4592 (11)	
Total	10.1025 (4)	11 (8)		

We have

Cell (1) value: $\left[\dfrac{(39.6)^2 + (41.1)^2 + (40.5)^2 + (39.9)^2}{3} \right] - \dfrac{(161.1)^2}{4 \cdot 3}$

$$= 2163.21 - 2162.7675 = 0.4425$$

Cell (2) value: $\left[\dfrac{(54.8)^2 + (54.1)^2 + (52.2)^2}{4} \right] - \dfrac{(161.1)^2}{4 \cdot 3}$

$$= 2163.6725 - 2162.7675 = 0.905$$

Cell (3) value: $[(13.4)^2 + (11.8)^2 + \cdots + (12.1)^2]$

$$- \left[\frac{(39.6)^2 + (41.1)^2 + (40.5)^2 + (39.9)^2}{3} \right]$$

$$- \left[\frac{(54.8)^2 + (54.1)^2 + (52.2)^2}{4} \right] + \frac{(161.1)^2}{4 \cdot 3}$$

$$= 2172.87 - 2163.21 - 2163.6725 + 2162.7675$$

$$= 8.755$$

Cell (4) value: $= [(13.4)^2 + (11.8)^2 + \cdots + (12.1)^2] - \dfrac{(161.1)^2}{4 \cdot 3}$

$$= 2172.87 - 2162.7675 = 10.1025$$

Cell (5) value: $4 - 1 = 3$

Cell (6) value: $3 - 1 = 2$

Cell (7) value: $(4 - 1)(3 - 1) = 6$

Cell (8) value: $4 \cdot 3 - 1 = 12 - 1 = 11$

Cell (9) value: $\dfrac{0.4425}{3} = 0.1475$

Cell (10) value: $\dfrac{0.905}{2} = 0.4525$

Cell (11) value: $\dfrac{8.755}{6} = 1.4592$

Cell (12) value: $\dfrac{0.1475}{1.4592} = 0.1011$

Cell (13) value: $\dfrac{0.4525}{1.4592} = 0.3101$

From Table IX in the Appendix we find that the F-value with 3 degrees of freedom for the numerator and 6 degrees of freedom for the denominator is 4.76 at the 5% level of significance. Similarly, the F-value with 2 degrees of freedom for the numerator and 6 degrees of freedom for the denominator is 5.14. In both cases we obtained values of F equal to 0.1011 and 0.3101, respectively, that are less than 4.76 and 5.14. Thus, we do not reject the null hypothesis that the average drying time for the several brands of paint is significantly different nor do we reject the null hypothesis that the different testing conditions were a factor in determining average drying time.

COMMENT   Our analysis of two-way ANOVA is by no means a complete discussion of the topic. A thorough two-way ANOVA problem and solution depends on how the treatments were assigned to the experimental units (i.e., how the randomization was conducted) as well as on the number of replications of each experimental condition. Our intention was merely to introduce you to the idea of two-way ANOVA. A more detailed analysis can be found by consulting any book on the analysis of variance.

COMPARISON OF NEW YORK STATE AND NATIONAL AVERAGES ON COLLEGE BOARD ACHIEVEMENT TEST, 1990

	Average Score	
Achievement Test	NYS	National
English Composition	524	512
Mathematics Level 1	580	539
Biology	588	546
Chemistry	601	571
American History	544	508
Spanish	556	529
Mathematics Level 2	669	654
Physics	611	595
French	561	546
Literature	529	517
European History	591	544
German	587	551
Hebrew	641	602
Latin	579	548
Russian	728	642
All Tests Combined	558	532

Source: New York State Department of Education

An analysis of this article indicates that the average scores of students on many different tests are given. How do we determine if there is any significant difference between the average scores of New York students and the national average on the various tests? A test-by-test comparison is very time-consuming. This is true even if we know the sample sizes.

EXERCISES FOR SECTION 13.4

1. Recent data from the National Health and Nutrition Examination Survey[*] showed that about 31% of the adult U.S. population had total cholesterol levels between 200 and 239 mg/dl—borderline high—while in another 20%, total cholesterol was ≥ 240 mg/dl—high. Often drug therapy is prescribed to lower cholesterol levels.

[*]*Source:* JAMA 269(23):3009–14, June 16, 1993.

To compare the effectiveness of four new pills in lowering blood serum cholesterol levels, 20 people with high cholesterol levels were carefully selected so as to make them as comparable as possible. The 20 people were divided into four groups of five each. Each person in a subgroup was then given a pill. The decrease in the blood cholesterol level of each person taking the pill is indicated below:

	Decrease in Blood Cholesterol Level (mg)				
Pill A	22	26	28	23	25
Pill B	20	25	32	21	27
Pill C	16	13	19	22	20
Pill D	27	29	21	24	24

Using a 1% level of significance, test the null hypothesis that the average decrease in the blood serum cholesterol levels of people taking each of these types of pills is the same.

Often a researcher must compare the effect of different diets on the reduction of blood serum cholesterol levels. What analysis is needed?   (© *Herb C. Ohlmeyer, Fran Heyl Associates*)

MORE HOMES
CONTAMINATED WITH RADON

Bergen, *Sept. 12*: A new study released today by the state's environmental commission reveals that the number of homes containing excessive amounts of deadly radon gas is larger than expected. The radon gas is entering many homes on the East Coast (particularly those in the states of New Jersey and Pennsylvania) through holes or cracks in the foundation. It is believed that the gas is coming from deep within the earth.

The Commission urged the governer to appropriate additional funds to enable it to expand its monitoring activities.

September 12, 1994

2. Consider the accompanying newspaper article. The EPA recommends that corrective measures be taken if levels reach 4 or more picocuries per liter (pc/l). A government official took a sample of six readings at each of four day-care centers. The following results, in pc/l, were obtained:

	Day-Care Center 1	Day-Care Center 2	Day-Care Center 3	Day-Care Center 4
	1.8	6.2	3.2	1.9
	1.2	5.8	2.8	2.1
Radon Level	0.9	2.3	3.1	2.3
	1.7	4.7	3.5	1.8
	0.8	3.9	2.7	1.6
	1.3	5.1	3.3	2.2

Using a 5% level of significance, test the null hypothesis that there is no significant difference in the average radon level at these four day-care centers.

3. Consider the newspaper article on the next page. Environmental pathologists carefully examined the catches of six returning fishing ships to pinpoint the cause of death. The pathologists found traces of four different chemicals in the bodies of those dead fish with the frequencies shown below:

Chemical Found

Pesticide A	Pesticide B	Pesticide C	Pesticide D
42	38	35	28
44	49	47	29
36	41	43	34
27	30	46	38
31	40	41	36
28	29	39	39

Using a 5% level of significance, test the null hypothesis that there is no significant difference in the average number of these different pesticides.

MORE DEAD FISH FOUND WASHING UP ON THE SHORE

Dec. 28: Health officials reported yesterday that more than 284 dead fish were found over the weekend on the city's shoreline, bringing to well over 2500 the number of dead fish that were found recently. A health official, who spoke on condition of anonymity, speculated that the fish might have died as a result of the accidental release of highly toxic chemicals into the waterways from a neighboring company.

December 28, 1994

4. Refer back to the previous exercise. A second group of pathologists decided to analyze whether the average number of pesticides was the same for the different species of fish caught. They obtained the following additional data.

Chemical Found

		Pesticide A	Pesticide B	Pesticide C	Pesticide D
	I	31	42	49	29
	II	27	41	48	27
Species	III	28	32	40	26
of Fish					
Examined	IV	26	33	41	25
	V	34	38	39	32
	VI	39	29	45	35

Using a two-way analysis of variance, test the null hypothesis that the average number of pesticides found is the same for the various species of fish and that the average is not significantly different for the pesticides. (Use a 5% level of significance.)

5. Carbon monoxide is an odorless gas. Nevertheless, it can be deadly. Many companies now market carbon monoxide detectors. An independent testing agency is analyzing the carbon monoxide detectors produced by four different companies to determine if there is any significant difference between the average amount of time elapsed before the detector emits an audible siren in response to a certain stimulus. The agency randomly purchases three detectors produced by each of the four companies and subjects them to the same conditions. The number of seconds that elapses before the carbon monoxide detectors emit an audible siren is as follows:

	Time Elapsed (seconds)		
Brand A	22	14	17
Brand B	16	18	11
Brand C	23	15	18
Brand D	19	13	16

Do the data indicate that there is a significant difference in the average time elapsed before the carbon monoxide detectors emit an audible sound among the different brands? (Use a 5% level of significance.)

6. The Energy Information Administration conducts surveys on the appliances used by U.S. households and publishes the results in *Residential Energy Consumption Survey: Housing Characteristics*. The following data represent the energy consumption of households from various regions of the United States (rounded to the nearest 10 million BTU).

Region of the United States			
East	West	South	North
12	11	7	13
19	14	8	11
17	19	6	14
16	22	9	16
18	10		10
	13		

Using a 5% level of significance, test the null hypothesis that there is no significant difference in the mean energy consumption for the regions mentioned.

7. Many colleges require all entering freshmen to take various tests administered by the American College Testing (ACT) program before they are accepted into the

college. The results of the tests are analyzed and appear in *High School Profile Report*. Some colleges combine the ACT results with their own test results before admitting students. The Commissioner of Education of a state is analyzing the average combined math scores of students from high schools located throughout the state on those tests. The Commissioner believes that the average math scores of students is significantly different for the various high schools and is related to the geographic region of the school district in which the high school is located. In support of this claim, the following data are available for students from four high schools located in different geographic regions within the state.

		Combined Math Scores			
	Northern	458	473	496	446
Geographic Location of High School Within The State	Southern	512	501	498	422
	Eastern	416	440	432	423
	Western	508	488	469	510

At the 5% level of significance, is the Commissioner's claim justified?

8. Accidental breakage of glass items by customers as they push their shopping carts through the aisles of a supermarket can be dangerous to other customers as well as costly to the management. A large supermarket chain reported the store locations and the following number of reported cases of the breakage of glass items on a weekly basis for a period of four weeks.

		Number of Cases of Glass Breakage			
	Riverside Mall	45	32	36	40
Store Location	Liberty Mall	39	34	46	36
	Apollo Mall	36	37	38	45
	Two-Towns Mall	42	39	31	40

Using a 5% level of significance, test the null hypothesis that the average number of accidental glass breakage cases is not significantly different for all the stores, irrespective of mall location.

9. Environmentalists took four water samples at each of three different locations of a river in an effort to measure whether the quantity of dissolved oxygen in the water varied from one location to another and from season to season. The quantity of dissolved oxygen (on an appropriate scale) in the water is used to determine the extent of the water pollution. The results of the survey are as follows:

		Quantity of Dissolved Oxygen		
		Location I	Location II	Location III
Season	Autumn	19.4	16.8	16.9
	Winter	18.6	18.3	17.8
	Spring	18.9	17.5	19.9
	Summer	19.3	17.6	18.8

Using a two-way analysis of variance, do the data indicate that the average quantity of dissolved oxygen is the same at all three locations and is the same for all the seasons of the year? (Use a 1% level of significance.)

10. Many drug companies claim that their products are superior in their ability to bring quick relief to suffering patients. A consumer's group decided to test the drugs manufactured by four different companies. The average time (in minutes) required for 20 patients to feel relief is shown in the accompanying chart:

		Average Time to Pain Relief (minutes)				
Drug Manufacturer	Company A	12	20	24	24	26
	Company B	22	24	22	26	19
	Company C	11	11	15	18	12
	Company D	12	16	18	16	22

Do the data indicate that there is a significant difference between the average amount of time that elapses when a patient takes the drugs produced by the different companies before relief is felt? (Use a 5% level of significance.)

11. Twenty moviegoers were asked to rate five different movies (on a certain scale). However, the air-conditioning system in the theater at the time the movies were shown was not functioning properly. As a result, it is suspected that the room temperature may have been an additional factor in the moviegoers' ratings of the movies, which are given below:

		Temperature in Theater			
		70–74°	75–79°	80–84°	85–90°
Movie Viewed	A	36	26	22	19
	B	33	23	27	29
	C	30	27	25	25
	D	30	29	32	36
	E	25	27	28	32

Using a two-way analysis of variance, test the null hypothesis that the average ratings for these movies is the same and that the temperature in the theater is not a factor. (Use a 5% level of significance.)

12. An experimenter is interested in knowing which method yields the greatest amount of usable oil from shale rock. The experimenter gathers shale rock from five different parts of the country and applies one of four conversion techniques. The quantity of oil obtained (on a certain scale) is given below:

		Region and Quantity of Oil Obtained				
		A	B	C	D	E
	I	38	34	44	92	41
Conversion	II	32	36	37	47	57
Technique Used	III	44	33	25	35	50
	IV	23	27	22	26	48

Using a two-way analysis of variance, test the null hypothesis that the average quantity of oil obtained is the same for all the conversion techniques and is also the same for shale rock obtained from each of the regions. (Use a 1% level of significance.)

13. Consider the newspaper article on the next page. After taking samples of the air on different days at the various tunnels and bridges, the following levels of pollutants were obtained:

		Day of the Week				
		Mon.	Tues.	Wed.	Thurs.	Fri.
	Bridge 1	2.9	3.4	2.8	3.9	3.8
Location	Bridge 2	2.4	2.8	2.9	3.2	2.9
Where	Bridge 3	3.9	3.9	3.8	3.4	2.5
Sample Was						
Obtained	Tunnel 1	2.9	2.9	3.4	3.5	3.6
	Tunnel 2	2.7	3.6	2.5	3.2	3.9

Using a two-way analysis of variance, test the null hypothesis that the average amount of sulfur-oxide pollutants is the same at all the bridges and tunnels and is the same for each day of the week. (Use a 5% level of significance.)

TOLL COLLECTORS PROTEST UNHEALTHY WORKING CONDITIONS

New Dorp, *May 17*: Toll collectors staged a 3-hour protest yesterday against the unhealthy working conditions occuring at different times of the day at the toll plazas of the city's seven tunnels. A spokesperson for the toll collectors indicated that the level of sulfur oxide pollutants in the air at the plazas was far in excess of the recommended safety level. Between the hours of 10 A.M. and 1 P.M., motorists were able to use the tunnels without paying any toll. In an effort to resolve the dispute, Bill Sigowsky has been appointed as mediator.

May 17, 1990

14. A lawyer has gathered the following statistics from several states for the length of a jail term for people who had the same prior record but were convicted of different types of crimes.

Jail Term by Crime (years)

		Burglary	Armed Robbery	Assault	Murder
	1	5	8	3	36
State	*2*	4	7	4	38
	3	5	9	2	39
	4	4	10	3	37

Using a two-way analysis of variance, test the null hypothesis that the average length of a jail term is the same for the various crimes mentioned and that this average is not significantly different for the several states. (Use a 1% level of significance.)

13.5 USING COMPUTER PACKAGES

MINITAB greatly simplifies the computations involved when working with ANOVA problems. We illustrate the output from MINITAB when applied to the data of Example 2 of Section 13.3 (page 712).

```
MTB  > READ THE FOLLOWING DATA INTO C1, C2, C3, C4
DATA > 64    73    61    63
DATA > 73    82    79    69
DATA > 69    71    71    68
DATA > 75    69    73    74
DATA > 78    74    66    75
DATA > END

MTB  > AOVONE WAY ON C1-C4

ANALYSIS OF VARIANCE
SOURCE    DF      SS       MS       F        p
FACTOR     3     52.2     17.4     0.56     0.652
ERROR     16    500.4     31.3
TOTAL     19    552.6
                                  INDIVIDUAL 95 PCT CI'S FOR MEAN
                                  BASED ON POOLED STDEV
LEVEL      N     MEAN     STDEV  --------+---------+---------+-------
C1         5    71.800    5.450          (-------------*------------)
C2         5    73.800    4.970              (-------------*------------)
C3         5    70.000    6.856    (-----------*------------)
C4         5    69.800    4.868  (-------------*------------)
        --------+---------+---------+-------
POOLED STDEV = 5.592           68.0        72.0              76.0

MTB > STOP
```

COMMENT   Although the MINITAB analysis of variance for this problem will produce additional output (involving confidence intervals), we will not analyze such information here, as our main objective is merely to indicate how MINITAB handles simple ANOVA problems.

EXERCISES FOR SECTION 13.5

Use MINITAB to solve each of the following exercises.

1. After a home computer has been purchased and removed from the packing material, the various components have to be assembled and connected for the machine to become functional. Many home computer stores provide this service at no charge. One large midwestern computer store employs five people who perform this function. The number of minutes required by each to complete the assembly task for 26 identical machines is shown at the top of the next page:

		Time to Complete Task (minutes)					
	Bill	68	50	56	75	65	84
	Mary	56	61	65	52	56	
Employee	Bob	64	60	52	61	60	67	75
	Sue	68	66	62	59		
	Sandra	52	61	54	48		

Using a 1% level of confidence, test the null hypothesis that the average time required to assemble the computer is the same for all of the employees.

2. The U.S. Department of Agriculture publishes figures on acreage and production yields by state in *Agricultural Statistics*. In one study of four different chemical fertilizers, the following bushel yield, per acre, was obtained.

No Chemical Fertilizer Used	*Chemical Fertilizer Used*			
	Brand A	Brand B	Brand C	Brand D
267	326	309	231	293
344	310	258	316	316
281	318	284	311	313
217	267	299	266	349
264	290	312	299	342
244	307	299	286	341

Is the production yield per acre of land dependent upon the type of chemical fertilizer used? (*USDA*)

Using a 5% level of significance, test the null hypothesis that average yield per acre is not significantly different when the different chemical fertilizers are used or when no chemical fertilizer is used.

3. A consumer's public interest research group sampled four different supermarket chains to determine if there is a significant difference in the average price charged for a quart of milk. The following results were obtained:

				Price Charged per Quart of Milk (in cents)				
	A	63	64	62	63	67	60	62	61
Supermarket	B	68	68	62	65	67	61	69	68
Chain	C	67	61	60	59	59	62	65	63
	D	62	66	67	63	64	69	67	66

Does the data indicate that there is a significant difference in the average price charged by these supermarkets for a quart of milk? (Use a 5% level of significance.)

4. A consumer's group wishes to verify the claim made by several paint manufacturers that the average gallon of a certain type of paint will cover a 420-square-foot wall when applied according to specifications. Fifteen walls are divided into groups of five each. Each wall in a subgroup is then painted with one of these brands of paint. The coverage (in square feet) by each of these brands of paint is as follows:

	Paint Coverage (square feet)				
Brand A	405	395	408	407	415
Brand B	428	418	428	418	423
Brand C	418	425	429	415	414

Using a 5% level of significance, test the null hypothesis that the average coverage by each of these brands is not significantly different.

13.6 SUMMARY

In this chapter we briefly introduced the important statistical technique known as analysis of variance or ANOVA. This technique is used when we wish to test a hypothesis about the equality of several means. We limited our discussion to normal populations with equal variances.

When applying the ANOVA technique, we do not test the differences between the means directly. Instead, we test the variances. If the variances are zero, there is no difference between the means. We therefore analyze the source of the variation. Is it due purely to chance or is the difference in the variation significant? When analyzing variation by means of analysis of variance techniques, we use an ANOVA table.

Although we discussed only single-factor (one-way) and two-way ANOVA, it is worth noting that ANOVA techniques can be applied to more complicated situations or when we have replications for each test condition. However, these are beyond the scope of this book.

Study Guide

The following is a chapter outline in capsule form. You should now be able to demonstrate your knowledge of the ideas mentioned by giving definitions, descriptions, or specific examples. Page references are given in parentheses.

We can analyze whether there is any significant difference between several means by using analysis of variance techniques. This is abbreviated as **ANOVA**. (page 701)

When setting up an ANOVA table the first number that we calculate is called the **Total Sum of Squares** and is abbreviated as SS(total). This number is obtained by squaring each number in the table and then adding these squares together. From this we subtract the square of the grand total divided by the number in the sample. (page 705)

The second number that we calculate is the **Sum of Squares Due to the Factor**. We square each of the row totals and divide the sum of these squares by the number in each row. From this result we subtract the square of the grand total divided by the number in the sample. (page 706)

Then we calculate the **Sum of Squares of the Error**, abbreviated as SS(error). We find the sum of the squares of each entry in the table. From this sum we subtract the sum of the squares of each row total divided by the number in each row. (page 706)

The **number of degrees for the factor**, df(factor), is one less than the number of possible levels at which the factor is tested. (page 707)

The **number of degrees for the error**, df(error), is one less than the number of repetitions of each condition multiplied by the number of possible levels of the factor. (page 707)

The **number of degrees for the total**, df(total), is one less than the total number in the sample. (page 707)

To determine the entries that belong in the **mean square** column, we divide the sum of squares for the row by the number of degrees of freedom for that row. (page 708)

If we are given two samples of size n_1 and n_2 for two populations that are roughly normally distributed with sample standard deviations of s_1 and s_2, respectively, then we base tests on the equality of the two population standard deviations on the ratio $\dfrac{s_1}{s_2}$ by using the F-distribution. (page 708)

The **F-distribution** depends on the number of degrees of freedom of both sample estimates, $n_1 - 1$ and $n_2 - 1$. We **reject the null hypotheses** of equal variances when the test statistic value of F exceeds the chart value of F at the appropriate level of significance. (page 709)

For a single-factor analysis of variance we have the following **ANOVA table**

Source of Variation	Source of Squares	Degrees of Freedom	Mean Square	F-Ratio
Factors of the experiment				
Error				
Total				

When performing an analysis of variance, if we are discussing only one factor, it is referred to as **treatments**. (*Note:* MINITAB uses "FACTOR" instead of "Treatments.") We then have the single-factor ANOVA table given above. If a second factor is involved, it is referred to as **blocks** and we have a two-way ANOVA. The two-way ANOVA table is slightly different. Instead of having a single line for factors, it has two lines, one for treatments and one for blocks. (page 716)

Formulas to Remember

The most important thing to remember when testing several means is how to set up an ANOVA table and how to find the appropriate values for each cell of the ANOVA table. Both of these ideas are summarized on pages 709–710 for single-factor ANOVA and on pages 716–717 for a two-way ANOVA.

Testing Your Understanding of This Chapter's Concepts

1. We cannot use one-way ANOVA techniques without first satisfying certain assumptions. Name three of them.
2. For an F-distribution, find the probability that $F \leq 3.37$ when the degrees of freedom for the numerator is 10 and the degrees of freedom for the denominator is 20.
3. The ANOVA techniques discussed in this chapter actually analyze whether the difference between several sample means is significant. Explain how variance plays a key role in such analyses.
4. When we are working with two populations ($k = 2$), then $MSE = s_p^2$, where s_p^2 is the pooled variance as defined in Section 10.7. What does $\sqrt{MSE}$ represent?
5. Suppose we perform a one-way ANOVA test where

$$H_0: \quad \mu_1 = \mu_2 = \cdots = \mu_n$$

$$H_1: \quad \text{Not all means are equal.}$$

If we reject the null hypothesis, does this mean that no two of the populations have the same mean? Explain your answer.

6. When performing a one-way ANOVA, which statistic measures the variation
   a. among the sample means?
   b. within the sample means?

Chapter Test

For Questions 1–10, use the following information: The number of children reported lost on a daily basis at the five city beaches during a five-day July 4th weekend are as follows:

Number of Children Reported Lost Daily, July 4th Weekend

Beach 1	48	39	45	49	42
Beach 2	46	38	36	35	48
Beach 3	40	32	29	40	35
Beach 4	24	59	29	45	42
Beach 5	23	35	40	28	41

In an effort to determine whether there is a significant difference in the average number of children lost on a daily basis at the various beaches, the following one-way ANOVA table has been set up:

ANOVA Table

Source of Variation	Sum of Squares	Degrees of Freedom	Mean Square	F-Ratio
Different Beaches	(1)	(4)	(7)	(9)
Error	(2)	(5)	(8)	
Total	(3)	(6)		

1. The appropriate entry for cell (1) is _____ .
2. The appropriate entry for cell (2) is _____ .
3. The appropriate entry for cell (3) is _____ .
4. The appropriate entry for cell (4) is _____ .

5. The appropriate entry for cell (5) is _____ .
6. The appropriate entry for cell (6) is _____ :
7. The appropriate entry for cell (7) is _____ .
8. The appropriate entry for cell (8) is _____ .
9. The appropriate entry for cell (9) is _____ .
10. Does the data indicate that there is a significant difference between the average number of reports processed daily at the five different locations? (Use a 5% level of significance.)

    a. yes   b. no   c. not enough information given   d. none of these

11. The Pleasantville Transit Authority services its busses at one of four garages located throughout the city. Transit authority officials would like to know if all of the garages service, on the average, the same number of busses per week. The research manager obtains the following statistics on the number of busses serviced per week by each of the four garages over a period of time.

Number of Busses Serviced (per week)

Ryder Ave. Garage	Scranton Ave. Garage	Clover St. Garage	Paradise Lane Garage
22	12	14	28
23	18	10	18
29	13	22	22
26	15	15	29
15	19	18	20
	16	19	

At the 5% level of significance, test the null hypothesis that the mean number of busses serviced per week at each of these garages is the same. (Assume that each garage performs the same type of servicing.)

12. Refer back to the U.S. General Accounting Office National Assessment of Educational progress report given in the "Did You Know That" section at the beginning of this chapter. The following math scores were obtained by third graders "on the move" from five different cities:

City A	City B	City C	City D	City E
180	206	199	186	203
203	212	186	193	187
212	208	195	199	196
194	201	206	202	208
197	179	214	212	205
	192	200	201	
		203	216	

At the 5% level of significance, test the null hypothesis that the average score obtained by third graders in each of these cities is about the same.

13. According to an alumni survey, the starting salaries (in thousands of dollars) and the area of specialization of recent graduates of Carlesville University are as follows:

Area of Specialization

Computer Science	Nursing	Business	Liberal Arts	Early Childhood Education
36.0	28.7	38.3	27.1	22.0
29.0	26.9	35.0	23.6	26.2
34.2	30.2	33.0	20.0	21.0
39.0	30.8	37.1		25.3
31.0	29.1	36.0		
32.6		30.3		
		32.4		

Using a 5% level of significance, test the null hypothesis that the average starting salary is approximately the same for all these graduates with these five areas of specialization.

14. Many people must restrict their intake of sodium (salt). The U.S. FDA requires all packers to label each can of salmon with its sodium content. An independent testing laboratory analyzed the sodium content in five cans sold by four leading companies. The following sodium content (in mg) per serving size (½ cup) was obtained.

Sodium Content per Can of Salmon (mg/½ cup)

Company A	Company B	Company C	Company D
480	460	493	453
462	449	487	464
493	458	482	470
475	471	476	463
493	462	473	477

Using a 1% level of significance, test the null hypothesis that the average sodium content in the serving sizes of salmon is the same for the cans packed by all of these companies.

15. A consumer rating agency tested nickel-cadmium rechargeable batteries produced by four different companies. The number of times that these batteries could be recharged before becoming useless was follows:

Brand A	Brand B	Brand C	Brand D
862	845	867	855
893	888	895	873

(*Table continues on page 736*)

Brand A	Brand B	Brand C	Brand D
764	819	834	808
791	902		777
	913		916
			868

Using a 5% level of significance, test the null hypothesis that the average number of times that these batteries can be recharged is approximately the same for the four brands of batteries.

16. A toy manufacturer is interested in determining whether different package wrappings for toys have any effect on sales. The manufacturer decides to package the toys in one of four possible package wrappings and to sell the toys with each of the possible wrappings at five large department stores. The price for the toys at each store is the same, irrespective of the packaging, and each package contains the same toy. The following are the sales reported for the toy.

		Package Wrapping			
		A	B	C	D
	I	79	80	74	98
	II	65	55	46	79
Store	III	68	86	50	86
	IV	46	34	59	67
	V	59	68	58	68

Using a two-way analysis of variance, do the data indicate that the average sales reported for the toy is the same for all four possible wrappings and also the same for all the stores? (Use a 5% level of significance.)

17. A medical researcher is interested in comparing the effects of four different diets on hypertension (high blood pressure). Twenty men with high blood pressure were randomly assigned (five each) to each of the four diet groups. After several months, the blood pressure of the participants was measured and is given below

			Blood Pressure		
Diet A	140	167	154	174	165
Diet B	152	159	146	149	146
Diet C	164	138	138	156	149
Diet D	158	150	149	159	160

Do the above data indicate that the type of diet used significantly affects the average blood pressure level? (Use a 1% level of significance.)

18. There are three hospitals with emergency rooms in Adams City. Thirty-three patients who had been treated at these emergency rooms are randomly selected and asked how long they had to wait before seeing a doctor. Their answers are as follows:

	Length of Wait (minutes)											
Lincoln Hospital	29	27	24	41	33	27	35	37	22	18	31	22
Mt. Sinai Hospital	19	26	42	31	31	27	22	28	29	34	27	
Columbia Hospital	23	22	26	27	23	22	34	21	28	35		

Do the data indicate that there is a significant difference between the average waiting time in the emergency rooms at these hospitals before seeing a doctor? (Use a 5% level of significance.)

19. The Acme Medical Corp. is interested in buying a new laser jet computer printer and is considering three equally rated brands. The purchasing agent has obtained the following prices for the cost of a laser jet computer printer for each of these brands:

	Cost of Laser Jet Computer Printer				
Brand A	$529	$536	$553	$502	$544
Brand B	509	528	519	537	503
Brand C	499	537	488	519	523

Do the data indicate that there is a significant difference in the cost of a laser jet printer for each of these brands? (Use a 1% level of significance.)

20. The number of arrests for illegal drug use on a daily basis at the three city beaches during the five-day July 4th weekend is as follows:

	Number of Arrests				
Beach 1	21	26	30	27	16
Beach 2	34	27	20	33	25
Beach 3	14	22	27	26	32

Do the data indicate that the average number of arrests for illegal drug use is not significantly different at these beaches? (Use a 1% level of significance.)

21. The following table gives the gains in weight of four different rations:

		Weight Gain (pounds)			
		Hog Type I	Hog Type II	Hog Type III	Hog Type IV
	A	13	20	16	19
Method	B	22	23	29	26
	C	12	20	13	18

Using a two-way analysis of variance, test the null hypothesis that there is no significant difference in the average weight gain of the different types of hogs and in the different rations used. (Use a 1% level of significance.)

22. Sixteen music students were divided into four groups of four students each. Each group was taught how to play a new song on a particular instrument, each by a different method. The following chart indicates the number of minutes needed by each student to learn to play the song.

		Time to Learn New Song (minutes)			
	A	50	43	44	53
	B	37	34	54	45
Method	C	41	27	46	43
	D	33	42	49	34

Do the data indicate that there is a significant difference in the average time needed by each student to learn to play the song by the different methods? (Use a 5% level of significance.)

23. It is alleged that some stores discriminate against the poor by charging different prices for similar items. To check out this allegation, a consumer's group obtained the following prices charged by stores of the same supermarket chain for several brands of a half-gallon of orange juice.

		Type of Neighborhood		
		Upper Class	Middle Class	Lower Class
	A	$1.69	$1.75	$1.78
Brand of	B	1.73	1.79	1.81
Orange Juice	C	1.81	1.84	1.87
Purchased	D	1.75	1.79	1.81
	E	1.72	1.76	1.79

Using a two-way analysis of variance, do the data indicate that the average price of a half-gallon of orange juice is not significantly different for the brands of orange juice and for all the neighborhoods? (Use a 5% level of significance.)

24. A large automobile insurance company operates drive-in claim centers where drivers involved in automobile accidents can bring their cars for immediate claims processing. The company is interested in determining whether the particular claims examiner involved in a settlement has any effect on the amount of money that the company must pay. The company decides to sample identical cases resolved by five different claims examiners in each of the four drive-in centers that it operates in a large northeastern state. The car damage is identical in all claims examined. The following are the reported settlement amounts.

		Drive-In Center			
		A	B	C	D
Claim Examiner	I	$1525	$1540	$1475	$1505
	II	$1720	$1695	$1405	$1595
	III	$1395	$1900	$1545	$1565
	IV	$1825	$1605	$1720	$1450
	V	$1540	$1610	$1590	$1730

Using a two-way analysis of variance, do the data indicate that the average amount of money in a settlement claim is not significantly different for all the centers and is not significantly different for all the employees? (Use a 5% level of significance.)

25. All students at Staten Island University take the same departmental final at the end of their Calculus I course to make sure that they are adequately prepared for the Calculus II course. The final exam grades for all the students in four different sections of the Calculus I were as follows:

Exam Grades from Four Professors

Smith	Carter	Jones	Kennedy
76	72	53	88
61	84	74	72
63	64	97	91
87	93	88	69
96	77	79	59
90	96	76	
80		85	
		88	

Do the preceding data indicate that the students in each of these teacher's classes are not significantly different in their performance on the Calculus I final exam? (Use a 5% level of significance.)

> ▶ **Thinking Critically**

1. A researcher designed an experiment that utilizes one qualitative factor with five levels (A, B, C, D, and E). What are the treatments?

2. If the Total Sum of Squares for a completely randomized experiment with five treatments and $n = 35$ measurements (five per treatment) is 900 and with error sum of squares equal to 450, perform an F-test of the null hypothesis that the five sample means are equal. (Use a 5% level of significance.)

3. If our analysis of variance leads us to conclude that the sample means are not all the same, then a confidence interval estimate of the difference between the two treatment means is given by

$$\bar{x}_1 - \bar{x}_2 \pm t_{\alpha/2}\sqrt{MSE\left(\frac{1}{n_1} + \frac{1}{n_2}\right)}$$

where $\bar{x}_1$ and $\bar{x}_2$ are the sample means for the first and second treatments respectively, n_1 and n_2 are the sample sizes, $MSE =$ mean square error as given in the ANOVA table, and $t_{\alpha/2} = t$-value (using the t-distribution) based on the number of degrees of freedom for the error as given in the ANOVA table. Using a 5% level of significance ($\alpha = 0.05$), find a confidence interval for the difference between the average tar content of Brand X and Y cigarettes as given in Example 1 of Section 13.3. on page 710).

4. Refer to Question 3. Find a confidence interval for the difference between the average tar content of
   a. Brand X and Z cigarettes.
   b. Brand Y and Z cigarettes.

5. A medical research team is investigating the effectiveness of three different cholesterol-lowering drugs. Twelve people who have a high blood serum cholesterol level are randomly selected. Four are given a medication produced by Company A, four are given a medication produced by Company B, and four are given a medication produced by Company C. All other factors such as diet, exercise, and so forth are the same for all of the subjects. After one month of medication, the blood serum cholesterol of each of these subjects is again tested. The following one-way ANOVA MINITAB output shows the mean cholesterol decrease by these subjects.

ANALYSIS OF VARIANCE

Source	DF	SS	MS	F	p
Factor				0.82	0.326
Error			62.83		
Total					

Fill in the missing entries.

Case Study

Are the prices charged for college textbooks outrageously high? Recently, a public research interest group (NYPIRG) conducted a survey of the prices charged for randomly selected college math textbooks in algebra and trigonometry, statistics, calculus, and liberal arts mathematics. The following prices were obtained:

Prices Charged For Textbooks

Algebra and Trigonometry	Statistics	Calculus	Liberal Arts Mathematics
$42.75	$49.95	$69.95	$46.23
44.82	47.68	72.36	42.86
43.79	48.50	71.42	45.12
47.53	48.88	65.23	
	49.35	68.69	
		74.49	

1. Using a 5% level of significance, test the null hypothesis that the average prices of a college textbook in algebra and trigonometry, statistics, calculus, and liberal arts mathematics are not significantly different.

14

NONPARAMETRIC STATISTICS

Chapter Outline

DID YOU KNOW THAT

a while prison capacity grew by 60% between 1984 and 1990, the number of prisoners grew faster, increasing by nearly 70%? (*Source: ABA Journal*, April 1993)

b a public health campaign urging parents to put newborns to sleep on their back has reduced the nationwide incidence of sudden infant death syndrome 12% in just six months? (*Source: Archives of Pediatrics and Adolescent Medicine*, 1994)

c the level of radon gas present in the air in some public buildings exceeds the recommended safety level of 4 picocuries per liter of air? (*Source:* Environment Protection Agency, 1995)

Often an animal trainer uses different techniques to teach lions to perform tricks. How do we determine which technique is superior? See discussion on page 769. (*Courtesy of Ringling Bros. and Barnum & Bailey Combined Shows, Inc.*)

I n recent years much emphasis has been placed on nonparametric statistics. Since the concepts behind nonparametric statistics (or distribution-free methods as they are called) are based on elementary probability theory, many of the formulas and tests used can be derived quite easily by employing only a little algebra. Once we develop a table of critical values for a given situation, the nonparametric techniques are quite easy to use.

In this chapter we discuss some of the more popular nonparametric techniques.

Chapter Objectives

- **To analyze** several tests that can be used when assumptions about normally distributed populations or sample size cannot be satisfied. These are called nonparametric statistics or distribution-free methods. (Sections 14.1 and 14.2)

- **To apply** the sign test that is used in the "before and after" type study. We test whether or not $\mu_1 = \mu_2$ when we know that the samples are not independent. We can also use the Wilcoxon signed-rank test. (Sections 14.3 and 14.4)

- **To discuss** an alternative to the standard significance tests for the difference between two sample means that is used when we have nonnormally distributed populations. This is the rank-sum test. We can also use this test when the population variances are not equal. (Section 14.5)

- **To understand** the Spearman rank correlation test. (Section 14.6)

- **To point out** how we use the runs test when we wish to test for randomness. (Section 14.7)

- **To work with** the Kruskal-Wallis H-test as a nonparametric test that can be used to test whether the difference between numerous sample means is significant. (Section 14.8)

S T A T I S T I C S   I N   A C T I O N

Referring to the article below, we can conclude that a technique is needed for measuring the effects of a new policy like the one described in the article. Such "be- fore and after" types of situations often occur when new products or policies are introduced or when a new technique is tried. Are such statistical tests available?

PRIORITIES WRONG?

Chicago, *April* 5: According to an American Bar Association report on the state of our criminal justice system, more attention is being directed toward drug offenses and less to violent crime even though drug use is down and violent crime is up.

According to the ABA report, drug arrests between 1986 and 1991 increased by 25 percent, and the number of persons imprisoned for drug offenses increased 327 percent. The following data are available for one county jail.

Number of Persons Imprisoned for Drug Offenses

	1986	1991
Jan	121	293
Feb	132	301
Mar	127	288
Apr	152	317
May	142	309

"The criminal justice system is on the track to collapse" declared Neal Sonnett, chair of the ABA's Criminal Justice section.

April 5, 1994

14.1   INTRODUCTION

In previous chapters we discussed procedures for testing various hypotheses involving means, proportions, variances, and the like. In almost all the cases discussed we assumed that the populations from which the samples were taken were approximately normally distributed. Only when we applied the chi-square distribution in comparing observed frequencies with expected frequencies did we not specify the normal distribution.

Nonparametric statistics
Distribution-free methods
Standard methods

Since there are many situations where this requirement cannot be satisfied, statisticians have developed techniques to be used in such cases. These techniques are known as **nonparametric statistics** or **distribution-free methods.** As the names imply, these methods are not dependent upon the distribution or parameters involved.

There are advantages and disadvantages associated with using nonparametric statistics. The advantages in using these methods as opposed to the **standard methods** are as follows:

1. They are easier to understand.
2. They often involve much less computation.
3. They are less demanding in their assumptions about the nature of the sampled populations.

For these reasons, many people often refer to nonparametric statistics as shortcut statistics. The disadvantages associated with nonparametric statistics are that they usually waste information, as we will see shortly, and that they tend to result in the acceptance of null hypotheses more often than they should.

Standard tests

Nonparametric statistical methods are frequently used when samples are small since most of the **standard tests** require that the sample sizes be reasonably large.

In this chapter we discuss only briefly some of the more commonly used nonparametric tests. A complete discussion of all these methods would require many chapters or perhaps even several volumes.

14.2   THE ONE-SAMPLE SIGN TEST

All the standard sample tests involving means that we have discussed so far in this book are based on the assumption that the populations are approximately normally distributed. This often may not be the case.

One-sample sign test

When the preceding assumption is not necessarily true, then we can replace the standard tests by one of the numerous nonparametric tests. By far the simplest of all the nonparametric tests is the **one-sample sign test**. It is used as an alternative to the *t*-test with one mean for symmetric populations that was discussed in Section 10.5. We apply this test when the hypothesis we are testing concerns the value of the mean or median of the population and we are sampling a continuous population in the vicinity of the unknown mean or median M. Under these assumptions the probability that a sample value is less than the mean or median and the probability that a sample value is greater than the mean or median are both $\frac{1}{2}$.

If the sample size is small, we can perform the sign test by referring to a table of binomial probabilities (Table III in the Appendix). When the sample size is large we can perform the sign test by using the normal curve approximation to the binomial. We illustrate the techniques with several examples.

EXAMPLE   1   A trucking industry spokesperson claims that the median weight of a load carried by a truck traveling on State Highway No. 7 is 15,000 pounds (lb). A transportation official believes that this figure is much too low. To verify the trucking industry claim, a random survey of 15 trucks is taken and the weight of each truck's load is determined. It is found that 5 trucks have loads with weights below 15,000 lbs, 2 trucks have loads equal to 15,000 lbs, and 8 trucks have loads above 15,000 lbs. Using the one-sample sign test, test the null hypothesis that the median weight of a load carried by a truck is 15,000 lbs against the alternative hypothesis that the median weight is more than 15,000 lbs. (Use a 5% level of significance.)

SOLUTION   We replace each truck's load weight above 15,000 lbs with a plus sign and each truck's load weight below 15,000 lbs with a minus sign. We discard those trucks whose load weight equals 15,000 lbs. We then test the null hypothesis that the plus signs and minus signs are values of a random variable having the binomial distribution with $p = \frac{1}{2}$. In our case we have 5 minus signs and 8 plus signs. Here $n = 13$ and not 15 since we disregard those trucks whose load weight is exactly 15,000 lbs. We then determine whether 8 plus signs in 13 observed trials agrees with the null hypothesis that $p = \frac{1}{2}$ or with the alternative hypothesis that $p > \frac{1}{2}$. Using Table III in the Appendix, we find that for $n = 13$ and $p = \frac{1}{2}$ the probability of obtaining 8 or more successes is

$$\text{Prob}\ \ (8\ \text{successes}) = 0.157$$

$$\text{Prob}\ \ (9\ \text{successes}) = 0.087$$

$$\text{Prob}\ (10\ \text{successes}) = 0.035$$

$$\text{Prob}\ (11\ \text{successes}) = 0.010$$

$$\text{Prob}\ (12\ \text{successes}) = 0.002$$

$$\text{Prob}\ (13\ \text{successes}) = \text{------}$$

$$\text{Prob}\ (8\ \text{or more successes}) = 0.291$$

Since this value is more than $\alpha = 0.05$, we do not reject the null hypothesis. The sample data do not contradict the trucking industry's claim.

The previous example was a one-sided hypothesis problem. However, the same procedure can be used when testing two-sided alternative hypotheses problems. In essence, what we do is calculate the number of plus signs and the number of minus signs. We then determine whether a random variable having a binomial distribution can have the calculated number of plus signs or minus signs and still have $p = \frac{1}{2}$. By referring to the binomial probability chart (Table III in the Appendix), we then specify the critical rejection region.

Since applications using the sign test occur often, statisticians have constructed a chart that enables us to determine whether the number of plus signs or minus signs is significant. This eliminates the need to use the binomial distribution probability chart and to add probabilities. One such chart is given in Table 14.1.

TABLE 14.1   Critical Values for Sign Test

n	Level of Significance ($\alpha =$) 0.01	0.05	0.10	0.25	n	Level of Significance ($\alpha =$) 0.01	0.05	0.10	0.25
1					21	4	5	6	7
2					22	4	5	6	7
3				0	23	4	6	7	8
4				0	24	5	6	7	8
5			0	0	25	5	7	7	9
6		0	0	1	26	6	7	8	9
7		0	0	1	27	6	7	8	10
8	0	0	1	1	28	6	8	9	10
9	0	1	1	2	29	7	8	9	10
10	0	1	1	2	30	7	9	10	11
11	0	1	2	3	31	7	9	10	11
12	1	2	2	3	32	8	9	10	12
13	1	2	3	3	33	8	10	11	12
14	1	2	3	4	34	9	10	11	13
15	2	3	3	4	35	9	11	12	13
16	2	3	4	5	36	9	11	12	14
17	2	4	4	5	37	10	12	13	14
18	3	4	5	6	38	10	12	13	14
19	3	4	5	6	39	11	12	13	15
20	3	5	5	6	40	11	13	14	15

RULE

When using Table 14.1 for a two-sided test, the following applies:

1. n represents the total number of plus signs and negative signs, disregarding any zeros.
2. The test statistic is the number of the less frequent sign. This means that we first determine the number of plus signs and the number of minus signs. We select the smaller number between the number of plus signs or minus signs. This represents the test statistic.
3. We reject the null hypothesis if the test statistic value is less than or equal to the chart value. If the test statistic is larger than the chart value, we do not reject the null hypothesis.

COMMENT   Table 14.1 gives us the critical values for a two-sided test. When working with a one-sided test for a lower tail test, it is the number of plus signs. For an upper tail test, it is the number of minus signs.

Use of the preceding rule is illustrated in the following example.

E X A M P L E   2   A doctor suspects that the median annual cost for malpractice insurance in her specialty is approximately $18,000. Her nurse believes that the median annual cost is not equal to $18,000. She samples 13 insurance companies and obtains the following quotes for the identical cost of malpractice insurance.

$14,350	$17,010	$13,936	$17,073	$17,985	$18,840	$17,240
$18,000	$19,420	$17,840	$16,090	$17,360	$17,053	

Using the one-sample sign test, test the null hypothesis that the median cost for malpractice insurance is $18,000 against the alternative hypothesis that the median cost is not 18,000. (Use a 5% level of significance.)

SOLUTION   We replace each price quote with a minus sign if it is less than $18,000 and with a plus sign if it is above $18,000. We neglect those quotes that equal $18,000. We then have

$$- \ - \ - \ - \ - \ + \ - \ + \ - \ - \ - \ -$$

or a total of 10 minus signs and 2 plus signs in 12 trials. Here we are testing whether the number of plus signs (two in our case) supports the null hypothesis that $p = \dfrac{1}{2}$ or the alternative hypothesis that $p \neq \dfrac{1}{2}$. The number of plus signs is 2 and the number of minus signs is 10. The smaller of these two numbers is 2. This represents the test statistic. Now we look in Table 14.1 to find the appropriate critical value for $n = 12$ and

$\alpha = 0.05$. The chart value is 2. Since the test statistic value that we obtained is less than or equal to the chart value of 2 (in our case it equals 2), we reject the null hypothesis and conclude that the median cost of malpractice insurance is not $18,000.

14.3   THE PAIRED-SAMPLE SIGN TEST

Paired data
Two dependent
samples
Before and after
type study

The sign test can also be used when working with **paired data** that occur when we deal with **two dependent samples.** This often happens when we measure the same sample twice, as is done in the **before and after type study.** We proceed in a manner similar to what was done in Section 14.2.

To illustrate, suppose a college administrator is interested in knowing how a particular 3-week math minicourse affects a student's grade. Twenty students are selected and are given a math test. Then these students attend the minicourse and are retested. The administration would like to use the results of these two tests to determine whether the minicourse actually improves a student's score.

Table 14.2 contains the scores of the 20 students on the precourse test and the postcourse test. In this table we have taken each student's precourse test score and

TABLE 14.2   Scores of Twenty Students on Precourse and Postcourse Tests

Student	Precourse Score	Postcourse Score	Sign of Difference
1	68	71	+
2	63	65	+
3	82	88	+
4	70	79	+
5	65	57	−
6	66	77	+
7	64	62	−
8	69	73	+
9	72	70	−
10	74	76	+
11	71	68	−
12	80	80	
13	59	71	+
14	85	80	−
15	57	65	+
16	83	87	+
17	43	48	+
18	94	94	
19	82	93	+
20	91	94	+

subtracted it from the student's postcourse test score to obtain the change score. An increase in score is assigned a plus ($+$) sign and a decrease in score is assigned a minus ($-$) sign. No sign is indicated when the two scores are identical.

The null hypothesis in this case is $\mu_1 = \mu_2$. This means that the minicourse does not significantly affect a student's score. Under this assumption we would expect an equal number of plus signs and minus signs. Thus, if $\hat{p}$ is the proportion of plus signs, we would expect $\hat{p}$ to be around 0.5 (subject only to chance error).

Since we think that the minicourse does increase a student's score, the alternative hypothesis would be $\mu_2 > \mu_1$. Thus,

$$H_0: \quad \mu_1 = \mu_2$$

$$H_1: \quad \mu_2 > \mu_1$$

We can now apply the methods discussed in Section 10.8 (pages 566–569) for testing a proportion. Recall that for testing a proportion we use the test statistic

$$z = \frac{\hat{p} - p}{\sqrt{\dfrac{p(1 - p)}{n}}}$$

In applying this test statistic, we let p be the true proportion of plus signs as specified in the null hypothesis. Thus, if the minicourse does not affect a student's scores, we would expect as many plus signs as minus signs. There should be no more students obtaining higher scores than students obtaining lower scores as a result of this minicourse. Therefore, $p = 0.50$. We now count the number of plus signs. There are 13 plus signs out of a possible 18 sign changes. We ignore the cases that involve no change. So, $\hat{p} = \dfrac{13}{18}$, or 0.72, $n = 18$, and $p = 0.50$. Applying the test statistic, we have

$$z = \frac{\hat{p} - p}{\sqrt{\dfrac{p(1 - p)}{n}}}$$

$$= \frac{0.72 - 0.50}{\sqrt{\dfrac{(0.5)(1 - 0.5)}{18}}}$$

$$= \frac{0.22}{\sqrt{0.013889}} = \frac{0.22}{0.118}$$

$$= 1.86$$

We use the one-tail rejection region of Figure 14.1. Since the value of $z = 1.86$ falls in the critical region, we reject the null hypothesis at the 5% level of significance. Thus, the minicourse seems to have improved a student's score.

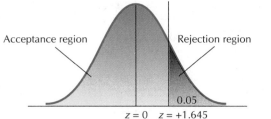

FIGURE 14.1

The following example further illustrates the paired-sign test technique.

EXAMPLE   1   A new weight-reducing pill is given to 15 people once a week for 3 months to determine its effectiveness in reducing weight. The following data indicate the before and after weights (in pounds) of these 15 people.

Weight Before Taking Pill	Weight After Taking Pill
131	125
127	128
116	118
153	155
178	179
202	200
192	195
183	180
171	180
182	180
169	174
155	150
163	169
171	172
208	200

Using a 5% level of significance, test the null hypothesis that the pill is not effective in reducing weight.

SOLUTION   We arrange the data as follows.

Weight Before	Weight After	Sign of Difference
131	125	−
127	128	+
116	118	+

(*Table continues on page 752*)

Weight Before	Weight After	Sign of Difference
153	155	+
178	179	+
202	200	−
192	195	+
183	180	−
171	180	+
182	180	−
169	174	+
155	150	−
163	169	+
171	172	+
208	200	−

For each person we determine the change in weight. A plus sign indicates a gain and a minus sign indicates a loss. If the weight-reducing pill is not effective, the average weight should be the same before and after taking this pill. Since we are testing whether a person's weight remains the same or is reduced, we have

$$H_0: \quad \mu_1 = \mu_2$$

$$H_1: \quad \mu_2 < \mu_1$$

Out of the 15 sign changes 6 are minus so that

$$\hat{p} = \frac{6}{15} = 0.4 \quad \text{and} \quad n = 15$$

Also, $p = 0.50$. Applying the test statistic, we get

$$z = \frac{\hat{p} - p}{\sqrt{\dfrac{p(1 - p)}{n}}}$$

$$= \frac{0.4 - 0.5}{\sqrt{\dfrac{(0.5)(1 - 0.5)}{15}}}$$

$$= \frac{-0.1}{\sqrt{0.01667}} = \frac{-0.1}{0.129}$$

$$= -0.78$$

We use the one-tail rejection region of Figure 14.2. Since the value of $z = -0.78$ falls in the acceptance region, we cannot reject the null hypothesis. The weight-reducing pill does not seem to be effective in reducing weight. It may even cause an increase in weight.

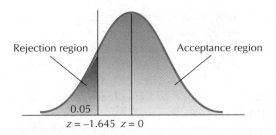

Rejection region   Acceptance region

0.05

$z = -1.645$ $z = 0$ **FIGURE 14.2**

COMMENT   We mentioned earlier that nonparametric methods are wasteful. A plus or minus sign merely tells us that a person gained or lost weight. It does not specify whether the gain was 1 pound, 10 pounds, or even 100 pounds. The same is true for the minus signs.

COMMENT   Since the sign test is so easy to use, many people use it even when the *standard* tests can be used.

COMMENT   The paired-sample sign test is actually a nonparametric alternative to the paired difference *t*-test.

EXERCISES FOR SECTION 14.3

1. The median U.S. income in 1992 as reported in *Statistical Abstract of the United States* was $32,073. The income of 12 randomly selected people of Middletown was $30,000, $28,638, $35,360, $31,350, $32,430, $33,000, $32,000, $33,500, $31,000, $32,850, $31,800, $31,400. Using the one-sample sign test, test the null hypothesis that the median income for the people of Middletown was $32,073 against the alternate hypothesis that the median income of the people of Middletown is lower than $32,073. (Use a 5% level of significance.)

2. According to the U.S. Census Bureau report *Marriage, Divorce and Remarriage in the 1990s*, the median duration of first marriages that end in divorce is 6.3 years. A random survey of 15 first marriages that ended in divorce in Bloomburg found that the duration of the marriage was 5, 8, 6, 7, 3, 9, 12, 1, 2, 4, 3, 5, 6, 7, and 2 years. Using the one-sample sign test, test the null hypothesis that the median duration of first marriages that end in divorce in Bloomburg is 6.3 years against the alternative hypothesis that the median duration is less than 6.3 years for the people in Bloomburg. (Use a 5% level of significance.)

3. A merchant has been accused of price gouging by charging higher prices for a quart of milk in a poorer neighborhood than the price charged by competitors. The merchant claims that the median cost of a container of milk in the neighborhood is 64 cents. A random survey of the prices charged by 12 competitors for a quart of milk disclosed prices of 61, 59, 58, 62, 64, 60, 58, 63, 65, 67, 58, and 63 cents, respectively. Use a one-sample sign test to test the null hypothesis that the merchant's claim is correct against the alternative hypothesis that the merchant is price gouging. (Use a 5% level of significance.)

4. The administrator of Lincoln Hospital claims that the median waiting time for a patient in the Emergency Room before being examined by a physician is 14 minutes. Several patient representatives claim that the median waiting time is considerably higher. A random sample of 13 patients in the Emergency Room is taken. The patients waited 15, 11, 21, 17, 16, 15, 20, 22, 18, 14, 19, 17, and 12 minutes, respectively, before being examined by a physician. Use the one-sample sign test to test the null hypothesis that the administrator's claim is correct against the alternative hypothesis that the median waiting time is more than 14 minutes. (Use a 5% level of significance.)

SUBWAYS ARE MUCH SAFER NOW

New York, *Dec. 22*: Following a rash of muggings on the subways, transit authority officials announced yesterday that the additional transit police assigned to each train had resulted in a decrease in the number of reported muggings in the subways.

December 22, 1993

5. Consider the accompanying newspaper article. The following table gives the number of mugging arrests per day for comparable 12-week periods both before and after the additional police were assigned to each train.

Number of Daily Arrests for Mugging

Week	Before New Police Assigned	After New Police Assigned
1	10	9
2	9	8
3	8	8
4	6	7
5	5	3
6	9	8
7	9	7
8	9	9
9	5	4
10	8	6
11	4	3
12	7	2

Using a 5% level of significance, test the null hypothesis that the additional police assigned to each train have no effect on the daily number of arrests for mugging against the alternate hypothesis that the number of daily arrests has decreased.

6. In recent years, as a result of difficult economic times and threatened by foreclosures, more and more workers are opting to purchase their financially unstable company from management. Does an employee-owned company result in increased productivity? The following data are available for 11 workers:

Number of Products Completed Daily by Each Worker

		Before Company Became Employee Owned	After Company Became Employee Owned
	A	38	47
	B	34	39
	C	36	36
	D	28	28
	E	49	47
Worker	F	27	37
	G	17	21
	H	18	24
	I	16	19
	J	10	11
	K	14	15

Using a 5% level of significance, test the null hypothesis that the number of products completed daily is not affected by the worker becoming an owner-employee against the alternative hypothesis that the number of products completed has increased.

7. Several home owners in the Northeast who recently converted from oil heat to gas heat in order to save money are comparing their current heating bills with their heating bills of years before converting. All of the homes are insulated in a comparable manner. The heating bills for a full season are as follows:

Heating Bill

		Oil	Gas
	A	$2200	$2400
	B	2000	1900
	C	2700	2800
Home Owner	D	2600	2480
	E	3000	3100
	F	2500	2540

(Table continues on page 756)

		Heating Bill	
		Oil	Gas
	G	2300	2380
Home Owner	H	2100	2000
	I	2375	2345

Use a 5% level of significance, test the null hypothesis that there is no significant difference in the annual heating bill for homes heated by oil or by gas against the alternative hypothesis that gas heating is cheaper.

8. Consider the accompanying newspaper article. Using a 5% level of significance, test the null hypothesis that the new fares have no effect on the commuter ridership

NEW STATISTICS INVOLVING COMMUTER RIDERSHIP RELEASED YESTERDAY

Long Island, *Feb. 18*: Officials of the regional Transportation Commission released the following statistics indicating how commuter ridership has been affected by the 75% increase in fares that took effect Febuuary 4. The data are for riders who use the 8:05 A.M. train daily between Pleasure Point and The City.

Number of Riders (in thousands) Using the 8:05 A.M. Daily

	Before Feb. 4	After Feb. 4
Monday	183	179
Tuesday	177	175
Wednesday	169	168
Thursday	180	179
Friday	179	179

A spokesperson for the railroad said that the effects of the rate increase on commuter ridership was under study and that no comments would be forthcomming at this time.

Monday –Frbruary 18, 1991

It is often claimed that railroad ridership decreases as fares are raised. How do we determine whether the effects of fare increases are significant? *(© Barbara Rios/Photo Researchers, Inc.)*

against the alternative hypothesis that the number of commuters using the 8:05 A.M. train has decreased.

9. Refer to the newspaper article on the next page. The number of felony arrests over a comparable period for ten precincts both before and after the one-police-officer-per-patrol-car policy was instituted is as follows:

		Number of Felony Arrests	
		Before One-Police-Officer-per-Car Policy	After One-Police-Officer-per-Car Policy
	7	12	8
	9	6	4
	12	10	9
	13	8	8
	16	10	6
Precinct	18	14	9
	22	11	5
	47	10	6
	66	9	10
	68	13	12

Using a 5% level of significance, test the null hypothesis that the daily number of felony arrests is not affected by the number of police officers in the patrol car against the alternative hypothesis that the number of felony arrests has decreased.

ONE OFFICER PATROL CARS A FIASCO

New York, *May 10*: Police department officials have refused to comment on surveys conducted by civic groups which indicate that the number of police arrests for various crimes including robberies, felonies, burglaries, etc. has dropped sharply ever since the department's new one police officer per patrol car policy went into effect. Although this new policy was started as an economy measure, its effectiveness is being questioned.

The department officials steadfastly refuse to admit that this new policy has any effect on the number af police arrests.

A 16-year-old driver getting a "talking-to" instead of a ticket from a police officer after a minor traffic violation. (© *Spencer Grant/Photo Researchers, Inc.*)

10. A large Wall Street brokerage firm has instituted a new policy whereby it randomly "listens in" and records all broker conversations with clients. The number of complaints about incorrect orders executed by ten sales brokers both before and after the new policy was instituted is shown in the accompanying table.

		Before New Policy	After New Policy
	1	47	38
	2	39	34
	3	28	16
	4	31	26
	5	30	30
Broker	6	53	47
	7	23	39
	8	42	38
	9	23	21
	10	14	10

Using a 5% level of significance, test the null hypothesis that the new policy has no effect on the number of incorrect orders against the alternative hypothesis that the number of complaints has decreased.

11. The number of different times that dizziness was reported per month by patients prior to and after taking a certain medicine is shown below:

		Reported Dizziness	
		Before Taking Medicine	After Taking Medicine
	1	19	17
	2	18	24
	3	9	12
	4	8	4
	5	7	7
Patient	6	12	15
	7	16	19
	8	22	25
	9	19	16
	10	18	24

Using a 5% level of significance, test the null hypothesis that the medicine has no effect on the number of times that dizziness is reported against the alternative hypothesis that the medicine increases the number of times that dizziness is reported.

14.4   THE WILCOXON SIGNED-RANK TEST

Paired-sample sign test

Wilcoxon signed-rank test

As we indicated in the last section, the **paired-sample sign test** merely utilizes information concerning whether the differences between pairs of numbers are positive or negative. Quite often the sign test will accept a null hypothesis simply because too much information is "thrown away." The **Wilcoxon signed-rank test** is less likely to accept a null hypothesis since it considers both the *magnitude* as well as the *direction* (positive or negative) of the differences between pairs.

To illustrate the use of the Wilcoxon signed-rank test, let us consider the following exam scores of 10 students who were tested both before and after receiving special instruction.

Student	Pre-instruction Score X_B	Post-instruction Score X_A	Difference $D = X_B - X_A$	Absolute Value of Difference $\lvert D \rvert$	Rank of $\lvert D \rvert$	Signed Rank
1	46	81	$46 - 81 = -35$	35	9	-9
2	58	73	$58 - 73 = -15$	15	8	-8
3	69	72	$69 - 72 = -3$	3	2	-2
4	72	77	$72 - 77 = -5$	5	4	-4
5	82	82	$82 - 82 = 0$	—	—	—
6	65	72	$65 - 72 = -7$	7	6	-6
7	69	63	$69 - 63 = 6$	6	5	$+5$
8	72	68	$72 - 68 = 4$	4	3	$+3$
9	73	85	$73 - 85 = -12$	12	7	-7
10	87	88	$87 - 88 = -1$	1	1	-1

Notice that we have added several new columns.

To apply the Wilcoxon signed-rank test we proceed as follows:

1. Find the entry for the difference column by subtracting the new value from the corresponding old value. These differences may be positive, negative, or zero.
2. Form the absolute value of these differences.
3. Rank these absolute values in order from the lowest (1) to the highest.
4. Give each rank a plus (+) sign or a minus (−) sign, which is the same sign as in the column for D.
5. The test statistic is the sum of the ranks with the smaller sum. If the null hypothesis is correct, then we would expect the sum of the positive ranks and the sum of the negative ranks to balance each other. If the sum of the ranks is considerably more positive or considerably more negative, then we would be more likely to reject the null hypothesis.

In our case the sum of the ranks is

$$\text{Positive sign ranks} = (+5) + (+3) = +8$$

$$\text{Negative sign ranks} = (-9) + (-8) + (-2) + (-4)$$
$$+ (-6) + (-7) + (-1) = -37$$

We select +8 since this is the sum of the ranks with the smaller sum. We now compare this test statistic value with the critical value given in Table 14.3 using $n = 9$ and $\alpha = 0.05$. The chart value for a two-tailed test is 5. Since our test statistic value of +8 is larger than the chart value, we do not reject the null hypothesis. If, on the other hand, the test statistic value is less than or equal to the chart value, then we reject the null hypothesis.

TABLE 14.3　　　　Critical Values for Wilcoxon Signed-Rank Test

	One-Tailed Test Level of Significance ($\alpha = $)					Two-Tailed Test Level of Significance ($\alpha = $)			
n	0.005	0.01	0.025	0.05	n	0.01	0.02	0.05	0.10
5	—	—	—	0	5	—	—	—	0
6	—	—	0	2	6	—	—	0	2
7	—	0	2	3	7	—	0	2	3
8	0	1	3	5	8	0	1	3	5
9	1	3	5	8	9	1	3	5	8
10	3	5	8	10	10	3	5	8	10
11	5	7	10	13	11	5	7	10	13
12	7	9	13	17	12	7	9	13	17
13	9	12	17	21	13	9	12	17	21
14	12	15	21	25	14	12	15	21	25
15	15	19	25	30	15	15	19	25	30
16	19	23	29	35	16	19	23	29	35
17	23	27	34	41	17	23	27	34	41
18	27	32	40	47	18	27	32	40	47
19	32	37	46	53	19	32	37	46	53
20	37	43	52	60	20	37	43	52	60
21	42	49	58	67	21	42	49	58	67
22	48	55	65	75	22	48	55	65	75
23	54	62	73	83	23	54	62	73	83
24	61	69	81	91	24	61	69	81	91
25	68	76	89	100	25	68	76	89	100
26	75	84	98	110	26	75	84	98	110
27	83	92	107	119	27	83	92	107	119
28	91	101	116	130	28	91	101	116	130
29	100	110	126	140	29	100	110	126	140
30	109	120	137	151	30	109	120	137	151

COMMENT   In performing the calculations in the previous example we did not use the fifth student's scores since, as with the sign test, a difference of zero is not considered as positive or negative for our purposes.

COMMENT   All entries given in Table 14.3 are for absolute values of the test statistic.

COMMENT   Occasionally, two differences will have the same rank. For example, if two differences are tied for the fourth place, then each is assigned a rank of 4.5. We then assign the next value a rank of 6. The same procedure is followed when there is a tie for any rank.

We illustrate the use of the Wilcoxon signed-rank test with another example.

E X A M P L E   1   A new cholesterol-lowering pill is given to 15 people once a day for 2 months to determine its effectiveness in lowering blood serum cholesterol levels. The following data indicate the before and after cholesterol levels (in milligrams) of these 15 people.

Cholesterol Level Before Taking Pill (in milligrams)	Cholesterol Level After Taking Pill (in milligrams)
240	227
261	238
283	257
276	276
220	208
186	193
195	198
198	199
247	233
238	227
220	210
250	241
263	255
298	276
317	269

Using a 5% level of significance, test the null hypothesis that the pill has no effect on a person's blood serum cholesterol level; it neither lowers nor raises it.

SOLUTION   We arrange the data as follows.

Cholesterol Level Before X_B	Cholesterol Level After X_A	Difference $D = X_R - X_A$	Absolute Value of Difference $|D|$	Rank of $|D|$	Signed Rank
240	227	$240 - 227 = 13$	13	9	+9
261	238	$261 - 238 = 23$	23	12	+12
283	257	$283 - 257 = 26$	26	13	+13
276	276	$276 - 276 = 0$	—	—	—
220	208	$220 - 208 = 12$	12	8	+8
186	193	$186 - 193 = -7$	7	3	-3
195	198	$195 - 198 = -3$	3	2	-2
198	199	$198 - 199 = -1$	1	1	-1
247	233	$247 - 233 = 14$	14	10	+10
238	227	$238 - 227 = 11$	11	7	+7
220	210	$220 - 210 = 10$	10	6	+6
250	241	$250 - 241 = 9$	9	5	+5
263	255	$263 - 255 = 8$	8	4	+4
298	276	$298 - 276 = 22$	22	11	+11
317	269	$317 - 269 = 48$	48	14	+14

For each person we determine the difference in the cholesterol level and the absolute value of the difference. Then we assign ranks from the lowest to the highest and assign a plus sign or a minus sign to each of these ranks. The sign is the same as the sign in the column for D. Now we calculate the sum of the ranks. We have

$$\text{Positive sign ranks} = (+9) + (+12) + (+13) + (+8) + (+10) + (+7)$$
$$+ (+6) + (+5) + (+4) + (+11) + (+14) = +99$$

$$\text{Negative sign ranks} = (-3) + (-2) + (-1) = -6$$

We select -6 since this is the sum of the ranks with the smaller sum. The absolute value of -6 is 6. We now compare this test statistic value with the critical value given in Table 14.3 using $n = 14$ and $\alpha = 0.05$. The chart value for a two-tailed test is 21. Since our test statistic value of 6 is less than the chart value, we reject the null hypothesis and conclude that the pill does affect the blood serum cholesterol level of an individual.

COMMENT   When using the Wilcoxon signed-rank test, it is assumed that the sample data are continuous and that the sampled population is symmetric. Furthermore, the data results can be arranged into relationships of "greater than" or "less than." In an applied example it is often difficult to verify that the sampled population is symmetric.

EXERCISES FOR SECTION 14.4

1. Many schools claim that a lot of food is wasted simply because students do not eat the lunches that have been prepared for them. At one school, a new caterer has been hired. The caterer is supposed to make the food served more visually appealing. After a ten-week trial period the following data are obtained:

		Number of Students Completing Their Lunch	
		Before New Caterer Began Preparing Lunches	**After New Caterer Began Preparing Lunches**
	1	241	256
	2	233	249
	3	262	258
	4	256	265
Week	*5*	271	278
	6	293	289
	7	312	310
	8	270	315
	9	275	305
	10	285	342

Using the Wilcoxon signed-rank test, test the null hypothesis that the number of students completing their lunch has not changed even after the new caterer was hired. Use a 5% level of significance.

2. Consider the newspaper article on the next page. The number of reported cases of car theft for a specific time period both before and after the use of the extensive computer tracking in several counties of the state is as follows:

		Number of Reported Cases of Car Theft	
		Before New Policy	**After New Policy**
	A	36	32
	B	40	34
	C	41	41
	D	42	32
County	*E*	39	40
	F	33	38
	G	54	46
	H	48	47

> ### INCIDENCE OF CAR THEFT BEGINING TO DECLINE
>
> Trenton, *Dec. 12*: Police department records as well as insurance company statistics indicate that the number of reported claims have been decreasing in the state ever since the state instituted a computer tracking system as well as computer cross-checks of all reported car thefts and vigorous prosecution of all false car claim thefts. This represents the first time that the reported number of car thefts showed any decline. Over the past years car theft had become a burgeoning crime.

		Number of Reported Cases of Car Theft	
		Before New Policy	After New Policy
County	*I*	56	54
	J	47	44

Using the Wilcoxon signed-rank test at a 5% level of significance, test the null hypothesis that the new policy has no effect on reducing car theft.

3. A radio talk show has been experimenting with having guest celebrities on the air. The estimated audience (in hundreds of thousands of people) both before and after instituting this policy for eight sections of a large city over a given time period is shown below:

		Radio Audience (in hundred thousands)	
		Before New Policy	After New Policy
Section of City	*Brighton*	24	28
	Queens	23	27
	Boro Hall	20	22

(Table continues on page 766)

		Radio Audience (in hundred thousands)	
		Before New Policy	After New Policy
	Gravesend	26	25
	Togo	19	26
Section of City	Braverly	25	25
	Beaver	18	26
	Kensington	27	27

Using the Wilcoxon signed-rank test at a 5% level of significance, test the null hypothesis that the new policy has no effect on radio audience.

4. Cholesterol-free and low-fat and/or no-fat bakery products are appearing more and more on supermarket shelves as people are more health conscious. In one study, 12 people were randomly selected and asked to rate a popular cake (on a certain scale) both before and after a certain fat-free ingredient was added. Their ratings are shown below:

		Cake Rating	
		Before Ingredient Added	After Ingredient Added
	1	111	108
	2	119	120
	3	138	140
	4	116	115
	5	120	120
Volunteer Tester	6	137	140
	7	128	136
	8	117	129
	9	120	104
	10	118	129
	11	137	124
	12	126	149

Using the Wilcoxon signed-rank test, do the data indicate that there is a significant difference in the ratings of the popular cake before and after the fat-free ingredient was added? (Use a 5% level of significance.)

14.5   THE MANN-WHITNEY TEST

Wilcoxon rank-sum test Mann-Whitney test

An important nonparametric test that is used as an alternative to the standard significance tests for the difference between two sample means is the **Wilcoxon rank-sum test** or the **Mann-Whitney Test**. We can use this test when the assumption about normality is not satisfied.

To illustrate how this test is used, we consider the following data on the number of minutes needed by two independent groups of music students to learn to play a particular song. Group A received special instruction whereas Group B did not.

										Average
Group A	35	39	51	63	48	31	29	41	55	43.56
Group B	85	28	42	37	61	54	36	57		50

The means of these two samples are 43.56 and 50. In this case we wish to decide whether the difference between the means is significant.

The two samples are arranged jointly, as if they were one sample, in order of increasing time. We get

Time	Group	Rank
28	B	1
29	A	2
31	A	3
35	A	4
36	B	5
37	B	6
39	A	7
41	A	8
42	B	9
48	A	10
51	A	11
54	B	12
55	A	13
57	B	14
61	B	15
63	A	16
85	B	17

We indicate each value, whether it belongs to Group A or to Group B. Then we assign the ranks 1, 2, 3, 4, . . . , 17 to the scores, in this order, as shown.

Notice that the Group A scores occupy the ranks of 2, 3, 4, 7, 8, 10, 11, 13, and 16. The Group B scores occupy the ranks of 1, 5, 6, 9, 12, 14, 15, and 17. Now we sum the ranks of the group with the *smaller* sample size, in this case Group B, getting

$$1 + 5 + 6 + 9 + 12 + 14 + 15 + 17 = 79$$

The sum of the ranks is denoted by R. In this case $R = 79$.

We always let n_1 and n_2 denote the sizes of the two samples where n_1 represents the smaller of the two sample sizes. Thus, R represents the sum of the ranks of this smaller group. (If both groups are of equal sizes, then either one is called n_1, and R represents the sum of the ranks of this group.) Statistical theory tells us that if both n_1 and n_2 are large enough, each equal to 8 or more, then the distribution of R can be approximated by a normal distribution. The test statistic is given by Formula 14.1.

FORMULA 14.1

$$z = \frac{R - \mu_R}{\sigma_R}$$

where

$$\mu_R = \frac{n_1(n_1 + n_2 + 1)}{2}$$

and

$$\sigma_R = \sqrt{\frac{n_1 n_2(n_1 + n_2 + 1)}{12}}$$

Using a 5% level of significance, we reject the null hypothesis of equal means if $z > 1.96$ or $z < -1.96$.

In our case $R = 79$, $n_1 = 8$, and $n_2 = 9$ so that

$$\mu_R = \frac{n_1(n_1 + n_2 + 1)}{2}$$

$$= \frac{8(8 + 9 + 1)}{2}$$

$$= 72$$

and

$$\sigma_R = \sqrt{\frac{n_1 n_2(n_1 + n_2 + 1)}{12}}$$

$$= \sqrt{\frac{8(9)(8 + 9 + 1)}{12}} = \sqrt{108}$$

$$= 10.39$$

The test statistic then becomes

$$z = \frac{R - \mu_R}{\sigma_R} = \frac{79 - 72}{10.39} = 0.67$$

Since the value of $z = 0.67$ falls in the acceptance region of Figure 14.3, we *do not* reject the null hypothesis. There is no significant difference between the means of these two groups.

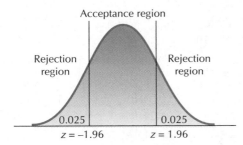

FIGURE 14.3

COMMENT   The test that we have just described is the Wilcoxon rank-sum test with the normal approximation of the test statistic. **Mann-Whitney's test** is equivalent, but the test statistic is calculated in a slightly different way. Statisticians have constructed tables that give the appropriate critical values when both sample sizes, n_1 and n_2, are smaller than 8. The interested reader can find such tables in many books on nonparametric statistics. The corresponding exact statistic is called the **Mann-Whitney U test**.

Mann-Whitney U test

E X A M P L E   1   An animal trainer in a circus is teaching 20 lions to perform a special trick. The lions have been divided into two groups, A and B. Group A gets positive reinforcement of food and favorable comments during the learning session whereas Group B does not. The following table indicates the number of days needed by each lion to learn the trick.

Group A	78	95	82	69	111	65	73	84	92	110
Group B	121	132	101	79	94	88	102	93	98	127

Using a 5% level of significance, test the null hypothesis that the mean time for both groups is the same.

SOLUTION   The two samples are first arranged jointly, as if they were one large sample, in order of increasing size. We get

Days	Group	Rank
65	A	1
69	A	2
73	A	3
78	A	4

(*Table continues on page 770*)

Days	Group	Rank
79	B	5
82	A	6
84	A	7
88	B	8
92	A	9
93	B	10
94	B	11
95	A	12
98	B	13
101	B	14
102	B	15
110	A	16
111	A	17
121	B	18
127	B	19
132	B	20

Since both groups are of equal size, we will work with Group A. The sum of the ranks of Group A is

$$1 + 2 + 3 + 4 + 6 + 7 + 9 + 12 + 16 + 17 = 77$$

Thus, $R = 77$. Now we apply Formula 14.1. We have $R = 77$, $n_1 = 10$, and $n_2 = 10$ so that

$$\mu_R = \frac{n_1(n_1 + n_2 + 1)}{2}$$

$$= \frac{10(10 + 10 + 1)}{2}$$

$$= 105$$

and

$$\sigma_R = \sqrt{\frac{n_1 n_2 (n_1 + n_2 + 1)}{12}}$$

$$= \sqrt{\frac{10 \cdot 10(10 + 10 + 1)}{12}}$$

$$= \sqrt{175}$$

$$\approx 13.23$$

The test statistic then becomes

$$z = \frac{R - \mu_R}{\sigma_R}$$

$$= \frac{77 - 105}{13.23}$$

$$= -2.12$$

We use the two-tail rejection of Figure 14.3. Since the value of $z = -2.12$ falls in the rejection region, we reject the null hypothesis and conclude that the number of minutes needed by each group is not the same. Positive reinforcement affects learning time.

EXAMPLE 2 Many airlines are forced to cancel scheduled flights for a variety of reasons. Records submitted to aviation officials indicate that the weekly number of canceled flights reported by two large airline companies over a consecutive period of weeks is as follows.

Airline A	Airline B
7	13
9	5
12	4
14	8
8	11
6	3
11	17
	15

Using a 5% level of significance, test the null hypothesis that the average number of canceled flights is the same for both airlines.

SOLUTION   The null hypothesis is that the average number of canceled flights is the same for both airlines. We arrange both samples jointly, as if they were one large sample, in order of increasing size. We get

Number of Canceled Flights	Airline	Rank
3	B	1
4	B	2
5	B	3
6	A	4
7	A	5
8	A	6.5
8	B	6.5
9	A	8
11	A	9.5
11	B	9.5
12	A	11
13	B	12
14	A	13
15	B	14
17	B	15

You will notice that there are some ties for several ranks. Whenever a tie comes up, each of the tied observations is assigned the mean of the ranks they occupy. Thus, we assign a rank of 6.5 when the number of canceled flights was 8, and a rank of 9.5 when the number of canceled flights was 11. For Airline A we have 7 weeks of data and for Airline B we have 8 weeks of data. Thus, we will work with the smaller group, which is A. The sum of the ranks for Airline A is

$$4 + 5 + 6.5 + 8 + 9.5 + 11 + 13 = 57$$

Thus, $R = 57$. Applying Formula 14.1 with $n_1 = 7$ and $n_2 = 8$ gives

$$\mu_R = \frac{n_1(n_1 + n_2 + 1)}{2}$$

$$= \frac{7(7 + 8 + 1)}{2}$$

$$= 56$$

and

$$\sigma_R = \sqrt{\frac{n_1 n_2(n_1 + n_2 + 1)}{12}}$$

$$= \sqrt{\frac{7(8)(7 + 8 + 1)}{12}}$$

$$= \sqrt{74.6667}$$

$$\approx 8.641$$

The test statistic then becomes

$$z = \frac{R - \mu_R}{\sigma_R}$$

$$= \frac{57 - 56}{8.641}$$

$$= -0.116$$

The value of $z = -0.116$ falls in the acceptance region of Figure 14.3. Thus, we cannot reject the null hypothesis. There are not sufficient data to conclude that the average number of cancellations of scheduled flights is significantly different for both airlines.

EXERCISES FOR SECTION 14.5

1. Two groups of students have been given a special math competency exam. Group A students are from schools located in the northern part of the city, an area that is predominantly middle class. Group B students are from schools located in the south-

ern part of the city, an area that is predominantly working class. The accompanying data indicate the results of the exam:

Group A	Group B
44	51
49	40
47	29
53	41
46	42
57	37
52	46
43	42
48	36
54	41
56	43
	44
	40
	39
	45

Using the rank-sum test, test the null hypothesis that the mean score for both groups is the same. (Use a 5% level of significance.)

2. *The Effects of Alcohol.* Eight younger volunteers aged 20–30 years and ten older volunteers aged 40–50 years were each given the same amount of alcohol and then asked to perform a special task correctly. The number of trials required by each volunteer before completing the task is shown below:

Younger Volunteer	Older Volunteer
19	33
27	35
31	32
16	27
23	18
24	28
20	23
22	26
	22
	28

Using the rank-sum test, test the null hypothesis that the average number of trials required by a member of each group to complete the task after being given alcohol is the same. (Use a 5% level of significance.)

3. The following data indicate the number of minutes of practice needed by two groups of people at a health club to learn a particular aerobic exercise.

Group A	56	62	54	51	65	66	59	50	60	55	
Group B	47	67	63	55	59	58	64	71	56	61	52	57

Using the rank-sum test, test the hypothesis that the mean time needed by both groups to learn the aerobic exercise is the same. (Use a 1% level of significance.)

4. The number of airplanes arriving at least 15 minutes after their scheduled arrival time daily for two large carriers over a period of time is as follows:

Airline A	5	16	9	18	14	15	10	8	11	2	6	28	
Airline B	17	6	11	21	16	13	4	8	12	3	7	15	4

Using a 5% level of significance, test the null hypothesis that the average number of late arrivals is the same for both airlines.

5. Ten male and 13 female drivers were asked to drive over a road that had many obstacles. The number of minutes needed by each to complete the trip is as follows:

Female	56	61	59	65	66	73	69	45	56	62	54	69	63
Male	47	71	63	63	59	57	67	74	54	49			

Using the rank-sum test, test the hypothesis that the average time needed by both groups is the same. (Use a 5% level of significance.)

6. The number of incidences of burglary reported to police officials in Monticello for the first eight days of January and to police officials in Boonton for the first nine days of January are as follows:

Monticello	157	160	172	149	155	182	169	132	
Boonton	161	146	183	195	103	113	188	170	153

Using a 5% level of significance, test the null hypothesis that the average number of incidence of burglary reported to police officials in both cities is the same.

7. Environmentalists took samples of the air in eight classrooms of the Brighton Day-Care Center and 12 samples of the air in classrooms of the Bergenville Day-Care Center to determine the level of radon gas in the air in picocuries per liter (pc/l). The following results were obtained:

Radon Levels (in pc/l)

Brighton Day-Care Center	Bergenville Day-Care Center
1.8	3.4
1.4	4.2
0.6	3.9
1.8	3.6
1.9	4.1
2.8	3.5
3.2	3.6
2.3	2.8
	2.7
	2.1
	2.7
	3.3

Using a 5% level of significance, test the null hypothesis that the average radon levels in the air at both day-care centers is approximately the same.

14.6   THE SPEARMAN RANK CORRELATION TEST

For many years the most widely used nonparametric statistical test was the rank correlation test developed by C. Spearman in the early 1900s. Although originally devised as a shortcut method for computing the coefficient of correlation discussed in Chapter 11, the Spearman rank correlation test has the advantage that it uses rankings only, it makes no assumptions about the distribution of the underlying populations, and it does not assume normality. We simply arrange some data in rank order and then apply the following formula.

FORMULA 14.2

The Spearman Rank Correlation Coefficient

$$R = 1 - \frac{6 \, \Sigma(x - y)^2}{n(n^2 - 1)}$$

where $\Sigma(x - y)^2$ represents the sum of the squares of the difference in ranks and n stands for the number of individuals who have been ranked.

COMMENT   When using Formula 14.2 the value of R will be between -1 and $+1$. It is used in much the same way that we used the correlation coefficient in Chapter 11.

COMMENT   When using Formula 14.2 the null hypothesis to be tested is that there is no significant correlation between the two rankings as opposed to the alternative hypothesis, which assumes that there is a significant correlation between the rankings.

We illustrate the use of this formula with several examples.

EXAMPLE 1

Beer Tasting

Two judges are testing five different brands of beer for their taste appeal, and the judges rate the beers as follows.

	Judge 1	Judge 2
Brand A	3	4
Brand B	4	3
Brand C	5	1
Brand D	2	2
Brand E	1	5

Using a 5% level of significance, test the null hypothesis that there is no significant (linear) correlation between the two judge's ratings.

SOLUTION We first rewrite the preceding rankings by letting x represent Judge 1's rankings and by letting y represent Judge 2's rankings. We then have the following.

x	y	$x - y$	$(x - y)^2$
3	4	-1	1
4	3	1	1
5	1	4	16
2	2	0	0
1	5	-4	16
			34

$$\Sigma(x - y)^2 = 34$$

Now we apply Formula 14.2. We have

$$R = 1 - \frac{6\,\Sigma(x - y)^2}{n(n^2 - 1)}$$

$$= 1 - \frac{6(34)}{5(25 - 1)}$$

$$= 1 - 1.7 = -0.7$$

Table XI in the Appendix allows us to interpret the value of the Spearman rank coefficient correctly. We use this table in the following way:

1. First compute the value of R using Formula 14.2.
2. Then look in the chart for the appropriate R-value corresponding to some given value of n where n is the number of pairs of scores.
3. The value of R is *not* statistically significant if it is between $-R_{0.025}$ and $R_{0.025}$ for a particular value of n at the 5% level of significance.

Returning to the previous example, we conclude that since the value of R is -0.7, it is not significant at the 5% level of significance. We cannot conclude that there is a significant correlation between the two rankings.

EXAMPLE   2   Several 9-year-old children recently competed in the Brown Bowling League play-offs and the Southview Bowling League games. The scores of these contestants were as follows.

	Score	
Contestant	Brown League Play-Off Game	Southview Bowling League Game
Kim	180	162
Heather	176	157
Jason	198	176
Cassandra	197	183
Wilfredo	171	188

Using a 5% level of significance, test the null hypothesis that there is no significant correlation between the Brown League play-off game scores and the Southview scores.

SOLUTION   We first rewrite each of the scores in terms of rankings from the highest score, which is assigned a ranking of 1 to the lowest score, which is assigned a ranking of 5. Now we let x represent the Brown score rankings and let y represent the Southview score rankings. We then have the following.

x	y	$x - y$	$(x - y)^2$
3	4	-1	1
4	5	-1	1
1	3	-2	4
2	2	0	0
5	1	4	16
			22
			$\Sigma(x - y)^2 = 22$

Applying Formula 14.2 gives

$$R = 1 - \frac{6 \, \Sigma(x - y)^2}{n(n^2 - 1)}$$

$$= 1 - \frac{6(22)}{5(25 - 1)}$$

$$= 1 - 1.1 = -0.1$$

By referring to Table XI in the Appendix, we conclude that since our value of R is -0.1, it is not significant at the 5% level of significance. We cannot conclude that there is a significant correlation between the two bowling league scores.

14.7 THE RUNS TEST

Runs test

All the samples discussed so far in this book were assumed to be random samples. How does one test for randomness?

In recent years mathematicians have developed a **runs test** for determining the randomness of samples. This test is based on the order in which the observations are made. For example, suppose 25 people are waiting in line for admission to a theater and they are arranged as follows where m denotes male and f denotes female:

f, f, f, f, m, f, m, m, m, m, f, f, f, f, f, f, m, m, m, m, f, m, m, f, m

Is this a random arrangement of the m's and f's?

Theory of runs

In order to answer this question, statisticians use the **theory of runs.** We first have the following definition.

Definition 14.1
Run

A **run** is a succession of identical letters or symbols that is followed and preceded by a different letter or by no letter at all.

There are ten runs in the preceding sequence of m's and f's. These are

Run	Letters
1	ffff
2	m
3	f
4	mmmm
5	ffffff
6	mmmm
7	f
8	mm
9	f
10	m

Many runs would indicate that the data occur in definite cycles according to some pattern. The same is true for data with few runs. In either case we do not have a random sample. We still need some way of determining when the number of runs is reasonable.

When using the runs test note that the length of each individual run is not important. What is important is the number of times that each letter appears in the entire sequence of letters. Thus, in our example there are 25 people waiting in line; 13 are female and 12 are male so that f appears 13 times and m appears 12 times. We have n_1 samples of one kind and n_2 samples of another kind. We now wish to test whether this sample is random.

Table X in the Appendix gives us the critical values for the total number of runs. To use this table we first determine the larger of n_1 and n_2. In our case there are 13 f's and 12 m's so that the larger is 13 and the smaller is 12. We now move across the top of the chart until we reach 13. Then we move down until we get to the 12 row. Notice that there are two numbers in the box corresponding to larger 13, smaller 12. These are the numbers 8 and 19. These are also the critical values. If the number of runs is between 8 and 19, we do not reject the null hypothesis. This would mean that we have a random sample. If the number of runs is less than 8 or more than 19, we no longer have a random sample. In our case since we had ten runs we do not reject the null hypotheses of randomness.

Let us further illustrate the runs test with several examples.

EXAMPLE 1   Twenty people are waiting on line in a bank. These people will either deposit or withdraw money. Let d represent a customer who makes a deposit and let w represent a customer who is making a withdrawal. If the people are arranged in the following order, test for randomness. (Use a 5% level of significance.)

d, d, d, w, w, w, w, w, d, d, d, d, d, d, d, w, w, w, d, d

SOLUTION   There are five runs as shown in the following chart:

Run	Letters
1	ddd
2	wwwww
3	ddddddd
4	www
5	dd

There are also 12 d's and 8 w's in this succession of letters, so that the larger is 12 and the smaller is 8. From Table X in the Appendix we find that the critical values, where the larger number is 12 and the smaller number is 8, are 6 and 16 runs. Since we obtained only 5 runs, we reject the null hypothesis and conclude that these people are not arranged in random order.

E X A M P L E   2   Thirty dresses are arranged on a rack as follows, where r represents a red dress and b represents a blue dress. Using a 5% level of significance, determine if the dresses are arranged in a random order.

r, b, r, b, r, r, b, r, b, b, r, r, r, r, r, b, r, b, b, r, b,
r, b, r, b, b, b, b, r, r

SOLUTION   There are 16 r's and 14 b's so that the larger number is 16 and the smaller number is 14. There are also 19 runs as shown.

Run	Letters
1	r
2	b
3	r
4	b
5	rr
6	b
7	r
8	bb
9	rrrr
10	b
11	r
12	bb
13	r
14	b
15	r
16	b
17	r
18	bbbb
19	rr

From Table X we note that the critical values, where the larger number is 16 and the smaller number is 14, are 10 and 22. Since we obtained 19 runs, we do not reject the null hypothesis of randomness.

The theory of runs can also be applied to any set of numbers to determine whether or not these numbers appear in a random order. In such cases we first calculate the median of these numbers (see page 133) since approximately one-half the numbers are below the median and one-half the numbers are above the median. We then go through the sequence of numbers and replace each number with the letter a if it is above the median and with the letter b if it is below the median. We omit any values that equal the median. Once we have a sequence of a's and b's, we proceed in exactly the same way as we did in Examples 1 and 2 of this section.

EXAMPLE   3   The number of defective items produced by a machine per day over a period of a month is

13, 17, 14, 20, 18, 16, 14, 19, 21, 20, 14, 17, 12, 14, 19,
20, 17, 18, 14, 20, 17, 19, 17, 14, 19, 21, 16, 12, 15, 22

Using a 5% level of significance, test for randomness.

SOLUTION   We first arrange the numbers in order, from the smallest to the largest, to determine the median. We get 12, 12, 13, 14, 14, 14, 14, 14, 14, 15, 16, 16, 17, 17, 17, 17, 17, 18, 18, 19, 19, 19, 19, 20, 20, 20, 20, 21, 21, 22. The median of these numbers is 17. We now replace all the numbers of the original sequence with the letter b if the number is below 17 and with the letter a if the number is above 17. We do not replace the 17 with any letter. The new sequence then becomes

b, b, a, a, b, b, a, a, a, b, b, b, a, a, a, b, a, a, b, a,
a, b, b, b, a

There are now 13 a's and 12 b's, and there are 12 runs as follows:

Run	Letters
1	bb
2	aa
3	bb
4	aaa
5	bbb
6	aaa
7	b
8	aa
9	b
10	aa
11	bbb
12	a

From Table X we note that the critical values, where the larger number is 13 and the smaller number is 12, are 8 and 19. Since we obtained 12 runs, we do not reject the null hypothesis.

COMMENT   Table X can be used only when n_1 and n_2 are not greater than 20. If either is larger than 20, we use the normal curve approximation. In this case the test statistic is

$$z = \frac{X - \mu_R}{\sigma_R}$$

where

$$\mu_R = \frac{2n_1 n_2}{n_1 + n_2} + 1$$

and

$$\sigma_R = \sqrt{\frac{2n_1 n_2 (2n_1 n_2 - n_1 - n_2)}{(n_1 + n_2)^2 (n_1 + n_2 - 1)}}$$

EXERCISES FOR SECTION 14.7

1. Many large companies send recruiters to college campuses across the country in an effort to identify the most talented prospective employees. These recruiters then rate these candidates. Four recruiters for one company have rated five potential prospects (Kimberly, Jonathan, Heather, Thomas, and Jason) in terms of the prospect's potential according to the following rankings:

Prospective Rankings

Recruiter 1	Recruiter 2	Recruiter 3	Recruiter 4
Kimberly	Kimberly	Jonathan	Jonathan
Jason	Jonathan	Heather	Jason
Heather	Thomas	Thomas	Thomas
Thomas	Jason	Jason	Heather
Jonathan	Heather	Kimberly	Kimberly

Using a 5% level of significance, test the null hypothesis that there is no significant correlation between the rankings of
a. recruiters 1 and 2.
b. recruiters 1 and 3.
c. recruiters 2 and 4.

2. All entering freshmen at Unity College must take the SAT (Scholastic Aptitude Test) as well as the college's own math placement exam. The following list gives the test scores of ten entering freshmen on both of these tests:

Student	Math Score on SAT Test	College Math Placement Exam
Margaret	585	46
Gina	582	86
Lorraine	565	70
Donna	577	74

Student	Math Score on SAT Test	College Math Placement Exam
Lisa	573	66
Joseph	581	64
Gail	589	62
Christine	559	71
Renée	569	75
Marie	599	90

Using a 5% level of significance, test the null hypothesis that there is no correlation between the SAT test scores and the college test scores, that is, that they are independent.

3. The ages of the 19 people who applied for unemployment insurance on Monday (in the order that they applied) are as follows:

20, 25, 59, 64, 43, 40, 32, 29, 23, 68, 45, 38, 54, 33, 30, 19, 23, 35, 47

Test for randomness.

4. The number of child abuse cases per month that were classified as serious by the welfare department of a city over the past two years is as follows:

70, 60, 90, 103, 80, 91, 57, 73, 80, 75, 90, 78, 54, 65, 103, 109, 81, 76, 64, 79, 68, 79, 95, 103

Test for randomness.

5. The number of homeless individuals in the United States has been a continuing problem. A newspaper reporter assigned to investigate the situation observed 20 homeless people with the following sequence of males (M) and females (F):

M, F, M, M, F, M, M, F, F, F, M, M, F, F, F, M, M, F, F, M

Test for randomness.

6. The first 25 cars that enter a parking lot are classified as either foreign made (F) or American made (A) as follows:

F, F, F, F, A, A, A, F, F, F, F, F, A, F, A, F, A, F, F, A, A, A, A, F, F

Test for randomness.

7. Patients who have been hospitalized for a long period of time often complain about the indifference and lack of emotional support on the part of the nursing staff. In a recent study, five patients who had been hospitalized at Maimomides General

Hospital were asked to rate seven nurses on their bedside manners. Their rankings were as follows:

Nurse	Patient 1	Patient 2	Patient 3	Patient 4	Patient 5
Ashley	7	7	2	6	4
Maureen	5	1	1	3	6
Dorothy	3	2	7	1	5
Agnes	4	3	5	5	1
Heather	6	4	3	7	7
Priscilla	1	6	6	4	3
Jacqueline	2	5	4	2	2

Using a 5% level of significance, test the null hypothesis that there is no significant correlation between the patient's ratings of the nurses.

8. Many colleges poll their recent graduates to gather information on their jobs, salaries, and so forth. A survey of 22 recent business graduates of Victory College revealed the following starting salaries (in the order that the information was received) for these graduates:

$37,693	$48,840	$39,055	$26,000	$29,970	$41,236	$33,792
31,003	33,007	39,057	43,068	24,769	54,103	48,897
45,692	40,038	36,092	37,504	38,113	40,153	36,045
36,421						

Test for randomness.

9. The number of cases of credit card complaints resulting from telemarketing fraudulent schemes received by authorities over the past 19 days is as follows:

11,   18,   15,   19,   16,   17,   15,   25,   24,   27,   28,   16,   19,   27,   21,
18,   16,   14,   13

Test for randomness.

10. There are seven sports counselors in Camp Shafne. Two campers were asked to rank those counselors from the best to the worst. Their rankings were as follows:

Counselor	Ratings by Camper 1	Ratings by Camper 2
Meredith	7	5
José	2	4
John	1	6
Vasilious	3	3
Vicky	6	7
Pete	4	1
Andrew	5	2

Using a 5% level of significance, test the null hypothesis that there is no correlation between the campers' rankings of the counselors.

11. A gas station attendant observes that the first 25 customers who bought gas paid for it either by cash (C) or by credit card (R) as follows:

    C, R, C, C, R, R, R, C, R, R, R, R, C, C, R, R, R,

    C, C, C, C, C, R, R, C

    Test for randomness.

12. The sex of the first 25 runners who have registered to run in a marathon is as follows (where M = male and F = female):

    M, F, F, F, M, M, F, M, F, F, F, M, F, F, M, M, M,

    F, F, F, M, M, F, F, F

    Test for randomness.

14.8   THE KRUSKAL-WALLIS H-TEST

Kruskal-Wallis H-test

In Chapter 13 we discussed ANOVA techniques for determining whether the sample means of several populations are equal. In order to apply the ANOVA techniques and the F-distribution we assumed that the samples were randomly selected from independent populations that were approximately normally distributed.

A nonparametric statistical test that does not require that these assumptions be satisfied is the **Kruskal-Wallis H-test.** This rank-sum test is used when we wish to test the null hypothesis that r independent random samples were obtained from identical (not necessarily normal) populations. Of course, the alternative hypothesis is that the means of these populations are not the same. The only requirement is that each sample have at least five observations.

When using the Kruskal-Wallis H-test we combine all the data and rank them jointly as if they represent one single sample. If the null hypothesis is true, then the sampling distribution of these numbers can be approximated by a chi-square distribution with $r - 1$ degrees of freedom. We accept the null hypothesis that the sample means are equal whenever the test statistic value is less than the χ^2 value for $r - 1$ degrees of freedom at a level of significance of α. Otherwise we reject the null hypothesis. The test statistic value is calculated using the following formula.

FORMULA 14.3

Test Statistic When Using the Kruskal-Wallis H-Test

$$\frac{12}{n(n + 1)} \sum_{i=1}^{r} \frac{R_i^2}{n_i} - 3(n + 1)$$

When using the preceding test statistic, we have

n = the total number in the entire sample
  (when the data are joined together)

that is,

$$n = n_1 + n_2 + \cdots + n_r$$

R_i = the sum of the ranks assigned to n_i values of the ith sample.

We illustrate the use of this formula with several examples.

EXAMPLE   1   A consumer's group tested numerous cans of paint from three different companies to determine whether there is any significant difference between these brands in average drying time. The number of minutes needed for each of these brands of paint to dry when applied to identical walls was as follows:

Brand A	Brand B	Brand C
38	52	47
32	48	30
27	39	37
29	42	41
23	46	44
43	53	49

Use the Kruskal-Wallis H-test to determine if there is any significant difference in the average drying time for these brands of paint. (Use a 5% level of significance.)

SOLUTION   The null hypothesis is that the means of these different brands are the same. The three samples are first arranged jointly, as if they were one large sample, in order of increasing size. We get

Drying Time	Brand	Rank
23	A	1
27	A	2
29	A	3
30	C	4
32	A	5
37	C	6
38	A	7
39	B	8
41	C	9
42	B	10
43	A	11
44	C	12
46	B	13
47	C	14
48	B	15
49	C	16
52	B	17
53	B	18

The sum of the rankings for Brand A is

$$R_1 = 1 + 2 + 3 + 5 + 7 + 11 = 29$$

The sum of the rankings for Brand B is

$$R_2 = 8 + 10 + 13 + 15 + 17 + 18 = 81$$

The sum of the rankings for Brand C is

$$R_3 = 4 + 6 + 9 + 12 + 14 + 16 = 61$$

In this case the number of cans of paint tested for each brand is 6 so that $n_1 = n_2 = n_3 = 6$. Also, the total number in the entire joined sample is $n = 18$. Substituting these values into the test statistic formula gives

$$\text{Test statistic} = \frac{12}{18(18 + 1)} \left(\frac{29^2}{6} + \frac{81^2}{6} + \frac{61^2}{6} \right) - 3(18 + 1)$$

$$= 8.047$$

From Table VIII in the Appendix the χ^2 value with $3 - 1$, or 2, degrees of freedom at the 5% level of significance is 5.991. Since our test statistic value of 8.047 exceeds this value, we reject the null hypothesis and conclude that the average drying time for these brands of paint is not the same.

EXAMPLE   2   An independent agency is interested in determining how long (in minutes) it takes a teller to complete the identical transaction at each of three banks. The following results were obtained when numerous tellers at the three banks were surveyed.

Time to Transaction Completion
(minutes)

Bank 1	Bank 2	Bank 3
12	23	15
14	8	20
11	17	21
16	9	7
10	13	22
18	24	
19		

Do the data indicate that there is a significant difference in the average time required to complete the transaction at these banks? (Use the Kruskal-Wallis H-test at the 5% level of significance.)

SOLUTION   The null hypothesis is that the means for the three banks are the same. The three samples are first arranged jointly, as if they constitute one large sample, in order of increasing size. We get

Time	Bank	Rank
7	3	1
8	2	2
9	2	3
10	1	4
11	1	5
12	1	6
13	2	7
14	1	8
15	3	9
16	1	10
17	2	11
18	1	12
19	1	13
20	3	14
21	3	15
22	3	16
23	2	17
24	2	18

The sum of the rankings for Bank 1 is

$$R_1 = 4 + 5 + 6 + 8 + 10 + 12 + 13 = 58$$

The sum of the rankings for Bank 2 is

$$R_2 = 2 + 3 + 7 + 11 + 17 + 18 = 58$$

The sum of the rankings for Bank 3 is

$$R_3 = 1 + 9 + 14 + 15 + 16 = 55$$

In this case the number of tellers from Bank 1 is 7, so $n_1 = 7$; the number of tellers from Bank 2 is 6, so $n_2 = 6$; and the number of tellers from Bank 3 is 5, so $n_3 = 5$. Also, the total number in the entire joined sample is $n = 18$. Substituting these values into the test statistic formula gives

$$\text{Test statistic} = \frac{12}{18(18 + 1)} \left(\frac{58^2}{7} + \frac{58^2}{6} + \frac{55^2}{5} \right) - 3(18 + 1)$$

$$= 0.7627$$

From Table VIII in the Appendix the χ^2 value with $3 - 1$, or 2, degrees of freedom at the 5% level of significance is 5.991. Since our test statistic value of 0.7627 is less than this value, we do not reject the null hypothesis. We conclude that there is no significant difference in the average time required to complete the transaction at these banks.

EXERCISES FOR SECTION 14.8

1. The number of summonses issued on a daily basis by each of five police officers during a five-day period is as follows:

Number of Summonses by Police Officer

Smith	Eskey	Kien	Walsh	Nuzzo
14	16	13	17	10
26	12	18	27	33
22	25	23	19	20
11	29	30	8	24
15	32	28	31	21

Do the above data indicate that there is a significant difference in the average number of summonses issued by each of these police officers? (Use a 5% level of significance.)

2. *Water pollution.* Environmentalists have accused a chemical company of polluting the waters of a particular river by dumping untreated chemical wastes in it. To test this charge, a judge orders that water samples from four different locations on the river be taken and the quantity of dissolved oxygen contained in the river at each location be determined. (The quantity of dissolved oxygen contained in water is often used to determine the extent of water pollution. The lower the dissolved oxygen content in the water, the higher level of water pollution.) The locations to be used are

1. upstream above the chemical plant.
2. adjacent to the chemical plant's discharge pipe.
3. one-half mile downstream from the chemical plant.
4. a considerable distance downstream from the chemical plant.

It is decided that at least five samples will be taken at each location. The results of the experiment are as follows:

Average Quantity of Oxygen in Water

Location								
I	4.3	3.8	6.0	3.7	4.7	6.2		
II	5.8	5.9	3.4	5.3	5.7	5.0	3.5	4.2
III	4.8	5.4	5.5	5.1	5.2	6.9		
IV	6.1	5.6	4.9	4.4	3.3			

Do the above data indicate that there is a significant difference in the average dissolved oxygen content at these four locations? (Use a 1% level of significance.)

14.9   COMPARISON OF PARAMETRIC AND NONPARAMETRIC STATISTICS

In this chapter we have discussed only some of the nonparametric statistical techniques. Actually, there are many other such techniques that can be used. The question that is often asked is: Why should one bother using the standard statistical techniques discussed in the remainder of this book when the nonparametric statistical techniques discussed in this chapter are easier to use? As a matter of fact, many statisticians actually recommend the use of such nonparametric techniques in many different situations.

The decision as to which statistical technique to use depends on the particular situation. Generally speaking, if you are sure that the data come from a population that is approximately normally distributed, you should use a parametric test. On the other hand, if you are not sure, then use the appropriate nonparametric test. Other factors, such as the possible error generated by using one test as opposed to another, have to be considered. While nonparametric calculations are easier, the widespread availability and use of the computer today for parametric calculations should make such techniques easier to use.

For the benefit of the reader, we summarize some of the nonparametric and parametric statistical techniques discussed and the cases in which each is used.

When Performing Tests Involving	Parametric Test to Use	Nonparametric Test to Use
One mean (median)	t-test (page 549)	One-sample or paired-sample sign test (pages 745 and 749)
Two means (independent samples)	t-test (page 561)	Mann-Whitney test (page 767)
Two means (paired samples)	Paired t-test (page 561)	Sign test or Wilcoxon signed-rank test (page 760)
More than two means (independent samples)	ANOVA (page 704)	Kruskal-Wallis H-test (page 785)
Correlation	Linear correlation (page 594)	Spearman's rank correlation test (page 775)
Randomness		Runs test (page 778)

14.10   USING COMPUTER PACKAGES

The MINITAB statistical package is well suited to perform many of the nonparametric statistical tests discussed in this chapter. We illustrate the use of the MINITAB for one such nonparametric test. Let us return to Example 1 of Section 14.6 (page 776). In that example two judges were asked to rate five different brands of beer on the basis of taste appeal. The actual test results were

	Judge A	Judge B
Brand 1	72	68
Brand 2	68	70
Brand 3	62	79
Brand 4	81	77
Brand 5	87	63

Although we deleted the actual test results and presented only the judge's rankings, when using MINITAB we do not have to do this. MINITAB does the ranking and then computes the Spearman rank coefficient. This is shown in the following printout:

```
MTB > READ JUDGE A INTO C1, JUDGE B INTO C3
DATA > 72    68
DATA > 68    70
DATA > 62    79
DATA > 81    77
DATA > 87    63
DATA > END
       5 ROWS READ
MTB > RANK THE VALUES IN C1, PUT RANKS INTO C2
MTB > RANK THE VALUES IN C3, PUT THE RANKS INTO C4
MTB > PRINT C1 - C4

ROW    C1    C2    C3    C4

 1     72     3    68     2
 2     68     2    70     3
 3     62     1    79     5
 4     81     4    77     4
 5     87     5    63     1

MTB > CORRELATION COEFFICIENT BETWEEN RANKS IN C2 AND C4

CORRELATION OF C2 AND C4 = -0.700

MTB > STOP
```

In addition to the above application, MINITAB can also be applied when we wish to test for randomness (the runs test of Section 14.7) or when wish to use the Kruskal-Wallis H-test of Section 14.8. These applications are illustrated in the following examples.

EXAMPLE 1 Refer back to the data of Example 3 on page 781. Use MINITAB to test for randomness.

SOLUTION We first enter the data into C1 and then use the MEDIAN command to determine the median of the data as shown.

```
MTB > SET THE FOLLOWING IN C1
DATA> 13 17 14 20 18 16 14 19 21 20 14 17 12 14 19
DATA> 20 17 18 14 20 17 19 17 14 19 21 16 12 15 22
DATA> END
MTB > MEDIAN C1
   MEDIAN =       17.000


MTB > RUNS 17.000 C1
C1


K =     17.0000

THE OBSERVED NO. OF RUNS = 16
THE EXPECTED NO. OF RUNS = 15.7333
13 OBSERVATIONS ABOVE K     17 BELOW
          THE TEST IS SIGNIFICANT AT 0.9196
          CANNOT REJECT AT ALPHA = 0.05
```

←This includes those observations that are the same as the median or less than the median.

Thus, we do not reject the null hypothesis. This is the same result that we obtained earlier.

EXAMPLE 2 Refer back to the data of Example 1 on page 786. Use MINITAB to determine if there is a significant difference in the average drying time for these brands of paint.

SOLUTION To apply the MINITAB command KRUSKAL-WALLIS or simply the MINITAB command KRUS, we must enter the data in a slightly different way. We enter in column C1 all the values for Brand A, followed by the values for Brand B, and finally the values for Brand C. In column C2 we indicate which sample each entry in column C1 is associated with. We proceed as follows:

```
MTB > SET THE FOLLOWING IN C1
DATA> 38 32 27 29 23 43
DATA> 52 48 39 42 46 53
DATA> 47 30 37 41 44 49
DATA> END
MTB > SET THE FOLLOWING IN C2
DATA> 1 1 1 1 1 1
DATA> 2 2 2 2 2 2
```

```
DATA>  3  3  3  3  3  3
DATA>  END
MTB >  KRUS  C1  C2

LEVEL        NOBS      MEDIAN     AVE RANK      Z VALUE
   1           6        30.50          4.8        -2.62
   2           6        47.00         13.5         2.25
   3           6        42.50         10.2         0.37
OVERALL       18                       9.5

H = 8.05      d.f. = 2  p = 0.018
```

The value of the test statistic, H = 8.05, agrees with the value that we obtained earlier. The above output indicates that the p-value is 0.018, which tells us that the null hypothesis of equal medians must be rejected.

EXERCISES FOR SECTION 14.10

Use MINITAB to solve each of the following:

1. A computer-literacy test was given to students in a non-air-conditioned room on a humid day. A comparable test was then given to these students in an air-conditioned room. The following list gives the results of the two tests.

		Test Results	
		Non-Air-Conditioned Room	Air-Conditioned Room
	A	77	80
	B	39	65
	C	85	76
	D	84	83
	E	35	44
	F	13	29
Student	G	68	59
	H	86	86
	I	56	79
	J	79	84
	K	81	72
	L	80	73
	M	65	76

Using the Spearman rank correlation test, test the null hypothesis that an air-conditioned room on a humid day has no effect on the performance on this literacy test against the alternative hypothesis that an air-conditioned room on a humid day increases performance on the literacy test. (Use a 5% level of significance.)

2. Exercise 8 of Section 14.7 on page 784.
3. Exercise 2 of Section 14.8 on page 789.

14.11   SUMMARY

In this chapter we discussed several of the nonparametric statistical methods that are often used when we cannot use the standard tests. By far the easiest and most popular of these methods is the sign test. This test is used when we wish to compare two sample means and we know that the samples are not independent. Because of its simplicity, the sign test is used by many people even when a standard test can be used. However, this method wastes much information.

Another important nonparametric test is the rank-sum test (Mann-Whitney test), which is used when the normality assumption is not satisfied or when the variances are not equal. The sum of the ranks is normally distributed when the sample size is large enough, in which case we can use an appropriate z statistic.

We also mentioned the Spearman rank coefficient test, which was originally devised as a shortcut method for computing the coefficient of correlation.

Then we discussed the runs test, which is used to test for randomness or a lack of it. In determining whether or not we have a random sample, we use Table X in the Appendix to find the appropriate critical values.

Finally, we presented the Kruskal-Wallis H-test, which we can use instead of the ANOVA techniques discussed in Chapter 13.

Study Guide

The following is a chapter summary in capsule form. You should now be able to demonstrate your knowledge of the ideas mentioned by giving definitions, descriptions, or specific examples. Page references are given in parentheses.

Statistical techniques that are not dependent on the distribution or parameters involved are known as **nonparametric statistics** or **distribution-free methods** as opposed to **standard methods.** (page 745)

Nonparametric tests are often used in place of the **standard tests** because their assumptions are less demanding. (page 745)

The **one-sample sign test** can be used as an alternative to the t-test with one mean when the assumption about normally distributed populations is not satisfied. (page 745)

The sign test can also be used when working with **paired data** that occur when we deal with **two dependent samples.** This can happen when we measure the same sample twice, as is done in the **before and after type study.** (page 749)

The **paired-sample sign test** is a nonparametric alternative to the paired difference *t*-test. (page 749)

The **Wilcoxon signed-rank test** utilizes information concerning whether the differences between pairs of numbers is positive or negative. It considers both the magnitude as well as the direction of the differences between pairs. (page 760)

The **Wilcoxon rank-sum test** or the **Mann-Whitney test** is a nonparametric test that can be used to test whether the difference between two observed means is significant when the assumption about normality is not satisfied. The two samples are arranged jointly as if they were one sample. (page 767)

The **Spearman rank correlation test** is a nonparametric shortcut method that can be used to compute the coefficient of correlation between data. It uses rankings only. (page 775)

A **run** is a succession of identical letters or symbols that is followed and preceded by a different letter or no letter at all. (page 778)

The **theory of runs** allows us to test for randomness. (page 778)

The **Kruskal-Wallis H-test** is a nonparametric test that can be used to test whether the sample means of several populations are equal. It is the counterpart to the ANOVA techniques discussed in Chapter 13. When using the Kruskal-Wallis H-test, we combine all the data and rank them jointly as if they represent one single sample. (page 785)

Formulas to Remember

You should be able to identify each symbol in the following formulas, understand the relationship among the symbols expressed in the formulas, understand the significance of the formulas, and use the formulas in solving problems.

1. When using the Mann-Whitney test (rank-sum test):

$$z = \frac{R - \mu_R}{\sigma_R}$$

where

$$\mu_R = \frac{n_1(n_1 + n_2 + 1)}{2}$$

and

$$\sigma_R = \sqrt{\frac{n_1 n_2 (n_1 + n_2 + 1)}{12}}$$

and n_1 is the smaller sample size.

2. Spearman rank coefficient test:

$$R = 1 - \frac{6\,\Sigma(x-y)^2}{n(n^2-1)}$$

3. Normal curve approximation for the runs test:

$$z = \frac{X - \mu_R}{\sigma_R}$$

$$\mu_R = \frac{2n_1 n_2}{n_1 + n_2} + 1$$

$$\sigma_R = \sqrt{\frac{2n_1 n_2(2n_1 n_2 - n_1 - n_2)}{(n_1 + n_2)^2(n_1 + n_2 - 1)}}$$

4. Kruskal-Wallis H-test statistic:

$$\frac{12}{n(n+1)} \sum_{i=1}^{r} \frac{R_i^2}{n_i} - 3(n+1)$$

where n = the total number in the entire sample (when the data are joined together) and R_i = the sum of the ranks assigned to n_i values of the ith sample.

Testing Your Understanding of This Chapter's Concepts

1. Assume that we are performing a Wilcoxon rank-sum test for independent random variables. Is there any difference between a one-tailed and a two-tailed test? Explain your answer.
2. In order to apply the Wilcoxon signed-rank test, we assume that the probability distribution of the differences is continuous. Explain why this is necessary.
3. When using the Kruskal-Wallis H-test only large values of H will lead to rejection of the null hypothesis that the samples come from identical populations. Can we say that the Kruskal-Wallis H-test is a right-tailed test? Explain your answer.
4. When using the Kruskal-Wallis H-test, if we have three samples, each of size 5, what are the smallest and the largest values of H? Explain your answer.

Chapter Test

Multiple Choice Questions

1. The ages of the 18 people applying for unemployment insurance benefits on a particular day in a city (in the order in which they applied) are as follows:

    19, 23, 32, 27, 35, 26, 31, 44, 57, 48, 53, 22, 24, 29, 34, 52, 19, 25

Test for randomness.
a. random   b. not random
2. There are 25 people waiting in line to board an airplane. The order in which they are lined up (where M = male and F = female) is as follows:

F   F   F   F   M   M   F   F   M   F   M   M   M   M   F   M   F   F
M   M   M   F   M   F   F

Test for randomness.
a. random   b. not random
3. Eleven male students and twelve female students were asked to assemble a puzzle. The number of minutes needed by each student to assemble the puzzle is as follows:

Male	38	44	29	57	41	36	53	42	40	54	50	
Female	43	37	55	45	48	59	34	39	51	49	56	58

Using the rank-sum test, test the null hypothesis (at the 5% level of significance) that the mean number of minutes needed by both groups is not significantly different.
a. reject null hypothesis   b. do not reject null hypothesis   c. not enough information given
4. A circus performer trained two different groups of animals to perform a particular act. The performer used different teaching techniques. The following chart indicates the number of practice sessions needed by each of the animals to learn the act:

Group A	36	28	19	25	42	33	21	40	35	30	
Group B	27	29	31	26	39	34	32	43	18	37	38

Using the rank-sum test, test the null hypothesis (at the 5% level of significance) that the mean number of practice sessions is not significantly different for both groups.
a. reject null hypothesis   b. do not reject null hypothesis
c. not enough information given
5. A certain weight-reducing technique produced the following results for ten people.

Weight Before	132	147	153	179	182	158	147	180	193	175
Weight After	128	146	154	179	169	150	149	175	185	170

Using the sign test, test the null hypothesis (at the 5% level of significance) that there is a significant difference in weight changes as a result of this weight-reducing technique.
a. reject null hypothesis   b. do not reject null hypothesis
c. not enough information given

6. The blood serum cholesterol level of ten men both before and after taking an experimental drug was as follows:

Before Taking Drug	280	275	291	338	261	240	223	257	298	302
After Taking Drug	263	260	273	281	263	251	247	257	276	295

Using the sign test, test the null hypothesis (at the 5% level of significance) that the experimental drug does not significantly affect the blood serum cholesterol level of an individual.
   a. reject null hypothesis   b. do not reject null hypothesis
   c. not enough information given

7. Three movie reviewers were asked to rate six different movies that they had seen during the year. Their ratings are as follows:

	Reviewer 1	Reviewer 2	Reviewer 3
Movie A	1	3	5
Movie B	4	4	1
Movie C	3	1	6
Movie D	6	5	2
Movie E	2	6	4
Movie F	5	2	3

Using a 5% level of significance, test the null hypothesis that there is no significant (linear) correlation between the ratings of reviewer 1 and reviewer 2.
   a. reject null hypothesis   b. do not reject null hypothesis
   c. not enough information given

8. Refer back to the previous question. Test the null hypothesis (at the 5% level of significance) that there is no significant (linear) correlation between the ratings of reviewer 1 and reviewer 3.
   a. reject null hypothesis   b. do not reject null hypothesis
   c. not enough information given

Supplementary Exercises

9. The ages of the first 25 people registering their car at the Motor Vehicle Bureau on January 25, 1988 (in the order in which they registered) are as follows:

   31,   23,   27,   41,   33,   59,   47,   35,   46,   40,   19,   21,   22,   30,   29,
   18,   24,   36,   38,   29,   28,   25,   32,   24,   37

Test for randomness.

10. Refer back to the previous question. The sex of the 25 people registering their car was as follows (where M = male and F = female):

F F F M F M F F F F M M M M M F M M M

F M M F F M

Test for randomness.

11. The following data are available on the number of delayed flights (on a monthly basis) for two airlines. For Airline A, data are available for 11 months only.

Airline A	26	30	31	35	27	28	38	24	32	37	39
Airline B	34	23	14	21	33	16	25	29	36	22	19	20

Using the rank-sum test, test the null hypothesis (at the 5% level of significance) that the mean number of delayed flights is not significantly different for both airlines.

12. Two groups of mechanics were trained by the manufacturer using different teaching techniques on how to satisfactorily service a particular auto part. The number of minutes needed by each mechanic (after training) to service the part satisfactorily is as follows:

Group A	56	50	42	39	53	54	47	38	48	43
Group B	35	55	51	57	41	46	52	59	44	49	40	45

Using the Kruskal-Wallis H-test, test the null hypothesis (at the 5% level of significance) that the mean number of minutes needed to service the part satisfactorily is not significantly different for both groups.

13. An experimental triglyceride-reducing pill was given to 10 people. The following triglyceride levels were obtained:

Triglyceride Level Before	132	147	120	131	99	84	103	109	123	148
Triglyceride Level After	130	140	123	139	99	81	104	100	110	128

Using the sign test, test the null hypothesis (at the 5% level of significance) that there is no significant difference in the triglyceride level of a person as a result of using this pill.

14. The pulse rate of ten people both before and after seeing a particularly gruesome scene was as follows:

Before	74	68	71	75	70	79	76	73	81	80
After	76	62	62	68	70	72	71	70	70	81

Using the sign test, test the null hypothesis (at the 5% level of significance) that the gruesome scene does not significantly affect the person's pulse rate.

15. Six candidates (Bob, Paul, Sue, Kim, Dave, and Dale) at a college are being ranked by the President, Vice-President, and Dean for promotion opportunities. Their rankings are as follows:

President	Vice-President	Dean
Bob	Dave	Dale
Sue	Kim	Kim
Kim	Sue	Sue
Dale	Paul	Paul
Paul	Bob	Dave
Dave	Dale	Bob

Using a 5% level of significance, test the null hypothesis that there is a significant (linear) correlation between the President's and the Vice-President's rankings.

16. The following data are available on the number of canceled appointments (on a monthly basis) for two dentists. For Dr. Kok, data are available for 11 months only.

Dr. Kok	20	18	21	30	23	28	34	24	29	37	31.	
Dr. Blau	26	16	40	35	22	19	17	25	33	39	36	38

Using the rank-sum test, test the null hypothesis (at the 5% level of significance) that the mean number of canceled appointments is not significantly different for both dentists.

17. Two groups of dogs were trained by two different trainers to perform a difficult task. The following chart indicates the number of practice sessions needed by each of the dogs before mastering the task.

Group A	31	40	27	19	21	30	38	25	20	34	41
Group B	35	28	39	32	33	42	26	24	29	36	37	43

Using the Kruskal-Wallis H-test, test the null hypothesis (at the 5% level of significance) that the mean number of practice sessions is significantly different for both groups.

18. A certain muscle-building technique produced the following results (on a particular scale) for ten people:

Before	73	64	82	69	71	75	78	68	75	81
After	70	71	82	75	77	78	79	69	76	84

Using the sign test, test the null hypothesis (at the 5% level of significance) that there is no significant difference in the muscle strength of an individual as a result of using this technique.

19. The blood pressure of 11 people both before and after viewing a particular horror movie was as follows:

Before	130	140	132	128	156	129	132	137	120	127	128
After	137	141	132	124	148	135	134	139	152	135	142

Using the sign test, test the null hypothesis (at the 5% level of significance) that seeing the horror movie does not affect the person's blood pressure.

20. Six candidates (Pete, Debbie, Jane, Sherry, Jean, and Roy) in a bank are being ranked by the President, Vice-President, and Treasurer for possible advancement. Their rankings are as follows:

President	Vice-President	Treasurer
Debbie	Sherry	Jean
Jane	Jane	Roy
Jean	Jean	Debbie
Roy	Pete	Jane
Pete	Debbie	Pete
Sherry	Roy	Sherry

Using a 5% level of significance, test the null hypothesis that there is no significant (linear) correlation between the President's and Vice-President's rankings.

21. Consider the accompanying newspaper article. In an effort to combat the increasing number of cab robberies, many cabbies are installing two-way radios in their taxis. These radios keep the cab drivers in direct contact with the police department and

ANOTHER CAB DRIVER ROBBED AND SHOT

New York, *March 12*: Police report that José Rodriguez was shot by a fare that he had picked up in the Times Square area. The shooting of a cab driver, the 21st this year, occurred on Broadway after the passenger pulled a gun on Rodriguez and robbed him. The driver was then ordered to drive on Broadway where he was shot. The assailant escaped on foot.

March 12, 1991

their dispatcher. The following statistics are available on the number of cab robberies for one city.

Number of Weekly Robberies of Cab Drivers

Before Installation of Two-Way Radios	After Installation of Two-Way Radios
19	12
15	14
17	17
20	21
23	17
16	19
18	18
10	10
16	9
14	12

Using a 5% level of significance, test the null hypothesis that the two-way radios have no effect on the number of robberies of cab drivers.

Thinking Critically

1. When you are testing for randomness, the null hypothesis is that the sampling distribution is random so that each sequence position has the same prior chance of being assigned an a as any other. The following expression can be used as a test statistic:

$$z = \frac{R_a - \left(\dfrac{2n_a n_b}{n_a + n_b} + 1 \right)}{\sqrt{\dfrac{2n_a n_b(2n_a n_b - n_a - n_b)}{(n_a + n_b)^2(n_a + n_b - 1)}}} \qquad R_a = \text{number of runs for category a.}$$

where n_a and n_b represent the number of a's and b's in the sample. Applying this formula, test the data of Example 3 on page 781 for randomness.

2. Refer back to the newspaper clipping given on page 221 concerning the 1970 draft lottery. The U.S. Selective Service was vehemently criticized about the way in which the capsules were selected due to the wide differences in the draft priority numbers. People whose birthdays were later in the year had a higher priority. Apply the runs test to justify the null hypothesis of randomness.

3. Is it true that for the case of two samples the Kruskal-Wallis H-test is equivalent to the Wilcoxon rank-sum test? (*Hint:* For the case of two samples the Kruskal-Wallis test statistic equals the square of the test statistic used in the Wilcoxon rank-sum test.

Also, note that when we have 1 degree of freedom the critical values of χ^2 correspond to the square of the z-score.)

Case Study Medicines Can Make You Fat

Many people take medications for a variety of illnesses. According to Dr. Michael Steelman of the American Society of Bariatric Physicians (doctors who specialize in weight problems), some medications can make your body retain water, some increase your appetite, and some affect the way your body burns fat; a few do all three. Of course, not everyone who takes a drug is affected, but people who tend to gain weight easily are likely candidates. Drugs such as Deltasone, Medrol, and other corticosteroids commonly taken for asthma, lung disease, and inflammatory conditions can cause weight gain in at least 75 percent of those who take them.

1. Over a period of one year, Dr. George Criffasi monitored the weight changes of 10 patients who were prescribed a common asthma medication. These patients did not alter their diet and exercise regimen over the one-year period. The weights of these patients at the beginning and the end of this monitoring period are shown below.

Patient	Weight at Beginning of Year	Weight at End of Year
Meg	126	132
Martha	134	125
Christopher	176	175
Juanita	145	164
John	193	195
Jean	153	162
Vivian	137	146
Lisa	117	125
Donna	112	113
Jim	162	170

Source: Good Housekeeping, July 1996, p. 125.

Using the sign test, test the null hypothesis (at the 5% level of significance) that the asthma medication has no effect on a person's weight.

Appendix

STATISTICAL TABLES

TABLE I   Factorials

n	$n!$
0	1
1	1
2	2
3	6
4	24
5	120
6	720
7	5,040
8	40,320
9	362,880
10	3,628,800
11	39,916,800
12	479,001,600
13	6,227,020,800
14	87,178,291,200
15	1,307,674,368,000
16	20,922,789,888,000
17	355,687,428,096,000
18	6,402,373,705,728,000
19	121,645,100,408,832,000
20	2,432,902,008,176,640,000

TABLE II BINOMIAL COEFFICIENTS   **A-3**

TABLE II   Binomial Coefficients $\dfrac{n!}{x!(n - x!)}$

n \ x	2	3	4	5	6	7	8	9	10
2	1								
3	3	1							
4	6	4	1						
5	10	10	5	1					
6	15	20	15	6	1				
7	21	35	35	21	7	1			
8	28	56	70	56	28	8	1		
9	36	84	126	126	84	36	9	1	
10	45	120	210	252	210	120	45	10	1
11	55	165	330	462	462	330	165	55	11
12	66	220	495	792	924	792	495	220	66
13	78	286	715	1,287	1,716	1,716	1,287	715	286
14	91	364	1,001	2,002	3,003	3,432	3,003	2,002	1,001
15	105	455	1,365	3,003	5,005	6,435	6,435	5,005	3,003
16	120	560	1,820	4,368	8,008	11,440	12,870	11,440	8,008
17	136	680	2,380	6,188	12,376	19,448	24,310	24,310	19,448
18	153	816	3,060	8,568	18,564	31,824	43,758	48,620	43,758
19	171	969	3,876	11,628	27,132	50,388	75,582	92,378	92,378
20	190	1,140	4,845	15,504	38,760	77,520	125,970	167,960	184,756

TABLE III   Binomial Probabilities

							p					
n	x	0.05	0.1	0.2	0.3	0.4	0.5	0.6	0.7	0.8	0.9	0.95
2	0	0.902	0.810	0.640	0.490	0.360	0.250	0.160	0.090	0.040	0.010	0.002
	1	0.095	0.180	0.320	0.420	0.480	0.500	0.480	0.420	0.320	0.180	0.095
	2	0.002	0.010	0.040	0.090	0.160	0.250	0.360	0.490	0.640	0.810	0.902
3	0	0.857	0.729	0.512	0.343	0.216	0.125	0.064	0.027	0.008	0.001	
	1	0.135	0.243	0.384	0.441	0.432	0.375	0.288	0.189	0.096	0.027	0.007
	2	0.007	0.027	0.096	0.189	0.288	0.375	0.432	0.441	0.384	0.243	0.135
	3		0.001	0.008	0.027	0.064	0.125	0.216	0.343	0.512	0.729	0.857
4	0	0.815	0.656	0.410	0.240	0.130	0.062	0.026	0.008	0.002		
	1	0.171	0.292	0.410	0.412	0.346	0.250	0.154	0.076	0.026	0.004	
	2	0.014	0.049	0.154	0.265	0.346	0.375	0.346	0.265	0.154	0.049	0.014
	3		0.004	0.026	0.076	0.154	0.250	0.346	0.412	0.410	0.292	0.171
	4			0.002	0.008	0.026	0.062	0.130	0.240	0.410	0.656	0.815
5	0	0.774	0.590	0.328	0.168	0.078	0.031	0.010	0.002			
	1	0.204	0.328	0.410	0.360	0.259	0.156	0.077	0.028	0.006		
	2	0.021	0.073	0.205	0.309	0.346	0.312	0.230	0.132	0.051	0.008	0.001
	3	0.001	0.008	0.051	0.132	0.230	0.312	0.346	0.309	0.205	0.073	0.021
	4			0.006	0.028	0.077	0.156	0.259	0.360	0.410	0.328	0.204
	5				0.002	0.010	0.031	0.078	0.168	0.328	0.590	0.774
6	0	0.735	0.531	0.262	0.118	0.047	0.016	0.004	0.001			
	1	0.232	0.354	0.393	0.303	0.187	0.094	0.037	0.010	0.002		
	2	0.031	0.098	0.246	0.324	0.311	0.234	0.138	0.060	0.015	0.001	
	3	0.002	0.015	0.082	0.185	0.276	0.312	0.276	0.185	0.082	0.015	0.002
	4		0.001	0.015	0.060	0.138	0.234	0.311	0.324	0.246	0.098	0.031
	5			0.002	0.010	0.037	0.094	0.187	0.303	0.393	0.354	0.232
	6				0.001	0.004	0.016	0.047	0.118	0.262	0.531	0.735
7	0	0.698	0.478	0.210	0.082	0.028	0.008	0.002				
	1	0.257	0.372	0.367	0.247	0.131	0.055	0.017	0.004			
	2	0.041	0.124	0.275	0.318	0.261	0.164	0.077	0.025	0.004		
	3	0.004	0.023	0.115	0.227	0.290	0.273	0.194	0.097	0.029	0.003	
	4		0.003	0.029	0.097	0.194	0.273	0.290	0.227	0.115	0.023	0.004
	5			0.004	0.025	0.077	0.164	0.261	0.318	0.275	0.124	0.041
	6				0.004	0.017	0.055	0.131	0.247	0.367	0.372	0.257
	7					0.002	0.008	0.028	0.082	0.210	0.478	0.698
8	0	0.663	0.430	0.168	0.058	0.017	0.004	0.001				
	1	0.279	0.383	0.336	0.198	0.090	0.031	0.008	0.001			
	2	0.051	0.149	0.294	0.296	0.209	0.109	0.041	0.010	0.001		
	3	0.005	0.033	0.147	0.254	0.279	0.219	0.124	0.047	0.009		
	4		0.005	0.046	0.136	0.232	0.273	0.232	0.136	0.046	0.005	
	5			0.009	0.047	0.124	0.219	0.279	0.254	0.147	0.033	0.005

TABLE III BIONOMIAL PROBABILITIES   **A-5**

TABLE III   Binomial Probabilities (*continued*)

							p					
n	x	0.05	0.1	0.2	0.3	0.4	0.5	0.6	0.7	0.8	0.9	0.95
	6			0.001	0.010	0.041	0.109	0.209	0.296	0.294	0.149	0.051
	7				0.001	0.008	0.031	0.090	0.198	0.336	0.383	0.279
	8					0.001	0.004	0.017	0.058	0.168	0.430	0.663
9	0	0.630	0.387	0.134	0.040	0.010	0.002					
	1	0.299	0.387	0.302	0.156	0.060	0.018	0.004				
	2	0.063	0.172	0.302	0.267	0.161	0.070	0.021	0.004			
	3	0.008	0.045	0.176	0.267	0.251	0.164	0.074	0.021	0.003		
	4	0.001	0.007	0.066	0.172	0.251	0.246	0.167	0.074	0.017	0.001	
	5		0.001	0.017	0.074	0.167	0.246	0.251	0.172	0.066	0.007	0.001
	6			0.003	0.021	0.074	0.164	0.251	0.267	0.176	0.045	0.008
	7				0.004	0.021	0.070	0.161	0.267	0.302	0.172	0.063
	8					0.004	0.018	0.060	0.156	0.302	0.387	0.299
	9						0.002	0.010	0.040	0.134	0.387	0.630
10	0	0.599	0.349	0.107	0.028	0.006	0.001					
	1	0.315	0.387	0.268	0.121	0.040	0.010	0.002				
	2	0.075	0.194	0.302	0.233	0.121	0.044	0.011	0.001			
	3	0.010	0.057	0.201	0.267	0.215	0.117	0.042	0.009	0.001		
	4	0.001	0.011	0.088	0.200	0.251	0.205	0.111	0.037	0.006		
	5		0.001	0.026	0.103	0.201	0.246	0.201	0.103	0.026	0.001	
	6			0.006	0.037	0.111	0.205	0.251	0.200	0.088	0.011	0.001
	7			0.001	0.009	0.042	0.117	0.215	0.267	0.201	0.057	0.010
	8				0.001	0.011	0.044	0.121	0.233	0.302	0.194	0.075
	9					0.002	0.010	0.040	0.121	0.268	0.387	0.315
	10						0.001	0.006	0.028	0.107	0.349	0.599
11	0	0.569	0.314	0.086	0.020	0.004						
	1	0.329	0.384	0.236	0.093	0.027	0.005	0.001				
	2	0.087	0.213	0.295	0.200	0.089	0.027	0.005	0.001			
	3	0.014	0.071	0.221	0.257	0.177	0.081	0.023	0.004			
	4	0.001	0.016	0.111	0.220	0.236	0.161	0.070	0.017	0.002		
	5		0.002	0.039	0.132	0.221	0.226	0.147	0.057	0.010		
	6			0.010	0.057	0.147	0.226	0.221	0.132	0.039	0.002	
	7			0.002	0.017	0.070	0.161	0.236	0.220	0.111	0.016	0.001
	8				0.004	0.023	0.081	0.177	0.257	0.221	0.071	0.014
	9				0.001	0.005	0.027	0.089	0.200	0.295	0.213	0.087
	10					0.001	0.005	0.027	0.093	0.236	0.384	0.329
	11							0.004	0.020	0.086	0.314	0.569
12	0	0.540	0.282	0.069	0.014	0.002						
	1	0.341	0.377	0.206	0.071	0.017	0.003					
	2	0.099	0.230	0.283	0.168	0.064	0.016	0.002				
	3	0.017	0.085	0.236	0.240	0.142	0.054	0.012	0.001			

TABLE III Binomial Probabilities (*continued*)

n	x	0.05	0.1	0.2	0.3	0.4	0.5	0.6	0.7	0.8	0.9	0.95
	4	0.002	0.021	0.133	0.231	0.213	0.121	0.042	0.008	0.001		
	5		0.004	0.053	0.158	0.227	0.193	0.101	0.029	0.003		
	6			0.016	0.079	0.177	0.226	0.177	0.079	0.016		
	7			0.003	0.029	0.101	0.193	0.227	0.158	0.053	0.004	
	8			0.001	0.008	0.042	0.121	0.213	0.231	0.133	0.021	0.002
	9				0.001	0.012	0.054	0.142	0.240	0.236	0.085	0.017
	10					0.002	0.016	0.064	0.168	0.283	0.230	0.099
	11						0.003	0.017	0.071	0.206	0.377	0.341
	12							0.002	0.014	0.069	0.282	0.540
13	0	0.513	0.254	0.055	0.010	0.001						
	1	0.351	0.367	0.179	0.054	0.011	0.002					
	2	0.111	0.245	0.268	0.139	0.045	0.010	0.001				
	3	0.021	0.100	0.246	0.218	0.111	0.035	0.006	0.001			
	4	0.003	0.028	0.154	0.234	0.184	0.087	0.024	0.003			
	5		0.006	0.069	0.180	0.221	0.157	0.066	0.014	0.001		
	6		0.001	0.023	0.103	0.197	0.209	0.131	0.044	0.006		
	7			0.006	0.044	0.131	0.209	0.197	0.103	0.023	0.001	
	8			0.001	0.014	0.066	0.157	0.221	0.180	0.069	0.006	
	9				0.003	0.024	0.087	0.184	0.234	0.154	0.028	0.003
	10				0.001	0.006	0.035	0.111	0.218	0.246	0.100	0.021
	11					0.001	0.010	0.045	0.139	0.268	0.245	0.111
	12						0.002	0.011	0.054	0.179	0.367	0.351
	13							0.001	0.010	0.055	0.254	0.513
14	0	0.488	0.229	0.044	0.007	0.001						
	1	0.359	0.356	0.154	0.041	0.007	0.001					
	2	0.123	0.257	0.250	0.113	0.032	0.006	0.001				
	3	0.026	0.114	0.250	0.194	0.085	0.022	0.003				
	4	0.004	0.035	0.172	0.229	0.155	0.061	0.014	0.001			
	5		0.008	0.086	0.196	0.207	0.122	0.041	0.007			
	6		0.001	0.032	0.126	0.207	0.183	0.092	0.023	0.002		
	7			0.009	0.062	0.157	0.209	0.157	0.062	0.009		
	8			0.002	0.023	0.092	0.183	0.207	0.126	0.032	0.001	
	9				0.007	0.041	0.122	0.207	0.196	0.086	0.008	
	10				0.001	0.014	0.061	0.155	0.229	0.172	0.035	0.004
	11					0.003	0.022	0.085	0.194	0.250	0.114	0.026
	12					0.001	0.006	0.032	0.113	0.250	0.257	0.123
	13						0.001	0.007	0.041	0.154	0.356	0.359
	14							0.001	0.007	0.044	0.229	0.488
15	0	0.463	0.206	0.035	0.005							
	1	0.366	0.343	0.132	0.031	0.005						
	2	0.135	0.267	0.231	0.092	0.022	0.003					

TABLE III BIONOMIAL PROBABILITIES   **A-7**

TABLE III Binomial Probabilities (*continued*)

							p					
n	x	0.05	0.1	0.2	0.3	0.4	0.5	0.6	0.7	0.8	0.9	0.95
15	3	0.031	0.129	0.250	0.170	0.063	0.014	0.002				
	4	0.005	0.043	0.188	0.219	0.127	0.042	0.007	0.001			
	5	0.001	0.010	0.103	0.206	0.186	0.092	0.024	0.003			
	6		0.002	0.043	0.147	0.207	0.153	0.061	0.012	0.001		
	7			0.014	0.081	0.177	0.196	0.118	0.035	0.003		
	8			0.003	0.035	0.118	0.196	0.177	0.081	0.014		
	9			0.001	0.012	0.061	0.153	0.207	0.147	0.043	0.002	
	10				0.003	0.024	0.092	0.186	0.026	0.103	0.010	0.001
	11				0.001	0.007	0.042	0.127	0.219	0.188	0.043	0.005
	12					0.002	0.014	0.063	0.170	0.250	0.129	0.031
	13						0.003	0.022	0.092	0.231	0.267	0.135
	14							0.005	0.031	0.132	0.343	0.366
	15								0.005	0.035	0.206	0.463

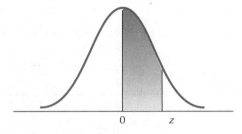

The entries in Table IV are the probabilities that a random variable having the standard normal distribution takes on a value between 0 and z; they are given by the area under the curve shaded in the diagram.

TABLE IV   Areas Under the Standard Normal Curve

					Second Decimal Place in z					
z	0.00	0.01	0.02	0.03	0.04	0.05	0.06	0.07	0.08	0.09
0.0	0.0000	0.0040	0.0080	0.0120	0.0160	0.0199	0.0239	0.0279	0.0319	0.0359
0.1	0.0398	0.0438	0.0478	0.0517	0.0557	0.0596	0.0636	0.0675	0.0714	0.0753
0.2	0.0793	0.0832	0.0871	0.0910	0.0948	0.0987	0.1026	0.1064	0.1103	0.1141
0.3	0.1179	0.1217	0.1255	0.1293	0.1331	0.1368	0.1406	0.1443	0.1480	0.1517
0.4	0.1554	0.1591	0.1628	0.1664	0.1700	0.1736	0.1772	0.1808	0.1844	0.1879
0.5	0.1915	0.1950	0.1985	0.2019	0.2054	0.2088	0.2123	0.2157	0.2190	0.2224
0.6	0.2257	0.2291	0.2324	0.2357	0.2389	0.2422	0.2454	0.2486	0.2517	0.2549
0.7	0.2580	0.2611	0.2642	0.2673	0.2704	0.2734	0.2764	0.2794	0.2823	0.2852
0.8	0.2881	0.2910	0.2939	0.2967	0.2995	0.3023	0.3051	0.3078	0.3106	0.3133
0.9	0.3159	0.3186	0.3212	0.3238	0.3264	0.3289	0.3315	0.3340	0.3365	0.3389
1.0	0.3413	0.3438	0.3461	0.3485	0.3508	0.3531	0.3554	0.3577	0.3599	0.3621
1.1	0.3643	0.3665	0.3686	0.3708	0.3729	0.3749	0.3770	0.3790	0.3810	0.3830
1.2	0.3849	0.3869	0.3888	0.3907	0.3925	0.3944	0.3962	0.3980	0.3997	0.4015
1.3	0.4032	0.4049	0.4066	0.4082	0.4099	0.4115	0.4131	0.4147	0.4162	0.4177
1.4	0.4192	0.4207	0.4222	0.4236	0.4251	0.4265	0.4279	0.4292	0.4306	0.4319
1.5	0.4332	0.4345	0.4357	0.4370	0.4382	0.4394	0.4406	0.4418	0.4429	0.4441
1.6	0.4452	0.4463	0.4474	0.4484	0.4495	0.4505	0.4515	0.4525	0.4535	0.4545
1.7	0.4554	0.4564	0.4573	0.4582	0.4591	0.4599	0.4608	0.4616	0.4625	0.4633
1.8	0.4641	0.4649	0.4656	0.4664	0.4671	0.4678	0.4686	0.4693	0.4669	0.4706
1.9	0.4713	0.4719	0.4726	0.4732	0.4738	0.4744	0.4750	0.4756	0.4761	0.4767
2.0	0.4772	0.4778	0.4783	0.4788	0.4793	0.4798	0.4803	0.4808	0.4812	0.4817
2.1	0.4821	0.4826	0.4830	0.4834	0.4838	0.4842	0.4846	0.4850	0.4854	0.4857
2.2	0.4861	0.4864	0.4868	0.4871	0.4875	0.4878	0.4881	0.4884	0.4887	0.4890
2.3	0.4893	0.4896	0.4898	0.4901	0.4904	0.4906	0.4909	0.4911	0.4913	0.4916
2.4	0.4918	0.4920	0.4922	0.4925	0.4927	0.4929	0.4931	0.4932	0.4934	0.4936

TABLE IV AREAS UNDER THE STANDARD NORMAL CURVE   **A-9**

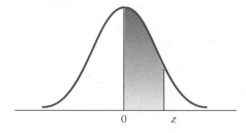

The entries in Table IV are the probabilities that a random variable having the standard normal distribution takes on a value between 0 and z; they are given by the area under the curve shaded in the diagram.

TABLE IV   Areas Under the Standard Normal Curve (*continued*)

| | | | | Second Decimal Place in z | | | | | |
z	0.00	0.01	0.02	0.03	0.04	0.05	0.06	0.07	0.08	0.09
2.5	0.4938	0.4940	0.4941	0.4943	0.4945	0.4946	0.4948	0.4949	0.4951	0.4952
2.6	0.4953	0.4955	0.4956	0.4957	0.4959	0.4960	0.4961	0.4962	0.4963	0.4964
2.7	0.4965	0.4966	0.4967	0.4968	0.4969	0.4970	0.4971	0.4972	0.4973	0.4974
2.8	0.4974	0.4975	0.4976	0.4977	0.4977	0.4978	0.4979	0.4979	0.4980	0.4981
2.9	0.4981	0.4982	0.4982	0.4983	0.4984	0.4984	0.4985	0.4985	0.4986	0.4986
3.0	0.4987	0.4987	0.4987	0.4988	0.4988	0.4989	0.4989	0.4989	0.4990	0.4990
3.1	0.4990	0.4991	0.4991	0.4991	0.4992	0.4992	0.4992	0.4992	0.4993	0.4993
3.2	0.4993	0.4993	0.4994	0.4994	0.4994	0.4994	0.4994	0.4995	0.4995	0.4995
3.3	0.4995	0.4995	0.4995	0.4996	0.4996	0.4996	0.4996	0.4996	0.4996	0.4997
3.4	0.4997	0.4997	0.4997	0.4997	0.4997	0.4997	0.4997	0.4997	0.4997	0.4998
3.5	0.4998	0.4998	0.4998	0.4998	0.4998	0.4998	0.4998	0.4998	0.4998	0.4998
3.6	0.4998	0.4998	0.4999	0.4999	0.4999	0.4999	0.4999	0.4999	0.4999	0.4999
3.7	0.4999	0.4999	0.4999	0.4999	0.4999	0.4999	0.4999	0.4999	0.4999	0.4999
3.8	0.4999	0.4999	0.4999	0.4999	0.4999	0.4999	0.4999	0.4999	0.4999	0.4999
3.9	0.5000°									

°For $z \geq 3.90$, the areas are 0.5000 to four decimal places.

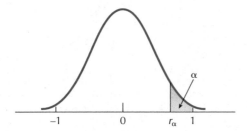

TABLE V   Critical Values of r

n	$r_{0.025}$	$r_{0.005}$	n	$r_{0.025}$	$r_{0.005}$
3	0.997		18	0.468	0.590
4	0.950	0.999	19	0.456	0.575
5	0.878	0.959	20	0.444	0.561
6	0.811	0.917	21	0.433	0.549
7	0.754	0.875	22	0.423	0.537
8	0.707	0.834	27	0.381	0.487
9	0.666	0.798	32	0.349	0.449
10	0.632	0.765	37	0.325	0.418
11	0.602	0.735	42	0.304	0.393
12	0.576	0.708	47	0.288	0.372
13	0.553	0.684	52	0.273	0.354
14	0.532	0.661	62	0.250	0.325
15	0.514	0.641	72	0.232	0.302
16	0.497	0.623	82	0.217	0.283
17	0.482	0.606	92	0.205	0.267

This table is abridged from Table VI of R. A. Fisher and F. Yates: *Statistical Tables for Biological, Agricultural, and Medical Research*, published by Longman Group, Ltd., London (previously published by Oliver & Boyd, Edinburgh), by permission of the authors and publishers.

TABLE VI TABLE OF RANDOM DIGITS   **A-11**

TABLE VI   Table of Random Digits

Col / Line	(1)	(2)	(3)	(4)	(5)	(6)	(7)	(8)	(9)	(10)	(11)	(12)	(13)	(14)
1	10480	15011	01536	02011	81647	91646	69179	14194	62590	36207	20969	99570	91291	90700
2	22368	46573	25595	85393	30995	89198	27982	53402	93965	34095	52666	19174	39615	99505
3	24130	48360	22527	97265	76393	64809	15179	24830	49340	32081	30680	19655	63348	58629
4	42167	93093	06243	61680	07856	16376	39440	53537	71341	57004	00849	74917	97758	16379
5	37570	39975	81837	16656	06121	91782	60468	81305	49684	60672	14110	06927	01263	54613
6	77921	06907	11008	42751	27756	53498	18602	70659	90655	15053	21916	81825	44394	42880
7	99562	72905	56420	69994	98872	31016	71194	18738	44013	48840	63213	21069	10634	12952
8	96301	91977	05463	07972	18876	20922	94595	56869	69014	60045	18425	84903	42508	32307
9	89579	14342	63661	10281	17453	18103	57740	84378	25331	12566	58678	44947	05585	56941
10	85475	36857	53342	53988	53060	59533	38867	62300	08158	17983	16439	11458	18593	64952
11	28918	69578	88231	33276	70997	79936	56865	05859	90106	31595	01547	85590	91610	78188
12	63553	40961	48235	03427	49626	69445	18663	72695	52180	20847	12234	90511	33703	90322
13	09429	93969	52636	92737	88974	33488	36320	17617	30015	08272	84115	27156	30613	74952
14	10365	61129	87529	85689	48237	52267	67689	93394	01511	26358	85104	20285	29975	89868
15	07119	97336	71048	08178	77233	13916	47564	81056	97735	85977	29372	74461	28551	90707
16	51085	12765	51821	51259	77452	16308	60756	92144	49442	53900	70960	63990	75601	40719
17	02368	21382	52404	60268	89368	19885	55322	44819	01188	65255	64835	44919	05944	55157
18	01011	54092	33362	94904	31273	04146	18594	29852	71585	85030	51132	01915	92747	64951
19	52162	53916	46369	58586	23216	14513	83149	98736	23495	64350	94738	17752	35156	35749
20	07056	97628	33787	09998	42698	06691	76988	13602	51851	46104	88916	19509	25625	58104
21	48663	91245	85828	14346	09172	30168	90229	04734	59193	22178	30421	61666	99904	32812
22	54164	58492	22421	74103	47070	25306	76468	26384	58151	06646	21524	15227	96909	44592
23	32639	32363	05597	24200	13363	38005	94342	28728	35806	06912	17012	64161	18296	22851
24	29334	27001	87637	87308	58731	00256	45834	15398	46557	41135	10367	07684	36188	18510
25	02488	33062	28834	07351	19731	92420	60952	61280	50001	67658	32586	86679	50720	94953
26	81525	72295	04839	96423	24878	82651	66566	14778	76797	14780	13300	87074	79666	95725
27	29676	20591	68086	26432	46901	20849	89768	81536	86645	12659	92259	57102	80428	25280
28	00742	57392	39064	66432	84673	40027	32832	61362	98947	96067	64760	64584	96096	98253
29	05366	04213	25669	26422	44407	44048	37937	63904	45766	66134	75470	66520	34693	90449
30	91921	26418	64117	94305	26766	25940	39972	22209	71500	64568	91402	42416	07844	69618
31	00582	04711	87917	77341	42206	35126	74087	99547	81817	42607	43808	76655	62028	76630
32	00725	69884	62797	56170	86324	88072	76222	36086	84637	93161	76038	65855	77979	88006
33	69011	65795	95876	55293	18988	27354	26575	08625	40801	59920	29841	80150	12777	48501
34	25976	57948	29888	88604	67917	48708	18912	82271	65424	69774	33611	54262	85963	03547
35	09763	83473	73577	12908	30883	18317	28290	35797	05998	41688	34952	37888	38917	88050
36	91567	42595	27958	30134	04024	86385	29880	99730	55536	84855	29080	09250	79656	73211
37	17955	56349	90999	49127	20044	59931	06115	20542	18059	02008	73708	83517	36103	42791

TABLE VI   Table of Random Digits (*continued*)

Col Line	(1)	(2)	(3)	(4)	(5)	(6)	(7)	(8)	(9)	(10)	(11)	(12)	(13)	(14)
38	46503	18584	18845	49618	02304	51038	20655	58727	28168	15475	56942	53389	20562	87338
39	92157	89634	94824	78171	84610	82834	09922	25417	44137	48413	25555	21246	35509	20468
40	14577	62765	35605	81263	39667	47358	56873	56307	61607	49518	89656	20103	77490	18062
41	98427	07523	33362	64270	01638	92477	66969	98420	04880	45585	46565	04102	46880	45709
42	34914	63976	88720	82765	34476	17032	87589	40836	32427	70002	70663	88863	77775	69348
43	70060	28277	39475	46473	23219	53416	94970	25832	69975	94884	19661	72828	00102	66794
44	53976	54914	06990	67245	68350	82948	11398	42878	80287	88267	47363	46634	06541	97809
45	76072	29515	40980	07391	58745	25774	22987	80059	39911	96189	41151	14222	60697	59583
46	90725	52210	83974	29992	65831	38857	50490	83765	55657	14361	31720	57375	56228	41546
47	64364	67412	33339	31926	14883	24413	59744	92351	97473	89286	35931	04110	23726	51900
48	08962	00358	31662	25388	61642	34072	81249	35648	56891	69352	48373	45578	78547	81788
49	95012	68379	93526	70765	10592	04542	76463	54328	02349	17247	28865	14777	62730	92277
50	15664	10493	20492	38391	91132	21999	59516	81652	27195	48223	46751	22923	32261	85653

Page 1 of *Table of 105,000 Random Decimal Digits*, Statement No. 4914, May 1949, File No. 261-A-1, Interstate Commerce Commission, Washington, D.C.

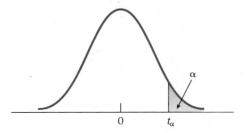

TABLE VII   The t-Distribution

df	$t_{0.050}$	$t_{0.025}$	$t_{0.010}$	$t_{0.005}$	df
1	6.314	12.706	31.821	63.657	1
2	2.920	4.303	6.965	9.925	2
3	2.353	3.182	4.541	5.841	3
4	2.132	2.776	3.747	4.604	4
5	2.015	2.571	3.365	4.032	5
6	1.943	2.447	3.143	3.707	6
7	1.895	2.365	2.998	3.499	7
8	1.860	2.306	2.896	3.355	8
9	1.833	2.262	2.821	3.250	9
10	1.812	2.228	2.764	3.169	10
11	1.796	2.201	2.718	3.106	11
12	1.782	2.179	2.681	3.055	12
13	1.771	2.160	2.650	3.012	13
14	1.761	2.145	2.624	2.977	14
15	1.753	2.131	2.602	2.947	15
16	1.746	2.120	2.583	2.921	16
17	1.740	2.110	2.567	2.898	17
18	1.734	2.101	2.552	2.878	18
19	1.729	2.093	2.539	2.861	19
20	1.725	2.086	2.528	2.845	20
21	1.721	2.080	2.518	2.831	21
22	1.717	2.074	2.508	2.819	22
23	1.714	2.069	2.500	2.807	23
24	1.711	2.064	2.492	2.797	24
25	1.708	2.060	2.485	2.787	25
26	1.706	2.056	2.479	2.779	26
27	1.703	2.052	2.473	2.771	27
28	1.701	2.048	2.467	2.763	28
29	1.699	2.045	2.462	2.756	29
inf.	1.645	1.960	2.326	2.576	inf.

This table is abridged from Table IV of R. A. Fisher and F. Yates: *Statistical Tables for Biological, Agricultural, and Medical Research*, published by Longman Group, Ltd., London (previously published by Oliver & Boyd, Edinburgh), by permission of the authors and publishers.

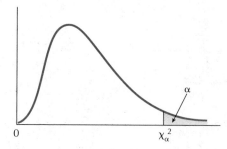

TABLE VIII   The χ^2 Distribution

df	$\chi^2_{0.05}$	$\chi^2_{0.01}$	df
1	3.841	6.635	1
2	5.991	9.210	2
3	7.815	11.345	3
4	9.488	13.277	4
5	11.070	15.086	5
6	12.592	16.812	6
7	14.067	18.475	7
8	15.507	20.090	8
9	16.919	21.666	9
10	18.307	23.209	10
11	19.675	24.725	11
12	21.026	26.217	12
13	22.362	27.688	13
14	23.685	29.141	14
15	24.996	30.578	15
16	26.296	32.000	16
17	27.587	33.409	17
18	28.869	34.805	18
19	30.144	36.191	19
20	31.410	37.566	20
21	32.671	38.932	21
22	33.924	40.289	22
23	35.172	41.638	23
24	36.415	42.980	24
25	37.652	44.314	25
26	38.885	45.642	26
27	40.113	46.963	27
28	41.337	48.278	28
29	42.557	49.588	29
30	43.773	50.892	30

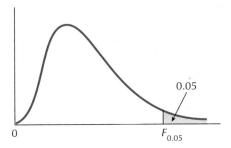

0.05

0    $F_{0.05}$

TABLE IX   Critical Values of the F-Distribution ($\alpha = 0.05$)

		Degrees of Freedom for Numerator									
		1	2	3	4	5	6	7	8	9	10
	1	161	200	216	225	230	234	237	239	241	242
	2	18.5	19.0	19.2	19.2	19.3	19.3	19.4	19.4	19.4	19.4
	3	10.1	9.55	9.28	9.12	9.01	8.94	8.89	8.85	8.81	8.79
	4	7.71	6.94	6.59	6.39	6.26	6.16	6.09	6.04	6.00	5.96
	5	6.61	5.79	5.41	5.19	5.05	4.95	4.88	4.82	4.77	4.74
	6	5.99	5.14	4.76	4.53	4.39	4.28	4.21	4.15	4.10	4.06
	7	5.59	4.74	4.35	4.12	3.97	3.87	3.79	3.73	3.68	3.64
	8	5.32	4.46	4.07	3.84	3.69	3.58	3.50	3.44	3.39	3.35
	9	5.12	4.26	3.86	3.63	3.48	3.37	3.29	3.23	3.18	3.14
	10	4.96	4.10	3.71	3.48	3.33	3.22	3.14	3.07	3.02	2.98
	11	4.84	3.98	3.59	3.36	3.20	3.09	3.01	2.95	2.90	2.85
	12	4.75	3.89	3.49	3.26	3.11	3.00	2.91	2.85	2.80	2.75
	13	4.67	3.81	3.41	3.18	3.03	2.92	2.83	2.77	2.71	2.67
Degrees	14	4.60	3.74	3.34	3.11	2.96	2.85	2.76	2.70	2.65	2.60
of	15	4.54	3.68	3.29	3.06	2.90	2.79	2.71	2.64	2.59	2.54
Freedom	16	4.49	3.63	3.24	3.01	2.85	2.74	2.66	2.59	2.54	2.49
for	17	4.45	3.59	3.20	2.96	2.81	2.70	2.61	2.55	2.49	2.45
Denominator	18	4.41	3.55	3.16	2.93	2.77	2.66	2.58	2.51	2.46	2.41
	19	4.38	3.52	3.13	2.90	2.74	2.63	2.54	2.48	2.42	2.38
	20	4.35	3.49	3.10	2.87	2.71	2.60	2.51	2.45	2.39	2.35
	21	4.32	3.47	3.07	2.84	2.68	2.57	2.49	2.42	2.37	2.32
	22	4.30	3.44	3.05	2.82	2.66	2.55	2.46	2.40	2.34	2.30
	23	4.28	3.42	3.03	2.80	2.64	2.53	2.44	2.37	2.32	2.27
	24	4.26	3.40	3.01	2.78	2.62	2.51	2.42	2.36	2.30	2.25
	25	4.24	3.39	2.99	2.76	2.60	2.49	2.40	2.34	2.28	2.24
	30	4.17	3.32	2.92	2.69	2.53	2.42	2.33	2.27	2.21	2.16
	40	4.08	3.23	2.84	2.61	2.45	2.34	2.25	2.18	2.12	2.08
	60	4.00	3.15	2.76	2.53	2.37	2.25	2.17	2.10	2.04	1.99
	120	3.92	3.07	2.68	2.45	2.29	2.18	2.09	2.02	1.96	1.91
	∞	3.84	3.00	2.60	2.37	2.21	2.10	2.01	1.94	1.88	1.83

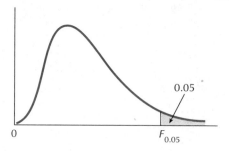

0.05

$F_{0.05}$

0

TABLE IX   Critical Values of the F-Distribution ($\alpha = 0.05$) (*continued*)

		12	15	20	24	30	40	60	120	∞
					Degrees of Freedom for Numerator					
	1	6,106	6,157	6,209	6,235	6,261	6,287	6,313	6,339	6,366
	2	99.4	99.4	99.4	99.5	99.5	99.5	99.5	99.5	99.5
	3	27.1	26.9	26.7	26.6	26.5	26.4	26.3	26.2	26.1
	4	14.4	14.2	14.0	13.9	13.8	13.7	13.7	13.6	13.5
	5	9.89	9.72	9.55	9.47	9.38	9.29	9.20	9.11	9.02
	6	7.72	7.56	7.40	7.31	7.23	7.14	7.06	6.97	6.88
	7	6.47	6.31	6.16	6.07	5.99	5.91	5.82	5.74	5.65
	8	5.67	5.52	5.36	5.28	5.20	5.12	5.03	4.95	4.86
	9	5.11	4.96	4.81	4.73	4.65	4.57	4.48	4.40	4.31
	10	4.71	4.56	4.41	4.33	4.25	4.17	4.08	4.00	3.91
	11	4.40	4.25	4.10	4.02	3.94	3.86	3.78	3.69	3.60
	12	4.16	4.01	3.86	3.78	3.70	3.62	3.54	3.45	3.36
Degrees	13	3.96	3.82	3.66	3.59	3.51	3.43	3.34	3.25	3.17
of	14	3.80	3.66	3.51	3.43	3.35	3.27	3.18	3.09	3.00
Freedom	15	3.67	3.52	3.37	3.29	3.21	3.13	3.05	2.96	2.87
for										
Denominator	16	3.55	3.41	3.26	3.18	3.10	3.02	2.93	2.84	2.75
	17	3.46	3.31	3.16	3.08	3.00	2.92	2.83	2.75	2.65
	18	3.37	3.23	3.08	3.00	2.92	2.84	2.75	2.66	2.57
	19	3.30	3.15	3.00	2.92	2.84	2.76	2.67	2.58	2.49
	20	3.23	3.09	2.94	2.86	2.78	2.69	2.61	2.52	2.42
	21	3.17	3.03	2.88	2.80	2.72	2.64	2.55	2.46	2.36
	22	3.12	2.98	2.83	2.75	2.67	2.58	2.50	2.40	2.31
	23	3.07	2.93	2.78	2.70	2.62	2.54	2.45	2.35	2.26
	24	3.03	2.89	2.74	2.66	2.58	2.49	2.40	2.31	2.21
	25	2.99	2.85	2.70	2.62	2.53	2.45	2.36	2.27	2.17
	30	2.84	2.70	2.55	2.47	2.39	2.30	2.21	2.11	2.01
	40	2.66	2.52	2.37	2.29	2.20	2.11	2.02	1.92	1.80
	60	2.50	2.35	2.20	2.12	2.03	1.94	1.84	1.73	1.60
	120	2.34	2.19	2.03	1.95	1.86	1.76	1.66	1.53	1.38
	∞	2.18	2.04	1.88	1.79	1.70	1.59	1.47	1.32	1.00

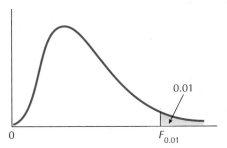

0.01

0                    $F_{0.01}$

TABLE IX   Critical Values of the F-Distribution ($\alpha = 0.01$) (*continued*)

		1	2	3	4	5	6	7	8	9	10
		Degrees of Freedom for Numerator									
	1	4,052	5,000	5,403	5,625	5,764	5,859	5,928	5,982	6,023	6,056
	2	98.5	99.0	99.2	99.2	99.3	99.3	99.4	99.4	99.4	99.4
	3	34.1	30.8	29.5	28.7	28.2	27.9	27.7	27.5	27.3	27.2
	4	21.2	18.0	16.7	16.0	15.5	15.2	15.0	14.8	14.7	14.5
	5	16.3	13.3	12.1	11.4	11.0	10.7	10.5	10.3	10.2	10.1
	6	13.7	10.9	9.78	9.15	8.75	8.47	8.26	8.10	7.98	7.87
	7	12.2	9.55	8.45	7.85	7.46	7.19	6.99	6.84	6.72	6.62
	8	11.3	8.65	7.59	7.01	6.63	6.37	6.18	6.03	5.91	5.81
	9	10.6	8.02	6.99	6.42	6.06	5.80	5.61	5.47	5.35	5.26
	10	10.0	7.56	6.55	5.99	5.64	5.39	5.20	5.06	4.94	4.85
	11	9.65	7.21	6.22	5.67	5.32	5.07	4.89	4.74	4.63	4.54
	12	9.33	6.93	5.95	5.41	5.06	4.82	4.64	4.50	4.39	4.30
	13	9.07	6.70	5.74	5.21	4.86	4.62	4.44	4.30	4.19	4.10
Degrees	*14*	8.86	6.51	5.56	5.04	4.70	4.46	4.28	4.14	4.03	3.94
of	*15*	8.68	6.36	5.42	4.89	4.56	4.32	4.14	4.00	3.89	3.80
Freedom											
for	*16*	8.53	6.23	5.29	4.77	4.44	4.20	4.03	3.89	3.78	3.69
Denominator	*17*	8.40	6.11	5.19	4.67	4.34	4.10	3.93	3.79	3.68	3.59
	18	8.29	6.01	5.09	4.58	4.25	4.01	3.84	3.71	3.60	3.51
	19	8.19	5.93	5.01	4.50	4.17	3.94	3.77	3.63	3.52	3.43
	20	8.10	5.85	4.94	4.43	4.10	3.87	3.70	3.56	3.46	3.37
	21	8.02	5.78	4.87	4.37	4.04	3.81	3.64	3.51	3.40	3.31
	22	7.95	5.72	4.82	4.31	3.99	3.76	3.59	3.45	3.35	3.26
	23	7.88	5.66	4.76	4.26	3.94	3.71	3.54	3.41	3.30	3.21
	24	7.82	5.61	4.72	4.22	3.90	3.67	3.50	3.36	3.26	3.17
	25	7.77	5.57	4.68	4.18	3.86	3.63	3.46	3.32	3.22	3.13
	30	7.56	5.39	4.51	4.02	3.70	3.47	3.30	3.17	3.07	2.98
	40	7.31	5.18	4.31	3.83	3.51	3.29	3.12	2.99	2.89	2.80
	60	7.08	4.98	4.13	3.65	3.34	3.12	2.95	2.82	2.72	2.63
	120	6.85	4.79	3.95	3.48	3.17	2.96	2.79	2.66	2.56	2.47
	∞	6.63	4.61	3.78	3.32	3.02	2.80	2.64	2.51	2.41	2.32

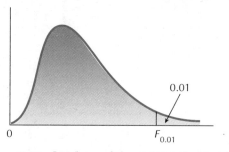

TABLE IX Critical Values of the *F*-Distribution ($\alpha = 0.01$) (*continued*)

		12	15	20	24	30	40	60	120	∞
	1	244	246	248	249	250	251	252	253	254
	2	19.4	19.4	19.4	19.5	19.5	19.5	19.5	19.5	19.5
	3	8.74	8.70	8.66	8.64	8.62	8.59	8.57	8.55	8.53
	4	5.91	5.86	5.80	5.77	5.75	5.72	5.69	5.66	5.63
	5	4.68	4.62	4.56	4.53	4.50	4.46	4.43	4.40	4.37
	6	4.00	3.94	3.87	3.84	3.81	3.77	3.74	3.70	3.67
	7	3.57	3.51	3.44	3.41	3.38	3.34	3.30	3.27	3.23
	8	3.28	3.22	3.15	3.12	3.08	3.04	3.01	2.97	2.93
	9	3.07	3.01	2.94	2.90	2.86	2.83	2.79	2.75	2.71
	10	2.91	2.85	2.77	2.74	2.70	2.66	2.62	2.58	2.54
	11	2.79	2.72	2.65	2.61	2.57	2.53	2.49	2.45	2.40
	12	2.69	2.62	2.54	2.51	2.47	2.43	2.38	2.34	2.30
	13	2.60	2.53	2.46	2.42	2.38	2.34	2.30	2.25	2.21
Degrees of Freedom for Denominator	14	2.53	2.46	2.39	2.35	2.31	2.27	2.22	2.18	2.13
	15	2.48	2.40	2.33	2.29	2.25	2.20	2.16	2.11	2.07
	16	2.42	2.35	2.28	2.24	2.19	2.15	2.11	2.06	2.01
	17	2.38	2.31	2.23	2.19	2.15	2.10	2.06	2.01	1.96
	18	2.34	2.27	2.19	2.15	2.11	2.06	2.02	1.97	1.92
	19	2.31	2.23	2.16	2.11	2.07	2.03	1.98	1.93	1.88
	20	2.28	2.20	2.12	2.08	2.04	1.99	1.95	1.90	1.84
	21	2.25	2.18	2.10	2.05	2.01	1.96	1.92	1.87	1.81
	22	2.23	2.15	2.07	2.03	1.98	1.94	1.89	1.84	1.78
	23	2.20	2.13	2.05	2.01	1.96	1.91	1.86	1.81	1.76
	24	2.18	2.11	2.03	1.98	1.94	1.89	1.84	1.79	1.73
	25	2.16	2.09	2.01	1.96	1.92	1.87	1.82	1.77	1.71
	30	2.09	2.01	1.93	1.89	1.84	1.79	1.74	1.68	1.62
	40	2.00	1.92	1.84	1.79	1.74	1.69	1.64	1.58	1.51
	60	1.92	1.84	1.75	1.70	1.65	1.59	1.53	1.47	1.39
	120	1.83	1.75	1.66	1.61	1.55	1.50	1.43	1.35	1.25
	∞	1.75	1.67	1.57	1.52	1.46	1.39	1.32	1.22	1.00

Degrees of Freedom for Numerator

From E. S. Pearson and H. O. Hartley, *Biometrika Tables for Statisticians*, 1(1958), 159–163. Reprinted by permission of the Biometrika Trustees.

TABLE X   Critical Values for Total Number of Runs (Table Shows Critical Values for Two-Tailed Test at $\alpha = 0.05$)

The Larger of n_1 and n_2

Smaller	5	6	7	8	9	10	11	12	13	14	15	16	17	18	19	20
2								2/6	2/6	2/6	2/6	2/6	2/6	2/6	2/6	2/6
3		2/8	2/8	2/8	2/8	2/8	2/8	2/8	2/8	2/8	3/8	3/8	3/8	3/8	3/8	3/8
4	2/9	2/9	2/10	3/10	3/10	3/10	3/10	3/10	3/10	3/10	3/10	4/10	4/10	4/10	4/10	4/10
5	2/10	3/10	3/11	3/11	3/12	3/12	4/12	4/12	4/12	4/12	4/12	4/12	4/12	5/12	5/12	5/12
6		3/11	3/12	3/12	4/13	4/13	4/13	4/13	5/14	5/14	5/14	5/14	5/14	5/14	6/14	6/14
7			3/13	4/13	4/14	5/14	5/14	5/14	5/15	5/15	6/15	6/16	6/16	6/16	6/16	6/16
8				4/14	5/14	5/15	5/15	6/16	6/16	6/16	6/16	6/17	7/17	7/17	7/17	7/17
9					5/15	5/16	6/16	6/16	6/17	7/17	7/18	7/18	7/18	8/18	8/18	8/18
10						6/16	6/17	7/17	7/18	7/18	7/18	8/19	8/19	8/19	8/20	9/20
11							7/17	7/18	7/19	8/19	8/19	8/20	9/20	9/20	9/21	9/21
12								7/19	8/19	8/20	8/20	9/21	9/21	9/21	10/22	10/22
13									8/20	9/20	9/21	9/21	10/22	10/22	10/23	10/23
14										9/21	9/22	10/22	10/23	10/23	11/23	11/24
15											10/22	10/23	11/23	11/24	11/24	12/25
16												11/23	11/24	11/25	12/25	12/25
17													11/25	12/25	12/26	13/26
18														12/26	13/26	13/27
19															13/27	13/27
20																14/28

(Row labels under "The Smaller of n_1 and n_2"; each cell shows the lower critical value over the upper critical value.)

From C. Eisenhart and F. Swed, "Tables for testing randomness of grouping in a sequence of alternatives," *The Annals of Statistics*, 14(1943), 66–87. Reprinted by permission.

TABLE XI   Critical Value of Spearman's Rank Correlation Coefficient

	Level of Significance for One-Tailed Test			
	0.05	0.025	0.01	0.005
	Level of Significance for Two-Tailed Test			
n	*0.10*	*0.05*	*0.02*	*0.01*
5	0.900	1.000	1.000	—
6	0.829	0.886	0.943	1.000
7	0.714	0.786	0.893	0.929
8	0.643	0.738	0.833	0.881
9	0.600	0.683	0.783	0.833
10	0.564	0.648	0.745	0.794
11	0.523	0.623	0.736	0.818
12	0.497	0.591	0.703	0.780
13	0.475	0.566	0.673	0.745
14	0.457	0.545	0.646	0.716
15	0.441	0.525	0.623	0.689
16	0.425	0.507	0.601	0.666
17	0.412	0.490	0.582	0.645
18	0.399	0.476	0.564	0.625
19	0.388	0.462	0.549	0.608
20	0.377	0.450	0.534	0.591
21	0.368	0.438	0.521	0.576
22	0.359	0.428	0.508	0.562
23	0.351	0.418	0.496	0.549
24	0.343	0.409	0.485	0.537
25	0.336	0.400	0.475	0.526
26	0.329	0.392	0.465	0.515
27	0.323	0.385	0.456	0.505
28	0.317	0.377	0.448	0.496
29	0.311	0.370	0.440	0.487
30	0.305	0.364	0.432	0.478

TABLE XII THE EXPONENTIAL FUNCTION   A-21

TABLE XII   The Exponential Function

x	e^{-x}	x	e^{-x}	x	e^{-x}	x	e^{-x}
0.00	1.00000	**0.40**	0.67032	**0.80**	0.44933	**1.20**	0.30119
0.01	0.99005	0.41	0.66365	0.81	0.44486	1.21	0.29820
0.02	0.98020	0.42	0.65705	0.82	0.44043	1.22	0.29523
0.03	0.97045	0.43	0.65051	0.83	0.43605	1.23	0.29229
0.04	0.96079	0.44	0.64404	0.84	0.43171	1.24	0.28938
0.05	0.95123	0.45	0.63763	0.85	0.42741	1.25	0.28650
0.06	0.94176	0.46	0.63128	0.86	0.42316	1.26	0.28365
0.07	0.93239	0.47	0.62500	0.87	0.41895	1.27	0.28083
0.08	0.92312	0.48	0.61878	0.88	0.41478	1.28	0.27804
0.09	0.91393	0.49	0.61263	0.89	0.41066	1.29	0.25727
0.10	0.90484	**0.50**	0.60653	**0.90**	0.40657	**1.30**	0.27253
0.11	0.89583	0.51	0.60050	0.91	0.40252	1.31	0.26982
0.12	0.88692	0.52	0.59452	0.92	0.39852	1.32	0.26714
0.13	0.87810	0.53	0.58860	0.93	0.39455	1.33	0.26448
0.14	0.86936	0.54	0.58275	0.94	0.39063	1.34	0.26185
0.15	0.86071	0.55	0.57695	0.95	0.38674	1.35	0.25924
0.16	0.85214	0.56	0.57121	0.96	0.38289	1.36	0.25666
0.17	0.84366	0.57	0.56533	0.97	0.37908	1.37	0.25411
0.18	0.83527	0.58	0.55990	0.98	0.37531	1.38	0.25158
0.19	0.82696	0.59	0.55433	0.99	0.37158	1.39	0.24908
0.20	0.81873	**0.60**	0.54881	**1.00**	0.36788	**1.40**	0.24660
0.21	0.81058	0.61	0.54335	1.01	0.36422	1.41	0.24414
0.22	0.80252	0.62	0.53794	1.02	0.36059	1.42	0.24171
0.23	0.79453	0.63	0.53259	1.03	0.35701	1.43	0.23931
0.24	0.78663	0.64	0.52729	1.04	0.35345	1.44	0.23693
0.25	0.77880	0.65	0.52205	1.05	0.34994	1.45	0.23457
0.26	0.77105	0.66	0.51685	1.06	0.34646	1.46	0.23224
0.27	0.76338	0.67	0.51171	1.07	0.34301	1.47	0.22993
0.28	0.75578	0.68	0.50662	1.08	0.33960	1.48	0.22764
0.29	0.74826	0.69	0.51058	1.09	0.33622	1.49	0.22537
0.30	0.74082	**0.70**	0.49659	**1.10**	0.33287	**1.50**	0.22313
0.31	0.73345	0.71	0.49164	1.11	0.32956	1.51	0.22091
0.32	0.72615	0.72	0.48675	1.12	0.32628	1.52	0.21871
0.33	0.71892	0.73	0.48191	1.13	0.32303	1.53	0.21654
0.34	0.71177	0.74	0.47711	1.14	0.31982	1.54	0.21438
0.35	0.70469	0.75	0.47237	1.15	0.31664	1.55	0.21225
0.36	0.69768	0.76	0.46767	1.16	0.31349	1.56	0.21014
0.37	0.69073	0.77	0.46301	1.17	0.31037	1.57	0.20805
0.38	0.68386	0.78	0.45841	1.18	0.30728	1.58	0.20598
0.39	0.67706	0.79	0.45384	1.19	0.30422	1.59	0.20393

TABLE XII   The Exponential Function (*continued*)

x	e^{-x}	x	e^{-x}	x	e^{-x}	x	e^{-x}
1.60	0.20190	**2.00**	0.13534	**2.40**	0.09072	**2.80**	0.06081
1.61	0.19989	2.01	0.13399	2.41	0.08982	2.81	0.06020
1.62	0.19790	2.02	0.13266	2.42	0.08892	2.82	0.05961
1.63	0.19593	2.03	0.13134	2.43	0.08804	2.83	0.05901
1.64	0.19398	2.04	0.13003	2.44	0.08716	2.84	0.05843
1.65	0.19205	2.05	0.12873	2.45	0.08629	2.85	0.05783
1.66	0.19014	2.06	0.12745	2.46	0.08543	2.86	0.05727
1.67	0.18825	2.07	0.12619	2.47	0.08458	2.87	0.05670
1.68	0.18637	2.08	0.12493	2.48	0.08374	2.88	0.05613
1.69	0.18452	2.09	0.12369	2.49	0.08291	2.89	0.05558
1.70	0.18268	**2.10**	0.12246	**2.50**	0.08208	**2.90**	0.05502
1.71	0.18087	2.11	0.12124	2.51	0.08127	2.91	0.05448
1.72	0.17907	2.12	0.12003	2.52	0.08046	2.92	0.05393
1.73	0.17728	2.13	0.11884	2.53	0.07966	2.93	0.05340
1.74	0.17552	2.14	0.11765	2.54	0.07887	2.94	0.05287
1.75	0.17377	2.15	0.11648	2.55	0.07808	2.95	0.05234
1.76	0.17204	2.16	0.11533	2.56	0.07730	2.96	0.05182
1.77	0.17033	2.17	0.11418	2.57	0.07654	2.97	0.05130
1.78	0.16864	2.18	0.11304	2.58	0.07577	2.98	0.05079
1.79	0.16696	2.19	0.11192	2.59	0.07502	2.99	0.05029
1.80	0.16530	**2.20**	0.11080	**2.60**	0.07427	**3.00**	0.04979
1.81	0.16365	2.21	0.10970	2.61	0.07353	3.01	0.04929
1.82	0.16203	2.22	0.10861	2.62	0.07280	3.02	0.04880
1.83	0.16041	2.23	0.10753	2.63	0.07208	3.03	0.04832
1.84	0.15882	2.24	0.10646	2.64	0.07136	3.04	0.04783
1.85	0.15724	2.25	0.10540	2.65	0.07065	3.05	0.04736
1.86	0.15567	2.26	0.10435	2.66	0.06995	3.06	0.04689
1.87	0.15412	2.27	0.10331	2.67	0.06925	3.07	0.04642
1.88	0.15259	2.28	0.10228	2.68	0.06856	3.08	0.04596
1.89	0.15107	2.29	0.10127	2.69	0.06788	3.09	0.04550
1.90	0.14957	**2.30**	0.10026	**2.70**	0.06721	**3.10**	0.04505
1.91	0.14808	2.31	0.09926	2.71	0.06654	3.11	0.04460
1.92	0.14661	2.32	0.09827	2.72	0.06587	3.12	0.04416
1.93	0.14515	2.33	0.09730	2.73	0.06522	3.13	0.04372
1.94	0.14370	2.34	0.09633	2.74	0.06457	3.14	0.04328
1.95	0.14227	2.35	0.09537	2.75	0.06393	3.15	0.04285
1.96	0.14086	2.36	0.09442	2.76	0.06329	3.16	0.04243
1.97	0.13946	2.37	0.09348	2.77	0.06266	3.17	0.04200
1.98	0.13807	2.38	0.09255	2.78	0.06204	3.18	0.04159
1.99	0.13670	2.39	0.09163	2.79	0.06142	3.19	0.04117

TABLE XII THE EXPONENTIAL FUNCTION   **A-23**

TABLE XII   The Exponential Function (*continued*)

x	e^{-x}	x	e^{-x}	x	e^{-x}	x	e^{-x}
3.20	0.04076	**3.60**	0.02732	**4.00**	0.01832	**4.40**	0.01228
3.21	0.04036	3.61	0.02705	4.01	0.01813	4.41	0.01216
3.22	0.03996	3.62	0.02678	4.02	0.01795	4.42	0.01203
3.23	0.03956	3.63	0.02652	4.03	0.01777	4.43	0.01191
3.24	0.03916	3.64	0.02625	4.04	0.01760	4.44	0.01180
3.25	0.03877	3.65	0.02599	4.05	0.01742	4.45	0.01168
3.26	0.03839	3.66	0.02573	4.06	0.01725	4.46	0.01156
3.27	0.03801	3.67	0.02548	4.07	0.01708	4.47	0.01145
3.28	0.03763	3.68	0.02522	4.08	0.01691	4.48	0.01133
3.29	0.03725	3.69	0.02497	4.09	0.01674	4.49	0.01122
3.30	0.03688	**3.70**	0.02472	**4.10**	0.01657	**4.50**	0.01111
3.31	0.03652	3.71	0.02448	4.11	0.01641	4.51	0.01100
3.32	0.03615	3.72	0.02423	4.12	0.01624	4.52	0.01089
3.33	0.03579	3.73	0.02399	4.13	0.01608	4.53	0.01078
3.34	0.03544	3.74	0.02375	4.14	0.01592	4.54	0.01067
3.35	0.03508	3.75	0.02352	4.15	0.01576	4.55	0.01057
3.36	0.03474	3.76	0.02328	4.16	0.01561	4.56	0.01046
3.37	0.03439	3.77	0.02305	4.17	0.01545	4.57	0.01036
3.38	0.03405	3.78	0.02282	4.18	0.01530	4.58	0.01025
3.39	0.03371	3.79	0.02260	4.19	0.01515	4.59	0.01015
3.40	0.03337	**3.80**	0.02237	**4.20**	0.01500	**4.60**	0.01005
3.41	0.03304	3.81	0.02215	4.21	0.01485	4.61	0.00995
3.42	0.03271	3.82	0.02193	4.22	0.01470	4.62	0.00985
3.43	0.03239	3.83	0.02171	4.23	0.01455	4.63	0.00975
3.44	0.03206	3.84	0.02149	4.24	0.01441	4.64	0.00966
3.45	0.03175	3.85	0.02128	4.25	0.01426	4.65	0.00956
3.46	0.03143	3.86	0.02107	4.26	0.01412	4.66	0.00947
3.47	0.03112	3.87	0.02086	4.27	0.01398	4.67	0.00937
3.48	0.03081	3.88	0.02065	4.28	0.01384	4.68	0.00928
3.49	0.03050	3.89	0.02045	4.29	0.01370	4.69	0.00919
3.50	0.03020	**3.90**	0.02024	**4.30**	0.01357	**4.70**	0.00910
3.51	0.02990	3.91	0.02004	4.31	0.01343	4.71	0.00900
3.52	0.02960	3.92	0.01984	4.32	0.01330	4.72	0.00892
3.53	0.02930	3.93	0.01964	4.33	0.01317	4.73	0.00883
3.54	0.02901	3.94	0.01945	4.34	0.01304	4.74	0.00874
3.55	0.02872	3.95	0.01925	4.35	0.01291	4.75	0.00865
3.56	0.02844	3.96	0.01906	4.36	0.01278	4.76	0.00857
3.57	0.02816	3.97	0.01887	4.37	0.01265	4.77	0.00848
3.58	0.02788	3.98	0.01869	4.38	0.01253	4.78	0.00840
3.59	0.02760	3.99	0.01850	4.39	0.01240	4.79	0.00831

TABLE XII   The Exponential Function (*continued*)

x	e^{-x}	x	e^{-x}	x	e^{-x}
4.80	0.00823	**6.00**	0.00248	**9.00**	0.00012
4.81	0.00815	6.10	0.00244	9.10	0.00011
4.82	0.00807	6.20	0.00203	9.20	0.00010
4.83	0.00799	6.30	0.00184	9.30	0.00009
4.84	0.00791	6.40	0.00166	9.40	0.00008
4.85	0.00783	6.50	0.00150	9.50	0.00007
4.86	0.00775	6.60	0.00136	9.60	0.00007
4.87	0.00767	6.70	0.00123	9.70	0.00006
4.88	0.00760	6.80	0.00111	9.80	0.00006
4.89	0.00752	6.90	0.00101	9.90	0.00005
4.90	0.00745	**7.00**	0.00091	**10.00**	0.00005
4.91	0.00737	7.10	0.00083	10.10	0.00004
4.92	0.00730	7.20	0.00075	10.20	0.00004
4.93	0.00723	7.30	0.00068	10.30	0.00003
4.94	0.00715	7.40	0.00061	10.40	0.00003
4.95	0.00708	7.50	0.00055	10.50	0.00003
4.96	0.00701	7.60	0.00050	10.60	0.00020
4.97	0.00694	7.70	0.00045	10.70	0.00002
4.98	0.00687	7.80	0.00041	10.80	0.00002
4.99	0.00681	7.90	0.00037	10.90	0.00002
5.00	0.00674	**8.00**	0.00034	**11.00**	0.00002
5.10	0.00610	8.10	0.00030	11.10	0.00002
5.20	0.00552	8.20	0.00027	11.20	0.00001
5.30	0.00499	8.30	0.00025	11.30	0.00001
5.40	0.00452	8.40	0.00022	11.40	0.00001
5.50	0.00409	8.50	0.00020	11.50	0.00001
5.60	0.00370	8.60	0.00018	11.60	0.00001
5.70	0.00335	8.70	0.00017	11.70	0.00001
5.80	0.00303	8.80	0.00015	11.80	0.00001
5.90	0.00274	8.90	0.00014	11.90	0.00001

ANSWERS TO SELECTED EXERCISES

CHAPTER 1

Section 1.4 (pages 18–22)

1. Although it should be descriptive only, in reality it is used as both descriptive and inferential statistics.
3. a. Sex of student, eye color, marital status, religion, etc.
   b. Number of brothers or sisters that each college student has, etc.
   c. Weight or height of each college student, etc.
5. a. The part that indicates that the number of Americans over 65 years of age increased by 6.3% over last year at this time.
   b. The part that predicts that by the end of the century approximately 25% of the American population will be over 65 years of age.
7. Choice (d)   9. It should be descriptive only.   11. Descriptive statistics
13. Descriptive statistics   15. No. It probably will not be a random sample.
21. No. The 10% raise in salary is based on a lower salary.   23. No
25. Technically yes, but I would not recommend it.   27. No

Section 1.5 (pages 26–27)

1–6.

ROW	JUNE	SALES	TOTAL
1	1	157802	1002919
2	2	181362	
3	3	159576	
4	4	172887	
5	5	193216	
6	6	138076	

7. This command names the data entered in C1 as JUNE.

9. This command adds the entries in column C2 and places the sum in C3.
11. This command prints the contents of columns C1, C2, and C3.
13. This command saves the data file as ASSIGN1 on a floppy disk in drive A.
15. This command exits you from MINITAB.

Testing Your Understanding of This Chapter's Concepts (page 28)

1. Probably
2. It is descriptive because it indicates the student's past performance, and it is inferential in that it is used to predict a student's likelihood of success in college.
3. Not necessarily   4. Not necessarily
5. The tobacco industry claims that people may develop lung cancer because of other factors such as air pollution, water pollution, etc.
6. Not necessarily; people may be more nutrition oriented.
7. Not necessarily. The 90% is only an average.
8. Not necessarily. The 8 out of every 11 is only a ratio in reduced form.

Chapter Test (pages 28–32)

1. Choice (c)   2. Choice (b)   3. Choice (b)   4. Choice (a)   5. Choice (d)
6. The part that indicates that the blood levels stood at 20% of their normal level and that only 837 pints of blood were donated last week.
7. The part that indicates that the findings have far-reaching implications for those planning to have major surgery in the future and for the general health of our citizens.
8. Choice (d)
9. a. Statistical inference   b. Descriptive statistics
   c. Statistical inference   d. Statistical inference
10. True   11. True
12. The part that indicates that 14 of the 527 houses were contaminated with radon.
13. The part that estimates that about 15% of the homes in the region are contaminated with radon gas.
14. a. Qualitative   b. Quantitative and continuous   c. Qualitative
    d. Quantitative and continuous   e. Quantitative and continuous
    f. Quantitative and continuous   g. Qualitative   h. Qualitative
15. Not necessarily
17. A discrete random variable is a variable that represents observations or measurements of something that results from a count. A continuous random variable is a variable that represents observations or measurements of something that results from a measure of quantity.

Thinking Critically (pages 32–34)

1. It is possible that the number of dolphins in the sea has already been reduced from earlier catches.
2. No. Not a random sample.   3. No. Not all women drive.   4. No. Not a random sample.

5. No. Less cars on the road.
6. No. Medical advances (among other things) may be a contributing factor.   7. No
8. No. There may be less highway patrol officers conducting sobriety tests.
9. No. There was a general increase in population.

Case Study (pages 34–35)

1. and 2. The claims are unsubstantiated.

CHAPTER 2

Section 2.2 (pages 50–55)

1.

Miles per Gallon	Tally	Frequency
25–26	|	1
27–28	||||	4
29–30	||| ||||	9
31–32	|||| |||| |	11
33–34	|||| ||||	9
35–36	|||| |	6
37–38	|||| |	6
39–40	||	2
41–42	|	1
43–44	|	1
		50

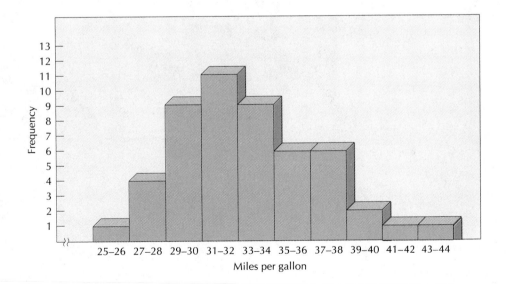

3.

Premium Cost	Tally	Frequency
$1500–1699	||||	4
1700–1899	ЖЖ |||	8
1900–2099	ЖЖ ЖЖ ЖЖ	15
2100–2299	ЖЖ |	6
2300–2499	|	1
2500–2699		0
2700–2899		0
2900–3099		0
3100–3299		0
3300–3504	|	1
		$\overline{35}$

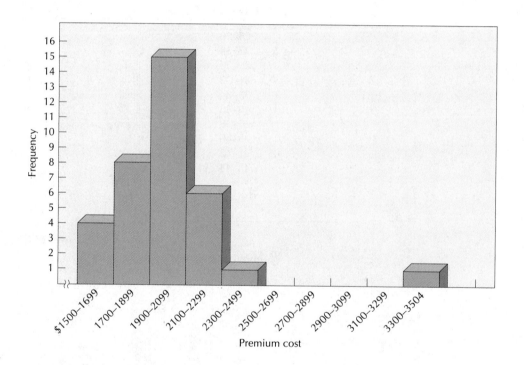

5.

Monthly Bill	Tally	Frequency
$14.69–15.70	|||	3
15.71–16.72			1
16.73–17.74	|||||	5
17.75–18.76	||||| |||	8
18.77–19.78	||||| ||	7
19.79–20.80	||||| ||||	9
20.81–21.82	||||	4
21.83–22.84			1
22.85–23.86			1
23.87–24.88			1
		40

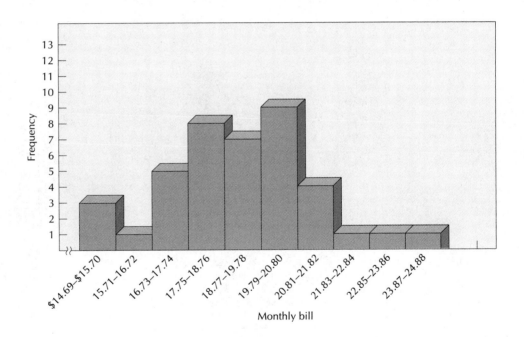

7. a. The frequency distribution using 10 classes.

Number of Pounds Collected	Tally	Frequency
120–308	||||	4
309–497	||||	4
498–686	৸৸ ||||	9
687–875	৸৸ ৸৸ |	11
876–1064	৸৸ ৸৸	10
1065–1253	৸৸ |||	8
1254–1442	||||	4
1443–1631	৸৸ |	6
1632–1820	||	2
1821–2009	||	2
		60

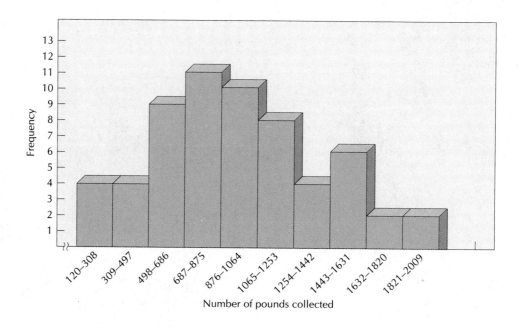

b. The frequency distribution using 5 classes.

Number of Pounds Collected	Tally	Frequency
120–497	||||| |||	8
498–875	||||| ||||| ||||| |||||	20
876–1253	||||| ||||| ||||| |||	18
1254–1631	||||| |||||	10
1632–2009	||||	4
		60

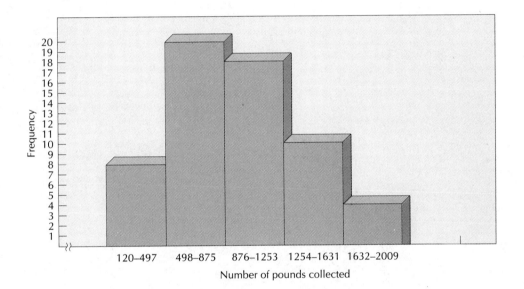

c. The frequency distribution using 15 classes.

Number of Pounds Collected	Tally	Frequency
120–245	||	2
246–371	|||	3
372–497	|||	3
498–623	||||	4
624–749	⩘ ||	7
750–875	⩘ ||||	9
876–1001	⩘ |||	8
1002–1127	⩘ |	6
1128–1253	||||	4
1254–1379	||	2
1380–1505	⩘	5
1506–1631	|||	3
1632–1757	|	1
1758–1883	|	1
1884–2009	||	2
		60

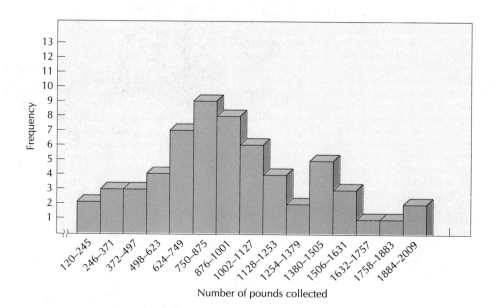

Number of pounds collected

9. a.

Number of Products Completed	Tally	Frequency
11–13	|	1
14–16	||||	4
17–19	||||	4
20–22	||||	4
23–25	|	1
26–28	|||	3
29–31	|	1
32–34	||	2
		20

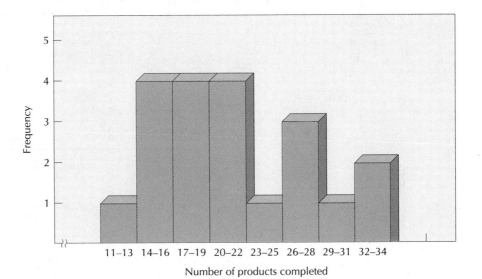

11. To avoid misinterpreting the data.

13.

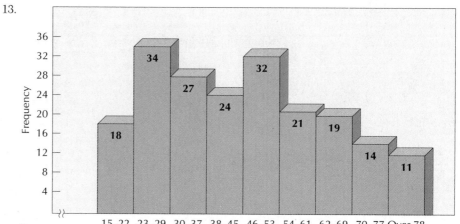

a. $\dfrac{103}{200} = 51.5\%$ b. $\dfrac{121}{200} = 60.5\%$ c. $\dfrac{83}{200} = 41.5\%$

d. $\dfrac{14}{200} = \dfrac{7}{100} = 7\%$ e. $\dfrac{186}{200} = \dfrac{93}{100} = 93\%$

15. a. (i)

Number of Stocks	Tally	Frequency
1–6	||||| |||	8
7–12	|||||	5
13–18	||||| ||||| |||	13
19–24	||||| ||||| ||||| ||||	19
25–30	|||||	5
		50

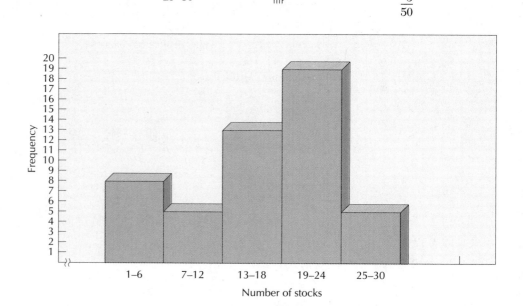

(ii)

Number of Stocks	Tally	Frequency
1–2	||	2
3–4	|||	3
5–6	|||	3
7–8	|||	3
9–10	||	2
11–12		0
13–14	|||	3
15–16	⫴⫴ ||	7
17–18	|||	3
19–20	||	2
21–22	⫴⫴ ||	7
23–24	⫴⫴ ⫴⫴	10
25–26		0
27–28	||	2
29–30	|||	3
		50

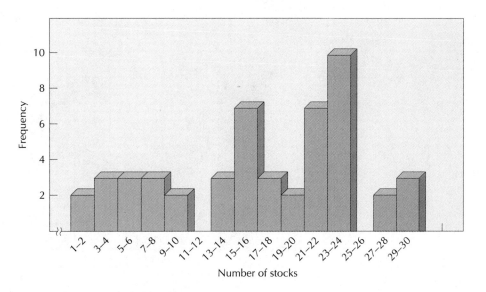

b. When too few or too many classes are used, much information is lost, or the information that is presented is hard to interpret.

Section 2.3 (pages 74–84)

1.

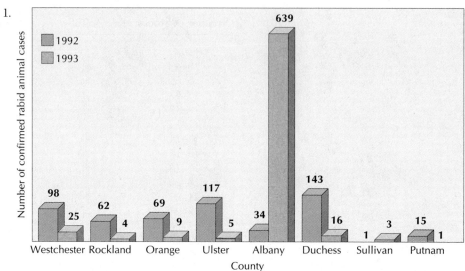

3.

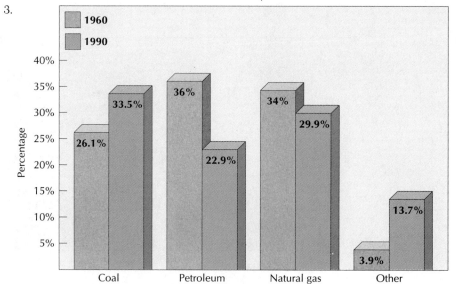

5. a.

Device	1980	1990
Telephone Service	82%	84%
Television	85%	98%
Radio Sets	85%	85%
Cable TV	18%	62%
VCR	2%	77%

b. From 2% to 77% or by 75%   c. No change

7. a. During the periods March 1–15, March 16–30, and May 30–June 12   b. March 1–15

9.

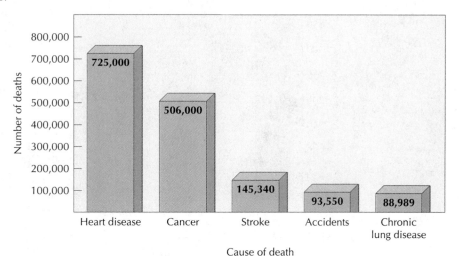

11. a. $0.15 \times 312 = 46.8$ b. $0.22 \times 312 = 68.64$ c. $0.43 \times 312 = 134.16$

13. a.

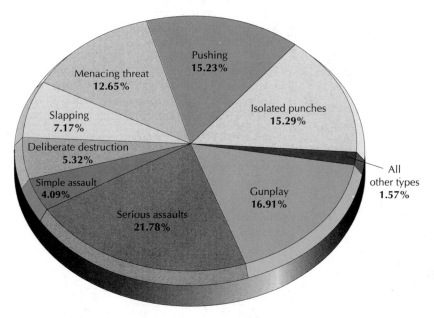

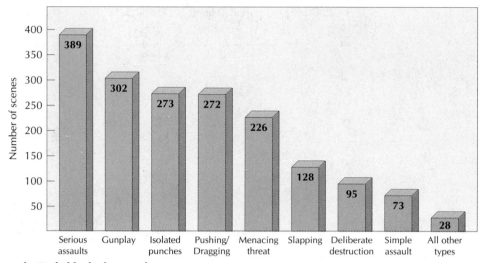

b. Probably the bar graph

15. a. $0.16 \times 4276 = 684.16$ customers   b. $(0.15 + 0.19) \times 4276 = 1453.84$ customers
    c. $0.79 \times 4276 = 3378.04$ customers

17. a. China: 850 million
       India: 625 million
       Soviet Union: 250 million
       United States: 225 million
    b. $850 - 250 = 600$ million

19. a. Not normally distributed   b. Not normally distributed   c. Normally distributed
    d. Not normally distributed   e. Normally distributed   f. Normally distributed

21. a. Assuming that the 1158 samples were selected from all of the 6 counties together, we have
       $1158/6 = 193$ samples per county.

       Nassau: $0.015 \times 193 = 2.895$ samples
       Rockland: $0 \times 193 = 0$
       Suffolk: $0.026 \times 193 = 5.018$ samples
       Westchester: $0 \times 193 = 0$
       New York City: $0.013 \times 193 = 2.509$ samples

    b. The bars of the bar graph are not drawn to scale. For example, consider the height of the
       rectangles for Rockland and Westchester (0%).

Section 2.4 (pages 89–91)

1. a.

Stem	Leaves
2	4 3 4 4 4 4
2	9 6 6 9 5 7 8 9 7 7 7 6 8 7 7 8 8 6 5 8 6 9 7 8 8 9 6 7 8
3	1 0 1 1 2 2 0 3 3 0 0 1 4 2 3

b.

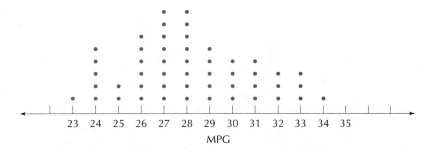

MPG

3.

Stem	Leaves
13	7000
14	0000 2000 3000
15	8000
16	0000 9000 5000
17	2000 0000 5000 5000 6000
18	5000 9000 0000 0000 5000 1000
19	8000 4000 8000
20	5000 8000 0000
21	5000 5000
22	9000 0000 0000 5000 5000
23	8000
24	8000 5000
25	4000 1000 0000
26	8000 5000
27	0000 0000

5.

Stem	Leaves
18 (1850–1899)	75
19 (1900–1949)	45 45 01 45
19 (1950–1999)	98 89 76 87 98 84
20 (2000–2049)	00 00 04 49
20 (2050–2099)	50
21 (2100–2149)	49 49 45 00
21 (2150–2199)	98 50
22 (2200–2249)	49 49 00

7. a.

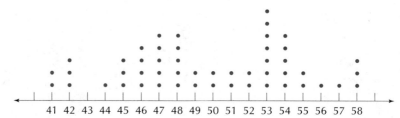

b.

Stem	Leaves
4	2 1 4 1 2 2
4	7 7 5 6 5 8 8 6 7 8 8 7 8 5 7 6 6 9 9
5	3 1 4 4 0 3 3 0 4 3 1 3 3 4 2 3 2 4
5	7 5 8 6 8 8 5

c. Probably the dotplot

Section 2.5 (pages 94–98)

1. Both graphs are statistically correct, but because different spacings are used on the vertical scale, the graphs appear different.
3. The graph is truncated. Also, the scale is misleading.
5. It should be $\dfrac{26.6}{19.3}$ or 1.378 times as tall; i.e., different spacings are used on the vertical scale.

   The conclusion may be true but not based upon the information obtained from this graph.

Section 2.6 (pages 101–102)

1. $\dfrac{1.38}{1.26} \times 100 = 109.52$. Price increased by approximately 9.52% when compared with 1989.
3. $\dfrac{1.38}{1.38} \times 100 = 100$. Base year price.
5.

Year	Cost Index	Interpretation
1990	$\dfrac{3300}{3300} \times 100 = 100$	Base year price
1991	$\dfrac{3700}{3300} \times 100 = 112.12$	Price increased by approximately 12.12% when compared with 1990
1992	$\dfrac{4000}{3300} \times 100 = 121.21$	Price increased by approximately 21.21% when compared with 1990
1993	$\dfrac{4400}{3300} \times 100 = 133.33$	Price increased by approximately 33.33% when compared with 1990
1994	$\dfrac{4600}{3300} \times 100 = 139.39$	Price increased by approximately 39.39% when compared with 1990
1995	$\dfrac{5000}{3300} \times 100 = 151.52$	Price increased by approximately 51.52% when compared with 1990

7.

Year	Cost Index	Interpretation
1989	$\dfrac{10,200}{13,000} \times 100 = 78.46$	Prices were approximately $78.46 - 100 = 21.54\%$ cheaper in 1989 when compared with 1992.
1990	$\dfrac{11,100}{13,000} \times 100 = 85.38$	Prices were approximately $85.38 - 100 = 14.62\%$ cheaper in 1990 when compared with 1992.
1991	$\dfrac{12,000}{13,000} \times 100 = 92.31$	Prices were approximately $92.31 - 100 = 7.69\%$ cheaper in 1991 when compared with 1992.
1992	$\dfrac{13,000}{13,000} \times 100 = 100$	Base year price
1993	$\dfrac{14,500}{13,000} \times 100 = 111.54$	Prices increased approximately 11.54% when compared with 1992.
1994	$\dfrac{17,000}{13,000} \times 100 = 130.77$	Prices increased approximately 30.77% when compared with 1992.
1995	$\dfrac{20,000}{13,000} \times 100 = 153.85$	Prices increased approximately 53.85% when compared with 1992.

9. Pension costs in January 1994 were 6% less when compared with January 1990.

Section 2.7 (pages 106–107)

1. a.
```
MTB   > SET THE FOLLOWING DATA INTO C1
DATA  > 430 441 460 420 437 450 426 460 438 456
DATA  > 400 425 409 428 416 442 422 475 453 443
DATA  > 450 410 422 406 404 460 429 420 446 405
DATA  > 375 408 430 411 417 420 435 403 422 412
DATA  > 493 430 401 429 406 409 442 427 409 459
DATA  > END
MTB   > GSTD
MTB   > HISTOGRAM OF C1;
SUBC  > START 381;
SUBC  > INCREMENT 12.


Histogram of C1    N = 50

              Midpoint       Count
                 381.0          1        *
                 393.0          0
                 405.0         12        ***********
                 417.0         10        **********
                 429.0          9        *********
                 441.0          8        ********
                 453.0          4        ****
                 465.0          4        ****
                 477.0          1        *
                 489.0          1        *
```

b. MTB > STEM AND LEAF OF C1

```
Stem and leaf of C1     N = 50
Leaf Unit = 1.0

         1       37      5
         1       38
         1       39
        12       40      01345668999
        17       41      01267
       (12)      42      000222567899
        21       43      000578
        15       44      12236
        10       45      00369
         5       46      000
         2       47      5
         1       48
         1       49      3
```

3. a. MTB > SET THE FOLLOWING IN C1
   DATA > 31 29 26 27 46 37 43 37 30 30 29 25 25
   DATA > 25 32 28 26 45 38 41 36 32 32 26 29
   DATA > 27 34 40 31 49 36 42 38 33 33 28 26
   DATA > 26 33 35 38 47 39 43 39 32 31 27 27
   DATA > 28 27 38 36 42 32 45 34 31 25 25 25
   DATA > END
   MTB > GSTD
   MTB > HISTOGRAM OF C1;
   MTB > START 26;
   MTB > INCREMENT 3.

```
Histogram of C1     N = 61

         Midpoint       Count
           26.00         16     ****************
           29.00          8     ********
           32.00         12     ************
           35.00          6     ******
           38.00          8     ********
           41.00          4     ****
           44.00          4     ****
           47.00          2     **
           50.00          1     *
```

b. **MTB   > STEM AND LEAF OF C1**

```
Stem and leaf of C1     N = 61
Leaf Unit = 1.0
                 6       2       555555
                16       2       6666677777
                22       2       888999
                28       3       001111
               (8)       3       22222333
                25       3       445
                22       3       66677
                17       3       888899
                11       4       01
                 9       4       2233
                 5       4       55
                 3       4       67
                 1       4       9
```

c. **MTB   > DOTPLOT OF C1**

Character Dotplot

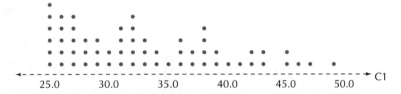

5. **MTB   > DOTPLOT OF C1**

Character Dotplot

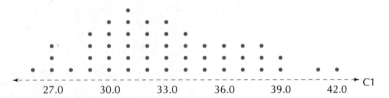

Testing Your Understanding of This Chapter's Concepts (pages 110–111)

1. Yes   2. True
3. a. $4 + 6 + 9 + 8 + 4 + 3 + 7 + 5 + 1 = 47$ b. Frequency histogram
4. $\dfrac{112 - 0}{12} = 9.33$. Use 10 as the class width.

Chapter Test (pages 111–118)

1. Choice (d)   2. Choice (c)   3. Choice (b)   4. Choice (d)   5. Choice (b)   6. Choice (c)
7. Choice (e)   8. Choice (d). It decreased during this time period.
9.

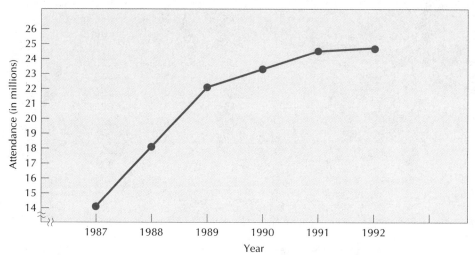

10.

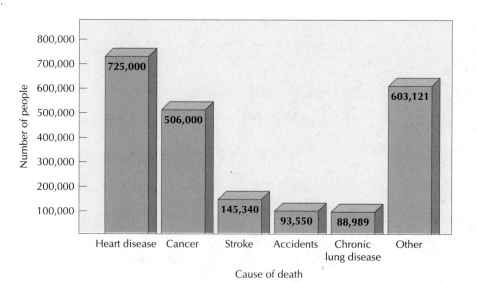

11.

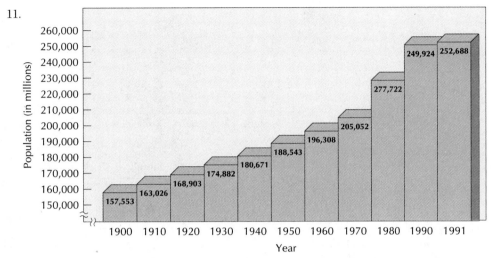

12.

Fee Charged	Tally	Frequency
$4500–4,599	||	2
4600–4699	~~||||~~ ~~||||~~ ~~||||~~	15
4700–4799	|	1
4800–4899	~~||||~~ |	6
4900–4999	~~||||~~	5
5000–5099	||||	4
5100–5199	~~||||~~ |||	8
5200–5299	|||	3
5300–5399	||	2
5400–5499	||||	4
		$\overline{50}$

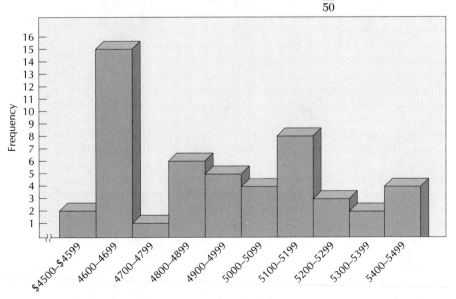

13.

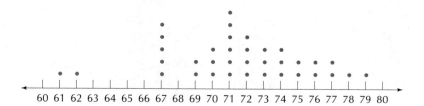

14.

Stem	Leaves
4	.4 .9
5	.0
6	.6 .8
7	.9 .2
8	.4 .6
9	.5
10	.0 .3
11	.9
12	.3 .9
13	.8

15.

Year	Cost Index	Interpretation
1988	$\dfrac{2500}{2500} \times 100 = 100$	Base year cost
1989	$\dfrac{2600}{2500} \times 100 = 104$	Prices increased by approximately 4% when compared with 1988.
1990	$\dfrac{2750}{2500} \times 100 = 110$	Prices increased by approximately 10% when compared with 1988.
1991	$\dfrac{3000}{2500} \times 100 = 120$	Prices increased by approximately 20% when compared with 1988.
1992	$\dfrac{3200}{2500} \times 100 = 128$	Prices increased by approximately 28% when compared with 1988.
1993	$\dfrac{3500}{2500} \times 100 = 140$	Prices increased by approximately 40% when compared with 1988.

16. The coins should be of the same size, and there should be twice as many coins for 1995 when compared with the number of coins stacked vertically for 1985.

17. a.

Cholesterol Level	Tally	Frequency
224–234	ⅢⅠ	5
235–245	ⅢⅠ ⅢⅠ Ⅲ	14
246–256	ⅢⅠ Ⅰ	6
257–267	ⅢⅠ ⅢⅠ Ⅲ	14
268–278	Ⅲ	4
279–289	ⅢⅠ Ⅲ	8
290–300	ⅠⅠ	2
301–311	ⅠⅠ	2
312–322	Ⅲ	3
323–333	ⅠⅠ	2
		60

b. and c.

Cholesterol Level	Percent Frequency Distribution	Cumulative Percent Frequency Distribution
224–234	$\frac{5}{60} = 8.33\%$	$\frac{5}{60} = 8.33\%$
235–245	$\frac{14}{60} = 23.33\%$	$\frac{19}{60} = 31.67\%$
246–256	$\frac{6}{60} = 10.00\%$	$\frac{25}{60} = 41.67\%$
257–267	$\frac{14}{60} = 23.33\%$	$\frac{39}{60} = 65.00\%$
268–278	$\frac{4}{60} = 6.67\%$	$\frac{43}{60} = 71.67\%$
279–289	$\frac{8}{60} = 13.33\%$	$\frac{51}{60} = 85.00\%$
290–300	$\frac{2}{60} = 3.33\%$	$\frac{53}{60} = 88.33\%$
301–311	$\frac{2}{60} = 3.33\%$	$\frac{55}{60} = 91.67\%$
312–322	$\frac{3}{60} = 5.00\%$	$\frac{58}{60} = 96.67\%$
323–333	$\frac{2}{60} = 3.33\%$	$\frac{60}{60} = 100.00\%$

d.

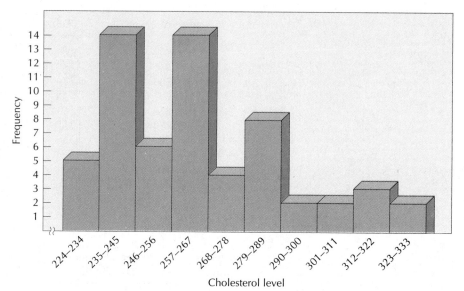

e.

Stem	Leaves
22	9 8 9 9 4
23	9 7 7 8 9 6
24	0 4 5 1 2 3 5 1 8 8
25	1 9 8 6 8 6 1
26	7 7 2 1 2 2 1 2 1 7 1
27	8 9 9 2 8 2
28	7 1 1 9 9 5
29	6 1
30	3 9
31	2 2
32	4 7 1
33	

18. $3 + 5 + 5 + 6 + 3 + 5 + 2 = 29$

19.

Year	Cost Index
1990	$\dfrac{12.45}{12.45} \times 100 = 100$
1991	$\dfrac{13.16}{12.45} \times 100 = 105.70$
1992	$\dfrac{15.67}{12.45} \times 100 = 125.86$
1993	$\dfrac{16.49}{12.45} \times 100 = 132.45$
1994	$\dfrac{18.11}{12.45} \times 100 = 145.46$
1995	$\dfrac{19.95}{12.45} \times 100 = 160.24$

20.

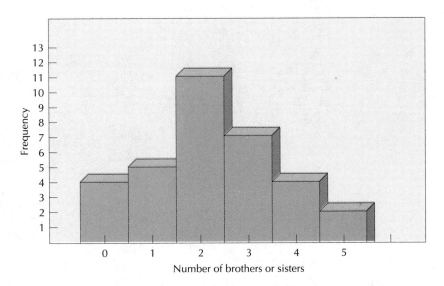

21. a. Normally distributed   b. Probably not normally distributed
    c. Probably not normally distributed   d. Normally distributed

22. a.

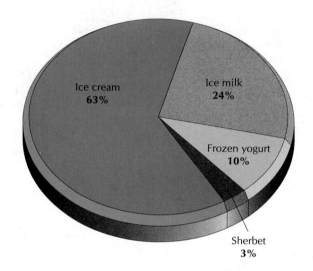

b.

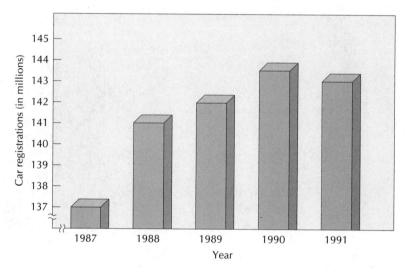

c.

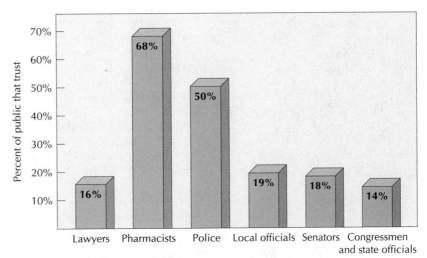

23. a. $0.17 \times 473{,}977 = 80{,}576.09$ or 80,577 immigrants
    b. $(0.24 + 0.12) \times 473{,}977 = 170{,}631.72$ or 170,632 immigrants
24.

Where do American children live?

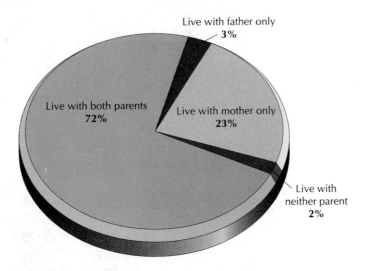

Thinking Critically (pages 118–120)

1. Yes. The next interval is supposed to begin where the previous interval ends. No gaps. Also, the interval lengths are not the same.
2. The interval endpoints are overlapping. In which category should we put a monthly bill of $15? Also, the interval lengths are not the same.

3.

Stem	Leaves
4 (400–419)	19
4 (420–439)	28 33
4 (440–459)	49
4 (460–479)	78 63
4 (480–499)	83 93 88 86
5 (500–519)	12 12 01 13 10 12 03 17
5 (520–539)	31 32 23 26 28
5 (540–559)	
5 (560–579)	76 76 73 72
5 (580–599)	86 93 86

4. a.

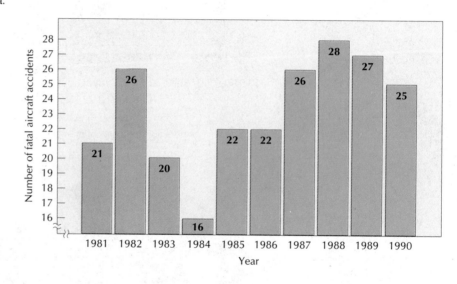

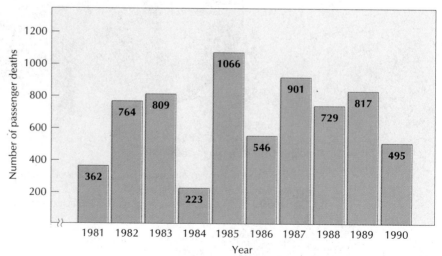

b.

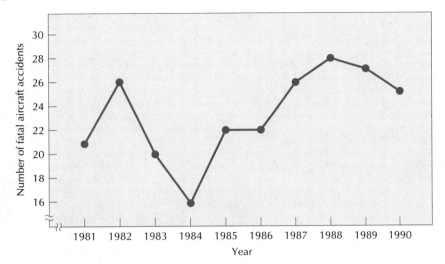

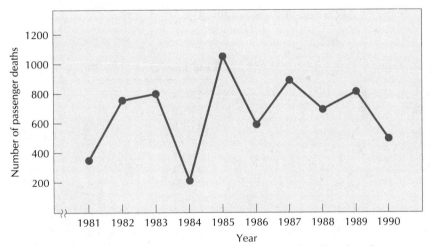

c. Both are useful, although some people may prefer one rather than the other.

5.

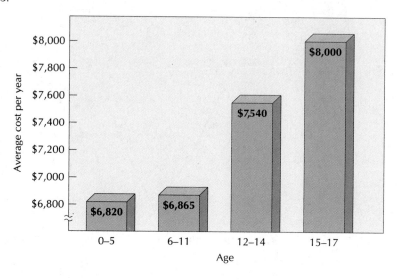

Case Studies (pages 120–121)

1. a.

Lead Levels	Tally	Frequency
13–14	|	1
15–16	||||	4
17–18	₩ ₩ |	11
19–20	₩ ₩ |	11
21–22	₩ ₩ ||	12
23–24	₩ |||	8
25–26	||||	4
27–28	||||	4
29–30	|||	3
31–32	||	2
		60

b.

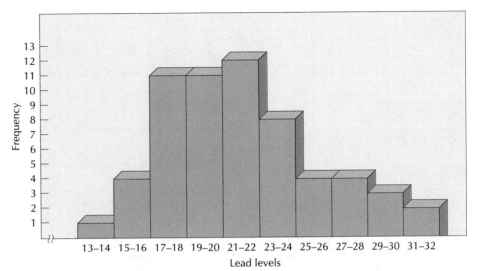

c.

Stem	Leaves
1	5 5 4
1	9 7 9 8 8 9 8 8 8 9 7 6 8 9 8 9 8 6 7
2	2 1 4 3 2 3 0 2 2 3 0 2 0 4 3 4 2 1 0 3 2 0 1 1 1
2	9 6 7 8 5 9 8 6 7 5
3	1 2 0

d.

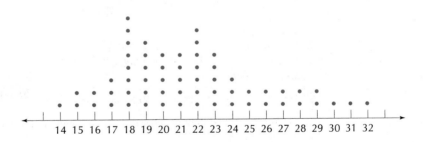

CHAPTER 3

Section 3.2 (pages 131–132)

1. a. $\displaystyle\sum_{i=1}^{7} x_i$ b. $\displaystyle\sum_{i=1}^{7} x_i^2$ c. $\displaystyle\sum_{i=1}^{7} x_i f_i$ d. $\displaystyle\sum_{i=1}^{7} 10x_i$ or $10\displaystyle\sum_{i=1}^{7} x_i$ e. $\displaystyle\sum_{i=1}^{n} (2y_i + x_i)$

3. a. 68 b. 4624 c. 1156 d. 93 e. 1961 f. 151 g. 5485

5. a. $\displaystyle\sum_{i=1}^{10} (x_i + 8) = \sum_{i=1}^{10} x_i + \sum_{i=1}^{10} 8 = 18 + 8 \cdot 10 = 98$

   b. $\displaystyle\sum_{i=1}^{10} (x_i + 8)^2 = \sum_{i=1}^{10} (x_i^2 + 16x_i + 64)$

   $\displaystyle = \sum_{i=1}^{10} x_i^2 + 16 \sum_{i=1}^{10} x_i + \sum_{i=1}^{10} 64 = 37 + 16(18) + 10 \cdot 64 = 965$

7. a. The total number of students submitting essays was 36.

   b. The total number of essays (of all types) that she received was 97.

9. a. 50.2 b. 2520.04 c. 252.34

Section 3.3 (pages 142–148)

1. Mean = 23.2 years

   Median = $\dfrac{23 + 24}{2} = 23.5$

   Mode = none

3. Mean = $46,200

   Median = $\dfrac{\$45,500 + 47,000}{2} = \$46,250$

   Mode = $49,000

   Extreme scores have an effect on the mean.

5. Mean = $127,900

   Median = $\dfrac{\$115,000 + 125,000}{2} = \$120,000$

   Mode = none

7. New mean = 7257.17 + 500 = 7757.17 copies

   New mode = 7589.50 + 500 = 8089.50 copies

   Mode = none

9. Average cost = $50.11   11. 159 female inmates   13. $\dfrac{\$1,716,000}{22} = \$78,000$

15. No. Both classes do not necessarily have the same number of students.

17. Harmonic mean = 6.6359

    Geometric mean = 6.8173

19. a. Midrange = $\dfrac{2 + 14}{2} = 8$ b. Midrange = $\dfrac{498 + 611}{2} = 554.5$

21. $\displaystyle\sum_{i=1}^{n} (x_i - \bar{x}) = \sum_{i=1}^{n} x_i - \sum_{i=1}^{n} \bar{x}$

$\displaystyle = \sum_{i=1}^{n} x_i - n\bar{x}$

$\displaystyle = \sum_{i=1}^{n} x_i - n \frac{\displaystyle\sum_{i=1}^{n} x_i}{n}$

$\displaystyle = \sum_{i=1}^{n} x_i - \sum_{i+1}^{n} x_i = 0$

Section 3.5 (pages 157–161)

1. Mean $\mu = 12$
   Range $= 9$
   Population variance, $\sigma^2 = 8.33333$
   Population standard deviation, $\sigma = \sqrt{8.33333} \approx 2.88675$
3. Sample mean $= 177$
   Range $= 46$
   Sample variance, $s^2 = 261.5556$
   Sample standard deviation, $s = \sqrt{261.5556} \approx 16.1727$
   Average deviation $= 13.8$
5. Sample mean $= 25$
   Sample variance, $s^2 = 33$
   Sample standard deviation, $s = \sqrt{33} \approx 5.7446$
   Average deviation $= 4.6667$
7. Sample mean $= 12.7857$
   Sample variance, $s^2 = 71.62597$
   Sample standard deviation, $s = \sqrt{71.62597} \approx 8.46321$
9. New sample mean $= 20$
   New sample variance, $s^2 = 180$
   New sample standard deviation, $s = \sqrt{180} \approx 13.4164$
   New average deviation $= 11.5556$
   The sample variance is 16 times as great as it was originally.
   The sample standard deviation and average deviation are four times as great as they were originally.
11. The new mean will be increased by 20%. It will be $308 + 0.20(308)$ or \$369.60. The new standard deviation will be increased by 20%. It will be $27 + 0.20(27)$ or \$32.40.
13. Probably Brand B.
15. a. $s^2 = 40.6783$
       $s = \sqrt{40.6783} \approx 6.37796$
    b. $s^2 = 38.1505$
       $s = \sqrt{38.1505} \approx 6.1766$
    c. The standard deviation for the males is larger.

Section 3.6 (pages 164–165)

1. $\bar{x} = 568$
   $s = \sqrt{4223.6} \approx 64.989$

When $k = 2$, then at least $\frac{3}{4}$ of the terms fall between $568 \pm 2(64.989)$ or between 438.022 and 697.978.

When $k = 3$, then at least $\frac{8}{9}$ of the terms fall between $568 \pm 3(64.989)$ or between 373.033 and 762.967.

3. $k = 1.48$. At least $1 - \dfrac{1}{(1.48)^2} = 0.5435$ or 54.35% of the pilots.

5. a. $\bar{x} = 3.315$, $s^2 = 1.7319$, $s = \sqrt{1.7319} \approx 1.3160$

   b. When $k = 1.75$, then at least $1 - \dfrac{1}{(1.75)^2} = 0.67$ or 67% of the terms fall within the interval $\bar{x} \pm 1.75s = 3.315 \pm 1.75(1.3160)$ or between 1.012 and 5.618.

   c. All the measurements fall within the specified intervals.

   d. Chebyshev's Theorem is valid.

Section 3.7 (pages 176–179)

1. Percentile rank of Michele = 45 percentile.
   Percentile rank of Ricardo = 87.5 percentile.
3. Ed's percentile rank = 71.43 percentile.
5. a.

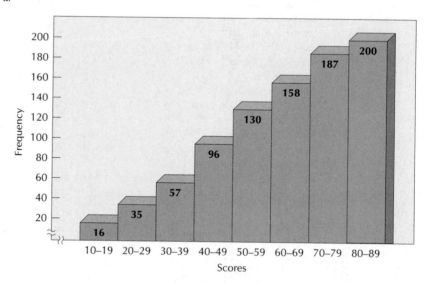

b.

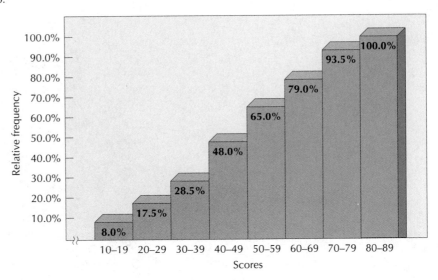

7. When arranged in increasing order, we have

    16 18 19 19 20 21 22 23 23 24 25 25 26 27 27 28 29 30 31 31

    The median is between 24 and 25. It is 24.5.
    a. The lower quartile is the median of 16 18 19 19 20 21 22 23 23 24, or between 20 and 21. It is 20.5.
    b. The upper quartile is the median of 25 25 26 27 27 28 29 30 31 31, or between 27 and 28. It is 27.5.
    c. The interquartile range $= Q_3 - Q_1 = 27.5 - 20.5 = 7$.
    d.

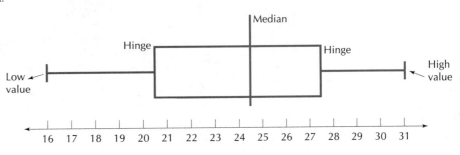

9. a.

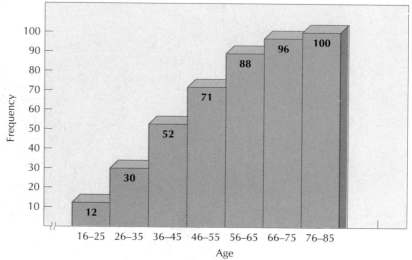

b.

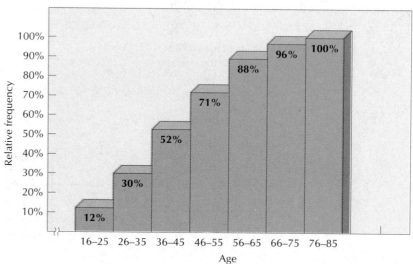

11. a. Percentile rank of Dawn = 45.45 percentile.
    b. Percentile rank of Spector = 97.73 percentile.
    c. D_{10} is approximately 92.

Section 3.8 (pages 183–186)

1. a. $z = 2$ b. $z = -1.5$ c. $z = 3$ d. $z = 0$
3. a. $x = 42.817$ b. $x = 58.562$ c. $x = 29.187$

5.

x	x^2
-1.52	2.3104
-0.54	0.2916
0.00	0.0000
0.65	0.4225
1.41	1.9881
0.00	5.0126

$\sigma = \sqrt{1.00252} \approx 1$
The slight difference from 1 is due to rounding.

and

x	x^2
-1.32	1.7424
-0.66	0.4356
0.00	0.0000
0.33	0.1089
1.65	2.7225
0.00	5.0094

$\sigma = \sqrt{1.00188} \approx 1$
The slight difference from 1 is due to rounding.

7. Claudette's actual salary is $31,000 + (2.6)2700 = \$38,020$.
   Tom's actual salary is $31,000 + (-0.98)2700 = \$28,354$.
   Gina's actual salary is $31,000 + (1.37)2700 = \$34,699$.

9. New York Lakes: $\sigma = \sqrt{50.48438} \approx 7.11$
   $\mu = 25.625$
   New Jersey Lakes: $\sigma = \sqrt{24.3333} \approx 4.933$
   $\mu = 70$
   a. New York Lake G since it has the highest z-value ($z = 1.60$).
   b. New Jersey Lake S since it has the smallest z-value ($z = -1.82$).

Section 3.9 (pages 187–189)

1. a.
```
MTB  > SET THE FOLLOWING IN C1
DATA > 29500 28700 25500 20500 23000 29750
DATA > 23000 22000 22000 26100 24000 21000
DATA > 28500 20500 21000 23000 21750 23000
DATA > 27400 22750 27500 22750 25000 28000
DATA > 26000 19700 26750 22000 26000 21500
DATA > 25500 27500 25000 24000
DATA > END
MTB  > DESCRIBE C1
```

Descriptive Statistics

Variable	N	Mean	Median	TrMean	StDev	SEMean
C1	34	24416	24000	24357	2859	490

Variable	Min	Max	Q1	Q3
C1	19700	29750	22000	26913

b. **MTB > BOXPLOT OF C1**

Character Boxplot

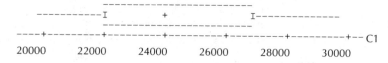

```
                          ----------------------------
         ------------I              +              I---------------
                          ----------------------------
         ----+---------+---------+---------+---------+---------+-- C1
         20000      22000      24000      26000      28000      30000
```

c. Interquartile range $= Q_3 - Q_1 = 26{,}913 - 22{,}000 = 4913$.

3. a. **MTB   > SET THE FOLLOWING IN C3**
   DATA > 65 62 73 70 73
   DATA > 74 54 76 73 73
   DATA > 46 73 60 64 63
   DATA > 71 72 41 75 72
   DATA > 60 47 48 61 70
   DATA > 73 46 71 57 39
   DATA > END
   MTB > DESCRIBE C3

Descriptive Statistics

Variable	N	Mean	Median	TrMean	StDev	SEMean
C3	30	63.40	67.50	64.27	11.27	2.06

Variable	Min	Max	Q1	Q3
C3	39.00	76.00	56.25	73.00

b. **MTB   > BOXPLOT OF C3**

Character Boxplot

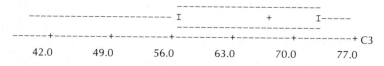

```
                                   ----------------------------
         -------------------------- I            +         I-----
                                   ----------------------------
         ------+---------+---------+---------+---------+---------+ C3
         42.0       49.0       56.0       63.0       70.0       77.0
```

c. Interquartile range $= Q_3 - Q_1 = 73.00 - 56.25 = 16.75$.

Testing Your Understanding of This Chapter's Concepts (pages 194–195)

1. b.   c.

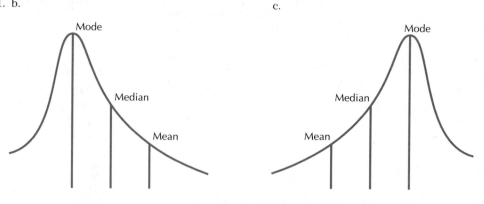

2. 75%

3. No. The average depth may be 3 feet. At some points, the depth may exceed 6 feet.

4. We cannot compute the z-score since $\sigma = 0$ and we would have to divide by 0. All the terms are the same.

5. No. Both schools do not necessarily have the same number of students.

6. He may be denied admission if his high school average of 93 has a percentile rank below 85.

7. Yes. Especially when you have extreme scores.   8. The second student.

9. 69° since this represents the mode.

Chapter Test (pages 195–200)

1. Choice (b)   2. Population mean = 137. Choice (a).

3. The median is between 137 and 145. It is 141. Choice (c).   4. Choice (b)   5. Choice (d)

6. Choice (b)   7. Choice (c)   8. Choice (d)   9. Choice (a)   10. Choice (d)

11. a. $\Sigma x = 550$

 b. Variance $= \dfrac{\Sigma x^2}{n} - \dfrac{(\Sigma x)^2}{n^2}$

 $25 = \dfrac{\Sigma x^2}{10} - \dfrac{(550)^2}{100}$

 $30{,}500 = \Sigma x^2$

12. Using Chebyshev's Theorem, the proportion of terms is at least

$$1 - \frac{1}{(2.3)^2} \approx 81\%$$

13. a. Median ≈ 22

 b. Upper quartile ≈ 23
 Lower quartile ≈ 16

 c. Interquartile range $= Q_3 - Q_1 = 23 - 16 = 7$

14.

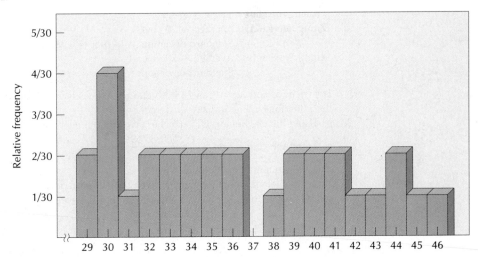

b. Midrange $\dfrac{29 + 46}{2} = 37.5$

15. When arranged in increasing order we have

29 29 30 30 30 30 31 32 32 33 33 34 34 35 35 36 36 38 39 39 40 40
41 41 42 43 44 44 45 46

The median is between 35 and 36. It is 35.5.

The lower quartile is the median of 29 29 30 30 30 30 31 32 32 33 33 34 34 35 35. It is 32.

The upper quartile is the median of 36 36 38 39 39 40 40 41 41 42 43 44 44 45 46. It is 41.

16. 29.914 years   17. 39.20 percentile

18. a. Marketing/sales: $z = 1.28$

Plant maintenance: $z = -0.32$

Inventory/shipping $z = 1.29$

Personnel: $z = -0.34$

b. Inventory/shipping c. Personnel

19. 360 children

a. $Q_1 = P_{25} = \dfrac{25}{100} \cdot 360 = 90$. Three children (by cumulative frequencies).

b. $Q_3 = P_{75} = \dfrac{75}{100} \cdot 360 = 270$. Four children.

c. $P_{80} = \dfrac{80}{100} \cdot 360 = 288$. Five children. d. $P_{95} = \dfrac{95}{100} \cdot 360 = 342$. Six children.

20. Lori: 71.78 minutes
Coty: 39.541 minutes
Walter: 64.261 minutes

21. a. Sample mean $= 4.2253$
Sample standard deviation $= \sqrt{12.4944} \approx 3.5347$

b. $P_{55} = \dfrac{55}{100} \cdot 415 = 228.25$. Using cumulative frequencies, this falls in the 4–5 category.

Using the class mark, $P_{55} = 4.5$.

22. Dating Service A: Average $= 20$ years
                        Standard deviation $= \sqrt{19.2} \approx 4.3818$
Dating Service B: Average $= 20$ years
                        Standard deviation: $= \sqrt{2} \approx 1.4142$

His friend is correct. Although both dating services have the same average age, the standard deviation for Dating Service A is larger.

23. a. Using the class mark, $P_{60} = 524.5$. (We use the cumulative frequencies.)

b. Using the class mark, $P_{80} = \dfrac{550 + 599}{2} = 574.5$.

24. Average amount contributed is $8.00.

25. Probably the one from Company B since the standard deviation is smaller. Others may prefer the one from Company A since the average life is greater.

26. When $k = 2$, then at least $1 - \dfrac{1}{2^2} = 75\%$ or more of the female smokers will be within $31.6 \pm 2(8.6)$ or between 14.4 and 48.8 years.

When $k = 3$, then at least $1 - \dfrac{1}{3^2} = 89\%$ or more of the female smokers will be within $31.6 \pm 3(8.6)$ or between 5.8 and 57.4 years.

27. z-score $= 0.70$. His grade is slightly above average.

Using Chebyshev's Theorem when $k = 2$, then at least $1 - \dfrac{1}{2^2} = 75\%$ or more of the grades will be within $82 \pm 2(4.3)$ or between 73.4 and 90.6.

When $k = 3$, then at least $1 - \dfrac{1}{3^2} = 89\%$ or more of the grades will be within $82 \pm 3(4.3)$ or between 69.1 and 94.9.

Thinking Critically (pages 200–201)

1. Yes, if all the numbers are the same.
2. $\Sigma(x - \mu) = \Sigma x - \Sigma \mu$

$\qquad = \Sigma x - n\mu$

$\qquad = \Sigma x - n\left(\dfrac{\Sigma x}{n}\right) \qquad$ since $\mu = \dfrac{\Sigma x}{n}$

$\qquad = \Sigma x - \Sigma x$

$\qquad = 0$

3. $\sqrt{\dfrac{\Sigma(x - \bar{x})^2}{n - 1}} = \sqrt{\dfrac{\Sigma(x^2 - 2x\bar{x} + \bar{x}^2)}{n - 1}}$

$\qquad = \sqrt{\dfrac{\Sigma x^2 - \Sigma 2x\bar{x} + \Sigma \bar{x}^2}{n - 1}}$

$\qquad = \sqrt{\dfrac{\Sigma x^2}{n - 1} - \dfrac{2\bar{x}\,\Sigma x}{n - 1} + \dfrac{n(\bar{x}^2)}{n - 1}} \qquad$ Note: $\bar{x} = \dfrac{\Sigma n}{n}$.

$\qquad = \sqrt{\dfrac{\Sigma x^2}{n - 1} - \dfrac{2n(\bar{x}^2)}{n - 1} + \dfrac{n(\bar{x}^2)}{n - 1}}$

$\qquad = \sqrt{\dfrac{\Sigma x^2}{n - 1} - \dfrac{n(\bar{x}^2)}{n - 1}}$

$\qquad = \sqrt{\dfrac{\Sigma x^2}{n - 1} - \dfrac{(\Sigma x)(\Sigma x)}{n(n - 1)}}$

$\qquad = \sqrt{\dfrac{n(\Sigma x^2) - (\Sigma x)^2}{n(n - 1)}}$

4. Zero 5. Not necessarily true. See the data given in Exercise 4 of Section 3.7.
6. False. Positive z-scores will occur for any terms (positive or negative) that are above the mean.
7. You will have 99.7% of the terms of the distribution falling between $z = -3$ and $z = +3$.
8. Each one is using a different measure of central tendency.

Case Studies (pages 201–203)

1. Sample mean = $68.51
   Sample standard deviation = $\sqrt{24.065315} \approx 4.9056$
2. a. Population mean = 42.9048
      Population standard deviation = $\sqrt{125.5147} \approx 11.2033$
   b. **Character Boxplot**

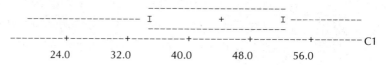

3. a. Sample mean = 31.33333
      Sample standard deviation = $\sqrt{73.06666} \approx 8.5479$
   b. **Character Boxplot**

4. Poorer Neighborhoods: Sample mean = 60 cents
                          Sample standard deviation = $\sqrt{56} \approx 7.4833$

   Middle Class Neighborhoods: Sample mean = 60
                                Sample standard deviation = $\sqrt{2.5} \approx 1.5811$

   Although the average cost in both neighborhoods is the same, the standard deviation for the prices charged in poorer neighborhoods is considerably larger.
5. Sample mean = 724
   Sample variance = 82,650.505
   Sample standard deviation = $\sqrt{82,650.505} \approx 287.49$

CHAPTER 4

Section 4.2 (pages 220–226)

1. a. $\dfrac{209}{516}$ b. $\dfrac{181}{516}$ c. $\dfrac{335}{516}$ d. $\dfrac{73}{516}$

3. $\dfrac{4}{16} = \dfrac{1}{4}$

5. a. $\dfrac{32}{75}$ b. $\dfrac{17}{75}$ c. $\dfrac{9}{100}$ d. 0

7. a. $\dfrac{9}{100}$ b. $\dfrac{39}{100}$

9. a. $\dfrac{3}{17}$ b. $\dfrac{38}{85}$ c. $\dfrac{7}{85}$

11. a. $\dfrac{627}{702} = \dfrac{209}{234}$ b. $\dfrac{540}{702} = \dfrac{10}{13}$ c. $\dfrac{5}{702}$

13. a. $\dfrac{7}{20}$ b. $\dfrac{1}{2}$ c. $\dfrac{13}{20}$ d. $\dfrac{7}{20}$

15. $\dfrac{19,400}{58,911}$ 17. $\dfrac{3}{8}$ 19. $\dfrac{1}{4}$ 21. $\dfrac{1}{24}$

Section 4.3 (pages 230–233)

1. 42 possible setups

3.

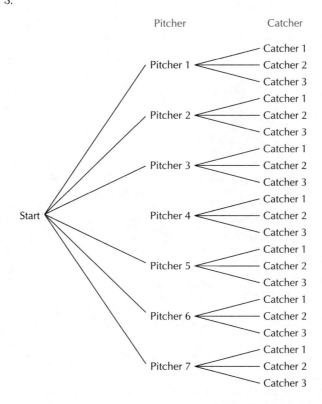

5. 132,600 possible ways   7. 720 possible ways   9. 8,000,000 possible numbers

11. 720 possible ways

13. a. 192 possible numbers   b. 18 possible numbers

15. a. 5040 possible ways   b. 3600 possible ways

17.

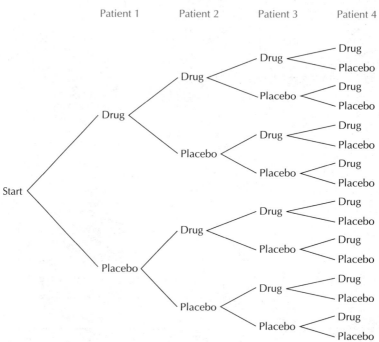

16 different ways

Section 4.4 (pages 240–242)

1. a. $6! = 720$ b. $7! = 5040$ c. $2! = 2$ d. $\dfrac{7!}{6!} = 7$ e. $\dfrac{0!}{3} = \dfrac{1}{3}$ f. $\dfrac{6!}{4!2!} = 15$ g. $\dfrac{8!}{6!2!} = 28$

h. $\dfrac{6!}{3!3!} = 20$ i. $_7P_5 = 2520$ j. $_6P_4 = 360$ k. $_7P_3 = 210$ l. $_5P_4 = 120$ m. $_6P_6 = 720$

n. $_4P_4 = 24$ o. $_0P_0 = 1$ p. $_5P_0 = 1$

3. $_6P_6 = 720$ possibilities 5. $_7P_3 = 210$ possible ways, assuming order counts

7. a. $_{12}P_{12} = 479{,}001{,}600$

   b. $_4P_4 \cdot _3P_3 \cdot _2P_2 \cdot _3P_3 = 24 \cdot 6 \cdot 2 \cdot 6 = 1728$. However, since books on the same subject can be permuted as a group, we must multiply this by $_4P_4$ or 24. Thus, we have $24(1728) = 41{,}472$ possible permutations.

9. a. $_{12}P_{12} = 479{,}001{,}600$

   b. The men can be seated in $_6P_6 = 720$ possible ways and the women can also be seated in $_6P_6$ or 720 possible ways. However, the men can be seated on either the left side or right side, so that there are $720 \times 720 \times 2 = 1{,}036{,}800$ possible ways.

   c. The couples can be seated in $_6P_6 = 6!$ ways. However, each man can sit on the left or right side of his wife. Thus, there are $2^6 \times 720 = 46{,}080$ possible ways.

11. a. $_5P_5 = 5! = 120$ possible ways b. $\dfrac{1}{120}$

13. The letters T, R, and O are each repeated twice. If the first letter is an O, then there are $\dfrac{11!}{2!2!}$ or 9,979,200 possible ways. However, since either of the O's can be the first letter, we have $2 \times 9{,}979{,}200 = 19{,}958{,}400$ possible ways. If the first letter is an I, there are $\dfrac{11!}{2!2!2!} = 4{,}989{,}600$ possible ways. If the first letter is an E, there are $\dfrac{11!}{2!2!2!} = 4{,}989{,}600$ possible ways. Thus, there are $19{,}958{,}400 + 4{,}989{,}600 + 4{,}989{,}600 = 29{,}937{,}600$ possible ways.

15. a. $_9P_9 = 9! = 362{,}880$ possible orders   b. $\dfrac{1}{9}$ c. No

17. $_7P_7 = 5040$ possible ways

19. a. Assuming there is no "head" of the circle, we can think of any one as being arbitrarily designated as the "head" and the other $n - 1$ people then can be arranged in $(n - 1)!$ ways.
    b. 120 possible ways

Section 4.5 (pages 252–254)

1. a. $_7C_6 = 7$ b. $_6C_4 = 15$ c. $_8C_3 = 56$ d. $_9C_0 = 1$ e. $_6C_1 = 6$ f. $_8C_5 = 56$

   g. $\begin{pmatrix} 9 \\ 6 \end{pmatrix} = 84$ h. $\begin{pmatrix} 8 \\ 6 \end{pmatrix} = 28$ i. $\begin{pmatrix} 7 \\ 8 \end{pmatrix} = {_7C_8} =$ impossible   j. $\begin{pmatrix} 6 \\ 6 \end{pmatrix} = 1$

3. $_9C_3 = 84$ 5. $_{10}C_4 = 210$ 7. $_{14}C_6 = 3003$

9. a. $\dfrac{_{13}C_2 \cdot {_{11}C_0}}{_{24}C_2} = \dfrac{13}{46}$ b. $\dfrac{_{13}C_0 \cdot {_{11}C_2}}{_{24}C_2} = \dfrac{55}{276}$ c. $\dfrac{_{13}C_1 \cdot {_{11}C_1}}{_{24}C_2} = \dfrac{143}{276}$

11. a. $_{13}C_4 \cdot {_{11}C_3} = 715 \cdot 165 = 117{,}975$ different ways
    b. The new lawyers can be

    | 4 women | 5 women | 6 women | 7 women |
    | 3 men | + 2 men | + 1 man | + 0 men |

    $_{11}C_4 \cdot {_{13}C_3} + {_{11}C_5} \cdot {_{13}C_2} + {_{11}C_6} \cdot {_{13}C_1} + {_{11}C_7} \cdot {_{13}C_0}$
    $= 136{,}752$ different ways

13. a. $_{18}C_3 \cdot {_2C_0} = 816$ b. $_{18}C_2 \cdot {_2C_1} = 306$ c. $_{18}C_1 \cdot {_2C_2} = 18$

15. a. $_8C_3 \cdot {_{10}C_3} = 6720$
    b. Unit can consist of

    | 5 males | 4 males | 3 males | 2 males | 1 male | 0 males |
    | 1 female | + 2 females | + 3 females | + 4 females | + 5 females | + 6 females |

    $_8C_5 \cdot {_{10}C_1} + {_8C_4} \cdot {_{10}C_2} + {_8C_3} \cdot {_{10}C_3} + {_8C_2} \cdot {_{10}C_4} + {_8C_1} \cdot {_{10}C_5} + {_8C_0} \cdot {_{10}C_6}$
    $= 18{,}536$

    c. Unit can consist of

    | 5 males | 4 males | 3 males | 2 males | 1 male |
    | 1 female | + 2 females | + 3 females | + 4 females | + 5 females |

    $_8C_5 \cdot {_{10}C_1} + {_8C_4} \cdot {_{10}C_2} + {_8C_3} \cdot {_{10}C_3} + {_8C_2} \cdot {_{10}C_4} + {_8C_1} \cdot {_{10}C_5}$
    $= 18{,}326$

17. a. $\dfrac{_{36}C_0 \cdot {_{14}C_4}}{_{50}C_4} = \dfrac{1 \cdot 1001}{230{,}300} = \dfrac{1001}{230{,}300}$

b. Committee can consist of

2 Republicans		3 Republicans		4 Republicans
2 Democrats	+	1 Democrat	+	0 Democrats

$$\frac{_{36}C_2 \cdot _{14}C_2}{_{50}C_4} + \frac{_{36}C_1 \cdot _{14}C_3}{_{50}C_4} + \frac{_{36}C_0 \cdot _{14}C_4}{_{50}C_4}$$

$$= \frac{2041}{6580}$$

c. $\dfrac{_{36}C_4 \cdot _{14}C_0}{_{50}C_4} = \dfrac{11{,}781}{46{,}060}$

Section 4.6 (pages 258–261)

1. $17.64   3. $-$0.50   5. $-$1.92   7. 6:7
9. a. 0.88 b. 88:12 or 22:3
11. Visit accounts I and II.   13. 79:21

Section 4.7 (page 263)

1. a.
```
MTB  > RANDOM 180 C1;
SUBC > INTEGER 1 TO 6.
MTB  > PRINT C1
MTB  > TALLY C1
```
Answers will vary.
  b.
```
MTB  > RANDOM 300 C1;
SUBC > INTEGER 1 TO 6.
MTB  > PRINT C1
MTB  > TALLY C1
```
Answers will vary.
  c.
```
MTB  > RANDOM 600 C1;
SUBC > INTEGER 1 TO 6.
MTB  > PRINT C1
MTB  > TALLY C1
```
Answers will vary.
  d.
```
MTB  > RANDOM 6000 C1;
SUBC > INTEGER 1 TO 6.
MTB  > PRINT C1
MTB  > TALLY C1
```
Answers will vary.
  e. Yes
3. Yes, as the observed frequencies do not seem to correspond to theoretical frequencies.

Testing Your Understanding of This Chapter's Concepts (pages 266–267)

1. $\dfrac{1}{120}$ 2. 100,000,000
3. It should really be called a permutation lock.

4. $2^6(6 - 1)! = 64 \cdot 120 = 7680$, as the wife can sit on the left side or the right side of her husband.

5. $10 \cdot 10 \cdot 10 \cdot 10 \cdot 10 \cdot 10 \cdot 10 \cdot 10 \cdot 10 = 1,000,000,000$ possible numbers. If a tenth digit is added, there will be $10 \cdot 10 \cdot 10 \cdot 10 \cdot 10 \cdot 10 \cdot 10 \cdot 10 \cdot 10 \cdot 10 = 10,000,000,000$ possible numbers.

6. a. $40 \cdot 40 \cdot 40 \cdot 40 \cdot 40 \cdot 40 \cdot 40 \cdot 40 \cdot 40 \cdot 40 = 40^{10}$, assuming repetition of numbers is allowed. If repetition is not allowed, then there are $40 \cdot 39 \cdot 38 \cdot 37 \cdot 36 \cdot 35 \cdot 34 \cdot 33 \cdot 32 \cdot 31$ possible lottery tickets.

   b. $1:(40^{10} - 1)$, assuming repetition of numbers is allowed.

7. $\dfrac{7!}{3!2!2!} = 210$

Chapter Test (pages 267–272)

1. Choice (b)   2. Choice (d)   3. Choice (c)   4. Choice (a)   5. Choice (a)
6. Choice (d)   7. Choice (b)   8. Choice (c)   9. Choice (a)   10. Choice (b)
11. 120   12. 120   13. 27,000   14. 29:71
15. a. $-\$0.66$ b. 49,999:1   c. 49,994:6 or 24,997:3
16. a. 10,000,000,000   b. $\dfrac{1}{10,000,000,000}$

17. 80

18.

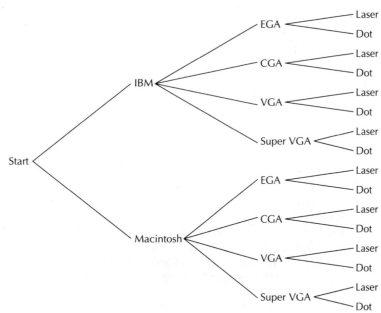

16 possible computer systems

19. a. $\dfrac{13!}{2!2!2!2!} = 389,188,800$ b. $\dfrac{9!}{2!2!} = 90,720$ c. $\dfrac{11!}{3!} = 6,652,800$

d. $\dfrac{12!}{2!2!2!2!} = 29{,}937{,}600$

20. a. $\dfrac{600}{1000} = 0.6$ b. $\dfrac{479}{1000}$ c. $\dfrac{293}{1000}$ d. $\dfrac{521}{1000}$

21. 20 to 80 or 1:4   22. 180   23. $\dfrac{1}{4}$ 24. 6,760,000   25. 60   26. $_{10}C_3 = 120$

27. $_{11}P_4 = 7920$ 28. $\dfrac{7!}{4!3!} = 35$ 29. $\dfrac{3}{5}$

30. Four letters followed by three numbers yields $26 \cdot 26 \cdot 26 \cdot 26 \cdot 10 \cdot 10 \cdot 10 = 456{,}976{,}000$ possible codes. Four numbers followed by three letters yields $10 \cdot 10 \cdot 10 \cdot 10 \cdot 26 \cdot 26 \cdot 26 = 175{,}760{,}000$ possible codes. Use four letters followed by three numbers.

31. $\dfrac{1}{6}$

32.

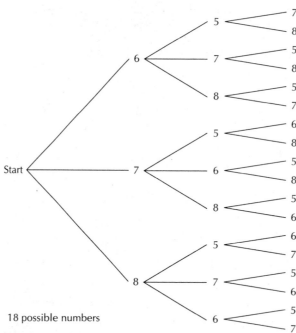

18 possible numbers

33. 85:15 or 17:3

34. a. $10 \cdot 10 \cdot 10 = 1000$. No, as there are 17,225 students.
     b. $26 \cdot 26 \cdot 26 = 17{,}576$. Yes, as there are only 17,225 students.

35. a. 1,000,000,000 possible numbers   b. 10,000,000,000 possible numbers
     c. 26,000,000,000 possible numbers

36. 5184 possible codes

37. a. 3:3 or 1:1   b. $2.67

Thinking Critically (pages 272–274)

1. No

2. a. $\dfrac{1}{8}$ b. $\dfrac{1}{4}$ c. $\dfrac{5}{16}$ d. $\dfrac{5}{16}$

3. Yes. Since the sample space has been reduced, the player should switch. Others may disagree.

4. a. ${}_nP_1 = \dfrac{n!}{(n-1)!}$ ${}_nC_1 = \dfrac{n!}{1!(n-1)!}$ They are equal.

   b. ${}_nP_n = n!$ ${}_nP_{n-1} = \dfrac{n!}{[n-(n-1)]!} = n!$ They are equal.

6. $$\binom{n+1}{r} = \binom{n}{r} + \binom{n}{r-1}$$

   $$\frac{(n+1)!}{r!(n+1-r)!} = \frac{n!}{r!(n-r)!} + \frac{n!}{(r-1)!(n-r+1)!}$$

   $$= \frac{(n-r+1)n! + n!(r)}{r!(n-r+1)!} = \frac{(n+1)!}{r!(n-r+1)!}$$

7. $$\binom{n}{r} = \binom{n}{n-r}$$

   $$\frac{n!}{r!(n-r)!} = \frac{n!}{(n-r)![n-(n-r)]!}$$

   $$= \frac{n!}{(n-r)!r!}$$

   $$= \frac{n!}{r!(n-r)!}$$

   If we choose any positive integer values for n and r, then the above results can be verified by using Pascal's triangle.

8. For row 0, the only entry is 1. Sum is $2^0 = 1$.
   For row 1, the entries are 1 and 1. Sum is $2^1 = 2$.
   For row 2, the entries are 1, 2, and 1. Sum is $2^2 = 4$.
   For row 3, the entries are 1, 3, 3, and 1. Sum is $2^3 = 8$.
   For row 4, the entries are 1, 4, 6, 4, and 1. Sum is $2^4 = 16$.
   For row 5, the entries are 1, 5, 10, 10, 5, and 1. Sum is $2^5 = 32$.
   etc. . . .

9. No, the events do not have the same probability.   10. 2^{70} possible coding schemes

Case Study (pages 274–275)

1. a. The probability of winning $= \dfrac{1}{24{,}975}$, assuming there are no duplications and that only one Dodge Caravan was awarded.

   b. 24,975 to 1

2. a. $\dfrac{66}{100} = 0.66$ b. $\dfrac{2}{100} = 0.02$

CHAPTER 5

Section 5.2 (pages 289–292)

1. a. Not mutually exclusive b. Mutually exclusive
   c. Mutually exclusive (assuming that a person buys only one ticket)
   d. Not mutually exclusive e. Mutually exclusive f. Mutually exclusive
   g. Not mutually exclusive h. Mutually exclusive
3. 0.96 5. 0.62 7. 0.14 9. 0.26 11. 0.55 13. 1.00

Section 5.3 (pages 298–301)

1. a. $\dfrac{281}{433}$ b. $\dfrac{59}{281}$ c. $\dfrac{59}{96}$

3. $\dfrac{35}{86}$ 5. $\dfrac{23}{77}$ 7. 0.0171 9. 0.66

11. a. $\dfrac{41}{137}$ b. $\dfrac{41}{92}$ c. $\dfrac{274}{819}$

Section 5.4 (pages 305–308)

1. 0.0507 3. 0.7885 5. 0.6603 7. 0.2270 9. 0.0026 11. 0.3144 13. 0.2538

Section 5.5 (pages 316–318)

1. $\dfrac{2}{19}$ 3. $\dfrac{1}{4}$ 5. $\dfrac{40}{103}$ 7. $\dfrac{115}{214}$ 9. $\dfrac{11}{29}$ 11. $\dfrac{99}{167}$

Section 5.6 (page 319)

```
1. MTB  < READ C1   C2
   DATA < 1         0.35
   DATA < 2         0.25
   Data < 3         0.22
   DATA < 4         0.18
   DATA < END

        4 ROWS READ

   MTB  < RANDOM 100 VALUES AND PLACE IN C3;
   SUBC < DISCRETE SAMPLE VALUES FROM C1 WITH PROBABILITIES IN C2.
   MTB  < HISTOGRAM OF C3;
   SUBC < START 1;
   SUBC < INCREMENT 1.

   HISTOGRAM OF C3  N = 100
```

In this case, 1 represents Math, 2 represents Physics, 3 represents Chemistry, and 4 represents Biology. The actual histogram will vary as each computer is generating numbers randomly.

```
3. MTB  < READ C1   C2
   DATA < 1          0.45
   DATA < 2          0.35
   DATA < 3          0.20
   DATA < END

        4 ROWS READ

   MTB  < RANDOM 100 VALUES AND PLACE IN C3;
   SUBC < DISCRETE SAMPLE VALUES FROM C1 WITH PROBABILITIES IN C2.
   MTB  < HISTOGRAM OF C3;
   SUBC < START 1;
   SUBC < INCREMENT 1.

   HISTOGRAM OF C3  N = 100
```

Testing Your Understanding of This Chapter's Concepts (pages 321–322)

1. Mutually exclusive events are two events that cannot occur at the same time. Independent events are two events where the likelihood of the occurrence of one event is in no way affected by the occurrence or nonoccurrence of the other event.
2. False. Let A = event of selecting a picture card from a deck of cards and let B = the event of selecting an ace from a deck of cards. Then $p(A) = \dfrac{12}{52}$ and $p(B) = \dfrac{4}{52}$. Events A and B are mutually exclusive, but $p(A) + p(B) \neq 1$
3. No. Use same illustration as in the previous question.
4. No. Events are not necessarily independent.
5. No   6. No   7. No

Chapter Test (pages 322–326)

1. Choice (b)   2. Choice (d)   3. Choice (a)   4. Choice (c)   5. Choice (d)   6. Choice (a)
7. Choice (e)   8. Choices (a) and (b)
9. 0.3672   10. $\dfrac{52}{83}$ 11. 0.0359   12. $\dfrac{192}{317}$ 13. $\dfrac{3}{68}$ 14. $\dfrac{89}{359}$ 15. 0.0001   16. $\dfrac{7}{12}$ 17. $\dfrac{9}{26}$
18. 9% are involved in Teaching and Consultation.
19. a. $\dfrac{4}{5}$ b. $\dfrac{20}{99}$ c. $\dfrac{316}{495}$
20. 0.6750   21. 0.57   22. 0.90   23. 0.485
24. a. 0.1808   b. 0.0224
25. a. $\dfrac{97}{289}$ b. $\dfrac{2208}{2211}$
26. $(0.50)^{50}$ or $\left(\dfrac{1}{2}\right)^{50}$

Thinking Critically (pages 326–327)

1. $\dfrac{364}{365}$, assuming no leap year

2. $p(\text{at least two have same birthday}) = 1 - p(\text{all birthdays are different})$
$$= 1 - 0.9918$$
$$= 0.0082$$

3. $p(\text{all four are different}) = p(\text{second different from first}) \cdot p(\text{third different from first 2} \mid \text{first and second are different}) \cdot (\text{fourth different from first 3} \mid \text{first, second, and third are different})$
$$= \left(\frac{364}{365}\right)\left(\frac{363}{365}\right)\left(\frac{362}{365}\right) = 0.9836, \text{ assuming no leap year.}$$

4. $\dfrac{31}{365}$, assuming no leap year   5. $\dfrac{1}{5}$ 6. $\dfrac{57}{63}$

CHAPTER 6

Section 6.2 (pages 339–342)

1. a. 0, 1, 2, . . .   b. 0, 1, 2, . . .   c. 0, 1, 2, . . .   d. Any nonnegative real number
   e. Any real number   f. 0, 1, 2, . . .
   g. 0, 1, 2, ... (It actually depends upon the Arab country.)

3. 0, 1, 2, 3, 4, 5, 6, 7, 8, 9, 10   5. 0.20

7.

Number of Bull's-Eye Hits, x	$p(x)$
0	$\dfrac{1}{16}$
1	$\dfrac{4}{16}$
2	$\dfrac{6}{16}$
3	$\dfrac{4}{16}$
4	$\dfrac{1}{16}$
	$\dfrac{16}{16} = 1$

9.

Number of Drivers Checking Oil, x	$p(x)$
0	$(0.94)(0.94)(0.94) = 0.8306$
1	$3(0.06)(0.94)(0.94) = 0.1590$
2	$3(0.06)(0.06)(0.94) = 0.0102$
3	$(0.06)(0.06)(0.06) = \underline{0.0002}$
	1.0000

11.

Number of Parents, x	$p(x)$
0	$(0.70)(0.70)(0.70) = 0.3430$
1	$3(0.30)(0.70)(0.70) = 0.4410$
2	$3(0.30)(0.30)(0.70) = 0.1890$
3	$(0.30)(0.30)(0.30) = \underline{0.0270}$
	1.0000

13. a.

Number of Boys in Family, x	$p(x)$
0	$\dfrac{1}{16}$
1	$\dfrac{4}{16}$
2	$\dfrac{6}{16}$
3	$\dfrac{4}{16}$
4	$\dfrac{1}{16}$
	$\dfrac{16}{16} = 1$

b. $p(x$ has a value of at least $1) = \dfrac{15}{16}$

15.

x	$p(x)$
0	$\dfrac{7}{28}$
1	$\dfrac{6}{28}$
2	$\dfrac{5}{28}$
3	$\dfrac{4}{28}$
4	$\dfrac{3}{28}$
5	$\dfrac{2}{28}$
6	$\dfrac{1}{28}$
7	0
	$\dfrac{28}{28} = 1$

Since the sum of the probabilities is 1, this represents the probability function of some random variable x.

17. a. $p(x = 4) = 0.38$ b. $p(x = 7) = 0$ c. $p(x \leq 3) = 0.36$ d. $p(x \leq 6) = 1$
    e. $p(x \leq 4 \text{ or } x > 5) = 0.83$
19. a. $p(x \leq 3) = 0.62$ b. $p(x > 4) = 0.15$ c. $p(x < 7) = 1$ d. $p(1 \leq x \leq 6) = 0.94$

Section 6.4 (pages 349–354)

1. Mean = $\mu = 38.7$
   Variance = $\sigma^2 = 209.31$
   Standard deviation = $\sigma = \sqrt{209.31} \approx 14.4675$
3. Mean = $\mu = 1.97$
   Variance = $\sigma^2 = 1.5491$
   Standard deviation = $\sigma = \sqrt{1.5491} \approx 1.2446$
5. Mean = $\mu = 597$
   Variance = $\sigma^2 = 29491$
   Standard deviation = $\sigma = \sqrt{29491} \approx 171.7294$
7. Mean = $\mu = 1.99$
   Variance = $\sigma^2 = 1.6699$
   Standard deviation = $\sigma = \sqrt{1.6699} \approx 1.2922$
9. Mean = $\mu = 4.45$
   Variance = $\sigma^2 = 2.6275$
   Standard deviation = $\sigma = \sqrt{2.6275} \approx 1.6210$
11. a. Mean = $\mu = 4.615$ b. Variance = $\sigma^2 = 8.3488$
    c. Standard deviation = $\sigma = \sqrt{8.3488} \approx 2.8894$
13. a. Expected income = $12,770
    b. Yes. The expected income minus the cost for rain insurance is still a positive number.

Section 6.5 (pages 368–370)

1. 0.2765   3. 0.0394
5. a. 0.2734   b. $0.0078 + 0.0547 + 0.1641 + 0.2734 = 0.5000$
   c. $0.2734 + 0.2734 + 0.1641 + 0.0547 + 0.0078 = 0.7734$
7. 0.9780
9. a. 0.0404   b. 0.0735
11. a. 0.2903   b. 0.0280   c. 0.0016
13. 0.2397   15. 0.9929

Section 6.6 (pages 373–375)

1. $\mu = 270$, $\sigma \approx 15.6748$ 3. $\mu = 152.46$. Yes, since this exceeds 151.
5. $\mu = 1797.40$, $\sigma \approx 33.3825$
7. $\mu = 2420$
   2420 changes @ $35 charge per change will yield $2420(35) = \$84,700$.
   $\sigma = \sqrt{22000(0.11)(0.89)} \approx 46.4091$ $(46.4091)(35) = \$1624.32$
9. $\mu = 1584$, $\sigma \approx 13.7870$ 11. $\mu = 270$, $\sigma \approx 16.0156$

Section 6.7 (pages 377–379)

 1. 0.2381
 3. a. 0.1246   b. 0.9596   c. 0.0842
 5. 0.4422
 7. a. 0.3733   b. 0.1600
 9. 0.2942
11. a. 0.8754   b. 0.2650   c. 0.1404   d. 0.1404

Section 6.8 (pages 382–384)

1. a. 0.2919   b. 0.8295

3. $\dfrac{\binom{6}{5}\binom{6}{0}}{\binom{12}{5}} = \dfrac{1}{132}$ 5. $\dfrac{\binom{15}{2}\binom{10}{3}}{\binom{25}{5}} = \dfrac{1260}{5313}$

7. a. $\dfrac{\binom{5}{0}\binom{8}{4}}{\binom{13}{4}} = \dfrac{14}{143}$ b. $\dfrac{\binom{5}{1}\binom{8}{3}}{\binom{13}{4}} = \dfrac{56}{143}$ c. $\dfrac{\binom{5}{2}\binom{8}{2}}{\binom{13}{4}} = \dfrac{56}{143}$

   d. $\dfrac{\binom{5}{3}\binom{8}{1}}{\binom{13}{4}} = \dfrac{16}{143}$ e. $\dfrac{\binom{5}{4}\binom{8}{0}}{\binom{13}{4}} = \dfrac{1}{143}$

9. $\dfrac{\binom{470}{25}\binom{30}{5}}{\binom{500}{30}} = 0.0211$

Section 6.9 (pages 386–387)

1.
```
MTB  > PDF;
SUBC > BINOMIAL WITH N = 10 p = 0.60.

BINOMIAL WITH N = 10, p = 0.600000
                x             P(X = x)
                0               0.0001
                1               0.0016
                2               0.0106
                3               0.0425
                4               0.1115
                5               0.2007
                6               0.2508
                7               0.2150
                8               0.1209
                9               0.0403
               10               0.0060
```

3. a. 0.0105 b. 0.9884 c. 0.4002

5. **MTB > PDF;**

   SUBC > POISSON 11.4.

   POISSON WITH MU = 11.4000

x	$P(X = x)$
0	0.0000
1	0.0001
2	0.0007
3	0.0028
4	0.0079
5	0.0180
6	0.0341
7	0.0556
8	0.0792
9	0.1003
10	0.1144
11	0.1185
12	0.1126
13	0.0987
14	0.0804
15	0.0611
16	0.0435
17	0.0292
18	0.0185
19	0.0111
20	0.0063
21	0.0034
22	0.0018
23	0.0009
24	0.0004
25	0.0002
26	0.0001
27	0.0000

7. **MTB > PDF;**

   SUBC > POISSON 6.4.

   POISSON WITH MU = 6.4000

x	$P(X = x)$
1	0.0017
2	0.0106
3	0.0340
4	0.0726
5	0.1162
6	0.1487
7	0.1586
8	0.1450
9	0.1160
10	0.0825

7. (*continued*)

```
POISSON WITH MU = 6.4000
      x        P(X = x)
     11          0.0528
     12          0.0307
     13          0.0164
     14          0.0081
     15          0.0037
     16          0.0016
     17          0.0006
     18          0.0002
     19          0.0000
```

Testing Your Understanding of This Chapter's Concepts (pages 391–392)

1. a.

x	$p(x)$	$x \cdot p(x)$	x^2	$x^2 \cdot p(x)$
4	$\dfrac{4}{36}$	$\dfrac{16}{36}$	16	$\dfrac{64}{36}$
5	$\dfrac{4}{36}$	$\dfrac{20}{36}$	25	$\dfrac{100}{36}$
6	$\dfrac{5}{36}$	$\dfrac{30}{36}$	36	$\dfrac{180}{36}$
7	$\dfrac{6}{36}$	$\dfrac{42}{36}$	49	$\dfrac{294}{36}$
8	$\dfrac{7}{36}$	$\dfrac{56}{36}$	64	$\dfrac{448}{36}$
9	$\dfrac{4}{36}$	$\dfrac{36}{36}$	81	$\dfrac{324}{36}$
10	$\dfrac{3}{36}$	$\dfrac{30}{36}$	100	$\dfrac{300}{36}$
11	$\dfrac{2}{36}$	$\dfrac{22}{36}$	121	$\dfrac{242}{36}$
12	$\dfrac{1}{36}$	$\dfrac{12}{36}$	144	$\dfrac{144}{36}$
		$\dfrac{264}{36}$		$\dfrac{2096}{36}$

b. $\mu = \dfrac{22}{3}$, $\sigma^2 = 4.4444$, $\sigma = \sqrt{4.4444} \approx 2.108$

2. 0.9287

3.

x	$p(x)$	x^2	$x \cdot p(x)$	$x^2 \cdot p(x)$
0	$\dfrac{9}{45}$	0	0	0
1	$\dfrac{8}{45}$	1	$\dfrac{8}{45}$	$\dfrac{8}{45}$
2	$\dfrac{7}{45}$	4	$\dfrac{14}{45}$	$\dfrac{28}{45}$
3	$\dfrac{6}{45}$	9	$\dfrac{18}{45}$	$\dfrac{54}{45}$
4	$\dfrac{5}{45}$	16	$\dfrac{20}{45}$	$\dfrac{80}{45}$
5	$\dfrac{4}{45}$	25	$\dfrac{20}{45}$	$\dfrac{100}{45}$
6	$\dfrac{3}{45}$	36	$\dfrac{18}{45}$	$\dfrac{108}{45}$
7	$\dfrac{2}{45}$	49	$\dfrac{14}{45}$	$\dfrac{98}{45}$
8	$\dfrac{1}{45}$	64	$\dfrac{8}{45}$	$\dfrac{64}{45}$
			$\dfrac{120}{45}$	$\dfrac{540}{45} = 12$

$$\mu = \frac{8}{3}, \; \sigma = \sqrt{4.8889} \approx 2.21108$$

4. a.

x	$p(x)$	$x \cdot p(x)$	x^2	$x^2 \cdot p(x)$
3	$\dfrac{35}{210}$	$\dfrac{105}{210}$	9	$\dfrac{315}{210}$
4	$\dfrac{105}{210}$	$\dfrac{420}{210}$	16	$\dfrac{1680}{210}$
5	$\dfrac{63}{210}$	$\dfrac{315}{210}$	25	$\dfrac{1575}{210}$
6	$\dfrac{7}{210}$	$\dfrac{42}{210}$	36	$\dfrac{252}{210}$
		$\dfrac{882}{210} = 4.2$		$\dfrac{3822}{210} = 18.2$

b. $\mu = 4.2$, $\sigma^2 = 0.56$, $\sigma = \sqrt{0.56} \approx 0.7483$

c. The interval $\mu - 2\sigma = 4.2 - 2(0.7483) = 2.7034$.
   The interval $\mu + 2\sigma = 4.2 + 2(0.7483) = 5.6966$.
   In this case, $k = 2$. Using Chebyshev's Theorem, the proportion of terms that lie within

2 standard deviations of the mean is at least $1 - \dfrac{1}{2^2} = 1 - \dfrac{1}{4} = \dfrac{3}{4}$. Therefore, the probability that x will fall within the interval 2.7034 to 5.6966 is at least 0.7500.

5.

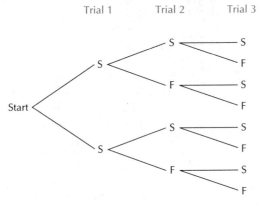

Where S = Success and F = Failure

6. a.

x(sum)	p	$x \cdot p(x)$	x^2	$x^2 \cdot p(x)$
2¢	$\left(\dfrac{5}{15}\right)\left(\dfrac{5}{15}\right) = \dfrac{25}{225}$	$\dfrac{50}{225}$	4	$\dfrac{100}{225}$
6	$2\left(\dfrac{5}{15}\right)\left(\dfrac{3}{15}\right) = \dfrac{30}{225}$	$\dfrac{180}{225}$	36	$\dfrac{1080}{225}$
10	$\left(\dfrac{3}{15}\right)\left(\dfrac{3}{15}\right) = \dfrac{9}{225}$	$\dfrac{90}{225}$	100	$\dfrac{900}{225}$
11	$2\left(\dfrac{5}{15}\right)\left(\dfrac{4}{15}\right) = \dfrac{40}{225}$	$\dfrac{440}{225}$	121	$\dfrac{4840}{225}$
15	$2\left(\dfrac{3}{15}\right)\left(\dfrac{4}{15}\right) = \dfrac{24}{225}$	$\dfrac{360}{225}$	225	$\dfrac{5400}{225}$
20	$\left(\dfrac{4}{15}\right)\left(\dfrac{4}{15}\right) = \dfrac{16}{225}$	$\dfrac{320}{225}$	400	$\dfrac{6400}{225}$
26	$2\left(\dfrac{5}{15}\right)\left(\dfrac{2}{15}\right) = \dfrac{20}{225}$	$\dfrac{520}{225}$	676	$\dfrac{13520}{225}$
30	$2\left(\dfrac{3}{15}\right)\left(\dfrac{2}{15}\right) = \dfrac{12}{225}$	$\dfrac{360}{225}$	900	$\dfrac{10800}{225}$
35	$2\left(\dfrac{4}{15}\right)\left(\dfrac{2}{15}\right) = \dfrac{16}{225}$	$\dfrac{560}{225}$	1225	$\dfrac{19600}{225}$
50	$\left(\dfrac{2}{15}\right)\left(\dfrac{2}{15}\right) = \dfrac{4}{225}$	$\dfrac{200}{225}$	2500	$\dfrac{10000}{225}$
51	$2\left(\dfrac{1}{15}\right)\left(\dfrac{5}{15}\right) = \dfrac{10}{225}$	$\dfrac{510}{225}$	2601	$\dfrac{26010}{225}$

continued

6. a. (*continued*)

x(sum)	p	$x \cdot p(x)$	x^2	$x^2 \cdot p(x)$
55	$2\left(\dfrac{3}{15}\right)\left(\dfrac{1}{15}\right) = \dfrac{6}{225}$	$\dfrac{330}{225}$	3025	$\dfrac{18150}{225}$
60	$2\left(\dfrac{1}{15}\right)\left(\dfrac{4}{15}\right) = \dfrac{8}{225}$	$\dfrac{480}{225}$	3600	$\dfrac{28800}{225}$
75	$2\left(\dfrac{1}{15}\right)\left(\dfrac{2}{15}\right) = \dfrac{4}{225}$	$\dfrac{300}{225}$	5625	$\dfrac{22500}{225}$
100	$\left(\dfrac{1}{15}\right)\left(\dfrac{1}{15}\right) = \dfrac{1}{225}$	$\dfrac{100}{225}$	10,000	$\dfrac{10000}{225}$
		$\dfrac{4800}{225} = \dfrac{192}{9}$		$\dfrac{178100}{225}$

b. $\mu = \dfrac{192}{9}$ c. Standard deviation $\sigma = \sqrt{336.4444} \approx 18.3424$

Chapter Test (pages 392–396)

1. Choice (d)   2. Choice (b)   3. Choice (b)   4. Choice (b)   5. Choice (d)
6. Choice (d)   7. Choice (b)   8. Choice (c)
9. 0.0011   10. 0.1404   11. 0.2075   12. 0.0154   13. 0.3006   14. 0.9807   15. 0.0316
16. $\mu = 88$, $\sigma = \sqrt{68.64} \approx 8.2849$ 17. 0.5921
18. The expected number of people who will show up = $220(0.91) = 200.2$. Since the airline has 210 seats, the airline will have a seat for each passenger who has a reservation and who shows up, assuming that the past show rate of 91% is applicable for future flights.
19. $\Sigma \dfrac{e^{-6}6^x}{x!} \geq 0.90$. By trial and error, the owner should have 9 or more on hand.
20. Approximately 0   21. 0.9437
22. a. 0.0286   b. 0.1912
23. $\mu = 954$, $\sigma = \sqrt{448.38} \approx 21.1750$ 24. 0.1981   25. $\mu = 20$, $\sigma = \sqrt{15} \approx 3.8730$
26. $\mu = 304$, $\sigma = \sqrt{188.48} \approx 13.7288$
27.

x	$p(x)$	$x \cdot p(x)$	x^2	$x^2 \cdot p(x)$
4	0.05	0.20	16	0.80
5	0.43	2.15	25	10.75
6	0.25	1.50	36	9.00
7	0.17	1.19	49	8.33
8	0.06	0.48	64	3.84
9	0.03	0.27	81	2.43
10	0.01	0.10	100	1.00
		5.89		36.15

$\mu = 5.89$, $\sigma = \sqrt{1.4579} \approx 1.2074$

Thinking Critically (pages 397–398)

1.

x	$p(x)$
0	$\left(\dfrac{3}{8}\right)\left(\dfrac{3}{8}\right) = \dfrac{9}{64}$
1	$2\left(\dfrac{5}{8}\right)\left(\dfrac{3}{8}\right) = \dfrac{30}{64}$
2	$\left(\dfrac{5}{8}\right)\left(\dfrac{5}{8}\right) = \dfrac{25}{64}$

2.

x	$p(x)$
0	$\left(\dfrac{11}{18}\right)\left(\dfrac{11}{18}\right)\left(\dfrac{11}{18}\right) = \dfrac{1331}{5832}$
1	$3\left(\dfrac{7}{18}\right)\left(\dfrac{11}{18}\right)\left(\dfrac{11}{18}\right) = \dfrac{2541}{5832}$
2	$3\left(\dfrac{7}{18}\right)\left(\dfrac{7}{18}\right)\left(\dfrac{11}{18}\right) = \dfrac{1617}{5832}$
3	$\left(\dfrac{7}{18}\right)\left(\dfrac{7}{18}\right)\left(\dfrac{7}{18}\right) = \dfrac{343}{5832}$

3. $(p + q) = p^3 + 3p^2q + 3pq^2 + q^3$

The first term, p^3, is the probability of successes on all three trials when a binomial experiment is performed three times. The second term, $3p^2q$, is the probability of two successes in the three trials. The third term, $3pq^2$, is the probability of one success in the three trials, and the last term, q^3, is the probability of no successes in the three trials.

4. a.

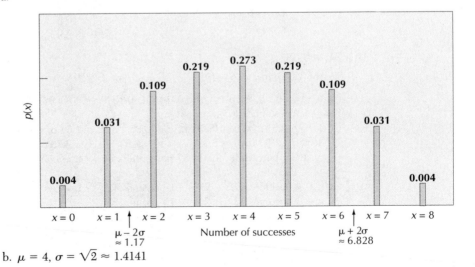

b. $\mu = 4,\ \sigma = \sqrt{2} \approx 1.4141$

5.

$$\sigma = \Sigma(x - \mu)^2 \cdot p(x)$$
$$= \Sigma(x^2 - 2\mu x + \mu^2) \cdot p(x)$$
$$= \Sigma x^2 \cdot p(x) - 2\mu \, \Sigma x \cdot p(x) + \mu^2 \, \Sigma p(x) \quad \text{Note: } \mu = \Sigma x \cdot p(x) \text{ and } \Sigma p(x) = 1.$$
$$= \Sigma x^2 \cdot p(x) - 2\mu\mu + \mu^2$$
$$= \Sigma x^2 \cdot p(x) - \mu^2$$
$$= \Sigma x^2 \cdot p(x) - [\Sigma x \cdot p(x)]^2$$

6. a.

x	$p(x)$	$x \cdot p(x)$	x^2	$x^2 \cdot p(x)$
0	0.03	0.00	0	0.00
1	0.62	0.62	1	0.62
2	0.19	0.38	4	0.76
3	0.09	0.27	9	0.81
4	0.07	0.28	16	1.12
		1.55		3.31

$\mu = 1.55$, $\sigma = \sqrt{0.9075} \approx 0.9526$

b. Using Chebyshev's theorem, the probability that x will fall within $k = 2$ standard deviations of the mean is at least $1 - \dfrac{1}{2^2} = 0.75$. We can also calculate this probability directly. Two standard deviations of the mean represents the interval $1.55 \pm 2(0.9526)$, or between -0.3552 and 3.4552. In our case, this represents

$$p(x = 0) + p(x = 1) + p(x = 2) + p(x = 3) = 0.03 + 0.62 + 0.19 + 0.09 = 0.93$$

7. We start with the basic formula $\sigma = \sqrt{\Sigma x^2 \cdot p(x) - [\Sigma x \cdot p(x)]^2}$.

Case Study (pages 398–399)

1. 0.9977 2. 0.9197

CHAPTER 7

Section 7.3 (pages 416–419)

1. a. 0.4656 b. 0.1628 c. 0.0314 d. 0.1949 e. 0.0084 f. 0.9194
   g. 0.9593 h. 0.2652
3. a. $z = 0.58$ b. $z = 0.95$ c. $z = 0.39$ d. $z = 1.64$ or $z = 1.65$
5. a. $z = 2.07$ b. $z = 2.18$ c. $z = -0.83$ d. $z = 2.68$ e. $z = 2.27$ f. $z = 2.93$
7. a. 12.71 percentile b. 71.57 percentile c. 3.14 percentile
9. $\sigma = 15.3846$
11. $z = 0$ becomes 47. $z = -1$ becomes 41.
    $z = 1$ becomes 53. $z = -2$ becomes 35.
    $z = 2$ becomes 59. $z = -3$ becomes 29.
    $z = 3$ becomes 65.

13. Area of region I is 0.0215.
    Area of region II is 0.1359.
    Area of region III is 0.3413.
    Area of region V is 0.1359.
    Area of region VI is 0.0215.

Section 7.4 (pages 422–424)

1. 0.0668 3. 0.2946 5. 0.2206
7. a. 0.0228 b. 0.5155
9. 3.1905 years 11. $\mu = 84.3232$ 13. $x = 206$ minutes
15. a. 0.5934 b. 0.6628 c. 0.0062
17. $\sigma = 9.6154$

Section 7.5 (pages 431–434)

1. 0.0207 3. 0.9998 5. 0.5714 7. 0.3936 9. 0.0159 11. 0.5557 13. 0.0239
15. 0.9713 17. 0.4761 19. 0.0066

Section 7.7 (pages 437–438)

1. a. 0.022750 b. 0.158655 c. 0.839995 d. 0.9759
3.
```
MTB  > SET C1
DATA > 45.5   100
DATA > END
MTB  > CDF C1;
SUBC > NORMAL MU = 50   SIGMA = 5.

NORMAL WITH MEAN = 50.0000 AND STANDARD DEVIATION = 5.00000

        x         P(X <= x)
  45.5000           0.1841
 100.0000           1.0000

MTB  > SET C2
DATA > 59.5   100
DATA > END
MTB  > CDF C2;
SUBC > NORMAL MU = 50   SIGMA = 5.

NORMAL WITH MEAN = 50.0000 AND STANDARD DEVIATION = 5.00000

        x         P(X <= x)
  59.5000           0.9713
 100.0000           1.0000
```

Thus, the probability that there are between 45 and 60 boys is $0.9713 - 0.1841 = 0.7872$

Testing Your Understanding of This Chapter's Concepts (page 441)

1. The one with mean $\mu = 2$ and standard deviation $\sigma = 4$.
2. $x = 12.9 + (-0.67)(2.9) = 10.957$
   $x = 12.9 + 0(2.9) = 12.9$
   $x = 12.9 + (0.67)(2.9) = 14.843$
3. $x = Q_3 = 40 + (0.67)5 = 43.35$
4. $z = +1.50$ gives a probability of 0.4332.
   $z = -1.50$ gives a probability of 0.4332.

   Adding gives $0.4332 + 0.4332$ or 0.8664 as the probability that the random variable z will fall within 1.5 standard deviations of the mean.

   Using Chebyshev's theorem, we get $1 - \dfrac{1}{1.5^2} = 0.5556$ as the minimum probability that the random variable z will fall within 1.5 standard deviations of the mean.

5. For all parts the probability is $0.3888 + 0.4788 = 0.8676$ because for a continuous distribution the probability that z equals a particular value is 0.

6. a.

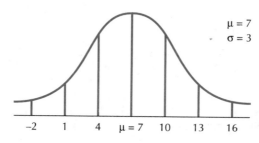

$\mu = 7$
$\sigma = 3$

b.

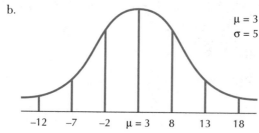

$\mu = 3$
$\sigma = 5$

c.

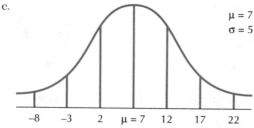

$\mu = 7$
$\sigma = 5$

7. No, as $np = 6\left(\dfrac{1}{2}\right)$ is not greater than 5.

8. Yes, the mean has no effect on shape. Different values of σ will make the curve wider or narrower; μ merely locates the center of the distribution.

Chapter Test (pages 441–446)

1. Choice (d)   2. Choice (b)   3. Choice (d)   4. Choice (a)   5. Choice (c)
6. Choice (b)   7. Choice (c)
8. 0.0784   9. 0.6340   10. 82.89 percentile   11. 0.0398   12. 0.0409   13. 0.1736
14. 0.3121   15. 0.7842   16. 0.7549   17. 0.0116   18. 0.1357   19. 0.9332   20. 0.3015
21. 0.4013   22. 0.4373   23. 0.2207   24. 0.7704   25. 0.2119
26. a. 0.4207   b. 0.3557   c. 0.1112   d. 0.5186
27. 0.3594

Thinking Critically (page 446)

1. $29 = \mu - 1.00\sigma$
   $67 = \mu + 2.00\sigma$

Solving simultaneously for μ and σ gives $\mu = 41.6667$ and $\sigma = 12.6667$.
2. $31.25 = \sigma$ 3. $k \geq 18.2574$
4. $15.2152 = \mu - (0.89)\sigma$
   $15.948 = \mu + 1.40\sigma$

Solving simultaneously gives $\mu = 15.5$ and $\sigma = 0.32$.
5. When $n = 12$ and $p = 0.05$, we can use the binomial distribution formula.

$$p(\text{at most 1 success}) = p(0 \text{ successes}) + p(1 \text{ success})$$
$$= 0.54036 + 0.34128 = 0.88164$$

Using normal approximation, we have

$$\mu = np\,12(0.05) = 0.6 \quad z = \frac{1.5 - 0.6}{0.7550} = 1.19 \quad \begin{array}{r} 0.5000 \\ +0.3830 \\ \hline 0.8830 \end{array}$$

$$\sigma = \sqrt{12(0.05)(0.95)} \approx 0.7550$$

The answers are slightly different since np and nq are not greater than 5.
6. $z = \dfrac{3}{4}$

Case Study (page 447)

1. 0.0037

CHAPTER 8

Section 8.2 (pages 456–458)

1. Those requests whose numbers are 691, 279, 151, 394, 604, 186, 711, 577, 388, 568, 186, 363, 676, 475, 607, 553, 185, 458, 609, and 665.
3. Those bonds whose numbers are 07856, 06121, 09172, 04024, 02304, 06138, 04146, 06691, 00256, 04542, 06115, 09922, 05859, 04734, and 08625.
5. Those customers whose numbers are 14194, 53402, 24830, 53537, 81305, 70659, 18738, 56869, 84378, 62300, 05859, 72695, 17617, 81056, 92144, 44819, 29852, 13602, 04734, 26384, 28728, 15398, 61280, 14778, 81536, 61362, 63904, 22209, 36086, 08625, 82271, 35797, 20542, 58727, and 25417.

7. Those returns whose numbers are 1048, 2236, 2413, 4216, 3757, 7792, 9956, 9630, 8957, 8547, 2891, 6355, 0942, 1036, 0711, 5108, 0236, 0101, 5216, 0705, 4866, 5416, 3263, 2933, 0248, 8152, 2967, 0074, 0536, and 9192.

9. Those volunteers whose numbers are 0201, 1665, 0797, 1028, 0342, 0817, 0999, 1434, 2420, 0735, 2643, 2642, 1290, 3013, 0739, 2999, 3192, 2538, 3099, and 0785.

Section 8.5 (pages 470–471)

1.

Number of Patients x	Sample Means $\bar{x}$	$\bar{x} - \mu_{\bar{x}}$	$(\bar{x} - \mu_{\bar{x}})^2$
6 and 10	8	$8 - 7.3333 = 0.6667$	0.4445
6 and 4	5	$5 - 7.3333 = -2.3333$	5.4443
6 and 7	6.5	$6.5 - 7.3333 = -0.8333$	0.6944
6 and 9	7.5	$7.5 - 7.3333 = 0.1667$	0.0278
6 and 8	7	$7 - 7.3333 = -0.3333$	0.1111
10 and 4	7	$7 - 7.3333 = -0.3333$	0.1111
10 and 7	8.5	$8.5 - 7.3333 = 1.1667$	1.3612
10 and 9	9.5	$9.5 - 7.3333 = 2.1667$	4.6946
10 and 8	9	$9 - 7.3333 = 1.6667$	2.7779
4 and 7	5.5	$5.5 - 7.3333 = -1.8333$	3.3610
4 and 9	6.5	$6.5 - 7.3333 = -0.8333$	0.6944
4 and 8	6	$6 - 7.3333 = -1.3333$	1.7777
7 and 9	8	$8 - 7.3333 = 0.6667$	0.4445
7 and 8	7.5	$7.5 - 7.3333 = 0.1667$	0.0278
9 and 8	8.5	$8.5 - 7.3333 = 1.1667$	1.3612
	110.0		23.3335

c. $\mu_{\bar{x}} = 7.3333$ d. $\sigma_{\bar{x}} = \sqrt{1.5556} \approx 1.2472$

3.

Number of Alarms x	Sample Means $\bar{x}$	$\bar{x} - \mu_{\bar{x}}$	$(\bar{x} - \mu_{\bar{x}})^2$
8, 5, and 12	8.3333	$8.3333 - 7.8333 = 0.5$	0.2500
8, 5, and 9	7.3333	$7.3333 - 7.8333 = -0.5$	0.2500
8, 5, and 6	6.3333	$6.3333 - 7.8333 = -1.5$	2.2500
8, 5, and 7	6.6667	$6.6667 - 7.8333 = -1.1666$	1.3610
5, 12, and 9	8.6667	$8.6667 - 7.8333 = 0.8334$	0.6946
5, 12, and 6	7.6667	$7.6667 - 7.8333 = -0.1666$	0.0278
5, 12, and 7	8.0000	$8.0000 - 7.8333 = 0.1667$	0.0278
12, 9, and 6	9.0000	$9.0000 - 7.8333 = 1.1667$	1.3612
12, 9, and 7	9.3333	$9.3333 - 7.8333 = 1.5$	2.2500

continued

3. (*continued*)

Number of Alarms x	Sample Means $\bar{x}$	$\bar{x} - \mu_{\bar{x}}$	$(\bar{x} - \mu_{\bar{x}})^2$
8, 12, and 9	9.6667	$9.6667 - 7.8333 = 1.8334$	3.3614
8, 12, and 6	8.6667	$8.6667 - 7.8333 = 0.8334$	0.6946
8, 12, and 7	9.0000	$9.0000 - 7.8333 = 1.1667$	1.3612
8, 9, and 6	7.6667	$7.6667 - 7.8333 = -0.1666$	0.0278
8, 9, and 7	8.0000	$8.0000 - 7.8333 = 0.1667$	0.0278
8, 6, and 7	7.0000	$7.0000 - 7.8333 = -0.8333$	0.6944
5, 9, and 6	6.6667	$6.6667 - 7.8333 = -1.1666$	1.3610
5, 9, and 7	7.0000	$7.0000 - 7.8333 = -0.8333$	0.6944
5, 6, and 7	6.0000	$6.0000 - 7.8333 = -1.8333$	3.3614
12, 6, and 7	8.3333	$8.3333 - 7.8333 = 0.5$	0.2500
9, 6, and 7	7.3333	$7.3333 - 7.8333 = -0.5$	0.2500
	156.6667		20.5564

c. $\mu_{\bar{x}} = 7.8333$ d. $\sigma_{\bar{x}} = \sqrt{1.0278} \approx 1.0138$

5. a.

x	$x - \mu$	$(x - \mu)^2$
12	$12 - 18 = -6$	36
21	$21 - 18 = 3$	9
16	$16 - 18 = -2$	4
15	$15 - 18 = -3$	9
26	$26 - 18 = 8$	64
90	0	122

$$\mu = \frac{90}{5} = 18$$

$$\sigma = \sqrt{\frac{122}{5}} = \sqrt{24.4} \approx 4.9396$$

b. Size 2

Number of Cases, x	Sample Means, $\bar{x}$	$\bar{x} - \mu_{\bar{x}}$	$(\bar{x} - \mu_{\bar{x}})^2$
12 and 21	16.5	-1.5	2.25
12 and 16	14.0	-4.0	16.00
12 and 15	13.5	-4.5	20.25
12 and 26	19.0	1.0	1.00
21 and 16	18.5	0.5	0.25
21 and 15	18.0	0.0	0.00
21 and 26	23.5	5.5	30.25
16 and 15	15.5	-2.5	6.25
16 and 26	21.0	3.0	9.00
15 and 26	20.5	2.5	6.25
	180.0		91.50

Size 3

Number of Cases, x	Sample Means, $\bar{x}$	$\bar{x} - \mu_{\bar{x}}$	$(\bar{x} - \mu_{\bar{x}})^2$
12, 21, and 16	16.3333	−1.6667	2.7779
12, 21, and 15	16.0000	−2.0000	4.0000
12, 21, and 26	19.6667	1.6667	2.7779
12, 16, and 15	14.3333	−3.6667	13.4447
12, 16, and 26	18.0000	0.0000	0.0000
12, 15, and 26	17.6667	−0.3333	0.1111
21, 16, and 15	17.3333	−0.6667	0.4445
21, 16, and 26	21.0000	3.0000	9.0000
21, 15, and 26	20.6667	2.6667	7.1113
16, 15, and 26	19.0000	1.0000	1.0000
	180.0000		40.6674

c.   Size 2   Size 3

$$\mu_{\bar{x}} = \frac{180}{10} = 18 \qquad\qquad \mu_{\bar{x}} = \frac{180}{10} = 18$$

$$\sigma_{\bar{x}} = \sqrt{\frac{91.5}{10}} = \sqrt{9.15} \approx 3.0249 \qquad \sigma_{\bar{x}} = \sqrt{\frac{40.6674}{10}} = \sqrt{4.06674} \approx 2.0166$$

d. We use $\sigma_{\bar{x}} = \dfrac{\sigma}{\sqrt{n}} \sqrt{\dfrac{N-n}{N-1}}$. In both cases $N = 5$.

$$\sigma_{\bar{x}} = \frac{4.9396}{\sqrt{2}} \sqrt{\frac{5-2}{5-1}} \quad \text{and} \quad \sigma_{\bar{x}} = \frac{4.9396}{\sqrt{3}} \sqrt{\frac{5-3}{5-1}}$$

$$\approx 3.0249 \qquad\qquad\qquad \approx 2.0166$$

7. a. $\mu_{\bar{x}} = \$964$ and $\sigma_{\bar{x}} = \dfrac{102}{\sqrt{49}} \approx \$14.5714$ b. $\mu_{\bar{x}} = \$964$ and $\sigma_{\bar{x}} = \dfrac{102}{\sqrt{100}} = \$10.20$

Section 8.7 (pages 477–479)

1. 0.9818 3. 0.0344 5. 0.9909 7. Between 19.7435 and 20.4565 ounces 9. 0.9686
11. 0.6278

Section 8.8 (page 481)

1.
```
MTB  > RANDOM 10 OBS, INTO C1;
SUBC > NORMAL MU = 75, SIGMA = 10.
MTB  > AVERAGE THE OBSERVATIONS IN C1
```

Repeat the above procedure 50 times to obtain 50 sample means. These are then entered in C51.

```
MTB  > SET THE FOLLOWING DATA INTO C51
DATA > ...
```

```
DATA > END
MTB  > DESCRIBE C51
```

The results will vary depending upon the numbers generated. For an alternate and time-saving way to generate all 50 samples at the same time, see the last comment at the end of the section.

3.
```
MTB  > RANDOM 50 C1;
SUBC > INTEGER 1  100.
. . .
```

(Note: If an integer is repeated, disregard it.)

5. Same as Exercise 1 except that we now have

```
MTB  > RANDOM 15 OBS, INTO C1;
SUBC > NORMAL MU = 18, SIGMA = 4.
```

Repeat above procedure 80 times. (See last comment at end of section.)

Testing Your Understanding of This Chapter's Concepts (pages 484–485)

1. The relationship is expressed in the formula $\sigma_{\bar{x}} = \dfrac{\sigma}{\sqrt{n}}$ when the sample is less than 5% of the population size.

2. Yes, since in this case a sample of any size will be less than 5% of the entire infinite population.

3. a.

Number of Alarms, x	$x - \mu$	$(x - \mu)^2$
23	3	9
18	−2	4
16	−4	16
31	11	121
12	−8	64
100		214

$\mu = 20$, $\sigma = \sqrt{42.8} \approx 6.5422$

b. Size 2

Number of Alarms, x	Sample Mean, $\bar{x}$	$\bar{x} - \mu_{\bar{x}}$	$(\bar{x} - \mu_{\bar{x}})^2$
23 and 18	20.5	0.5	0.25
23 and 16	19.5	−0.5	0.25
23 and 31	27.0	7.0	49.00
23 and 12	17.5	−2.5	6.25
18 and 16	17.0	−3.0	9.00
18 and 31	24.5	4.5	20.25
18 and 12	15.0	−5.0	25.00
16 and 31	23.5	3.5	12.25
16 and 12	14.0	−6.0	36.00
31 and 12	21.5	1.5	2.25
	200.0		160.50

Size 3

Number of Alarms, x	Sample Mean, $\bar{x}$	$\bar{x} - \mu_{\bar{x}}$	$(\bar{x} - \mu_{\bar{x}})^2$
23, 18, and 16	19.0000	−1.0	1.0000
23, 18, and 31	24.0000	4.0	16.0000
23, 18, and 12	17.6667	−2.3333	5.4443
23, 16, and 31	23.3333	3.3333	11.1109
23, 16, and 12	17.0000	−3.0	9.0000
23, 31, and 12	22.0000	2.0	4.0000
18, 16, and 31	21.6667	1.6667	2.7779
18, 16, and 12	15.3333	−4.6667	21.7781
18, 31, and 12	20.3333	0.3333	0.1111
16, 31, and 12	19.6667	−0.3333	0.1111
	200.0000		71.3334

c. Size 2 $\mu_{\bar{x}} = 20, \sigma_{\bar{x}} = \sqrt{16.05} \approx 4.0062$
   Size 3 $\mu_{\bar{x}} = 20, \sigma_{\bar{x}} = \sqrt{7.13334} \approx 2.6708$

d. Size 2 $\sigma_{\bar{x}} = \dfrac{6.5422}{\sqrt{2}} \sqrt{\dfrac{5-2}{5-1}} \approx 4.0063$

   Size 3 $\sigma_{\bar{x}} = \dfrac{6.5422}{\sqrt{3}} \sqrt{\dfrac{5-3}{5-1}} \approx 2.6708$

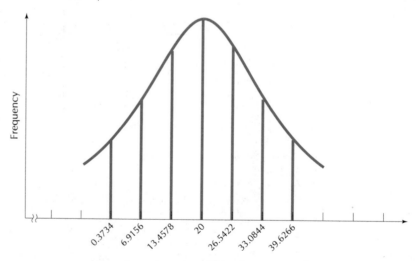

Average number of false alarms reported

Chapter Test (pages 485–487)

1. Choice (a)   2. Choice (b)   3. Choice (a)   4. Choice (a)   5. Choice (c)   6. Choice (a)
7. Those nurses whose numbers are 816, 309, 763, 078, 061, 277, 188, 174, 530, and 709.
8. 0.8767   9. 0.0027   10. 0.0021   11. Between 64.2409 and 65.7591
12. Those officers whose numbers are 648, 163, 534, 310, 209, 181, 595, 694, 334, and 522.

13. 0.9801   14. 0.0107   15. 0.8365   16. $\mu_{\bar{x}} = 758$ 17. $\sigma_{\bar{x}} = \sqrt{976} \approx 31.2410$ 18. 0.8776
19. Between 9.3987 and 10.0014 years   20. Between 42.0083 and 43.9917 years
21. Those doctors whose numbers are 691, 279, 151, 394, 604, 186, 711, 577, 388, and 568.
22. 0.9962   23. 0.0012   24. 0.0188   25. $\mu_{\bar{x}} = 478$ 26. $\sigma_{\bar{x}} = \sqrt{1256} \approx 35.4401$ 27. 0.9600
28. Between 11.1799 and 11.8201 years
29. $\mu_{\bar{x}} = 193$

$$\sigma_{\bar{x}} = \frac{28}{\sqrt{100}} = 2.8$$

Thinking Critically (pages 487–488)

1. The larger the sample size the better the estimate of the population mean since it is based on more data that encompasses more of the population.
2. σ_x represents the standard deviation of the x's, whereas $\sigma_{\bar{x}}$ represents the standard deviation of the sampling distribution of the mean. It is the standard error of the mean. Except for $n = 1$, $\sigma_{\bar{x}}$ will be smaller than σ.
3. Probable error of the mean $= (0.06745)\left(\dfrac{1281}{\sqrt{400}}\right) = 4.3202.$
4. Since in this case, the sample constitutes at least 5% of the population.   5. Probably yes.
6. a. 0.9949   b. 0.9999   c. The larger sample size produces greater probability.

Case Study (pages 488–489)

1. a. Between 4.2079 and 4.6521 hours   b. Between 1.8363 and 2.1237 hours
   c. The interval for the average number of hours spent by 95% of urban young people listening to music is considerably larger than the interval for the average number of hours spent by 95% of suburban teenagers.

CHAPTER 9

Section 9.3 (pages 500–502)

1. 90% confidence interval: $127,276.19 to $130,063.81
3. 95% confidence interval: $115,516.35 to $118,465.66
5. 95% confidence interval: $11.34 to $12.66
7. 95% confidence interval: 9.7933 to 12.2067 hours
9. 90% confidence interval: 2.0736 to 2.3264 smoke detectors
11. a. 95% confidence interval: 18.2493 to 20.6395
    b. 99% confidence interval: 17.8713 to 21.0175

Section 9.4 (pages 506–510)

1. 95% confidence interval: Between $37.57 and $42.33
3. 90% confidence interval: Between 3.5543 and 4.9457 pounds
5. 90% confidence interval: Between 132.9470 and 161.0530 calories
7. 99% confidence interval: Between $3.58 and $4.20
9. 90% confidence interval: Between 5.0675 and 9.9325 hours
11. 90% confidence interval: Between 15.0404 and 20.9596

Section 9.6 (pages 513–514)

1. 95% confidence interval for σ: Between $8.15 and $12.13
3. 95% confidence interval for σ: Between $0.89 and $1.23
5. Sample size = 70   7. Sample size = 28   9. Sample size = 6

Section 9.7 (pages 519–521)

1. a. 95% confidence interval: Between 0.4155 and 0.4845
   b. 95% confidence interval: Between 0.6032 and 0.6968
3. 95% confidence interval: Between 0.4586 and 0.7508
5. 90% confidence interval: Between 0.0430 and 0.1570
7. 0.2578   9. 0.1446   11. 0.1867

Section 9.8 (pages 521–522)

```
1. MTB  > SET THE FOLLOWING DATA IN C1
   DATA > 23 17 16 8 31 14 12 17 15 17 15 13
   DATA > END
   MTB  > TINTERVAL WITH 90 PERCENT CONFIDENCE FOR DATA IN C1
```

Confidence Intervals

Variable	N	Mean	StDev	SE Mean	90.0% C.I.
C1	12	16.50	5.79	1.67	(13.50, 19.50)

```
3. MTB  > SET THE FOLLOWING DATA IN C1
   DATA > 3 4 13 31 33 16 23 9 17 13
   DATA > 12 7 8 23 16 14 12 16 16 19
   DATA > 5 14 24 17 14 12 14 12 14 11
   DATA > 6 12 21 19 22 11 13 13 10 16
   DATA > 4 19 17 18 19 19 2 10 8 10
   DATA > 10 21 16 23 12 17 1 11 6 11
   DATA > END
   MTB  > TINTERVAL WITH 99 PERCENT CONFIDENCE FOR DATA IN C1
```

Confidence Intervals

Variable	N	Mean	StDev	SE Mean	99.0% C.I.
C1	60	13.983	6.474	0.836	(11.758, 16.208)

Testing Your Understanding of This Chapter's Concepts (page 525)

1. Choice (d)
2. a. 90% confidence interval: Between 0.4123 and 0.4543
   b. Sample size = 1695
3. 99.7% confidence interval: Between 0.6977 and 0.8023
4. a. 95% confidence interval for μ: Between 5.2754 and 9.3246
   b. 95% confidence interval for σ: Between 1.9679 and 5.0387
      (Although the formulas given are for large samples and 10 does not qualify as large, our computations are based on the given information of past experience with large sample sizes.)

Chapter Test (pages 525–528)

1. Choice (a)   2. Choice (b)   3. Choice (c)   4. Choice (a)   5. Choice (b)   6. Choice (c)
7. Choice (c)   8. 0.2578
9. 95% confidence interval: Between 0.4985 and 0.6215   10. Sample size = 16
11. 95% confidence interval: Between $34.09 and $35.91
12. 95% confidence interval: Between $46.65 and $53.25
13. 90% confidence interval: Between $113.18 and $126.82
14. 95% confidence interval for σ: Between $32.32 and $49.16
15. 99% confidence interval for σ: Between $6.29 and $10.99
16. Sample size = 40   17. 0.2005   18. 0.2389
19. 95% confidence interval: Between 0.5487 and 0.6513   20. Sample size = 28
21. 95% confidence interval: Between $21.29 and $22.61
22. 95% confidence interval: Between $30.74 and $33.16
23. 90% confidence interval: Between $79.16 and $90.84
24. 95% confidence interval for σ: Between 2.3746 and 3.5323 hours

Thinking Critically (page 528)

1. True. To double accuracy, we cut e in half and the formula indicates that we need to quadruple the sample size.
2. $n = \left(\dfrac{1.645\sigma}{e} \right)^2$
3. 95% confidence interval: Between 8.6123 and 11.3877
4. 95% confidence interval for σ: Between 13.2422 and 15.2141
5. Large sample size
6. $df = 10 - 1 = 9$, $\quad y^2 = \left[1.645 \left(\dfrac{8(9) + 3}{8(9) + 1} \right) \right]^2 = 2.8563$,

$$t = \sqrt{9 \cdot \left(2.718^{\frac{2.8563}{9}} - 1 \right)}$$

$\quad\quad = 1.833$
The $t_{0.05}$-value with 9 degrees of freedom is 1.833.

Case Study (pages 528–529)

1. 95% confidence interval: Between 0.1142 and 0.1698

CHAPTER 10

Section 10.4 (pages 547–549)

1. $z = -2.95$. Reject manufacturer's claim.
3. $z = -1.39$. No. We cannot reject engineer's claim.
5. $z = 2.73$. Yes. We reject the claim that appears in *Statistical Abstract of the United States*. Average cost has risen.
7. $z = -4.43$. Yes. Reject mayor's claim.
9. $z = 1.87$. Yes. We cannot reject Motor Vehicle Bureau claim.
11. $z = 8.88$. Yes. Reject banking industry official's claim.

Section 10.5 (pages 552–553)

1. $t = 47.43$. Yes. The charge for electricity by these utilities is significantly above average.
3. $t = -7.37$. Yes. Reject "waterproof" claim.
5. $t = 5.37$. Yes. Reject claim that the average fare is $26.20.
7. $t = -4.33$. Yes. Reject claim that average weight is 32 ounces.
9. $t = -1.44$. Do not reject null hypothesis.

Section 10.6 (pages 559–561)

1. $z = 5.036$. Reject null hypothesis. There is a significant difference.
3. $z = -1.09$. Do not reject null hypothesis. There is no significant difference.
5. $z = 3.25$. Reject null hypothesis. There is a significant difference.
7. $z = -5.66$. Reject null hypothesis. There is a significant difference.
9. $z = -2.89$. Reject null hypothesis. There is a significant difference.

Section 10.7 (pages 564–566)

1. $s_p \approx 6.2843$, $t = -4.08$. Reject null hypothesis. There is evidence to indicate a significant difference.
3. $s_p \approx 4001.4315$, $t = -0.56$. Do not reject null hypothesis. There is not enough information to indicate a significant difference.
5. $s_p \approx 1.9845$, $t = 0.8322$. Do not reject null hypothesis. There is not enough information to indicate a significant difference.
7. $s_p \approx 1.6554$, $t = -1.91$. Do not reject null hypothesis. There is not enough information to indicate a significant difference.
9. $s_p \approx 729.3968$, $t = 2.54$. Reject null hypothesis. There is evidence to indicate a significant difference.

Section 10.8 (pages 570–572)

1. $\hat{p} = \dfrac{10}{84} = 0.119$, $z = 0.58$. No. Do not reject null hypothesis and the study.
3. $\hat{p} = \dfrac{35}{65} = 0.5385$, $z = 1.43$. No. Do not reject null hypothesis and the claim.
5. $\hat{p} = 0.30$, $z = 1.03$. No. Do not reject null hypothesis and the claim.
7. $\hat{p} = \dfrac{46}{258} = 0.1783$, $z = 1.27$. No. Do not reject null hypothesis and the claim.
9. $\hat{p} = \dfrac{35}{50} = 0.70$, $z = 0.5971$, $z = -0.1558$. No. Results do not differ from IIHS or *Prevention* claim. Do not reject null hypothesis.
11. $\hat{p} = \dfrac{41}{80} = 0.5125$, $z = -0.67$. Yes. We cannot reject the null hypothesis and the 55% claim.

Section 10.9 (pages 574–577)

```
1. MTB   > SET THE FOLLOWING IN C1
   DATA > 36 34 39 43 29 32 35 36
   DATA > END
   MTB   > TTEST, MU = 44, C1
```

```
TEST OF MU = 44.000 VS NOT = MU 44.000

        N     MEAN    STDEV   SE MEAN        T    P VALUE
C1      8    35.50     4.24      1.50    -5.67     0.0008
```

Since the *P*-value of 0.0008 is less than 0.01, we must reject null hypothesis. There is a significant difference between the average amount of pollutant in the air found by health officials and the union claim.

3. **MTB > SET THE FOLLOWING IN C1**
   DATA > 42 38 59 46 31 39 42 55 69 51 50 22 36 39 21 19 26
   DATA > 22 29 34 36 52 40 21 29 37 62 50 46 54 71 23 39 26
   DATA > 34 23 38
   DATA > END
   MTB > DESCRIBE C1

```
VARIABLE      N      MEAN    MEDIAN   TRMEAN    STDEV    SEMEAN
C1           37     39.22     38.00    38.52    13.76      2.26

VARIABLE     MIN       MAX        Q1       Q3
C1         19.00     71.00     27.50    50.00
```

   MTB > ZTEST, MU = 31, SIGMA = 13.76, C1;
   SUBC > ALTERNATIVE = +1.

```
TEST OF MU = 31.000 VS MU > 31.000
THE ASSUMED SIGMA = 13.8

        N      MEAN     STDEV   SE MEAN        Z    P VALUE
C1     37    39.216    13.762    22.262     3.63     0.0001
```

Since the *P*-value of 0.0001 is less than 0.05, we reject the hospital official's claim that the average waiting time is 31 minutes before being examined by a doctor.

5. **MTB > SET THE FOLLOWING IN C1**
   DATA > 6 8 4 2 3 7 8 1 3 2 7 9
   DATA > END
   MTB > TTEST, MU = 8.62, C1;
   SUBC > ALTERNATIVE = -1.

```
TEST OF MU = 8.620 VS MU < 8.620

        N     MEAN    STDEV   SE MEAN        T    P VALUE
C1     12    5.000    2.796     0.807    -4.48     0.0005
```

Since the *P*-value of 0.0005 is less than 0.05, we reject the null hypothesis and the welfare department official's claim.

7. **MTB > SET THE FOLLOWING IN C1**
   DATA > 10 12 5 17 21 18 19 10 11 6 7 9 10 3 18 16 17
   DATA > 12 5 7 6 8 9 3 12 15 17 16 10 12 4 11 17 15 13
   DATA > END
   MTB > DESCRIBE C1

```
VARIABLE      N      MEAN    MEDIAN   TR MEAN    STDEV    SEMEAN
C1           35    11.457    11.000    11.452    4.973     0.841
```

```
VARIABLE        MIN        MAX        Q1        Q3
C1             3.000     21.000     7.000    16.000

MTB  > ZTEST, MU = 8.4, SIGMA = 4.973, C1;
SUBC > ALTERNATIVE = +1.

TEST OF MU = 8.40 VS MU > 8.40
THE ASSUMED SIGMA = 4.97

         N      MEAN     STDEV    SE MEAN      Z     P VALUE
C1      35     11.457    4.973     0.841     3.64     0.0001
```

Since the P-value is less than 0.05, we reject the null hypothesis and the union claim.

9.
```
MTB  > SET THE FOLLOWING IN C1
DATA > 24.6 25.3 28.6 26.1 23.7 26.4 25.9 25.1 24.3
DATA > END
MTB  > SET THE FOLLOWING IN C2
DATA > 26.4 27.3 25.4 26.1 24.2 23.8 24.0 22.1
DATA > END
MTB  > TWOSAMPLE T C1 C2;
SUBC > ALTERNATIVE = +1.

TWOSAMPLE T FOR C1 VS C2

       N      MEAN     STDEV    SE MEAN
C1     9     25.56     1.44       0.48
C2     8     24.91     1.69       0.60

95% C.I. FOR MU C1 - MU C2: (-1.01, 2.30)
T-TEST MU C1 = MU C2 (VS >): T = 0.84 P = 0.21 DF = 13
```

Since the P-value of 0.21 is greater than 0.05, we do not reject the null hypothesis and conclude that there is not sufficient evidence to say that the new chemical additive significantly increases gasoline mileage.

Testing Your Understanding of This Chapter's Concepts (pages 579–580)

1. $s_p \approx 3.0356$. The endpoints of the confidence interval for $\mu_1 - \mu_2$ are

$$(\bar{x}_1 - \bar{x}_2) \pm t_{\frac{\alpha}{2}} \cdot s_p \sqrt{\frac{1}{n_1} + \frac{1}{n_2}} \qquad df = n_1 + n_2 - 2 = 12 + 14 - 2 = 24$$

90% confidence interval: Between 11.96 cents and 16.04 cents

2. a.

Jonathan Williams Houses	Martin Luther King Houses
Mean, $\bar{x} = 16491$	Mean, $\bar{x} = 15942$
Standard deviation, $s = 1844$	Standard deviation, $s = 2171$
$n = 10$	$n = 12$

b. $s_p \approx 2030.3777$. The endpoints of the confidence interval for $\mu_1 - \mu_2$ are

$$(\bar{x}_1 - \bar{x}_2) \pm t_{\frac{\alpha}{2}} \cdot s_p \sqrt{\frac{1}{n_1} + \frac{1}{n_2}} \qquad df = n_1 + n_2 - 2 = 10 + 12 - 2 = 20$$

95% confidence interval: Between $-1264.48 and $2362.48

c. $t \approx 0.6315$. Do not reject null hypothesis. The mean family incomes are not significantly different.

Chapter Test (pages 580–584)

1. $z = -3.95$. Reject null hypothesis. Company is paying lower wages. Choice (a).
2. $z = -1.59$. Do not reject null hypothesis. Choice (b).
3. $t = -2.89$. Reject null hypothesis. Choice (a).
4. $t = 2.33$. Reject null hypothesis. Choice (a).
5. $z = -23.029$. Reject null hypothesis. Choice (a).
6. $z = -6.27$. Reject null hypothesis. The difference in waiting time is significant.
7. $s_p \approx 6.4996$, $t = 1.9867$. Do not reject null hypothesis. The difference in registration time at both campuses is not statistically significant.
8. $\hat{p} = \dfrac{50}{70} = 0.7143$, $z = -4.2662$. Reject null hypothesis and the mayor's claim.
9. $\hat{p} = \dfrac{48}{58} = 0.8276$, $z = -0.48$. Do not reject null hypothesis. Do not reject chemist's claim.
10. $s_p \approx 2.1506$, $t = -1.89$. Reject null hypothesis. The average time spent by the workers in the advertising department is more than the average time spent by the workers of the production department.
11. $t = 1.15$. Do not reject null hypothesis. The average charge at these hospitals does not significantly exceed the state average.
12. $z = -9.32$. Reject null hypothesis. The average age at these jails is significantly lower than the average age at all of the state's jails.
13. $t = 2.57$. Do not reject null hypothesis. The special mechanism does not significantly affect the average life of a transmission.
14. $t = -3.18$. Reject null hypothesis. The average height of police officers in Buscane is significantly less than 69 inches.
15. $z = 3.796$. Reject null hypothesis. The difference between the average time for both groups is significant.
16. $z = -9.75$. Reject the null hypothesis. The difference between the waiting time is significant.
17. $s_p \approx 5.5587$, $t = -1.78$. Do not reject null hypothesis. The difference in the average amount of coliform bacteria in both states is not statistically significant.
18. $s_p \approx 4.7655$, $t = -1.90$. Reject null hypothesis. The average number of practice days required by dolphins in the first park is significantly less than the average number of days required by the dolphins in the second park.
19. $\hat{p} = \dfrac{69}{80} = 0.8625$, $z = -0.4817$. Do not reject null hypothesis. There is not enough evidence to reject the researcher's claim.
20. $\hat{p} = \dfrac{80}{168} = 0.4762$, $z = -1.40$. Do not reject null hypothesis. Do not reject researcher's claim.

21. $z = 0.95$. Do not reject null hypothesis. The difference between the average lasting time of the paints is not statistically significant.

22. $s_p \approx 849.2214$, $t = -3.82$. Reject null hypothesis. There is a significant difference between the average annual salaries at these schools.

23. $s_p \approx 0.5290$, $t = -1.90$. Reject null hypothesis. The average price charged by pharmacies in Baltimore is significantly less than the average price charged by pharmacies in Chicago.

24. $\hat{p} = \dfrac{68}{80} = 0.85$, $z = -1.49$. Do not reject null hypothesis. Do not reject the store owner's claim.

25. $\hat{p} = \dfrac{200}{300} = 0.6667$, $z = -1.26$. Do not reject null hypothesis. There is not sufficient evidence to reject the politician's claim.

Thinking Critically (page 584)

1. No. When the null hypothesis is rejected, either a type-I error is made or a correct decision is made. When the null hypothesis is not rejected, either a type-II error is made or a correct decision is made. Since it is not possible to simultaneously reject and accept the null hypothesis, it is not possible to make both errors at the same time.

2. a. Not for any reasonable level of significance ($\alpha < 0.50$). The rejection region for such a level involves positive values of the test statistic; the given sample mean and hypothesized value for the mean yield a negative value for the test statistic. Therefore, the null hypothesis would not be rejected.

   b. Possibly a type-II error if H_0 is false.

3. If both sample sizes n_1 and n_2 are equal call them both n. Then,

$$s_p = \sqrt{\frac{(n-1)s_1^2 + (n-1)s_2^2}{n + n - 2}}$$

$$= \sqrt{\frac{(n-1)s_1^2 + (n-1)s_2^2}{2n - 2}}$$

$$= \sqrt{\frac{(n-1)(s_1^2 + s_2^2)}{2(n-1)}}$$

$$= \sqrt{\frac{s_1^2 + s_2^2}{2}}$$

Thus, s_p^2 is the mean of s_1^2 and s_2^2.

4. P(type-II error) could be 1.

Case Study (pages 584–585)

1. $\hat{p} = \dfrac{33}{98} = 0.3367$, $z = 2.51$. Reject null hypothesis. The number of births by Caesarean section is significantly more than the claimed 23%.

CHAPTER 11

Section 11.3 (pages 599–604)

1. a. Positive correlation   b. Zero correlation
    c. Probably positive correlation. Others may disagree.   d. Positive correlation.
    e. Positive correlation   f. Negative correlation   g. Positive correlation
    h. Positive correlation

3. a.

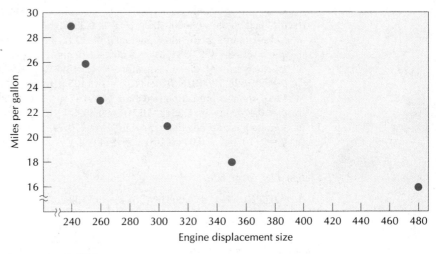

$r = -0.8831$

5.

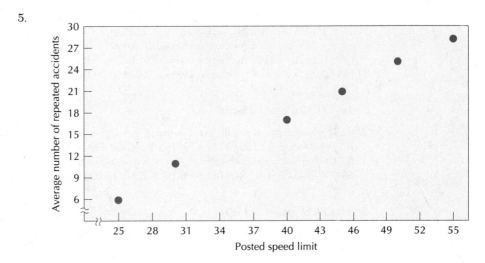

$r = 0.9981$

7. $r = -0.9734$ 9. $r = 0.9843$ 11. $r = 0.9959$

Section 11.4 (page 605)

1. Significant   3. Significant   5. Significant   7. Significant   9. Significant
11. Significant

Section 11.6 (pages 616–620)

1. a. Least-squares prediction equation: $\hat{y} = 12.023 + 0.84238x$
   b. $\hat{y} = 12.023 + 0.84238(74) = 74.359$ inches
3. a. Least-squares prediction equation: $\hat{y} = 0.285630 + 0.00004068x$
   b. $\hat{y} = 0.285630 + (0.00004068)(875) = 0.32123$
5. a. Least-squares prediction equation: $\hat{y} = -11.845 + 0.3884x$
   b. $\hat{y} = -11.845 + 0.3884(48) = 6.800$ accidents
7. a. Least-squares prediction equation: $\hat{y} = 261526.3 - 9263.1579x$
   b. $\hat{y} = 261526.3 - (9263.1579)(13) = 141{,}105$ people
9. a. Least-squares prediction equation: $\hat{y} = 0.47494 + 0.23322x$
   b. $\hat{y} = 0.47494 + (0.023322)(16.50) = 0.8598$
11. a. Least-squares prediction equation: $\hat{y} = -10.794 + 0.048591x$
    b. $\hat{y} = -10.794 + 0.048591(512) = \$14.084$

Section 11.7 (page 626)

1. $s_e \approx 0.5019$ 3. $s_e \approx 0.008457$ 5. $s_e \approx 0.3499$ 7. $s_e \approx 9003$ 9. $s_e \approx 0.1462$
11. $s_e \approx 2.203$

Section 11.8 (pages 632–633)

1. $t = 18.1537$,   95% prediction interval: 73.044 to 75.674
3. $t = 4.5836$,   95% prediction interval: 0.29903 to 0.34342
5. $t = 14.3387$,   95% prediction interval: 5.748 to 7.853
7. $t = -7.0912$,   95% prediction interval: 113,556 to 168,655
9. $t = 6.4949$,   95% prediction interval: 0.5296 to 1.1899
11. $t = 5.3205$,   95% prediction interval: 8.014 to 20.154

Section 11.10 (pages 635–636)

1. a. The least-squares prediction equation is $\hat{y} = 3.648 + 0.1139x_1 + 0.7171x_2$.
   b. $\hat{y} = 3.648 + 0.1139(67) + 0.7171(205) = 158.290$
3. a. The least-squares prediction equation is $\hat{y} = 39.0005 + 0.3845x_1 - 0.1918x_2$.
   b. $\hat{y} = 39.0005 + 0.3845(123) - 0.1918(12) = 83.9924$

Section 11.11 (pages 641–643)

```
1. MTB  > READ AGE IN C1, PERCENT REDUCTION IN C2
   DATA > 52 60
   DATA > 53 54
   DATA > 55 46
   DATA > 56 40
   DATA > 57 35
```

```
DATA > 58 23
DATA > 59 18
DATA > 60 12
DATA > 61  4
DATA > END
MTB   NAME C1 'x' C2 'y'
MTB   PLOT 'y' vs 'x'
```

Character Plot

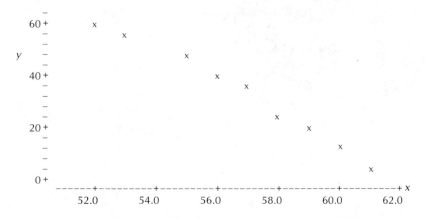

```
MTB  > CORRELATION 'y' 'x'
```

Correlations (Pearson)

Correlation of y and x = -0.992

```
MTB  > REGRESS 'y' 1 'x';
SUBC > PREDICT 54.
```

Regression Analysis

The regression equation is
$y = 387 - 6.25 x$

Predictor	Coef	Stdev	t-ratio	P
Constant	387.22	16.81	23.04	0.000
x	-6.2485	0.2956	-21.14	0.000

s = 2.570   R-sq = 98.5%   R-sq(adj) = 98.2%

Analysis of Variance

Source	DF	SS	MS	F	p
Regression	1	2950.0	2950.0	446.76	0.000
Error	7	46.2	6.6		
Total	8	2996.2			

Fit	Stdev Fit	95.0% C.I.	95.0% P.I.
49.801	1.187	(46.995, 52.608)	(43.107, 56.496)

3. MTB > READ HEIGHT IN C1, STARTING SALARY IN C2
   DATA > 64 21000
   DATA > 67 24000
   DATA > 66 23000
   DATA > 68 26000
   DATA > 69 26000
   DATA > 71 27000
   DATA > 65 25000
   DATA > END
   MTB > NAME C1 'x' C2 'y'
   MTB > PLOT 'y' vs 'x'

Character Plot

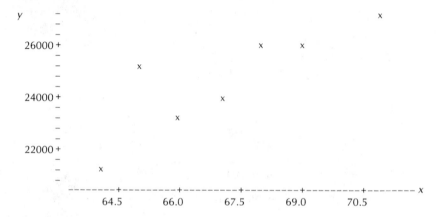

MTB > CORRELATION 'y' 'x'

Correlations (Pearson)

Correlation of y and x = 0.849

MTB > REGRESS 'y' 1 'x';
SUBC > PREDICT 70.

Regression Analysis

The regression equation is
$y = -24410 + 730x$

Predictor	Coef	Stdev	t-ratio	P
Constant	−24410	13620	−1.79	0.133
x	729.5	202.7	3.60	0.016

$s = 1197$ R-sq = 72.1%   R-sq(adj) = 66.6%

Analysis of Variance

Source	DF	SS	MS	F	P
Regression	1	18550352	18550352	12.95	0.016

```
Error           5      7163934      1432787
Total           6     25714286

  Fit    Stdev. Fit      95.0% C.I.           95.0% P.I.
26656           735    ( 24766, 28546)     ( 23044, 30267)
```

Testing Your Understanding of This Chapter's Concepts (page 647)

1. Most likely choice (d). However, it really can be large, small, or anything in between as it will be approximately equal to the standard deviation of the y's.
2. Choice (a)   3. Yes   4. 0
5. a. Least-squares prediction equation: $\hat{y} = 37.3211 + 4.2844x$
   b. $75 = 37.3211 + 4.2844x$. Solving for x gives $x = 8.7944$ units of the drug.

Chapter Test (pages 647–651)

1. Choice (a)   2. Choice (a)   3. Choice (b)   4. Choice (c)   5. Choice (d)   6. Choice (c)
7. Choice (a)   8. Choice (a)   9. Choice (c)   10. Choice (b)
11. $r = -.9838$ 12. Least-squares prediction equation: $\hat{y} = 69.2143 - 16.2286x$
13. 32.69995   14. 1.7864   15. 95% prediction interval: From 26.832 to 38.568   16. 0.9679
17. Yes
18. **Character Plot**

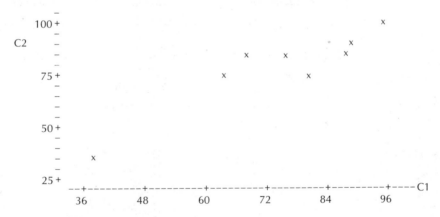

   Probably linear correlation
19. a. $r = 0.9402$
    b. Least-squares prediction: $\hat{y} = -18.3996 + 8.9673x$
    c. $s_e \approx 10.3839$ $t = 5.5183$ Reject H_0: $\beta_1 = 0$
    d. 95% prediction interval: 46.57 to 113.91
20. a. Least-squares prediction equation: $\hat{y} = 11102 - 1130x_1 - 0.008711x_2$
    b. Predicted selling price is $6,788.40

21. a. **Character Plot**

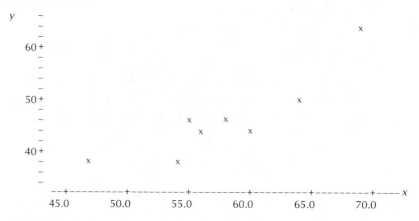

b. $r = 0.8943$ c. Least-squares prediction equation: $\hat{y} = -17.6023 + 1.0925x$
d. $\hat{y} = 50.1327$ e. $s_e \approx 3.934$ 95% prediction interval: 39.6768 to 60.5886.

Thinking Critically (pages 651–652)

1. Start with the formula for b_1 in Formula 11.2. Divide numerator and denominator by $n - 1$.
Using the fact that $s_x^2 = \dfrac{n \sum x^2 - (\sum x)^2}{n(n - 1)}$, we get the equation for b_1 given in Formula 11.3.
Also the formulas for b_0 are equivalent since $\dfrac{1}{n} \sum x = \bar{x}$ and $\dfrac{1}{n} \sum y = \bar{y}$.

2. $s_x = \sqrt{\dfrac{n \sum x^2 - (\sum x)^2}{n(n - 1)}}$ and $\sqrt{\dfrac{n(\sum y^2) - (\sum y)^2}{n(n - 1)}}$

Also, $s_{xy} = \dfrac{\sum(x - \bar{x})(y - \bar{y})}{n - 1}$

Thus, $\dfrac{\sum(x - \bar{x})(y - \bar{y})/(n - 1)}{\sqrt{\dfrac{n \sum x^2 - (\sum x)^2}{n(n - 1)}} \sqrt{\dfrac{n \sum y^2 - (\sum y)^2}{n(n - 1)}}}$

$= \dfrac{n(\sum xy) - (\sum x)(\sum y)}{\sqrt{n \sum x^2 - (\sum x)^2} \sqrt{n \sum y^2 - (\sum y)^2}} = \dfrac{s_{xy}}{s_x s_y}$

This is the formula for r. Thus, $r = \dfrac{s_{xy}}{s_x s_y}$.

3. Correlation analysis determines a correlation coefficient for measuring the strength of a relationship between the variables. Regression analysis involves finding an equation connecting the variables. A regression equation is determined once we know that there is a significant correlation between the variables.

Case Studies (pages 652–653)

1. a. **Character Plot**

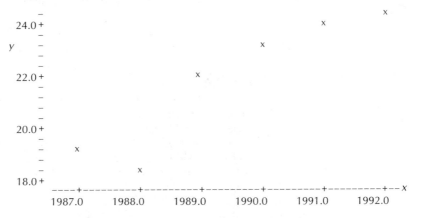

b. Least-squares prediction equation: $\hat{y} = -1725.8855 + 0.8786x$ c. Technically yes, but realistically no.

2. a. Regression equation for men: $\hat{y}_1 = 419.6071 - 0.16029x$

   b. Regression equation for women: $\hat{y}_2 = 1316.3691 - 0.60943x$

CHAPTER 12

Section 12.2 (pages 664–669)

1. $\chi^2 = 9.0431$. Reject null hypothesis. There is a significant difference between the corresponding proportion of women in the various age-groups who apply sunscreen.

3. $\chi^2 = 155.0999$. Reject null hypothesis. The percentage of students from one-parent homes is not the same for these students.

5. $\chi^2 = 5.420$. Do not reject null hypothesis. There is no significant difference in the percentage of constituents who are in favor of the new legislation, regardless of frequency of visits.

7. $\chi^2 = 1.070$. Do not reject null hypothesis. There is no significant difference in the percentage of policy holders who use seat belts in all the geographical locations within the state.

9. $\chi^2 = 4.201$. Do not reject null hypothesis. There is no significant difference in the proportion of residents who would allow new jails and/or drug rehabilitation centers to be built in their neighborhoods.

11. $\chi^2 = 5.623$. Do not reject null hypothesis. There is no significant difference in the principal computer usage among the uses listed by those business executives.

13. $\chi^2 = 0.292$. Do not reject null hypothesis. There is no significant difference between the corresponding proportion of male or female visitors who are left-handed.

Section 12.3 (pages 673–675)

1. $\chi^2 = 22.3468$. Reject null hypothesis. The proportion of babies delivered during the various seasons is not independent of the season of the year.

3. $\chi^2 = 5.4138$. Do not reject null hypothesis and the manufacturer's claim.

5. $\chi^2 = 4.7841$. Do not reject null hypothesis that people in some age-groups are more likely to be afraid of heights than people in other age-groups.

7. $\chi^2 = 22.6525$. Reject null hypothesis. The frequencies of these characteristics are not in the ratio of $8:6:3:2$.

9. $\chi^2 = 96.8754$. Reject null hypothesis. The proportions of the drinks dispensed are not in the ratio $7:5:4:3:1$.

11. $\chi^2 = 4.4575$. Do not reject null hypothesis and the claim that the amount of money spent per day per tourist is independent of the tourist's home country.

Section 12.4 (pages 680–683)

1. $\chi^2 = 41.450$. Reject null hypothesis. The region in which a respondent lives is not independent of the resident's opinion on legalizing mercy killings.

3. $\chi^2 = 0.7568$. Do not reject null hypothesis that the number of males or females who will buy a shirt for a man is not significantly different at the various locations.

5. $\chi^2 = 3.151$. Do not reject null hypothesis that the number of defective items produced on the various production lines is independent of the day of the week.

7. $\chi^2 = 30.874$. Reject null hypothesis. Age of customer is not independent of the customer's views as to how the tours should be arranged.

9. $\chi^2 = 12.105$. Do not reject null hypothesis that the educational level attained by at least one of the partners is independent of the number of years that a marriage will last.

11. $\chi^2 = 0.965$. Do not reject the null hypothesis that the principal use of a computer by a computer science major is independent of the class standing of the individual.

Section 12.5 (pages 685–686)

1. $\text{Expected} = \dfrac{212 + 276 + 198 + 246 + 253}{5} = 237$

```
MTB  > READ THE FOLLOWING IN C1, C2
DATA > 212 237
DATA > 276 237
DATA > 198 237
DATA > 246 237
DATA > 253 237
DATA > END
         5 rows read

MTB  > LET C3 = (C1 - C2)**2/C2
MTB  > PRINT C1 C2 C3
```

Data Display

ROW	C1	C2	C2
1	212	237	2.63713
2	276	237	6.41772
3	198	237	6.41772
4	246	237	0.31477
5	253	237	1.08017

```
MTB  > SUM C3
```

Column Sum

Sum of C3 $= 16.8945$

Reject null hypothesis. The number of calls received is not independent of the day of the week.

3. **MTB > READ THE FOLLOWING INTO C1, C2, C3**
   DATA > 71 101 52
   DATA > 82 92 57
   DATA > END
         2 rows read

   MTB > CHISQUARE ON C1 - C3

   Chi-Square Test

   Expected counts are printed below observed counts

	C1	C2	C3	Total
1	71	101	52	224
	75.32	95.02	53.66	
2	82	92	57	231
	77.68	97.98	55.34	
Total	153	193	109	455

   Chisq =   0.248 +   0.377 + 0.051 +
              0.241 +   0.366 + 0.050 =   1.333
   df = 2, p = 0.514

   Do not reject null hypothesis that the sex of the viewer is independent of the viewer's opinion.

Testing Your Understanding of This Chapter's Concepts (page 687–688)

1. Since the null hypothesis is rejected only when the value of the test statistic is too large, the rejection region is always on the right, i.e., it is always right-tailed.
2. No
3. b. Since this is a contingency table and not a simple chi-square problem, the value of p will be different for each cell.

Chapter Test (pages 689–695)

1. Choice (c) 2. Choice (a) 3. Choice (c) 4. Choice (b) 5. Choice (b) 6. Choice (b)
7. Choice (a) 8. Choice (c) 9. Choice (a) 10. Choice (d)
11. $\chi^2 = 0.907$. Do not reject null hypothesis that the obesity of a parent is not a factor in the obesity of the child.
12. $\chi^2 = 1.369$. Do not reject null hypothesis that the financial status of parents is independent of the frequency of the use of drugs by teenagers.
13. $\chi^2 = 3.5460$. Do not reject null hypothesis that the proportion of people in favor of legalizing gambling casinos is the same for all socioeconomic classes.
14. $\chi^2 = 102.4701$. Reject null hypothesis. The vending machine operator's claim is incorrect.
15. $\chi^2 = 147.402$. Reject null hypothesis. The number of children that a woman has is not independent of the size of the family from which she comes.
16. $\chi^2 = 44.049$. Reject null hypothesis. Grade level is a factor in determining which parent will come to school to discuss the child's academic progress with the teacher on open school day.
17. $\chi^2 = 7.810$. Do not reject null hypothesis that the gender of the respondent is independent of the factor considered important by a potential buyer.

18. Choice (e)

19. $\chi^2 = 1.3939$. Do not reject null hypothesis that the number of calls received is independent of the day of the week.

20. $\chi^2 = 2.474$. Do not reject null hypothesis that the length of hospital stay is independent of the extent of hospital insurance reimbursement.

21. $\chi^2 = 14.0349$. Reject null hypothesis. The given percentages are incorrect.

22. $\chi^2 = 2.257$. Do not reject null hypothesis that the level of pollution is independent of the river involved.

23. $\chi^2 = 9.177$. Do not reject null hypothesis that the proportion of Americans who think that our government is doing enough for the homeless is the same for the cities mentioned.

24. $\chi^2 = 3.366$. Do not reject null hypothesis that the time of day when an accident occurs is independent of the day of the week that the accident occurs.

25. $\chi^2 = 1.4140$. Do not reject null hypothesis that the data are consistent with the theory.

Thinking Critically (pages 695–696)

1. $\chi^2 = 7.471$. Do not reject null hypothesis.

2. $\chi^2 = 1.297$. Do not reject null hypothesis that the type of antitheft device used is independent of the size of the car.

3. $\chi^2 = 23.152$. Reject null hypothesis that the proportion of people in each age-group who fear heights is the same.

4.

	1	2	Total
1	a	b	$a + b$
2	c	d	$c + d$
Total	$a + c$	$b + d$	$a + b + c + d$

Column one, row 1 expected value: $\dfrac{(a + b)(a + c)}{a + b + c + d}$

Column two, row 1 expected value: $\dfrac{(a + b)(b + d)}{a + b + c + d}$

Column one, row 2 expected value: $\dfrac{(a + c)(c + d)}{a + b + c + d}$

Column two, row 2 expected value: $\dfrac{(c + d)(b + d)}{a + b + c + d}$

$$\chi^2 = \frac{\left[a - \dfrac{(a + b)(a + c)}{a + b + c + d}\right]^2}{\dfrac{(a + b)(a + c)}{a + b + c + d}} + \frac{\left[b - \dfrac{(a + b)(b + d)}{a + b + c + d}\right]^2}{\dfrac{(a + b)(b + d)}{a + b + c + d}}$$

$$+ \frac{\left[c - \dfrac{(a + c)(c + d)}{a + b + c + d}\right]^2}{\dfrac{(a + c)(c + d)}{a + b + c + d}} + \frac{\left[d - \dfrac{(c + d)(b + d)}{a + b + c + d}\right]^2}{\dfrac{(c + d)(b + d)}{a + b + c + d}}$$

This simplifies to $\chi^2 = \dfrac{(a + b + c + d)(ad - bc)^2}{(a + b)(c + d)(b + d)(a + c)}$

5. a. $\chi^2 = 0.114$. Do not reject null hypothesis.

b. $z = \dfrac{p_1 - p_2}{\sigma_{p_1 - p_2}} = \dfrac{\dfrac{32}{39} - \dfrac{38}{48}}{0.085479} = 0.33746$

Do not reject null hypothesis.

c. In both cases, we conclude that we should not reject the null hypothesis.

Case Study (page 697)

1. $\chi^2 = 9.043$. Reject null hypothesis. There is a significant difference between the corresponding proportion of vacationers who use sunscreen.

CHAPTER 13

Section 13.4 (pages 719–727)

1. Do not reject null hypothesis that the average decrease in the blood serum cholesterol levels of people taking each of the different types of pills is the same $(F = 4.70)$.
3. Do not reject null hypothesis that there is no significant difference in the average number of these different pesticides $(F = 2.067)$.
5. Do not reject null hypothesis $(F = 0.591)$.
7. Reject null hypothesis. The commissioner's claim is justified $(F = 5.066)$.
9. Do not reject null hypothesis that the average quantity of dissolved oxygen is the same at all three locations $(F = 2.9433)$ and is the same for all the seasons of the year $(F = 0.8519)$.
11. Do not reject null hypothesis that the average ratings for these movies is the same $(F = 1.0839)$ and that the temperature in the theater is not a factor $(F = 1.0371)$.
13. Do not reject null hypothesis that the average amount of sulfur oxide pollutants is the same at all the bridges and tunnels $(F = 1.289)$ and is the same for each day of the week $(F = 0.8392)$.

Section 13.5 (pages 728–730)

1.
```
MTB  > READ THE FOLLOWING DATA INTO C1, C2, C3, C4, C5
DATA > 68 56 64 68 52
DATA > 50 61 60 66 61
DATA > 56 65 52 62 54
DATA > 75 52 61 59 48
DATA > 65 56 60 67 84
DATA > END
MTB  > AOVONE WAY ON C1-C5
```

One-Way Analysis of Variance

```
Analysis of Variance
```

Source	DF	SS	MS	F	P
Factor	4	472.1	118.0	1.89	0.149
Error	21	1308.3	62.3		
Total	25	1780.3			

```
                                             Individual 95% CIs For Mean
                                             Based on Pooled StDev

 Level        N      Mean      StDev     ---+---------+---------+---------+--

 C1           6     66.333    12.372                      (--------*--------)

 C2           5     58.000     5.050       (--------*---------)
 C3           7     62.714     7.111         (------*-------)
 C4           4     63.750     4.031          (---------*----------)

 C5           4     53.750     5.439     (---------*----------)

                                         ---+---------+---------+---------+--
 Pooled StDev =    7.893                 48.0      56.0      64.0      72.0
```

Do not reject null hypothesis that the average time required to assemble the computer is the same for all of the employees.

3. **MTB   > READ THE FOLLOWING DATA INTO C1, C2, C3, C4**
   DATA > 63 68 67 62
   DATA > 64 68 61 66
   DATA > 62 62 60 67
   DATA > 63 65 59 63
   DATA > 67 67 59 64
   DATA > 60 61 62 69
   DATA > 62 69 65 67
   DATA > 61 68 63 66
   DATA > END
   MTB   > AOVONE WAY ON C1-C4

One-Way Analysis of Variance

```
Analysis of Variance

Source     DF        SS       MS       F        P
Factor      3      94.38    31.46    4.60    0.010
Error      28     191.50     6.84
Total      31     285.87
```

```
                                             Individual 95% CIs For Mean
                                             Based on Pooled StDev

 Level        N      Mean      StDev     ----------+---------+---------+--------

 C1           8     62.750     2.141     (----------*------------)
 C2           8     66.000     3.024                (----------*------------)
 C3           8     62.000     2.878     (---------*----------)
 C4           8     65.500     2.330               (---------*----------)

                                         ----------+---------+---------+--------
 Pooled StDev =    2.615                         62.5      65.0      67.5
```

Reject null hypothesis. There is a significant difference in the average price charged by these supermarkets for a quart of milk.

Testing Your Understanding of This Chapter's Concepts (page 732–733)

1. Assumptions for one-way ANOVA:
   a. *Independent samples:* The samples taken from the various populations are independent of one another.
   b. *Normal populations:* The populations from which the samples are obtained are (approximately) normally distributed.
   c. *Equal standard deviations:* The populations from which the samples are obtained all have the same (often unknown) variance, σ^2.
2. $F_{20}^{10}(0.01) = 3.37$ so that the probability that $F \leq 3.37$ is approximately 0.99.
3. ANOVA techniques tell us whether the variation among the sample means is too large to be attributed to chance. This, of course, implies that the differences among the sample means are significant.
4. $\sqrt{MSE}$ represents the pooled sample standard deviation, s_p.
5. No. We can only conclude that they are not *all* the same.
6. a. Treatment mean square
   b. Error mean square, *MSE*

Chapter Test (pages 733–739)

1. Cell (1): 399.8   2. Cell (2): 1315.2   3. Cell (3): 1715.0   4. Cell (4): 4   5. Cell (5): 20
6. Cell (6): 24   7. Cell (7): 99.95   8. Cell (8): 65.76   9. Cell (9): 1.5199
10. Choice (b)
11. Reject null hypothesis. The mean number of busses serviced per week at each of these garages is not the same $(F = 5.27)$.
12. Do not reject null hypothesis that the average score obtained by third graders in each of the cities is the same $(F = 0.1210)$.
13. Reject null hypothesis. The average starting salary is not the same for all these graduates with the five areas of specialization $(F = 15.5949)$.
14. Reject null hypothesis. The average sodium content in the serving sizes of salmon is not the same for the cans packed by all of these companies $(F = 6.4337)$.
15. Do not reject null hypothesis that the average number of times that these batteries can be recharged is approximately the same for the four brands of batteries $(F = 0.7724)$.
16. Reject null hypothesis. Both the stores and the wrappings have an effect on sales $(F = 5.5855$ and $F = 4.4222)$.
17. Do not reject null hypothesis that the type of diet used significantly affects the average blood pressure level $(F = 1.3778)$.
18. Do not reject null hypothesis that there is a significant difference between the average waiting time in the emergency rooms at these hospitals before seeing a doctor $(F = 0.6792)$.
19. Do not reject null hypothesis that there is a significant difference in the cost of a laser jet printer for each of these brands $(F = 1.5932)$.
20. Do not reject null hypothesis that the average number of arrests for illegal drug use is not significantly different at these beaches $(F = 0.6271)$.
21. Reject null hypothesis about the different rations used. There is a significant difference in the ration method used $(F = 12.6377)$. Do not reject null hypothesis that there is a significant difference in the type of hog used $(F = 2.3798)$.
22. Do not reject the null hypothesis that there is a significant difference in the average time needed by each student to learn to play the song by the different methods $(F = 1.0217)$.

23. Reject null hypothesis. The average price of a half-gallon of orange juice is significantly different for the brands of orange juice ($F = 68.26$) and for the different neighborhoods ($F = 104.95$).

24. Do not reject null hypothesis that the average amount of money in a settlement claim is not significantly different for all the centers ($F = 0.6796$) and is not significantly different for all the employees ($F = 0.5031$).

25. Do not reject null hypothesis that the students in each of these teacher's classes are not significantly different in their performance on the calculus I final exam ($F = 0.1604$).

Thinking Critically (page 740)

1. A, B, C, D, and E.   2. Reject null hypothesis that the sample means are equal ($F = 7.5$).

3. The 95% confidence interval for the difference between average tar content of Brand X and Brand Y cigarettes is between -5.438 and 3.772.

4. a. The 95% confidence interval for the difference between average tar content of Brand X and Brand Z cigarettes is between -11.438 and -2.228.

   b. The 95% confidence interval for the difference between average tar content of Brand Y and Brand Z cigarettes is between -10.605 and -1.395.

5.

```
Analysis of Variance
```

Source	SS	DF	MS	F	p
Factor	103.0412	2	51.5206	0.82	0.326
Error	565.4700	9	62.83		
Total	668.5112	11			

Level	N	Mean	StDev
C1	4	11.82	7.641
C2	4	12.69	9.123
C3	4	8.93	8.569

```
Pooled StDev
```

Case Study (page 741)

1. Reject null hypothesis that the average prices of college textbooks in algebra and trigonometry, statistics, calculus, and liberal arts mathematics are not significantly different ($F = 149.3868$).

CHAPTER 14

Section 14.3 (pages 753–760)

1. $n = 12$ and the number of plus signs is 5. Do not reject the null hypothesis that the median salary is $32,073.

3. $n = 11$, the number of plus signs is 2, and the number of minus signs is 9. Do not reject the claim that the median price is 64 cents.

5. There are 9 minus signs out of a possible 10 sign changes. Reject null hypothesis. The number of daily arrests has decreased.

7. There are 4 minus signs out of a possible 9 sign changes. Do not reject null hypothesis. There is no significant difference (at least for these homeowners) in the annual heating bill.
9. There are 8 minus signs out of a possible 9 sign changes. Reject null hypothesis. The daily number of felony arrests has decreased.
11. There are 6 plus signs out of a possible 9 sign changes. Do not reject null hypothesis. The data do not indicate that the medicine significantly affects the number of times that dizziness is reported.

Section 14.4 (pages 764–766)

1. Reject the null hypothesis. The number of students completing their lunch has changed significantly after the new caterer was hired.
3. Do not reject null hypothesis. We cannot conclude that the new policy significantly affects the radio audience.

Section 14.5 (pages 772–775)

1. Reject null hypothesis. The mean score for both groups is not the same ($z = 3.63$).
3. Do not reject null hypothesis that the mean time needed by both groups to learn the aerobic exercise is the same ($z = -0.56$).
5. Do not reject null hypothesis that the average time needed by both groups is the same ($z = -0.19$).
7. Reject null hypothesis. The average radon levels in the air at both day-care centers is not the same ($z = -3.05$).

Section 14.7 (pages 782–785)

1. a. $R = 0.1$ is not significant. We cannot conclude that there is a significant correlation between the ratings of recruiters 1 and 2.
   b. $R = -0.9$ is not significant. We cannot conclude that there is a significant correlation between the ratings of recruiters 1 and 3.
   c. $R = -0.5$ is not significant. We cannot conclude that there is a significant correlation between the ratings of recruiters 2 and 4.
3. There are 6 runs where we have 9 a's and 9 b's. Do not reject the null hypothesis of randomness.
5. There are 11 runs where we have 10 M's and 10 F's. Do not reject the null hypothesis of randomness.
7. Between patient 1 and patient 2: $R = 0$
   Between patient 1 and patient 3: $R = -0.714$
   Between patient 1 and patient 4: $R = 0.643$
   Between patient 1 and patient 5: $R = 0.5$
   Between patient 2 and patient 3: $R = 0$
   Between patient 2 and patient 4: $R = 0.429$
   Between patient 2 and patient 5: $R = -0.321$
   Between patient 3 and patient 4: $R = -0.429$
   Between patient 3 and patient 5: $R = -0.393$
   Between patient 4 and patient 5: $R = 0.179$
   There is no significant correlation between the patient's ratings of the nurses.

9. There are 7 runs where we have 8 a's and 9 b's. Do not reject null hypothesis of randomness.

11. There are 11 runs where we have 12 C's and 13 R's. Do not reject null hypothesis of randomness.

Section 14.8 (page 789)

1. Do not reject the null hypothesis. The data do not indicate that there is a significant difference in the average number of summonses issued by each of these police officers (test statistic = 1.5951).

Section 14.10 (pages 793–794)

```
1. MTB  > READ NON AIR-CONDITIONED INTO C1, AIR-CONDITIONED INTO C2
   DATA > 77 80
   DATA > 39 65
   DATA > 85 76
   DATA > 84 83
   DATA > 35 44
   DATA > 13 29
   DATA > 68 59
   DATA > 86 86
   DATA > 56 79
   DATA > 79 84
   DATA > 81 72
   DATA > 80 73
   DATA > 65 76
   DATA > END
         13 rows read.

   MTB  > RANK THE VALUES IN C1, PUT RANKS IN C3
   MTB  > RANK THE VALUES IN C2, PUT RANKS IN C4
   MTB  > PRINT C1-C4
```

Data Display

Row	C1	C2	C3	C4
1	77	80	7	10.0
2	39	65	3	4.0
3	85	76	12	7.5
4	84	83	11	11.0
5	35	44	2	2.0
6	13	29	1	1.0
7	68	59	6	3.0
8	86	86	13	13.0
9	56	79	4	9.0
10	79	84	8	12.0
11	81	72	10	5.0
12	80	73	9	6.0
13	65	76	5	7.5

```
MTB > CORRELATION COEFFICIENT BETWEEN RANKS IN C3 AND C4
```

Correlation (Pearson)

```
Correlation of C3 and C4 = 0.669
```

Reject null hypothesis. An air-conditioned room on a humid day significantly affects the performance on this literacy test.

```
3. MTB  > SET THE FOLLOWING IN C1
   DATA > 4.3 3.8 6.0 3.7 4.7 6.2
   DATA > 5.8 5.9 3.4 5.3 5.7 5.0 3.5 4.2
   DATA > 4.8 5.4 5.5 5.1 5.2 6.9
   DATA > 6.1 5.6 4.9 4.4 3.3
   DATA > END
   MTB  > SET THE FOLLOWING IN C2
   DATA > 1 1 1 1 1 1
   DATA > 2 2 2 2 2 2 2 2
   DATA > 3 3 3 3 3 3
   DATA > 4 4 4 4 4
   DATA > END
   MTB  > KRUS C1 C2
```

Kruskal-Wallis Test

LEVEL	NOBS	MEDIAN	AVE. RANK	Z VALUE
1	6	4.500	11.8	-0.45
2	8	5.150	12.3	-0.35
3	6	5.300	15.8	1.08
4	5	4.900	12.2	-0.27
OVERALL	25		13.0	

$H = 1.18$ d.f. $= 3$ $p = 0.757$

Do not reject null hypothesis. The data do not indicate that there is a significant difference in the average dissolved oxygen content at the four locations.

Testing Your Understanding of This Chapter's Concepts (page 796)

1. Yes. In the one-tailed test, the alternate hypothesis is that the probability distribution for population A is shifted to the right of that for B. In the two-tailed test, the alternate hypothesis is that the probability distribution for population A is shifted to the left or to the right of that for B.
2. The distributions are assumed to be continuous so that the probability of tied measurements is 0, and each measurement can be assigned a unique rank.
3. Yes. Because large values of the test statistic support the alternative hypothesis that the populations have different probability distributions, and that the rejection region is located in the upper tail of the χ^2 distribution.
4. The lowest value H of the test statistic is 0 and largest value is 12.5 and occurs when the ranks for sample 1 are 1, 2, 3, 4, and 5, the ranks for sample 2 are 6, 7, 8, 9, and 10 and the ranks for sample 3 are 11, 12, 13, 14, and 15.

Chapter Test (pages 796–802)

1. Choice (a) 2. Choice (a) 3. Choice (b) 4. Choice (b) 5. Choice (a)
6. Choice (b) 7. Choice (b) 8. Choice (b)
9. There are 9 runs where we have 12 a's and 12 b's. Do not reject null hypothesis of randomness.
10. There are 12 runs where we have 12 F's and 13 M's. Do not reject null hypothesis of randomness.
11. Reject null hypothesis. The mean number of delayed flights for both airlines is significantly different ($z = 2.40$).
12. Do not reject the null hypothesis. The data does not indicate that the mean number of minutes needed to service the part satisfactorily is not significantly different for both groups (test statistic = 0.1087).
13. There are 6 minus signs out of a possible 9 sign changes. Do not reject null hypothesis that there is a significant difference in the triglyceride level of a person as a result of using this pill ($z = 1.00$).
14. There are 7 minus signs out of a possible 9 sign changes. Do not reject null hypothesis that the gruesome scene does not significantly affect the person's pulse rate ($z = 1.67$).
15. $R = -0.37$. Do not reject null hypothesis that there is a significant correlation between the President's and the Vice-President's rankings.
16. Do not reject null hypothesis that the mean number of canceled appointments is not significantly different for both dentists ($z = -0.615$).
17. Do not reject the null hypothesis that the mean number of practice sessions is significantly different for both groups (test statistic = 1.515).
18. There is 1 minus sign out of a possible 9 sign changes. Reject null hypothesis. There is a significant difference in the muscle strength of an individual as a result of using this technique ($z = -2.33$).
19. There are 2 minus signs out of a possible 10 sign changes. Do not reject null hypothesis. It would seem that seeing the horror movie does not significantly affect the person's blood pressure (assuming a two-tailed test) ($z = -1.898$).
20. $R = -0.3143$. Do not reject null hypothesis that there is no significant correlation between the President's and the Vice-President's rankings.
21. There are 5 minus signs out of a possible 7 sign changes. Do not reject null hypothesis. The two-way radios have no significant effect on the number of robberies of cab drivers ($z = 1.13$).

Thinking Critically (pages 802–803)

1. Do not reject null hypothesis ($z = -0.61$). 3. Yes

Case Study (page 803)

1. There are 8 plus signs out of a possible 10 sign changes. Reject null hypothesis. The asthma medication significantly affects a person's weight ($z = 1.898$).

INDEX

Frequently Used Formulas

Relative frequency: $\dfrac{f_i}{n}$

Mean: $\dfrac{\Sigma x}{n} = \dfrac{x_1 + x_2 + \cdots + x_n}{n}$

Weighted mean: $\bar{x}_w = \dfrac{\Sigma xw}{\Sigma w}$

Variance: $\sigma^2 = \dfrac{\Sigma(x - \mu)^2}{n}$ or $\dfrac{\Sigma x^2}{n} - \dfrac{(\Sigma x)^2}{n^2}$

Standard deviation: $\sigma = \sqrt{\dfrac{\Sigma(x - \mu)^2}{n}}$

Average deviation: $\dfrac{\Sigma|x - \mu|}{n}$

Sample standard deviation: $\sqrt{\dfrac{\Sigma(x - \bar{x})^2}{n - 1}}$

Percentile rank of x: $\dfrac{B + \frac{1}{2}E}{n} \cdot 100$

z-Score: $z = \dfrac{x - \mu}{\sigma}$ Original score: $x = \mu + z\sigma$

Probability, p: $p = \dfrac{f}{n}$

$_nP_r = \dfrac{n!}{(n - r)!}$ $_nP_n = n!$ $_nC_r = \dfrac{n!}{r!(n - r)!}$

Number of permutations with repetitions: $\dfrac{n!}{p!q!r! \cdots}$

Mathematical expectation: $m_1p_1 + m_2p_2 + m_3p_3 + \cdots$

Addition rule (for mutually exclusive events): $p(A \text{ or } B) = p(A) + p(B)$

Addition rule (general case): $p(A \text{ or } B) = p(A) + p(B) - p(A \text{ and } B)$

Complement of event A: $p(A') = 1 - p(A)$

Conditional probability formula: $p(A|B) = \dfrac{p(A \text{ and } B)}{p(B)}$

Multiplication rule: $p(A \text{ and } B) = p(A|B) \cdot p(B)$

Multiplication rule (for independent events): $p(A \text{ and } B) = p(A) \cdot p(B)$

Bayes' rule: $\dfrac{p(B|A_n)p(A_n)}{p(B|A_1)p(A_1) + p(B|A_2)p(A_2) + \cdots + p(B|A_n)p(A_n)}$

Mean of probability distribution: $\mu = \Sigma xp(x)$

Variance of a probability distribution: $\sigma^2 = \Sigma(x - \mu)^2p(x)$ or $\sigma^2 = \Sigma x^2p(x) - [\Sigma(x)p(x)]^2$